An Introduction to Computer Simulation Methods

Applications to Physical Systems

Second Edition

Harvey Gould

Department of Physics
Clark University

Jan Tobochnik

Department of Physics
Kalamazoo College

Addison-Wesley Publishing Company

Reading, Massachusetts • Menlo Park, California • New York
Don Mills, Ontario • Wokingham, England • Amsterdam • Bonn
Sydney • Singapore • Tokyo • Madrid • San Juan • Milan • Paris

Sponsoring Editor: Julia Berrisford
Associate Editor: Jennifer Albanese
Production Supervisor: Helen Wythe
Cover Designer: Marshall Henrichs
Senior Manufacturing Supervisor: Hugh Crawford
Composition: Windfall Software (Paul C. Anagnostopoulos, Joe Snowden), using ZzTEX

Library of Congress Cataloging-in-Publication Data

Gould, Harvey
    An introduction to computer simulation methods / Harvey Gould, Jan Tobochnik.
        p.   cm.
    ISBN 0-201-50604-1
    1. Physics—Data processing.   2. Physics—Simulation methods.
    3. Physics—Computer simulation.   I. Tobochnik.
    II. Title.
    QC52-G68   1996
    530'.01'13—dc20   94-13897
                                                                       CIP

1 2 3 4 5 6 7 8 9 10-MA-98979695

To the memory of
Sheng-keng Ma

P R E F A C E

Computer simulations are an integral part of contemporary basic and applied physics and computation is becoming as important as theory and experiment. The ability "to compute" is now part of the essential repertoire of research scientists. It is time for computation to play a similar role in science education.

Since writing the first edition of our text, more courses devoted to the study of physics using computers have been introduced into the physics curriculum, and many more traditional courses are incorporating numerical examples. We are gratified to see that our text has helped shape these innovations. The purpose of our book includes the following:

- To provide a means for students to *do* physics.
- To give students an opportunity to gain a deeper understanding of the physics they have learned in other courses.
- To encourage students to "discover" physics in a way similar to how physicists learn in the context of research.
- To introduce numerical methods and new areas of physics that can be studied with these methods.
- To give examples of how physics can be applied in a much broader context than it is discussed in the traditional physics undergraduate curriculum.
- To teach structured programming in the context of doing science.

Our overall goal is to encourage students to learn about science through experience and by asking questions. Our objective always is understanding, not the generation of numbers.

Many of the problems in the text are open-ended, and students are encouraged to develop the skills necessary to determine when they are satisfied with the results they have obtained. There are no answers in the back of the book. We want students to think critically about their work, and not to rely on instructors to tell them when they have the correct answer.

Our experience has been that the best format for achieving the above goals is a laboratory based course. Such a format allows students to proceed at their own pace, encourages students to learn from each other, and makes it possible for students to choose topics of special interest to them. A separate course also reduces the necessity of introducing new computational methods into already overburdened courses. At Clark University and Kalamazoo College, first-year undergraduates to graduate students have taken courses based on drafts of this text. Student users of our text include majors in a variety of disciplines, including students outside the physical sciences.

At first glance, the second edition might seem formidable because it is longer than the first edition and is published in one volume. There are over 700 pages and many of the topics might be unfamiliar. However, most of the text can be understood at some level by students with a minimal background in physics and calculus. No background in programming is assumed. The first five chapters, and parts of all the remaining chapters (except Chapter 18, which requires a background in quantum mechanics) are accessible to students with only introductory physics and calculus. We have included a few exercises to provide more practice in writing simple programs. Most of the more advanced parts of the text are located in the starred sections, the starred problems, and the project sections. They are not prerequisites to the rest of the text, and can be used by advanced students who wish to use this text for independent or guided research projects.

Most of the chapters are relatively independent, but some necessary background is established in certain key chapters. Chapter 2 provides an introduction to programming and to the finite difference solution of a differential equation. Chapter 3 introduces several approaches for the numerical solution of Newton's equations of motion. Similar methods are used in Chapters 4, 5, 8, 9, and the latter half of Chapter 6. Chapters 7 and 11 introduce the ideas of probability and Monte Carlo methods. The latter is used in a variety of contexts in the remaining chapters. An introductory course for students with a limited background might use Chapters 1–3, 7, 11, and selections from other chapters. Parts of Chapters 9 and 10 on waves and electrodynamics require more theoretical background than most of the other chapters.

Many improvements have been made to the text. The most obvious is that the text is no longer divided into two separate parts. Unifying the text under a single cover has made it possible for us to emphasize the unity of physics by providing more connections between the topics. A single volume also allows beginners to see some of the more interesting topics covered in later chapters, and the more advanced students to review the techniques they need from the introductory chapters. Specific changes include the following. We have introduced the more powerful, but still simple Euler-Richardson algorithm early in the text to simulate particle motion. We have significantly enhanced our chapters on chaos, molecular dynamics, normal modes and waves, electrodynamics, and quantum mechanics. We also have introduced an elementary chapter on random processes (Chapter 7) early in the text so that we can use Monte Carlo methods in later chapters. Finally, we have expanded the original chapter on fractals and cellular automata into two chapters, Chapters 14 and 15. We also have made the graphical output of many programs more interesting and colorful.

Each chapter contains a brief discussion of the important physical concepts, and is followed by program listings, problems, and relevant questions. The discussion, programs, and problems are interrelated, and in general, the discussion will be more readily understood after the problems are completed. All of the problems should be read whether or not they are attempted. The programs are designed to be simple and easy to read, rather than elegant or efficient. We regard the program listings as text for the reader rather than source code for the computer. To do most of the problems, the reader must understand the logic of the programs, and hence the logic of the underlying physical system. Most of the problems require at least some modification of the programs. The problems are organized so that the earlier ones in a chapter provide the basis for the later problems in the same chapter and in succeeding chapters.

We have added projects to the end of many chapters. These projects are similar in scope to the starred problems, but are frequently more open-ended and are appropriate for independent research projects. The recommended readings at the end of each chapter have been selected for their pedagogical value rather than for completeness or for historical accuracy. We apologize to our colleagues whose work has been inadvertently omitted, and we would appreciate suggestions for new and additional references.

We have decided to continue using True BASIC as the programming language discussed in the body of the text. Our motivation for this choice is that True BASIC is easy to learn and use, has true subroutines and excellent graphics capabilities, is inexpensive, and is identical on computers using DOS, Apple Macintosh, and the UNIX operating systems. Readers familiar with other dialects of BASIC should have few problems adapting the programs in the text. The syntax of True BASIC is very similar to Fortran, and hence learning True BASIC provides an easy route to that important scientific language. We discuss the basic syntax of Fortran and C in the appendices. For our purposes, the similarities of BASIC, Fortran, and C are much greater than their differences. Students and instructors may use any language they wish, while treating our True BASIC program listings as pseudocode.

We believe that typing in the programs line-by-line will help students learn programming more easily. The programs have been carefully checked for errors and typos, and the programs have been incorporated into the text directly from the source code. Nevertheless, there may be subtle errors of which we are not aware. It is our experience that few programs are perfect, and we cannot guarantee that our programs are free of error.

We began the revision of our text less than five years after the publication of the first edition. The rate of change in computational science is rapid, and we expect that further revisions will be necessary. To encourage reader feedback and to provide a forum for incorporating changes, we have set up a World Wide Web server devoted to users of our text and to faculty and students using computer simulations in an educational context. Our plan is to continuously update our text, so that everyone can benefit from the experience of others. All the programs in the text can be downloaded for noncommercial use from the server. We hope that our home page (`http://www.clarku.edu/~sip` or `http://www.kzoo.edu/~sip`) will serve as a forum for faculty and students to share ideas and programs and to make comments, corrections, and suggestions. In particular, we have posted suggestions for the use of the various chapters by instructors.

We acknowledge generous support from the National Science Foundation for an ILI grant to develop curriculum materials for a laboratory-based course on computer simulations at the lower division level, and a grant to develop new computer simulation materials at the upper division level. This support has helped us to find time to write our text and to provide workshops and develop other curricular materials essential to expanding the number of institutions that can offer courses similar to ours.

The text was produced using TEXtures. We thank Paul C. Anagnostopoulos of Windfall Software who wrote the macros for the book and who patiently explained the vagaries of TEX. Almost all the figures were written in PostScript by Louis Colonna-Romano. In some cases they were edited using Illustrator 5.5. It is difficult to imagine how we could have produced the figures without his help. Lou also read every chapter, ran many of the programs, and made numerous helpful comments and suggestions. We enjoyed working with our former editor, Stuart Johnson, who encouraged us to write a second edition and waited patiently for us to finish it while never losing his sense of humor. We also thank Jennifer Albanese, Associate Editor for Physics, who coordinated our project, and Mona Zeftel for her help with the technical aspects of book publishing. We look forward to working with our new editor at Addison-Wesley, Julia Berrisford, on future projects.

Many colleagues and students have generously commented on preliminary drafts of the second edition and have given encouragement and advice. We thank the students in our classes at Clark and Kalamazoo for their patience and good humor while they suffered through various drafts, and in particular mention students Jeremy Carfi, Aaron Isabelle, Faina Ryvkin, and Jonathan Wall for their helpful suggestions. We thank our physics colleagues at Clark and Kalamazoo for their encouragement and willingness to allow us to take some extra time to complete this text, and our secretaries Sujata Davis and Peggy Cauchy for their assistance. We also acknowledge Bill Klein, Richard Brower, and Roscoe Giles and their colleagues at Boston University for their encouragement and suggestions. We wish to thank Kenneth Basye, Marc Bourzutschy, Alyce Brady, Denis Donnelly, Robert Ehrlich, Allan Ferrenberg, Roger Kohin, Jon Machta, Andrew Mel'cuk, Mark Novotny, George Phillies, Dennis Rapaport, and Dietrich Stauffer for reviewing sections of the text. We are specially indebted to Peter Reynolds and Fred Harris who reviewed large portions of the text. Of course, the responsibility for the remaining errors, typos, and unclear passages is ours. We also wish to acknowledge the work of many people who have contributed to the computer simulation column in Computers in Physics. Their contributions have allowed us to learn about new developments, much of which we have incorporated into our text. One of us thanks Donna and Mark Mogul, the proprietors of Sweetreats, for providing the coffee and bagels that continue to sustain members of the Clark physics department.

Finally, we are grateful to our wives, Patti Gould and Andrea Moll Tobochnik, and to our children, Joshua, Emily, and Evan Gould and Steven and Howard Tobochnik, for their patience and understanding during the course of this work.

CONTENTS

C H A P T E R

1

Introduction

The importance of computers in physics and the nature of computer simulation is discussed.

1.1   IMPORTANCE OF COMPUTERS IN PHYSICS

Computers are becoming increasingly important in how our society is organized. Like any new technology they are affecting how we learn and how we think. Physicists are in the vanguard of the development of new computer hardware and software and computation has become an integral part of contemporary science. For the purposes of discussion, we divide the use of computers in physics into the following categories:

- numerical analysis
- symbolic manipulation
- simulation
- collection and analysis of data
- visualization

In the *numerical analysis* mode, the simplifying physical principles are discovered prior to the computation. For example, we know that the solution of many problems in physics can be reduced to the solution of a set of simultaneous linear equations. Consider the equations

$$2x + 3y = 18$$
$$x - y = 4.$$

It is easy to find the analytical solution $x = 6$, $y = 2$ using the method of substitution and pencil and paper. Suppose we have to solve a set of four simultaneous equations. We again can find an analytical solution, perhaps using a more sophisticated method. If the number of variables becomes much larger, we would have to use numerical methods and a computer to find a numerical solution. In this mode, the computer is a tool of numerical analysis, and the essential physical principles, e.g., the reduction of the problem to the inversion of a matrix, are included in the computer program. Because it is often necessary to compute a multidimensional integral, manipulate large matrices, or solve a complex differential equation, we know that this use of the computer is important in physics.

An increasingly important use is *symbolic manipulation*. As an example, suppose we want to know the solution to the quadratic equation, $ax^2 + bx + c = 0$. A symbolic manipulation program can give us the solution as $x = \left[-b \pm \sqrt{b^2 - 4ac} \right]/2a$. In addition, such a program can give us the usual numerical solutions for specific values of a, b, and c. Mathematical operations such as differentiation, integration, matrix inversion, and power series expansion can be performed using most symbolic manipulation programs. What will happen to education when such programs become readily available on personal computers?

In the *simulation* mode, the essential elements of the model are included with a minimum of analysis. As an example, suppose a teacher gives $10 to each student in a class of 100. The teacher, who also begins with $10 in her pocket, chooses a student at random and flips a coin. If the coin is "tails," the teacher gives $0.50 to the student; otherwise, the student gives $0.50 to the teacher. If either the teacher or the student would go into debt by this transaction, the transaction is not allowed. After many exchanges, we would like to know the probability that a student has s

dollars and the probability that the teacher has t dollars. Are these two probabilities the same? One way to find the answers to such questions is to do a classroom experiment. However, such an experiment would be difficult to arrange, and it would be tedious to do a sufficient number of transactions. Although this particular problem can be solved exactly by analytical methods, many problems of this nature cannot be solved in this way. Another way to proceed is to incorporate the rules of the game into a computer program, simulate many exchanges, and compute the probabilities. After we compute the probabilities of interest, we might gain a better understanding of their relation to the exchanges of money, and a better idea how to find an analytical solution if one exists. We also can modify the rules and ask "what if?" questions. For example, how would the probabilities change if the exchanges were \$1.00 rather than \$0.50? What would happen if the teacher were allowed to go into debt?

In all these approaches, the main goal of the computation is generally insight rather than simply numbers. Computation has had a profound effect on the way we do physics, on the nature of the important questions in physics, and on the physical systems we choose to study. All these approaches require the use of at least some simplifying approximations to make the problem computationally feasible. However, because the simulation mode requires a minimum of analysis and emphasizes an exploratory mode of learning, we stress this approach in this text. For example, if we change the names of the players in the above example, e.g., let money $\rightarrow$ energy, then the above questions would be applicable to problems in magnetism and particle physics.

Computers often are involved in all phases of a laboratory experiment, from the design of the apparatus to the collection and analysis of data. This involvement of the computer has not only permitted experimentalists to sleep better at night, but has made possible experiments that would otherwise be impossible. Some of these tasks are similar to those encountered in theoretical computation. However, the tasks involved in control and interactive data analysis are qualitatively different and involve real-time programming and the interfacing of computer hardware to various types of instrumentation. For these reasons we do not discuss the application of computers in experimental physics.

1.2   THE NATURE OF COMPUTER SIMULATION

Why is computation becoming so important in physics? One reason is that most of our analytical tools such as differential calculus are best suited to the analysis of *linear* problems. For example, you probably have learned to analyze the motion of a particle attached to a spring by assuming a linear restoring force and solving Newton's second law of motion. In this case, a small change in the displacement of the particle leads to a small change in the force. However, many natural phenomena are *nonlinear*, and a small change in a variable might produce a large change in another. Because relatively few nonlinear problems can be solved by analytical methods, the computer gives us a new tool to explore nonlinear phenomena. Another reason for the importance of computation is the increased interest in systems with many variables or with many degrees of freedom. The money exchange example described in Section 1.1 is an example of such a problem.

Laboratory experiment	Computer simulation
sample	model
physical apparatus	computer program
calibration	testing of program
measurement	computation
data analysis	data analysis

Table 1.1   Analogies between a computer simulation and a laboratory experiment.

Computer simulations are sometimes referred to as *computer experiments* because they share much in common with laboratory experiments. Some of the analogies are shown in Table 1.1. The starting point of a computer simulation is the development of an idealized model of a physical system of interest. We then need to specify a procedure or *algorithm* for implementing the model on a computer. The computer program simulates the physical system and defines the computer experiment. Such a computer experiment serves as a bridge between laboratory experiments and theoretical calculations. For example, we can obtain essentially exact results by simulating an idealized model that has no laboratory counterpart. The comparison of the simulation results with an approximate theoretical calculation serves as a stimulus to the development of methods of calculation. On the other hand, a simulation can be done on a realistic model in order to make a more direct comparison with laboratory experiments.

Computer simulations, like laboratory experiments, are not substitutes for thinking, but are tools that we can use to understand complex phenomena. But the goal of all our investigations of fundamental phenomena is to seek explanations of physical phenomena that fit on the back of an envelope or that can be made by the wave of a hand.

Developments in computer technology are leading to new ways of thinking about physical systems. Asking the question "How can I formulate the problem on a computer?" has led to new formulations of physical laws and to the realization that it is both practical and natural to express scientific laws as rules for a computer rather than in terms of differential equations. This new way of thinking about physical processes is leading some physicists to consider the computer as a physical system and to develop novel computer architectures that can more efficiently model physical systems in nature.

1.3   IMPORTANCE OF GRAPHICS

Because computers are changing the way we do physics, they will change the way we learn physics. For example, as the computer plays an increasing role in our understanding of physical phenomena, the visual representation of complex numerical results will become even more important. The human eye in conjunction with the visual processing capacity of the brain is a very sophisticated device for analyzing visual information. Most of us can draw a good approximation to the best straight line through a sequence

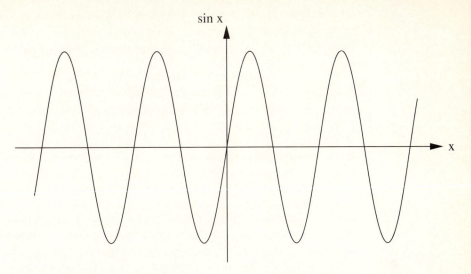

Fig. 1.1   Plot of the sine function.

of data points very quickly. And such a straight line is more meaningful to us than the "best fit" line drawn by a statistical package that we do not understand. Our eye can determine patterns and trends that might not be evident from tables of data and can observe changes with time that can lead to insight into the important mechanisms underlying a system's behavior.

At the same time, the use of graphics will increase our understanding of the nature of analytical solutions. For example, what does a sine function mean to you? We suspect that your answer is not the series, $\sin x = x - x^3/3! + x^5/5! + \ldots$, but rather a periodic, constant amplitude graph (see Fig. 1.1). What is important is the visualization of the form of the function.

An increasing application of computers is the visualization of large amounts of data. Traditional modes of presenting data include two- and three-dimensional plots including contour and field line plots. Frequently, more than three variables are needed to understand the behavior of a system, and new methods of using color and texture are being developed to help researchers gain greater insight about their data.

1.4   PROGRAMMING LANGUAGES

We will consider four programming languages in the text: BASIC, C, Fortran, and Pascal. The BASIC programming language was developed in 1965 by John G. Kemeny and Thomas E. Kurtz of Dartmouth College as a language for introductory courses in computer science. Pascal was designed in 1970 by Niklaus Wirth of the Swiss Federal Institute of Technology and also was intended for teaching purposes. Both languages have since been adapted to many other purposes. Fortran was developed by John Backus and his colleagues at IBM between 1954 and 1957 and is the most common language used in scientific applications. The C programming language evolved from a succession of programming languages at Bell Laboratories and was originally intended as a language for systems programming.

There is no single best programming language any more than there is a best natural language. Programming languages are not static, but continue to change with developments in hardware and theories of computation. As with any language, a working knowledge of one makes it easier to learn another. BASIC is easy to learn, but some of its features are hardware dependent, and in many of its implementations, it lacks modular programming features. Pascal is well structured and allows a rich set of data structures. Pascal also imposes requirements that make it difficult to avoid writing modular programs. One of the disadvantages of Pascal is that these requirements make it more difficult to write simple programs quickly. Another disadvantage is that Pascal was not really intended for scientific programming. Fortran allows structured programming, and much of its syntax is simpler than Pascal. Many of the major disadvantages of Fortran 77 have been eliminated by Fortran 90, but this new standard is not yet widely available. C has become the language of choice for writing application programs for workstations and personal computers.

BASIC, C, Fortran, and Pascal are examples of procedural languages. Procedural languages change the state or memory of the machine by a sequence of statements. In a *functional* language such as LISP, the focus of the language is not on changing the data in specific memory locations. Instead, a program consists of a function that takes an input and produces an output. Mathematica and Maple are other examples of functional languages. For most of the problems discussed in this text functional languages are slower than procedural languages. This disadvantage exists in part because most computers have been designed with procedural languages in mind. Another important class of programming languages are *object-oriented* languages such as the C++ extension of C and Smalltalk. The key idea is that functions and data are not treated separately, but are grouped together in an *object*. A program is a structured collection of objects that communicate with each other causing the data within a given object to change. Object oriented programming is particularly useful for creating icon based user interfaces. One goal of the object oriented approach is to produce large libraries of objects that can be easily modified and extended for individual problems.

In most of the text we will use a version of BASIC developed by Kemeny and Kurtz, the original developers of BASIC. This version, known as True BASIC, has subroutines similar to Fortran and has excellent graphics capabilities that are hardware independent. Identical True BASIC programs can run on personal computers such as MS-DOS compatibles and the Apple Macintosh, as well as on most UNIX workstations. Because of the importance of graphics, we have adopted True BASIC as the programming language in the body of the text. Readers who already know another programming language should consider the True BASIC program listings to be pseudocode that easily can be translated into a language of their choice. A comparison of the core syntax of True BASIC and Fortran is given in Appendix A, and a similar comparison with C is given in Appendix B. These appendices also contain sample programs in Fortran and C.

1.5   LEARNING TO PROGRAM

If you already know how to program, try reading a program that you wrote several years, or even several weeks, ago. Many people would not be able to follow the logic of their program and consequently would have to rewrite it. And your program would probably be of little use to a friend who needs to solve a similar problem. If you are learning programming for the first time, it is important to learn good programming habits and to avoid this problem. The programs in this text employ *structured* programming techniques such as the IF THEN ELSE constructs and avoid the use of GO TO statements. The programs also are written in *modules*, which are subprograms that perform specific tasks.

Because of their background, students of science have special advantages in learning how to program. We know that our mistakes cannot harm the computer (except for spilling coffee or soda on the keyboard). More importantly, we have an existing context in which to learn programming. The past several decades of doing physics research with computers has given us numerous examples that we can use to learn physics, programming, and data analysis. Hence, we encourage you to learn programming in the context of the examples in each chapter.

Our experience is that the single most important criterion of program quality is readability. If a program is easy to read and follow, it is probably a good program. The analogies between a good program and a well-written paper are strong. Few programs come out perfectly on their first draft, regardless of the techniques and rules we use to write it. Rewriting is an important part of programming.

1.6   HOW TO USE THIS BOOK

In general, each chapter begins with a short background summary of the nature of the system and the important physical questions. We then introduce the computer algorithms, True BASIC syntax if necessary, and discuss a sample program. The programs are meant to be read as text on an equal basis with the discussions and the problems that are interspersed throughout the text. It is strongly recommended that all the problems be read, because many concepts are introduced after the simulation of a physical process.

It is a good idea to maintain a laboratory notebook to record your programs, results, graphical output, and analysis of the data. This practice will help you develop good habits for future research projects, prevent duplication, organize your thoughts, and save time. After a while, you will find that most of your new programs will use parts of your earlier programs. Ideally, you will use your notebook to write a laboratory report or mini-research paper on your work. Guidelines for writing such a paper are given in Appendix 1A.

Many of the problems in the text are open ended and do not lend themselves to simple "back of the book" answers. So how will you know if your results are correct? How will you know when you have done enough? There are no simple answers to either question, but we can give some guidelines. First, you always should compare the results of your program to known results. The known results might come from an analytical solution that exists in certain limits or from published results. You also should look at your numbers and graphs, and determine if they make sense. Do the numbers have the right sign? Are they the right order of magnitude? Do the trends make sense as you change the parameters? What is the statistical error in the data? What is the systematic error? Some of the problems explicitly ask you to do these checks, but you should make it a habit to do as many as you can whenever possible.

How do you know when you are finished? The main guideline is that if you can tell a coherent story about your system of interest, then you are probably done. If you only have a few numbers and do not know their significance, then you need to do more. Let your curiosity lead you to more explorations. Do not let the questions asked in the problems limit what you do. They are only starting points, and frequently you will be able to think of your own questions.

Appendix 1A   LABORATORY REPORT

Laboratory reports should reflect clear writing style and obey proper rules of grammar and correct spelling. Write in a manner that can be understood by another person who has not done the assignment. In the following, we give a suggested format for your reports.

1. *Introduction*. Briefly summarize the nature of the physical system, the basic numerical method or algorithm, and the interesting or relevant questions.

2. *Method*. Describe the algorithm and how it is implemented in the program. In some cases this explanation can be given in the program itself. Give a typical listing of your program. Simple modifications of the program can be included in the appendix if necessary. The program should include your name, date, and title of your file, and be annotated in a way that is as self-explanatory as possible. Be sure to discuss any important features of your program.

3. *Verification of program*. Confirm that your program is not incorrect by considering special cases and by giving at least one comparison to a hand calculation or known result.

4. *Data*. Show the results of some typical runs in graphical or tabular form. Additional runs can be included in an appendix. All runs should be labeled, and all tables and figures must be referred to in the body of the text. Each figure and table should have a caption with complete information, e.g., the value of the time step.

5. *Analysis*. In general, the analysis of your results will include a determination of qualitative and quantitative relationships between variables, and an estimation of numerical accuracy.

6. *Interpretation*. Summarize your results and explain them in simple physical terms whenever possible. Specific questions that were raised in the assignment should be addressed here. Also give suggestions for future work or possible extensions. It is not necessary to answer every part of each question in the text.

7. *Critique*. Summarize the important physical concepts for which you gained a better understanding and discuss the numerical or computer techniques you learned. Make specific comments on the assignment and your suggestions for improvements or alternatives.

8. *Log*. Keep a log of the time spent on each assignment and include it with your report.

References and Suggestions for Further Reading

Programming

We recommend that you learn programming the same way you learned English — with practice and with a little help from your friends and manuals. We list some of our favorite True BASIC programming books here. The True BASIC reference manual is well written and essential.

William S. Davis, *True BASIC Primer*, Addison-Wesley (1986).

Larry Joel Goldstein, C. Edward Moore, and Peter J. Welcher, *Structured Programming with True BASIC*, Prentice Hall (1986).

Brian D. Hahn, *True BASIC by Problem Solving*, VCH Publishers (1988).

Stewart M. Venit and Sandra Schleiffers, *Programming in True BASIC*, West Publishing Co. (1992).

John G. Kemeny and Thomas E. Kurtz, *True BASIC*, Addison-Wesley (1985).

General References on Physics and Computers

Richard E. Crandall, *Projects in Scientific Computation*, Springer-Verlag (1994). See also Richard E. Crandall, *Pascal Applications for the Sciences*, John Wiley & Sons (1984). This book contains many examples of Pascal programs and a discussion of the nature of Pascal.

Marvin L. De Jong, *Introduction to Computational Physics*, Addison-Wesley (1991).

Paul L. DeVries, *A First Course in Computational Physics*, John Wiley & Sons (1994). Fortran is used in this text.

Alejandro L. Garcia, *Numerical Methods for Physics*, Prentice Hall (1994). MatLab and Fortran are used.

Dieter W. Heermann, *Computer Simulation Methods in Theoretical Physics*, second edition, Springer-Verlag (1990). A discussion of molecular dynamics and Monte Carlo methods directed toward advanced undergraduate and beginning graduate students.

Steven E. Koonin and Dawn C. Meredith, *Computational Physics*, Addison-Wesley (1990). The emphasis of this book is on applications of numerical methods. Many nontrivial problems are given.

P. K. MacKeown and D. J. Newman, *Computational Techniques in Physics*, Adam Hilger (1987).

Dietrich Stauffer, Friedrich W. Hehl, Volker Winkelmann, and John G. Zabolitsky, *Computer Simulation and Computer Algebra*, Springer-Verlag (1988).

William J. Thompson, *Computing for Scientists and Engineers*, John Wiley & Sons (1992). This text contains many examples of C programs and a discussion of the nature of C.

Samuel S. M. Wong, *Computational Methods in Physics and Engineering*, Prentice Hall (1992).

2

The Coffee Cooling Problem

We introduce a simple numerical method for the solution of first-order differential equations and discuss several simple programming and graphical techniques.

2.1   BACKGROUND

Before we become acquainted with computer programming and numerical methods, it is a good idea to sit down and enjoy a cup of hot coffee (or tea). Suppose that when we take our first sip, we find that the coffee is too hot. If we are in a hurry, we can add cream. But if the coffee is still too hot, we must wait for the coffee to cool. If we want the coffee to cool as soon as possible, is it better to add the cream immediately after the coffee is made, or should we wait for a while before we add the cream? Although you might already know the answer, this type of "what if?" question will serve as our introduction to computer simulation.

The nature of the heat flow from the hot water to the surrounding air is complicated and involves the mechanisms of convection, radiation, evaporation, and conduction. However, if the temperature difference between the water and its surroundings is not too large, the rate of change of the temperature of the water is proportional to this temperature difference. We can formulate this statement more precisely in terms of a differential equation:

$$\frac{dT}{dt} = -r\,(T - T_s),\tag{2.1}$$

where T is the temperature of the water, T_s is the temperature of its surroundings, t is the time, and r is the "cooling constant." The cooling constant r depends on the heat transfer mechanism, the contact area with the surroundings, and the thermal properties of the water. The minus sign in (2.1) implies that if $T > T_s$, the temperature of the water will decrease with time. The relation (2.1) is known as *Newton's law of cooling*, even though the relation is only approximate and is not a law. Moreover, Newton did not express the rate of cooling in this form. However, we shall refer to (2.1) as Newton's law of cooling to conform to common usage.

An analytical solution of (2.1) for $T(t)$, the temperature as a function of the time, can be found by writing (2.1) as $dT/(T - T_s) = -r\,dt$ and integrating both sides (see (2.7)). Equation (2.1) is an example of a *first-order* differential equation because only the first derivative of the unknown function $T(t)$ appears. Let us consider first-order differential equations of the form

$$\frac{dy}{dx} = f(x, y).\tag{2.2}$$

Because many types of systems can be modeled by differential equations of this form, it is important to know how to solve such equations. In general, an *analytical* solution of (2.2), that is, a solution in terms of well-known functions, does not exist. Moreover, even if an analytical solution exists, our understanding will be aided by a visual display of the dependence of the solution on the relevant parameters. For these reasons, we are motivated to find numerical solutions of differential equations and to learn simple ways of representing the solutions visually.

2.2   THE EULER ALGORITHM

The standard technique for numerically solving a differential equation is to convert the differential equation to a *finite difference* equation. Let us analyze the meaning of (2.2). Suppose that at $x = x_0$, y has the value y_0. Because (2.2) tells us how y changes at (x_0, y_0), we can find the *approximate* value of y at the neighboring point $x_1 = x_0 + \Delta x$, if Δx is small. The simplest approximation is to assume that $f(x, y)$, the rate of change of y with respect to x, is constant over the interval x_0 to x_1. Then the approximate value of y at $x_1 = x_0 + \Delta x$ is given by

$$y_1 = y(x_0) + \Delta y \approx y(x_0) + f(x_0, y_0)\Delta x. \tag{2.3}$$

We can repeat this procedure to find the value of y at the point $x_2 = x_1 + \Delta x$:

$$y_2 = y(x_1 + \Delta x) \approx y(x_1) + f(x_1, y_1)\Delta x. \tag{2.4}$$

This procedure can be generalized to calculate the approximate value of y at any point $x_{n+1} = x_n + \Delta x$ by the iterative formula

$$y_{n+1} = y_n + f(x_n, y_n)\Delta x. \tag{2.5}$$

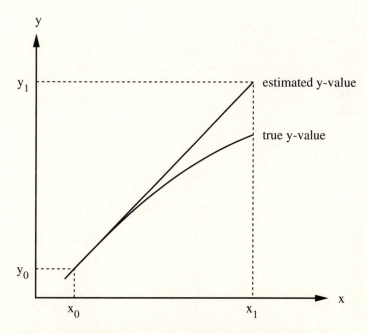

Fig. 2.1   Graphical illustration of the Euler method. The slope is evaluated at the beginning of the interval. The Euler approximation and the true function are represented by a straight line and a curve respectively.

This procedure is called the constant slope or *Euler* method. We expect that (2.5) will yield a good approximation to the exact value of y if Δx is sufficiently small; the degree of "smallness" of Δx depends on our requirements and must be left vague until we consider specific applications.

The Euler method assumes that the rate of change of y is constant over the interval x_n to x_{n+1}, and that the rate of change can be evaluated at the *beginning* of the interval. The graphical interpretation of (2.5) is shown in Fig. 2.1. We see that if the slope changes during an interval, a discrepancy occurs between the numerical solution and the exact solution. Nonetheless, the discrepancy can be made smaller if we choose a smaller value of Δx.

2.3 A SIMPLE EXAMPLE

We first use the Euler method to compute the numerical solution of the differential equation $dy/dx = 2x$ with the initial condition $y_0 = 1$ at $x_0 = 1$. Suppose that we wish to find the approximate value of y at $x = 2$. We choose $\Delta x = 0.1$, so that the number of steps is $n = (x_n - x_0)/\Delta x = (2-1)/\Delta x = 10$.

The calculation can be arranged as in Table 2.1. The initial condition $x_0 = 1$ determines the initial slope $f(x_0) = 2x_0 = 2$. The value of y at the end of the interval, y_1, is obtained from y_0, the value of y at the beginning of the interval, by the relation

$$y_1 = y_0 + \text{slope} \times \Delta x = 1 + 2 \times 0.1 = 1.2. \tag{2.6}$$

This value of y is then transferred to the second line of Table 2.1 and the process is repeated. In this way we find that $y = 3.90$ at $x = 2$. For comparison, the true solution is $y = x^2 = 4$; the error is 2.5%. Convince yourself that a smaller value of Δx improves the accuracy of the solution by redoing Table 2.1 using $\Delta x = 0.05$.

n	x_n	y_n	$g(x) = 2x$	$y_n + \text{slope} \times 0.10$
0	1.00	1.00	2.00	$1.00 + 2.00 \times 0.10 = 1.20$
1	1.10	1.20	2.20	$1.20 + 2.20 \times 0.10 = 1.42$
2	1.20	1.42	2.40	$1.42 + 2.40 \times 0.10 = 1.66$
3	1.30	1.66	2.60	$1.66 + 2.60 \times 0.10 = 1.92$
4	1.40	1.92	2.80	$1.92 + 2.80 \times 0.10 = 2.20$
5	1.50	2.20	3.00	$2.20 + 3.00 \times 0.10 = 2.50$
6	1.60	2.50	3.20	$2.50 + 3.20 \times 0.10 = 2.82$
7	1.70	2.82	3.40	$2.82 + 3.40 \times 0.10 = 3.16$
8	1.80	3.16	3.60	$3.16 + 3.60 \times 0.10 = 3.52$
9	1.90	3.52	3.80	$3.52 + 3.80 \times 0.10 = 3.90$
10	2.00	3.90		

Table 2.1 Iterated solution of the differential equation $dy/dx = 2x$ with $y = 1$ at $x = 1$. The step size is $\Delta x = 0.1$. Three significant figures are shown.

2.4   SOME TRUE BASIC PROGRAMS

To ease our introduction to programming, we first write a program that multiplies two numbers and prints the result:

```
PROGRAM product
LET m = 2                            ! mass in kilograms
LET a = 4                            ! acceleration in mks units
LET force = m*a                      ! force in Newtons
PRINT force
END
```

The first *statement* of the program is a *program header*. LET statements are used to *assign* values to variables and to compute formulas. The end of a program is denoted by the END statement. PROGRAM, LET, PRINT, and END are *keywords* that are part of the syntax of True BASIC. We will write keywords in uppercase letters, but True BASIC is not case sensitive and also ignores extra spaces. Comment statements begin with ! and can be included anywhere in a program. Use them freely to make your programs readable. Comment statements and the program header are ignored by the computer.

We introduce more syntax by modifying *Program product* so that we can enter the desired values of m and a from the keyboard and print the force in a more readable fashion.

```
PROGRAM product2
INPUT m
INPUT prompt "acceleration a (mks units) = ": a
LET force = m*a                      ! force in Newton's
PRINT "force (in Newtons) ="; force
END
```

Note the use of the INPUT and INPUT prompt statements and the simple modification of the PRINT statement. What happens if you replace the semicolon after the expression in the PRINT statement by a comma? You also can modify the program so that it adds, subtracts, and divides two numbers.

Of course, we would not bother to write a program that only multiplies or adds two numbers. However, if we need to do a series of mathematical operations, a simple program would be very convenient. Suppose that we want to know the sum $\sum_{n=1}^{\infty} 1/n^2 = 1 + 1/4 + \cdots$. (The exact answer is $\pi^2/6$ if we sum the infinite series.) One way to compute the sum for a finite number of terms is shown in *Program series*.

```
PROGRAM series
! add the first 100 terms of a simple series
! True BASIC automatically initializes variables to zero,
! but other languages might not.
LET sum = 0
FOR n = 1 to 100
    LET sum = sum + 1/(n*n)
    PRINT n,sum
NEXT n
END
```

Note the use of the FOR NEXT loop structure to execute a set of statements a predetermined number of times. The index or control variable monitors the number of times the loop has been executed. The FOR statement specifies the first and last value of the index, and the amount that the index is incremented each time the NEXT statement is reached. Unless otherwise specified, the index is increased by 1 until the index is greater than its last value in which case the program goes to the statement after the NEXT statement. In *Program series*, the loop index is n and assumes the values 1 through 100. The block of statements inside the loop is indented for clarity.

True BASIC initializes all numeric variables to zero each time a subroutine is entered and at the beginning of the main program. If you want to save the value of the variable after leaving a subroutine, then the variable must be passed back to the calling program. All computer languages do not work this way. For example, in standard Pascal an error message will occur if a variable is used before it is initialized. In standard Fortran, an uninitialized variable might be assigned a meaningless value determined by whatever happens to be in the storage location assigned to the variable.

There are more general situations where the number of repetitions is not predetermined. Suppose that we wish to compute the series until the next term changes the sum by less than 0.01%. One way to do so is illustrated in *Program series_test*:

```
PROGRAM series_test
! illustrate use of DO LOOP strucrure
LET sum = 0                          ! initialize variables
LET n = 0
LET relative_change = 1              ! choose large value
DO while relative_change > 0.0001
   LET n = n + 1
   LET newterm = 1/(n*n)
   LET sum = sum + newterm
   LET relative_change = newterm/sum
   PRINT n,relative_change,sum
LOOP
END
```

Note the use of DO while LOOP structure to repeat the sum until the specified condition is no longer satisfied. An example of the DO until loop structure is illustrated in Problem 2.6.

So far, our programs have performed a simple task, e.g., the addition of a series of numbers. In general, we require programs that perform a series of tasks. In this case it is convenient to divide the program into separate tasks, each of which can be written and tested separately. The program then consists of a main program and a series of units that implement each task. As an example, the following program adds and multiplies two numbers that are entered from the keyboard.

```
PROGRAM tasks            ! illustrate use of subroutines
! note how variables are passed
CALL initial(x,y)                   ! initialize variables
```

```
CALL add(x,y,sum)                    ! add two variables
CALL multiply(x,y,product)
PRINT "sum ="; sum, "product ="; product
END                                  ! end of main program

SUB initial(x,y)
    INPUT x
    INPUT y
END SUB

SUB add(x,y,sum)
    LET sum = x + y
END SUB

SUB multiply(x,y,product)
    LET product = x*y
END SUB
```

Program tasks is an example of a *modular* program, i.e., a program that is divided into separate tasks each of which can be written and tested separately. A complete program contains at least a *main* program consisting of a series of invocations or *calls* to subprograms. These subprograms or tasks include two types, *subroutines* and *functions*. Subroutines are invoked by the CALL statement as shown in *Program tasks*. Examples of functions are given in Exercise 2.3 and Section 2.8.

Whenever possible, we will use the features of True BASIC that emphasize the similarities between C, Fortran, and Pascal. For this reason we will use only *external* subroutines and functions in True BASIC. External program units are defined in any order *after* the END statement of the main program. A subroutine is defined by a SUB statement and the end of a subroutine is denoted by the END SUB statement.

A subroutine is a separate program unit with its own *local* variables, that is, the variables in the main program and in each external subroutine and function are available only to the program subunit. A variable name represents a memory location in the computer. If the same variable name is used in two program units, the name represents two different memory locations. How then do the various program units communicate with one another? In True BASIC the most common method for subroutines to pass information to the main program and to other subroutines is via arguments in the subroutine calls. In *Program tasks* the variables x, y, sum, and product are passed in this way. In *Program no_pass* the variable name x is used in the main program and in SUB add_one, but is not passed. Convince yourself that the result printed for x in the main program is 10. What is the value of x in SUB add_one?

```
PROGRAM no_pass
LET x = 10              ! local variable name x defined in main program
CALL add_one
PRINT x
END
```

```
SUB add_one
    ! variable name x not passed
    LET x = x + 1      ! local variable name defined in subroutine
END SUB
```

It is actually a little misleading to say that variable names are passed to a subroutine. More precisely, the variable names are not passed, but rather the memory location in the computer is passed. The variable name is simply a label that identifies the memory location. What are the values of x and y if SUB add_one in *Program pass* is written in the following form?

```
PROGRAM pass
LET x = 10
LET y = 3
CALL add_one(x)
CALL add_one(y)
PRINT x,y
END

SUB add_one(s)                  ! example of the use of dummy arguments
    LET s = s + 1
END SUB
```

Note the variable name in the subroutine declaration need not match the name used in calling the subroutine. We call the declaration variable name a "dummy variable," because it may not actually be used in the program execution. It is only necessary that the number of variables in the declaration equal the number in the calling of a subroutine. Fortran treats variable passing in much the same way as True BASIC. Another way of passing information in True BASIC is discussed in Appendix 3C. A discussion of how Fortran and C pass information is discussed in Appendices A and B, respectively.

Why do we care about subroutines? One of the important practical reasons is that they simplify programming. If we think of a program as a sequence of tasks, then many tasks appear again and again in different programs. Hence we can use previously written subroutines developed either by ourselves or by others. Because the subroutines are separate programs, we do not have to read through the listings and see if the same variable name has been used in different parts of our program. Another important reason is that the use of subroutines encourages the development of modular programs whose parts are as decoupled as possible. In this way, a change in one part does not affect other parts. The separation of a program into parts lets us concentrate on one part at a time, and makes it easier for us to follow the logic of the program and to ensure that each part works as it should.

Exercise 2.1   Function evaluation

a.   Write a program to print the values of $f(x) = a e^{-bx}$, where a and b are parameters. The program should use a subroutine that inputs the parameters a and b, the minimum and maximum values of x, and the interval dx between the output of x and $f(x)$. A second subroutine should loop through the values of x, compute $f(x)$, and print x and $f(x)$ on the screen.

b. Modify your program to calculate and print on the same line the following functions: $f_1(x) = x$, $f_2(x) = x^2$, $f_3(x) = 2^x - 1$, and $f_4(x) = 1 + \ln x$ for $1 \le x \le 20$.

Exercise 2.2   Swap

Write a subroutine that takes two inputs x and y and swaps their values. For example, if $x = 5$ and $y = 10$ before the call to the subroutine, then $x = 10$ and $y = 5$ after the call to the subroutine. Write a program to test your subroutine.

2.5   A COMPUTER PROGRAM FOR THE EULER METHOD

Now that we know how to find a numerical solution of a first-order differential equation using the Euler method and are familiar with some of the elements of programming, we are ready to restate the Euler method as a computer *algorithm*, that is, as a finite sequence of precise steps or rules that solve a problem. We will then develop a computer program to implement the algorithm. The Euler method for the numerical solution of the first-order differential equation (2.2) is specified by the following algorithm:

1. Choose the initial conditions, the step size Δx, and the maximum value of x.
2. Determine y and the slope at the beginning of the interval.
3. Calculate y at the end of the interval by adding the change, the slope times the step size, to the value of y at the beginning of the interval and print the result.
4. Repeat steps 2 and 3 until the desired value of x is reached.

The first step is to write the above algorithm as a series of separate tasks. So the outline of our main program might look something like the following:

```
PROGRAM example
CALL initial              ! define initial conditions and parameters
DO while x <= xmax        ! repeat while x less than desired value
   CALL Euler             ! do one iteration
LOOP
END
```

We use True BASIC to implement each task and consider the solution of the same differential equation, $dy/dx = 2x$, as discussed in Section 2.3. The initial conditions and the numerical parameters are specified in SUB initial.

```
SUB initial(y,x,delta_x,xmax)
   LET x = 1                 ! initial value of x
   LET y = 1                 ! initial value of y
   LET xmax = 2              ! maximum value of x
```

```
        LET delta_x = 0.1            ! step size
        PRINT " x "," y "            ! print heading
        PRINT                        ! skip line
        PRINT x,y                    ! print initial values
    END SUB
```

Note that the variables y, x, delta_x, and xmax are passed. Our next task implements
the Euler method for one iteration:

```
    SUB Euler(y,x,delta_x)
        ! variables slope and change introduced for sake of clarity
        LET slope = 2*x              ! slope at beginning of interval
        LET change = slope*delta_x   ! estimate of change during interval
        LET y = y + change           ! new value of y (at end of interval)
        LET x = x + delta_x          ! value of x at end of interval
    END SUB
```

The complete program is listed in the following:

```
    PROGRAM example                  ! beginning of main program
    ! specify initial values and parameters
    CALL initial(y,x,delta_x,xmax)
    DO while x <= xmax
        CALL Euler(y,x,delta_x)      ! use simple Euler algorithm
        PRINT x,y
    LOOP
    END                              ! end of main program

    SUB initial(y,x,delta_x,xmax)
        LET x = 1                    ! initial value of x
        LET y = 1                    ! initial value of y
        LET xmax = 2                 ! maximum value of x
        LET delta_x = 0.1            ! step size
        PRINT " x "," y "            ! print heading
        PRINT                        ! skip line
        PRINT x,y                    ! print initial values
    END SUB

    SUB Euler(y,x,delta_x)
        ! variables slope and change introduced for sake of clarity only
        LET slope = 2*x              ! slope at beginning of interval
        LET change = slope*delta_x   ! estimate of change during interval
        LET y = y + change           ! new value of y
        LET x = x + delta_x          ! value of x at end of interval
    END SUB
```

The subroutines SUB initial and SUB Euler are listed after the END statement of the main program. They are separate program units and do not share variables with the main program unless the variables are passed to and from the main program. Hence the variables slope and change in SUB Euler are local variables. If you are not familiar with a programming language, run the program and see how it works. The procedure we have followed in developing the complete program is sometimes called "top-down programming," because we began by thinking about the overall structure of the program and then completed the detailed tasks using subroutines and functions.

2.6 THE COFFEE COOLING PROBLEM

We now return to our cup of coffee and develop a computer program to obtain a numerical solution of Newton's law of cooling. To determine if our model of cooling is reasonable, we give some experimental data in Table 2.2 for the cooling of a cup of coffee.

Program cool, which implements the Euler algorithm for the numerical solution of Newton's law of cooling, is similar in structure to *Program example*, but the printing of the results has been assigned to SUB output. This modular approach will be helpful when we wish to plot rather than to print our results. Note that the name of a *numeric* variable, a variable that represents a number, must start with a letter, but may contain digits and the underscore character. The latter allows us to form hyphenated names.

time (min)	T°C black	T°C cream	time (min)	T°C black	T°C cream
0	82.3	68.8	24	51.2	45.9
2	78.5	64.8	26	49.9	44.8
4	74.3	62.1	28	48.6	43.7
6	70.7	59.9	30	47.2	42.6
8	67.6	57.7	32	46.1	41.7
10	65.0	55.9	34	45.0	40.8
12	62.5	53.9	36	43.9	39.9
14	60.1	52.3	38	43.0	39.3
16	58.1	50.8	40	41.9	38.6
18	56.1	49.5	42	41.0	37.7
20	54.3	48.1	44	40.1	37.0
22	52.8	46.8	46	39.5	36.4

Table 2.2   The temperature of coffee in a glass. The temperature was recorded with an estimated accuracy of 0.1°C. The air temperature was 17°C. The second column corresponds to black coffee and the third column corresponds to coffee with heavy cream.

```
PROGRAM cool
! numerical solution of Newton's law of cooling
CALL initial(t,T_coffee,T_room,r,delta_t,tmax,nshow)
LET counter = 0                      ! initialize counter for clarity
DO while t <= tmax
   CALL Euler(t,T_coffee,T_room,r,delta_t)
   LET counter = counter + 1      ! number of iterations
   IF mod(counter,nshow) = 0 then
      CALL output(t,T_coffee,T_room)
   END IF
LOOP
END

SUB initial(t,T_init,T_room,r,delta_t,tmax,nshow)
   ! time is keyword in True BASIC so cannot be used as a variable
   LET t = 0                        ! time
   LET T_init = 82.3                ! initial coffee temperature (C)
   LET T_room = 17                  ! room temperature (C)
   INPUT prompt "cooling constant r = ": r
   INPUT prompt "time step dt = ": delta_t
   INPUT prompt "duration of run = ": tmax        ! minutes
   INPUT prompt "time between output of data = ": tshow
   LET nshow = round(tshow/delta_t)
   CALL output(t,T_init,T_room)
END SUB

SUB Euler(t,T_coffee,T_room,r,delta_t)
   LET change = -r*(T_coffee - T_room)*delta_t
   LET T_coffee = T_coffee + change
   LET t = t + delta_t
END SUB

SUB output(t,T_coffee,T_room)
   IF t = 0 then
      PRINT
      PRINT "time (min)","T_coffee","T_coffee - T_room"
      PRINT
   END IF
   PRINT t,T_coffee,T_coffee - T_room        ! show results on screen
END SUB
```

Note that we have used the mod function to test whether one number is a multiple of another and have used the IF ··· then ··· END IF control structure to print the results at the desired time intervals. To ensure that you understand how the mod function works, write a little program to compute mod(n,10) for various values of n. It would be a good idea to also test the round function used in SUB initial. The reason for using the variable counter rather than the time t as the control variable is discussed in Appendix 2A.

Problem 2.1   Coffee cooling program

a.   After you have entered `Program cool` and have corrected any syntax errors, how do you know whether or not the program correctly implements the desired algorithm? For example, you might have typed a minus sign rather than a plus sign. If possible, you should compare the results of your program to limiting cases for which an analytical solution or hand calculation is available. Use a calculator to obtain a numerical solution of Newton's law of cooling for the Euler method. It is easier to do the hand calculation if you choose $r = 1$ and $\Delta t = 0.1$. Compare your hand calculation to the output of `Program cool` and verify your program in this case.

b.   Because time is measured in minutes, the unit of the cooling constant r is min^{-1}. You might have noticed that the value $r = 0.1\,\text{min}^{-1}$ yields a cooling curve $T(t)$ that does not correspond to the data given in Table 2.2. Find an approximate value of r that describes the experimental results shown in Table 2.2. What is your implicit criterion for determining the best value of r? Can you fit the empirical data for all t?

c.   Use the value of r found in part (b) and make a graph showing the dependence of temperature on time. Plot the data given in Table 2.2 on the same graph and compare your results. Although we will learn in Section 2.8 how to write a computer program to plot data, it is a good idea to make preliminary plots by hand to obtain a feel for the nature of your data.

d.   Does the time step Δt have any physical significance? Make sure that your value of Δt is sufficiently small so that it does not affect your results.

e.   Is the value of r the same for black coffee as it is for coffee with cream? Based on what you know about coffee and cream, do you expect r to be greater or less for black coffee?

f.   Is Newton's law of cooling applicable to the cooling of a cup of coffee?

g.   The initial difference in temperature between the black coffee and its surroundings is approximately 65°C. How long does it take for the coffee to cool so that the difference is $65/2 = 32.5°C$? How long does it take the difference to become 65/4 and 65/8? Try to understand your results in simple terms without first using a computer.

Problem 2.2   Is it faster to add cream first or later?

Suppose that the initial temperature of the coffee is 90°C, but that the coffee can be sipped comfortably only when its temperature is below 75°C. Assume that the addition of cream cools the coffee by 5°C. If you are in a hurry and want to wait the shortest possible time, should the cream be added first and the coffee be allowed to cool, or should you wait until the coffee has cooled to 80°C before adding the cream? Use your program to "simulate" the two cases. Choose a value of r so that Newton's Law approximates the behavior of a real cup of coffee. Does your general conclusion depend on using a different value of r when the cream is added? This type of "what if" simulation of a dynamical system is an important technique in policy analysis in the social sciences (see for example Roberts et al.).

Exercise 2.3   Evaluating lists

a. To gain more experience in programming, write a subroutine that has two parameters x and xmax. The subroutine changes the value of xmax to the larger of x and xmax. Write a program to test your subroutine so that the user can input a series of numbers from the keyboard and the program prints the maximum value.

b. Write your subroutine so that it also computes the minimum value. Modify your program so that the minimum and maximum values of a series of numbers are shown.

c. Modify your program so that the user can input a number y and then search a list of additional numbers for the number y. The output of the program should indicate whether the number y was found.

d. Modify your program so that the subroutines that you wrote in parts (a) and (b) are replaced by functions. Because a function returns a value, you might implement the task of finding the maximum as LET xmax = findmax(x,xmax), where findmax is the name of a function. This function can be declared as follows:

```
FUNCTION findmax(x,xmax)
    IF x > xmax then
       LET findmax = x
    ELSE
       LET findmax = xmax
    END IF
END DEF
```

2.7   ACCURACY AND STABILITY

Now that we have learned how to use the Euler method to find a numerical solution to a first-order differential equation, we need to develop some practical guidelines to help us estimate the accuracy of the method. Because we have replaced a differential equation by a difference equation, our numerical solution cannot be identically equal, in general, to the "true" solution of the original differential equation. In general, the discrepancy between the two solutions has two causes. One cause is that computers do not store numbers with infinite precision, but rather to a maximum number of digits that is hardware and software dependent. Most programming languages allow the programmer to distinguish between *floating point* or *real* numbers, i.e., numbers with decimal points, and *integer* or *fixed point* numbers. Arithmetic with numbers represented by integers is exact. However in general, we cannot solve a differential equation using integer arithmetic. Hence, arithmetic operations such as addition and division, which involve real numbers, can introduce an error, called the *roundoff error*. For example, if a computer only stored real numbers to two significant figures, the product 2.1×3.2 would be stored as 6.7 rather than the exact value 6.72. The significance of roundoff errors is that they accumulate as the number of mathematical operations increases. Ideally we should choose algorithms that do not significantly magnify the roundoff error, e.g., we should avoid subtracting numbers that are nearly the same in magnitude.

The other source of the discrepancy between the true answer and the computed answer is the error associated with the choice of algorithm. This error is called the *truncation error*. A truncation error would exist even on an idealized computer that stored floating point numbers with infinite precision and hence had no roundoff error. Because this error depends on the choice of algorithm and hence can be controlled by the programmer, you should be motivated to learn more about numerical analysis and the estimation of truncation errors. However, there is no general prescription for the best method for obtaining numerical solutions of differential equations. We will find in later chapters that each method has advantages and disadvantages and the proper selection depends on the nature of the solution, which you might not know in advance, and on your objectives. How accurate must the answer be? Over how large an interval do you need the solution? What kind of computer are you using? How much computer time and personal time do you have?

In practice, we usually can determine the accuracy of our numerical solution by reducing the value of Δt until the numerical solution is unchanged at the desired level of accuracy. Of course, we have to be careful not to choose Δt too small, because too many steps would be required and the total computation time and roundoff error would increase.

In addition to accuracy, another important consideration is the stability of an algorithm. For example, it might happen that the numerical results are very good for short times, but diverge from the true solution for longer times. This divergence might occur if small errors in the algorithm are multiplied many times, causing the error to grow geometrically. Such an algorithm is said to be *unstable* for the particular problem. We consider the accuracy and the stability of the Euler method in Problem 2.3.

Problem 2.3   Accuracy and stability of the Euler method

a. Show that the analytical solution of (2.1) can be written in the form

$$T(t) = T_s - (T_s - T_0)e^{-rt}. \tag{2.7}$$

Note that $T(t = 0) = T_s - (T_s - T_0) = T_0$ and that $T(t \to \infty) = T_s$.

b. Use *Program cool* to compute the temperature at $t = 1$ min with $\Delta t = 0.1$, 0.05, 0.025, 0.01, and 0.005. Choose a value of r corresponding to a real cup of coffee. Make a table showing the difference between the exact solution (2.7) and your numerical solution as a function of Δt. Is the difference a decreasing function of Δt? If Δt is decreased by a factor of two, how does the difference change? Plot the difference as a function of Δt. If your points fall approximately on a straight line, then the difference is proportional to Δt (for $\Delta t \ll 1$). A numerical method is called nth order, if the difference between the analytical solution and the numerical solution is proportional to $(\Delta t)^n$ at a fixed value of t. What is the order of the Euler method?

c. One way to determine the accuracy of a numerical solution is to repeat the calculation with a smaller step size and compare the results. If the two calculations agree to p decimal places, we can reasonably assume that the results are correct to p decimal places. What value of Δt is necessary for 0.1% accuracy at $t = 1$? What value of Δt is necessary for 0.1% accuracy at $t = 4$?

d. Consider the differential equation

$$R\frac{dQ}{dt} = V - \frac{Q}{C}, \tag{2.8}$$

with $Q = 0$ at $t = 0$. This equation represents the charging of a capacitor in an RC circuit with an applied voltage V. Measure t in seconds and choose $R = 2000\,\Omega$, $C = 10^{-6}$ farads, and $V = 10$ volts. Do you expect $Q(t)$ to increase with t? Does $Q(t)$ increase indefinitely or does it reach a steady-state value? Write a program to solve (2.8) numerically using the Euler method. What value of Δt is necessary to obtain three decimal accuracy at $t = 0.005$ s?

e. What is the nature of your numerical solution to (2.8) at $t = 0.005$ s for $\Delta t = 0.005$, 0.0025, and 0.001? Does a small change in Δt lead to a large change in the computed value of Q? Is the Euler method stable in this calculation for any value of Δt?

2.8 SIMPLE PLOTS

Most computers can generate visual displays quickly and easily. The most common use of a visual display is to plot a set of data. We also might wish to find the best straight line through the data points. Another common use is animation. These two uses are related to the rapidly growing fields of computer graphics and visualization. How can we represent large amounts of data so that the representation is most meaningful? What is the appropriate use of color and shading? How can we use sound? Of course, we can only begin to discuss these questions in this introductory chapter.

We first distinguish between two levels of plotting. In the "quick and dirty" mode, we want to plot the results produced by our program to help check the program and to gain a preliminary understanding of how the system behaves. In the "presentation" mode, we want to do a careful graphical analysis of the results and make plots that we can present to others. In the latter mode we want to optimize the range of each axis, label the axes, have tick marks at convenient spacings, etc. For this second mode we recommend that you port your data to a separate program that is optimized for making graphs and fits to data. There are many public domain, shareware, and commercial programs available with these capabilities. These programs allow you to manipulate the data in many ways, do simple statistics, and to set up the plot in many different formats. Of course, you also can write a separate True BASIC program to draw presentation quality graphs. It is a good idea to keep your programs as simple as possible and not to try to write one program that does everything.

We can use the graphics statements in True BASIC to make simple plots. Although a standard graphics language does not yet exist, the core graphics statements in True BASIC are similar to the statements in more sophisticated graphics packages. Hence an understanding of the True BASIC graphics statements and associated concepts will allow you to learn another graphics language easily.

A graphics screen is covered by a grid of pixels. The number of pixels is hardware-dependent. In True BASIC the number of pixels is irrelevant. That is, the mapping of the absolute values of the coordinates to the device coordinates or pixels is done by True BASIC. The first step in using a graphics display is to specify the range of coordinates that are to be plotted. The statement

```
SET WINDOW xmin, xmax, ymin, ymax
```

determines the minimum and maximum x (horizontal) and y (vertical) coordinates. The statement

```
PLOT POINTS: x,y;
```

draws a point at (x, y) in the current window coordinates. The statement

```
PLOT LINES: x1,y1; x2,y2;
```

draws a line between $(x1, y1)$ and $(x2, y2)$. The following program uses these statements to draw a set of axes and plots the function in (2.7).

```
PROGRAM plot_function                ! plot analytical solution
CALL initial(t,tmax,r,Ts,T0)
CALL set_up_window(t,tmax,Ts,T0)
CALL show_plot(t,tmax,r,Ts,T0)
END

SUB initial(t,tmax,r,Ts,T0)
    LET t = 0
    LET T0 = 83                      ! initial coffee temperature (C)
    LET Ts = 22                      ! room temperature (C)
    INPUT prompt "cooling constant r = ": r      ! cooling constant
    INPUT prompt "duration = ": tmax   ! time (minutes)
    CLEAR
END SUB

SUB set_up_window(xmin,xmax,ymin,ymax)
    LET mx = 0.01*(xmax - xmin)    ! margin
    LET my = 0.01*(ymax - ymin)
    SET WINDOW xmin - mx,xmax + mx,ymin - my,ymax + my
    ! default background color black on all computers except Macintosh
    PLOT xmin,ymin; xmax,ymin       ! abbreviation for PLOT LINES:
    PLOT xmin,ymin; xmin,ymax
END SUB
```

```
SUB show_plot(t,tmax,r,Ts,T0)
    DECLARE DEF f                        ! declare use of external function
    SET COLOR "red"
    DO while t <= tmax
        PLOT t,f(t,r,Ts,t0);             ! abbreviation for PLOT LINES
        LET t = t + 0.1
    LOOP
END SUB

DEF f(t,r,Ts,T0)
    LET f = Ts - (Ts - T0)*exp(-r*t)
END DEF
```

Note the use of a separate subroutine to set up the screen and another subroutine to plot the function. The function f is defined in a separate subprogram called a *function*. The main distinction between functions and subroutines is that functions return a single value. In True BASIC, unlike other languages, functions do not change the value of arguments passed to them. In *Program plot_function* the function is so simple that there really is no need to define it separately, and our reason for doing so is to introduce the idea of a function in a simple context. Modify *Program plot_function* so that the sine function is plotted. Why is it convenient to have the function defined separately?

Problem 2.4   Cooling time

a. Incorporate the graphics statements in *Program plot_function* into *Program cool* so that the numerical solution for the coffee temperature is plotted rather than printed. An example of a subroutine for simple plots is shown below in SUB plot_output. This subroutine can be used directly in *Program cool* by replacing the calls to SUB output by calls to SUB plot_output. What are the appropriate maximum and minimum values for the temperature? What is a reasonable range of times? The straightforward replacement of one output subroutine by another illustrates one reason for creating modular programs. In later chapters we often will write the output statements as two subroutines, where one subroutine defines the output format and the other outputs the results.

b. Find the time it takes for the coffee to reach $1/e \approx 0.37$ of the difference between the initial coffee temperature and the room temperature. This time is called the *relaxation time*. Does the relaxation time depend on the initial temperature or the room temperature? Try different values of r and determine the qualitative dependence of the relaxation time on r.

```
SUB plot_output(t,T_coffee,T_room)
    IF t = 0 then
        SET WINDOW -2,22,-5,105
        PLOT LINES: 0,100;0,0;20,0       ! axes
```

```
                  PLOT LINES: 0,T_room;20,T_room  ! equilibrium
            END IF
            PLOT t,T_coffee              ! plot results on screen
      END SUB
```

2.9   VISUALIZATION

Another example of the use of graphics illustrates how we might visualize a physical system changing with time. Suppose that a rigid wheel rolls without slipping on a surface. Is energy lost due to friction? We know that the force of friction always is parallel to the surface. There will be work done by friction only if there is a component of the displacement of a point on the wheel parallel to this force when the point touches the surface. The equation of motion for a point on the wheel is given by

$$x(t) = -r \cos \omega t + v_{cm} t \tag{2.9a}$$
$$y(t) = r \sin \omega t. \tag{2.9b}$$

Program wheel draws the point x, y as a function of t.

```
PROGRAM wheel
CALL initial(r,omega,vcm,t,dt,xmax)
DO while x <= xmax
    LET x = -r*cos(omega*t) + vcm*t
    LET y = r*sin(omega*t)
    ! note the semicolon after PLOT x,y
    PLOT x,y;                        ! abbreviation of PLOT POINT:
    LET t = t + dt
LOOP
END

SUB initial(r,omega,vcm,t,dt,xmax)
    LET t = 0
    LET r = 1                        ! radius of wheel
    LET omega = 1                    ! angular frequency of wheel
    LET vcm = r*omega                ! velocity of center of mass
    LET dt = 0.01                    ! time increment
    LET xmin = -1
    LET xmax = 50
    LET mx = 0.01*(xmax - xmin)
    SET WINDOW xmin-mx,xmax+mx,-4,4
    ! draw ground at y = -r (bottom of wheel)
    PLOT LINES: xmin,-r;xmax,-r
END SUB
```

Run *Program wheel* and explain why friction does no work. Now you know why we use wheels.

A more abstract example of the use of graphics is given by the dynamical system specified by the two coupled difference equations:

$$x_{n+1} = 1 - y_n + |x_n| \tag{2.10a}$$
$$y_{n+1} = x_n. \tag{2.10b}$$

For initial values x_0 and y_0, we can find the values x_1, y_1, x_2, y_2, etc. from (2.10). The set of equations (2.10) represents a dynamical system since x and y are changing with n, a discrete time. Use a calculator to find the first several values of x_n, y_n for $n = 1, 2, \cdots$ starting from $x_0 = 0$, $y_0 = 0$.

How can we obtain a qualitative idea of how the values of x_n, y_n are changing with n? One way is to make a table of their values. Another way is to draw the points x_n, y_n and see if they show a pattern. The following program draws the points:

```
PROGRAM simple_map
CALL initial(x,y,a,b)
FOR i = 1 to 100000
    LET xnew = 1 - y + abs(x)
    LET y = x
    LET x = xnew
    PLOT x,y
NEXT i
END

SUB initial(x0,y0,a,b)
    INPUT prompt "initial value of x = ": x0
    INPUT prompt "initial value of y = ": y0
    CLEAR
    LET xmin = -3
    LET xmax = 8
    LET ymin = -3
    LET ymax = 8
    LET mx = 0.01*(xmax - xmin)    ! margin
    LET my = 0.01*(ymax - ymin)
    SET WINDOW xmin - mx,xmax + mx,ymin - my,ymax + my
    BOX LINES xmin,xmax,ymin,ymax       ! draw border
    ASK MAX COLOR mc
    IF mc > 2 then SET COLOR "blue"
END SUB
```

Note the use of the FOR NEXT loop in the main program. The variable i starts with the value 1 and increases in steps of 1 until 100000 is reached. The two new graphics statements, BOX LINES and ASK MAX COLOR, are easy to understand from the context.

Of course, we do not expect you to understand the significance of dynamical maps and (2.10) at this stage. Our main reason for introducing these equations here is to introduce some additional programming statements and to motivate you to look at other chapters to see what topics might interest you. The following problem explores some of the properties of (2.10).

Problem 2.5   A two-dimensional map

Use *Program* `simple_map` and iterate (2.10) starting from the initial condition $x_0 = -0.1$, $y_0 = 0$. Note that the program computes the new value of y using the old value of x and not the new value of x. Choose the SET WINDOW statement

such that all values of x_n and y_n within the box drawn by the statement BOX
LINES -3,8,-3,8 are plotted.

Problem 2.6 Lissajous figures

A computer screen can be used to simulate the output seen on an oscilloscope.
Imagine that the vertical and horizontal inputs to an oscilloscope are sinusoidal
in time, i.e., $x = A \sin(\omega_x t + \phi_x)$ and $y = B \sin(\omega_y t + \phi_y)$. If the curve that is
drawn repeats itself, such a curve is called a *Lissajous figure*. For what values of
the angular frequencies ω_x and ω_y do you obtain a Lissajous figure? How do the
phase factors ϕ_x and ϕ_y and the amplitudes A and B affect the curves? Choose
$A = B = 1$, $\omega_x = 2$, $\omega_y = 3$, $\phi_x = \pi/6$, and $\phi_y = \pi/4$ for starters. Write a program
to plot y versus x, as t advances from $t = 0$. One way to stop the program is to let
the program run until the user hits any key. The following code illustrates how to
use the key input statement to repeat a loop a desired number of times.

```
PROGRAM do_loop
LET x = 0
DO
    LET x = x + 1
    PRINT x
LOOP until key input
PRINT "loop finished"
END
```

A summary of the core True BASIC graphics statements is given in Table 2.3. Other
commands will be introduced in later chapters as they are needed. It would be a good

```
SET BACKGROUND COLOR "black"
SET WINDOW xmin,xmax,ymin,ymax      ! default window coordinates are 0,1,0,1
PLOT POINTS: x,y                    ! abbreviation is PLOT x,y
PLOT LINES: x1,y1; x2,y2;
PLOT                                ! start a new curve
PLOT AREA: 1,1;2,1;2,2              ! draw filled triangle
BOX LINES xmin,xmax,ymin,ymax       ! draw rectangle
BOX AREA xmin,xmax,ymin,ymax        ! draw filled rectangle
BOX CLEAR xmin,xmax,ymin,ymax       ! erase rectangle
BOX CIRCLE xmin,xmax,ymin,ymax      ! draw inscribed circle (or ellipse)
ASK MAX COLOR mc                    ! mc is number of foreground colors
! other names include green, blue, brown, magenta, and white
SET COLOR "red"
CLEAR                               ! erase contents of current window
! fill enclosed graphical region with current foreground color
FLOOD x,y
```

Table 2.3 Summary of core True BASIC graphics statements.

idea to make your own summary of the True BASIC syntax you have learned in this chapter.

*2.10 NUCLEAR DECAY

The power of mathematics when applied to physics comes from the fact that frequently unrelated problems have the same mathematical formulation. Hence, if we can solve one problem, we can solve other problems that might appear to be unrelated. For example, the cooling of a cup of hot water is equivalent to the discharge of a capacitor in an RC circuit. In the following, we will find that nuclear decay can be formulated in a similar way.

Consider a large number of radioactive nuclei. Although the number of nuclei is discrete, we can treat this number as a continuous variable because the number of nuclei is large. (We will treat this number as a discrete variable in Chapter 7.) The fundamental law of radioactive decay is that the rate of decay is proportional to the number of nuclei. Thus we can write

$$\frac{dN}{dt} = -\lambda N, \tag{2.11}$$

where N is the number of nuclei and λ is the decay constant. Note that (2.11) is similar in form to (2.1) with the constant T_s in (2.1) replaced by zero. Hence, we can use the coffee cooling program to solve simple nuclear decay problems. Of course, we do not need to use a computer to solve either problem because both problems are easy to solve analytically. A more interesting case occurs when the decay product of a radioactive nucleus is itself radioactive. In this case we have

$$\frac{dN_1}{dt} = -\lambda_1 N_1 \tag{2.12a}$$

and

$$\frac{dN_2}{dt} = \lambda_1 N_1 - \lambda_2 N_2. \tag{2.12b}$$

The interpretation of (2.12b) is that the second species (the daughter) is formed at the rate its parent decays, $\lambda_1 N_1$, and decays at the rate $\lambda_2 N_2$. The set of equations (2.12) also can be solved analytically and the resulting expressions are not too complicated. But if there are further decays, then it is more transparent to solve the resulting equations numerically.

Problem 2.7   Single nuclear species decay

a. Modify *Program cool* so that nuclear decay notation is used in your program. For $\lambda = 1$, compute the difference between the analytical result and the result of the Euler method for $N(t)/N(0)$ at $t = 1$ and $t = 2$. For each value of t show that the difference is proportional to Δt.

b.   The rate of radioactive decay is more conveniently given in units comparable to the half-life $T_{1/2}$ rather than λ. The half-life is the time it takes for one-half of the original nuclei to decay and is given by $T_{1/2} = \ln 2/\lambda$. Use your modified program to verify this relation. Determine the decay constant for $^{238}\text{U} \rightarrow {}^{234}\text{Th}$ if the half-life is 4.5×10^9 years.

c.   Because it is awkward to treat very large or very small numbers on the computer, it is convenient to measure time in terms of the half-life. Modify your program so that time is measured in this way. Use your program to determine the time (in years) for $100\,\mu\text{g}$ of ^{238}U to decay to 20% of its original amount.

* **d.**   As we saw in part (a), the Euler method is only first-order in accuracy. We can increase the accuracy to second-order by the following reasoning. Expand $N(t)$ in a Taylor series:

$$N(t + \Delta t) = N(t) + \frac{dN(t)}{dt}\Delta t + \frac{1}{2}\frac{d^2 N(t)}{dt^2}(\Delta t)^2 + \cdots, \tag{2.13}$$

and use (2.11) to write $d^2 N(t)/dt^2 = -\lambda\, dN(t)/dt$. We also use (2.11) to write $dN(t)/dt$ in terms of $N(t)$ and express (2.13) as

$$N(t + \Delta t) = N(t) - \lambda N(t)\Delta t + \frac{\lambda^2}{2}N(t)(\Delta t)^2 + \cdots \tag{2.14}$$

If we write the Euler estimate as $N_e = N(t) - \lambda N(t)\Delta t$, we can rewrite (2.14) as

$$N(t + \Delta t) = N(t) - \frac{\lambda}{2}(N(t) + N_e)\Delta t. \tag{2.15}$$

From the form of (2.15), we see that a second-order algorithm can be obtained by using the average of the Euler estimate for the value of N at $t + \Delta t$ and the value of N at time t in an Euler-like expression. Modify your program so that the second-order algorithm (2.15) is used and repeat parts (a) and (c).

Problem 2.8   Nuclear decays

a. ^{76}Kr decays to ^{76}Br via electron capture with a half-life of 14.8 h; ^{76}Br decays to ^{76}Se via electron capture and positron emission with a half-life of 16.1 h. Suppose that the sample initially contains 1 gm of pure ^{76}Kr. In this case there are two half-lives, and it is convenient to measure time in units comparable to the smallest half-life. Modify your program to compute the time dependence of the amount of ^{76}Kr and ^{76}Se over an interval of one week.

b. ^{28}Mn decays via beta emission to ^{28}Al with a half-life of 21 h; ^{28}Al decays by positron emission to ^{28}Si with a half-life of 2.31 min. If we were to use minutes as the unit of time, our program would have to do many iterations before we would see a significant decay of the ^{28}Mn. What simplifying assumption can you make to speed up the computation?

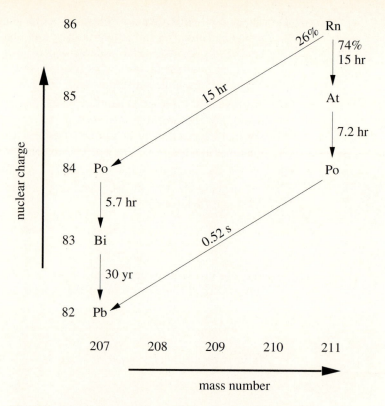

Fig. 2.2   The decay scheme of ^{211}Rn. Note that ^{211}Rn decays via two
branches, and the final product is the stable isotope ^{207}Pb.
All vertical transitions are by electron capture (a form of beta
decay), and all diagonal transitions are by alpha decay. The
times represent half-lives.

c. ^{211}Rn decays via two branches as shown in Fig. 2.2. Make any necessary
approximations and compute the amount of each isotope as a function of time,
assuming that the sample initially consists of 1 μg of ^{211}Rn.

2.11   OVERVIEW

Although this chapter is introductory in nature, you might feel that you have been
introduced to many new ideas and techniques. If you are not familiar with computer
programming, the first use of a computer and the introduction to a new language might
be a little bewildering. But take heart! There is not much more syntax to learn. Look at
the program listings in later chapters, and you will find that you already recognize most
of the syntax. The reason that we can use relatively simple program statements in most
computer simulations is that nature at its most fundamental level can be described in

terms of simple physical laws. You probably are familiar with Newton's laws of motion and how compactly they can be written if we use the language of differential vector calculus. You also probably realize that the consequences of these seemingly simple laws are far-reaching and in many cases counterintuitive. So you will not be surprised when we find in later chapters that simple algorithms can yield complex behavior.

In this chapter we have emphasized that seemingly unrelated physical problems can have the same mathematical formulation in the context of simple differential equations. In later chapters we will find that our thinking about physical systems has expanded so that we can think about models of physical systems not only in terms of differential equations, but also in terms of an algorithm, that is, as a series of instructions to a computer. You might not be surprised to learn that seemingly unrelated physical systems can have the same formulation in terms of a computer algorithm. But we are getting ahead of ourselves. So we will discuss in Chapters 3–5 how our understanding of Newton's laws of motion can be enhanced by doing some simple simulations.

Appendix 2A   INTEGER AND REAL VARIABLES

Unlike most programming languages, True BASIC does not allow the user to distinguish explicitly between integer variables and real variables. However, if we make an assignment such as LET i = 1, then True BASIC implicitly treats the variable i as a 16-bit integer variable whenever possible (a bit is either 0 or 1). For integers outside the range -32768 to 32767 floating point representation is used. The fact that real variables are not represented with infinite precision means that we have to be careful when comparing two real variables. Run the following program to see what happens.

```
PROGRAM compare
LET i = 0
LET compare = 1000
DO while i <= 10000
   LET i = i + 1
   IF i = compare then
      PRINT "first test satisfied"
   END IF
LOOP
PRINT "finished first loop"
LET t = 0
LET dt = 0.1
LET finished = 100
DO while t < 1000
   LET t = t + dt
   IF t = finished then
      PRINT "second test satisfied"
   END IF
LOOP
PRINT "t =";t
END
```

On our computer, the comparison of the two integer variables is successful, but the comparison of the two real variables is not. Your results will depend on the accuracy of the floating point representation on your computer. Note that operations between integer variables are done with integer arithmetic, but operations between two reals or a real and an integer are done with real arithmetic. Because arithmetic involving variables in floating point representation is not exact, control structures based on tests of equality between real numbers should be avoided.

In Fortran, variables that begin with the letters I,J,K,L,M,N (upper or lower case) are implicitly assumed to be integer variables. There is no such convention in other languages.

References and Suggestions for Further Reading

George B. Arfken, David F. Griffing, Donald C. Kelly, and Joseph Priest, *University Physics*, second edition, Harcourt Brace Jovanovich (1989). Chapter 20 on heat transfer discusses Newton's law of cooling.

Craig F. Bohren, "Comment on Newton's law of cooling," *Am. J. Phys.* **59**, 1044 (1991). See also Craig F. Bohren, *Clouds in a Glass of Beer*, John Wiley & Sons (1987); Colm T. O'Sullivan, "Newton's law of cooling–a critical assessment," *Am. J. Phys.* **58**, 958 (1990); W. G. Rees and C. Viney, "On cooling tea and coffee," *Am. J. Phys.* **56**, 434 (1988).

Nancy Roberts, David Andersen, Ralph Deal, Michael Jaret, and William Shaffer, *Introduction to Computer Simulation: The System Dynamics Approach,* Addison-Wesley (1983). A book on computer simulation in the social sciences.

Emilio Segré, *Nuclei and Particles*, second edition, W. A. Benjamin (1977). Chapter 5 discusses decay cascades. The decay schemes described briefly in Problem 2.8 are taken from C. M. Lederer, J. M. Hollander, and I. Perlman, *Table of Isotopes*, sixth edition, John Wiley & Sons (1967).

It is impossible to list all the excellent books on numerical analysis. We list a few of our favorites.

Forman S. Acton, *Numerical Methods That Work*, Harper & Row (1970); corrected edition, Mathematical Association of America (1990). A somewhat advanced, but clearly written text.

Paul DeVries, *A First Course in Computational Physics*, John Wiley (1994).

William H. Press, Saul A. Teukolsky, William T. Vetterling, and Brian P. Flannery, *Numerical Recipes*, second edition, Cambridge University Press (1992).

CHAPTER

3
The Motion of Falling Objects

We introduce finite difference methods for obtaining numerical solutions to Newton's equations of motion and discuss the qualitative and quantitative behavior of bodies falling near the earth's surface.

3.1   BACKGROUND

A common example of motion is an object falling near the earth's surface. The simplest description of the motion of the object ignores rotational and internal motion and describes an idealized object called a *particle*, an object without internal structure. Of course, objects such as planets, rocks, baseballs, and atoms have internal structure. Nonetheless, for many purposes their internal motion can be neglected and we can regard them as particles.

Our initial discussion emphasizes one-dimensional motion with only one spatial coordinate needed to determine the position of a particle. The instantaneous position $y(t)$, velocity $v(t)$, and acceleration $a(t)$ of a particle can be defined using the language of differential calculus:

$$v(t) = \frac{dy(t)}{dt} \tag{3.1}$$

and

$$a(t) = \frac{dv(t)}{dt}. \tag{3.2}$$

These quantities are known as *kinematical* quantities, because they describe the motion without regard to its cause.

Why do we need the concept of acceleration in kinematics? The answer can be found only *a posteriori*. Thanks to Newton, we know that the net force acting on a particle determines its acceleration. Newton's second law of motion tells us that

$$a(t) = \frac{F(y, v, t)}{m}, \tag{3.3}$$

where F is the net *force* and m is the *inertial mass*. In general, the force depends on position, velocity, and time. Note that Newton's law implies that the motion of a particle does not depend on d^2v/dt^2 or on any higher derivative of the velocity. It is a property of nature, not of mathematics, that we can find simple explanations for motion.

The description of motion of a particle requires the solution of two coupled first-order differential equations (3.1) and (3.3). The two first-order equations (3.1) and (3.3) can be combined to obtain a *second-order* differential equation for the position:

$$\frac{d^2y(t)}{dt^2} = \frac{F}{m}. \tag{3.4}$$

3.2   THE FORCE ON A FALLING OBJECT

In the absence of air resistance, all objects, regardless of size, mass, or composition, have the same acceleration at the same point near the earth's surface. This idealized motion is called "free fall." According to (3.3), constant acceleration implies that the force per unit mass (called the gravitational field), F/m, is a constant. This constant commonly is denoted by the symbol g. Near the earth's surface, the magnitude of g

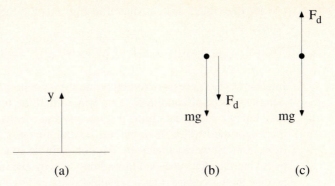

Fig. 3.1 (a) Coordinate system with y measured positive upwards from the ground. (b) The force diagram for upward motion. (c) The force diagram for downward motion.

is approximately $9.8\,\text{N/kg}$ or $9.8\,\text{m/s}^2$. Let us adopt the coordinate system shown in Fig. 3.1 with the positive direction upward. In this case $a = -g$ and the solution of (3.4) can be written as

$$v(t) = v_0 - gt \tag{3.5a}$$

and

$$y(t) = y_0 + v_0 t - \frac{1}{2} g t^2, \tag{3.5b}$$

where y_0 and v_0 are the initial position and velocity of the particle respectively. Note that two initial conditions are necessary to specify the motion.

For free fall near the earth's surface, the analytical solution (3.5) is so simple that further analysis is not necessary. However, it is not difficult to think of more realistic models of the motion near the earth's surface for which the equations of motion do not have simple analytical solutions. For example, if we take into account the variation of the earth's gravitational field with the distance from the center of the earth, then the force on a particle is not constant. According to Newton's law of gravitation, the force due to the earth on a particle of mass m is given by

$$F = \frac{GMm}{(R+y)^2} = \frac{GMm}{R^2(1+y/R)^2} = mg\left(1 - 2\frac{y}{R} + \cdots\right), \tag{3.6}$$

where y is measured from the earth's surface, R is the radius of the earth, M is the mass of the earth, G is the gravitational constant, and $g = GM/R^2$.

For objects near the earth's surface, a more important modification of the free fall problem is the retarding drag force due to air resistance. The direction of the drag force is opposite to the velocity of the object. We first discuss the motion of a falling particle. The direction of the drag force F_d is upward as shown in Fig. 3.1c. If we use the coordinate system shown in Fig. 3.1, the total force F on the particle can be written as

$$F = -mg + F_d. \tag{3.7}$$

In general, it is necessary to determine the velocity dependence of F_d empirically over a limited range of conditions. One way to find the form of $F_d(v)$ is to measure y as a function of t and to determine the velocity and acceleration as a function of t. From this information it is possible to find the acceleration as a function of v and to extract $F_d(v)$ from (3.7). However, this method is not useful in general, because errors are introduced by taking the slopes needed to find the velocity and acceleration. A better method is to reverse the procedure, that is, assume an explicit form for the v dependence of $F_d(v)$, and use it to solve for $y(t)$. If the calculated $y(t)$ is consistent with the experimental values of $y(t)$, the assumed v dependence of $F_d(v)$ is justified empirically.

The two most commonly assumed forms of the velocity dependence of $F_d(v)$ are

$$F_d(v) = k_1 v \tag{3.8a}$$

and

$$F_d(v) = k_2 v^2, \tag{3.8b}$$

where the parameters k_1 and k_2 depend on the properties of the medium and the shape of the object. We stress that the forms (3.10) are not exact laws of physics, but instead are useful *phenomenological* expressions that yield approximate results for $F_d(v)$ over a limited range of v.

Because $F_d(v)$ increases as v increases, there is a limiting or *terminal velocity* (speed) at which $F_d = mg$ and the acceleration equals zero. This speed can be found from (3.7) and (3.8) and is given by

$$v_t = \frac{mg}{k_1} \tag{3.9a}$$

or

$$v_t = \left(\frac{mg}{k_2}\right)^{1/2} \tag{3.9b}$$

for the linear and quadratic cases, respectively. It frequently is convenient to measure velocities in terms of the terminal velocity. We can use (3.10) and (3.9) to write F_d in the linear and quadratic cases as

$$F_{1,d} = k_1 v_t \left(\frac{v}{v_t}\right) = mg \frac{v}{v_t} \tag{3.10a}$$

and

$$F_{2,d} = k_2 v_t^2 \left(\frac{v}{v_t}\right)^2 = mg \left(\frac{v}{v_t}\right)^2. \tag{3.10b}$$

Hence, we can write the net force on a falling object in the form

$$F_1(v) = -mg\left(1 - \frac{v}{v_t}\right) \tag{3.11a}$$

and

$$F_2(v) = -mg\left(1 - \frac{v^2}{v_t^2}\right). \tag{3.11b}$$

To determine if the effects of air resistance are important in the fall of ordinary objects, consider the fall of a spherical pebble of mass $m = 10^{-2}$ kg. To a good approximation, the drag force is proportional to v^2. For a pebble of radius 0.01 m, k_2 is found empirically to be approximately 10^{-4} kg/m. From (3.9b) we find the terminal velocity is about 30 m/s. Because this speed would be achieved by a freely falling body in a vertical fall of approximately 50 m in a time of about 3 s, we expect that the effects of air resistance would be appreciable for comparable times and distances. Hence, many of the textbook problems we encounter in elementary mechanics courses are not realistic.

3.3 THE EULER METHOD FOR NEWTON'S LAWS OF MOTION

The analytical solution to the equation of motion (3.4) with the net force (3.11a) is straightforward, and there is no need to use a computer. However, to ease our introduction to numerical methods, we will use a computer in a familiar context.

The first step in applying the Euler method to Newton's equation of motion is to write (3.4) as two coupled first-order differential equations, (3.1) and (3.3). We let Δt be the time interval between successive steps and a_n, v_n, and y_n be the values of a, v, and y at time $t_n = t_0 + n\Delta t$, e.g., $a_n = a_n(y_n, v_n, t_n)$. A straightforward generalization of the Euler method introduced in Chapter 2 is

$$v_{n+1} = v_n + a_n \Delta t \tag{3.12a}$$

and

$$y_{n+1} = y_n + v_n \Delta t. \qquad \text{(Euler)} \tag{3.12b}$$

Note that v_{n+1}, the velocity at the end of an interval, is determined by a_n, the rate of change of the velocity at the *beginning* of the interval. In the same spirit, y_{n+1}, the position at the end of an interval, is determined by v_n, the rate of change of the position at the *beginning* of the interval.

The algorithm for obtaining a numerical solution of a differential equation is not unique, and there are many algorithms that reduce to the same differential equation in the limit $\Delta t \to 0$. For example, a simple variation of (3.12) is to determine y_{n+1} using v_{n+1}, the velocity at the *end* of the interval rather than at the beginning. We write this modified Euler method as

$$v_{n+1} = v_n + a_n \Delta t \tag{3.13a}$$

and

$$y_{n+1} = y_n + v_{n+1}\Delta t. \qquad \text{(Euler-Cromer)} \qquad (3.13b)$$

We refer to (3.13) as the Euler-Cromer method. Is there any *a priori* reason to prefer one method over the other?

It might occur to you that it would be better to compute the velocity at the middle of the interval rather than at the beginning or at the end of the interval. The Euler-Richardson algorithm is based on this idea and the simple Euler method. This algorithm is particularly useful for velocity-dependent forces, but does as well as other simple algorithms for forces that do not depend on the velocity. The algorithm consists of using the Euler method to find the intermediate position y_{mid} and velocity v_{mid} at a time $t_{mid} = t + \Delta t/2$. Then we compute the force, $F(y_{mid}, v_{mid}, t_{mid})$ and the acceleration a_{mid} at t_{mid}. The new position x_{n+1} and velocity v_{n+1} at time t_{n+1} is found using v_{mid} and a_{mid}. We summarize the Euler-Richardson algorithm as follows:

$$a_n = F(y_n, v_n, t_n)/m \qquad (3.14a)$$

$$v_{mid} = v_n + \frac{1}{2}a_n\Delta t, \qquad (3.14b)$$

$$y_{mid} = y_n + \frac{1}{2}v_n\Delta t, \qquad (3.14c)$$

$$a_{mid} = F(y_m, v_m, t + \frac{1}{2}\Delta t)/m, \qquad (3.14d)$$

and

$$v_{n+1} = v_n + a_{mid}\Delta t. \qquad (3.15a)$$
$$y_{n+1} = y_n + v_{mid}\Delta t. \qquad \text{(Euler-Richardson)} \qquad (3.15b)$$

Even though we need to do twice as many computations per time step, the Euler-Richardson algorithm is much faster because we can make the time step larger and still obtain better accuracy than with either the Euler or Euler-Cromer algorithms. A "derivation" of the Euler-Richardson algorithm is given in Appendix 3D.

3.4   A PROGRAM FOR ONE-DIMENSIONAL MOTION

A simple implementation of the Euler method for the motion of a freely falling object is given in *Program free_fall*. The structure of the program is similar to *Program cool*.

```
PROGRAM free_fall                ! no air resistance
CALL initial(y,v,a,g,t,dt)    ! initial conditions and parameters
CALL print_table(y,v,a,t,nshow)   ! print initial conditions
LET counter = 0
DO while y >= 0
    CALL Euler(y,v,a,g,t,dt)
    LET counter = counter + 1
```

```
            IF mod(counter,nshow) = 0 then
                CALL print_table(y,v,a,t,nshow)
            END IF
        LOOP
        CALL print_table(y,v,a,t,nshow)    ! print values at surface
        END

        SUB initial(y,v,a,g,t,dt)
            LET t = 0                 ! initial time (sec)
            LET y = 10                ! initial height (m)
            LET v = 0                 ! initial velocity
            LET g = 9.8               ! (magnitude) of accel due to gravity
            LET a = -g
            INPUT prompt "time step dt = ": dt
        END SUB

        SUB Euler(y,v,a,g,t,dt)
            LET y = y + v*dt          ! use velocity at beginning of interval
            ! following included to remind us that acceleration is constant
            LET a = -g                ! y positive upward
            LET v = v + a*dt
            LET t = t + dt
        END SUB

        SUB print_table(y,v,a,t,nshow)
            IF t = 0 then
                INPUT prompt "number of time steps between output = ": nshow
                CLEAR      ! might want to omit this graphics statement
                PRINT "time (s)", "y (m)", "velocity (m/s)", "accel (m/s^2)"
                PRINT
            END IF
            PRINT t,y,v,a
        END SUB
```

In *Program free_fall* we use the notation dt for the finite time step Δt. The reader should not confuse the finite time step dt with the mathematical notation *dt* representing an infinitesimal interval.

It is instructive to see how *Program free_fall* can be modified so that the results can be given in different forms. For example, a simple way to "observe" the trajectory of the falling object is to replace SUB print_table by the subroutine:

```
        SUB show_particle(y,t,xmid,r,nshow)
            IF t = 0 then
                INPUT prompt "number of time steps between output = ": nshow
                CLEAR
                SET WINDOW 0,11,-1,y + 1
                PLOT LINES: 0,0;11,0        ! surface of earth
                LET xmid = 6                ! place line in middle of screen
```

```
        LET r = 0.1
     END IF
     BOX AREA xmid - r,xmid + r,y - r,y + r  ! show particle
END SUB
```

Note the use of the BOX AREA statement (see Table 2.3) to represent the particle.

We also might wish to make a graph of the position and velocity of the falling body as a function of time. In this case we can replace the output subroutine by SUB show_graph:

```
SUB show_graph(y,t,nshow)
    IF t = 0 then
        LET nshow = 10
        LET tmin = t
        LET tmax = 2                        ! estimate of time to reach surface
        LET ymin = 0
        LET ymax = y
        CALL draw_axes(tmin,tmax,ymin,ymax)  ! see Appendix 3A
    END IF
    PLOT LINES: t,y;
END SUB
```

SUB show_graph calls SUB draw_axes which we list and discuss in Appendix 3A. In Appendix 3B we discuss how to save our numerical results in a file so that we can tabulate or plot them using a separate program.

In the following problems, we use simple modifications of *Program free_fall* to study the vertical motion of simple objects with and without the effects of air resistance.

Problem 3.1   Comparison of algorithms

a.   Use *Program free_fall* to determine the time dependence of the velocity and position of a freely falling body near the earth's surface. Assume the values $y = 20$ and $v = 0$ at $t = 0$. The coordinate system used in *Program free_fall* is shown in Fig. 3.1. What is a suitable value of Δt? Compare your output to the exact results given in (3.5).

b.   Modify *Program free_fall* to implement the Euler-Cromer algorithm. (All that is necessary is the interchange of two lines in SUB Euler.) Is there any reason to prefer one algorithm over the other? Think of a simple modification of either algorithm that yields exact results for the case of a freely falling body (no air resistance).

c.   Modify *Program free_fall* to implement the Euler-Richardson algorithm. Adopt the Euler-Richardson algorithm for the remaining problems in this section.

d.   Use SUB show_particle to observe the position of the falling object at equal time intervals. Is the distance the object has traveled over equal time intervals equal?

e. Use SUB show_graph to plot y, v, and a as functions of time.

Now that you have tested your programs in the absence of air resistance, you are ready to consider several more realistic problems. In Table 3.1, we show the observed position of a falling styrofoam ball as a function of time. Are the effects of air resistance important? *Program drag* incorporates the quadratic drag force and is useful for Problem 3.2.

```
PROGRAM drag
! assume drag force proportional to v^2
CALL initial(y,y0,v,t,g,vt2,dt,dt_2)
CALL print_table(y,y0,t,dt,nshow)
DO until t >= 0
    CALL Euler_Richardson(y,v,t,g,vt2,dt,dt_2)
LOOP
CALL print_table(y,y0,t,dt,nshow)          ! values at t = 0
LET counter = 0
DO while t <= 0.80
    CALL Euler_Richardson(y,v,t,g,vt2,dt,dt_2)
    LET counter = counter + 1
    IF mod(counter,nshow) = 0 then
        CALL print_table(y,y0,t,dt,nshow)
    END IF
LOOP
END

SUB initial(y,y0,v,t0,g,vt2,dt,dt_2)
    LET t0 = -0.132                    ! initial time (see Table 3.1)
    LET y0 = 0
    LET y = y0
    LET v = 0                          ! initial velocity
    INPUT prompt "terminal velocity (m/s) = ": vt
    LET vt2 = vt*vt
    LET dt = 0.001
    LET dt_2 = 0.5*dt
    LET g = 9.8
END SUB

SUB Euler_Richardson(y,v,t,g,vt2,dt,dt_2)    ! Euler-Richardson method
    LET v2 = v*v
    LET a = -g*(1 - v2/vt2)
    LET vmid = v + a*dt_2              ! velocity at midpoint
    LET ymid = y + v*dt_2              ! position at midpoint
    LET vmid2 = vmid*vmid
    LET amid = -g*(1 - vmid2/vt2)         ! acceleration at midpoint
    LET v = v + amid*dt
    LET y = y + vmid*dt
    LET t = t + dt
END SUB
```

```
SUB print_table(y,y0,t,dt,nshow)
    IF t < 0 then
        LET show_time = 0.1           ! time interval between output
        ! choice of dt = 0.001 convenient
        LET nshow = int(show_time/dt)
        CLEAR
        PRINT "time","displacement"
        PRINT
    END IF
    LET distance_fallen = y0 - y
    PRINT t,distance_fallen
END SUB
```

Problem 3.2   The fall of a styrofoam ball

a.   Use the empirical data for the displacement $y(t)$ in Table 3.1 to estimate the velocity $v(t)$ and the acceleration $a(t)$. To estimate these derivatives, recall that the derivative of the function $y(t)$ is defined as

$$v(t) = \frac{dy(t)}{dt} = \lim_{\Delta t \to 0} \frac{y(t + \Delta t) - y(t)}{\Delta t}. \tag{3.16}$$

Hence if Δt is small enough, the derivative is approximately given by

$$v(t) \approx \frac{y(t + \Delta t) - y(t)}{\Delta t}. \tag{3.17}$$

Equation (3.17) is known as the *forward difference* approximation. The *central difference* approximation for the derivative is given by

$$v(t) \approx \frac{y(t + \Delta t) - y(t - \Delta t)}{2\Delta t}. \tag{3.18}$$

t (s)	position (m)
-0.132	0.0
0.0	0.075
0.1	0.260
0.2	0.525
0.3	0.870
0.4	1.27
0.5	1.73
0.6	2.23
0.7	2.77
0.8	3.35

Table 3.1   Results by Greenwood, Hanna, and Milton (see references) for the vertical fall of a styrofoam ball of mass 0.254 gm and radius 2.54 cm. Note that the initial time is negative and not an integer multiple of 0.1.

Show that if we write the acceleration as $a(t) \approx (v(t + \Delta t) - v(t))/\Delta t$ and use (3.17) for the velocity, we can express the acceleration as

$$a(t) \approx \frac{y(t + \Delta t) - 2y(t) + y(t - \Delta t)}{(\Delta t)^2}. \tag{3.19}$$

Use (3.18) and (3.19) to estimate the velocity and acceleration. Estimate the terminal velocity from the data given in Table 3.1. This estimation is nontrivial, in part because the terminal velocity has not yet been reached during the time interval shown in the table. Explain why your results for the acceleration do not seem to be very accurate.

b. Use *Program* drag with the net force given by either (3.11a) or (3.11b) and the terminal velocity found in part (a). Make sure that your computed results for the displacement of the particle, $y(t) - y(0)$, do not depend on Δt. Compare your plot of the computed values of $y(t) - y(0)$ for the linear and quadratic form with the empirical values of $y(t) - y(0)$ and visually determine which form of the drag force yields the best overall fit. What is your criteria for the "best" fit? Should you match your results with the experimental data at early times or at later times? Or should you adopt another strategy? What are the qualitative differences between the two computed forms of $y(t) - y(0)$?

Problem 3.3   Effect of air resistance on the ascent and descent of a pebble

a. Let us verify the claim made in Section 3.2 that the effects of air resistance on a falling pebble can be appreciable. Compute the speed at which a pebble reaches the ground if it is dropped from rest at a height of 50 m. Compare this speed to that of a freely falling object under the same conditions. Assume that the drag force is proportional to v^2 and that the terminal velocity is 30 m/s.

b. Suppose a pebble is thrown vertically upward with an initial velocity v_0. In the absence of air resistance, we know that the maximum height reached by the pebble is $v_0^2/2g$, its velocity upon return to the earth equals v_0, the time of ascent equals the time of descent, and the total time in the air is $2v_0/g$. Before performing a numerical simulation, give a simple qualitative explanation of how you think these quantities will be affected by air resistance. Then perform a simulation to determine if your qualitative answers are correct. Assume that the drag force is proportional to v^2. From (3.10) we see that we can characterize the magnitude of the drag force by a terminal velocity even if the pebble never attains this velocity. Choose the terminal velocity $v_t = 30$ m/s. Suggestions: It is a good idea to choose an initial velocity that allows the pebble to remain in the air for a total time such that the time of ascent differs appreciably from the time of descent. A reasonable choice is $v(t = 0) = 50$ m/s. Choose the coordinate system shown in Fig. 3.1 with y positive upward. What is the net force for $v > 0$ and $v < 0$? You might find it convenient to use the sgn function in True BASIC, where sgn(v) returns the sign of v. Or you can write the drag force in the form $F_d = -v*\mathtt{abs}(v)$. One way to determine the maximum height of the pebble is to use the statement

```
IF v*vold < 0 then PRINT "maximum height = "; y
```

where $v = v_{n+1}$ and vold $= v_n$.

Problem 3.4 Position-dependent force

Modify *Program free_fall* to simulate the fall of a particle with the position-dependent force law (3.6). Assume that a particle is dropped from a height h with zero initial velocity and compute its speed when it hits the ground. Determine the value of h for which this impact velocity differs by one percent from its value with a constant acceleration $g = 9.8 \, \mathrm{m/s^2}$. Take the radius of the earth to be 6.37×10^6 m.

3.5 TWO-DIMENSIONAL TRAJECTORIES

You are probably familiar with two-dimensional trajectory problems in the absence of air resistance. For example, if a ball is thrown in the air with an initial velocity v_0 at an angle θ_0 with respect to the ground, how far will the ball travel in the horizontal direction, and what is its maximum height and time of flight? Suppose that a ball is released at a nonzero height h above the ground. What is the launch angle for the maximum range? Are your answers still applicable if air resistance is taken into account? We consider these and similar questions in the following.

Consider an object of mass m whose initial velocity $\mathbf{v}_0$ is directed at an angle θ_0 above the horizontal (see Fig. 3.2a). The particle is subjected to gravitational and drag forces of magnitude mg and F_d; the direction of the drag force is opposite to $\mathbf{v}$ (see Fig. 3.2b). Newton's equations of motion for the x and y components of the motion can be written as

$$m\frac{dv_x}{dt} = -F_d \cos\theta \tag{3.20a}$$

$$m\frac{dv_y}{dt} = -mg - F_d \sin\theta. \tag{3.20b}$$

As an example, let us maximize the range of a round steel ball of radius 4 cm. A reasonable assumption for an object of this size (a "shot") and typical speed is that $F_d = k_2 v^2$. Because $v_x = v\cos\theta$ and $v_y = v\sin\theta$, we can rewrite (3.20) as

$$\frac{dv_x}{dt} = -Cvv_x \tag{3.21a}$$

$$\frac{dv_y}{dt} = -g - Cvv_y, \tag{3.21b}$$

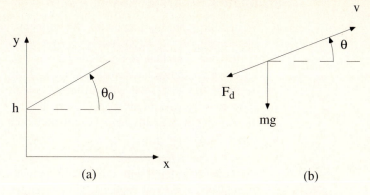

Fig. 3.2 (a) A ball is thrown from a height h at an launch angle θ_0 measured with respect to the horizontal. The initial velocity is $\mathbf{v}_0$. (b) The gravitational and drag forces on a particle.

where $C = k_2/m$. Note that $-Cvv_x$ and $-Cvv_y$ are the x and y components of the drag force $-Cv^2$. Because (3.21a) and (3.21b) for the change in v_x and v_y involve the square of the velocity, $v^2 = v_x{}^2 + v_y{}^2$, we cannot calculate the vertical motion of a falling body without reference to the horizontal component, that is, the motion in the x and y direction is coupled.

A simple modification of SUB Euler in *Program drag* that computes the trajectory in two dimensions is given below:

```
SUB Euler_Richardson(x,y,vx,vy,t,C,g,dt,dt_2)
    LET v2 = vx*vx + vy*vy
    LET v = sqr(v2)
    LET ax = -C*v*vx
    LET ay = -g - C*v*vy
    LET vxmid = vx + ax*dt_2          ! velocity at midpoint
    LET vymid = vy + ay*dt_2
    LET xmid = x + vx*dt_2            ! position at midpoint
    LET ymid = y + vy*dt_2
    LET vmid2 = vxmid*vxmid + vymid*vymid
    LET vmid = sqr(vmid2)
    LET axmid = -C*vmid*vxmid         ! acceleration at midpoint
    LET aymid = -g - C*vmid*vymid
    LET vx = vx + axmid*dt
    LET vy = vy + aymid*dt
    LET x = x + vxmid*dt
    LET y = y + vymid*dt
    LET t = t + dt
END SUB
```

Problem 3.5   Trajectory of a shot

a. Write a program to compute the two-dimensional trajectory of a ball moving in air and plot y as a function of x. As a check on your program, first neglect air resistance so that you can compare your computed results with the exact results. For example, assume that a ball is thrown from ground level at an angle θ_0 above the horizontal with an initial velocity of $v_0 = 15\,\text{m/s}$. Vary θ_0 and show that the maximum range occurs at $\theta_0 = \theta_{max} = 45°$. What is R_{max}, the maximum range, at this angle? Compare your numerical value to the analytical result $R_{max} = v_0^2/g$.

b. Suppose that a ball ("shot") is thrown ("put") from a height h at an angle θ_0 above the horizontal with the same initial speed as in part (a). If you neglect air resistance, do you expect θ_{max} to be larger or smaller than 45°? Although this problem can be solved analytically, you can determine the numerical value of θ_{max} without changing your program. What is θ_{max} for $h = 2\,\text{m}$? By what percent is the range R changed if θ is varied by 2% from θ_{max}?

c. Consider the effects of air resistance on the range and optimum angle of a shot put. For a typical shot put (mass $\approx 7\,\text{kg}$ and cross-sectional area $\approx 0.01\,\text{m}^2$), the parameter k_2 defined by (3.8b) is approximately 0.01. What are the units of k_2? It is convenient to exaggerate the effects of air resistance so that you can more easily determine the qualitative nature of the effects. Compute the optimum angle for $h = 2\,\text{m}$, $v_0 = 30\,\text{m/s}$, and $C = k_2/m = 0.1$, and compare your answer to the value found in part (b). Is R more or less sensitive to changes in θ_0 from θ_{max} than in part (b)? Determine the optimum launch angle and the corresponding range for the more realistic value of $C = 0.001$. A detailed discussion of the maximum range of the shot has been given by Lichtenberg and Wills (see references).

Problem 3.6   Coupled motion

Consider the motion of two identical objects that both start from a height h. One object is dropped vertically from rest and the other is thrown with a horizontal velocity v_0. Which object reaches the ground first?

a. Give reasons for your answer assuming that air resistance can be neglected.

b. Assume that air resistance cannot be neglected and that the drag force is proportional to v^2. Give reasons for your anticipated answer for this case. Then perform numerical simulations using, for example, $C = k_2/m = 0.1$, $h = 10\,\text{m}$, and $v_0 = 30\,\text{m/s}$. Are your qualitative results consistent with your anticipated answer? If they are not, the source of the discrepancy might be an error in your program. Or the discrepancy might be due to your failure to anticipate the effects of the coupling between the vertical and horizontal motion.

c. Suppose that the drag force is proportional to v rather than to v^2. Is your anticipated answer similar to that in part (b)? Do a numerical simulation to test your intuition.

3.6   LEVELS OF SIMULATION

So far we have done simulations of models in which the microscopic complexity of the system has been simplified considerably. Consider for example, the motion of a pebble falling through the air. First we reduced the complexity by representing the pebble as a particle. Then we reduced the number of degrees of freedom even more by representing the collisions of the pebble with the billions of molecules in the air by a velocity-dependent friction term. It is remarkable that the resultant phenomenological model is a fairly accurate representation of realistic physical systems. However, what we gain in simplicity, we lose in range of applicability.

In a detailed model, the individual physical processes are represented in a more microscopic manner. For example, we can imagine doing a simulation in which the effects of the air are represented by a fluid of particles that collide with one another and with the falling particle. How accurately do we need to represent the potential energy of interaction between the fluid particles? Clearly the level of detail that is needed in a model depends on the accuracy of the corresponding experimental data and the type of information in which we are interested. For example, we do not need to take into account the influence of the moon on a pebble falling near the earth's surface. And the level of detail that we can simulate depends in part on the available computer resources.

The words *simulation* and *modeling* are frequently used interchangeably and their precise meaning is not important, especially since people who work with models and who do simulations do not use them precisely. Most practitioners would say that in Chapters 2 and 3 we have solved several mathematical models numerically. Beginning with Chapter 4, we will be able to say that we actually are doing simulations. The difference is that our models will represent physical systems in more detail, and we will need to give more attention to what physical quantities we should measure. In other words, our simulations will become more analogous to laboratory experiments.

3.7   FURTHER APPLICATIONS

Applications of the ideas and methods that we have discussed are important in the physics of clouds. For example, to understand the behavior of falling water droplets, it is necessary to take into account drag resistance as well as droplet growth by condensation and other mechanisms. Because of the variety and complexity of the mechanisms, simulation plays an essential role.

Another area of interest is the trajectory of balls of various shapes through the air. Of particular interest to sports fans is the curve of balls in flight due to their rotation and the effect of air resistance on the range and speed of table tennis balls.

Appendix 3A   SUBROUTINE FOR DRAWING AXES

To make simple plots we need to define the coordinates of our window, draw horizontal and vertical axes, and put "tick" marks in the appropriate place to show the scale. If you are ambitious, you can add statements (using PLOT TEXT) to label the axes and make a title for the plot.

```
SUB draw_axes(xmin,xmax,ymin,ymax)
    ! number of tick marks
    LET ntick = 10
    ! dx distance between tick marks on x axis
    LET dx = (xmax - xmin)/ntick
    ! dy distance between tick marks on y axis
    LET dy = (ymax - ymin)/ntick
    ! include margin in window statement
    SET WINDOW xmin - dx,xmax + dx,ymin - dy,ymax + dy
    LET x0 = max(0,xmin)
    LET y0 = max(0,ymin)
    IF ymin*ymax < 0 then
       LET y0 = 0
    ELSE
       LET y0 = ymin
    END IF
    PLOT LINES: xmin,y0;xmax,y0      ! horizontal axis
    PLOT LINES: x0,ymin;x0,ymax      ! vertical axis
    LET tx = 0.1*dy                  ! size of tick mark
    LET ty = 0.1*dx
    FOR itick = 0 to ntick
        LET x = xmin + itick*dx
        PLOT LINES: x,y0 - tx; x,y0 + tx      ! draw ticks on x axis
        LET y = ymin + itick*dy
        PLOT LINES: x0 - ty,y; x0 + ty,y      ! draw ticks on y axis
    NEXT itick
END SUB
```

Because we will use SUB draw_axes in many programs, it is convenient to place the subroutine in a separate file that may be used by any program. In True BASIC, such a file is called a **library**. The only modification we need to make is to start the library with the statement EXTERNAL. (Add this statement before the statement SUB draw_axes.) Suppose we name the library file as csgraphics. Then if we add the statement

```
LIBRARY "csgraphics"
```

to the main program, True BASIC will find all the subroutines (and functions) that are included in the library file and invoked by the program. (It is simplest to keep the library file in the same directory as the main program. Alternatives are discussed in the True BASIC manual.) Programming languages such as C, Fortran, and Pascal have analogous methods for incorporating external subroutines and functions.

Appendix 3B   DATA FILES

The following True BASIC program illustrates how to open a text file on a disk, write to it, close the file, and how to read the file. The details of writing and reading files depend strongly on the programming language (see Appendices A and B).

```
PROGRAM single_column
! save data in a single column
! file$ example of string variable
INPUT prompt "name of file for data? ": file$
! channel number #1 associated with file and can be passed to
! subroutines
! various options may be specified in OPEN statement
! access output: write to file only
! create newold: open file if exists, else create new file
OPEN #1: name file$, access output, create newold
! True BASIC does not overwrite data in a text file
! ERASE #1                                ! erase contents of file
! RESET #1: end    ! allows data to be added to end of file
FOR i = 1 to 4
    LET x = i*i
    PRINT #1: x                           ! print column of data
NEXT i
CLOSE #1
! channel # irrelevant if only one open at a time
OPEN #2: name file$, access input
FOR i = 1 to 4
    INPUT #2: y                           ! print column of data
    PRINT y
NEXT i
! files automatically closed when program terminates
CLOSE #2                                  ! not necessary but good practice
END
```

Note that *Program single_column* uses a *string* or character variable for the name of the file. String variables end in a dollar sign ($). An example of an assignment of a string variable is LET name$ = "hello". The writing of files with multiple columns is more complicated and is illustrated in Chapter 8 (see SUB save_config).

Appendix 3C   STRONG TYPING AND DEBUGGING

A common mistake while entering a program is to make a typo, e.g., typing dr instead of dt. Such a mistake is difficult to find in a True BASIC and Fortran program because the program will compile and run. One of the advantages of PASCAL as a teaching language is that it requires the programmer to declare all variables before they can be used. In this way, the above typo would lead to a compile error since the variable dr would not have been declared. The enforced declaration of variables is called *strong typing*. Although more work is required to declare all the variables, much time can be saved in finding errors in long programs.

Fortunately, True BASIC (and Fortran) allow us to "turn on" strong typing. An example of strong typing in True BASIC is given in the following modified version of *Program free_fall*:

```
PROGRAM free_fall
! example of strong typing
OPTION TYPO
PUBLIC y,v,a,g,t,dt,nshow      ! declare variables
LOCAL counter
CALL initial
CALL print_table
LET counter = 0
DO while y >= 0
    CALL Euler
    LET counter = counter + 1
    IF mod(counter,nshow) = 0 then
        CALL print_table
    END IF
LOOP
CALL print_table
END

SUB initial
    DECLARE PUBLIC y,v,a,g,t,dt
    LET t = 0
    LET y = 10
    LET v = 0
    LET g = 9.8
    LET a = -g
    LET dt = 0.01
END SUB

SUB Euler
    DECLARE PUBLIC y,v,a,g,t,dt
    LET y = y + v*dt
    LET a = -g
    LET v = v + a*dt
    LET t = t + dt
END SUB

SUB print_table
    DECLARE PUBLIC y,v,a,t,nshow
    IF t = 0 then
        INPUT prompt "number of time steps between output = ": nshow
        CLEAR        ! might want to omit this graphics statement
        PRINT "time (s)", "y (m)", "velocity (m/s)", "accel (m/s*s))"
        PRINT
    END IF
    PRINT t,y,v,a
END SUB
```

OPTION TYPO requires that all variables be declared. In *Program free_fall*, the variables are declared to be either LOCAL or PUBLIC (global). A program subunit that uses a public variable must have a DECLARE PUBLIC statement. Note that public variables are not passed in the arguments of the subroutines. Of course, variables also can be declared without using OPTION TYPO.

It is rare to write a program that is free of errors on the first attempt. The process of debugging (finding errors) can be very frustrating and can be more time consuming than the actual writing of the program. Although there is no foolproof way of avoiding errors, there is one general rule that even the most experienced programmer needs to remember: "keep it simple." Some specific rules that help reduce the number of errors include using a structured programming style with subroutines and functions with well defined tasks, self-explanatory variable names, and well documented code. Another rule to remember is that in general, your time is more valuable than the computer's, and hence clarity is more important than efficiency. Two other general rules that are usually learned the hard way is to save your program and other files early and often (especially before running a program) and to make major changes in a copy of your program rather than in the original.

Typical programming errors include misspelled variable names (easily checked with OPTION TYPO), logical errors (e.g., typing > instead of <), and failure to pass variables. Problems also can arise if IF statements are based on the equality of two real numbers. Why? In general, the best way of finding errors is to check that each subroutine and function is working separately and to compare your results with known results whenever possible. Programming aids such as the Trace command in True BASIC let you go through the program step-by-step and "trace" the values of your variables. The liberal inclusion of temporary PRINT statements to examine variables that are not normally printed also is very valuable. For convenience, these print statements can be commented out and left in the program for future use if necessary.

Although the above rules will save you time, they are not sufficient. Our experience is that most programming problems are not errors of programming, but errors in understanding or implementing the desired algorithm. There is no substitute for thinking!

Exercise 3.1   Communication between program units

In the above version of *Program free_fall*, all the variables that are used by the subroutines are declared public. It is not usually a good practice to declare all variables public, because sometimes we would like to see how program units are directly communicating, particularly when there are many levels of subroutines. For example, suppose that within SUB top there is a call to SUB bottom. If x is a public variable declared in the main program, and is listed in a DECLARE PUBLIC statement in SUB bottom, we do not know by looking only at SUB bottom whether x has been changed in SUB top. How do we determine which variables should be public and which should be passed as arguments of subroutines? Two kinds of variables that can safely be declared public include (1) variables such as g, nshow, and dt which are never changed, and (2) variables such as x, v, a, and t which are used by every program unit.

Modify *Program drag* so that OPTION TYPO is used and variables are declared public only if they meet the above two criteria. Note that if a variable is passed in a subroutine, that variable is declared as a local variable in the calling program unit, but is not declared within the subroutine.

Appendix 3D   THE EULER-RICHARDSON METHOD

We motivate the Euler-Richardson method (3.14) in the following. We write $y(t + \Delta t)$ as a Taylor series to second-order in $(\Delta t)^2$:

$$y_1(t + \Delta t) = y(t) + v(t)\Delta t + \frac{1}{2}a(t)(\Delta t)^2, \tag{3.22}$$

where $a(t) = a(y(t), v(t), t)$. The notation y_1 implies that $y(t + \Delta t)$ is related to $y(t)$ by one time step. We also divide the step Δt into half steps and write the first half step, $y(t + \frac{1}{2}\Delta t)$, as

$$y(t + \frac{1}{2}\Delta t) = y(t) + v(t)\frac{\Delta t}{2} + \frac{1}{2}a(t)\left(\frac{\Delta t}{2}\right)^2. \tag{3.23}$$

The second half step, $y_2(t + \Delta t)$, can be written as

$$y_2(t + \Delta t) = y(t + \frac{1}{2}\Delta t) + v(t + \frac{1}{2}\Delta t)\frac{\Delta t}{2} + \frac{1}{2}a(t + \frac{1}{2}\Delta t)\left(\frac{\Delta t}{2}\right)^2. \tag{3.24}$$

We substitute (3.23) into (3.24) and obtain

$$y_2(t + \Delta t) = y(t) + \frac{1}{2}\left[v(t) + v(t + \frac{1}{2}\Delta t)\right]\Delta t$$
$$+ \frac{1}{2}\left[a(t) + a(t + \frac{1}{2}\Delta t)\right]\left(\frac{1}{2}\Delta t\right)^2. \tag{3.25}$$

Now $a(t + \frac{1}{2}\Delta t) = a(t) + \frac{1}{2}a'(t)\Delta t + \ldots$ Hence to order $(\Delta t)^2$, (3.25) becomes

$$y_2(t + \Delta t) = y(t) + \frac{1}{2}\left[v(t) + v(t + \frac{1}{2}\Delta t)\right]\Delta t + \frac{1}{2}(2a(t))\left(\frac{1}{2}\Delta t\right)^2. \tag{3.26}$$

We can create an approximation that is accurate to order $(\Delta t)^3$ by combining (3.22) and (3.26) so that the terms to order $(\Delta t)^2$ cancel. The combination that works is $2y_2 - y_1$, which gives the Euler-Richardson result:

$$y_{er} = 2y_2(t + \Delta t) - y_1(t + \Delta t) = y(t) + v(t + \frac{1}{2}\Delta t)\Delta t. \tag{3.27}$$

The same reasoning leads to an approximation for the velocity accurate to $(\Delta t)^3$ giving

$$v_{er} = 2v_2(t + \Delta t) - v_1(t + \Delta t) = v(t) + a(t + \frac{1}{2}\Delta t)\Delta t. \tag{3.28}$$

A bonus of the Euler-Richardson method is that the quantities $|y_2 - y_1|$ and $|v_2 - v_1|$ give an estimate for the error in the procedure. We can use these estimates to change the time step so that the error is always within some desired level of precision.

References and Suggestions for Further Reading

William R. Bennett, *Scientific and Engineering Problem-Solving with the Computer*, Prentice Hall (1976). One of the first and best books that incorporates computer problem solving. Many one- and two-dimensional falling body problems are considered in Chapter 5.

Byron L. Coulter and Carl G. Adler, "Can a body pass a body falling through the air?," *Am. J. Phys.* **47**, 841 (1979). The authors discuss the limiting conditions for which the drag force is linear or quadratic in the velocity.

Alan Cromer, "Stable solutions using the Euler approximation," *Am. J. Phys.* **49**, 455 (1981). The author shows that a minor modification of the usual Euler approximation yields stable solutions for oscillatory systems including planetary motion and the harmonic oscillator.

Paul L. DeVries, *A First Course in Computational Physics*, John Wiley & Sons (1994). Chapter 1 includes some excellent guidelines for good programming style.

R. M. Eisberg, *Applied Mathematical Physics with Programmable Pocket Calculators*, McGraw-Hill (1976). Chapter 3 of this handy paperback is similar in spirit to the present discussion.

Richard P. Feynman, Robert B. Leighton and Matthew Sands, *The Feynman Lectures on Physics, Vol. 1*, Addison-Wesley (1963). Feynman discusses the numerical solution of Newton's equations in Chapter 9.

A. P. French, *Newtonian Mechanics,* W. W. Norton & Company (1971). Chapter 7 has an excellent discussion of air resistance and does a detailed analysis of motion in the presence of drag resistance.

Ian R. Gatland, "Numerical integration of Newton's equations including velocity-dependent forces," *Am J. Phys.* **62**, 259 (1994). The author discusses the Euler-Richardson algorithm.

Margaret Greenwood, Charles Hanna, and John Milton, "Air resistance acting on a sphere: numerical analysis, strobe photographs, and videotapes," *Phys. Teacher* **24**, 153 (1986). More experimental data and theoretical analysis are given for the fall of ping-pong and styrofoam balls. Also see Mark Peastrel, Rosemary Lynch, and Angelo Armenti, "Terminal velocity of a shuttlecock in vertical fall," *Am. J. Phys.* **48**, 511 (1980).

K. S. Krane, "The falling raindrop: variations on a theme of Newton," *Am. J. Phys.* **49**, 113 (1981). The author discusses the problem of mass accretion by a drop falling through a cloud of droplets.

D. B. Lichtenberg and J. G. Wills, "Maximizing the range of the shot put," *Am. J. Phys.* **46**, 546 (1978). Problem 3.5 is based in part on the discussion in this paper.

Rabindra Mehta, "Aerodynamics of sports balls" in *Ann. Rev. Fluid Mech.* **17**, 151 (1985).

Neville de Mestre, *The Mathematics of Projectiles in Sport*, Cambridge University Press (1990). The emphasis of this text is on solving many problems in projectile motion, e.g., baseball, basketball, and golf, in the context of mathematical modeling. Many references to the relevant literature are given.

R. R. Rogers, *A Short Course in Cloud Physics*, Pergamon Press (1976).

C H A P T E R

4
The Two-Body Problem

We apply Newton's laws of motion to planetary motion and emphasize some of the counter-intuitive consequences of Newton's laws.

4.1   INTRODUCTION

Planetary motion is of special significance since it played an important role in the conceptual history of the mechanical view of the universe. Few theories have affected Western civilization as much as Newton's laws of motion and the law of gravitation, which together relate the motion of the heavens to the motion of terrestrial bodies.

Much of our knowledge of planetary motion is summarized by Kepler's three laws, which can be stated as:

1. Each planet moves in an elliptical orbit with the sun located at one of the foci of the ellipse.

2. The speed of a planet increases as its distance from the sun decreases such that the line from the sun to the planet sweeps out equal areas in equal times.

3. The ratio T^2/a^3 is the same for all planets that orbit the sun, where T is the period of the planet and a is the semimajor axis of the ellipse.

Kepler obtained these laws by a careful analysis of the observational data collected over many years by Tycho Brahe.

Kepler's first and third laws describe the shape of the orbit rather than the time dependence of the position and velocity of a planet. Because it is not possible to obtain this time dependence in terms of elementary functions, we are motivated to discuss the numerical solution of the equations of motion of planets and satellites in orbit. In addition, we consider the effects of perturbing forces on the orbit and problems that challenge our intuitive understanding of Newton's laws of motion.

4.2   THE EQUATIONS OF MOTION

The motion of the sun and earth is an example of a *two-body problem*. We can reduce this problem to a one-body problem in one of two ways. The easiest way is to use the fact that the mass of the sun is much greater than the mass of the earth. Hence we can assume that, to a good approximation, the sun is stationary and is a convenient choice of the origin of our coordinate system. If you are familiar with the concept of a *reduced mass,* you know that this reduction is more general. That is, the motion of two objects of mass m and M whose total potential energy is a function only of their relative separation can be reduced to an equivalent one-body problem for the motion of an object of reduced mass μ given by

$$\mu = \frac{Mm}{m+M}. \tag{4.1}$$

Because the mass of the earth $m = 5.99 \times 10^{24}$ kg and the mass of the sun $M = 1.99 \times 10^{30}$ kg, we find that for most practical purposes, the reduced mass of the sun and the earth is that of the earth alone. In the following, we consider the problem of a single

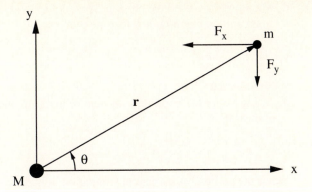

Fig. 4.1 An object of mass m moves under the influence of a central force F. Note that $\cos\theta = x/r$ and $\sin\theta = y/r$, providing useful relations for writing the equations of motion in component form suitable for numerical simulation.

particle of mass m moving about a fixed center of force, which we take as the origin of the coordinate system.

Newton's universal law of gravitation states that a particle of mass M attracts another particle of mass m with a force given by

$$\mathbf{F} = -\frac{GMm}{r^2}\hat{\mathbf{r}} = -\frac{GMm}{r^3}\mathbf{r}, \tag{4.2}$$

where the vector $\mathbf{r}$ is directed from M to m (see Fig. 4.1). The negative sign in (4.2) implies that the gravitational force is attractive, i.e., it tends to decrease the separation r. The gravitational constant G is determined experimentally to be

$$G = 6.67 \times 10^{-11}\,\frac{\text{m}^3}{\text{kg} \cdot \text{s}^2}. \tag{4.3}$$

The force law (4.2) applies to objects of negligible spatial extent. Newton delayed publication of his law of gravitation for twenty years while he invented integral calculus and showed that (4.2) also applies to any uniform sphere or spherical shell of matter if the distance r is measured from the center of each mass.

The gravitational force has two general properties: its magnitude depends only on the separation of the particles, and its direction is along the line joining the particles. Such a force is called a *central force*. The assumption of a central force implies that the orbit of the earth is restricted to a plane (x-y), and the *angular momentum* $\mathbf{L}$ is conserved and lies in the third (z) direction. We write L_z in the form

$$L_z = (\mathbf{r} \times m\mathbf{v})_z = m(xv_y - yv_x), \tag{4.4}$$

where we have used the cross-product definition $\mathbf{L} = \mathbf{r} \times \mathbf{p}$ and $\mathbf{p} = m\mathbf{v}$. An additional constraint on the motion is that the total energy E given by

$$E = \frac{1}{2}mv^2 - \frac{GMm}{r} \tag{4.5}$$

is conserved.

If we fix the coordinate system at mass M, the equation of motion of the particle of mass m is

$$m\frac{d^2\mathbf{r}}{dt^2} = -\frac{GMm}{r^3}\mathbf{r}. \tag{4.6}$$

For the purpose of numerical simulation, it is convenient to write the force in Cartesian coordinates (see Fig. 4.1):

$$F_x = -\frac{GMm}{r^2}\cos\theta = -\frac{GMm}{r^3}x \tag{4.7a}$$

$$F_y = -\frac{GMm}{r^2}\sin\theta = -\frac{GMm}{r^3}y. \tag{4.7b}$$

Hence, the equations of motion in Cartesian coordinates are

$$\frac{d^2x}{dt^2} = -\frac{GM}{r^3}x \tag{4.8a}$$

$$\frac{d^2y}{dt^2} = -\frac{GM}{r^3}y, \tag{4.8b}$$

where $r^2 = x^2 + y^2$. Equations (4.8a) and (4.8b) are examples of "coupled differential equations" because each differential equation contains both x and y.

4.3   CIRCULAR AND ELLIPTICAL ORBITS

Because many planetary orbits are nearly circular, it is useful to obtain the condition for a circular orbit. The magnitude of the acceleration $\mathbf{a}$ is related to the radius r of the circular orbit by

$$a = \frac{v^2}{r}, \tag{4.9}$$

where v is the speed of the object. The acceleration always is directed towards the center and is due to the gravitational force. Hence we have

$$\frac{mv^2}{r} = \frac{GMm}{r^2} \tag{4.10}$$

and

$$v = \left(\frac{GM}{r}\right)^{1/2}. \tag{4.11}$$

The relation (4.11) between the radius and the speed is the general condition for a circular orbit.

We also can find the dependence of the period T on the radius of a circular orbit using the relation

$$T = \frac{2\pi r}{v}$$

(4.12)

in combination with (4.11) to obtain

$$T^2 = \frac{4\pi^2}{GM} r^3.$$

(4.13)

The relation (4.13) is a special case of Kepler's third law where the radius r corresponds to the semimajor axis of an ellipse.

Because the most general orbit is an ellipse, we summarize the description of an elliptical orbit. A simple geometrical characterization of an elliptical orbit is shown in Fig. 4.2. The two *foci* of an ellipse, F_1 and F_2, have the property that for any point P, the distance $F_1 P + F_2 P$ is a constant. In general, an ellipse has two perpendicular axes of unequal length. The longer axis is the major axis; half of this axis is the *semimajor axis a*. The shorter axis is the minor axis; the *semiminor axis b* is half of this distance.

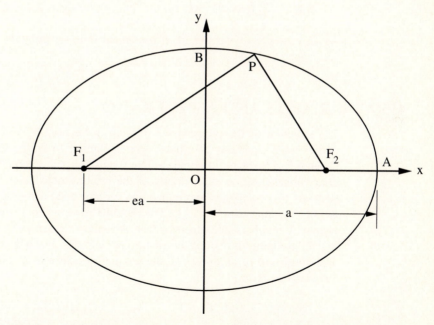

Fig. 4.2 The characterization of an ellipse in terms of the semimajor axis a and the eccentricity e. The semiminor axis b is the distance OB. The origin O in Cartesian coordinates is at the center of the ellipse.

It is common in astronomy to specify an elliptical orbit by a and by the *eccentricity* e, where e is the ratio of the distance between the foci to the length of the major axis. Because $F_1P + F_2P = 2a$, it is easy to show that

$$e = \sqrt{1 - \frac{b^2}{a^2}} \tag{4.14}$$

with $0 < e < 1$. (Consider P at $x = 0$, $y = b$.) A special case is $b = a$ for which the ellipse reduces to a circle and $e = 0$.

4.4   ASTRONOMICAL UNITS

It is convenient to choose a system of units in which the magnitude of the product GM is not too large and not too small. To describe the earth's motion, the convention is to choose the length of the earth's semimajor axis as the unit of length. This unit of length is called the *astronomical unit* (AU) and is

$$1\,\text{AU} = 1.496 \times 10^{11}\,\text{m}. \tag{4.15}$$

The unit of time is assumed to be one year or 3.15×10^7 s. In these units, the period of the earth is $T = 1$ yr and its semimajor axis is $a = 1$ AU. Hence

$$GM = \frac{4\pi^2 a^3}{T^2} = 4\pi^2 \text{AU}^3/\text{yr}^2. \qquad \text{(astronomical units)} \tag{4.16}$$

As an example of the use of astronomical units, a program distance of 1.5 would correspond to 2.244×10^{11} m.

4.5   ARRAY VARIABLES AND ASPECT RATIO

The structure of the program we develop in Section 4.7 for the numerical solution of Newton's equations of motion (4.8) is similar to the programs we considered in Chapter 3. The main difference is that we use an array variable to represent the two-dimensional position and velocity of the orbiting object. An array variable is a data structure that consists of more than one variable and is an ordered set of elements that are of the same data type. (In True BASIC there are only two data types for array elements, numbers and strings.) One advantage of arrays is that they allow for the logical grouping of data of the same type, e.g., the x and y coordinates of a particle. The *dimension* of an array and the passing of arrays to a subroutine is illustrated in *Program vector*.

```
PROGRAM vector              ! illustrate use of arrays
DIM a(3),b(3)               ! arrays defined in DIM statement
CALL initial(a(),b())
CALL dot(a(),b())
CALL cross(a(),b())
END
```

```
SUB initial(a(),b())
    LET a(1) = 2
    LET a(2) = -3
    LET a(3) = -4
    LET b(1) = 6
    LET b(2) = 5
    LET b(3) = 1
END SUB

SUB dot(a(),b())
    LET dot_product = 0
    FOR i = 1 to 3
        LET dot_product = dot_product + a(i)*b(i)
    NEXT i
    PRINT "scalar product = "; dot_product
END SUB

SUB cross(r(),s())
    ! arrays can be defined in main program or subroutine
    ! note use of dummy variables
    DIM cross_product(3)
    FOR component = 1 to 3
        LET i = mod(component,3) + 1
        LET j = mod(i,3) + 1
        LET cross_product(component) = r(i)*s(j) - s(i)*r(j)
    NEXT component
    PRINT
    PRINT "three components of the vector product:"
    PRINT " x "," y "," z "
    FOR component = 1 to 3
        PRINT cross_product(component),
    NEXT component
END SUB
```

The properties of arrays in True BASIC include:

1. Arrays are defined in a DIM statement and the total number of elements of an array is given in parentheses. The array variables a and b in the main program and the array variables r and s in SUB cross are examples of one-dimensional arrays. Higher dimensional arrays are first considered in Chapter 9.

2. The lower and upper limit of each subscript in an array can be specified; the default lower limit is 1. Examples of other limits are DIM r(0 to 2) and DIM s(-3 to 3). A colon may be used instead of TO in the DIM statement, e.g., DIM r(0:2) and DIM s(-3:3). The arguments in a DIM statement must be numbers, not variables.

3. An element of an array is specified by its subscript value. Arrays can be passed to a subroutine or a function, with empty parentheses and commas used to indicate the dimension of the array.

4. The same name cannot be used for both an array variable and for another type of variable.

Note that the entire array is not actually passed. Rather the address of the first element of the array is passed, and there is no memory or speed penalty when arrays are passed to a subroutine.

Exercise 4.1   Vector products

Consider the three-dimensional vectors $\mathbf{a} = \hat{\mathbf{x}} + 2\hat{\mathbf{y}} + 3\hat{\mathbf{z}}$, $\mathbf{b} = 2\hat{\mathbf{x}} - \hat{\mathbf{y}} + 4\hat{\mathbf{z}}$, and $\mathbf{c} = \hat{\mathbf{x}} - \hat{\mathbf{y}} + \hat{\mathbf{z}}$. Modify *Program* vector_product to answer the following. What is the magnitude of $\mathbf{a}$ and $\mathbf{b}$? What is the angle between them? What is the component of $\mathbf{a}$ in the direction of $\mathbf{b}$? Compute $\mathbf{a} \times (\mathbf{b} \times \mathbf{c})$ and confirm that it equals $\mathbf{b}\,(\mathbf{a} \cdot \mathbf{c}) - \mathbf{c}\,(\mathbf{a} \cdot \mathbf{b})$.

Because we will be interested in the shape of the orbit, we need to write a subroutine that draws a circle. However, the following simple program probably does not do the job:

```
PROGRAM no_circle
LET r = 1                                ! radius of circle
SET WINDOW -r,r,-r,r
SET COLOR "blue"
BOX CIRCLE -r,r,-r,r
! FLOOD 0,0: start from the point 0,0 and color continuous pixels
! until boundary of region is reached
FLOOD 0,0
END
```

Note the use of the FLOOD statement. What happens if you try to draw a square with a BOX AREA or BOX LINES statement?

The problem is that few screens are square and have the same number of pixels in the horizontal and vertical direction. The shape of your window also can vary. Because we want a circular orbit to appear circular on the screen, we must correct for the *aspect ratio* of your screen. The following program computes the aspect ratio and sets the window so that a circle really looks like a circle.

```
PROGRAM circle
LET r = 1                                ! radius of circle
CALL compute_aspect_ratio(r,xwin,ywin)
SET WINDOW -xwin,xwin,-ywin,ywin
SET COLOR "blue"
BOX CIRCLE -r,r,-r,r                      ! draw circle
END

SUB compute_aspect_ratio(r,x,y)
    LET m = 0.1*r                        ! margin
    LET size = r + m
```

```
              ! px, py: # pixels in horizonal and vertical direction
              ASK PIXELS px,py
              IF px > py then
                 LET aspect_ratio = px/py
                 LET x = aspect_ratio*size
                 LET y = size
              ELSE
                 LET aspect_ratio = py/px
                 LET x = size
                 LET y = aspect_ratio*size
              END IF
          END SUB
```

It would be a good idea to add SUB compute_aspect_ratio to your graphics library (see Appendix 3A).

Exercise 4.2   Aspect ratio

How many pixels are there in the horizontal and vertical directions for the screen you are using? What is its aspect ratio? What is the maximum number of colors your video card can support (use ASK MAX COLOR mc)?

Exercise 4.3   Sorting a list of numbers

Write a program that places a list of numbers entered by the user into an array list, and then places the numbers in ascending order into a second array order. Finally, the program prints the sorted list of numbers. Each of these steps should be performed in a separate subroutine. To sort the numbers in list, use an algorithm known as *insertion sort*. First place list(1) into order(1). Then decide where list(2) should go. If list(2) < list(1), then move order(1) to order(2), and place list(2) in order(1); otherwise, place list(2) in order(2). Continue this process until all the numbers are sorted. If there are N numbers to be sorted, then the CPU time required for an insertion sort is approximately proportional to N^2. More sophisticated algorithms can reduce this time to $N \log N$ by sorting parts of the list first (see Problem 14.7).

4.6   LOG-LOG AND SEMILOG PLOTS

In Problem 4.3, we will obtain numerical results for the period T and the semimajor axis a of several elliptical orbits. The values of T and a for the solar system are given in Table 4.1. We first analyze these values and determine if T and a satisfy a simple mathematical relationship.

Suppose we wish to test whether two variables y and x satisfy a functional relationship, $y = f(x)$. To simplify the analysis, we ignore possible errors in the measurements of y and x. The simplest relation between y and x is linear, i.e., $y = mx + b$. The existence of such a relation can be seen by plotting y versus x on simple graph paper.

planet	T (earth years)	a (AU)
Mercury	0.241	0.387
Venus	0.615	0.723
Earth	1.0	1.0
Mars	1.88	1.523
Jupiter	11.86	5.202
Saturn	29.5	9.539
Uranus	84.0	19.18
Neptune	165	30.06
Pluto	248	39.44

Table 4.1   The period T and semimajor axis a of the planets. The unit of length is the astronomical unit (AU). The unit of time is one (earth) year.

From Table 4.1 we see that T is not a linear function of a. For example, an increase in T from 0.24 to 1, an increase of approximately 4, yields an increase in T of approximately 2.5. Depending on the nature of the problem, it might be reasonable to assume the general form

$$y = C\, e^{rx} \tag{4.17}$$

or

$$y = C\, x^n, \tag{4.18}$$

where C, r, and n are unknown parameters.

If we assume the form (4.17), we can take the natural logarithm of both sides to find

$$\ln y = \ln C + rx. \tag{4.19}$$

Hence if (4.17) and (4.19) are applicable, a plot of $\ln y$ versus x would yield a straight line with slope r and intercept $\ln C$. The traditional way to plot the relationship (4.19) is to use semi-log graph paper with the vertical axis ruled logarithmically and the horizontal axis ruled linearly. An alternative procedure is to plot $\ln y$ versus x on regular graph paper.

If we assume the form (4.18), we can take the natural logarithm of both sides and obtain

$$\ln y = \ln C + n \ln x. \tag{4.20}$$

If (4.18) and (4.20) applies, a plot of y versus x on log-log paper yields the exponent n, which is the usual quantity of physical interest.

We illustrate a simple analysis of the data in Table 4.1. Because we expect that the relation between T and a has the form $T = Ca^n$, we plot $\ln T$ versus $\ln a$ (see Fig. 4.3). Inspection of the plot indicates that a linear relationship between $\ln T$ and $\ln a$ is reasonable. The measured slope is found to be approximately 1.50 in agreement with

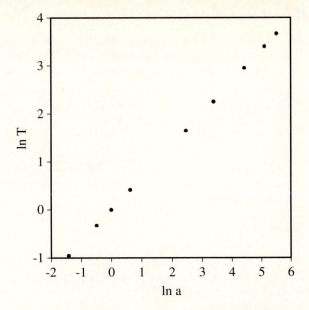

Fig. 4.3   Plot of ln T versus ln a using the astronomical
data in Table 4.1. Verify that the slope is 1.50.

Kepler's second law. In Chapter 7, we discuss the "least squares" method for fitting a
straight line through a number of data points. With a little practice you can do a visual
analysis that is nearly as good.

Problem 4.1   Simple functional forms

Consider the three sets of measurements shown in Table 4.2 of the quantity y as
a function of the variable x. What is the functional form of $y(x)$ that best fits
each set?

x	$y_1(x)$	$y_2(x)$	$y_3(x)$
0	0.00	0.00	2.00
0.5	0.75	1.59	5.44
1.0	3.00	2.00	14.78
1.5	6.75	2.29	40.17
2.0	12.00	2.52	109.20
2.5	18.75	2.71	296.83

Table 4.2   Determine the functional forms of $y(x)$ for the three sets of numerical data. There
are no measurement errors, but there are roundoff errors.

4.7   SIMULATION OF THE ORBIT

Program planet uses three one-dimensional arrays of two elements each to represent the position, velocity, and acceleration of the orbiting object. The use of an array allows us to write the equation of motion (4.8) in a symmetrical form:

```
LET accel(i) = -GM*pos(i)/(r*r*r)
LET vel(i) = vel(i) + accel(i)*dt
LET pos(i) = pos(i) + vel(i)*dt
```

The x and y-coordinates of the object are represented by pos(1) and pos(2), respectively. The program uses the Euler-Cromer algorithm. Note that we have used a separate subroutine to plot the orbit.

```
PROGRAM planet                          ! planetary motion
LIBRARY "csgraphics"
DIM pos(2),vel(2)                       ! define dimension of arrays
CALL initial(pos(),vel(),t,GM,dt,nshow)
CALL output(pos(),t)
LET counter = 0
DO
    CALL Euler(pos(),vel(),t,GM,dt)
    LET counter = counter + 1
    IF mod(counter,nshow) = 0 then
        CALL output(pos(),t)            ! draw position of "earth"
    END IF
LOOP until key input
END

SUB initial(pos(),vel(),t,GM,dt,nshow)
    LET GM = 4.0*pi^2                    ! GM constant in astronomical units
    LET t = 0
    INPUT prompt "time step = ": dt
    INPUT prompt "number of time steps between plots = ": nshow
    INPUT prompt "initial x position = ": pos(1)
    LET pos(2) = 0                       ! initial y-position
    LET vel(1) = 0                       ! initial x-velocity
    INPUT prompt "initial y-velocity = ": vel(2)
    ! assumed maximum value of semi-major axis
    LET r = 2*pos(1)
    CALL compute_aspect_ratio(r,xwin,ywin)  ! in library
    SET WINDOW -xwin,xwin,-ywin,ywin
END SUB

SUB Euler(pos(),vel(),t,GM,dt)      ! Euler-Cromer algorithm
    DIM accel(2)
    LET r2 = pos(1)*pos(1) + pos(2)*pos(2)
    LET r3 = r2*sqr(r2)
```

```
        FOR i = 1 to 2
            LET accel(i) = -GM*pos(i)/r3
            LET vel(i) = vel(i) + accel(i)*dt
            LET pos(i) = pos(i) + vel(i)*dt
        NEXT i
        LET t = t + dt
    END SUB

    SUB output(pos(),t)
        IF t = 0 then
            CLEAR
            LET radius = 0.05              ! radius of "sun" (not to scale)
            ASK MAX COLOR mc
            IF mc > 2 then SET COLOR "red"
            BOX CIRCLE -radius,radius,-radius,radius   ! "sun" at origin
            FLOOD 0,0                      ! paint sun in foreground color
            IF mc > 2 then SET COLOR "green"
        END IF
        PLOT pos(1),pos(2)
    END SUB
```

Problem 4.2   Verification of Program planet for circular orbits

a. Verify *Program planet* by considering the special case of a circular orbit. For example, choose (in astronomical units) $x(t = 0) = 1$, $y(t = 0) = 0$, and $v_x(t = 0) = 0$. Use the relation (4.11) to find the numerical value of $v_y(t = 0)$ that yields a circular orbit. How small a value of Δt is needed so that a circular orbit is repeated over many periods? Modify the program so that the Euler-Richardson algorithm is incorporated. Compare the orbits for the same choice of initial conditions and Δt. Which algorithm appears to give a circular orbit that repeats over the most periods?

b. Write a subroutine to compute the total energy (see (4.5)). (It is sufficient to calculate E/m.) For the same initial conditions as in part (a), choose Δt so that the total energy E is conserved to a good approximation. Is this value of Δt small enough to yield a circular orbit reproducible over many periods? Is it possible to choose a value of Δt that conserves the energy exactly? For a given value of Δt, which algorithm conserves the total energy better?

c. Think of a simple condition that allows you to find the numerical value of the period and implement this condition in a separate subroutine. Run *Program planet* for different sets of values of $x(t = 0)$ and $v_y(t = 0)$, consistent with the condition for a circular orbit. Set $y(t = 0) = 0$ and $v_x(t = 0) = 0$. For each orbit, measure the radius and the period and verify Kepler's third law.

d. Show that the simple Euler method does not yield stable orbits for the same choice of Δt as you used in parts (a)–(c). Is it possible to simply choose a smaller value of Δt or is the Euler method not stable for this dynamical system?

Problem 4.3   Verification of Kepler's second and third law

a. Set $y(t = 0) = 0$ and $v_x(t = 0) = 0$ and find by trial and error several values
of $x(t = 0)$ and $v_y(t = 0)$ that yield elliptical orbits of a convenient size. Compute the total energy, angular momentum, semimajor and semiminor axes,
eccentricity, and period for each orbit. What is the significance of the negative sign for the total energy? Plot your data for the dependence of the period
T on the semimajor axis a and verify Kepler's third law. Given the ratio of
T^2/a^3 that you found in your program, determine the numerical value of this
ratio in SI units for our solar system.

b. Choose initial conditions such that a convenient elliptical orbit is obtained.
Because the distance between adjacent "points" of an orbit is a measure of the
speed of the planet, determine qualitatively the variation of the planet's speed.
Where is the speed a maximum (minimum)?

c. The force center is at $(x, y) = (0, 0)$ and is one focus. Find the second focus
by symmetry. Compute the sum of the distances from each point on the orbit
to the two foci and verify that the orbit is an ellipse.

d. According to Kepler's second law, the orbiting object sweeps out equal areas
in equal times. Because the time step Δt is fixed, it is sufficient to compute the
area of the triangle swept in each time step. This area equals one-half the base
of the triangle times its height, or $\frac{1}{2}\Delta t\,(\mathbf{r} \times \mathbf{v}) = \frac{1}{2}\Delta t\,(xv_y - yv_x)$. Is this area
a constant?

*e. You probably have noticed that the Euler-Cromer and the Euler-Richardson
algorithms with a fixed value of Δt break down if the "planet" is too close to
the sun. How can you visually confirm the breakdown of the algorithms? What
is the cause of the failure of the method? One possibility is to use the Euler-
Richardson algorithm with a variable time step to obtain better numerical
results (see Project 4.3).

Problem 4.4   Non-inverse square forces

a. Consider the dynamical effects of a small change in the attractive inverse-
square force law, e.g., let the magnitude of the force equal $Cm/r^{2+\delta}$, where
$\delta << 1$. Take the numerical value of the constant C to be $4\pi^2$ as before.
Consider the initial conditions $x(t = 0) = 1$, $y(t = 0) = 0$, $v_x(t = 0) = 0$, and
$v_y(t = 0) = 5$. Choose $\delta = 0.05$ and determine the nature of the orbit. Does
the orbit of the planet retrace itself? Verify that your result is not due to your
choice of Δt. Does the planet spiral away from or toward the sun? The path of
the planet can be described as an elliptical orbit that slowly rotates or *precesses*
in the same sense as the motion of the planet. A convenient measure of the
precession is the angle between successive orientations of the semimajor axis
of the ellipse. This angle is the rate of precession per revolution. Estimate the
magnitude of this angle for your choice of initial conditions and δ. What is the
effect of decreasing the semimajor axis? What is the effect of changing δ?

b. Einstein's theory of classical gravitation (the general theory of relativity) predicts corrections to Newton's equations of motion in the weak gravitational field limit. The result is that the equation of motion for the trajectory of a particle can be written as

$$\frac{d^2\mathbf{r}}{dt^2} = -\frac{GM}{r^2}\left[1 - \frac{\alpha GM}{c^2 r}\right]\hat{\mathbf{r}}. \tag{4.21}$$

The constant α is dimensionless. Choose units such that $GM = 4\pi^2$ and assume $\alpha/c^2 = 10^{-3}$. Determine the nature of the orbit for this potential. (For our solar system the constant α/c^2 is a maximum for the planet Mercury, but is much smaller than 10^{-3}.)

c. Suppose that the attractive gravitational force law depends on the inverse-cube of the distance, e.g., Cm/r^3. What are the units of C? Take the numerical value of C to be $4\pi^2$. Consider the initial condition $x(t=0) = 1$, $y(t=0) = 0$, $v_x(t=0) = 0$, and determine analytically the value of $v_y(t=0)$ required for a circular orbit. How small a value of Δt is needed so that the numerical simulation yields a circular orbit over several periods? How does this value of Δt compare with the value needed for the inverse-square force law?

d. Vary $v_y(t=0)$ by approximately 2% from the circular orbit condition that you determined in part (c). What is the nature of the new orbit? What is the sign of the total energy? Is the orbit bound? Is it closed?

Problem 4.5 Effect of drag resistance on satellite orbit

Consider a satellite in orbit about the earth. In this case it is convenient to measure distances in terms of the radius of the earth, $R = 6.37 \times 10^6$ m, and time in terms of hours. In these earth units (EU), the value of the gravitational constant G is

$$G = 6.67 \times 10^{-11} \frac{m^3}{kg \cdot s^2}\left(\frac{1\,EU}{6.37 \times 10^6\,m}\right)^3\left(3.6 \times 10^3 s/h\right)^2$$

$$= 3.34 \times 10^{-24} \frac{EU^3}{kg \cdot h^2}. \tag{4.22}$$

Because the force on the satellite is proportional to Gm, where m is the mass of the earth, we need to evaluate the product Gm in earth units. We obtain

$$Gm = 3.34 \times 10^{-24} \frac{EU^3}{kg \cdot h^2} \times 5.99 \times 10^{24}\,kg$$

$$= 20.0\,EU^3/h^2. \qquad \text{(earth units)} \tag{4.23}$$

Modify *Program planet* to incorporate the effects of drag resistance on the motion of an orbiting earth satellite. Assume the drag force is proportional to the square of the speed of the satellite. To be able to observe the effects of air resistance in a reasonable time, take the magnitude of the drag force to be approximately one-tenth of the magnitude of the gravitational force. Choose initial conditions such that a circular orbit would be obtained in the absence of drag resistance and allow at least one revolution before "switching on" the drag resistance. Describe the qualitative change in the orbit due to drag resistance. How does the total energy and the speed of the satellite change with time?

4.8 PERTURBATIONS

We now challenge your intuitive understanding of Newton's laws of motion by considering perturbations on the motion of an orbiting object. In each case answer the questions before doing the simulation.

Problem 4.6   Radial perturbations

a. Suppose that a small radial "kick" or impulsive force is applied to a satellite in a circular orbit about the earth (see Fig. 4.4a.) How will the orbit change?

b. How does the change in the orbit depend on the strength of the kick and its duration?

c. After you have answered parts (a) and (b), modify your program and do the simulation (see *Program mouse*) and SUB Euler listed after Problem 4.7. Choose earth units so that the numerical value of the product *Gm* is given by (4.23). Is the orbit *stable*, e.g., does a small impulse lead to a small change in the orbit? Does the orbit retrace itself indefinitely if no further perturbations are applied?

d. Determine if the angular momentum and total energy are changed by a radial perturbation.

Problem 4.7   Tangential perturbations

a. Suppose that a small tangential kick is applied to a satellite in a circular orbit about the earth (see Fig. 4.4b). How will the orbit change?

b. How does the change in the orbit depend on the strength of the kick and its duration?

c. After you have answered parts (a) and (b), perform the simulation and determine the new orbit. Is the orbit stable?

d. Determine if the angular momentum and total energy are changed by a tangential perturbation.

e. Determine the stability of the inverse-cube force law (see Problem 4.4) to radial or tangential perturbations.

We now modify SUB Euler so that we can apply a kick. If we apply a vertical kick when the position of the satellite is as shown in Fig. 4.4a, the impulse would be tangential to the orbit. A radial kick can be applied when the satellite is as shown in

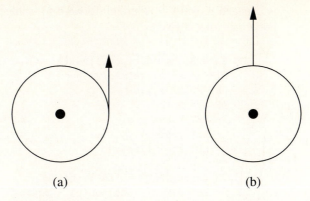

(a)   (b)

Fig. 4.4 (a) An impulse applied in the tangential direction. (b) An impulse applied in the radial direction.

Fig. 4.4b. It is convenient to use a mouse to determine where the impulsive force will be applied. *Program mouse* illustrates how the mouse works in True BASIC:

```
PROGRAM mouse
DO
    ! return current x and y window coordinates and state of mouse
    GET MOUSE x,y,s
    IF s = 1 then
       PRINT "button down",x,y
    ELSE IF s = 2 then
       PRINT "button clicked",x,y
    ELSE IF s = 3 then
       PRINT "button released",x,y
    END IF
LOOP
END
```

The variable s is the state of the mouse when the mouse cursor is at position (x, y) and has the meaning:

$s = 0$ Mouse button not pressed.
$s = 1$ Mouse button held down while dragging.
$s = 2$ Mouse button clicked at (x, y).
$s = 3$ Mouse button released at (x, y).

The modified SUB EulerRichardson listed below uses the GET MOUSE statement to determine the coordinates where the mouse is clicked. The impulse is applied the next time the orbit is in the neighborhood of these coordinates. Since the impulse changes the momentum, the velocity (momentum) of the orbiting object in the desired direction is changed directly. It is straightforward to change SUB EulerRichardson so that a horizontal kick is given.

```
SUB EulerRichardson(pos(),vel(),t,GM,dt,kick_flag$,xsave,ysave)
    DIM accel(2),velmid(2),posmid(2)
    GET MOUSE x,y,s
    IF s = 2 then
        LET xsave = x                    ! save position where mouse clicked
        LET ysave = y
        LET kick_flag$ = "on"
    END IF
    IF (kick_flag$ = "on") then
        LET diffx = abs(pos(1) - xsave)
        LET diffy = abs(pos(2) - ysave)
        ! wait until orbit reaches point where mouse clicked
        IF (diffx < 0.02) and (diffy < 0.02) then
            LET v2 = vel(1)*vel(1) + vel(2)*vel(2)
            LET v = sqr(v2)
            LET impulse = 0.1*v        ! magnitude of impulse/mass
            LET vel(2) = vel(2) + impulse      ! vertical kick
            LET kick_flag$ = "off"
        END IF
    END IF
    ! compute acceleration at beginning of interval
    CALL acceleration(pos(),accel(),GM)
    FOR i = 1 to 2
        LET velmid(i) = vel(i) + 0.5*accel(i)*dt
        LET posmid(i) = pos(i) + 0.5*vel(i)*dt
    NEXT i
    ! compute acceleration at middle of interval
    CALL acceleration(posmid(),accel(),GM)
    ! position and velocity at end of interval
    FOR i = 1 to 2
        LET vel(i) = vel(i) + accel(i)*dt
        LET pos(i) = pos(i) + velmid(i)*dt
    NEXT i
    LET t = t + dt
END SUB


SUB acceleration(position(),accel(),GM)
    LET r2 = position(1)*position(1) + position(2)*position(2)
    LET r3 = r2*sqr(r2)
    FOR i = 1 to 2
        LET accel(i) = -GM*position(i)/r3
    NEXT i
END SUB
```

Use the modified form of SUB EulerRichardson to simulate the effects of the perturbations in Problems 4.6 and 4.7.

4.9   VELOCITY SPACE

In Problems 4.6 and 4.7 your intuition might have been incorrect. For example, you might have thought that the orbit would elongate in the direction of the kick. In fact the orbit does elongate, but in a direction *perpendicular* to the kick. Do not worry, you are in good company! Few students have a good qualitative understanding of Newton's law of motion (cf. McCloskey). One qualitative way to state Newton's second law is

> *Forces act on the paths of particles by changing velocity not position.*

If we fail to take into account this property of Newton's second law, we will encounter physical situations that appear counter-intuitive.

Because force acts to change velocity, it is reasonable to consider both velocity and position on an equal basis. In fact position and momentum are treated in such a manner in advanced formulations of classical mechanics (see Chapter 5) and in quantum mechanics.

In Problem 4.8 we "discover" some of the properties of orbits in velocity space in the context of the bound motion of a particle in an inverse-square force. Modify your program so that the path in velocity space of the earth is plotted. That is, plot the point (v_x, v_y) the same way you plotted the point (x, y). The path in velocity space is a series of successive values of the object's velocity vector. If the position space orbit is an ellipse, what is the shape of the orbit in velocity space?

Problem 4.8   Properties of velocity space orbits

a.   Verify that the velocity space orbit is a circle, even if the orbit in position space is an ellipse. (All that is needed is to pass the array vel() to SUB output.) Does the center of this circle coincide with the origin $(v_x, v_y) = (0, 0)$ in velocity space? Consider both elliptical and circular orbits in position space. Choose the same initial conditions that you considered in Problems 4.2 and 4.3.

****b.**   It is instructive to observe the orbit in position space and in velocity space at the same time. One way to do so is to divide the screen into two windows. For example, the statement OPEN #1: screen 0,0.5,0,1 opens the left-half of the screen and the statement OPEN #2: screen 0.5,1,0,1 opens the right-half. A window is identified by a channel number, consisting of # followed by a number or numeric expression. A channel number can be passed as a parameter to a subroutine. For example, if we want to plot the velocity in window #2, we can use the statement CALL output(#2,vel(),"blue") with

```
SUB output(#9,plot(),color$)
    WINDOW #9
    SET COLOR color$
    PLOT plot(1),plot(2)
END SUB
```

A more detailed example of multiple windows is discussed in Section 5.4.

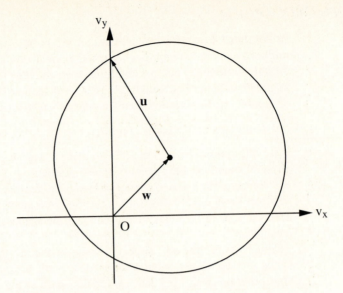

Fig. 4.5   The orbit of a particle in velocity space. The vector
w points from the origin in velocity space to the
center of the circular orbit. The vector **u** points from
the center of the orbit to the point (v_x, v_y).

* **c.**   Let **u** denote the radius vector of a point on the velocity circle, and **w** denote
the vector from the origin in velocity space to the center of the velocity circle
(see Fig. 4.5). Then the velocity of the particle can be written as

$$\mathbf{v} = \mathbf{u} + \mathbf{w}. \tag{4.24}$$

Compute **u** and verify that its magnitude is given by

$$u = GMm/L, \tag{4.25}$$

where L is the magnitude of the angular momentum. Note that L is proportional to m so that it is not necessary to know the magnitude of m.

* **d.**   Verify that at each moment in time, the planet's position vector **r** is perpendicular to **u**. Explain why this relation holds.

* Problem 4.9   Perturbations in velocity space

How does the velocity space orbit change when an impulsive kick is applied in the
tangential or in the radial direction? How does the magnitude and direction of **w**
change? From the observed change in the velocity orbit and the above considerations, explain the observed change of the orbit in position space.

4.10   A MINI-SOLAR SYSTEM

So far our study of planetary orbits has been restricted to two-body central forces. However, the solar system is not a two-body system, because the planets exert gravitational forces on one another. Although the interplanetary forces are small in magnitude in comparison to the gravitational force of the Sun, they can produce measurable effects. For example, the existence of the planets Neptune and Pluto was conjectured on the basis of a discrepancy between the experimentally measured orbit of Uranus and the predicted orbit calculated from the known forces.

The presence of other planets implies that the total force on a given planet is not a central force. Furthermore, because the orbits of the planets are not exactly in the same plane, an analysis of the solar system must be extended to three dimensions if accurate calculations are required. However for the sake of simplicity, we consider a model of a two-dimensional solar system with two planets in orbit about a fixed sun.

The equations of motion of two planets of mass m_1 and mass m_2 can be written in vector form as (see Fig. 4.6)

$$m_1 \frac{d^2\mathbf{r}_1}{dt^2} = -\frac{GMm_1}{r_1^3}\mathbf{r}_1 + \frac{Gm_1m_2}{r_{21}^3}\mathbf{r}_{21} \tag{4.26a}$$

and

$$m_2 \frac{d^2\mathbf{r}_2}{dt^2} = -\frac{GMm_2}{r_2^3}\mathbf{r}_2 - \frac{Gm_1m_2}{r_{21}^3}\mathbf{r}_{21}, \tag{4.26b}$$

where $\mathbf{r}_1$ and $\mathbf{r}_2$ are directed from the sun to planets 1 and 2 respectively, and $\mathbf{r}_{21} = \mathbf{r}_2 - \mathbf{r}_1$ is the vector from planet 1 to planet 2. It is convenient to divide (4.26) by the mass of each planet and to write the equations of motion as

$$\frac{d^2\mathbf{r}_1}{dt^2} = -\frac{GM}{r_1^3}\mathbf{r}_1 + \frac{Gm_2}{r_{21}^3}\mathbf{r}_{21} \tag{4.27a}$$

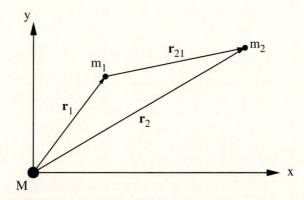

Fig. 4.6   The coordinate system used in (4.26). Planets of mass m_1 and m_2 orbit a sun of mass M.

and

$$\frac{d^2\mathbf{r}_2}{dt^2} = -\frac{GM}{r_2{}^3}\mathbf{r}_2 - \frac{Gm_1}{r_{21}{}^3}\mathbf{r}_{21}. \tag{4.27b}$$

A numerical solution of (4.27) can be obtained by a straightforward extension of our previous methods. To keep *Program planet2* simple, we have used separate one-dimensional arrays for the x and y position and the x and y velocity of the two planets. The array ratio is defined such that $\mathtt{ratio(1)} = +m_2G$ and $\mathtt{ratio(2)} = -m_1G$. If we assume that planet one corresponds to the Earth and planet two corresponds to Jupiter, then $m_1/M \approx 3 \times 10^{-6}$ and $m_2/M \approx 9.1 \times 10^{-4}$. For illustrative purposes, we adopt the numerical values $m_1/M = 10^{-3}$ and $m_2/M = 4 \times 10^{-2}$, and hence $\mathtt{ratio(1)} = (m_2/M)GM = 0.04 \times \mathrm{GM}$, and $\mathtt{ratio(2)} = -(m_1/M)GM = -0.001 \times \mathrm{GM}$.

```
PROGRAM planet2
! d = 2 solar system with major and minor planet
DIM x(2),y(2),vx(2),vy(2),ratio(2)
LIBRARY "csgraphics"
CALL initial(x(),y(),vx(),vy(),t,GM,ratio(),dt,nshow)
CALL output(x(),y(),t)
LET counter = 0
DO
    CALL Euler(x(),y(),vx(),vy(),t,GM,ratio(),dt)
    LET counter = counter + 1
    IF mod(counter,nshow) = 0 then
        CALL output(x(),y(),t)
    END IF
LOOP until key input
END

SUB initial(x(),y(),vx(),vy(),t,GM,ratio(),dt,nshow)
    LET t = 0
    LET GM = 4.0*pi*pi              ! gravitational constant
    LET dt = 0.001                  ! time step (yrs)
    LET nshow = 20                  ! # of iterations between output
    ! planet one initial conditions
    LET x(1) = 2.52
    LET y(1) = 0
    LET vx(1) = 0
    LET vy(1) = sqr(GM/x(1))
    ! mass of planet 2 is 0.04 mass of sun
    LET ratio(1) = 0.04*GM
    ! planet two initial conditions
    LET x(2) = 5.24
    LET y(2) = 0
    LET vx(2) = 0
    LET vy(2) = sqr(GM/x(2))
    ! mass of planet 1 is 0.001 mass of sun
    LET ratio(2) = -0.001*GM
```

```
                    ! set up window
                    CALL compute_aspect_ratio(2*x(2),xwin,ywin)
                    SET WINDOW -xwin,xwin,-ywin,ywin
                END SUB

                SUB Euler(x(),y(),vx(),vy(),t,GM,ratio(),dt)
                    ! compute distance between planets 1 and 2
                    LET dx = x(2) - x(1)
                    LET dy = y(2) - y(1)
                    LET dr2 = dx*dx + dy*dy
                    LET dr = sqr(dr2)
                    LET dr3 = dr2*dr
                    FOR i = 1 to 2                    ! sum over planets
                        LET r2 = x(i)*x(i) + y(i)*y(i)
                        LET r = sqr(r2)
                        LET r3 = r2*r
                        LET ax = -GM*x(i)/r3 + ratio(i)*dx/dr3
                        LET ay = -GM*y(i)/r3 + ratio(i)*dy/dr3
                        LET vx(i) = vx(i) + ax*dt
                        LET vy(i) = vy(i) + ay*dt
                        LET x(i) = x(i) + vx(i)*dt
                        LET y(i) = y(i) + vy(i)*dt
                    NEXT i
                    LET t = t + dt
                END SUB

                SUB output(x(),y(),t)                ! plot orbits
                    IF t = 0 then
                        SET COLOR "red"
                        BOX CIRCLE -0.1,0.1,-0.1,0.1    ! sun at origin
                        FLOOD 0,0
                    END IF
                    SET COLOR "blue"
                    PLOT POINTS: x(1),y(1)           ! planet one
                    SET COLOR "green"
                    PLOT POINTS: x(2),y(2)           ! planet two
                END SUB
```

Problem 4.10   Planetary perturbations

Use *Program planet2* with the initial conditions given in the program. What
would be the shape of the orbits and the periods of the two planets if they did not
mutually interact? What is the qualitative effect of their mutual interaction? Why
is one planet affected more by the mutual interaction than the other? Describe the
shape of the two orbits. Is the angular momentum and the total energy of planet
one conserved? Is the total energy and total angular momentum of the two planets
conserved? A related, but more time consuming problem is given in Project 4.2.

Problem 4.11   Double stars

Another interesting dynamical system is a planet orbiting about two fixed stars of equal mass. In this case there are no closed orbits, but the orbits can be classified as either stable or unstable. Stable orbits may be open loops that encircle both stars, figure eights, or Kepler-like orbits that encircle only one star. Unstable orbits will eventually collide with one of the stars. Modify *Program planet2* to simulate the double star system, with the first star located at the origin and the second star of equal mass located at $(2, 0)$. Place the planet at $(1.1, 1)$ and systematically vary the x and y components of the velocity to obtain different types of orbits. Try other initial positions.

4.11   TWO-BODY SCATTERING

Much of our current understanding of the structure of matter comes from scattering experiments. In this section we explore one of the more difficult concepts in the theory of scattering, the *differential cross section*.

A typical experiment involves a beam with many incident particles all with the same kinetic energy. The coordinate system is shown in Fig. 4.7. The incident particles come from the left with an initial velocity **v** in the $+x$ direction. We assume that the radius of the beam is R. The target contains many scattering centers, but for calculational purposes we may consider scattering off only one particle if the target is sufficiently thin. In general, the width of the beam is larger than the size of the target. We take the center of the beam and the center of the target to be on the x axis. The *impact parameter* b is the perpendicular distance from the initial trajectory to a parallel line through the center of the target (see Fig. 4.7).

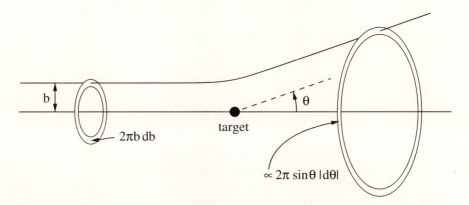

Fig. 4.7   The coordinate system used to define the differential scattering cross section. Particles passing through the beam area $2\pi b\,db$ are scattered into the solid angle $d\Omega$.

When an incident particle comes close to the target, it is deflected. In a typical experiment, the scattered particles are counted in a detector that is far from the target. The final velocity of the scattered particles is $\mathbf{v}'$, and the angle between $\mathbf{v}$ and $\mathbf{v}'$ is the scattering angle θ.

Let us assume that the scattering is elastic and that the target is much more massive than the beam particles so that the target can be considered to be fixed. (The latter condition can be relaxed by using center of mass coordinates.) We also assume that no incident particle is scattered more than once. These considerations imply that the initial speed and final speed of the incident particles are equal. The functional dependence of θ on b depends on the nature of the target, i.e., the force on the beam particles. In a typical experiment the number of particles in an angular region between θ and $\theta + d\theta$ is detected for many values of θ. These detectors measure the number of particles scattered into the solid angle $d\Omega = \sin\theta d\theta d\phi$ centered about θ. The *differential cross section* $\sigma(\theta)$ is defined by the relation

$$\frac{dN}{N} = n\,\sigma(\theta)\,d\Omega, \tag{4.28}$$

where dN is the number of particles scattered into the solid angle $d\Omega$ centered about θ and the azimuthal angle ϕ, N is the total number of particles in the beam, and n is the target density defined as the number of targets per unit area.

The interpretation of (4.28) is that the fraction of particles that is scattered into the solid angle $d\Omega$ is proportional to $d\Omega$ and the density of the target. We see from (4.28) that $\sigma(\theta)$ can be interpreted as the effective area of a target particle for the scattering of an incident particle into the element of solid angle $d\Omega$. Particles that are not scattered are ignored. Another way of thinking about $\sigma(\theta)$ is that it is the ratio of the area $b\,db\,d\phi$ to the solid angle $d\Omega = \sin\theta d\theta\,d\phi$, where $b\,db\,d\phi$ is the infinitesimal cross sectional area of the beam that scatters into the solid angle defined by θ to $\theta + d\theta$ and ϕ to $\phi + d\phi$. The alternative notation for the differential cross section, $d\sigma/d\Omega$, comes from this interpretation.

To do an analytical calculation of $\sigma(\theta)$, we write

$$\sigma(\theta) = \frac{d\sigma}{d\Omega} = \frac{b}{\sin\theta}\left|\frac{db}{d\theta}\right|. \tag{4.29}$$

We see from (4.29) that the analytical calculation of $\sigma(\theta)$ involves b as a function of θ, or more precisely, how b changes to give scattering through an infinitesimally larger angle $\theta + d\theta$.

In a real experiment, the values of the impact parameters and the trajectories of the incident beam particles are not known. Instead, particles enter from the left (see Fig. 4.7) with random values of the impact parameter b and azimuthal angle ϕ and the number of particles scattered into the various detectors is measured. In our simulation we know the value of b, and we can integrate Newton's equations of motion to find the angle at which the incident particle is scattered. Hence, in contrast to the analytical calculation, a simulation naturally yields θ as a function of b.

Because the differential cross section is usually independent of ϕ, we need to consider only beam particles at $\phi = 0$. However, we have to take into account the fact that in a real beam, there are more particles at some values of b than at others. That is, the number of particles in a real beam is proportional to $2\pi b \Delta b$, the area of the ring between b and $b + \Delta b$, where we have integrated over the values of ϕ. Here Δb is the interval between the values of b used in the program. The smaller the value of Δb, the more accurate are the results for the differential cross section. Because there is only one target in the beam, the target density is $n = 1/(\pi R^2)$.

Program scatter uses the Euler-Richardson algorithm to compute the trajectories of beam particles with equally spaced impact parameters between $b = 0$ and the beam radius R. The trajectory of the beam particle is computed until the particle is outside the scattering region. The possible values of the scattering angle between 0 and $180°$ are divided into bins of width dtheta. To compute the number of particles coming from a ring of the beam of radius b that are scattered into the angle θ, we introduce the array dN(theta) and accumulate the value of b associated with each bin or "detector" (as discussed above, the number of particles in a ring of radius b is proportional to b). If theta is the scattering angle for an incident particle with impact parameter b, we have

```
LET dN(theta) = dN(theta) + b
```

We count N, the total number of scattered particles, in the same way. Hence, we write (see SUB detector)

```
LET N = N + b
```

After the trajectories of all the beam particles have been computed, the differential cross section $\sigma(\theta)$ is computed using the relation (4.28) in SUB output.

```
PROGRAM scatter
! compute differential and total cross section
DIM dN(0 to 360)
LIBRARY "csgraphics"
CALL initial(x0,vx0,R,delta_b,dt,nbin,dtheta)
LET N = 0                              ! number of incident particles
LET b = 0
! make sure that b = R actually done, but b might exceed R slightly
DO while b <= R + delta_b/2
    CALL trajectory(vx,vy,vx0,x0,b,dt)
    CALL detector(dN(),vx,vy,b,dtheta,N)
    LET b = b + delta_b
LOOP
SET COLOR "black/white"
SET CURSOR 22,1
PRINT "Hit any key for the differential cross section."
DO
LOOP until key input
CALL output(dN(),N,nbin,dtheta,R)
END

SUB initial(x0,vx0,R,delta_b,dt,nbin,dtheta)
    INPUT prompt "kinetic energy of incident particles = ": E
```

```
            INPUT prompt "incremental change in impact parameter = ": delta_b
            LET R = 1.0                    ! beam radius
            LET vx0 = sqr(2*E)             ! mass = 1
            LET dt = 0.01                  ! time step
            LET nbin = 9                   ! number of detectors
            LET dtheta = 180/nbin          ! degrees
            ! initial position of beam relative to target
            LET x0 = -3
            CALL compute_aspect_ratio(abs(x0),xwin,ywin)
            SET WINDOW -xwin,xwin,-ywin,ywin
            SET COLOR "blue"
            LET Rs = 1.0                    ! scattering region radius
            BOX CIRCLE -Rs,Rs,-Rs,Rs
            PLOT 0,0
            SET COLOR "red"
        END SUB

        SUB trajectory(vx,vy,vx0,x0,b,dt)
            LET y = b
            LET x = x0
            LET vx = vx0
            LET vy = 0
            DO
                CALL EulerRichardson(x,y,vx,vy,dt)    ! one time step
                PLOT POINTS: x,y
            LOOP until x*x + y*y > x0*x0 + b*b
        END SUB

        SUB detector(dN(),vx,vy,b,dtheta,N)
            DECLARE DEF arctan
            LET theta = int(arctan(vx,vy)/dtheta)    ! scattering angle
            LET dN(theta) = dN(theta) + b
            LET N = N + b
        END SUB

        SUB EulerRichardson(x,y,vx,vy,dt)
            ! could use arrays to reduce # of statements
            CALL force(x,y,fx,fy)
            ! mass is unity in program units
            LET vxmid = vx + 0.5*fx*dt
            LET vymid = vy + 0.5*fy*dt
            LET xmid = x + 0.5*vx*dt
            LET ymid = y + 0.5*vy*dt
            CALL force(xmid,ymid,fxmid,fymid)
            LET vx = vx + fxmid*dt
            LET vy = vy + fymid*dt
            LET x = x + vxmid*dt
            LET y = y + vymid*dt
        END SUB
```

```
SUB force(x,y,fx,fy)
    LET r2 = x*x + y*y
    LET r = sqr(r2)
    ! force for simple model of Hydrogen atom
    ! range of force set equal to unity
    IF r2 <= 1 then
       LET f = 1/r2 - r
    ELSE
       LET fr = 0
    END IF
    LET fx = f*x/r
    LET fy = f*y/r
END SUB

SUB output(dN(),N,nbin,dtheta,R)
    CLEAR
    LET dtheta_rad = dtheta*pi/180      ! radians
    LET target_density = 1/(pi*R*R)
    PRINT "theta","dN/N","sigma(theta)"
    PRINT
    FOR i = 0 to nbin
        LET theta = (i + 0.5)*dtheta_rad
        IF dN(i) > 0 then
            LET d_Omega = 2*pi*sin(theta)*dtheta_rad
            LET diffcross = dN(i)/(d_Omega*N*target_density)
            LET totalcross = totalcross + diffcross*d_Omega
            PRINT (i+1)*dtheta,dN(i)/N,diffcross
        END IF
    NEXT i
    PRINT "total cross section ="; totalcross
END SUB

FUNCTION arctan(x,y)
    ! convert -pi/2 to pi/2 radians to 0 to 180 degrees
    IF x > 0  then
       LET theta = atn(abs(y/x))
    ELSE IF  y > 0 then
       LET theta = atn(y/x) + pi
    ELSE
       LET theta = atn(y/x) + pi/2
    END IF
    LET arctan = theta*180/pi
END DEF
```

In Problem 4.12, we consider a model of the hydrogen atom for which the force on a beam particle equals zero for $r > 1$. Because we do not count the beam particles that are not scattered, we set the beam radius $R = 1$. For forces that are not identically zero in any region, e.g., $F \sim 1/r^2$, we need to choose a minimum angle for θ and not count particles whose scattering angle is less than this minimum.

Problem 4.12   Scattering from a model hydrogen atom

a.   Consider a model of the hydrogen atom for which a positively charged nucleus
of charge $+e$ is surrounded by a uniformly distributed negative charge of
equal magnitude. The spherically symmetric negative charge distribution is
contained within a sphere of radius a. It is straightforward to show that the
force between a positron of charge $+e$ and this model hydrogen atom is given
by

$$f(r) = \begin{cases} 1/r^2 - r/a^3, & \text{if } r \le a \\ 0, & r > a. \end{cases} \tag{4.30}$$

We have chosen units such that $e^2/(4\pi\epsilon_0) = 1$, and the mass of the positron is
unity. What is the ionization energy in these units? *Program scatter* incorporates this force with $a = 1$. Use *Program scatter* to compute the trajectories for $b = 0.25, 0.5$, and 0.75 and describe the trajectories qualitatively. Use
$E = 0.125$ and $\Delta t = 0.01$ and set the beam radius $R = 1$. Is the force on the
positron from the model hydrogen atom purely repulsive?

b.   Determine the cross section for $E = 0.125$. Choose nine bins so that the angular width of a detector is `delta` $= 20$ degrees and let `delta_b` $= 0.1, 0.01$,
and 0.002. How does the accuracy of your results depend on the number of
bins? Determine the differential cross section for different energies and explain its qualitative energy dependence.

c.   The total cross section σ_T is defined as

$$\sigma_T = \int \sigma(\theta)\, d\Omega. \tag{4.31}$$

What is the value of σ_T for $E = 0.125$? Does σ_T depend on E? The total cross
section has units of area. To what area does it refer? What would you expect
the total cross section to be for scattering from a hard sphere?

d.   Change the sign of the force so that it corresponds to electron scattering. How
do the trajectories change? Discuss the change in $\sigma(\theta)$.

Problem 4.13   Rutherford scattering

a.   One of the most famous scattering experiments was performed by Geiger and
Marsden who scattered a beam of alpha particles on a thin gold foil. Based on
these experiments, Rutherford deduced that the positive charge of the atom is
concentrated in a small region at the center of the atom rather than distributed
uniformly over the entire atom. Modify the force law in *Program scatter* so
that the force law is $f(r) = 1/r^2$ for all r, corresponding to electron-electron
scattering. Compute the trajectories for $b = 0.25, 0.5$, and 0.75 and describe
the trajectories qualitatively. Choose $E = 5$ and $\Delta t = 0.01$. The default value
of x_0, the initial x-coordinate of the beam, is $x_0 = -5$. Is this value reasonable?

b.   With $E = 5$ determine the cross section with nbin $= 18$. Choose $R = 2$. Then vary nbin and delta_b and discuss the accuracy of your results. The analytical result is that $\sigma(\theta)$ varies as $[\sin(\theta/2)]^{-4}$. How do your computed results compare with this dependence on θ? If necessary, decrease delta_b to obtain better results. Are your results better or worse at small angles, intermediate angles, or large angles near $180°$? Explain.

c.   Because the Coulomb force is long range, there is scattering at all impact parameters. Increase the beam radius R and determine if your results for $\sigma(\theta)$ change. What happens to the total cross section as you increase R?

d.   Compute $\sigma(\theta)$ for different values of E and estimate the dependence of $\sigma(\theta)$ on E.

Problem 4.14   Scattering by other potentials

a.   A simple phenomenological form for the effective interaction between electrons in metals is the screened Coulomb (or Thomas-Fermi) potential given by

$$V(r) = \frac{e^2}{4\pi\epsilon_0 r} e^{-r/a}. \tag{4.32}$$

The range of the interaction a depends on the density and temperature of the electrons. The form (4.32) is known as the Yukawa potential in the context of the interaction between nuclear particles and as the Debye potential in the context of classical plasmas. Choose units such that $a = 1$ and $e^2/(4\pi\epsilon_0) = 1$. Recall that the force is given by $f(r) = -dV/dr$. Incorporate this force law into *Program scatter* and compute the dependence of $\sigma(\theta)$ on the energy of the incident particle. Choose $R = 3$. Compare your results for $\sigma(\theta)$ with your results from the Coulomb potential.

b.   Modify the force law in *Program scatter* so that $f(r) = 24(2/r^{13} - 1/r^7)$. This form for $f(r)$ is used to describe the interactions between simple molecules (see Chapter 8). Describe some typical trajectories and compute the differential cross section for several different energies. Let $R = 2$. What is the total cross section? How do your results change if you vary R? Choose a small angle as the minimum scattering angle. How sensitive is the total cross section to this minimum angle? Does the differential cross section vary for any other angles beside the smallest scattering angle?

c.   Write a program to compute the differential cross section for the elastic scattering of an incident hard sphere (billiard ball) of radius R_s from a target hard sphere of radius R_t. In this case there is no need to integrate the equations of motion. Rather you need to use geometry and set the angle of incidence equal to the angle of reflection when the target and projectile touch. Before you do a simulation, what result do you expect for the total cross section? Explain your results for $\sigma(\theta)$ and the total cross section.

4.12   PROJECTS

Project 4.1   Effect of a "solar wind"

a. Assume that a satellite is affected not only by the earth's gravitational force, but also by a weak uniform "solar wind" of magnitude W acting in the horizontal direction. The equations of motion can be written as

$$\frac{d^2x}{dt^2} = -\frac{GMx}{r^3} + W \tag{4.33a}$$

$$\frac{d^2y}{dt^2} = -\frac{GMy}{r^3}. \tag{4.33b}$$

Choose initial conditions so that a circular orbit would be obtained for $W = 0$. Then choose a value of W whose magnitude is about 3% of the acceleration due to the gravitational field and compute the orbit. How does the orbit change?

b. Determine the change in the velocity space orbit when the solar wind (4.33) is applied. How does the total angular momentum and energy change? Explain in simple terms the previously observed change in the position space orbit. See Luehrmann for further discussion of this problem.

Project 4.2   Resonances and the asteroid belt

a. A histogram of the number of asteroids versus their distance from the Sun shows some distinct gaps. These gaps, called the *Kirkwood gaps*, are believed to be due to resonance effects. That is, if asteroids were in these gaps, their periods would be simple fractions of the period of Jupiter. Modify *Program planet2* so that planet two has the mass of Jupiter by setting ratio(1) = 0.001 * GM. Because the asteroid masses are very small compared to that of Jupiter, the gravitational force on Jupiter due to the asteroids can be neglected. The initial conditions listed in *Program planet2* are approximately correct for Jupiter. The initial conditions for the asteroid (planet one in *Program planet2*) correspond to the 1/3 resonance (the period of the asteroid is one third that of Jupiter). Modify SUB Euler so that the Euler-Richardson algorithm is used. Run the program with these changes and describe the orbit of the asteroid.

b. Use Kepler's third law, $T^2/a^3 = $ const, to determine the values of a, the asteroid's semimajor axis, such that the ratio of its period of revolution about the Sun to that of Jupiter is 1/2, 3/7, 2/5, and 2/3. Set the initial value of x(1) equal to a for each of these ratios and choose the initial value of vy(1) so that the asteroid would have a circular orbit if Jupiter were not present. Describe the orbits you obtain.

c. It is instructive to plot a as a function of time. However, because it is not straightforward to measure a directly in the simulation, it is more convenient to plot the quantity $-2GMm/E$, where E is the total energy of the asteroid and m is the mass of the asteroid. Because E is proportional to m, the quantity $-2GMm/E$ is independent of m. If the interaction of the asteroid with Jupiter is ignored, it can be shown that $a = -2GMm/E$. Derive this result for circular orbits. Plot the quantity $-2GMm/E$ versus time for about thirty revolutions for the initial conditions in part (b). Take E to be the asteroid kinetic energy plus the asteroid-sun potential energy.

d.   Compare the time dependence of $-2GMm/E$ (a good approximation to a) for those asteroid orbits whose initial positions x(1) range from 2.0 to 5.0 in steps of 0.2. (Choose the initial values of vy(1) so that circular orbits would be obtained in the absence of Jupiter.) Are there any values of x(1) for which the time dependence of a is unusual?

e.   Make a histogram of the number of asteroids versus the value of $a \approx -2GMm/E$ at $t = 2000$. Assume that the initial value of x(1) ranges from 2.0 to 5.0 in steps of 0.02, and use a histogram bin width of 0.1. If you have time, repeat for $t = 5000$, and compare the histogram with your previous results. Is there any evidence for Kirkwood gaps? A resonance occurs when the periods of the asteroid and Jupiter are related by simple fractions. We expect the number of asteroids for values of a corresponding to resonances to be small.

f.   Repeat part (e) with the addition of initial velocities varying from their values for a circular orbit by 1, 3, and 5%.

Project 4.3   The classical helium atom

The classical helium atom is a relatively simple example of a three-body problem and is similar to the gravitational three-body problem of a heavy sun and two light planets (see Fig. 4.8). The significant difference is that the two electrons repel one another, unlike the planetary case where the intraplanetary interaction is attractive. If we ignore the small motion of the heavy nucleus, the equations of motion for the two electrons can be written as:

$$\mathbf{a}_1 = -2\frac{\mathbf{r}_1}{r_1^3} + \frac{\mathbf{r}_1 - \mathbf{r}_2}{r_{12}^3} \qquad\qquad (4.34a)$$

$$\mathbf{a}_2 = -2\frac{\mathbf{r}_2}{r_2^3} + \frac{\mathbf{r}_2 - \mathbf{r}_1}{r_{12}^3}, \qquad\qquad (4.34b)$$

where $\mathbf{r}_1$ and $\mathbf{r}_2$ are measured from the fixed nucleus, and r_{12} is the distance between the two electrons. We have chosen units such that the mass and charge of the electron are both unity. (The charge of the helium nucleus is two in these units.) Because the electrons are sometimes very close to the nucleus, the acceleration can become very large, and a very small time step Δt is required. It is not efficient to use the same small time step throughout the simulation and instead a variable time step or an *adaptive* step size algorithm is suggested. An adaptive step size algorithm can be implemented as follows:

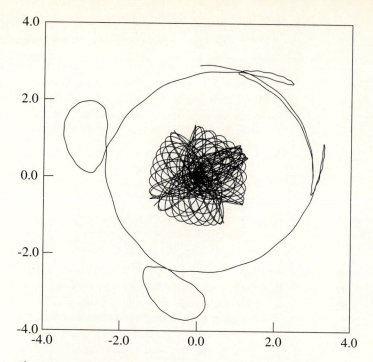

Fig. 4.8 Orbits of the two electrons using the initial condition
$\mathbf{r}_1 = (3, 0)$, $\mathbf{r}_2 = (1, 0)$, $\mathbf{v}_1 = (0, 0.4)$, and $\mathbf{v}_2 = (0, -1)$
(see Project 4.3c).

1. Given a set $\mathbf{f}(t)$ of positions and velocities at time t, use the current value of Δt to compute the set $\mathbf{f}_1(t + \Delta t)$ of positions and velocities at time $t + \Delta t$.

2. Given the same set $\mathbf{f}(t)$ of positions and velocities at time t as in step (1), use half the current value of Δt to compute the set $\mathbf{f}(t + \Delta t/2)$ of positions and velocities at time $t + \Delta t/2$. Then use the set $\mathbf{f}(t + \Delta t/2)$ and the time step $\Delta t/2$ to obtain a second set $\mathbf{f}_2(t + \Delta t)$ of positions and velocities.

3. Use $\mathbf{f}_1(t + \Delta t)$ and $\mathbf{f}_2(t + \Delta t)$ to compute the difference ΔE in the total energy. (It is possible to use the differences in the positions or velocities instead, but it is easier to compare just one dynamical quantity that depends on the positions and velocities.)

4. If ΔE is too large, e.g., greater than 10^{-5}, then reduce the time step by a factor of two and repeat steps (1)–(3). If ΔE is too small, e.g., less than 10^{-7}, then increase the time step by a factor of two. If ΔE is acceptable, e.g., $10^{-7} < \Delta E < 10^{-5}$, then retain the same time step. In the last two cases, use the results from step (2) to obtain the new positions and velocities.

An adaptive step size algorithm can be used with any standard numerical algorithm for solving differential equations. We suggest using the Euler-Richardson algorithm, although the fourth-order Runge Kutta method (see Appendix 5A) is more accurate.

a. For simplicity, we restrict our atom to two dimensions. Modify *Program planet2* to simulate the classical helium atom. Let the initial value of the time step be $\Delta t = 0.001$. Some of the possible orbits are similar to those we have seen in our mini-solar system. For example, try the initial condition $r_1 = (2, 0)$, $r_2 = (-1, 0)$, $v_1 = (0, 0.95)$, and $v_2 = (0, -1)$.

b. The initial condition $r_1 = (1.4, 0)$, $r_2 = (-1, 0)$, $v_1 = (0, 0.86)$, and $v_2 = (0, -1)$ gives "braiding" orbits. Most initial conditions result in unstable orbits in which one electron eventually leaves the atom (autoionization). Make small changes in this initial condition to observe autoionization.

c. The classical helium atom is capable of very complex orbits. Investigate the motion for the initial condition $r_1 = (3, 0)$, $r_2 = (1, 0)$, $v_1 = (0, 0.4)$, and $v_2 = (0, -1)$. Does the motion conserve the total angular momentum? Also try $r_1 = (2.5, 0)$, $r_2 = (1, 0)$, $v_1 = (0, 0.4)$, and $v_2 = (0, -1)$.

d. Choose the initial condition $r_1 = (2, 0)$, $r_2 = (-1, 0)$, and $v_2 = (0, -1)$. Then vary the initial value of v_1 from $(0.6, 0)$ to $(1.3, 0)$ in steps of $\Delta v = 0.02$. For each set of initial conditions calculate the time it takes for autoionization. Assume that ionization occurs when either electron exceeds a distance of six from the nucleus. Run each simulation for a maximum time equal to 2000. Plot the ionization time versus v_{1x}. Repeat for a smaller interval of Δv centered about one of the longer ionization times. (These calculations require much computer resources.) Do the two plots look similar? If so, such behavior is called "self-similar" and is characteristic of chaotic systems and the geometry of fractals (see Chapters 6 and 14). More discussion on the nature of the orbits can be found in Yamamoto and Kaneko.

References and Suggestions for Further Reading

Harold Abelson, Andrea diSessa, and Lee Rudolph, "Velocity space and the geometry of planetary orbits," *Am. J. Phys.* **43**, 579 (1975). See also Andrea diSessa, "Orbit: a mini-environment for exploring orbital mechanics," in 0. Lecarme and R. Lewis, editors, *Computers in Education* 359, North-Holland (1975). Detailed geometrical rather than calculus-based arguments on the origin of closed orbits for inverse-square forces are presented. Are geometrical arguments easier to understand than algebraic arguments? Sections 4.8 and 4.9 are based on these papers.

Ralph Baierlein, *Newtonian Dynamics*, McGraw-Hill (1983). An intermediate level text on mechanics. Of particular interest are the discussions on the stability of circular orbits and the effects of an oblate sun.

John J. Brehm and William J. Mullin, *Introduction to the Structure of Matter*, John Wiley & Sons (1989). See Section 3-4 for a discussion of Rutherford scattering.

R. P. Feynman, R. B. Leighton, M. Sands, *The Feynman Lectures in Physics, Vol. 1*, Addison-Wesley (1963). See Chapter 9.

A. P. French, *Newtonian Mechanics,* W. W. Norton & Company (1971). An introductory level text with more than a cursory treatment of planetary motion.

Ian R. Gatland, "Numerical integration of Newton's equations including velocity-dependent forces," *Am J. Phys.* **62**, 259 (1994). The author chooses a variable time step based on the difference in the calculation of the positions rather than the energy as we did in Project 4.3.

Herbert Goldstein, *Classical Mechanics,* second edition, Addison-Wesley (1980). Chapter 3 has an excellent discussion of the Kepler problem and the conditions for a closed orbit.

Myron Lecar and Fred A. Franklin, "On the original distribution of the asteroids. I," *Icarus* **20**, 422 (1973). The authors use simulations of the motions of asteroids and discuss the Kirkwood gaps.

Arthur W. Luehrmann, "Orbits in the solar wind—a mini-research problem," *Am. J. Phys.* **42**, 361 (1974). Luehrmann emphasizes the desirability of student problems requiring inductive rather than deductive reasoning.

Michael McCloskey, "Intuitive physics," *Sci. Amer.* **249**, 122 (April, 1983). A discussion of the counterintuitive nature of Newton's laws.

John R. Merrill and Richard A. Morrow, "An introductory scattering experiment by simulation," *Am. J. Phys.* **38**, 1104 (1970).

Tomomyuki Yamamoto and Kunihiko Kaneko, "Helium atom as a classical three-body problem," *Phys. Rev. Lett.* **70**, 1928 (1993).

5

Simple Linear and Nonlinear Systems

We explore the qualitative behavior of linear and nonlinear oscillatory systems in the context of classical mechanics and electronics, and introduce the concept of phase space.

5.1 SIMPLE HARMONIC MOTION

There are many physical systems that undergo regular, repeating motion. Motion that repeats itself at definite intervals, e.g., the motion of the earth about the sun, is said to be *periodic*. If an object undergoes periodic motion between two limits over the same path, we call the motion *oscillatory*. Examples of oscillatory motion that are familiar to us from our everyday experience include a plucked guitar string and the pendulum in a grandfather clock. Less obvious examples are microscopic phenomena such as the oscillations of the atoms in crystalline solids.

To illustrate the important concepts associated with oscillatory phenomena, consider an object of mass m connected to the free end of a spring. The object slides on a frictionless, horizontal surface (see Fig. 5.1). We specify the position of the object by x and take $x = 0$ to be the *equilibrium* position of the object, i.e., the position when the spring is relaxed. If the object is moved from $x = 0$ and then released, the object oscillates along a horizontal line. If the spring is not compressed or stretched too far from $x = 0$, the force on the object at position x is linearly related to x:

$$F = -kx. \tag{5.1}$$

The *force constant* k is a measure of the stiffness of the spring. The negative sign in (5.1) implies that the force acts to restore the object to its equilibrium position. Newton's equation of motion of the object can be written as

$$\frac{d^2x}{dt^2} = -\omega_0^2 x, \tag{5.2}$$

where the angular frequency ω_0 is defined by

$$\omega_0^2 = \frac{k}{m}. \tag{5.3}$$

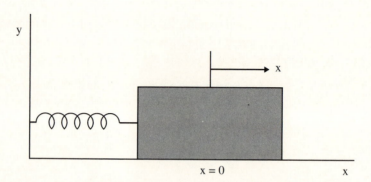

Fig. 5.1 A one-dimensional harmonic oscillator. The block slides horizontally on the frictionless surface.

The equation of motion (5.2) is an example of a *linear* differential equation because it involves only first powers of x and its derivatives. The dynamical behavior described by (5.2) is called *simple harmonic motion* and can be solved analytically in terms of sine and cosine functions. Because the form of the solution will help us introduce some of the terminology needed to discuss oscillatory motion, we include the solution here. One form of the solution is

$$x(t) = A\cos(\omega_0 t + \delta), \tag{5.4}$$

where A and δ are constants and the argument of the cosine is in radians. It is straightforward to check by substitution that (5.4) is a solution of (5.2). The constants A and δ are called the *amplitude* and the *phase* respectively, and can be determined by the initial conditions for x and the velocity $v = dx/dt$.

Because the cosine is a periodic function with period 2π, we know that $x(t)$ in (5.4) also is periodic. We define the *period* T as the smallest time for which the motion repeats itself, i.e.,

$$x(t + T) = x(t). \tag{5.5}$$

Because $\omega_0 T$ corresponds to one *cycle*, we have

$$T = \frac{2\pi}{\omega_0} = \frac{2\pi}{\sqrt{k/m}}. \tag{5.6}$$

The *frequency* ν of the motion is the number of cycles per second and is given by $\nu = 1/T$. Note that T depends on the ratio k/m and not on A and δ. Hence the period of simple harmonic motion is independent of the amplitude of the motion.

Although the position and velocity of the oscillator are continuously changing, the total energy E remains constant and is given by

$$E = \frac{1}{2}mv^2 + \frac{1}{2}kx^2 = \frac{1}{2}kA^2. \tag{5.7}$$

The two terms in (5.7) are the kinetic and potential energies, respectively.

5.2 NUMERICAL SIMULATION OF THE HARMONIC OSCILLATOR

Program sho computes the time dependence of the position and velocity for a linear harmonic oscillator using the Euler-Richardson algorithm. The results for the position x and velocity v are printed on the screen until any key is pressed. Examples of the use of the tab function and PRINT using functions are given in SUB initial and SUB output respectively.

```
PROGRAM sho                          ! simple harmonic oscillator
CALL initial(x,v,w2,t,dt,dt2,nshow)
LET counter = 0
```

```
DO
    CALL update(x,v,w2,t,dt,dt2)
    LET counter = counter + 1
    IF counter = nshow then
        CALL output(x,v,t)
        LET counter = 0
    END IF
LOOP until key input
END


SUB initial(x0,v,w2,t,dt,dt2,nshow)
    INPUT prompt "initial position = ": x0   ! meters
    LET t = 0
    LET v = 0                           ! initial velocity
    ! natural (angular) frequency
    INPUT prompt "ratio of k/m = ": w2
    INPUT prompt "time step = ": dt
    LET dt2 = 0.5*dt
    LET show_time = 0.1                 ! time interval between output
    LET nshow = int(show_time/dt)
    PRINT tab(7);"time";tab(17);"position";tab(28);"velocity"
    PRINT                               ! skip line
END SUB


SUB update(x,v,w2,t,dt,dt2)
    ! Euler-Richardson algorithm
    LET a = -w2*x
    LET vmid = v + a*dt2                ! velocity at midpoint
    LET xmid = x + v*dt2                ! position at midpoint
    LET amid = -w2*xmid                 ! acceleration at midpoint
    LET v = v + amid*dt
    LET x = x + vmid*dt
    LET t = t + dt
END SUB


SUB output(x,v,t)
    ! - print number with leading space or minus sign
    ! % print leading zeroes as '0'
    PRINT using "-----%.####": t,x,v
END SUB
```

In Problem 5.1 we verify that the Euler-Richardson algorithm can be applied to the dynamical motion of the simple harmonic oscillator using a reasonable choice of Δt.

Problem 5.1   Energy conservation

a.   Modify *Program sho* so that E_n, the total energy per unit mass, is computed at time $t_n = t_0 + n\Delta t$. Plot the difference $\Delta E_n = E_n - E_0$ as a function of

t_n for several complete cycles for a given value of Δt. (E_0 is the initial total energy.) Choose $x(t=0)=1$, $v(t=0)=0$ and $\omega_0{}^2 = k/m = 9$ and start with $\Delta t = 0.01$. Is the difference ΔE_n uniformly small throughout the cycle? Does ΔE_n drift, that is, become bigger with time? What is the optimum choice of Δt?

b. Modify the program so that the simple Euler algorithm is used. Compute ΔE_n and describe the qualitative difference between the time dependence of ΔE_n using the Euler and the Euler-Richardson methods. Which method is more consistent with the requirement of conservation of energy?

c. For fixed Δt, which algorithm yields better results for the position in comparison to the analytical solution (5.4)? Is the requirement of conservation of energy consistent with the relative accuracy of the computed positions?

d. Use the more appropriate algorithm and determine the values of Δt that are needed to conserve the total energy to within 0.1% over one cycle for $\omega_0 = 3$ and for $\omega_0 = 5$. Can you use the same value of Δt for both values of ω_0? If not, how do the values of Δt correspond to the relative values of the period in the two cases?

Problem 5.2 Analysis of simple harmonic motion

a. Modify *Program sho* so that the position and velocity of the oscillator are plotted as a function of the time t. Describe the qualitative behavior of the position and velocity.

b. Compute the period T for different values of ω_0. Assume that T is proportional to $(k/m)^\alpha$ and estimate the exponent α from a log-log plot of T versus k/m (see Section 4.6).

c. Plot the time dependence of the potential energy and the kinetic energy through one complete cycle. Where in the cycle is the kinetic energy a maximum?

d. Compute the average value of the kinetic energy and the potential energy during a complete cycle. Is there a relation between the two averages?

e. Compute $x(t)$ for different values of A and show that the shape of $x(t)$ is independent of A, that is, show that $x(t)/A$ is a *universal* function of t for a fixed value of k/m. In what units should the time be measured so that the ratio $x(t)/A$ is independent of A and k/m?

f. The state of motion of the one-dimensional oscillator is completely specified as a function of time by $x(t)$ and $v(t)$. These quantities may be interpreted as the coordinates of a point in a two-dimensional space known as *phase space*. As time increases, the point $(x(t), v(t))$ moves along a *trajectory in phase space*. Modify your program so that v is plotted as a function of x, that is, choose v and x as the vertical and horizontal axes respectively. Set $\omega_0 = 3$ and compute the phase space trajectories for the initial condition

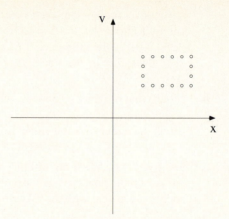

Fig. 5.2   What happens to a given area
in phase space for conservative
systems?

$x(t = 0) = 1$, $v(t = 0) = 0$. What is the shape of the trajectory in phase space? What is the shape for the initial conditions, $x(t = 0) = 0$, $v(t = 0) = 1$ and $x(t = 0) = 4$, $v(t = 0) = 0$? Do you find a different phase trajectory for each initial condition? What physical quantity distinguishes the phase trajectories? Is the motion of a representative point (x, v) always in the clockwise or counterclockwise direction?

*g.   Modify your program so that a set of initial conditions can be computed at one time. An easy way to do so is to use an array so that the array element x(i) represents the position of the oscillator corresponding to initial condition i. Write your program so that the phase space trajectory corresponding to each initial condition is shown simultaneously. Do the phase space trajectories for different initial conditions ever cross? Explain your answer in terms of the uniqueness of trajectories in a deterministic system.

*h.   Choose a set of initial conditions that form a rectangle (see Fig. 5.2). Does the shape of this area change with time? What happens to the total area?

5.3   THE SIMPLE PENDULUM

A common example of a mechanical system that exhibits oscillatory motion is the simple pendulum (see Fig. 5.3). A simple pendulum is an idealized system consisting of a particle or bob of mass m attached to the lower end of a rigid rod of length L and negligible mass; the upper end of the rod pivots without friction. If the bob is pulled to one side from its equilibrium position and released, the pendulum swings in a vertical plane.

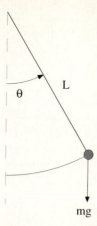

Fig. 5.3   Force diagram for a simple pendulum. The angle θ is
measured from the vertical direction and is positive if
the mass is to the right of the vertical and negative if it
is to the left.

Because the bob is constrained to move along the arc of a circle of radius L about
the center O, the bob's position is specified by the arc length or by the angle θ (see
Fig. 5.3). The linear velocity and acceleration of the bob as measured along the arc are
given by

$$v = L\frac{d\theta}{dt} \tag{5.8}$$

$$a = L\frac{d^2\theta}{dt^2}. \tag{5.9}$$

In the absence of friction, two forces act on the object: the force mg vertically down-
ward and the force of the rod which is directed inward to the center if $|\theta| < \pi/2$. Note
that the effect of the rigid rod is to constrain the motion of the bob along the arc. From
Fig. 5.3, we can see that the component of mg along the arc is $mg \sin \theta$ in the direction
of decreasing θ. Hence, the equation of motion can be written as

$$mL\frac{d^2\theta}{dt^2} = -mg \sin \theta \tag{5.10}$$

or

$$\frac{d^2\theta}{dt^2} = -\frac{g}{L} \sin \theta. \tag{5.11}$$

Equation (5.11) is an example of a nonlinear equation because $\sin \theta$ rather than θ
appears. Most nonlinear equations do not have analytical solutions in terms of well-

known functions, and (5.11) is no exception. However, if the pendulum undergoes oscillations of sufficiently small amplitude, then $\sin\theta \approx \theta$, and (5.11) reduces to

$$\frac{d^2\theta}{dt^2} \approx -\frac{g}{L}\theta. \qquad (\theta \ll 1) \qquad\qquad (5.12)$$

Remember that θ is measured in radians.

Part of the fun of studying physics comes from realizing that equations that appear in different areas (and different fields) are often identical. An example of this "crossover" effect can be seen from a comparison of (5.2) and (5.12). If we associate x with θ, we see that the two equations are identical in form, and we can immediately conclude that for $\theta \ll 1$, the period of a pendulum is given by

$$T = 2\pi\sqrt{L/g}. \qquad\qquad (5.13)$$

One way to understand the motion of a pendulum with large oscillations is to solve (5.11) numerically. Because we know that a numerical solution must be consistent with conservation of total energy, we derive its form here. The potential energy can be found from the following considerations. If the rod is deflected by the angle θ, then the bob is raised by the distance $h = L - L\cos\theta$ (see Fig. 5.3). Hence, the potential energy of the bob in the gravitational field of the earth can be expressed as

$$U = mgh = mgL(1 - \cos\theta), \qquad\qquad (5.14)$$

where the zero of the potential energy corresponds to $\theta = 0$. Because the kinetic energy of the pendulum is $\frac{1}{2}mv^2 = \frac{1}{2}mL^2(d\theta/dt)^2$, the total energy E is

$$E = \frac{1}{2}mL^2\left(\frac{d\theta}{dt}\right)^2 + mgL(1 - \cos\theta). \qquad\qquad (5.15)$$

In the following, we adopt the notation $\omega = d\theta/dt$ for the angular velocity.

Problem 5.3   Oscillations of a pendulum

a.   Assume the amplitude of the pendulum is sufficiently small so that $\sin\theta \approx \theta$. Modify your program for the linear oscillator so that the notation of a pendulum is used. Choose $g/L = 9$ and the initial condition $\theta(t = 0) = 0.1$, $\omega(t = 0) = 0$ and determine the period.

b.   Modify your program to simulate the large amplitude oscillations of a pendulum by replacing θ by $\sin\theta$. Set $g/L = 9$ and choose Δt so the numerical algorithm generates a stable solution. Check the stability of the solution by monitoring the total energy and ensuring that it does not drift from its initial value.

c.   Set $\omega(t = 0) = 0$ and make plots of $\theta(t)$ and $\omega(t)$ for the initial conditions $\theta(t = 0) = 0.1, 0.2, 0.4, 0.8$, and 1.0. Remember that θ is measured in radians. Describe the qualitative behavior of θ and ω. What is the period T and the amplitude θ_m in each case? Plot T versus θ_m and discuss the qualitative dependence of the period on the amplitude. How do your results for T compare in the linear and nonlinear cases, e.g., which period is larger? Explain the relative values of T in terms of the relative magnitudes of the restoring force in the two cases.

d.   Plot several phase space trajectories (ω versus θ) for the nonlinear pendulum for different values of the total energy. Are the phase space trajectories closed? Does the shape of the trajectory depend on the total energy?

* e.   Choose a set of initial conditions that form a rectangle in phase space (see Problem 5.2h). Does the shape of this area change with time? What happens to the total area?

The harmonic oscillator and the pendulum are examples of *conservative* systems, that is, systems for which the total energy is a constant. We have seen two general properties of conservative systems, the preservation of areas in phase space and the nonintersecting nature of the trajectories. These concepts will be important in Section 6.9 where we study the properties of conservative systems with more than one degree of freedom.

5.4   OUTPUT AND ANIMATION

So far we have printed our results on a screen in either tabular or graphical form. On most computers this screen output can be "dumped" to a printer by "printing the screen." However, frequently it is convenient to send the results directly to a printer. True BASIC treats the screen, keyboard, printer, and the disk drives as peripheral devices that are accessed by the computer using a path called a data channel. Before a device can be accessed, a channel to that device must be opened. The keyboard and display are assigned to channel #0 and are always open. Other channels are defined in an OPEN statement. For example, to access the printer we can use

```
OPEN #1: printer
```

where the channel number is arbitrary. We then can output the value of the variable x to the printer by using

```
PRINT #1: x
```

The last step is to close the channel number:

```
CLOSE #1
```

Program print accesses a printer, prints a result, and closes the printer. Note that a channel number can be passed as a parameter to a subroutine.

```
PROGRAM print
CALL initial(#1,x)
CALL add(x)
CALL output(#1,x)
CLOSE #1
END

SUB initial(#1,x)
    OPEN #1: printer
    LET x = 150
END SUB

SUB add(x)
    LET x = x + 200
END SUB

SUB output(#1,x)
    PRINT #1: x
END SUB
```

To make animated pictures, True BASIC can store screen images as a string variable and display them again without the need for additional calculation. The following True BASIC program illustrates the use of the BOX KEEP, BOX CLEAR, and BOX SHOW statements to create the illusion of motion across the screen. It might be useful to change the time of the PAUSE statement to slow the motion of the box across the screen.

```
PROGRAM animation
SET WINDOW 1,10,1,10
SET COLOR "red"
BOX AREA 1,2,1,2              ! draw shape
BOX KEEP 1,2,1,2 in box$      ! store shape in string variable box$
CLEAR
LET x = 1
DO while x < 10
    BOX CLEAR x,x+1,5,6       ! erase shape
    LET x = x + 0.1
    BOX SHOW box$ at x,5      ! redraw shape at different location
    PAUSE 0.01
LOOP
END
```

One way to gain more insight into the qualitative differences between linear and nonlinear restoring forces is to animate a linear and nonlinear pendulum under the same conditions. Program pendula shows an animation of a linear and nonlinear pendulum in two different windows. The "pendula" are represented by circles that are drawn on the screen at a position proportional to their angular displacement. The previous position of the pendula are erased and the circles move across the screen horizontally.

```
PROGRAM pendula
! animation of linear and nonlinear pendula
DIM ball$(2)
LIBRARY "csgraphics"
CALL initial(#1,#2,theta1,v1,theta2,v2,omega2,t,dt,dt_2,ball$(),r)
DO
   CALL linear(theta1,v1,omega2,dt,dt_2,x1old)
   CALL nonlinear(theta2,v2,omega2,dt,dt_2,x2old)
   LET t = t + dt
   CALL animation(#1,theta1,x1old,ball$(1),r)
   CALL animation(#2,theta2,x2old,ball$(2),r)
LOOP until key input
CLOSE #1
CLOSE #2
END

SUB initial(#1,#2,theta1,v1,theta2,v2,omega2,t,dt,dt_2,ball$(),r)
   LET t = 0
   LET theta1 = 0.25          ! position (radians) of linear oscillator
   LET theta2 = theta1        ! position of nonlinear oscillator
   LET v1 = 0
   LET v2 = v1
   LET omega2 = 9             ! g/L ratio
   LET dt = 0.01
   LET dt_2 = 0.5*dt
   LET xmax = theta1          ! dimension of window
   OPEN #1: screen 0,1,0,0.5     ! bottom half of screen
   CALL compute_aspect_ratio(xmax,xwin,ywin)
   SET WINDOW -xwin,xwin,-ywin,ywin
   PRINT "linear oscillator"
   ! draw circle
   SET COLOR "red"
   LET r = 0.1               ! radius of circle
   BOX CIRCLE theta1 - r,theta1 + r,- r,r
   ! store circle in string variable ball$
   BOX KEEP theta1 - r, theta1 + r,-r,r in ball$(1)
   OPEN #2: screen 0,1,0.5,1    ! top half of screen
   CALL compute_aspect_ratio(xmax,xwin,ymin)
   SET WINDOW -xwin,xwin,-ywin,ywin
   PRINT "nonlinear oscillator"
   SET COLOR "blue"
   BOX CIRCLE theta2 - r,theta2 + r,- r,r
   BOX KEEP theta2 - r, theta2 + r,-r,r in ball$(2)
END SUB

SUB linear(theta,v,omega2,dt,dt_2,x1old)
   LET x1old = theta
   LET a = -omega2*theta
   LET vmid = v + a*dt_2
```

```
        LET theta_mid = theta + v*dt_2
        LET amid = -omega2*theta_mid
        LET v = v + amid*dt
        LET theta = theta + vmid*dt
    END SUB

    SUB nonlinear(theta,v,omega2,dt,dt_2,x2old)
        LET x2old = theta
        LET a = -omega2*sin(theta)
        LET vmid = v + a*dt_2
        LET theta_mid = theta + v*dt_2
        LET amid = -omega2*sin(theta_mid)
        LET v = v + amid*dt
        LET theta = theta + vmid*dt
    END SUB

    SUB animation(#9,theta,theta_old,circle$,r)
        WINDOW #9
        LET x = theta_old
        ! erase old circle; using 0 same as box clear
        BOX SHOW circle$ at x - r,-r using 0
        LET x = theta
        BOX SHOW circle$ at x - r,-r  ! draw new circle
    END SUB
```

For convenience, SUB linear and SUB nonlinear are listed even though they are similar to SUB update in *Program sho*.

Problem 5.4   Animation of the motion of the pendulum

a.   Describe the qualitative nature of the motion of the pendula. Where do they move relatively quickly or slowly?

b.   Describe the qualitative features of the relative motion of the linear and non-linear pendulum.

* **c.**   Write a program to make the observations more interesting by replacing the circle by a more realistic looking pendulum. (One way is to use the PICTURE graphics subroutine in True BASIC.)

5.5   DISSIPATIVE SYSTEMS

We know from experience that most oscillatory motion in nature gradually decreases until the displacement becomes zero; such motion is said to be *damped* and the system is said to be *dissipative* rather than conservative. As an example of a damped harmonic oscillator, consider the motion of the block in Fig. 5.1 when a horizontal drag force is included. For small velocities, it is a reasonable approximation to assume that the drag force is proportional to the first power of the velocity. In this case the equation of motion can be written as

$$\frac{d^2x}{dt^2} = -\omega_0^2 x - \gamma \frac{dx}{dt}.$$ (5.16)

The *damping coefficient* γ is a measure of the magnitude of the drag term. Note that the drag force in (5.16) opposes the motion. What is the behavior of $x(t)$ if the linear restoring term in (5.16) is neglected? We simulate the behavior of the damped linear oscillator and the damped pendulum in Problems 5.5 and 5.6.

Problem 5.5 Damped linear oscillator

a. Incorporate the effects of damping into your program and plot the time dependence of the position and the velocity of the linear oscillator. Describe the qualitative behavior of $x(t)$ and $v(t)$ for $\omega_0 = 3$ and $\gamma = 0.5$ with $x(t = 0) = 1$, $v(t = 0) = 0$.

b. The period of the motion is the time between successive maxima of $x(t)$. Compute the period and corresponding angular frequency and compare their values to the undamped case. Is the period longer or shorter? Make additional runs for $\gamma = 1$, 2, and 3. Does the period increase or decrease with greater damping? Why?

c. The amplitude is the maximum value of x during one cycle. Compute the *relaxation time* τ, the time it takes for the amplitude of an oscillation to decrease by $1/e \approx 0.37$ from its maximum value. Is the value of τ constant throughout the motion? Compute τ for the values of γ considered in part (b) and discuss the qualitative dependence of τ on γ.

d. Plot the total energy as a function of time for the values of γ considered in part (b). If the decrease in energy is not monotonic, explain.

e. Compute the average value of the kinetic energy, the potential energy, and the total energy over a complete cycle. Plot these averages as a function of the number of cycles. Do these averages decrease monotonically? Characterize the time dependence of these averages in terms of τ for $\gamma = 0.5$ and $\gamma = 1$.

f. Compute the time dependence of $x(t)$ and $v(t)$ for $\gamma = 4$, 5, 6, 7, and 8. Is the motion oscillatory for all γ? How can you characterize the decay? For fixed ω_0, the oscillator is said to be *critically* damped at the smallest value of γ for which the decay to equilibrium is monotonic. For what value of γ does critical damping occur for $\omega_0 = 4$ and $\omega_0 = 2$? For each value of ω_0, compute the value of γ for which the system approaches equilibrium most quickly.

g. Compute the phase space diagram for $\omega_0 = 3$ and $\gamma = 0.5$, 2, 4, 6, and 8. Why does the phase space trajectory converge to the origin, $(x = 0, v = 0)$? This point is called an *attractor*. Convince yourself that points near the origin move toward the attractor; this attractor is said to be *stable*. Are these qualitative features of the phase space plot independent of γ?

Problem 5.6   Damped nonlinear pendulum

a.   Consider a damped pendulum with $g/L = 9$ and $\gamma = 1$ and the initial condition $\theta(t = 0) = 0.2$, $\omega(t = 0) = 0$. In what ways is the motion of the damped nonlinear pendulum similar to the damped linear oscillator? In what ways is it different? What is the shape of the phase space trajectory for the initial condition $\theta(t = 0) = 1$, $\omega(t = 0) = 0$? Do you find a different phase trajectory for other initial conditions? Remember that θ is restricted to be between $-\pi$ and $+\pi$.

*b.   As in Problems 5.2 and 5.3, choose a set of initial conditions that form a rectangle in phase space. Does the shape of this area change with time? What happens to the total area? Give reasons why you expect that your results are characteristic of dissipative systems in general.

5.6   RESPONSE TO EXTERNAL FORCES

How can we determine the period of a pendulum that is not already in motion? The obvious way is to disturb the system, for example, to displace the bob and observe its motion. We will find that the nature of the *response* of the system to the perturbation tells us something about the nature of the system in the absence of the perturbation.

Consider the *driven* damped linear oscillator with an external force $F(t)$ in addition to the linear restoring force and linear damping force. The equation of motion can be written as

$$\frac{d^2x}{dt^2} = -\omega_0{}^2x - \gamma v + \frac{1}{m}F(t). \tag{5.17}$$

It is customary to interpret the response of the system in terms of the displacement x rather than the velocity v.

The time dependence of $F(t)$ in (5.17) is arbitrary. Because many forces are periodic, we first consider the form

$$\frac{1}{m}F(t) = A_0 \cos \omega t, \tag{5.18}$$

where ω is the angular frequency of the driving force. In Problem 5.7 we consider the response of the damped linear oscillator to (5.18).

Problem 5.7   Motion of a driven damped linear oscillator

a.   Modify Program sho so that an external force of the form (5.18) is included. Include this force as an external function whose form can be modified easily without changing the main program. For example, we can write SUB update in the form:

```
SUB update(x,v,w2,gamma,omega,t,dt,dt2)
    DECLARE DEF f                        ! external function
    LET a = f(x,v,t,w2,gamma,omega)
    LET vmid = v + a*dt2
    LET xmid = x + v*dt2
    LET tmid = t + dt2
    LET amid = f(xmid,vmid,tmid,w2,gamma,omega)
    LET v = v + amid*dt
    LET x = x + vmid*dt
    LET t = tmid + dt2
END SUB

DEF f(x,v,t,w2,gamma,omega)
    LET f = -w2*x - gamma*v + cos(omega*t)
END DEF
```

The parameter w2 is the ratio $\omega_0{}^2 = k/m$. Both w2 and the angular frequency omega of the external force are defined in SUB initial.

b. Set $\omega_0 = 3$, $\gamma = 0.5$, $\omega = 2$ and the amplitude of the external force $A_0 = 1$ for all runs unless otherwise stated. We know that for these values of ω_0 and γ, the dynamical behavior in the absence of an external force corresponds to a underdamped oscillator. Plot $x(t)$ versus t in the presence of the external force with the initial condition, $x(t = 0) = 1$, $v(t = 0) = 0$. How does the qualitative behavior of $x(t)$ differ from the nonperturbed case? What is the period and angular frequency of $x(t)$ after several oscillations have occurred? Repeat the same observations for $x(t)$ with $x(t = 0) = 0$, $v(t = 0) = 1$. Does $x(t)$ approach a limiting behavior independently of the initial conditions? Does the short time behavior of $x(t)$ depend on the initial conditions? Identify a *transient* part of $x(t)$ that depends on the initial conditions and decays in time, and a *steady state* part that dominates at longer times and is independent of the initial conditions.

c. Compute $x(t)$ for several combinations of ω_0 and ω. What is the period and angular frequency of the steady state motion in each case? What parameters determine the frequency of the steady state behavior?

d. One measure of the long-term behavior of the driven harmonic oscillator is the amplitude of the steady state displacement $A(\omega)$. The following statements in SUB update allow $A(\omega)$ to be computed easily.

```
IF x > amplitude then
    LET amplitude = x
END IF
```

Be sure to pass the variable amplitude to the main program. Verify that the steady state behavior of $x(t)$ is given by

$$x(t) = A(\omega) \cos(\omega t + \delta). \tag{5.19}$$

The quantity δ is the phase difference between the applied force and the steady state motion. Compute $A(\omega)$ and $\delta(\omega)$ for $\omega_0 = 3$, $\gamma = 0.5$, and $\omega = 0$, 1.0, 2.0, 2.2, 2.4, 2.6, 2.8, 3.0, 3.2, and 3.4. Choose the initial condition, $x(t = 0) = 0$, $v(t = 0) = 0$. Repeat the simulation for $\gamma = 3.0$, and plot $A(\omega)$ and $\delta(\omega)$ versus ω for the two values of γ. Discuss the qualitative behavior of $A(\omega)$ and $\delta(\omega)$ for the two values of γ. If $A(\omega)$ has a maximum, determine the angular frequency ω_m at which the maximum of A occurs. Is the value of ω_m close to the natural angular frequency ω_0? Compare ω_m to ω_0 and to the frequency of the damped linear oscillator in the absence of an external force.

e. Compute $x(t)$ and $A(\omega)$ for a damped linear oscillator with the amplitude of the external force $A_0 = 4$. How do the steady state results for $x(t)$ and $A(\omega)$ compare to the case $A_0 = 1$? Does the transient behavior of $x(t)$ satisfy the same relation as the steady state behavior?

f. What is the shape of the phase space trajectory for the initial condition $x(t = 0) = 1$, $v(t = 0) = 0$? Do you find a different phase trajectory for other initial conditions?

* g. Describe the qualitative behavior of the steady state amplitude $A(\omega)$ near $\omega = 0$ and $\omega \gg \omega_0$. Why is $A(\omega = 0) < A(\omega)$ for small ω? Why does $A(\omega) \to 0$ for $\omega \gg \omega_0$?

* h. Does the mean kinetic energy resonate at the same frequency as does the amplitude? Compute the mean kinetic energy over one cycle once steady state conditions have been reached. Choose $\omega_0 = 3$ and $\gamma = 0.5$.

In Problem 5.7 we found that the response of the damped harmonic oscillator to an external driving force is linear. For example, if the magnitude of the external force is doubled, then the magnitude of the steady state motion also is doubled. This behavior is a consequence of the linear nature of the equation of motion. When a particle is subject to nonlinear forces, the response can be much more complicated (see Section 6.8).

For many problems, the sinusoidal driving force in (5.18) is not realistic. Another example of an external force can be found by observing someone pushing a child on a swing. Because the force is nonzero only for short intervals of time, this type of force is impulsive. In the following problem, we consider the response of a damped linear oscillator to an impulsive force.

* Problem 5.8   Response of a damped linear oscillator to nonsinusoidal external forces

a. Assume a swing can be modeled by a linear restoring force and a linear damping term. The effect of an impulse is to change the velocity. For simplicity, let the duration of the push equal the time step Δt. Introduce an integer variable for the number of time steps and use the mod function to ensure that the impulse is nonzero only at the time interval associated with the period of the external impulse.

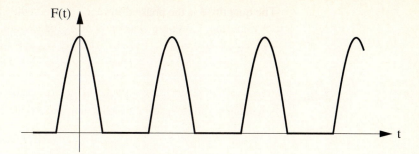

Fig. 5.4   A half-wave driving force corresponding to the positive part of a
cosine function.

b.   Determine the steady state amplitude $A(\omega)$ for $\omega = 1.0, 1.3, 1.4, 1.5, 1.6, 2.5$,
3.0, and 3.5. The corresponding period of the impulse is given by $T = 2\pi/\omega$.
Choose $\omega_0 = 3$ and $\gamma = 0.5$. Are your results consistent with your experience
in pushing a swing and with the comparable results of Problem 5.7?

c.   Consider the response to the half-wave external force consisting of the positive
part of a cosine function (see Fig. 5.4). Compute $A(\omega)$ for $\omega_0 = 3$ and $\gamma =
0.5$. At what values of ω does $A(\omega)$ have a relative maxima? Is the half-
wave cosine driving force equivalent to a sum of cosine functions of different
frequencies? For example, does $A(\omega)$ have more than one resonance?

*d.   Compute the steady state response $x(t)$ to the external force

$$\frac{1}{m}F(t) = \frac{1}{\pi} + \frac{1}{2}\cos t + \frac{2}{3\pi}\cos 2t - \frac{2}{15\pi}\cos 4t. \qquad (5.20)$$

How does a plot of $F(t)$ versus t compare to the half-wave cosine function?
Use your results to conjecture a principle of superposition for the solutions to
linear equations.

5.7   ELECTRICAL CIRCUIT OSCILLATIONS

In this section we discuss several electrical analogues of the mechanical systems we
have considered. Although the equations of motion are identical in form, it is convenient to consider electrical circuits separately, because the nature of the questions is
somewhat different.

   The starting point for electrical circuit theory is Kirchhoff's loop rule, which states
that the sum of the voltage drops around a closed path of an electrical circuit is zero.
This law is a consequence of conservation of energy, because a voltage drop represents
the amount of energy that is lost or gained when a unit charge passes through a circuit
element. The relationships for the voltage drops across each circuit element are summarized in Table 5.1.

element	voltage drop	units
resistor	$V_R = IR$	resistance R, ohms (Ω)
capacitor	$V_C = Q/C$	capacitance C, farads (F)
inductor	$V_L = L\,dI/dt$	inductance L, henries (H)

Table 5.1   The voltage drops across the basic electrical circuit elements. Q is the charge (coulombs) on one plate of the capacitor, and I is the current (amperes).

Imagine an electrical circuit with an alternating voltage source $V_s(t)$ attached in series to a resistor, inductor, and capacitor (see Fig. 5.5). The corresponding loop equation is

$$V_L + V_R + V_C = V_s(t). \qquad (5.21)$$

The voltage source term V_s in (5.21) is the *emf* and is measured in units of volts. If we substitute the relationships shown in Table 5.1, we find

$$L\frac{d^2Q}{dt^2} + R\frac{dQ}{dt} + \frac{Q}{C} = V_s(t), \qquad (5.22)$$

where we have used the definition of current $I = dQ/dt$. We see that (5.22) for the series RLC circuit is identical in form to the damped harmonic oscillator (5.17). The analogies between ideal electrical circuits and mechanical systems are summarized in Table 5.2.

Although we are already familiar with (5.22), we first consider the dynamical behavior of an RC circuit described by

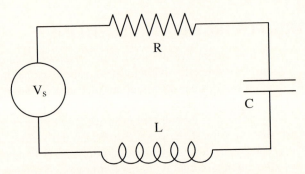

Fig. 5.5   A simple series RLC circuit with a voltage source V_s.

Electric circuit	Mechanical system
charge Q	displacement x
current $I = dQ/dt$	velocity $v = dx/dt$
voltage drop	force
inductance L	mass m
inverse capacitance $1/C$	spring constant k
resistance R	damping γ

Table 5.2 Analogies between electrical parameters and mechanical parameters.

$$R\frac{dQ}{dt} = RI(t) = V_s(t) - \frac{Q}{C}. \tag{5.23}$$

Two RC circuits corresponding to (5.23) are shown in Fig. 5.6. Although the loop equation (5.23) is identical regardless of the order of placement of the capacitor and resistor in Fig. 5.6, the output voltage measured by the oscilloscope in Fig. 5.6 is different. We will see in Problem 5.9 that these circuits act as *filters* that pass voltage components of certain frequencies while rejecting others.

An advantage of a computer simulation of an electrical circuit is that the measurement of a voltage drop across a circuit element does not affect the properties of the circuit. In fact, digital computers often are used to optimize the design of circuits for special applications. *Program rc* simulates an RC circuit with an alternating current (ac) voltage source of the form $V_s(t) = \cos \omega t$. The time dependencies of the voltage source and the voltage drop across the resistor are shown in separate windows. Because *Program rc* makes extensive use of windows, it must be extensively modified to be written in other programming languages.

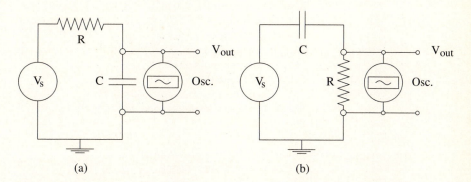

(a) (b)

Fig. 5.6 Examples of RC circuits used as low and high pass filters. Which circuit is which?

```
PROGRAM rc
! simulation of RC circuit with ac voltage source
LIBRARY "csgraphics"
CALL initial(Q,R,tau,V0,omega,tmax,t,dt)
CALL set_up_windows(#1,#2,V0,tmax)
DO while t <= tmax
   CALL scope(Q,I,R,tau,V0,omega,dt)
   LET t = t + dt
   CALL source_voltage(#1,V0,omega,t)
   CALL output_voltage(#2,I,R,t)
LOOP
CLOSE #1
CLOSE #2
END

SUB initial(Q,R,tau,V0,omega,tmax,t,dt)
   LET t = 0
   LET V0 = 1                        ! amplitude of external voltage
   INPUT prompt "external voltage frequency (Hertz) = ":f
   LET omega = 2*pi*f                ! angular frequency
   INPUT prompt "resistance (ohms) = ": R
   INPUT prompt "capacitance (farads) = ": C
   INPUT prompt "time step dt = ": dt
   LET Q = 0
   LET tau = R*C                     ! relaxation time
   LET period = 1/f                  ! period of external frequency
   IF period > tau then
      LET tmax = 2*period
   ELSE
      LET tmax = 2*tau
   END IF
END SUB

SUB set_up_windows(#1,#2,V0,tmax)
   OPEN #1: screen 0,1,0,0.5
   LET tmin = 0
   LET Vmin = -V0
   CALL draw_axes(tmin,tmax,Vmin,V0)
   PRINT "source voltage"
   OPEN #2: screen 0,1,.5,1
   CALL draw_axes(tmin,tmax,Vmin,V0)
   PRINT "voltage drop across resistor"
END SUB

SUB scope(Q,I,R,tau,V0,omega,dt)
   ! compute voltage drops
   DECLARE DEF V
   LET I = V(V0,omega,t)/R - Q/tau
   LET Q = Q + I*dt
END SUB
```

```
DEF V(V0,omega,t) = V0*cos(omega*t)

SUB source_voltage(#1,V0,omega,t)
    DECLARE DEF V
    WINDOW #1
    SET COLOR "blue"
    PLOT LINES: t,V(V0,omega,t);
END SUB

SUB output_voltage(#2,I,R,t)        ! voltage drop across the resistor
    WINDOW #2
    SET COLOR "red"
    PLOT LINES: t,I*R;
END SUB
```

Problem 5.9 Simple filter circuits

a. Use *Program rc* with $R = 1000\,\Omega$ and $C = 1.0\,\mu F\,(10^{-6}$ farads). Find the steady state amplitude of the voltage drops across the resistor and across the capacitor as a function of the angular frequency ω of the source voltage $V_s = \cos\omega t$. The program needs to be modified to plot the voltage drop across the capacitor. Consider the frequencies $f = 10$, 50, 100, 160, 200, 500, 1000, 5000, and 10000 Hz. (Remember that $\omega = 2\pi f$.) Choose Δt to be no more than $0.0001\,\text{s}$ for $f = 10$ Hz. What is a reasonable value of Δt for $f = 10000\,\text{Hz}$?

b. The output voltage depends on where the digital oscilloscope is connected. What is the output voltage of the oscilloscope in Fig. 5.6a? Plot the ratio of the amplitude of the output voltage to the amplitude of the input voltage as a function of ω. Use a logarithmic scale for ω. What range of frequencies is passed? Does this circuit act as a high pass or a low pass filter? Answer the same questions for the oscilloscope in Fig. 5.6b. Use your results to explain the operation of a high and low pass filter. Compute the value of the cutoff frequency for which the amplitude of the output voltage drops to $1/\sqrt{2}$ (half-power) of the input value. How is the cutoff frequency related to RC?

c. Plot the voltage drops across the capacitor and resistor as a function of time. The phase difference ϕ between each voltage drop and the source voltage can be found by finding the time t_m between the corresponding maxima of the voltages. Because ϕ is usually expressed in radians, we have the relation $\phi/2\pi = t_m/T$, where T is the period of the oscillation. What is the phase difference ϕ_C between the capacitor and the voltage source and the phase difference ϕ_R between the resistor and the voltage source? Do these phase differences depend on ω? Does the current lead or lag the voltage, i.e., does the maxima of $V_R(t)$ come before or after the maxima of $V_s(t)$? What is the phase difference between the capacitor and the resistor? Does the latter difference depend on ω?

d. Modify *Program rc* to find the steady state response of an LR circuit with a source voltage $V_s(t) = \cos \omega t$. Let $R = 100 \, \Omega$ and $L = 2 \times 10^{-3} \, \text{H}$. Because $L/R = 2 \times 10^{-5} \, \text{s}$, it is convenient to measure the time and frequency in units of $T_0 = L/R$. We write $t^* = t/T_0$, $\omega^* = \omega T_0$, and rewrite the equation for an LR circuit as

$$I(t^*) + \frac{dI(t^*)}{dt^*} = \frac{1}{R} \cos \omega^* t^*. \qquad (5.24)$$

Because it will be clear from the context, we now simply write t and ω rather than t^* and ω^*. What is a reasonable value of the step size Δt? Compute the steady state amplitude of the voltage drops across the inductor and the resistor for the input frequencies $f = 10, 20, 30, 35, 50, 100,$ and $200 \, \text{Hz}$. Use these results to explain how an LR circuit can be used as a low pass or a high pass filter. Plot the voltage drops across the inductor and resistor as a function of time and determine the phase differences ϕ_R and ϕ_L between the resistor and the voltage source and the inductor and the voltage source. Do these phase differences depend on ω? Does the current lead or lag the voltage? What is the phase difference between the inductor and the resistor? Does the latter difference depend on ω?

Problem 5.10   Square wave response of an RC circuit

Modify *Program rc* so that the voltage source is a periodic square wave as shown in Fig. 5.7. Use a $1.0 \, \mu\text{F}$ capacitor and a $3000 \, \Omega$ resistor. Plot the computed voltage drop across the capacitor as a function of time. Make sure the period of the square wave is long enough so that the capacitor is fully charged during one half-cycle. What is the approximate time dependence of $V_C(t)$ while the capacitor is charging (discharging)?

We now consider the steady state behavior of the series RLC circuit shown in Fig. 5.5 and represented by (5.22). The response of an electrical circuit is the current

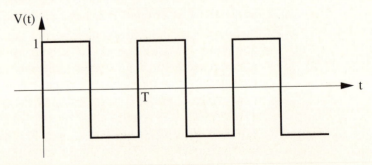

Fig. 5.7   Square wave voltage with period T and unit amplitude.

rather than the charge on the capacitor. Because we have simulated the analogous mechanical system, we already know much about the behavior of driven RLC circuits. Nonetheless, we will find several interesting features of ac electrical circuits in the following two problems.

Problem 5.11   Response of an RLC circuit

a. Consider an RLC series circuit with $R = 100\,\Omega$, $C = 3.0\,\mu F$, and $L = 2\,mH$. Modify `Program sho` or `Program rc` to simulate an RLC circuit and compute the voltage drops across the three circuit elements. Assume an ac voltage source of the form $V(t) = V_0 \cos \omega t$. Plot the current I as a function of time and determine the maximum steady state current I_m for different values of ω. Obtain the *resonance curve* by plotting $I_m(\omega)$ as a function of ω and compute the value of ω at which the resonance curve is a maximum. This value of ω is the *resonant frequency*.

b. The sharpness of the resonance curve of an ac circuit is related to the quality factor or Q value. (Q should not be confused with the charge on the capacitor.) The sharper the resonance, the larger the Q. Circuits with high Q (and hence a sharp resonance) are useful for tuning circuits in a radio so that only one station is heard at a time. We define $Q = \omega_0/\Delta\omega$, where the width $\Delta\omega$ is the frequency interval between points on the resonance curve $I_m(\omega)$ that are $\sqrt{2}/2$ of I_m at its maximum. Compute Q for the values of R, L, and C given in part (a). Change the value of R by 10% and compute the corresponding percentage change in Q. What is the corresponding change in Q if L or C is changed by 10%?

c. Compute the time dependence of the voltage drops across each circuit element for approximately fifteen frequencies ranging from 1/10 to 10 times the resonant frequency. Plot the time dependence of the voltage drops.

d. The ratio of the amplitude of the sinusoidal source voltage to the amplitude of the current is called the *impedance* Z of the circuit, i.e., $Z = V_m/I_m$. This definition of Z is a generalization of the resistance that is defined by the relation $V = IR$ for direct current circuits. Use the plots of part (c) to determine I_m and V_m for different frequencies and verify that the impedance is given by

$$Z(\omega) = \sqrt{R^2 + (\omega L - 1/\omega C)^2}. \tag{5.25}$$

For what value of ω is Z a minimum? Note that the relation $V = IZ$ holds only for the maximum values of I and V and not for I and V at any time.

e. Compute the phase difference ϕ_R between the voltage drop across the resistor and the voltage source. Consider $\omega \ll \omega_0$, $\omega = \omega_0$, and $\omega \gg \omega_0$. Does the current lead or lag the voltage in each case, i.e., does the current reach a maxima before or after the voltage? Also compute the phase differences ϕ_L and ϕ_C and describe their dependence on ω. Do the relative phase differences between V_C, V_R, and V_L depend on ω?

f. Compute the amplitude of the voltage drops across the inductor and the capacitor at the resonant frequency. How do these voltage drops compare to the voltage drop across the resistor and to the source voltage? Also compare the relative phases of V_C and V_L at resonance. Explain how an RLC circuit can be used to amplify the input voltage.

5.8 PROJECTS

Project 5.1 Chemical oscillations

The kinetics of chemical reactions can be modeled by a system of coupled first-order differential equations. As an example, consider the following reaction

$$A + 2B \rightarrow 3B + C, \tag{5.26}$$

where A, B, and C represent the concentrations of three different types of molecules. The corresponding rate equations for this reaction are

$$\frac{dA}{dt} = -kAB^2 \tag{5.27a}$$

$$\frac{dB}{dt} = kAB^2 \tag{5.27b}$$

$$\frac{dC}{dt} = kAB^2. \tag{5.27c}$$

The rate at which the reaction proceeds is determined by the reaction constant k. The terms on the right-hand side of (5.27) are positive if the concentration of the molecule increases in (5.26) as it does for B and C, and negative if the concentration decreases as it does for A. Note that the term $2B$ in the reaction (5.26) appears as B^2 in the rate equation (5.27).

Most chemical reactions proceed to equilibrium, where the mean concentrations of all molecules are constant. However, if the concentrations of some molecules are replenished, it is possible to observe other kinds of behavior, such as oscillations (see below) and chaotic behavior (see Project 6.10). In (5.27) we have assumed that the reactants are well stirred, so that there are no spatial inhomogeneities. In Section 12.4 we will discuss the effects of spatial inhomogeneities due to molecular diffusion.

To obtain chemical oscillations, it is essential to have a series of chemical reactions such that the products of some reactions are the reactants of others. In the following, we consider a simple set of reactions that can lead to oscillations under certain conditions (see Lefever and Nicolis):

$$A \rightarrow X \tag{5.28a}$$

$$B + X \rightarrow Y + D \tag{5.28b}$$

$$2X + Y \rightarrow 3X \tag{5.28c}$$

$$X \rightarrow C. \tag{5.28d}$$

If we assume that the reverse reactions are negligible and A and B are held constant by an external source, the corresponding rate equations are

$$\frac{dX}{dt} = A - (B + 1)X + X^2 Y \tag{5.29a}$$

$$\frac{dY}{dt} = BX - X^2 Y. \tag{5.29b}$$

For simplicity, we have chosen the rate constants to be unity.

a. The steady state solution of (5.29) can be found by setting the left-hand side equal to zero. Show that the steady state values for (X, Y) are $(A, B/A)$.

b. Write a program to solve numerically the rate equations given by (5.29). A simple Euler algorithm is sufficient. Your program should input the initial values of X and Y and the fixed concentrations A and B, and plot X versus Y as the reactions evolve.

c. Systematically vary the initial values of X and Y for given values of A and B. Are their steady state behaviors independent of the initial conditions?

d. Let the initial value of (X, Y) equal $(A + 0.001, B/A)$ for several different values of A and B, i.e., choose initial values close to the steady state values. Classify which initial values result in steady state behavior (stable) and which ones show periodic behavior (unstable). Find the relation between A and B that separates the two types of behavior.

Project 5.2 Comparison of algorithms

a. Consider a particle of unit mass moving in the Morse potential:

$$V(x) = e^{-2x} - 2e^{-x}. \tag{5.30}$$

The total energy $E = \frac{1}{2}p^2 + V(x) < 0$, and the force on the particle is given by

$$F(x) = -\frac{dV}{dx} = 2e^{-x}(e^{-x} - 1). \tag{5.31}$$

Plot $V(x)$ and $F(x)$ versus x. What is their qualitative dependence on x? What is the total energy for the initial condition $x_0 = 2$ and $v_0 = 0$? What type of motion do you expect?

b. Appendix 5A summarizes some of the more commonly used algorithms for the numerical solution of Newton's equations of motion. Compare the Euler, Euler-Richardson, Verlet, and fourth-order Runge Kutta algorithms by computing $x(t)$, $v(t)$, and E_n, where E_n is the total energy after the nth step. One measure of the error associated with the algorithm is ΔE_{max}, the maximum value of the difference $|E_n - E_0|$ over the time interval $t_n - t_0$. Compute this measure of the error at a given value of t_n for decreasing values of Δt with the initial condition $(x_0 = 2, v_0 = 0)$. If an algorithm is second-order, the error in the total energy should decrease as $(\Delta t)^2$. The absence of such a decrease might indicate that there is an error in your program or that roundoff errors are important. Which one of the above algorithms is the best trade-off between speed, accuracy, and simplicity?

    c.   Compare your results for $x(t)$ to the analytical result

$$x(t) = \ln\left(\alpha \cos \omega t + \beta \sin \omega t - E_0^{-1}\right) \tag{5.32}$$

with

$$\alpha = e^{x_0} + E_0^{-1}$$
$$\beta = \omega^{-1} v_0 \, e^{x_0}$$
$$\omega = (2|E_0|)^{1/2}.$$

    d.   Repeat the computations of part (a) with $x_0 = 3$, $v_0 = 1$. Is the motion periodic? Which algorithm is most suitable in this case?

Appendix 5A   NUMERICAL INTEGRATION OF NEWTON'S EQUATION OF MOTION

We summarize several of the common finite difference methods for the solution of Newton's equations of motion with continuous force functions. The variety of algorithms currently in use is evidence that no single method is superior under all conditions.

    To simplify the notation, we consider the motion of a particle in one dimension and write Newton's equations of motion in the form

$$\frac{dv}{dt} = a(t) \tag{5.33a}$$

and

$$\frac{dx}{dt} = v(t), \tag{5.33b}$$

where $a(t) \equiv a(x(t), v(t), t)$. The goal of finite difference methods is to determine the values of x_{n+1} and v_{n+1} at time $t_{n+1} = t_n + \Delta t$. We already have seen that Δt must be chosen so that the integration method generates a stable solution. If the system is conservative, Δt must be sufficiently small so that the total energy is conserved to the desired accuracy.

    The nature of many of the integration algorithms can be understood by expanding $v_{n+1} = v(t_n + \Delta t)$ and $x_{n+1} = x(t_n + \Delta t)$ in a Taylor series. We write

$$v_{n+1} = v_n + a_n \Delta t + O\big((\Delta t)^2\big) \tag{5.34a}$$

and

$$x_{n+1} = x_n + v_n \Delta t + \frac{1}{2} a_n (\Delta t)^2 + O\big((\Delta t)^3\big). \tag{5.34b}$$

The familiar Euler method is equivalent to retaining the $O(\Delta t)$ terms in (5.34):

$$v_{n+1} = v_n + a_n \Delta t \tag{5.35a}$$

and

$$x_{n+1} = x_n + v_n \Delta t. \qquad \text{(Euler)} \tag{5.35b}$$

Because order Δt terms are retained in (5.35), the local truncation error, the error in one time step, is order $(\Delta t)^2$. The global error, the total error over the time of interest, due to the accumulation of errors from step to step is order Δt. This estimate of the global error follows from the fact that the number of steps into which the total time is divided is proportional to $1/\Delta t$. Hence, the order of the global error is reduced by a factor of $1/\Delta t$ relative to the local error. We say that a method is nth order if its global error is order $(\Delta t)^n$. The Euler method is an example of a *first-order* method.

The Euler method is asymmetrical because it advances the solution by a time step Δt, but uses information about the derivative only at the beginning of the interval. We already have found that the accuracy of the Euler method is limited and that frequently its solutions are not stable. We also found that a simple modification of (5.35) yields solutions that are stable for oscillatory systems. For completeness, we repeat the Euler-Cromer algorithm or last-point approximation here:

$$v_{n+1} = v_n + a_n \Delta t, \tag{5.36a}$$

and

$$x_{n+1} = x_n + v_{n+1} \Delta t. \qquad \text{(Euler-Cromer)} \tag{5.36b}$$

Perhaps the most obvious way to improve the Euler method is to use the mean velocity during the interval to obtain the new position. The corresponding *midpoint* method can be written as

$$v_{n+1} = v_n + a_n \Delta t \tag{5.37a}$$

and

$$x_{n+1} = x_n + \frac{1}{2}(v_{n+1} + v_n)\Delta t. \qquad \text{(midpoint)} \tag{5.37b}$$

Note that if we substitute (5.37a) for v_{n+1} into (5.37b), we obtain

$$x_{n+1} = x_n + v_n \Delta t + \frac{1}{2}a_n \, \Delta t^2. \tag{5.38}$$

Hence, the midpoint method yields second-order accuracy for the position and first-order accuracy for the velocity. Although the midpoint approximation yields exact results for constant acceleration, it usually does not yield much better results than the Euler method. In fact, both methods are equally poor, because the errors increase with each time step.

A higher-order method whose error is bounded is the *half-step* method. In this method the average velocity during an interval is taken to be the velocity in the middle of the interval. The half-step method can be written as

$$v_{n+\frac{1}{2}} = v_{n-\frac{1}{2}} + a_n \Delta t,$$
(5.39a)

and

$$x_{n+1} = x_n + v_{n+\frac{1}{2}} \Delta t.$$
(half-step) (5.39b)

Note that the half-step method is not self-starting, i.e., (5.39a) does not allow us to calculate $v_{\frac{1}{2}}$. This problem can be overcome by adopting the Euler algorithm for the first half step:

$$v_{\frac{1}{2}} = v_0 + \frac{1}{2} a_0 \Delta t.$$
(5.39c)

Because the half-step method is stable, it is a common textbook method.

The *Euler-Richardson method* was introduced in Chapter 3, but we list it here for completeness.

$$v_m = v_n + \frac{1}{2} a_n \Delta t,$$
(5.40a)

$$y_m = y_n + \frac{1}{2} v_n \Delta t,$$
(5.40b)

$$t_m = t_n + \frac{1}{2} \Delta t$$
(5.40c)

$$a_m = F(y_m, v_m, t_m)/m,$$
(5.40d)

and

$$v_{n+1} = v_n + a_m \Delta t,$$
(5.41a)

$$y_{n+1} = y_n + v_m \Delta t.$$
(Euler-Richardson) (5.41b)

One of the most common drift-free higher-order algorithms is commonly attributed to Verlet. We write the Taylor series expansion for x_{n-1} in a form similar to (5.34b):

$$x_{n-1} = x_n - v_n \Delta t + \frac{1}{2} a_n (\Delta t)^2.$$
(5.42)

If we add the forward and reverse forms, (5.34b) and (5.42) respectively, we obtain

$$x_{n+1} + x_{n-1} = 2x_n + a_n (\Delta t)^2 + O((\Delta t)^4)$$
(5.43)

or

$$x_{n+1} = 2x_n - x_{n-1} + a_n (\Delta t)^2.$$
(5.44a)

Similarly, the subtraction of the Taylor series for x_{n+1} and x_{n-1} yields

$$v_n = \frac{x_{n+1} - x_{n-1}}{2\Delta t}.$$
(original Verlet) (5.44b)

Note that the global error associated with the *Verlet* algorithm (5.44) is third-order for the position and second-order for the velocity. However, the velocity plays no part in the integration of the equations of motion. In the numerical analysis literature, the Verlet algorithm is known as the "explicit central difference method."

Because the Verlet algorithm is not self-starting, another algorithm must be used to obtain the first few terms. An additional problem is that the new velocity (5.44b) is found by computing the difference between two quantities of the same order of magnitude. As we discussed in Chapter 2, such an operation results in a loss of numerical precision and may give rise to roundoff error.

A mathematically equivalent version of the original Verlet algorithm is given by

$$x_{n+1} = x_n + v_n \Delta t + \frac{1}{2} a_n (\Delta t)^2 \tag{5.45a}$$

and

$$v_{n+1} = v_n + \frac{1}{2} (a_{n+1} + a_n) \Delta t. \qquad \text{(velocity Verlet)} \tag{5.45b}$$

We see that (5.45), known as the *velocity* form of the Verlet algorithm, is self-starting and minimizes roundoff errors. Because we will make no use of (5.44) in the text, we will refer to (5.45) as the Verlet algorithm.

We can derive (5.45) from (5.44) by the following considerations. We first solve (5.44b) for x_{n-1} and write $x_{n-1} = x_{n+1} - 2v_n \Delta t$. If we substitute this expression for x_{n-1} into (5.44a) and solve for x_{n+1}, we find the form (5.45a). Then we use (5.44b) to write v_{n+1} as:

$$v_{n+1} = \frac{x_{n+2} - x_n}{2 \Delta t}, \tag{5.46}$$

and use (5.44a) to obtain $x_{n+2} = 2x_{n+1} - x_n + a_{n+1}(\Delta t)^2$. If we substitute this form for x_{n+2} into (5.46), we obtain

$$v_{n+1} = \frac{x_{n+1} - x_n}{\Delta t} + \frac{1}{2} a_{n+1} \Delta t. \tag{5.47}$$

Finally, we use (5.45a) for x_{n+1} to eliminate $x_{n+1} - x_n$ from (5.47); after some algebra we obtain the desired result (5.45b).

Another useful algorithm that avoids the roundoff error of the original Verlet algorithm is due to Beeman and Schofield. We write the *Beeman* algorithm in the form:

$$x_{n+1} = x_n + v_n \Delta t + \frac{1}{6} (4a_n - a_{n-1})(\Delta t)^2 \tag{5.48a}$$

and

$$v_{n+1} = v_n + \frac{1}{6} (2a_{n+1} + 5a_n - a_{n-1}) \Delta t. \tag{5.48b}$$

Note that (5.48) does not calculate particle trajectories more accurately than the Verlet algorithm. Its advantage is that in general it does a better job of maintaining energy conservation. However, the Beeman algorithm is not self-starting.

In general, the most common finite difference method for solving ordinary differential equations is the *Runge-Kutta* method. To explain the method, we first consider the solution of the first-order differential equation

$$\frac{dx}{dt} = f(x, t). \tag{5.49}$$

The *second-order* Runge-Kutta solution of (5.49) can be written using standard notation as:

$$k_1 = f(x_n, t_n)\Delta t \tag{5.50a}$$

$$k_2 = f(x_n + \frac{k_1}{2}, t_n + \frac{\Delta t}{2})\Delta t \tag{5.50b}$$

$$x_{n+1} = x_n + k_2 + O\big((\Delta t)^3\big). \tag{5.50c}$$

The interpretation of (5.50) is as follows. The Euler method assumes that the slope $f(x_n, t_n)$ at (x_n, t_n) can be used to extrapolate to the next step, i.e., $x_{n+1} = x_n + f(x_n, t_n)\Delta t$. A plausible way of making a better estimate of the slope is to use the Euler method to extrapolate to the midpoint of the interval and then to use the midpoint slope across the full width of the interval. Hence, the Runge-Kutta estimate for the slope is $f(x^*, t_n + \frac{1}{2}\Delta t)$, where $x^* = x_n + \frac{1}{2}f(x_n, t_n)\Delta t$ (see (5.50b)).

The application of the second-order Runge-Kutta method to Newton's equation of motion (5.33) yields

$$k_{1v} = a_n(x_n, v_n, t_n)\Delta t \tag{5.51a}$$

$$k_{1x} = v_n \Delta t \tag{5.51b}$$

$$k_{2v} = a(x_n + \frac{k_{1x}}{2}, v_n + \frac{k_{1v}}{2}, t + \frac{\Delta t}{2})\Delta t \tag{5.52a}$$

$$k_{2x} = (v_n + \frac{k_{1v}}{2})\Delta t \tag{5.52b}$$

$$v_{n+1} = v_n + k_{2v} \tag{5.53a}$$

$$x_{n+1} = x_n + k_{2x} \qquad \text{(second-order Runge Kutta)} \tag{5.53b}$$

In the *fourth-order* Runge-Kutta algorithm, the derivative is computed at the beginning of the time interval, in two different ways at the middle of the interval, and again at the end of the interval. The two estimates of the derivative at the middle of the interval are given twice the weight of the other two estimates. The algorithm for the solution of (5.49) can be written in standard notation as

$$k_1 = f(x_n, t_n)\Delta t \tag{5.54a}$$

$$k_2 = f(x_n + \frac{k_1}{2}, t_n + \frac{\Delta t}{2})\Delta t \tag{5.54b}$$

$$k_3 = f(x_n + \frac{k_2}{2}, t_n + \frac{\Delta t}{2})\Delta t \tag{5.54c}$$

$$k_4 = f(x_n + k_3, t_n + \Delta t)\Delta t \tag{5.54d}$$

and

$$x_{n+1} = x_n + \frac{1}{6}(k_1 + 2k_2 + 2k_3 + k_4). \tag{5.55}$$

The application of the fourth-order Runge-Kutta method to Newton's equation of motion (5.33) yields

$$k_{1v} = a(x_n, v_n, t_n)\Delta t \tag{5.56a}$$
$$k_{1x} = v_n \Delta t \tag{5.56b}$$

$$k_{2v} = a(x_n + \frac{k_{1x}}{2}, v_n + \frac{k_{1v}}{2}, t_n + \frac{\Delta t}{2})\Delta t \tag{5.57a}$$

$$k_{2x} = (v_n + \frac{k_{1v}}{2})\Delta t \tag{5.57b}$$

$$k_{3v} = a(x_n + \frac{k_{2x}}{2}, v_n + \frac{k_{2v}}{2}, t_n + \frac{\Delta t}{2})\Delta t \tag{5.58a}$$

$$k_{3x} = (v_n + \frac{k_{2v}}{2})\Delta t \tag{5.58b}$$

$$k_{4v} = a(x_n + k_{3x}, v_n + k_{3v}, t + \Delta t) \tag{5.59a}$$
$$k_{4x} = (v_n + k_{3x})\Delta t \tag{5.59b}$$

and

$$v_{n+1} = v_n + \frac{1}{6}(k_{1v} + 2k_{2v} + 2k_{3v} + k_{4v}) \tag{5.60a}$$

$$x_{n+1} = x_n + \frac{1}{6}(k_{1x} + 2k_{2x} + 2k_{3x} + k_{4x}). \quad \text{(fourth-order Runge-Kutta)} \tag{5.60b}$$

Because Runge-Kutta methods are self-starting, they are frequently used to obtain the first few iterations for an algorithm that is not self-starting.

Our last example is the *predictor-corrector* method. The idea is to first *predict* the value of the new position:

$$x_p = x_{n-1} + 2v_n \Delta t. \quad \text{(predictor)} \tag{5.61}$$

The predicted value of the position allows us to predict the acceleration a_p. Then using a_p, we obtain the *corrected* values of v_{n+1} and x_{n+1}:

$$v_{n+1} = v_n + \frac{1}{2}(a_p + a_n)\Delta t \tag{5.62a}$$

$$x_{n+1} = x_n + \frac{1}{2}(v_{n+1} + v_n)\Delta t. \quad \text{(corrected)} \tag{5.62b}$$

The corrected values of x_{n+1} and v_{n+1} are used to obtain a new predicted value of a_{n+1}, and hence a new predicted value of v_{n+1} and x_{n+1}. This process is repeated until the predicted and corrected values of x_{n+1} differ by less than the desired value. Note that the predictor-corrector method is not self-starting.

As we have emphasized, there is no single algorithm for solving Newton's equations of motion that is superior under all conditions. It is usually a good idea to start with a simple algorithm, and then to try a higher-order algorithm to see if any real improvement is obtained.

References and Suggestions for Further Reading

F. S. Acton, *Numerical Methods That Work*, The Mathematical Association of America (1990), Chapter 5.

G. L. Baker and J. P. Gollub, *Chaotic Dynamics: An Introduction*, Cambridge University Press (1990). A good introduction to the notion of phase space.

A. Douglas Davis, *Classical Mechanics*, Academic Press (1986). The author gives a "crash" course in conversational BASIC and Pascal and simple numerical solutions of Newton's equations of motion. Much emphasis is given to the harmonic oscillator problem.

S. Eubank, W. Miner, T. Tajima, and J. Wiley, "Interactive computer simulation and analysis of Newtonian dynamics," *Amer. J. Phys.* **57**, 457 (1989).

Richard P. Feynman, Robert B. Leighton, and Matthew Sands, *The Feynman Lectures on Physics*, Vol. 1, Addison-Wesley (1963). Chapters 21, 23, 24, and 25 are devoted to various aspects of harmonic motion.

Charles Kittel, Walter D. Knight, and Malvin A. Ruderman, *Mechanics*, second edition, revised by A. Carl Helmholz and Burton J. Moyer, McGraw-Hill (1973).

R. Lefever and G. Nicolis, "Chemical instabilities and sustained oscillations," *J. Theor. Biol.* **30**, 267 (1971).

Jerry B. Marion and Stephen T. Thornton, *Classical Dynamics*, fourth edition, Academic Press (1995). Excellent discussion of linear and nonlinear oscillators.

M. F. McInerney, "Computer-aided experiments with the damped harmonic oscillator," *Am. J. Phys.* **53**, 991 (1985).

William H. Press, Saul A. Teukolsky, William T. Vetterling, and Brian P. Flannery, *Numerical Recipes*, second edition, Cambridge University Press (1992). Chapter 16 discusses the integration of ordinary differential equations.

J. C. Sprott, *Introduction to Modern Electronics*, John Wiley & Sons (1981). The first five chapters treat the topics discussed in Section 5.7.

S. C. Zilio, "Measurement and analysis of large-angle pendulum motion," *Am. J. Phys.* **50**, 450 (1982).

6

The Chaotic Motion of Dynamical Systems

We study simple deterministic nonlinear models which exhibit complex behavior.

6.1   INTRODUCTION

Most natural phenomena are intrinsically nonlinear. Weather patterns and the turbulent motion of fluids are everyday examples. Although we have explored some of the properties of nonlinear physical systems in Chapter 5, it is easier to introduce some of the important concepts in the context of ecology. Our goal will be to analyze the one-dimensional difference equation

$$x_{n+1} = 4rx_n(1 - x_n), \qquad (6.1)$$

where x_n is the ratio of the population in the nth generation to a reference population. We shall see that the dynamical properties of (6.1) are surprisingly intricate and have important implications for the development of a more general description of nonlinear phenomena. The significance of the behavior of (6.1) is indicated by the following quote from the ecologist Robert May:

> " ... *Its study does not involve as much conceptual sophistication as does elementary calculus. Such study would greatly enrich the student's intuition about nonlinear systems. Not only in research but also in the everyday world of politics and economics we would all be better off if more people realized that simple nonlinear systems do not necessarily possess simple dynamical properties.*"

The study of chaos is currently very popular, but the phenomena is not new and has been of interest to astronomers and mathematicians for about one hundred years. Much of the current interest is due to the use of the computer as a tool for making empirical observations. We will use the computer in this spirit.

6.2   A SIMPLE ONE-DIMENSIONAL MAP

Many biological populations effectively consist of a single generation with no overlap between successive generations. We might imagine an island with an insect population that breeds in the summer and leaves eggs that hatch the following spring. Because the population growth occurs at discrete times, it is appropriate to model the population growth by *difference equations* rather than by differential equations. A simple model of density-independent growth that relates the population in generation $n + 1$ to the population in generation n is given by

$$P_{n+1} = aP_n, \qquad (6.2)$$

where P_n is the population in generation n and a is a constant. In the following, we assume that the time interval between generations is unity, and refer to n as the time.

If $a > 1$, each generation will be a times larger than the previous one. In this case (6.2) leads to geometrical growth and an unbounded population. Although the unbounded nature of geometrical growth is clear, it is remarkable that most of us do not integrate our understanding of geometrical growth into our everyday lives. Can a bank pay 4% interest each year indefinitely? Can the world's human population grow at a constant rate forever?

It is natural to formulate a more realistic model in which the population is bounded by the finite carrying capacity of its environment. A simple model of density-dependent growth is

$$P_{n+1} = P_n(a - bP_n). \tag{6.3}$$

Equation (6.3) is nonlinear due to the presence of the quadratic term in P_n. The linear term represents the natural growth of the population; the quadratic term represents a reduction of this natural growth caused, for example, by overcrowding or by the spread of disease.

It is convenient to rescale the population by letting $P_n = (a/b)x_n$ and rewriting (6.3) as

$$x_{n+1} = ax_n(1 - x_n). \tag{6.4}$$

The replacement of P_n by x_n changes the system of units used to define the various parameters. To write (6.4) in the form (6.1), we define the parameter $r = a/4$ and obtain

$$x_{n+1} = f(x_n) = 4rx_n(1 - x_n). \tag{6.5}$$

The rescaled form (6.5) has the desirable feature that its dynamics are determined by a single control parameter r. Note that if $x_n > 1$, x_{n+1} will be negative. To avoid this unphysical feature, we impose the conditions that x is restricted to the interval $0 \le x \le 1$ and $0 < r \le 1$.

Because the function $f(x)$ defined in (6.5) transforms any point on the one-dimensional interval [0, 1] into another point in the same interval, the function f is called a one-dimensional *map*. The form of $f(x)$ in (6.5) is known as the *logistic* map. The logistic map is a simple example of a *dynamical system*, that is, a deterministic, mathematical prescription for finding the future state of a system.

The sequence of values $x_0, x_1, x_2, \cdots$ is called the *trajectory* or the *orbit*. To check your understanding, suppose that the initial condition or *seed* is $x_0 = 0.5$ and $r = 0.2$. Use a calculator to show that the trajectory is $x_1 = 0.2$, $x_2 = 0.128$, $x_3 = 0.089293, \ldots$ In Fig. 6.1 the first thirty iterations of (6.5) are shown for two values of r.

Program iterate_map computes the trajectory for the logistic map (6.5). The trajectory is listed in window 1 and plotted in window 2.

```
PROGRAM iterate_map                    ! iterate logistic map
CALL set_up_windows(#1,#2)
DO
    CALL initial(x,r,#1,#2,flag$)
    CALL map(x,r,#1,#2,flag$)
LOOP until flag$ = "stop"
END
```

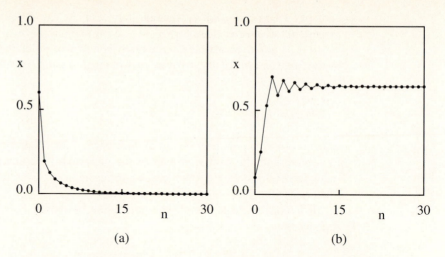

Fig. 6.1   (a) Time series for $r = 0.2$ and $x_0 = 0.6$. Note that the stable fixed point is
$x = 0$. (b) Time series for $r = 0.7$ and $x_0 = 0.1$. Note the initial transient
behavior. The lines between the points are a guide to the eye.

```
SUB initial(x0,r,#1,#2,flag$)
    WINDOW #2
    INPUT prompt "growth parameter (0 < r <= 1) = ": r
    LET x0 = 0.3
    CLEAR
    BOX LINES 0,1000,0,1
    SET CURSOR 1,2
    PRINT "r ="; r
    LET flag$ = ""
END SUB


SUB set_up_windows(#1,#2)
    OPEN #1: screen 0,1,0,0.5        ! text
    OPEN #2: screen 0,1,0.5,1        ! graphics
    LET nmax = 1000
    LET margin = 0.01*nmax
    SET WINDOW -margin,nmax+margin,-0.01,1.01
END SUB


SUB map(x,r,#1,#2,flag$)
    LET iterations = 0
    DO
        LET x = 4*r*x*(1 - x)        ! iterate map
        LET iterations = iterations + 1      ! number of iterations
        WINDOW #1
        SET COLOR "black/white"
```

```
        PRINT USING "#.######": x;
        ! period doubling implies convenient to start new line
        ! every 2^n iterations, where n = 2 or 3.
        IF mod(iterations,8) = 0 then PRINT   ! new line
        WINDOW #2
        SET COLOR "red"
        PLOT iterations,x
        IF key input then CALL change(#1,#2,flag$)
    LOOP until flag$ = "stop" or flag$ = "change"
    WINDOW #1
    PRINT
    PRINT "number of iterations = "; iterations
END SUB

SUB change(#1,#2,flag$)
    GET KEY k
    IF (k = ord("c")) or (k = ord("C")) then
        LET flag$ = "change"
        SET COLOR "black/white"
    ELSE IF (k = ord("s")) or (k = ord("S")) then
        LET flag$ = "stop"
    END IF
END SUB
```

In Problems 6.1 and 6.3 we use *Program map* to explore the dynamical properties of the logistic map (6.5). The program uses the GET key statement so that the key 'c' can be pressed to change the value of r and the key 's' can be pressed to stop the program.

Problem 6.1   Exploration of period-doubling

a. Explore the dynamical behavior of (6.5) with $r = 0.24$ for different values of x_0. Show that $x = 0$ is a *stable fixed point*. That is, for sufficiently small r, the iterated values of x converge to $x = 0$ independently of the value of x_0. If x represents the population of insects, describe the qualitative behavior of the population.

b. Explore the dynamical behavior of (6.5) for $r = 0.26, 0.5, 0.74$, and 0.748. A fixed point is *unstable* if for almost all x_0 near the fixed point, the trajectories diverge from it. Verify that $x = 0$ is an unstable fixed point for $r > 0.25$. Show that for the suggested values of r, the iterated values of x do not change after an initial *transient*, that is, the long time dynamical behavior is *period* 1. In Appendix 6A we show that for $r < 3/4$ and for x_0 in the interval $0 < x_0 < 1$, the trajectories approach the *attractor* at $x = 1 - 1/4r$. The set of initial points that iterate to the attractor is called the *basin* of the attractor. For the logistic map, the interval $0 < x < 1$ is the basin of attraction of the attractor $x = 1 - 1/4r$.

c.   Explore the dynamical properties of (6.5) for $r = 0.752, 0.76, 0.8$, and 0.862. For $r = 0.752$ and 0.862 approximately 1000 iterations are necessary to obtain convergent results. Show that if r is increased slightly beyond 0.75, x oscillates between two values after an initial transient behavior. That is, instead of a stable cycle of period 1 corresponding to one fixed point, the system has a stable cycle of period 2. The value of r at which the single fixed point x^* splits or *bifurcates* into two values x_1^* and x_2^* is $r = b_1 = 3/4$. The pair of x values, x_1^* and x_2^*, form a *stable attractor* of period 2.

d.   Describe an ecological scenario of an insect population that exhibits dynamical behavior similar to that observed in part (c).

e.   What are the stable attractors of (6.5) for $r = 0.863$ and 0.88? What is the corresponding period?

f.   What are the stable attractors and corresponding periods for $r = 0.89, 0.891$, and 0.8922?

Another way to determine the behavior of (6.5) is to plot the values of x as a function of r (see Fig. 6.2). The iterated values of x are plotted after the initial transient behavior is discarded. Such a plot is generated by *Program bifurcate*. For each value of r, the first ntransient values of x are computed but not plotted. Then the next nplot values of x are plotted, with the first half in red and the second half in blue. This process is repeated for a new value of r until the desired range of r values is reached. A typical value of ntransient should be in the range of 100–1000 iterations. The magnitude of nplot should be at least as large as the longest period that you wish to observe.

```
PROGRAM bifurcate
! plot values of x for different values of r
CALL initial(x,r,rmax,nvalues,dr,ntransient,nplot)
FOR ir = 0 to nvalues
    CALL output(x,r,ntransient,nplot)
    LET r = r + dr
NEXT ir
! maximum value of r done separately to avoid r > 1
CALL output(x,rmax,ntransient,nplot)
END

SUB initial(x0,r0,rmax,nvalues,dr,ntransient,nplot)
    INPUT prompt "initial value of control parameter r = ": r0
    ! important that r not be greater than 1
    INPUT prompt "maximum value of r = ": rmax
    ! suggest dr <= 0.01
    INPUT prompt "incremental change of r = ": dr
    INPUT prompt "number of iterations not plotted = ": ntransient
    INPUT prompt "number of iterations plotted = ": nplot
    LET nvalues = (rmax - r0)/dr  ! number of r values plotted
    LET nvalues = int(nvalues)
```

```
              LET x0 = 0.5                     ! initial value
              CLEAR
              LET xmax = 1                     ! maximum value of x
              LET mx = 0.05*xmax               ! margin
              SET WINDOW r0-dr,rmax+dr,-mx,xmax + mx
              BOX LINES r0,rmax,0,1
         END SUB

         SUB output(x,r,ntransient,nplot)
              DECLARE DEF f
              SET COLOR "black/white"
              SET CURSOR 1,1
              PRINT "             ";             ! erase previous output
              SET CURSOR 1,1
              PRINT "r ="; r
              FOR i = 1 to ntransient          ! x values not plotted
                  LET x = f(x,r)
              NEXT i
              SET COLOR "red"
              FOR i = 1 to 0.5*nplot
                  LET x = f(x,r)
                  ! show different x-values for given value of r
                  PLOT r,x
              NEXT i
              ! change color to see if values of x have converged
              SET COLOR "blue"
              FOR i = (0.5*nplot + 1) to nplot
                  LET x = f(x,r)
                  PLOT r,x
              NEXT i
         END SUB

         DEF f(x,r) = 4*r*x*(1 - x)
```

Problem 6.2   Qualitative features of the logistic map

a. Use *Program bifurcate* to identify period 2, period 4, and period 8 behavior as in Fig. 6.2. It might be necessary to "zoom in" on a portion of the plot. How many period-doublings can you find?

b. Change the scale so that you can follow the iterations of x from period 4 to period 16 behavior. How does the plot look on this scale in comparison to the original scale?

c. Describe the shape of the trajectory near the bifurcations from period 2 $\rightarrow$ period 4, period 4 $\rightarrow$ 8, etc. These bifurcations are frequently called *pitchfork bifurcations*.

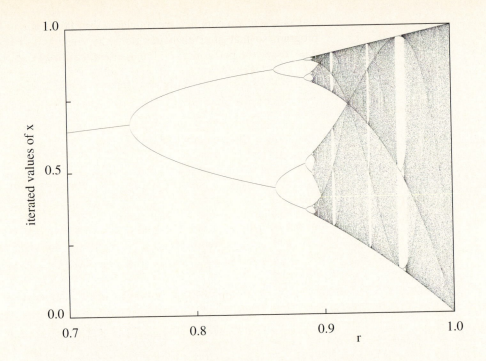

Fig. 6.2   Bifurcation diagram of the logistic map. For each value of r, the iterated values of
x_n are plotted after the first 1000 iterations are discarded. Note the transition from
periodic to chaotic behavior and the narrow windows of periodic behavior within
the region of chaos.

The final state or bifurcation diagram in Fig. 6.2 indicates that the period-doubling
behavior ends at $r \approx 0.892$. This value of r is known very precisely and is given by $r =
r_\infty = 0.892486417967\ldots$ At $r = r_\infty$, the sequence of period-doublings accumulate to
a trajectory of infinite period. In Problem 6.3 we explore the behavior of the trajectories
for $r > r_\infty$.

Problem 6.3   The chaotic regime

a.   For $r > r_\infty$, two initial conditions that are very close to one another can yield
very different trajectories after a small number of iterations. As an example,
choose $r = 0.91$ and consider $x_0 = 0.5$ and 0.5001. How many iterations are
necessary for the iterated values of x to differ by more than ten percent? What
happens for $r = 0.88$ for the same choice of seeds?

b.   The accuracy of floating point numbers retained on a digital computer is finite. To test the effect of the finite accuracy of your computer, choose $r = 0.91$

and $x_0 = 0.5$ and compute the trajectory for 200 iterations. Then modify your program so that after each iteration, the operations x = x/10 followed by x = 10 * x are performed. This combination of operations truncates the last digit that your computer retains. A similar effect can be obtained by using the True BASIC truncate(x,n) function, which truncates the variable x to n decimal places. Compute the trajectory again and compare your results. Do you find the same discrepancy for $r < r_\infty$?

c.   What are the dynamical properties for $r = 0.958$? Can you find other windows of periodic behavior in the interval $r_\infty < r < 1$?

6.3 PERIOD-DOUBLING

The results of the numerical experiments that we did in Section 6.2 have led us to adopt a new vocabulary to describe our observations and probably have convinced you that the dynamical properties of simple deterministic nonlinear systems can be quite complicated.

To gain more insight into how the dynamical behavior depends on r, we introduce a simple graphical method for iterating (6.5). In Fig. 6.3, we show a graph of $f(x)$ versus x for $r = 0.7$. A diagonal line corresponding to $y = x$ intersects the curve $y = f(x)$ at the two fixed points $x^* = 0$ and $x^* = 9/14 \approx 0.642857$. If x_0 is not one of the fixed points, we can find the trajectory in the following manner. Draw a vertical line from $(x = x_0, y = 0)$ to the intersection with the curve $y = f(x)$ at $(x_0, y_0 = f(x_0))$. Next draw a horizontal line from (x_0, y_0) to the intersection with the diagonal line at (y_0, y_0). On this diagonal line $y = x$, and hence the value of x at this intersection is the first iteration $x_1 = y_0$. The second iteration x_2 can be found in the same way. From the point (x_1, y_0), draw a vertical line to the intersection with the curve $y = f(x)$. Keep y fixed at $y = y_1 = f(x_1)$, and draw a horizontal line until it intersects the diagonal line; the value of x at this intersection is x_2. Further iterations can be found by repeating this process.

This graphical method is illustrated in Fig. 6.3 for $r = 0.7$ and $x_0 = 0.9$. If we begin with any x_0 (except $x_0 = 0$ and $x_0 = 1$), continued iterations will converge to the fixed point $x^* \approx 0.642857$. Repeat the procedure shown in Fig. 6.3 by hand and convince yourself that you understand the graphical solution of the iterated values of the map. For this value of r, the fixed point is stable (an attractor of period 1). In contrast, no matter how close x_0 is to the fixed point at $x = 0$, the iterates diverge away from it, and this fixed point is unstable.

How can we explain the qualitative difference between the fixed point at $x = 0$ and $x^* = 0.642857$ for $r = 0.7$? The local slope of the curve $y = f(x)$ determines the distance moved horizontally each time f is iterated. A slope steeper than 45° leads to a value of x further away from its initial value. Hence, the criterion for the stability of a fixed point is that the magnitude of the slope at the fixed point must be less than 45°. That is, if $|df(x)/dx|_{x=x^*}$ is less than unity, then x^* is stable; conversely, if $|df(x)/dx|_{x=x^*}$ is greater than unity, then x^* is unstable. Inspection of $f(x)$ in Fig. 6.3 shows that $x = 0$ is unstable because the slope of $f(x)$ at $x = 0$ is greater than unity. In

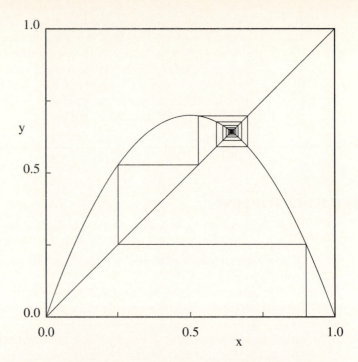

Fig. 6.3   Graphical representation of the iteration of the logistic
map (6.5) with $r = 0.7$ and $x_0 = 0.9$. Note that
the graphical solution converges to the fixed point
$x^* \approx 0.643$.

contrast, the magnitude of the slope of $f(x)$ at $x = x^*$ is less than unity and the fixed
point is stable. In Appendix 6A, we use similar analytical arguments to show that

$$x^* = 0 \text{ is stable for } 0 < r < 1/4 \tag{6.6a}$$

and

$$x^* = 1 - \frac{1}{4r} \text{ is stable for } 1/4 < r < 3/4. \tag{6.6b}$$

Thus for $0 < r < 3/4$, the eventual behavior after many iterations is known.

What happens if r is greater than 3/4? From our observations we have found
that if r is slightly greater than 3/4, the fixed point of f becomes unstable and gives
birth (bifurcates) to a cycle of period 2. Now x returns to the same value only
after every second iteration, and the fixed points of $f(f(x))$ are the attractors
of $f(x)$. In the following, we adopt the notation $f^{(2)}(x) = f(f(x))$, and write
$f^{(n)}(x)$ for the nth iterate of $f(x)$. (Do not confuse $f^{(n)}(x)$ with the nth derivative

of $f(x)$.) For example, the second iterate $f^{(2)}(x)$ is given by the fourth-order polynomial:

$$f^{(2)}(x) = 16r^2x(1-x)\left[1 - 4rx(1-x)\right]$$
$$= 16r^2x\left[-4rx^3 + 8rx^2 - (1+4r)x + 1\right]. \tag{6.7}$$

What happens if we increase r still further? Eventually the magnitude of the slope of the fixed points of $f^{(2)}(x)$ exceeds unity and the fixed points of $f^{(2)}(x)$ become unstable. Now the cycle of f is period 4, and we can study the stability of the fixed points of the fourth iterate $f^{(4)}(x) = f^{(2)}\big(f^{(2)}(x)\big) = f\big(f\big(f(f(x))\big)\big)$. These fixed points also eventually bifurcate, and we are led to the phenomena of *period-doubling* as we observed in Problem 6.2.

Program graph_sol implements the graphical analysis of $f(x)$. The nth order iterates are defined in DEF f(x,r,iterate) using *recursion*. (The quantity iterate is 1, 2, and 4 for the functions $f(x)$, $f^{(2)}(x)$, and $f^{(4)}(x)$ respectively.) Recursion is an idea that is simple once you understand it, but it can be difficult to grasp the idea initially. One way to understand how recursion works is to think of a stack, such as a stack of trays in a cafeteria. The first time a recursive function is called, the function is placed on the top of the stack. Each time the function calls itself, an exact copy of the function, with possibly different values of the input parameters, is placed on top of the stack. When a copy of the function is finished, this copy is popped off the top of the stack. To understand the function f(x,r,iterate), suppose we want to compute f(0.4,0.8,3). First we write f(0.4,0.8,3) on a piece of paper (see Fig. 6.4a). Follow the statements within the function until another call to f(0.4,0.8,iterate) occurs. In this case, the call is to f(0.4,0.8,iterate-1) which equals f(0.4,0.8,2). Write f(0.4,0.8,2) above f(0.4,0.8,3) (see Fig. 6.4b). When you come to the

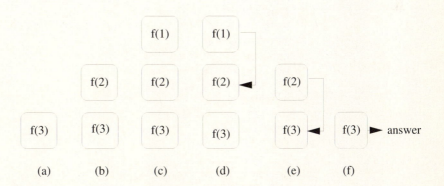

Fig. 6.4   Example of the calculation of f(0.4,0.8,3) using the recursive function defined in *Program graph_sol*. The number in each box is the value of the variable iterate. The values of $x = 0.4$ and $r = 0.8$ are not shown. The value of f(x, r, 3) = 0.7842.

end of the definition of the function, write down the value of f that is actually returned, and remove the function from the stack by crossing it out (see Fig. 6.4d). This returned value for f equals y if iterate > 1, or it is the output of the function for iterate $= 1$. Continue deleting copies of f as they are finished, until there are no copies left on the paper. The final value of f is the value returned by the computer. Write a miniprogram that defines f(x,r,iterate) and prints the value of f(0.4,0.8,3). Is the answer the same as your hand calculation?

```
PROGRAM graph_sol
! graphical solution for trajectory of logistic map
CALL initial(x,r,iterate)
CALL draw_function(r,iterate)
CALL trajectory(x,r,iterate)         ! press any key to stop
END

SUB initial(x0,r,iterate)
    INPUT prompt "control parameter r = ": r
    INPUT prompt "initial value of x = ": x0
    INPUT prompt "iterate of f(x) = ": iterate
    CLEAR
    PRINT "r ="; r
END SUB

SUB draw_function(r,iterate)
    DECLARE DEF f
    LET nplot = 200              ! # of points at which function computed
    LET delta = 1/nplot
    LET margin = 0.1
    SET WINDOW -margin,1 + margin,-margin,1 + margin
    PLOT LINES: 0,0;1,1          ! draw diagonal line y = x
    PLOT LINES: 0,1;0,0;1,0  ! draw axes
    PLOT                     ! left pen
    SET COLOR "red"
    LET x = 0
    FOR i = 0 to nplot
        LET y = f(x,r,iterate)
        PLOT x,y;
        LET x = x + delta
    NEXT i
END SUB

SUB trajectory(x,r,iterate)
    DECLARE DEF f
    LET y0 = 0
    LET x0 = x
    SET COLOR "blue"
    DO
        LET y = f(x,r,iterate)
        PLOT LINES: x0,y0; x0,y; y,y
```

```
                    LET x0 = y
                    LET y0 = y
                    LET x = y
                LOOP until key input
                GET KEY k
            END SUB

            DEF f(x,r,iterate)              ! f defined by recursive procedure
                IF iterate > 1 then
                    LET y = f(x,r,iterate - 1)
                    LET f = 4*r*y*(1 - y)
                ELSE
                    LET f = 4*r*x*(1 - x)
                END IF
            END DEF
```

Problem 6.4   Qualitative properties of the fixed points

a. Use *Program graph_sol* to show graphically that there is a single stable fixed point of $f(x)$ for $r < 3/4$. It would be instructive to insert a pause between each iteration of the map and to show the value of the slope at $y_n = f(x_n)$ in a separate window. At what value of r does the absolute value of this slope exceed unity? Let b_1 denote the value of r at which the fixed point of $f(x)$ bifurcates and becomes unstable. Verify that $b_1 = 0.75$.

b. Describe the trajectory of $f(x)$ for $r = 0.785$. What is the nature of the fixed point given by $x = 1 - 1/4r$? What is the nature of the trajectory if $x_0 = 1 - 1/4r$? What is the period of $f(x)$ for all other choices of x_0? What are the numerical values of the two-point attractor?

c. The function $f(x)$ is symmetrical about $x = \frac{1}{2}$ where $f(x)$ is a maximum. What are the qualitative features of the second iterate $f^{(2)}(x) = f\big(f(x)\big)$ for $r = 0.785$? Is $f^{(2)}(x)$ symmetrical about $x = \frac{1}{2}$? For what value of x does $f^{(2)}(x)$ have a minimum? Iterate $x_{n+1} = f^{(2)}(x_n)$ for $r = 0.785$ and find its two fixed points x_1^* and x_2^*. (Try $x_0 = 0.1$ and $x_0 = 0.3$.) Are the fixed points of $f^{(2)}(x)$ stable or unstable? How do these values of x_1^* and x_2^* compare with the values of the two-point attractor of $f(x)$? Verify that the slopes of $f^{(2)}(x)$ at x_1^* and x_2^* are equal.

d. Verify the following properties of the fixed points of $f^{(2)}(x)$. As r is increased, the fixed points of $f^{(2)}(x)$ move apart and the slope of $f^{(2)}(x)$ at the fixed points decreases. What is the value of $r = s_2$ at which one of the two fixed points of $f^{(2)}$ equals $\frac{1}{2}$? What is the value of the other fixed point? What is the slope of $f^{(2)}(x)$ at $x = \frac{1}{2}$? What is the slope at the other fixed point? As r is further increased, the slopes at the fixed points become negative. Finally at $r = b_2 \approx 0.8623$, the slopes at the two fixed points of $f^{(2)}(x)$ equal -1, and the two fixed points of $f^{(2)}$ become unstable. (It can be shown that the exact value of b_2 is $b_2 = (1 + \sqrt{6})/4$.)

e. Show that for r slightly greater than b_2, e.g., $r = 0.87$, there are four stable fixed points of the function $f^{(4)}(x)$. What is the value of $r = s_3$ when one of the fixed points equals $\frac{1}{2}$? What are the values of the three other fixed points at $r = s_3$?

f. Estimate the value of $r = b_3$ at which the four fixed points of $f^{(4)}$ become unstable.

g. Choose $r = s_3$ and estimate the number of iterations that are necessary for the trajectory to converge to period 4 behavior. How does this number of iterations change when neighboring values of r are considered? Choose several values of x_0 so that your results do not depend on the initial conditions.

Problem 6.5   Periodic windows in the chaotic regime

a. If you look closely at the bifurcation diagram in Fig. 6.2, you will see that the region of chaotic behavior for $r > r_\infty$ is interrupted by intervals of periodic behavior. Magnify your bifurcation diagram so that you can look at the interval $0.957107 \le r \le 0.960375$, where a periodic trajectory of period 3 occurs. (Period 3 behavior starts at $r = (1 + \sqrt{8})/4$.) What happens to the trajectory for slightly larger r, e.g., for $r = 0.9604$?

b. Plot the map $f^{(3)}(x)$ versus x at $r = 0.96$, a value of r in the period 3 window. Draw the line $y = x$ and determine the intersections with $f^{(3)}(x)$. (Use *Program graph_sol* without calling SUB trajectory.) The stable fixed points satisfy the condition $x^* = f^{(3)}(x^*)$. Because $f^{(3)}(x)$ is an eighth-order polynomial, there are eight solutions (including $x = 1$). Find the intersections of $f^{(3)}(x)$ with $y = x$ and identify the three stable fixed points. What are the slopes of $f^{(3)}(x)$ at these points? Then decrease r to $r = 0.957107$, the (approximate) value of r below which the system is chaotic. Draw the line $y = x$ and determine the number of intersections with $f^{(3)}(x)$. Note that at this value of r, the curve $y = f^{(3)}(x)$ is tangent to the diagonal line at the three stable fixed points. For this reason, this type of transition is called a *tangent bifurcation*. Note that there also is an unstable point at $x \approx 0.76$.

c. Plot $x_{n+1} = f^{(3)}(x_n)$ versus n for $r = 0.9571$, a value of r just below the onset of period 3 behavior. How would you describe the behavior of the trajectory? This type of chaotic motion is an example of *intermittency*, that is, nearly periodic behavior interrupted by occasional irregular bursts.

*d. Modify *Program graph_sol* so that you can study the graphical solution of $x_{n+1} = f^{(3)}(x_n)$ for the same value of r as in part (c). That is, "zoom in" on the values of x near the stable fixed points that you found in part (b) for r in the period 3 regime. Note the three narrow channels between the diagonal line $y = x$ and the plot of $f^{(3)}(x)$. The trajectory requires many iterations to squeeze through the channel, and we see period 3 behavior during this time. Eventually, the trajectory escapes from the channel and bounces around until it is sent into a channel at some unpredictable later time.

6.4   UNIVERSAL PROPERTIES AND SELF-SIMILARITY

In Sections 6.2 and 6.3 we found that the trajectory of the logistic map has remarkable properties as a function of the control parameter r. In particular, we found a sequence of period-doublings accumulating to a chaotic trajectory of infinite period at $r = r_\infty$. For most values of $r > r_\infty$, we saw that the trajectory is very sensitive to the initial conditions. We also found "windows" of period 3, 6, 12, ... embedded in the broad regions of chaotic behavior. How typical is this type of behavior? In the following, we will find further numerical evidence that the general behavior of the logistic map is independent of the details of the form (6.5) of $f(x)$.

You might have noticed that the range of r between successive bifurcations becomes smaller as the period increases (see Table 6.1). For example, $b_2 - b_1 = 0.112398$, $b_3 - b_2 = 0.023624$, and $b_4 - b_3 = 0.00508$. A good guess is that the decrease in $b_k - b_{k-1}$ is geometric, i.e., the ratio $(b_k - b_{k-1})/(b_{k+1} - b_k)$ is a constant. You can check that this ratio is not exactly constant, but converges to a constant with increasing k. This behavior suggests that the sequence of values of b_k has a limit and follows a geometrical progression:

$$b_k \approx r_\infty - \text{constant } \delta^{-k}, \tag{6.8}$$

where δ is known as the Feigenbaum number. From (6.8) it is easy to show that δ is given by the ratio

$$\delta = \lim_{k \to \infty} \frac{b_k - b_{k-1}}{b_{k+1} - b_k}. \tag{6.9}$$

Problem 6.6   Estimation of the Feigenbaum constant

a.   Plot $\delta_k = (b_k - b_{k-1})/(b_{k+1} - b_k)$ versus k using the values of b_k in Table 6.1 and estimate the value of δ. Are the number of decimal places given in

k	b_k
1	0.750 000
2	0.862 372
3	0.886 023
4	0.891 102
5	0.892 190
6	0.892 423
7	0.892 473
8	0.892 484

Table 6.1   Values of the control parameter b_k for the onset of the kth bifurcation. Six decimal places are shown.

Table 6.1 for b_k sufficient for all the values of k shown? The best estimate of δ is

$$\delta = 4.669\,201\,609\,102\,991\ldots \tag{6.10}$$

The number of decimal places in (6.10) is shown to indicate that δ is known precisely. Use (6.8) and (6.10) and the values of b_k to estimate the value of r_∞.

b.   In Problem 6.4 we found that one of the four fixed points of $f^{(4)}(x)$ is at $x^* = \frac{1}{2}$ for $r = s_3 \approx 0.87464$. We also found that the convergence to the fixed points of $f^{(4)}(x)$ is more rapid than at nearby values of r. In Appendix 6A we show that these *superstable* trajectories occur whenever one of the fixed points is at $x = \frac{1}{2}$. The values of $r = s_m$ that give superstable trajectories of period 2^{m-1} are much better defined than the points of bifurcation, $r = b_k$. The rapid convergence to the final trajectories also gives better numerical estimates, and we always know one member of the trajectory, namely $x = \frac{1}{2}$. It is reasonable that δ can be defined as in (6.9) with b_k replaced by s_m. Use the values of $s_1 = 0.5$, $s_2 \approx 0.809017$, and $s_3 = 0.874640$ to estimate δ. The numerical values of s_m are found in Project 6.1 by solving the equation $f^{(m)}(x = \frac{1}{2}) = \frac{1}{2}$ numerically; the first eight values of s_m are listed in Table 6.2.

We can associate another number with the series of pitchfork bifurcations. From Fig. 6.3 and Fig. 6.5 we see that each pitchfork bifurcation gives birth to "twins" with the new generation more densely packed than the previous generation. One measure of this density is the maximum distance M_k between the values of x describing the bifurcation (see Fig. 6.5). The disadvantage of using b_k is that the transient behavior of the trajectory is very long at the boundary between two different periodic behaviors. A more convenient measure of the density is the quantity $d_k = x_k^* - \frac{1}{2}$, where x_k^* is the value of the fixed point nearest to the fixed point $x^* = \frac{1}{2}$. The first two values of d_k are shown in Fig. 6.6 with $d_1 \approx 0.3090$ and $d_2 \approx -0.1164$. The next value is $d_3 \approx 0.0460$. Note that the fixed point nearest to $x = \frac{1}{2}$ alternates from one side of $x = \frac{1}{2}$ to the other. We define the quantity α by the ratio

$$\alpha = \lim_{k \to \infty} -\left(\frac{d_k}{d_{k+1}}\right). \tag{6.11}$$

The estimates $\alpha = 0.3090/0.1164 = 2.65$ for $k = 1$ and $\alpha = 0.1164/0.0460 = 2.53$ for $k = 2$ are consistent with the asymptotic limit $\alpha = 2.5029078750958928485\ldots$

We now give qualitative arguments that suggest that the general behavior of the logistic map in the period-doubling regime is independent of the detailed form of $f(x)$. As we have seen, period-doubling is characterized by self-similarities, e.g., the period-doublings look similar except for a change of scale. We can demonstrate these similarities by comparing $f(x)$ for $r = s_1 = 0.5$ for the superstable trajectory with period 1 to the function $f^{(2)}(x)$ for $r = s_2 \approx 0.809017$ for the superstable trajectory of period 2 (see Fig. 6.7). The function $f(x, r = s_1)$ has unstable fixed points at $x = 0$ and $x = 1$ and a stable fixed point at $x = \frac{1}{2}$. Similarly the function $f^{(2)}(x, r = s_2)$ has a stable fixed point at $x = \frac{1}{2}$ and an unstable fixed point at $x \approx 0.69098$. Note the similar shape,

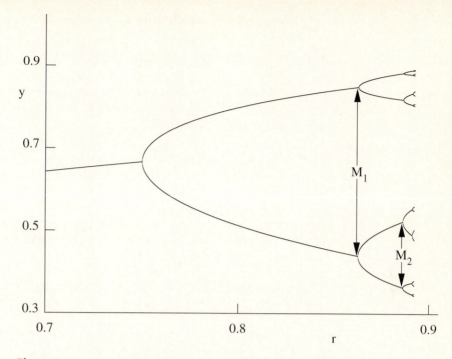

Fig. 6.5   The first few bifurcations of the logistic equation showing the scaling of the maximum distance M_k between the asymptotic values of x describing the bifurcation.

but different scale of the curves in the square box in part (a) and part (b) of Fig. 6.7. This similarity is an example of scaling. That is, if we scale $f^{(2)}$ and change *(renormalize)* the value of r, we can compare $f^{(2)}$ to f. (See Chapter 13 for a discussion of scaling and renormalization in another context.)

Our graphical comparison is meant only to be suggestive. A precise approach shows that if we continue the comparison of the higher-order iterates, e.g., $f^{(4)}(x)$ to $f^{(2)}(x)$, etc., the superposition of functions converges to a universal function that is independent of the form of the original function $f(x)$.

Problem 6.7   Further estimates of the exponents α and δ

a.   Write a subroutine to find the appropriate scaling factor and superimpose f and the rescaled form of $f^{(2)}$ found in Fig. 6.7.

b.   Use arguments similar to those discussed in the text in Fig. 6.7 and compare the behavior of $f^{(4)}(x, r = s_3)$ in the square about $x = \frac{1}{2}$ with $f^{(2)}(x, r = s_2)$ in its square about $x = \frac{1}{2}$. The size of the squares are determined by the unstable fixed point nearest to $x = \frac{1}{2}$. Find the appropriate scaling factor and superimpose $f^{(2)}$ and the rescaled form of $f^{(4)}$.

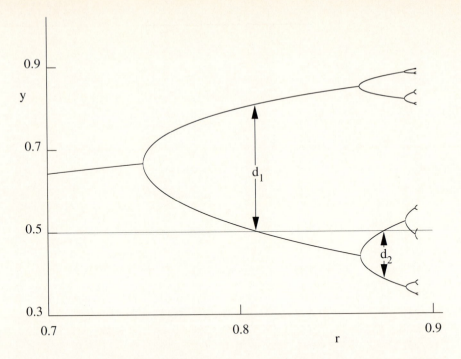

Fig. 6.6   The quantity d_k is the distance from $x^* = 1/2$ to the nearest element of the attractor of period 2^k. It is convenient to use this quantity to determine the exponent α.

It is easy to modify your programs to consider other one-dimensional maps. In Problem 6.8 we consider several one-dimensional maps and determine if they also exhibit the period-doubling route to chaos.

*Problem 6.8   Other one-dimensional maps

Determine the qualitative properties of the one-dimensional maps:

$$f(x) = xe^{r(1-x)} \tag{6.12}$$

$$f(x) = r \sin \pi x. \tag{6.13}$$

The map in (6.12) has been used by ecologists (cf. May) to study a population that is limited at high densities by the effect of epidemic disease. Although it is more complicated than (6.5), its advantage is that the population remains positive no matter what (positive) value is taken for the initial population. There are no restrictions on the maximum value of r, but if r becomes sufficiently large, x eventually becomes effectively zero, rendering the population extinct. What is the behavior of the time series of (6.12) for $r = 1.5, 2$, and 2.7? Describe the qualitative behavior of $f(x)$. Does it have a maximum?

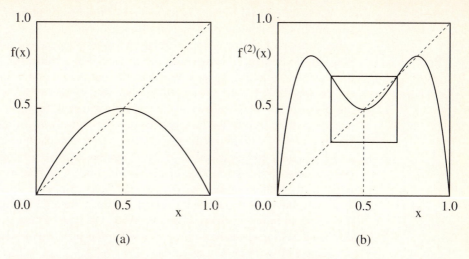

Fig. 6.7 Comparison of $f(x, r)$ for $r = s_1$ with the second iterate $f^{(2)}(x)$ for $r = s_2$. (a) The function $f(x, r = s_1)$ has unstable fixed points at $x = 0$ and $x = 1$ and a stable fixed point at $x = \frac{1}{2}$. (b) The function $f^{(2)}(x, r = s_1)$ has a stable fixed point at $x = \frac{1}{2}$. The unstable fixed point of $f^{(2)}(x)$ nearest to $x = \frac{1}{2}$ occurs at $x \approx 0.69098$, where the curve $f^{(2)}(x)$ intersects the line $y = x$. The upper right-hand corner of the square box in (b) is located at this point, and the center of the box is at $(\frac{1}{2}, \frac{1}{2})$. Note that if we reflect this square about the point $(\frac{1}{2}, \frac{1}{2})$, the shape of the reflected graph in the square box is nearly the same as it is in part (a), but on a smaller scale.

The sine map (6.13) with $0 < r \leq 1$ and $0 \leq x \leq 1$ has no special significance, except that it is nonlinear. If time permits, estimate the value of δ for both maps. What limits the accuracy of your determination of δ?

The above qualitative arguments and numerical results suggest that the quantities α and δ are *universal*, that is, independent of the detailed form of $f(x)$. In contrast, the values of the accumulation point r_∞ and the constant in (6.8) depend on the detailed form of $f(x)$. Feigenbaum has shown that the period-doubling route to chaos and the values of δ and α are universal property of maps that have a quadratic maximum, i.e., $f'(x)|_{x=x_m} = 0$ and $f''(x)|_{x=x_m} < 0$.

Why is the universality of period-doubling and the numbers δ and α more than a curiosity? The reason is that because this behavior is independent of the details, there might exist realistic systems whose underlying dynamics yield the same behavior as the logistic map. Of course, most physical systems are described by differential rather than difference equations. Can these systems exhibit period-doubling behavior? Several workers (cf. Testa et al.) have constructed nonlinear RLC circuits driven by an oscillatory source voltage. The output voltage shows bifurcations, and the measured values of the exponents δ and α are consistent with the predictions of the logistic map.

Of more general interest is the nature of turbulence in fluid systems. Consider a stream of water flowing past several obstacles. We know that at low flow speeds, the water flows past obstacles in a regular and time-independent fashion, called *laminar* flow. As the flow speed is increased (as measured by a dimensionless parameter called the Reynolds number), some swirls develop, but the motion is still time-independent. As the flow speed is increased still further, the swirls break away and start moving downstream. The flow pattern as viewed from the bank becomes time-dependent. For still larger flow speeds, the flow pattern becomes very complex and looks random. We say that the flow pattern has made a transition from laminar flow to *turbulent* flow.

This qualitative description of the transition to chaos in fluid systems is superficially similar to the description of the logistic map. Can fluid systems be analyzed in terms of the simple models of the type we have discussed here? In a few instances such as turbulent convection in a heated saucepan, period doubling and other types of transitions to turbulence have been observed. The type of theory and analysis we have discussed has suggested new concepts and approaches, and the study of turbulent flows is a subject of much current research.

6.5   MEASURING CHAOS

How do we know if a system is chaotic? The most important characteristic of chaos is *sensitivity to initial conditions*. In Problem 6.3 for example, we found that the trajectories starting from $x_0 = 0.5$ and $x_0 = 0.5001$ for $r = 0.91$ become very different after a small number of iterations. Because computers only store floating numbers to a certain number of digits, the implication of this result is that our numerical predictions of the trajectories are restricted to small time intervals. That is, sensitivity to initial conditions implies that even though the logistic map is deterministic, our ability to make numerical predictions is limited.

How can we quantify this lack of predictably? In general, if we start two identical dynamical systems from different initial conditions, we expect that the difference between the trajectories will change as a function of n. In Fig. 6.8 we show a plot of the difference $|\Delta x_n|$ versus n for the same conditions as in Problem 6.3a. We see that roughly speaking, $\ln |\Delta x_n|$ is a linearly increasing function of n. This result indicates that the separation between the trajectories grows exponentially if the system is chaotic. This divergence of the trajectories can be described by the *Lyapunov* exponent, which is defined by the relation:

$$|\Delta x_n| = |\Delta x_0|\, e^{\lambda n}, \tag{6.14}$$

where Δx_n is the difference between the trajectories at time n. If the Lyapunov exponent λ is positive, then nearby trajectories diverge exponentially. Chaotic behavior is characterized by exponential divergence of nearby trajectories.

A naive way of measuring the Lyapunov exponent λ is to run the same dynamical system twice with slightly different initial conditions and measure the difference of the trajectories as a function of n. We used this method to generate Fig. 6.8. Because the rate of separation of the trajectories might depend on the choice of x_0, a better method

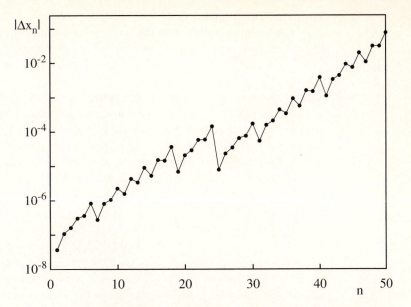

Fig. 6.8 The evolution of the difference Δx_n between the trajectories of the logistic map at $r = 0.91$ for $x_0 = 0.5$ and $x_0 = 0.5001$. The separation between the two trajectories increases with n, the number of iterations, if n is not too large. (Note that $|\Delta x_1| \sim 10^{-8}$ and that the trend is not monotonic.)

would be to compute the rate of separation for many values of x_0. This method would be tedious, because we would have to fit the separation to (6.14) for each value of x_0 and then determine an average value of λ.

A more important limitation of the naive method is that because the trajectory is restricted to the unit interval, the separation $|\Delta x_n|$ ceases to increase when n becomes sufficiently large. However, to make the computation of λ as accurate as possible, we would like to average over as many iterations as possible. Fortunately, there is a better procedure. To understand the procedure, we take the natural logarithm of both sides of (6.14) and write λ as

$$\lambda = \frac{1}{n} \ln \left| \frac{\Delta x_n}{\Delta x_0} \right|. \tag{6.15}$$

Because we want to use the data from the entire trajectory after the transient behavior has ended, we use the fact that

$$\frac{\Delta x_n}{\Delta x_0} = \frac{\Delta x_1}{\Delta x_0} \frac{\Delta x_2}{\Delta x_1} \cdots \frac{\Delta x_n}{\Delta x_{n-1}}. \tag{6.16}$$

Hence, we can express λ as

$$\lambda = \frac{1}{n}\sum_{i=0}^{n-1}\ln\left|\frac{\Delta x_{i+1}}{\Delta x_i}\right|. \tag{6.17}$$

The form (6.17) implies that we can consider x_i for any i as the initial condition.

We see from (6.17) that the problem of computing λ has been reduced to finding the ratio $\Delta x_{i+1}/\Delta x_i$. Because we want to make the initial difference between the two trajectories as small as possible, we are interested in the limit $\Delta x_i \to 0$. The idea of the more sophisticated procedure is to compute the differential dx_i from the equation of motion at the same time that the equation of motion is being iterated. We use the logistic map as an example. The differential of (6.5) can be written as

$$\frac{dx_{i+1}}{dx_i} = f'(x_i) = 4r(1-2x_i). \tag{6.18}$$

We can consider x_i for any i as the initial condition and the ratio dx_{i+1}/dx_i as a measure of the rate of change of x_i. Hence, we can iterate the logistic map as before and use the values of x_i and the relation (6.18) to compute dx_{i+1}/dx_i at each iteration. The Lyapunov exponent is given by

$$\lambda = \lim_{n\to\infty}\frac{1}{n}\sum_{i=0}^{n-1}\ln\left|f'(x_i)\right|, \tag{6.19}$$

where we begin the sum in (6.19) after the transient behavior is completed. We have included explicitly the limit $n \to \infty$ in (6.19) to remind ourselves to choose n sufficiently large. Note that this procedure weights the points on the attractor correctly, that is, if a particular region of the attractor is not visited often by the trajectory, it does not contribute much to the sum in (6.19).

Problem 6.9   Lyapunov exponent for the logistic map

a. Compute the Lyapunov exponent λ for the logistic map using the naive approach. Choose $r = 0.91$, $x_0 = 0.5$, and $\Delta x_0 = 10^{-6}$, and plot $\ln|\Delta x_n/\Delta x_0|$ versus n. What happens to $\ln|\Delta x_n/\Delta x_0|$ for large n? Estimate λ for $r = 0.91$, $r = 0.97$, and $r = 1.0$. Does your estimate of λ for each value of r depend significantly on your choice of x_0 or Δx_0?

b. Compute λ using the algorithm discussed in the text for $r = 0.76$ to $r = 1.0$ in steps of $\Delta r = 0.01$. What is the sign of λ if the system is not chaotic? Plot λ versus r, and explain your results in terms of behavior of the bifurcation diagram shown in Fig. 6.2. Compare your results for λ with those shown in Fig. 6.9. How does the sign of λ correlate with the behavior of the system as seen in the bifurcation diagram? If $\lambda < 0$, then the two trajectories converge and the system is not chaotic. If $\lambda = 0$, then the trajectories diverge algebraically, i.e., as a power of n. For what value of r is λ a maximum?

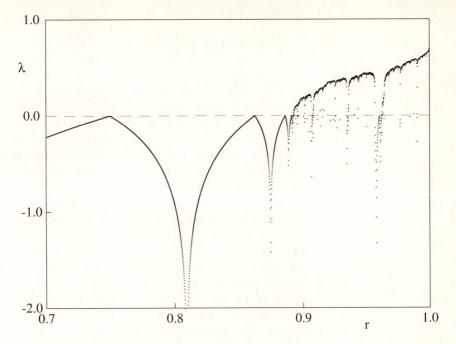

Fig. 6.9 The Lyapunov exponent calculated using the method in (6.19) as a function of the control parameter r. Compare the behavior of λ to the bifurcation diagram in Fig. 6.2. Note that $\lambda < 0$ for $r < 3/4$ and approaches zero at a period doubling bifurcation. A negative spike corresponds to a superstable trajectory. The onset of chaos is visible near $r = 0.892$, where λ first becomes positive. For $r > 0.892$, λ generally increases except for dips below zero whenever a periodic window occurs. Note the large dip due to the period 3 window near $r = 0.96$. For each value of r, the first 1000 iterations were discarded, and 10^5 values of $\ln |f'(x_n)|$ were used to determine λ.

c.   In Problem 6.3b we saw that roundoff errors in the chaotic regime make the computation of individual trajectories meaningless. That is, if the system's behavior is chaotic, then small roundoff errors are amplified exponentially in time, and the actual numbers we compute for the trajectory starting from a given initial value are not "real." Given this limitation, how meaningful is our computation of the Lyapunov exponent? Repeat your calculation of λ for $r = 1$ by changing the roundoff error as you did in Problem 6.3b. Does your computed value of λ change? We will encounter a similar question in Chapter 8 where we compute the trajectories of a system of many particles. The answer appears to be that although the trajectory we compute is not the one we thought we were trying to compute, the computed trajectory is close to a possible trajectory of the system. Quantities such as λ that are averaged over many possible trajectories are independent of the detailed behavior of an individual trajectory.

6.6 CONTROLLING CHAOS

The dream of classical physics has been that if the initial conditions and all the forces acting on a system are known, then we can predict the future with as much precision as we desire. The existence of chaos has shattered that dream, not just for some esoteric systems, but for the majority of dynamical systems observed in nature. However, even if a system is chaotic, we still might be able to control its behavior with small, but carefully chosen perturbations of the system. We will illustrate the method for the logistic map. The application of the method to other one-dimensional systems is straightforward, but the extension to higher dimensional systems is more complicated (cf. Ott, Lai).

Suppose that we want the logistic map to have periodic behavior even though the parameter r equals a value in the chaotic regime. How can we make the trajectory have periodic behavior without drastically changing r or imposing an external perturbation that is so large that the internal dynamics of the map become irrelevant? The key to the solution is that for any value of r in the chaotic regime, there is an infinite number of trajectories that have unstable periods. This property of the chaotic regime means that if we choose the value of the seed x_0 to be precisely equal to a point on an unstable trajectory with period p, the subsequent trajectory will have this period. However, if we choose a value of x_0 that differs ever so slightly from this special value, the trajectory will not be periodic. Our goal is to make slight perturbations to the system to keep it on the desired unstable periodic trajectory.

The first step is to find the values of $x(i)$, $i = 1$ to p, that constitute the unstable periodic trajectory. It is an interesting numerical problem to find the values of $x(i)$, and we consider this problem first. The trick is to find a fixed point of the map $f^{(p)}$. That is, we need to find the value of x^* such that

$$g^{(p)}(x^*) \equiv f^{(p)}(x^*) - x^* = 0. \tag{6.20}$$

The algorithms for finding the solution to (6.20) are called root finding algorithms. You might have heard of Newton's method, which we will describe in Chapter 10. Here we use the simplest root-finding algorithm, the *bisection* method. The algorithm works as follows:

1. Choose two values, x_{left} and x_{right}, with $x_{\text{left}} < x_{\text{right}}$, such that the product $g^{(p)}(x_{\text{left}})g^{(p)}(x_{\text{right}}) < 0$. There must be a value of x such that $g^{(p)}(x) = 0$ in the interval $[x_{\text{left}}, x_{\text{right}}]$.

2. Choose the midpoint, $x_{\text{mid}} = x_{\text{left}} + \frac{1}{2}(x_{\text{right}} - x_{\text{left}}) = \frac{1}{2}(x_{\text{left}} + x_{\text{right}})$, as the guess for x^*.

3. If $g^{(p)}(x_{\text{mid}})$ has the same sign as $g^{(p)}(x_{\text{left}})$, then replace x_{left} by x_{mid}; otherwise, replace x_{right} by x_{mid}. The interval for the location of the root is now reduced.

4. Repeat steps 2 and 3 until the desired level of precision is achieved.

The following program implements this algorithm for the logistic map. One possible problem is that some of the roots of $g^{(p)}(x) = 0$ also are roots of $g^{(p')}(x) = 0$ for p' equal to a factor of p. As p increases, it might become more difficult to find a root that is part of a period p trajectory and not part of a period p' trajectory.

```
PROGRAM period
! find fixed point of f(x) iterated p times
DECLARE DEF f
CALL initial(r,p,epsilon,xleft,xright,gleft,gright)
DO
    CALL bisection(r,p,xleft,xright,gleft,gright)
LOOP until abs(xleft - xright) < epsilon
LET x = 0.5*(xleft + xright)
PRINT "explicit demonstration of period"; p; "behavior"
PRINT
PRINT 0,x                         ! result
FOR i = 1 to 2*p + 1
    LET x = f(x,r,1)
    PRINT i,x
NEXT i
END

SUB initial(r,p,epsilon,xleft,xright,gleft,gright)
    DECLARE DEF f
    INPUT prompt "control parameter r = ": r
    INPUT prompt "period = " : p
    INPUT prompt "desired precision = ": epsilon
    LET done$ = "no"
    DO while done$ = "no"     ! do until zero between xleft and xright
       INPUT prompt "guess for xleft = ": xleft
       INPUT prompt "guess for xright = ": xright
       LET gleft = f(xleft,r,p) - xleft
       LET gright = f(xright,r,p) - xright
       IF gleft*gright < 0  then
          LET done$ = "yes"
       ELSE
          PRINT "range does not necessarily enclose a root"
       END IF
    LOOP
END SUB

SUB bisection(r,p,xleft,xright,gleft,gright)
    DECLARE DEF f
    ! midpoint between xleft and xright
    LET xmid = 0.5*(xleft + xright)
    LET gmid = f(xmid,r,p) - xmid
    IF gmid*gleft > 0 then
       LET xleft = xmid       ! change xleft
       LET gleft = gmid
```

```
        ELSE
           LET xright = xmid        ! change xright
           LET gright = gmid
        END IF
     END SUB


     DEF f(x,r,p)                    ! f defined by recursive procedure
        IF p > 1 then
           LET y = f(x,r,p-1)
           LET f = 4*r*y*(1-y)
        ELSE
           LET f = 4*r*x*(1-x)
        END IF
     END DEF
```

Problem 6.10   Unstable periodic trajectories for the logistic map

a.   Test *Program period* for values of r for which the logistic map has a stable period with $p = 1$ and $p = 2$. Set the desired precision ϵ equal to 10^{-7}. Initially use $x_{\text{left}} = 0.01$ and $x_{\text{right}} = 0.99$. Calculate the stable attractor analytically and compare the results of your program with the analytical results.

b.   Set $r = 0.95$ and find the periodic trajectories for $p = 1, 2, 5, 6, 7, 12, 13,$ and 19.

c.   Modify *Program period* so that n_b, the number of bisections needed to obtain the unstable trajectory, is listed. Choose three of the cases considered in part (c), and compute n_b for the precision ϵ equal to 0.01, 0.001, 0.0001, and 0.00001. Determine the functional dependence of n_b on ϵ.

Now that we know how to find the values of the unstable periodic trajectories, we discuss an algorithm for stabilizing this period. Suppose that we wish to stabilize the unstable trajectory of period p for a choice of $r = r_0$. The idea is to make small adjustments of $r = r_0 + \Delta r$ at each iteration so that the difference between the actual trajectory and the target periodic trajectory is small. If the actual trajectory is x_n and we wish the trajectory to be at $x(i)$, we make the next iterate x_{n+1} equal to $x(i + 1)$ by expanding the difference $x_{n+1} - x(i + 1)$ in a Taylor series and setting the difference to zero to first-order. We have $x_{n+1} - x(i + 1) = f(x_n, r) - f(x(i), r_0)$. If we expand $f(x_n, r)$ about $(x(i), r_0)$, we have to first-order:

$$x_{n+1} - x(i + 1) = \frac{\partial f(x, r)}{\partial x}\big[x_n - x(i)\big] + \frac{\partial f(x, r)}{\partial r}\Delta r = 0. \tag{6.21}$$

The partial derivatives in (6.21) are evaluated at $x = x(i)$ and $r = r_0$. The result can be expressed as

$$4r_0\big[1 - 2x(i)\big]\big[x_n - x(i)\big] + 4x(i)\big[1 - x(i)\big]\Delta r = 0. \tag{6.22}$$

The solution of (6.22) for Δr can be written as

$$\Delta r = -r_0 \frac{[1 - 2x(i)][x_n - x(i)]}{x(i)[1 - x(i)]}. \tag{6.23}$$

The procedure is to iterate the logistic map at $r = r_0$ until x_n is sufficiently close to $x(i)$. The nature of chaotic systems is that the trajectory is guaranteed to come close to the desired unstable trajectory eventually. Then we use (6.23) to change the value of r so that the next iteration is closer to $x(i + 1)$. We summarize the algorithm for controlling chaos as follows:

1. Find the unstable periodic trajectory $x(i)$, $i = 1$ to p, for the desired value of r_0.
2. Iterate the map with $r = r_0$ until x_n is within ϵ of $x(1)$. Then use (6.23) to determine r.
3. To turn off the control, set $r = r_0$.

Problem 6.11 Controlling chaos

a. Write a program that allows the user to turn the control on and off by hitting any key. The trajectory can be displayed either by listing the values of x_n or by plotting x_n versus n. The program should incorporate as input the desired unstable periodic trajectory $x(i)$, the period p, the value of r_0, and the parameter ϵ.

b. Test your program with $r_0 = 0.95$ and the periods $p = 1, 5$, and 13. Use $\epsilon = 0.02$.

c. Modify your program so that the values of r are shown as well as the values of x_n. How does r change if we vary ϵ? Try $\epsilon = 0.05, 0.01$, and 0.005.

*d. Add a subroutine to compute n_c, the number of iterations necessary for the trajectory x_n to be within ϵ of $x(1)$ when control is turned on. Find $<n_c>$, the average value of n_c, by starting with 100 random values of x_0. Compute $<n_c>$ as a function of ϵ for $\delta = 0.05, 0.005, 0.0005$, and 0.00005. What is the functional dependence of $<n_c>$ on ϵ?

*6.7 HIGHER-DIMENSIONAL MODELS

So far we have discussed the logistic map as a mathematical model that has some remarkable properties and produces some interesting computer graphics. In this section we discuss some two- and three-dimensional systems that also might seem to have little to do with realistic physical systems. However, as we will see in Sections 6.8 and 6.9, similar behavior is found in realistic physical systems under the appropriate conditions.

We begin with a two-dimensional map and consider the sequence of points (x_n, y_n) generated by

$$x_{n+1} = y_n + 1 - ax_n^2 \tag{6.24a}$$
$$y_{n+1} = bx_n. \tag{6.24b}$$

The map (6.24) was proposed by Hénon who was motivated by the relevance of this dynamical system to the behavior of asteroids and satellites.

Problem 6.12   The Hénon map

a.   Iterate (6.24) for $a = 1.4$ and $b = 0.3$ and plot 10^4 iterations starting from $x_0 = 0$, $y_0 = 0$. Make sure you compute the new value of y using the old value of x and not the new value of x. Do not plot the initial transient. Choose the SET window statement so that all values of the trajectory within the box drawn by the statement BOX LINES -1.5,1.5,-0.45,0.45 are plotted. Make a similar plot beginning from the second initial condition, $x_0 = 0.63135448$, $y_0 = 0.18940634$. Compare the shape of the two plots. Is the shape of the two curves independent of the initial conditions?

b.   Increase the scale of your plot so that all points with the box drawn by the statement BOX LINES 0.50,0.75,0.15,0.21 are shown. Begin from the second initial condition and increase the number of computed points to 10^5. Then make another plot showing all points within the box drawn by BOX LINES 0.62,0.64,0.185,0.191. If patience permits, make an additional enlargement and plot all points within the box drawn by BOX LINES 0.6305,0.6325,0.1889,0.1895. (You have to increase the number of computed points to order 10^6.) What is the structure of the curves within each box? Does the attractor appear to have a similar structure on smaller and smaller length scales? Is there a region in the plane from which the points cannot escape? The region of points that do not escape is the basin of the Hénon attractor. The attractor itself is the set of points to which all points in the basin are attracted. That is, two trajectories that begin from different conditions will soon lie on the attractor. We will find in Section 14.4 that the Hénon attractor is an example of a strange attractor.

c.   Determine if the system is chaotic, i.e., sensitive to initial conditions. Start two points very close to each other and watch their trajectories for a fixed time. Choose different colors for the two trajectories.

*d.   It is straightforward in principle to extend the method for computing the Lyapunov exponent that we used for a one-dimensional map to higher-dimensional maps. The idea is to linearize the difference (or differential) equations and replace dx_n by the corresponding vector quantity $d\mathbf{r}_n$. This generalization yields the Lyapunov exponent corresponding to the divergence along the fastest growing direction. If a system has f degrees of freedom, it has a set of f Lyapunov exponents. A method for computing all f exponents is discussed in Project 6.4.

One of the earliest indications of chaotic behavior was in an atmospheric model developed by Lorenz. His goal was to describe the motion of a fluid layer that is heated from below. The result is convective rolls, where the warm fluid at the bottom rises, cools off at the top, and then falls down later. Lorenz simplified the description by restricting the motion to two spatial dimensions. This situation has been modeled experimentally in the laboratory and is known as a Rayleigh-Benard cell. The equations that Lorenz obtained are

$$\frac{dx}{dt} = -\sigma x + \sigma y \tag{6.25a}$$

$$\frac{dy}{dt} = -xz + rx - y \tag{6.25b}$$

$$\frac{dz}{dt} = xy - bz, \tag{6.25c}$$

where x is a measure of the fluid flow velocity circulating around the cell, y is a measure of the temperature difference between the rising and falling fluid regions, and z is a measure of the difference in the temperature profile between the bottom and the top from the normal equilibrium temperature profile. The dimensionless parameters σ, r, and b are determined by various fluid properties, the size of the Raleigh-Benard cell, and the temperature difference in the cell. Note that the variables x, y, and z have nothing to do with the spatial coordinates, but are measures of the state of the system. Although it is not expected that you will understand the relation of the Lorenz equations to convection, we have included these equations here to reinforce the idea that simple sets of equations can exhibit chaotic behavior.

Problem 6.13   The Lorenz Model

a. Use either the simple Euler algorithm or one of the Runge-Kutta methods (see Appendix 5A) to obtain a numerical solution to the Lorenz equations (6.25). Generate plots of x versus y, y versus z, and x versus z as in Fig. 6.10 or use a separate graphics program to make three-dimensional plots. Explore the basin of the attractor with $\sigma = 10$, $b = 8/3$, and $r = 28$.

b. Determine qualitatively the sensitivity to initial conditions. Start two points very close to each other and watch their trajectories for approximately 10,000 time steps.

c. Let z_m denote the value of z where z is a relative maximum for the mth time. You can determine the value of z_m by finding the average of the two values of z when the right-hand side of (6.25c) changes sign. Plot z_{m+1} versus z_m and describe what you find. This procedure is one way that a continuous system can be mapped onto a discrete map. What is the slope of the z_{m+1} versus z_m curve? Is its magnitude always greater than unity? If so, then this behavior is an indication of chaos. Why?

The application of the Lorenz equations to weather prediction has led to a popular metaphor known as the *butterfly effect*. That is, if the conditions are such that the atmosphere displays chaotic behavior, then even a small effect such as the flapping of a butterfly's wings would make our long-term predictions of the weather meaningless. This metaphor is made even more meaningful by inspection of Fig. 6.10.

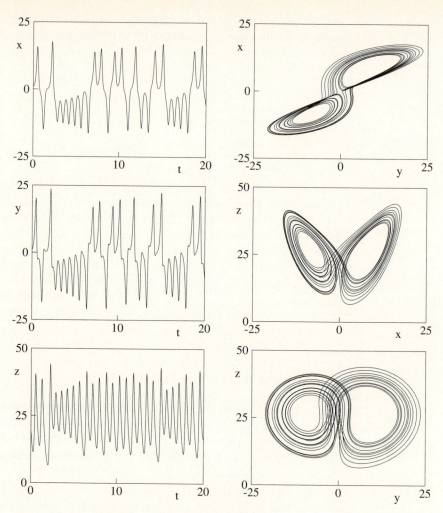

Fig. 6.10   A trajectory of the Lorenz model with $\sigma = 10$, $b = 8/3$, and $r = 28$ and the initial condition $x_0 = 1$, $y_0 = 1$, $z_0 = 20$. A time interval of $t = 20$ is shown with points plotted at intervals of 0.01. The fourth-order Runge-Kutta algorithm was used with $\Delta t = 0.0025$.

6.8   FORCED DAMPED PENDULUM

We now consider the dynamics of nonlinear mechanical systems described by classical mechanics. The general problem in classical mechanics is the determination of the positions and velocities of a system of particles subjected to certain forces. For example, we considered in Chapter 4 the celestial two-body problem and were able to predict the motion at any time. We will find that we cannot make long-time predictions for the trajectories of nonlinear classical systems when these systems exhibit chaos.

A familiar example of a nonlinear mechanical system is the simple pendulum (see Chapter 5). To make its dynamics more interesting, we assume that there is a linear damping term present and that the pivot is forced to move vertically up and down. Newton's second law for this system is (cf. McLaughlin or Percival and Richards)

$$\frac{d^2\theta}{dt^2} = -\gamma\frac{d\theta}{dt} - [\omega_0{}^2 + 2A\cos\omega t]\sin\theta, \tag{6.26}$$

where θ is the angle the pendulum makes with the vertical axis, γ is the damping coefficient, $\omega_0{}^2 = g/L$ is the natural frequency of the pendulum, and ω and A are the frequency and amplitude of the external force. Note that the effect of the vertical acceleration of the pivot is equivalent to a time-dependent gravitational field.

How do we expect the driven, damped simple pendulum to behave? Because there is damping present, we expect that if there is no external force, the pendulum would come to rest. That is, $(x = 0, v = 0)$ is a stable attractor. As A is increased from zero, this attractor remains stable for sufficiently small A. At a value of A equal to A_c, this attractor becomes unstable. How does the driven nonlinear oscillator behave as we increase the amplitude A? Because we are mainly interested in the (stable and unstable) fixed points of the motion, it is convenient to analyze the motion by plotting a point in phase space after every cycle of the external force. Such a phase space plot is called a *Poincaré map*. Hence, we will plot $d\theta/dt$ versus θ for values of t equal to nT for n equal to $1, 2, 3, \ldots$ If the system has a period T, then the Poincaré map consists of a single point. If the period of the system is nT, there will be n points.

Program poincare uses the fourth-order Runge-Kutta algorithm to compute $\theta(t)$ and the angular velocity $d\theta(t)/dt$ for the pendulum described by (6.26). A phase diagram for $d\theta(t)/dt$ versus $\theta(t)$ is shown in the left window. In the right window the Poincaré map is represented by drawing a small box at the point $(\theta, d\theta/dt)$ at time $t = nT$. If the system has period 1, i.e., if the same values of $(\theta, d\theta/dt)$ are drawn at $t = nT$, the box will flicker, indicating that there is already a box at that position. The keys 'i' and 'd' can be pressed to increase or decrease the amplitude A by an amount dA. Because the first few values of $(\theta, d\theta/dt)$ show the transient behavior, it is desirable to clear the screen and draw a new Poincaré map without changing A, θ, or $d\theta/dt$. Clearing the screen can be done by pressing any key other than i, d, or s. The latter key stops the execution of the program.

```
PROGRAM poincare
! plot phase diagram and Poincare map for damped, driven pendulum
CALL initial(theta,ang_vel,gamma,A,t,dt,nstep,dA,#1,#2,flag$)
CALL set_up_windows(A,bx,by,#1,#2)
CALL draw_axes(A,#1,#2)
DO
    FOR istep = 1 to nstep
        CALL RK4(theta,ang_vel,gamma,A,t,dt)
        CALL phase_space(theta,ang_vel,#1)
    NEXT istep
    CALL Poincare(theta,ang_vel,bx,by,#2)
    IF key input then CALL change_amplitude(A,dA,#1,#2,flag$)
LOOP UNTIL flag$ = "stop"
END
```

```
SUB initial(theta0,ang_vel0,gamma,A,t0,dt,nstep,dA,#1,#2,flag$)
    INPUT prompt "initial angle = ": theta0
    INPUT prompt "initial angular velocity = ": ang_vel0
    LET t0 = 0                    ! initial time
    INPUT prompt "damping constant = ": gamma
    INPUT prompt "amplitude of external force = ": A
    ! angular frequency of external force equals two and
    ! hence period of external force equals pi
    LET nstep = 100              ! # iterations between Poincare plot
    LET T = pi                   ! period of external force
    LET dt = T/nstep             ! time step
    LET dA = 0.01                ! increase in amplitude of external force
    ! left-side of screen for phase space plot
    OPEN #1: screen 0,0.49,0,1
    ! open second window for Poincare map
    OPEN #2: screen 0.51,1.0,0,1
    LET flag$ = ""
END SUB

SUB set_up_windows(A,bx,by,#1,#2)
    LET vmax = 4*A
    WINDOW #1
    SET WINDOW -pi,pi,-vmax,vmax
    WINDOW #2
    SET WINDOW -pi,pi,-vmax,vmax
    LET scale = 2*min(pi,vmax)
    LET bx = 0.005*scale       ! x-dimension of box
    ASK PIXELS px,py
    IF px > py then
        LET by = (px/py)*bx
    ELSE
        LET by = (py/px)*bx
    END IF
END SUB

SUB draw_axes(A,#1,#2)
    WINDOW #1
    CLEAR
    SET COLOR "black/white"
    LET vmax = 4*A
    PLOT LINES: 0,-vmax;0,vmax
    PLOT LINES: -pi,0;pi,0
    PRINT using "A = #.#####": A
    WINDOW #2
    CLEAR
    SET COLOR "black/white"
    PLOT LINES: -pi,0;pi,0
    PLOT LINES: 0,-vmax; 0,vmax
END SUB
```

```
SUB RK4(theta,ang_vel,gamma,A,t,dt)
    ! fourth-order Runge Kutta algorithm
    DECLARE DEF force
    LET k1v = force(theta,ang_vel,gamma,A,t)*dt
    LET k1x = ang_vel*dt
    LET t = t + 0.5*dt
    LET k2v = force(theta+0.5*k1x,ang_vel+0.5*k1v,gamma,A,t)*dt
    LET k2x = (ang_vel + 0.5*k1v)*dt
    LET k3v = force(theta+0.5*k2x,ang_vel+0.5*k2v,gamma,A,t)*dt
    LET k3x = (ang_vel + 0.5*k2v)*dt
    LET t = t + 0.5*dt
    LET k4v = force(theta+k3x,ang_vel+k3v,gamma,A,t)*dt
    LET k4x = (ang_vel + k3v)*dt
    LET ang_vel = ang_vel + (k1v + 2*k2v + 2*k3v + k4v)/6
    LET theta = theta + (k1x + 2*k2x + 2*k3x + k4x)/6
    IF theta > pi then LET theta = theta - 2*pi
    IF theta < -pi then LET theta = theta + 2*pi
END SUB

DEF force(theta,ang_vel,gamma,A,t)
    ! A is amplitude of driving force
    LET force = -gamma*ang_vel - (1 +2*A*cos(2*t))*sin(theta)
END DEF

SUB phase_space(theta,ang_vel,#9)
    WINDOW #9
    SET COLOR "red"
    PLOT theta,ang_vel
END SUB

SUB change_amplitude(A,dA,#1,#2,flag$)
    GET KEY k                ! press any key to clear screen
    IF k = ord("s") then LET flag$ = "stop"
    IF k = ord("i") then LET A = A + dA
    IF k = ord("d") then LET A = A - dA
    IF k = ord("i") or k = ord("d") then
        CALL set_up_windows(A,bx,by,#1,#2)
    END IF
    IF flag$ = "" then CALL draw_axes(A,#1,#2)
END SUB

SUB Poincare(theta,ang_vel,bx,by,#9)
    ! plot points and allow for change in amplitude of external force
    WINDOW #9
    SET COLOR "blue"
    BOX AREA theta-bx,theta+bx,ang_vel-by,ang_vel+by
END SUB
```

Problem 6.14 Dynamics of a driven, damped simple pendulum

a. Use `Program poincare` to simulate the driven, damped simple pendulum. In the program $\omega = 2$ so that the period T of the external force equals π. The program also assumes that $\omega_0 = 1$. Use $\gamma = 0.2$ and $A = 0.85$ and compute the phase space trajectory. After the initial transient, how many points do you see in the Poincaré plot? What is the period of the pendulum? Vary the initial values of θ and $d\theta/dt$. Is the attractor independent of the initial conditions? Remember to ignore the initial transient behavior.

b. Modify `Program poincare` so that it plots θ and $d\theta/dt$ as a function of t. Describe the qualitative relation between the Poincaré plot, the phase space plot, and the t dependence of θ and $d\theta/dt$.

c. The amplitude A plays the role of the control parameter for the dynamics of the system. Use the behavior of the Poincaré plot to find the value $A = A_c$ at which the $(0, 0)$ attractor becomes unstable. Start with $A = 0.1$ and continue increasing A until the $(0, 0)$ attractor becomes unstable.

d. Find the period for $A = 0.1, 0.25, 0.5, 0.7, 0.75, 0.85, 0.95, 1.00, 1.02, 1.031, 1.033, 1.036,$ and 1.05. Note that for small A, the period of the oscillator is twice that of the external force. The steady state period is 2π for $A_c < A < 0.71$, π for $0.72 < A < 0.79$, and then 2π again.

e. The first period-doubling occurs for $A \approx 0.79$. Find the approximate values of A for further period-doubling and use these values of A to compute the exponent δ defined by (6.10). Compare your result for δ with the result found for the one-dimensional logistic map. Are your results consistent with those that you found for the logistic map? An analysis of this system can be found in the article by McLaughlin.

f. Sometimes a trajectory does not approach a steady state even after a very long time, but a slight perturbation causes the trajectory to move quickly onto a steady state attractor. Consider $A = 0.62$ and the initial condition ($\theta = 0.3, d\theta/dt = 0.3$). Describe the behavior of the trajectory in phase space. During the simulation, change θ by 0.1. Does the trajectory move onto a steady state trajectory? Do similar simulations for other values of A and other initial conditions.

*g. Repeat the calculations of parts (b)–(d) for $\gamma = 0.05$. What can you conclude about the effect of damping?

*h. Replace the fourth-order Runge-Kutta algorithm by the lower-order Euler-Richardson algorithm. Which algorithm gives the better trade-off between accuracy and speed?

Problem 6.15 Basin of an attractor

a. For $\gamma = 0.2$ and $A > 0.79$ the pendulum rotates clockwise or counterclockwise in the steady state. Each of these two rotations is an attractor.

The set of initial conditions that lead to a particular attractor is called the basin of the attractor. Modify *Program poincare* so that the program draws the basin of the attractor with $d\theta/dt > 0$. For example, your program might simulate the motion for about 20 periods and then determine the sign of $d\theta/dt$. If $d\theta/dt > 0$ in the steady state, then the program plots a point in phase space at the coordinates of the initial condition. The program repeats this process for many initial conditions. Describe the basin of attraction for $A = 0.85$ and increments of the initial values of θ and $d\theta/dt$ equal to $\pi/10$.

b. Repeat part (a) using increments of the initial values of θ and $d\theta/dt$ equal to $\pi/20$ or as small as possible given your computer resources. Does the boundary of the basin of attraction appear smooth or rough? Is the basin of the attractor a single object or is it disconnected into more than one piece?

c. Repeat parts (a) and (b) for other values of A, including values near the onset of chaos and in the chaotic regime. Is there a qualitative difference between the basins of periodic and chaotic attractors? For example, can you always distinguish the boundaries of the basin?

*6.9 HAMILTONIAN CHAOS

Hamiltonian systems are a very important class of dynamical systems. The most familiar are mechanical systems without friction, and the most important of these is the solar system. The linear harmonic oscillator and the simple pendulum that we considered in Chapter 5 are two simple examples. Many other systems can be included in the Hamiltonian framework, e.g., the motion of charged particles in electric and magnetic fields, and ray optics. The Hamiltonian dynamics of charged particles is particularly relevant to confinement issues in particle accelerators, storage rings, and plasmas. In each case a function of all the coordinates and momenta called the Hamiltonian is formed. For many mechanical systems this function can be identified with the total energy. The Hamiltonian for a particle in a potential $V(x, y, z)$ is

$$H = \frac{1}{2m}(p_x^2 + p_y^2 + p_z^2) + V(x, y, z). \tag{6.27}$$

Typically we write (6.27) using the notation

$$H = \sum_i [\frac{p_i^2}{2m} + V(\{q_i\})], \tag{6.28}$$

where $p_1 \equiv p_x$, $q_1 \equiv x$, etc. This notation emphasizes that the p_i and the q_i are generalized coordinates. For example, in some systems p can represent the angular momentum and q can represent an angle. For a system of N particles in three dimensions, the sum in (6.28) runs from 1 to $3N$, where $3N$ is the number of degrees of freedom.

The methods for constructing the generalized momenta and the Hamiltonian are described in standard classical mechanics texts. The time dependence of the generalized momenta and coordinates is given by

$$\dot{p}_i \equiv \frac{dp_i}{dt} = -\frac{\partial H}{\partial q_i} \tag{6.29a}$$

$$\dot{q}_i \equiv \frac{dq_i}{dt} = \frac{\partial H}{\partial p_i}. \tag{6.29b}$$

Check that (6.29) leads to the usual form of Newton's second law by considering the simple example of a single particle in a potential.

As we found in Chapter 5, an important property of conservative systems is preservation of areas in phase space. Consider a set of initial conditions of a dynamical system that form a closed surface in phase space. For example, if phase space is two-dimensional, this surface would be a one-dimensional loop. As time evolves, this surface in phase space will typically change its shape. For Hamiltonian systems the volume enclosed by this surface remains constant in time. For dissipative systems this volume will decrease, and hence dissipative systems are not described by a Hamiltonian. One consequence of the constant phase space volume is that Hamiltonian systems do not have phase space attractors.

In general, the motion of Hamiltonian systems is very complex, but seems to fall into three classes. In the first class the motion is regular, and there is a constant of the motion (a quantity that does not change with time) for each degree of freedom. Such a system is said to be *integrable*. For time independent systems an obvious constant of the motion is the total energy. At the other end of the spectrum are fully chaotic systems. An important example which we will consider in Chapter 8 is a collection of molecules inside a box. Their chaotic motion is essential for the system to be described by the methods of statistical mechanics. For regular motion the change in shape of a closed surface in phase space would be rather uninteresting. For chaotic motion, nearby trajectories must exponentially diverge from each other, but be confined to a finite region of phase space. Hence, there will be local stretching of the surface accompanied by repeated folding to ensure confinement. There is another class of systems whose behavior is in between, that is, the system behaves regularly for some initial conditions, and chaotically for others. We will study these *mixed* systems in this section.

Consider the Hamiltonian for a system of N particles. If the system is integrable, there are $3N$ constants of the motion. It is natural to identify the generalized momenta with these constants. The coordinates that are associated with each of these constants will vary linearly with time. If the system is confined in phase space, then the coordinates must be periodic. If we have just one coordinate, we can think of the motion as being a point moving on a circle in phase space. In two dimensions the motion is a point moving in two circles at once, that is, a point moving on the surface of a torus. In three dimensions we can imagine a generalized torus with three circles, and so on. If the period of motion along each circle is a rational fraction of the period of all the other circles, then the torus is called a resonant torus, and the motion in phase space is

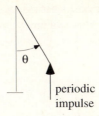

periodic
impulse

Fig. 6.11   A kicked rotor
consisting of a rigid
rod with moment
of inertia I. Gravity
and friction at the
pivot is ignored.

periodic. If the periods are not rational fractions of each other, then the torus is called
nonresonant.

  If we take an integrable Hamiltonian and change it slightly, what happens to these
tori? A partial answer is given by a theorem due to Kolmogorov, Arnold, and Moser
(KAM), which states that, under certain circumstances, these tori will remain. When the
perturbation of the Hamiltonian becomes large enough, these KAM tori are destroyed.

  To understand the basic ideas associated with mixed systems, we consider a simple
model known as the *standard* map. Consider the rotor shown in Fig. 6.11. The rod has
moment of inertia I and length L and is fastened at one end to a frictionless pivot. The
other end is subjected to a vertical periodic impulsive force of strength k/L applied at
time $t = 0, \tau, 2\tau, \ldots$ Gravity is ignored. The motion of the rotor can be described by the
angle θ and the corresponding angular momentum p_θ. The Hamiltonian for this system
can be written as

$$H(\theta, p_\theta, t) = \frac{p_\theta{}^2}{2I} + k \cos \theta \sum_n \delta(t - n\tau). \tag{6.30}$$

The term $\delta(t - n\tau)$ is zero everywhere except at $t = n\tau$; its integral over time is unity if
$t = n\tau$ is within the limits of integration. If we use (6.29) and (6.30), it is easy to show
that the corresponding equations of motion are given by

$$\frac{dp_\theta}{dt} = k \sin \theta \sum_n \delta(t - n\tau) \tag{6.31a}$$

$$\frac{d\theta}{dt} = \frac{p_\theta}{I}. \tag{6.31b}$$

From (6.31) we see that p_θ is constant between kicks (remember that gravity is assumed
to be absent), but changes discontinuously at each kick. The angle θ varies linearly with
t between kicks and is continuous at each kick.

  It is convenient to know the values of θ and p_θ at times just after the kick. We
let θ_n and p_n be the values of $\theta(t)$ and $p_\theta(t)$ at times $t = n\tau + 0^+$, where 0^+ is a

infinitesimally small positive number. If we integrate (6.31a) from $t = (n+1)\tau - 0^+$ to $t = (n+1)\tau + 0^+$, we obtain

$$p_{n+1} - p_n = k \sin \theta_{n+1}. \tag{6.32a}$$

(Remember that p is constant between kicks and the delta function contributes to the integral only when $t = (n+1)\tau$.) From (6.31b) we have

$$\theta_{n+1} - \theta_n = (\tau/I)p_n. \tag{6.32b}$$

If we choose units such that $\tau/I = 1$, we obtain the standard map

$$\theta_{n+1} = (\theta_n + p_n) \quad \text{modulo } 2\pi, \tag{6.33a}$$
$$p_{n+1} = p_n + k \sin \theta_{n+1}. \qquad \text{(standard map)} \tag{6.33b}$$

We have added the requirement in (6.33a) that the value of the angle θ is restricted to be between zero and 2π.

Before we iterate (6.33), let us check that (6.33) represents a Hamiltonian system, i.e., the area in q-p space is constant as n increases. (Here q refers to θ.) Suppose we start with a rectangle of points of length dq_n and dp_n. After one iteration, this rectangle will be deformed into a parallelogram of sides dq_{n+1} and dp_{n+1}. From (6.33) we have

$$dq_{n+1} = dq_n + dp_n \tag{6.34a}$$
$$dp_{n+1} = dp_n + k \cos q_{n+1} \, dq_{n+1}. \tag{6.34b}$$

If we substitute (6.34a) in (6.34b), we obtain

$$dp_{n+1} = (1 + k \cos q_{n+1}) \, dp_n + k \cos q_{n+1} \, dq_n. \tag{6.35}$$

To find the area of a parallelogram, we take the magnitude of the cross product of the vectors $d\mathbf{q}_{n+1} = (dq_n, dp_n)$ and $d\mathbf{p}_{n+1} = (1 + k \cos q_n dq_n, k \cos q_n dp_n)$. The result is $dq_n \, dp_n$, and hence the area in phase space has not changed. We say that the standard map is an example of an *area-preserving map*.

The qualitative properties of the standard map are explored in Problem 6.16. A summary of its properties follows. For $k = 0$, the rod rotates with a fixed angular velocity determined by the momentum $p_n = p_0 = \text{constant}$. If p_0 is a rational number times 2π, then the trajectory in phase space consists of a sequence of isolated points lying on a horizontal line (resonant tori). Can you see why? If p_0 is not a rational number times 2π or if your computer does not have sufficient precision, then after a long enough time, the trajectory will consist of a horizontal line in phase space. As we increase k, these horizontal lines are deformed into curves that run from $q = 0$ to $q = 2\pi$, and the isolated points of the resonant tori are converted into closed loops. For some initial conditions, the trajectories will become chaotic after the map has been iterated long enough so that the transient behavior has died out.

Problem 6.16   The standard map

a.   Write a program to iterate the standard map and plot its trajectory in phase space. Design the program so that more than one trajectory for the same value of the parameter k can be shown at the same time (using different colors). Choose a set of initial conditions that form a rectangle (see Problem 5.2). Does the shape of this area change with time? What happens to the total area?

b. Begin with $k = 0$ and choose an initial value of p that is a rational number times 2π. What types of trajectories do you obtain? If you obtain trajectories consisting of isolated points, do these points appear to shift due to numerical roundoff errors? How can you tell? (Use the True BASIC `truncate` function or equivalent.) What happens if p_0 is an irrational number times 2π? Remember that a computer can only approximate an irrational number.

c. Consider $k = 0.2$ and explore the nature of the phase space trajectories. What structures appear that do not appear at $k = 0$? Discuss the motion of the rod corresponding to some of the typical trajectories that you find.

d. Increase k until you first find some chaotic trajectories. How can you tell that they are chaotic? Do these chaotic trajectories fill all of phase space? If there is one trajectory that is chaotic at a particular value of k, are all trajectories chaotic? What is the approximate value for k_c above which chaotic trajectories appear?

We now discuss a discrete map that models the rings of the planet Saturn (see Fröyland). The assumption is that the rings of Saturn are due to perturbations produced by Mimas, one of Saturn's moons, which is a distance of $\sigma = 185.7 \times 10^3$ km from Saturn. There are two important forces acting on objects near Saturn. The force due to Saturn can be incorporated as follows. We know that each time Mimas completes an orbit, it traverses a total angle of 2π. Hence, the angle θ of any other moon of Saturn relative to Mimas can be expressed as

$$\theta_{n+1} = \theta_n + 2\pi \frac{\sigma^{3/2}}{r_n^{3/2}}, \tag{6.36}$$

where r_n is the radius of the orbit after n revolutions. The other important force is due to Mimas and causes the radial distance r_n to change. A discrete approximation to the radial acceleration dv_r/dt is

$$\begin{aligned}
\frac{\Delta v_r}{\Delta t} &\approx \frac{v_r(t + \Delta t) - v_r(t))}{\Delta t} \\
&\approx \frac{r(t + \Delta t) - r(t)}{(\Delta t)^2} - \frac{r(t) - r(t - \Delta t)}{(\Delta t)^2} \\
&= \frac{r(t + \Delta t) - 2r(t) + r(t - \Delta t)}{(\Delta t)^2}.
\end{aligned}$$

The acceleration equals the radial force due to Mimas. If we average over a complete period, then a reasonable approximation for the change in r_n due to Mimas is

$$r_{n+1} - 2r_n + r_{n-1} = f(r_n, \theta_n), \tag{6.37}$$

where $f(r_n, \theta_n)$ is the radial force. In general, the form of $f(r_n, \theta_n)$ is very complicated. We make a major simplifying assumption and take f to be proportional to $-(r_n - \sigma)^{-2}$ and to be periodic in θ_n. For simplicity, we express this periodicity in the simplest possible way, that is, as $\cos \theta_n$. We also want the map to be area conserving. These considerations lead us to the following two-dimensional map:

$$\theta_{n+1} = \theta_n + 2\pi \frac{\sigma^{3/2}}{r_n^{3/2}} \tag{6.38a}$$

$$r_{n+1} = 2r_n - r_{n-1} - a \frac{\cos \theta_n}{(r_n - \sigma)^2}. \tag{6.38b}$$

The constant a for Saturn's rings is approximately 2×10^{12} km^3. We can show, using a similar technique as before, that the volume in (r, θ) space is preserved, and hence (6.38) is a Hamiltonian map.

The purpose of the above discussion was only to motivate and not to derive the form of the map (6.38). In Problem 6.17 we investigate how the map (6.38) yields the qualitative structure of Saturn's rings. In particular, what happens to the values of r_n if the period of a moon is related to the period of Mimas by the ratio of two integers?

Problem 6.17   A simple model of the rings of Saturn

a. Write a program to implement the map (6.38). Be sure to save the last two values of r so that the values of r_n are updated correctly. The radius of Saturn is 60.4×10^3 km. Express all lengths in units of 10^3 km. In these units $a = 2000$. Plot the points $(r_n \cos \theta_n, r_n \sin \theta_n)$. Choose initial values for r between the radius of Saturn and σ, the distance of Mimas from Saturn, and find the bands of r_n values where stable trajectories are found.

b. What is the effect of changing the value of a? Try $a = 200$ and $a = 20000$ and compare your results with part (a).

c. Vary the force function. Replace $\cos \theta$ by other trigonometric functions. How do your results change? If the changes are small, does that give you some confidence that the model has something to do with Saturn's rings?

A more realistic dynamical system is the double pendulum, a system that can be demonstrated in the laboratory. This system consists of two equal point masses m, with one suspended from a fixed support by a rigid weightless rod of length L, and the other suspended from the first by a similar rod (see Fig. 6.12). Because there is no friction, this system is clearly an example of a Hamiltonian system. The four rectangular coordinates x_1, y_1, x_2, and y_2 of the two masses can be expressed in terms of two generalized coordinates θ_1, θ_2:

$$x_1 = L \sin \theta_1 \tag{6.39a}$$
$$y_1 = 2L - L \cos \theta_1 \tag{6.39b}$$
$$x_2 = L \sin \theta_1 + L \sin \theta_2 \tag{6.39c}$$
$$y_2 = 2L - L \cos \theta_1 - L \cos \theta_2. \tag{6.39d}$$

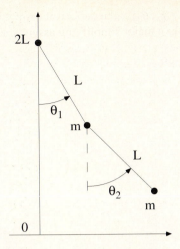

Fig. 6.12   The double pendulum.

The kinetic energy is given by

$$K = \frac{1}{2}m(\dot{x}_1^2 + \dot{x}_2^2 + \dot{y}_1^2 + \dot{y}_2^2) = \frac{1}{2}mL^2[2\dot{\theta}_1^2 + \dot{\theta}_2^2 + 2\dot{\theta}_1\dot{\theta}_2\cos(\theta_1 - \theta_2)], \qquad (6.40)$$

and the potential energy is given by

$$U = mgL(3 - 2\cos\theta_1 - \cos\theta_2). \qquad (6.41)$$

To use Hamilton's's equations of motion (6.29), we need to express the sum of the kinetic energy and potential energy in terms of the generalized momenta and coordinates. In rectangular coordinates we know that the momenta are equal to $p_i = \partial K/\partial \dot{q}_i$, where for example, $q_i = x_1$ and p_i is the x-component of $m\mathbf{v}_1$. This prescription works for generalized momenta as well, and the generalized momentum corresponding to θ_1 is given by $p_1 = \partial K/\partial \dot{\theta}_1$. If we calculate the appropriate derivatives, we can show that the generalized momenta can be written as

$$p_1 = mL^2\left[2\dot{\theta}_1 + \dot{\theta}_2\cos(\theta_1 - \theta_2)\right] \qquad (6.42a)$$

$$p_2 = mL^2\left[\dot{\theta}_2 + \dot{\theta}_1\cos(\theta_1 - \theta_2)\right]. \qquad (6.42b)$$

The Hamiltonian or total energy becomes

$$H = \frac{1}{2mL^2}\frac{p_1^2 + 2p_2^2 - 2p_1p_2\cos(q_1 - q_2)}{1 + \sin^2(q_1 - q_2)}$$
$$+ mgL(3 - 2\cos q_1 - \cos q_2), \qquad (6.43)$$

where $q_1 = \theta_1$ and $q_2 = \theta_2$. The equations of motion can be found by using (6.43) and (6.29).

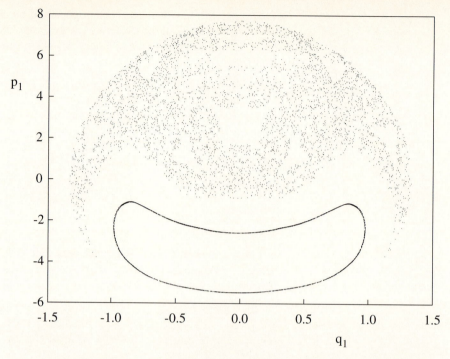

Fig. 6.13 Poincaré plot for the double pendulum with p_1 plotted versus q_1 at $q_2 = 0$ and $p_2 > 0$. Two sets of initial conditions, $(q_1, q_2, p_1) = (0, 0, 0)$ and $(1.1, 0, 0)$ respectively, were used to create the plot. The initial value of the coordinate p_2 is found from (6.43) by requiring that $E = 15$.

Fig. 6.13 shows a Poincaré map for the double pendulum. The coordinate p_1 is plotted versus q_1 for the same total energy $E = 15$, but for two different initial conditions. The map includes the points in the trajectory for which $q_2 = 0$ and $p_2 > 0$. Note the resemblance between Fig. 6.13 and plots for the standard map above the critical value of k, that is, there is a regular trajectory and a chaotic trajectory for the same parameters, but different initial conditions.

Problem 6.18 Double pendulum

a. Use either the fourth-order Runge-Kutta algorithm (with $\Delta t = 0.003$) or the second-order Euler-Richardson algorithm (with $\Delta t = 0.001$) to simulate the double pendulum. Choose $m = 1$, $L = 1$, and $g = 9.8$ (mks units). The input parameter is the total energy E. The initial values of q_1 and q_2 can be either chosen randomly within the interval $-\pi < q_i < \pi$ or by the user. Then set the initial $p_1 = 0$, and solve for p_2 using (6.43) with $H = E$. First explore the pendulum's behavior by plotting the generalized coordinates and momenta as a function of time in four windows. Consider the energies $E = 1, 5, 10, 15$, and 40. Try a few initial conditions for each energy. Determine visually whether the steady state behavior is regular or appears to be chaotic. Are there some values of E for which all the trajectories appear regular? Are there values of E for which all trajectories appear chaotic? Are there values of E for which both types of trajectories occur?

b. Repeat your investigations of part (a), but plot the phase space diagrams p_1 versus q_1 and p_2 versus q_2. Are these plots more useful for determining the nature of the trajectories than those drawn in part (a)?

c. Draw the Poincaré plot with p_1 plotted versus q_1 only when $q_2 = 0$ and $p_2 > 0$. Overlay trajectories from different initial conditions, but with the same total energy on the same plot. Duplicate the plot shown in Fig. 6.13. Then produce Poincaré plots for the values of E given in part (a), with at least five different initial conditions per plot. Describe the different types of behavior.

d. Is there a critical value of the total energy at which some chaotic trajectories first occur?

e. Write a subroutine to animate the double pendulum, showing the two masses moving back and forth. Describe how the motion of the pendulum is related to the behavior of the Poincaré plot.

Hamiltonian chaos has important applications in physical systems such as the solar system, the motion of the galaxies, and plasmas. It also has helped us understand the foundation for statistical mechanics. One of the most fascinating applications has been to quantum mechanics which has its roots in the Hamiltonian formulation of classical mechanics. A current area of much interest is the quantum analogue of classical Hamiltonian chaos. The meaning of this analogue is not obvious because well-defined trajectories do not exist in quantum mechanics. Moreover, Schrödinger's equation is linear and can be shown to have only periodic and quasiperiodic solutions. An introduction to problems in quantum chaos can be found in the text by Gutzwiller and in the series of articles by Srivastava, Kaufman, and Müller.

6.10 PERSPECTIVE

As the many recent books and review articles on chaos can attest, it is impossible to discuss all aspects of chaos in a single chapter. We will revisit chaotic systems in Chapter 14 where we introduce the concept of fractals. We will find that one of the characteristics of chaotic dynamics is that the resulting attractors often have an intricate geometrical structure.

The most general idea that we have discussed in this chapter is that *simple systems can exhibit complex behavior*. We also have learned that computers allow us to explore the behavior of dynamical systems and visualize the numerical output. However, the simulation of a system does not automatically lead to understanding. If you are interested in learning more about the phenomena of chaos and the associated theory, the suggested readings at the end of the chapter are a good place to start. We also invite you to explore chaotic phenomenon in more detail in the following projects.

6.11 PROJECTS

The first several projects are on various aspects of the logistic map. These projects do not exhaust the possible investigations of the properties of the logistic map.

Project 6.1   A more accurate determination of δ and α

We have seen that it is difficult to estimate δ accurately by finding the sequence of values of b_k at which the trajectory bifurcates for the kth time. A better way to estimate δ is to compute it from the sequence s_m of superstable trajectories of period 2^{m-1}. We already have found that $s_1 = 1/2$, $s_2 \approx 0.80902$, and $s_3 \approx 0.87464$. The parameters $s_1, s_2, \ldots$ can be computed directly from the equation

$$f^{(2^{m-1})}(x = \frac{1}{2}) = \frac{1}{2}. \tag{6.44}$$

For example, s_2 satisfies the relation $f^{(2)}(x = \frac{1}{2}) = \frac{1}{2}$. This relation, together with the analytical form for $f^{(2)}(x)$ given in (6.7), yields:

$$8r^2(1 - r) - 1 = 0. \tag{6.45}$$

Clearly $r = s_1 = \frac{1}{2}$ solves (6.45) with period 1. If we wish to solve (6.45) numerically for s_2, we need to be careful not to find the irrelevant solutions corresponding to a lower period. In this case we can factor out the solution $r = \frac{1}{2}$ and solve the resultant quadratic equation analytically to find $s_2 = (1 + \sqrt{5})/4$.

a. It is straightforward to adapt the bisection method discussed in Section 6.6. Adapt *Program period* to find the numerical solutions of (6.44). Good starting values for the left-most and right-most values of r are easy to obtain. The left-most value is $r = r_\infty \approx 0.8925$. If we already know the sequence s_1, $s_2, \ldots, s_m$, then we can estimate δ by

$$\delta_m = \frac{s_{m-1} - s_{m-2}}{s_m - s_{m-1}}. \tag{6.46}$$

We use this estimate for δ_m to estimate the right-most value of r:

$$r_{\text{right}}^{(m+1)} = \frac{s_m - s_{m-1}}{\delta_m}. \tag{6.47}$$

We choose the desired precision to be 10^{-16}, the number of decimal places stored by True BASIC on our computer. A summary of our results is given in Table 6.2. Verify these results and estimate δ.

b. Use your values of s_m to obtain a better estimate of α.

Project 6.2   Entropy of the logistic map

Another quantitative measure of chaos besides the Lyapunov exponent is based on the concept of entropy. Suppose that an experiment has M possible outcomes and

m	period	s_m
1	1	0.500 000 000
2	2	0.809 016 994
3	4	0.874 640 425
4	8	0.888 660 970
5	16	0.891 666 899
6	32	0.892 310 883
7	64	0.892 448 823
8	128	0.892 478 091

Table 6.2 Values of the control parameter s_m for the superstable trajectories of period 2^{m-1}. Nine decimal places are shown.

$p_1, p_2, \ldots, p_M$ are the probabilities of each outcome. The probabilities satisfy the normalization condition $\sum_{i=1}^{M} p_i = 1$. If we define the entropy S as

$$S = -\sum_{i=1}^{M} p_i \ln p_i, \tag{6.48}$$

then S is a measure of the uncertainty of the outcome of the experiment. For example, if $p_1 = 1$, there is no uncertainty since event 1 always occurs, and (6.48) gives $S = 0$, its minimum value. Maximum uncertainty corresponds to all M events being equally probable. In this case, $p_i = 1/M$ for all i, and $S = \ln M$.

We can determine S for the logistic map by dividing the interval $[0, 1]$ into M bins or subintervals and determining the relative number of times the trajectory falls into each bin. Write a program to compute S and determine its r dependence in the range $0.7 \leq r \leq 1$. Choose a bin size of $\Delta r = 0.01$. The following fragment of True BASIC code might suggest how to determine p_i.

```
DIM prob(1:100)
LET nbin = 100                      ! number of bins
LET delta_x = 1/nbin
LET n = 0                           ! number of iterations
LET x = 0.6
LET nmax = 10000
DO while n <= nmax
    LET x = 4*x*(1-x)
    LET n = n + 1
    LET ibin = x/delta_x
    LET ibin = truncate(ibin,0) + 1
    LET prob(ibin) = prob(ibin) + 1
LOOP
FOR ibin = 1 to nbin
    IF prob(ibin) > 0 then PRINT ibin,prob(ibin)/nmax
NEXT ibin
END
```

Plot the histogram p_i as a function of x for $r = 1$. For what value(s) of x is the histogram a maximum?

Project 6.3   From chaos to order

The bifurcation diagram of the logistic map (see Fig. 6.2) has many interesting features that we do not have space to explore. For example, you might have noticed that there are several smooth dark bands in the chaotic region for $r > r_\infty$. Use *Program bifurcate* to generate the bifurcation diagram for $r_\infty \leq r \leq 1$. Note that the points are not uniformly distributed in each vertical line. For example, if we start at $r = 1.0$ and decrease r, there is a band that narrows and eventually splits into two parts at $r \approx 0.9196$. If you look closely, you will see that the band splits into four parts at $r \approx 0.899$. In fact, if you look closely, you will see many more bands. What type of change occurs near the splitting (merging) of these bands)? Use *Program iterate_map* to look at the time series of $x_{n+1} = f(x_n)$ for $r = 0.9175$. You will notice that although the trajectory looks random, it oscillates back and forth between two bands. This behavior can be seen more clearly if you look at the time series of $x_{n+1} = f^{(2)}(x_n)$. A detailed discussion of the splitting of the bands can be found in Peitgen et al.

Project 6.4   Calculation of the Lyapunov spectrum

In Section 6.5 we discussed the calculation of the Lyapunov exponent for the logistic map. If a dynamical system has a multidimensional phase space, e.g., the Hénon map and the Lorenz model, there is a set of Lyapunov exponents, called the Lyapunov spectrum, that characterize the divergence of the trajectory. As an example, consider a set of initial conditions that forms a filled sphere in phase space for the (three-dimensional) Lorenz model. If we iterate the Lorenz equations, then the set of phase space points will deform into another shape. If the system has a fixed point, this shape contracts to a single point. If the system is chaotic, then, typically, the sphere will diverge in one direction, but become smaller in the other two directions. In this case we can define three Lyapunov exponents to measure the deformation in three mutually perpendicular directions. These three directions generally will not correspond to the axes of the original variables. Instead, we must use a Gram-Schmidt orthogonalization procedure.

The algorithm for finding the Lyapunov spectrum is as follows:

1. Linearize the dynamical equations. If **r** is the f-component vector containing the dynamical variables, then define $\Delta\mathbf{r}$ as the linearized difference vector. For example, the linearized Lorenz equations are

$$\frac{d\Delta x}{dt} = -\sigma\,\Delta x + \sigma\,\Delta y \tag{6.49a}$$

$$\frac{d\Delta y}{dt} = -x\,\Delta z - z\,\Delta x + r\,\Delta x - \Delta y \tag{6.49b}$$

$$\frac{d\Delta z}{dt} = x\,\Delta y + y\,\Delta x - b\,\Delta z. \tag{6.49c}$$

2. Define f orthonormal initial values for $\Delta\mathbf{r}$. For example, $\Delta\mathbf{r}_1(0) = (1, 0, 0)$, $\Delta\mathbf{r}_2(0) = (0, 1, 0)$, and $\Delta\mathbf{r}_3(0) = (0, 0, 1)$. Because these vectors appear in a linearized equation, they do not have to be small in magnitude.

3. Iterate the original and linearized equations of motion. One iteration yields a new vector from the original equation of motion and f new vectors $\Delta\mathbf{r}_\alpha$ from the linearized equations.

4. Find the orthonormal vectors $\Delta\mathbf{r}'_\alpha$ from the $\Delta\mathbf{r}_\alpha$ using the Gram-Schmidt procedure. That is,

$$\Delta\mathbf{r}'_1 = \frac{\Delta\mathbf{r}_1}{|\Delta\mathbf{r}_1|} \tag{6.50a}$$

$$\Delta\mathbf{r}'_2 = \frac{\Delta\mathbf{r}_2 - (\Delta\mathbf{r}'_1 \cdot \Delta\mathbf{r}_2)\Delta\mathbf{r}'_1}{|\Delta\mathbf{r}_2 - (\Delta\mathbf{r}'_1 \cdot \Delta\mathbf{r}_2)\Delta\mathbf{r}'_1|} \tag{6.50b}$$

$$\Delta\mathbf{r}'_3 = \frac{\Delta\mathbf{r}_3 - (\Delta\mathbf{r}'_1 \cdot \Delta\mathbf{r}_3)\Delta\mathbf{r}'_1 - (\Delta\mathbf{r}'_2 \cdot \Delta\mathbf{r}_3)\Delta\mathbf{r}'_2}{|\Delta\mathbf{r}_3 - (\Delta\mathbf{r}'_1 \cdot \Delta\mathbf{r}_3)\Delta\mathbf{r}'_1 - (\Delta\mathbf{r}'_2 \cdot \Delta\mathbf{r}_3)\Delta\mathbf{r}'_2|}. \tag{6.50c}$$

It is straightforward to generalize the method to higher dimensional models.

5. Set the $\Delta\mathbf{r}_\alpha(t)$ equal to the orthonormal vectors $\Delta\mathbf{r}'_\alpha(t)$.

6. Accumulate the running sum, S_α as $S_\alpha \to S_\alpha + \log|\Delta\mathbf{r}_\alpha(t)|$.

7. Repeat steps (3)–(6) and periodically output the estimate of the Lyapunov exponents $\lambda_\alpha = (1/n)S_\alpha$, where n is the number of iterations.

To obtain estimates for the Lyapunov spectrum that represent the steady state attractor, only include data after the transient behavior has died out.

a. Compute the Lyapunov spectrum for the Lorenz model with $\sigma = 16$, $b = 4$, and $r = 45.92$. Try other values of the parameters and compare your results.

b. Linearize the equations for the Hénon map and find the Lyapunov spectrum with $a = 1.4$ and $b = 0.3$ in (6.24).

Project 6.5   A spinning magnet

Consider a compass needle that is free to rotate in a periodically reversing magnetic field that is perpendicular to the axis of the needle. The equation of motion of the needle is given by

$$\frac{d^2\phi}{dt^2} = -\frac{\mu}{I}B_0 \cos \omega t \sin \phi, \tag{6.51}$$

where ϕ is the angle of the needle with respect to a fixed axis along the field, μ is the magnetic moment of the needle, I its moment of inertia, and B_0 and ω are the amplitude and the angular frequency of the magnetic field. Choose an appropriate numerical method for solving (6.51), and plot the Poincaré map at time $t = 2\pi n/\omega$. Verify that if the parameter $\lambda = \sqrt{2B_0\mu/I}/\omega > 1$, then the motion of the needle exhibits chaotic motion. Briggs (see references) discusses how to construct the corresponding laboratory system and other nonlinear physical systems.

Project 6.6   Billiard models

Consider a two-dimensional planar geometry in which a particle moves with constant velocity along straight line orbits until it elastically reflects off the boundary. This straight line motion occurs in various "billiard" systems. A simple example of such a system is a particle moving with fixed speed within a circle. For this geometry the angle between the particle's momentum and the tangent to the boundary at a reflection is the same for all points.

Suppose that we divide the circle into two equal parts and connect them by straight lines of length L as shown in Fig. 6.14a. This geometry is called a stadium billiard. How does the motion of a particle in the stadium compare to the motion in the circle? In both cases we can find the trajectory of the particle by geometrical considerations. The stadium billiard model and a similar geometry known as the Sinai billiard model (see Fig. 6.14b) have been used as model systems for exploring the foundations of statistical mechanics. There also is much interest in relating the behavior of a classical particle in various billiard models to the solution of Schrödinger's equation for the same geometries.

a.   Write a program to simulate the stadium billiard model. Use the radius r of the semicircles as the unit of length. The algorithm for determining the path of the particle is as follows:

    1. Begin with an initial position (x_0, y_0) and momentum (p_{x0}, p_{y0}) of the particle such that $|\mathbf{p}_0| = 1$.

    2. Determine which of the four sides the particle will hit. The possibilities are the top and bottom line segments and the right and left semicircles.

    3. Determine the next position of the particle from the intersection of the straight line defined by the current position and momentum, and the equation for the segment where the next reflection occurs.

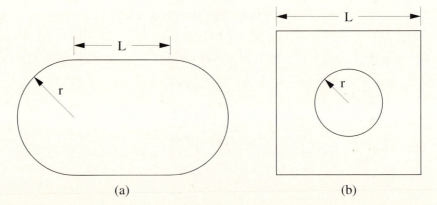

(a)   (b)

Fig. 6.14   (a) Geometry of the stadium billiard model. (b) Geometry of the Sinai billiard model.

4. Determine the new momentum, (p'_x, p'_y), of the particle after reflection such that the angle of incidence equals the angle of reflection. For reflection off the line segments we have $(p'_x, p'_y) = (p_x, -p_y)$. For reflection off a circle we have

$$p'_x = (y^2 - (x - x_c)^2) p_x - 2(x - x_c) y p_y \qquad (6.52a)$$

$$p'_y = -2(x - x_c) y p_x + ((x - x_c)^2 - y^2) p_y, \qquad (6.52b)$$

where $(x_c, 0)$ is the center of the circle.

5. Repeat steps (2)–(4).

b. Determine if the particle dynamics is chaotic by estimating the largest Lyapunov exponent. One way to do so is to start two particles with almost identical positions and/or momenta (varying by say 10^{-5}). Compute the difference Δs of the two phase space trajectories as a function of the number of reflections n, where Δs is defined by

$$\Delta s = \sqrt{|\mathbf{r}_1 - \mathbf{r}_2|^2 + |\mathbf{p}_1 - \mathbf{p}_2|^2}. \qquad (6.53)$$

Do the calculation for $L = 1$. The Lyapunov exponent can be found from a semilog plot of Δs versus n. Why does the exponential growth in Δs stop for sufficiently large n? Repeat your calculation for different initial conditions and average your values of Δs before plotting. Repeat the calculation for $L = 0.1, 0.5$, and 2.0 and determine if your results depend on L.

c. Another test for the existence of chaos is the reversibility of the motion. Reverse the momentum after the particle has made n reflections, and let the drawing color equal the background color so that the path can be erased. What limitation does roundoff error place on your results? Repeat this simulation for $L = 1$ and $L = 0$.

d. Place a small hole of diameter d in one of the circular sections of the stadium so that the particle can escape. Choose $L = 1$ and set $d = 0.02$. Give the particle a random position and momentum, and record the time when the particle escapes through the hole. Repeat for at least 10^4 particles and compute the fraction of particles $S(n)$ remaining after a given number of reflections n. The function $S(n)$ will decay with n. Determine the functional dependence of S on n, and calculate the characteristic decay time if $S(n)$ decays exponentially. Repeat for $L = 0.1, 0.5$, and 2.0. Is the decay time a function of L? Does $S(n)$ decays exponentially for the circular billiard model ($L = 0$) (see Bauer and Bertsch)?

e. Choose an arbitrary initial position for the particle in a stadium with $L = 1$, and a small hole as in part (d). Choose at least 5000 values of the initial value p_{x0} uniformly distributed between 0 and 1. Choose p_{y0} so that $|\mathbf{p}| = 1$. Plot the escape time versus p_{x0}, and describe the visual pattern of the trajectories. Then choose 5000 values of p_{x0} in a smaller interval centered about the value of p_{x0} for which the escape time was greatest. Plot these values of the escape time versus p_{x0}. Do you see any evidence of self-similarity?

f. Repeat steps (a)–(e) for the Sinai billiard geometry.

Project 6.7   Mode locking and the circle map

The driven, damped pendulum can be approximated by a one-dimensional difference equation for a range of amplitudes and frequencies of the driving force. This
difference equation is known as the *circle map* and is given by

$$\theta_{n+1} = \left(\theta_n + \Omega - \frac{K}{2\pi}\sin 2\pi\theta_n\right). \quad \text{modulo 1} \tag{6.54}$$

The variable θ represents an angle, and Ω represents a frequency ratio, the ratio
of the natural frequency of the pendulum to the frequency of the periodic driving
force. The parameter K is a measure of the strength of the nonlinear coupling of
the pendulum to the external force. An important quantity is the winding number
which is defined as

$$W = \lim_{m\to\infty} \frac{1}{m}\sum_{n=0}^{m-1}\Delta\theta_n, \tag{6.55}$$

where $\Delta\theta_n = \Omega - (K/2\pi)\sin 2\pi\theta_n$.

a.   Consider the linear case, $K = 0$. Choose $\Omega = 0.4$ and $\theta_0 = 0.2$ and determine
W. Verify that if Ω is a ratio of two integers, then $W = \Omega$ and the trajectory
is periodic. What is the value of W if $\Omega = \sqrt{2}/2$, an irrational number? Verify that $W = \Omega$ and that the trajectory comes arbitrarily close to any particular
value of θ. Does θ_n ever return exactly to its initial value? This type of behavior of the trajectory is termed *quasiperiodic*.

b.   For $K > 0$, we will find that $W \neq \Omega$ and "locks" into rational frequency ratios
for a range of values of K and Ω. This type of behavior is called *mode locking*.
For $K < 1$, the trajectory is either periodic or quasiperiodic. Estimate the value
of W for $K = 1/2$ and values of Ω in the range $O < \Omega \le 1$. The widths in
Ω of the various mode-locked regions where W is fixed increase with K. If
time permits, consider other values of K, and draw a diagram in the K-Ω
plane ($0 \le K, \Omega \le 1$) so that those areas corresponding to frequency locking
are shaded. These shaded regions are called *Arnold tongues*.

c.   For $K = 1$, all trajectories are frequency-locked periodic trajectories. Fix K at
$K = 1$ and determine the dependence of W on Ω. The plot of the W versus Ω
for $K = 1$ is called the *Devil's staircase*.

Project 6.8   Nonlinear ring laser cavity

Consider the ring laser in Fig. 6.15 consisting of four mirrors located at the corners
of a square and oriented at $45°$. A nonlinear dielectric medium (for which the index
of refraction depends on the intensity of the light) is placed between two of the
mirrors. Light is sent in from point A and is detected at point B. Mirrors 1 and 2
are partially reflective, and mirrors 3 and 4 are totally reflective.

The electric field can be represented by a plane wave modulated by a sinusoidal function of space and time. The electric field amplitude at B arises from

two sources – the light that arrives directly from A and the light that is reflected around the ring. The two sources do not have the same amplitude or phase, and the reflected field vector is rotated and diminished in amplitude relative to the input field. It can be shown (see Ikeda et al.) that the field at B at time t_{n+1} is related to the input field at A at time t_n by:

$$E_{n+1} = a + bE_n\, e^{ik - ip/(1+|E_n|^2)}. \tag{6.56}$$

We have written the electric field using complex notation, $\mathbf{E} = E_x + iE_y$, with $i = \sqrt{-1}$. The quantity a is the input amplitude of the laser at point A, b measures the loss due to the partially reflecting mirrors, and p measures the detuning due to the nonlinear dielectric medium; k would be the phase change if there were no dielectric medium present. Detuning refers to the relative rotation of the electric field due to the dielectric medium. The difference $[t_{n+1} - t_n]$ is the time for light to go once around the ring. Equation (6.56) can be written as a coupled two-dimensional map:

$$E_{x,n+1} = a + b\left(E_{x,n}\cos\theta_n - E_{y,n}\sin\theta_n\right) \tag{6.57a}$$

$$E_{y,n+1} = b\left(E_{x,n}\sin\theta_n + E_{y,n}\cos\theta_n\right), \tag{6.57b}$$

where $\theta_n = k - p/(1 + |E_n|^2)$, and $e^{i\theta} = \cos\theta + i\sin\theta$.

a.   Iterate the map (6.57) for $a = 0.8$, $b = 0.9$, and $k = 0.4$. Choose the initial condition, $(E_x, E_y) = (0, 0)$. Plot E_x versus E_y for different values of p in the range $0 < p \le 6$. Approximately determine the values of p for which the period of the system changes.

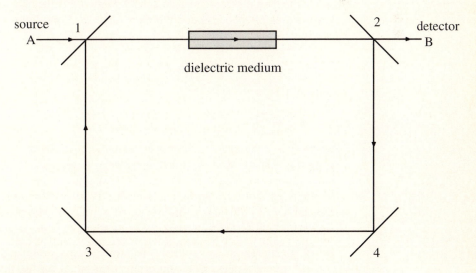

Fig. 6.15   The ring laser discussed in Project 6.8.

b. List the values of (E_x, E_y) and estimate as accurately as possible the value of p at which each period-doubling occurs. Use these values to estimate the Feigenbaum number δ. Compare your estimate to its value for the logistic map. How accurate is your calculation of δ? Does the map (6.57) exhibit behavior similar to the logistic map?

c. Is there a window of period three?

Project 6.9 Chaotic scattering

In Chapter 4 we discussed the classical scattering of particles off a fixed target, and found that the differential cross section for a variety of interactions is a smoothly varying function of the scattering angle. That is, a small change in the impact parameter b leads to a small change in the scattering angle θ. There is much current interest in cases where a small change in b leads to large changes in θ. Such a phenomenon is called *chaotic scattering*, because of the sensitivity to initial conditions that is characteristic of chaos. The study of chaotic scattering is relevant to the design of electronic nanostructures, because many experimental structures exhibit this type of scattering. In addition, the comparison of classical and quantum scattering is an active area of research in quantum chaos.

A typical scattering model consists of a target composed of a group of fixed hard disks and a scatterer consisting of a point particle. We trace the path of the scatterer as it bounces off the disks, and measure θ and the time of flight as a function of the impact parameter b. If a particle bounces inside the target region before leaving, the time of flight can be very long. There are even some trajectories for which the particle never leaves the target region.

Because it is difficult to monitor a trajectory that bounces back and forth between the hard disks, we instead consider a two-dimensional map that contains the key features of chaotic scattering (see Yalcinkaya and Lai for further discussion). The map is given by:

$$x_{n+1} = a\left[x_n - \frac{1}{4}(x_n + y_n)^2\right], \tag{6.58a}$$

and

$$y_{n+1} = \frac{1}{a}\left[y_n + \frac{1}{4}(x_n + y_n)^2\right], \tag{6.58b}$$

where a is a parameter. The target region is centered at the origin. In an actual scattering experiment, the relation between (x_{n+1}, y_{n+1}) and (x_n, y_n) would be much more complicated, but the map (6.58) captures most of the important features of realistic chaotic scattering experiments. The iteration number n is analogous to the number of collisions of the scattered particle off the disks. When x_n or y_n is significantly different from zero, the scatterer has left the target region.

a. Write a program to iterate the map (6.58). Let $a = 8.0$ and $y_0 = -0.3$. Choose 10^4 initial values of x_0 uniformly distributed on the interval $0 < x_0 < 0.1$. Determine the time $T(x_0)$, the number of iterations for which $x_n \leq -5.0$. After this time, x_n rapidly moves to $-\infty$. Plot $T(x_0)$ versus x_0. Then choose

10^4 initial values in a smaller interval centered about a value of x_0 for which $T(x_0) > 7$. Plot these values of $T(x_0)$ versus x_0. Do you see any evidence of self-similarity?

b. A trajectory is said to be *uncertain* if a small change ϵ in x_0 leads to a change in $T(x_0)$. We expect that the number of uncertain trajectories, N, will depend on a power of ϵ, i.e., $N \sim \epsilon^\alpha$. Determine $N(\epsilon)$ for $\epsilon = 10^{-p}$ with $p = 2$ to 7 using the values of x_0 in part (a). Then determine the uncertainty dimension $1 - \alpha$ from a log-log plot of N versus ϵ. Repeat these measurements for other values of a. Does α depend on a?

c. Choose 4×10^4 initial conditions in the same interval as in part (a) and determine the number of trajectories, $S(n)$, that have not yet reached $x_n = -5$ as a function of the number of iterations n. Plot $\ln S(n)$ versus n and determine if the decay is exponential. It is possible to obtain algebraic decay for values of a less than approximately 6.5.

d. Let $a = 4.1$ and choose 100 initial conditions uniformly distributed in the region $1.0 < x_0 < 1.05$ and $0.60 < y_0 < 0.65$. Are there any trajectories that are periodic and hence have infinite escape times? Due to the accumulation of roundoff error, it is possible to find only finite, but very long escape times. These periodic trajectories form closed curves, and the regions enclosed by them are called KAM surfaces.

Project 6.10   Chemical reactions

In Project 5.1 we discussed how chemical oscillations can occur when the reactants are continuously replenished. In this project we introduce a set of chemical reactions that exhibits the period doubling route to chaos. Consider the following reactions (see Peng et al.):

$$P \rightarrow A \tag{6.59a}$$

$$P + C \rightarrow A + C \tag{6.59b}$$

$$A \rightarrow B \tag{6.59c}$$

$$A + 2B \rightarrow 3B \tag{6.59d}$$

$$B \rightarrow C \tag{6.59e}$$

$$C \rightarrow D. \tag{6.59f}$$

Each of the above reactions has an associated rate constant. The time dependence of the concentrations of A, B, and C is given by:

$$\frac{dA}{dt} = k_1 P + k_2 PC - k_3 A - k_4 AB^2 \tag{6.60a}$$

$$\frac{dB}{dt} = k_3 A + k_4 AB^2 - k_5 B \tag{6.60b}$$

$$\frac{dC}{dt} = k_4 B - k_5 C. \tag{6.60c}$$

We assume that P is held constant by replenishment from an external source. We also assume the chemicals are well mixed so that there is no spatial dependence. In Section 12.4 we discuss the effects of spatial inhomogeneities due to molecular diffusion. Equations (6.59) can be written in a dimensionless form as

$$\frac{dX}{d\tau} = c_1 + c_2 Z - X - XY^2 \tag{6.61a}$$

$$c_3 \frac{dY}{d\tau} = X + XY^2 - Y \tag{6.61b}$$

$$c_4 \frac{dZ}{d\tau} = Y - Z, \tag{6.61c}$$

where the c_i are constants, $\tau = k_3 t$, and X, Y, and Z are proportional to A, B, and C, respectively.

a. Write a program to solve the coupled differential equations in (6.61). We suggest using a fourth-order Runge-Kutta algorithm with an adaptive step size. (See Project 4.3 for how to implement an adaptive step size algorithm.) Plot $\ln Y$ versus the time τ.

b. Set $c_1 = 10$, $c_3 = 0.005$, and $c_4 = 0.02$. The constant c_2 is the control parameter. Consider $c_2 = 0.10$ to 0.16 in steps of 0.005. What is the period of $\ln Y$ for each value of c_2?

c. Determine the values of c_2 at which the period doublings occur for as many period doublings as you can determine. Compute the constant δ (see (6.9)) and compare its value to the value of δ for the logistic map.

d. Make a bifurcation diagram by taking the values of $\ln Y$ from the Poincaré plot at $X = Z$, and plotting them versus the control parameter c_2. Do you see a sequence of period doublings?

e. If you have three-dimensional graphics capability, plot the trajectory of (6.61) with $\ln X$, $\ln Y$, and $\ln Z$ as the three axes. Describe the attractors for some of the cases considered in part (b).

Appendix 6A   STABILITY OF THE FIXED POINTS OF THE LOGISTIC MAP

In the following, we derive analytical expressions for the fixed points of the logistic map. The fixed-point condition is given by

$$x^* = f(x^*). \tag{6.62}$$

From (6.5) this condition yields the two fixed points

$$x^* = 0 \quad\text{and}\quad x^* = 1 - \frac{1}{4r}. \tag{6.63}$$

Because x is restricted to be positive, the only fixed point for $r < 1/4$ is $x = 0$. To determine the stability of x^*, we let

$$x_n = x^* + \epsilon_n \tag{6.64a}$$

and

$$x_{n+1} = x^* + \epsilon_{n+1}. \tag{6.64b}$$

Because $|\epsilon_n| \ll 1$, we have

$$\begin{aligned} x_{n+1} = f(x^* + \epsilon_n) &\approx f(x^*) + \epsilon_n f'(x^*) \\ &= x^* + \epsilon_n f'(x^*). \end{aligned} \tag{6.65}$$

If we compare (6.64b) and (6.65), we obtain

$$\epsilon_{n+1}/\epsilon_n = f'(x^*). \tag{6.66}$$

If $|f'(x^*)| > 1$, the trajectory will diverge from x^* since $|\epsilon_{n+1}| > |\epsilon_n|$. The opposite is true for $|f'(x^*)| < 1$. Hence, the local stability criteria for a fixed point x^* are

1. $|f'(x^*)| < 1$, x^* is stable;
2. $|f'(x^*)| = 1$, x^* is marginally stable;
3. $|f'(x^*)| > 1$, x^* is unstable.

If x^* is marginally stable, the second derivative $f''(x)$ must be considered, and the trajectory approaches x^* with deviations from x^* inversely proportional to the square root of the number of iterations.

For the logistic map the derivatives at the fixed points are respectively

$$f'(x = 0) = \frac{d}{dx}[4rx(1-x)]\bigg|_{x=0} = 4r \tag{6.67}$$

and

$$f'(x = x^*) = \frac{d}{dx}[4rx(1-x)]\bigg|_{x=1-1/4r} = 2 - 4r. \tag{6.68}$$

It is straightforward to use (6.67) and (6.68) to find the range of r for which $x^* = 0$ and $x^* = 1 - 1/4r$ are stable.

If a trajectory has period two, then $f^{(2)}(x) = f(f(x))$ has two fixed points. If you are interested, you can solve for these fixed points analytically. As we found in Problem 6.2, these two fixed points become unstable at the same value of r. We can derive this property of the fixed points using the chain rule of differentiation:

$$\frac{d}{dx}f^{(2)}(x)\bigg|_{x=x_0} = \frac{d}{dx}f(f(x))\bigg|_{x=x_0} = f'(f(x_0))f'(x)\bigg|_{x=x_0}.$$

If we substitute $x_1 = f(x_0)$, we can write

$$\frac{d}{dx} f(f(x))\Big|_{x=x_0} = f'(x_1) f'(x_0). \tag{6.69}$$

In the same way, we can show that

$$\frac{d}{dx} f^{(2)}(x)\Big|_{x=x_1} = f'(x_0) f'(x_1). \tag{6.70}$$

We see that if x_0 becomes unstable, then $|f^{(2)'}(x_0)| > 1$ as does $|f^{(2)'}(x_1)|$. Hence, x_1 also is unstable at the same value of r, and we conclude that both fixed points of $f^{(2)}(x)$ bifurcate at the same value of r, leading to an trajectory of period 4.

From (6.68) we see that $f'(x = x^*) = 0$ when $r = \frac{1}{2}$ and $x^* = \frac{1}{2}$. Such a fixed point is said to be *superstable*, because as we found in Problem 6.4, convergence to the fixed point is relatively rapid. In general, superstable trajectories occur whenever one of the fixed points is at $x^* = \frac{1}{2}$.

References and Suggestions for Further Reading

Books

Ralph H. Abraham and Christopher D. Shaw, *Dynamics–The Geometry of Behavior*, Addison-Wesley (1984). The authors use an abundance of visual representations.

Hao Bai-Lin, *Chaos II*, World Scientific (1990). A collection of reprints on chaotic phenomena. The following papers were cited in the text. James P. Crutchfield, J. Doyne Farmer, Norman H. Packhard, and Robert S. Shaw, "Chaos," *Sci. Amer.* **255**(6), 46–57 (1986); Mitchell J. Feigenbaum, "Quantitative universality for a class of nonlinear transformations," *J. Stat. Phys.* **19**, 25 (1978); M. Hénon, "A two-dimensional mapping with a strange attractor," *Commun. Math. Phys.* **50**, 50 (1976); Robert M. May, "Simple mathematical models with very complicated dynamics," *Nature* **261**, 459 (1976); Robert Van Buskirk and Carson Jeffries, "Observation of chaotic dynamics of coupled nonlinear oscillators," *Phys. Rev.* A **31**, 3332 (1985).

G. L. Baker and J. P. Gollub, *Chaotic Dynamics: An Introduction*, Cambridge University Press (1990). A good introduction to chaos with special emphasis on the forced damped nonlinear harmonic oscillator. Several programs are given in True BASIC.

Pedrag Cvitanovic, *Universality in Chaos*, second edition, Adam-Hilger (1989). A collection of reprints on chaotic phenomena including the articles by Hénon and May also reprinted in the Bai-Lin collection and the chaos classic, Mitchell J. Feigenbaum, "Universal behavior in nonlinear systems," *Los Alamos Sci.* **1**, 4 (1980).

Robert Devaney, *A First Course in Chaotic Dynamical Systems*, Addison-Wesley (1992). This text is a good introduction to the more mathematical ideas behind chaos and related topics.

Jan Fröyland, *Introduction to Chaos and Coherence*, Institute of Physics Publishing (1992). See Chapter 7 for a simple model of Saturn's rings.

Martin C. Gutzwiller, *Chaos in Classical and Quantum Mechanics*, Springer-Verlag (1990). A good introduction to problems in quantum chaos for the more advanced student.

Robert C. Hilborn, *Chaos and Nonlinear Dynamics*, Oxford University Press (1994). An excellent pedagogically oriented text.

Douglas R. Hofstadter, *Metamagical Themas*, Basic Books (1985). A shorter version is given in his article, "Metamagical themas," *Sci. Amer.* **245**(11), 22–43 (1981).

E. Atlee Jackson, *Perspectives of Nonlinear Dynamics*, Vols. 1 and 2., Cambridge University Press (1989, 1991). An advanced text that is a joy to read.

R. V. Jensen, "Chaotic scattering, unstable periodic orbits, and fluctuations in quantum transport," *Chaos* **1**, 101 (1991). This paper discusses the quantum version of systems similar to those discussed in Projects 6.9 and 6.6.

Francis C. Moon, *Chaotic and Fractal Dynamics, An Introduction for Applied Scientists and Engineers*, Wiley (1992). An engineering oriented text with a section on how to build devices that demonstrate chaotic dynamics.

Edward Ott, *Chaos in Dynamical Systems*, Cambridge University Press (1993). An excellent textbook on chaos at the upper undergraduate to graduate level. See also E. Ott, "Strange attractors and chaotic motions of dynamical systems," *Rev. Mod. Phys.* **53**, 655 (1981).

Edward Ott, Tim Sauer, James A. Yorke, editors, *Coping with Chaos*, John Wiley & Sons (1994). A reprint volume emphasizing the analysis of experimental time series from chaotic systems.

Heinz-Otto Peitgen, Hartmut Jürgens, and Dietmar Saupe, *Fractals for the Classroom*, Part II, Springer-Verlag (1992). A delightful book with many beautiful illustrations. Chapter 11 discusses the nature of the bifurcation diagram of the logistic map.

Ian Percival and Derek Richards, *Introduction to Dynamics*, Cambridge University Press (1982). An advanced undergraduate text that introduces phase trajectories and the theory of stability. A derivation of the Hamiltonian for the driven damped pendulum considered in Section 6.4 is given in Chapter 5, example 5.7.

Ivars Peterson, *Newton's Clock: Chaos in the Solar System*, W. H. Freeman (1993). An historical survey of our understanding of the motion of bodies within the solar system with a focus on chaotic motion.

Stuart L. Pimm, *The Balance of Nature*, The University of Chicago Press (1991). An introductory treatment of ecology with a chapter on applications of chaos to real biological systems. The author contends that much of the difficulty in assessing the importance of chaos is that ecological studies are too short.

Robert Shaw, *The Dripping Faucet as a Model Chaotic System*, Aerial Press, Santa Cruz, CA (1984).

Steven Strogatz, *Nonlinear Dynamics and Chaos with Applications to Physics, Biology, Chemistry and Engineering*, Addison-Wesley (1994). Another outstanding text.

Anastasios A. Tsonis, *Chaos: From Theory to Applications*, Plenum Press (1992). Of particular interest is the discussion of applications nonlinear time series forecasting.

Nicholas B. Tufillaro, Tyler Abbott, and Jeremiah Reilly, *Nonlinear Dynamics and Chaos*, Addison-Wesley (1992). See also, N. B. Tufillaro and A. M. Albano, "Chaotic dynamics of a bouncing ball," *Amer. J. Phys.* **54**, 939 (1986). The authors describe an undergraduate level experiment of a bouncing ball subject to repeated impacts with a vibrating table.

Articles

W. Bauer and G. F. Bertsch, "Decay of ordered and chaotic systems," *Phys. Rev. Lett.* **65**, 2213 (1990). See also the comment by Olivier Legrand and Didier Sornette, "First return, transient chaos, and decay in chaotic systems," *Phys. Rev. Lett.* **66**, 2172 (1991), and the reply by Wolfgang Bauer and George F. Bertsch on the following page. The authors of these papers point out that the dependence of decay laws on chaotic behavior is very general and has been considered in various contexts including room acoustics and the chaotic scattering of microwaves in an "elbow" cavity. They also mention that chaotic behavior is a sufficient, but not necessary condition for exponential decay.

Keith Briggs, "Simple experiments in chaotic dynamics," *Amer. J. Phys.* **55**, 1083 (1987).

J. P. Crutchfield, J. D. Farmer, and B. A. Huberman, "Fluctuations and simple chaotic dynamics," *Phys. Repts.* **92**, 45 (1982).

William L. Ditto and Louis M. Pecora, "Mastering chaos," *Sci. Amer.* **262**(8), 78 (1993).

J. C. Earnshaw and D. Haughey, "Lyapunov exponents for pedestrians," *Amer. J. Phys.* **61**, 401 (1993).

K. Ikeda, H. Daido, and O. Akimoto, "Optical turbulence: chaotic behavior of transmitted light from a ring cavity," *Phys. Rev. Lett.* **45**, 709 (1980). See also S. M. Hammel, C. K. R. T. Jones, and J. V. Maloney, "Global dynamical behavior of the optical field in a ring cavity," *J. Opt. Soc. Am.* B**2**, 552 (1985).

Ying-Cheng Lai, "Controlling chaos," *Computers in Physics* **8**, 62 (1994). Section 6.6 is based on this article.

J. B. McLaughlin, "Period-doubling bifurcations and chaotic motion for a parametrically forced pendulum," *J. Stat. Phys.* **24**, 375 (1981).

Bo Peng, Stephen K. Scott, and Kenneth Showalter, "Period doubling and chaos in a three-variable autocatalator," *J. Phys. Chem.* **94**, 5243 (1990).

Niraj Srivastava, Charles Kaufman, and Gerhard Müller, "Hamiltonian chaos," *Computers in Physics* **4**, 549 (1990); *ibid.* **5**, 239 (1991); *ibid.* **6**, 84 (1992).

Jan Tobochnik and Harvey Gould, "Quantifying chaos," *Computers in Physics* **3**(6), 86 (1989). Note that there is a typographical error in the equations for step (3) of the algorithm for computing the Lyapunov spectrum. The correct equations are given in Project 6.4.

Tolga Yalcinkaya and Ying-Cheng Lai, "Chaotic scattering," *Computers in Physics*, to be published (1995). Project 6.9 is based on a draft of this article. The map (6.58) is discussed in more detail in Yun-Tung Lau, John M. Finn, and Edward Ott, "Fractal dimension in nonhyperbolic chaotic scattering," *Phys. Rev. Lett.* **66**, 978 (1991).

7

Random Processes

Random processes are introduced in the context of several simple physical systems.

7.1 ORDER TO DISORDER

In Chapter 6 we saw several examples of how deterministic systems can generate chance. In this chapter we will see some examples of how chance can generate statistically predictable outcomes. For example, we know that if we bet often enough on the outcome of a game for which the outcome is determined by chance, we will lose money eventually if the probability of winning is less than 50%.

We first give an example that illustrates the tendency of many particle systems to evolve toward a well defined state. Imagine a closed box that is divided into two parts of equal volume (see Fig. 7.1). The left half contains a gas of N identical particles and the right half is empty. We then make a small hole in the partition between the two halves. What happens? We know that after some time, the system reaches equilibrium, and the average number of particles in each half of the box is $N/2$.

How can we simulate this process? One way is to give each particle an initial velocity and position and adopt a simple deterministic model of the motion of the particles. We could assume that each particle moves in a straight line until it hits a wall of the box and undergoes an elastic collision. We will consider similar deterministic models in Chapter 8, where we will find that this model is challenging to program. Instead, we consider a simpler approach based on the simulation of a *random process*.

The basic assumption underlying the probabilistic model is that the motion of the particles is so complex that we can assume random behavior. For simplicity, we assume that the particles do not interact with one another so that the probability per unit time that a particle goes through the hole is the same for all particles regardless of the number of particles in either half. We also assume that the size of the hole is such that only one particle can pass through it in one unit of time.

One way to implement this model is to choose a particle at random and move it to the other side. For example, we can use an array, LOC indexed by the particle label i, such that LOC(i) = -1 if particle i is in the left half, and LOC(i) = +1 if particle i is in the right half. We can simulate the random process by generating an integer i between 1 and N at random and changing the sign of LOC(i). The tool we need to do this simulation and other simulations of probabilistic systems is a random number generator.

It might seem strange that we can use a deterministic computer to generate sequences of random numbers. In Chapter 12 we discuss some of the methods for computing a set of numbers that appear statistically random, but are in fact generated by a deterministic algorithm. These algorithms are sometimes called pseudorandom number generators to distinguish their output from intrinsically random physical processes such as the time between clicks in a Geiger counter near a radioactive sample. We will not make such a distinction.

For the present we will be content to use the random number generator supplied with various programming languages, although these random number generators vary greatly in quality. In True BASIC the function rnd produces a random number r that is uniformly distributed in the interval $0 \le r < 1$. To generate a random integer i between 1 and N we write:

```
LET i = int(N*rnd) + 1
```

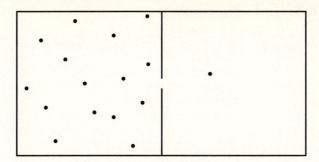

Fig. 7.1   A box is divided into two equal halves by a
partition. After a small hole is opened in the
partition, one particle can pass through the hole
per unit time.

Because the effect of the `int` function is to round the output of `rnd` down to its nearest integer, it is necessary to add 1. In True BASIC the same sequence of random numbers appears each time the program is run unless the function `RANDOMIZE` is called before the `rnd` function is used.

We now use the `rnd` statement to write a program to implement our probabilistic model for the particles in a box problem. However, the procedure we have specified is needlessly cumbersome, because our only interest is the number of particles on each side. Hence, we need to know only n, the number of particles on the left side; the number on the right side is $n' = N - n$. Because each particle has the same chance to go through the hole, the probability per unit time that a particle moves from left to right equals the number of particles on the left side divided by the total number of particles, that is, the probability of a move from left to right is n/N. The algorithm for simulating the evolution of the model can be summarized by the following steps:

1. Generate a random number r from a uniformly distributed set of random numbers in the interval $0 \leq r < 1$.

2. Compare r to the current value of the fraction of particles n/N on the left side of the box.

3. If $r \leq n/N$, move a particle from left to right, i.e., let $n \to n - 1$; otherwise, move a particle from right to left.

4. Increase the "time" by unity.

Program box implements this algorithm and plots the evolution of n. Note how straightforward it is to implement this probabilistic model.

```
PROGRAM box
! simulation of the particles in a box problem
CALL initial(N,tmax)
CALL move(N,tmax)                  ! move particles through hole
END
```

```
SUB initial(N,tmax)
    RANDOMIZE
    INPUT prompt "number of particles = ": N
    LET tmax = 20*N
    SET WINDOW -0.01*tmax,1.01*tmax,-0.01*N,1.01*N
    CLEAR
    BOX LINES 0,tmax,0,N
    SET COLOR "blue"
    PLOT 0,0.5*N;tmax,0.5*N   ! equilibrium value
END SUB

SUB move(N,tmax)
    LET nleft = N              ! initially all particles on left side
    SET COLOR "red"
    FOR t = 1 to tmax
        ! generate random number and move particle
        LET r = int(N*rnd) + 1
        IF r <= nleft then
           LET nleft = nleft - 1
        ELSE
           LET nleft = nleft + 1
        END IF
        PLOT t,nleft;
    NEXT t
END SUB
```

How long does it take for the system to reach equilibrium? How does this time depend on the number of particles? After the system is in equilibrium, what is the magnitude of the fluctuations? How do the fluctuations depend on the number of particles? Problems 7.1 and 7.2 address such questions.

Problem 7.1   Approach to equilibrium

a.   Run *Program box* and describe the time evolution of n, the number of particles on the left side of the box. Choose the total number of particles N to be $N = 8, 16, 64, 400, 800$, and 3600. Estimate the time for the system to reach equilibrium from the plots. How does this time depend on N? What criterion did you use for equilibrium? Does n change when the system is in equilibrium?

b.   For sufficiently large N, does the time dependence of n appear to be deterministic? Based on the shape of your plots of $n(t)$, what is the qualitative behavior of $n(t)$ before equilibrium is reached?

Problem 7.2   Equilibrium fluctuations

a.   Modify *Program box* so that averages are taken after equilibrium has been reached. What is the maximum deviation of $n(t)$ from $N/2$ for $N = 64, 400$,

800, and 3600? Run for a time that is long enough to yield meaningful results. How do you know if they are? How do your results for the maximum deviation depend on N?

b. A measure of the equilibrium fluctuations is the variance σ^2 defined as

$$\sigma^2 = \overline{(n - \bar{n})^2}$$
$$= \overline{n^2} - \bar{n}^2. \tag{7.1}$$

The bar denotes a time average taken after the system has reached equilibrium. The relative magnitude of the fluctuations is $\sigma/\bar{n}$. Compute the variance of n for the same values of N considered in part (a). How do the relative fluctuations, $\sigma/\bar{n}$, depend on N?

From Problem 7.1 we see that $n(t)$ decreases in time from its initial value to its equilibrium value in an almost deterministic manner if $N \gg 1$. It is instructive to derive the time dependence of $n(t)$ to show explicitly how chance can generate deterministic behavior. If there are $n(t)$ particles on the left side after t moves, then the change in $n(t)$ in the time interval Δt is given by

$$\Delta n = \left[\frac{-n(t)}{N} + \frac{N - n(t)}{N}\right]\Delta t. \tag{7.2}$$

(Recall that the time interval $\Delta t = 1$ in our simulations.) What is the meaning of the two terms in (7.2)? If we treat n and t as continuous variables and take the limit $\Delta t \to 0$, we have

$$\frac{dn}{dt} = 1 - \frac{2n(t)}{N}. \tag{7.3}$$

The solution of the differential equation (7.3) is

$$n(t) = \frac{N}{2}\left[1 + e^{-2t/N}\right], \tag{7.4}$$

where we have used the initial condition $n(t = 0) = N$. How does the exponential form (7.4) compare to your Monte Carlo results for various values of N?

*Problem 7.3 Variation on "particles in the box"

Modify Program box so that each side of the box is chosen with equal probability. A particle is then moved from the side chosen to the other side. If the side chosen does not have a particle in it, then no particle is moved during this time interval. Do you expect that the system behaves in the same way as before? Do the simulation starting with all the particles on the left side of the box and choose $N = 800$. Compare the behavior of $n(t)$ with your predicted behavior, and with the behavior of $n(t)$ found in Problem 7.2. How do the values of $\bar{n}$ and σ^2 compare? Is this variation of the model realistic?

The above probabilistic method for simulating the approach to equilibrium is an example of a *Monte Carlo* method, that is, the random sampling of the most probable outcomes. An alternative method is to use *exact enumeration* and to determine all the possibilities at each time interval. For example, suppose that at $t = 0$, $n = 8$, and $n' = 0$. At $t = 1$, the only possibility is $n = 7$ and $n' = 1$. Hence, $P(n = 7, t = 1) = 1$ and all other probabilities are zero. At $t = 2$, one of the seven particles on the left can move to the right, or the one particle on the right can move to the left. Because the first possibility can occur in seven different ways, we have the nonzero probabilities, $P(n = 6, t = 2) = 7/8$ and $P(n = 8, t = 2) = 1/8$. Hence at $t = 2$, the average number of particles on the left side of the box is

$$<n(t = 2)> = 6P(6, 2) + 8P(8, 2) = \frac{1}{8}[6 \times 7 + 8 \times 1] = 6.25.$$

Note that we have denoted the average by the brackets $< \ldots >$ to distinguish it from the time average that was computed in Problem 7.2. Is this exact result consistent with what you found in Problem 7.1? In this example N is small, and we can continue the enumeration of all the possibilities indefinitely. However for larger N, the number of possibilities becomes very large after a few time intervals, and we are forced to use Monte Carlo methods.

So far we have run each simulation only once. Clearly, running a Monte Carlo simulation only once cannot reproduce the exact enumeration results, because in general, each run gives somewhat different outcomes if we use a different sequence of random numbers. In general, we need to do a Monte Carlo simulation many times and average over the results to obtain meaningful averages. Each run is called a *trial*, or a *sample*, or sometimes an *experiment*. How do you know how many trials to use? The answer usually can be obtained empirically by averaging over more and more trials until the average results do not change within the desired level of accuracy.

7.2 THE POISSON DISTRIBUTION AND NUCLEAR DECAY

As we have seen, we often can change the names of several variables and do a seemingly different physical problem. Our goal in this section is to discuss the decay of unstable nuclei, but we first discuss a conceptually easier problem related to throwing darts. Related physical problems are the distribution of stars on the sky and the distribution of photons on a photographic plate.

Suppose we randomly throw $N = 100$ darts at a dart board that has been divided into $L = 1000$ equal size regions. The probability that a dart hits a given region for any one throw is $p = 1/1000$. If we count the number of darts in the different regions, we would find that most regions are empty, some regions have one dart, and other regions have more than one dart. What is the probability that a given region has a particular number of darts? The simulation of this problem is straightforward. Imagine that the L regions of the dart board are labeled and correspond to the L elements in the one-dimensional array B. Throwing a dart at random at the board is equivalent to choosing an integer at random between 1 and L. Randomly throw $N = 100$ darts and count how many elements in the array B have n darts. Call this number $H(n)$. Then repeat the simulation and add the results to the histogram $H(n)$. That is, if $H(n) = H_1(n)$ for the first simulation, and $H(n) = H_2(n)$ for the second simulation, then $H(n) = H_1(n) + H_2(n)$ for both simulations. Average $H(n)$ over many simulations (trials), and calculate the distribution

$$P(n) = \frac{H(n)}{\sum_n H(n)}. \tag{7.5}$$

Also calculate $<n>$, the average number of darts in a region, where

$$<n> = \sum_n n P(n). \tag{7.6}$$

We have denoted the average in (7.6) by the brackets $<\ldots>$ to distinguish it from the time average that was computed in Problem 7.2. In the latter case we considered *one* system and computed the time average of the number of particles on one side of the box. In contrast, in the darts problem, we average over *many* systems.

We usually choose N (the number of darts) to be much greater than unity, and p (the probability that a dart strikes a given region) to be much less then unity. The conditions $N \gg 1$ and $p \ll 1$ and the independence of the events (the landing of a dart in a particular region) satisfy the requirements for a *Poisson distribution*. The Poisson distribution, $P(n)$, is given by

$$P(n) = \frac{<n>^n}{n!} e^{-<n>}, \tag{7.7}$$

where n is the number of darts in a given region and $<n>$ is the mean number, $<n> = \sum_{n=0}^{\infty} n P(n)$. The upper limit of the sum should be N, but since $N \gg 1$, we can take the upper limit to be ∞ when it is convenient.

Problem 7.4   Darts and the Poisson distribution

a. Write a program to compute $\sum_{n=0}^{N} P(n)$, $\sum_{n=0}^{N} n P(n)$, and $\sum_{n=0}^{N} n^2 P(n)$ using the form (7.7) for $P(n)$. Choose a reasonable value for $<n¿$. Verify that $P(n)$ in (7.7) is normalized. What is the value of σ for the Poisson distribution?

b. Write a program to simulate the dart problem. Throw N darts at random in one trial. Do a total of ntrial "experiments" and determine $H(n)$, the number of regions that have n darts, and $<n>$, the mean number of darts in a region. Normalize $H(n)$ to determine the probability, and compare the measured probability with the Poisson distribution (7.7). Use your measured value of $<n>$ as input. Begin with $L = 1000$, $N = 50$, and ntrial $= 1000$. Increase the number of trials until you are confident that your data for $P(n)$ is reproducible. If time permits, use larger values of N.

c. Choose $L = 100$ and $N = 50$ and redo part (b). Are your results consistent with a Poisson distribution? What happens if $L = N = 50$?

Now that we are more familiar with the Poisson distribution, we consider the decay of radioactive nuclei. We know that a collection of radioactive nuclei will decay into other nuclei, and that there is no way to know a priori which nucleus will decay next. If all nuclei of a particular type are identical, why do they not all decay at the same time? The answer is based on the fundamental uncertainty inherent in the quantum description of matter at the microscopic level. In the following, we will see that a simple model of the decay process leads to an exponential decay law. This approach complements the continuum approach discussed in Section 2.10.

Because each nucleus is identical, we assume that during any time interval Δt, each nucleus has the same probability p of decaying. The basic algorithm is simple — choose an unstable nucleus and generate a random number r uniformly distributed in the unit interval $0 \leq r < 1$. If $r \leq p$, the unstable nucleus decays; otherwise, it does not. Every unstable nucleus is tested during each time interval. Note that for a system of unstable nuclei, there are many events that can happen during each time interval, e.g., $0, 1, 2, \ldots, n$ nuclei can decay. In contrast, for the particles in the box problem, there is a probability of unity that a particle is moved from one side to the other side at each time interval. Remember that once a nucleus decays, it is no longer in the group of unstable nuclei that is tested at each time interval. *Program nuclear_decay*, listed below, implements the nuclear decay algorithm.

```
PROGRAM nuclear_decay
! simulation of decay of unstable nuclei
DIM ncum(0 to 1000)
CALL initial(N0,p,ntrial,tmax,ncum())
CALL decay(N0,p,ntrial,tmax,ncum())
CALL output(ntrial,tmax,ncum())
END

SUB initial(N0,p,ntrial,tmax,ncum())
    RANDOMIZE
    INPUT prompt "initial number of unstable nuclei = ": N0
    INPUT prompt "decay probability for unstable nucleus = ": p
    INPUT prompt "number of time intervals per trial = ": tmax
    INPUT prompt "number of trials = ": ntrial
    FOR t = 1 to tmax
        LET ncum(t) = 0      ! accumulate number of unstable nuclei
    NEXT t
END SUB

SUB decay(N0,p,ntrial,tmax,ncum())
    DIM N(5000)
    FOR itrial = 1 to ntrial
        FOR i = 1 to N0
            LET N(i) = 1     ! nucleus = 1 if unstable
        NEXT i
        LET ncum(0) = ncum(0) + N0
```

```
                LET n_unstable = NO   ! # of unstable nuclei
                FOR t = 1 to tmax
                    FOR i = 1 to NO
                        IF N(i) = 1 then
                            IF rnd <= p then
                                LET N(i) = 0        ! nucleus decays
                                LET n_unstable = n_unstable - 1
                            END IF
                        END IF
                    NEXT i
                    LET ncum(t) = ncum(t) + n_unstable   ! accumulate data
                NEXT t
            NEXT itrial
        END SUB


        SUB output(ntrial,tmax,ncum())
            ! print data to file to be read by separate program
            OPEN #1: name "decay.dat", create new, access output
            PRINT #1: "time", "mean number of unstable nuclei"
            FOR t = 0 to tmax
                PRINT #1: t, ncum(t)/ntrial
            NEXT t
            CLOSE #1
        END SUB
```

Problem 7.5 Monte Carlo simulation of nuclear decay

a. To obtain better accuracy, Program nuclear_decay repeats the simulation ntrial times and averages the results. The time interval is assumed to be one second. Run Program nuclear_decay with NO = 100 (the initial number of unstable nuclei), $p = 0.01$, tmax = 100, and ntrial = 20. Use a separate graphics program to plot your results. Is your result for $N(t)$, the mean number of unstable nuclei at time t, consistent with the expected behavior, $N(t) = N(0) e^{-\lambda t}$ found in Section 2.10? What is the value of λ for this value of p?

b. There are a very large number of unstable nuclei in a typical radioactive source. We also know that over any reasonable time interval, only a relatively small number decay. Because $N \gg 1$ and $p \ll 1$, we expect that $P(n)$, the probability that n nuclei decay during a specified time interval, is a Poisson distribution. Modify Program nuclear_decay so that it outputs the probability that n unstable nuclei decay during the first time interval. Choose NO = 1000, $p = 0.001$, tmax = 1, and ntrial = 1000. What is the mean number $<n>$ of nuclei that decay during this interval? What is the associated variance? Plot $P(n)$ versus n and compare your results to the Poisson distribution (7.7) with your measured value of $<n>$ as input. Repeat the run for $p = 0.02$.

c. Modify Program nuclear_decay so that it outputs the probability that n unstable nuclei decay during two time intervals. Choose NO = 1000, $p = 0.001$,

$\texttt{tmax} = 2$, and $\texttt{ntrial} = 1000$. Compare the probability you obtain with your results from part (b). How do your results change as the time interval becomes larger?

d.   Increase p for fixed $N = 1000$ and determine $P(n)$ for a given time interval. Estimate the values of p and n for which the Poisson distribution is no longer applicable.

e.   Modify your program so that it flashes a small circle on the screen or makes a sound (like that of a Geiger counter) when a nucleus decays. Choose the location of the small circle at random. Do a single run and describe the qualitative differences between the visual and/or audio patterns for the situations in parts (a)–(d)? Choose $N \geq 5000$. Such a visualization might be somewhat misleading on a serial computer because only one nuclei can be considered at a time. In contrast, for a real system, nuclei can decay simultaneously. How can you improve the visualization to better approximate a real system?

7.3   INTRODUCTION TO RANDOM WALKS

In Section 7.1 we considered the random motion of many particles in a box, but we did not care about their trajectories—all we needed to know was the number of particles on each side. Now suppose that we want to characterize the motion of a dust particle in a glass of water. We know that a given dust particle collides with the water molecules, changing its direction frequently. Its motion appears so erratic that we cannot possibly predict its trajectory after even a small number of collisions. These considerations suggest that a simple model for the trajectory of a dust particle is that it moves in any direction with equal probability. Such a model is an example of a *random walk*.

The original statement of a random walk was formulated in the context of a "drunken sailor." If a drunkard begins at a lamp post and takes N steps of equal length in random directions, how far will the drunkard be from the lamp post? We will find that the mean square displacement of a random walker, e.g., a molecule or a drunkard, grows linearly with time. This result and its relation to diffusion leads to many applications that might seem to be unrelated to random walks.

We first consider an idealized one-dimensional example of a random walker that can move only along a line. Suppose that the walker begins at $x = 0$ and that each step is of equal length ℓ. At each interval of time the walker has a probability p of a step to the right and a probability $q = 1 - p$ of a step to the left. The direction of each step is independent of the preceding one. After N steps the displacement x of a walker is given by

$$x(N) = \sum_{i=1}^{N} s_i, \tag{7.8}$$

and the displacement squared x^2 is

$$x^2(N) = \left(\sum_{i=1}^{N} s_i \right)^2, \tag{7.9}$$

where $s_i = \pm\ell$. We can generate one walk of N steps by flipping a coin N times and increasing x by ℓ each time the coin is heads and decreasing x by ℓ each time the coin is tails. For simplicity, we first assume $p = q = \frac{1}{2}$. We expect that if we average over a sufficient number of walks of N steps, then the average of $x(N)$, denoted by $<x(N)>$, would be zero. That is, we expect to find as many steps in one direction as in the opposite direction. To find $<x^2(N)>$ analytically, we write (7.9) as two terms:

$$x^2(N) = \sum_{i=1}^{N} s_i^2 + \sum_{i\neq j=1}^{N} s_i s_j. \tag{7.10}$$

The first sum in (7.10) includes terms for which $i = j$; the second sum is over i and j such that $i \neq j$. The product $s_i s_j$ for $i \neq j$ equals $+\ell^2$ and $-\ell^2$ with equal probability, and hence the average of the second term in (7.10) is zero. Because $s_i^2 = \ell^2$ independently of the sign of s_i, the first term in (7.10) equals $\ell^2 N$ for all walks, and hence equals $\ell^2 N$ on the average. We conclude that

$$<x^2(N)> = \ell^2 N. \tag{7.11}$$

If the time interval for a step is Δt rather than unity, we should replace N in (7.11) by $N\Delta t$. The result (7.11) can be generalized to two- and three-dimensional walks where each step is a vector $\mathbf{s}$ of constant magnitude, but in a random direction.

The above derivation of the N dependence of $<x(N)>$ and $<x^2(N)>$ assumes that $p = \frac{1}{2}$. For general p, it is easy to show that $<x(N)> = (p-q)\ell N$. What is the meaning of this linear dependence on N? For $p \neq \frac{1}{2}$, it is convenient to consider the dispersion $<\Delta x^2(N)>$ defined as

$$<\Delta x^2(N)> \equiv <\big(x(N) - <x(N)>\big)^2> = <x^2(N)> - <x(N)>^2. \tag{7.12}$$

It is straightforward to show that for any value of p, the N dependence of $<\Delta x^2(N)>$ is given by

$$<\Delta x^2(N)> = 4pq\ell^2 N. \tag{7.13}$$

What is the N dependence of $<x^2(N)>$ for $p \neq q$?

We can gain more insight into the nature of random walks by doing a Monte Carlo simulation, i.e., by using a computer to "flip coins" and averaging over many trials. The implementation of the random walk algorithm is simple, e.g.,

```
IF rnd < p then
    LET x = x + 1
ELSE
    LET x = x - 1
END IF
```

The more difficult parts of the program are associated with bookkeeping. In *Program random_walk* we use the array element xcum(istep) to accumulate the value of x(istep), the displacement of the walker from the origin after istep steps. The walker takes a total of N steps in each trial and the average values of x and x^2 as a function of the number of steps are computed in SUB output.

```
PROGRAM random_walk
! simulation of a random walk in one dimension
DIM xcum(64),x2cum(64)
CALL initial(p,N,xcum(),x2cum(),ntrial)
FOR itrial = 1 to ntrial
    CALL walk(p,N,xcum(),x2cum())
NEXT itrial
CALL output(N,xcum(),x2cum(),ntrial)
END

SUB initial(p,N,xcum(),x2cum(),ntrial)
    RANDOMIZE
    INPUT prompt "maximum number of steps N = ": N
    LET p = 0.5
    LET ntrial = 1000                  ! number of trials
    FOR istep = 1 to N
        LET xcum(istep) = 0            ! not necessary in True BASIC
        LET x2cum(istep) = 0
    NEXT istep
END SUB

SUB walk(p,N,xcum(),x2cum())
    LET x = 0          ! initial position of walker for each trial
    FOR istep = 1 to N
        IF rnd <= p then
            LET x = x + 1
        ELSE
            LET x = x - 1
        END IF
        ! collect data after every step
        CALL data(x,xcum(),x2cum(),istep)
    NEXT istep
END SUB

SUB data(x,xcum(),x2cum(),istep)
    LET xcum(istep) = xcum(istep) + x
    LET x2cum(istep) = x2cum(istep) + x*x
END SUB

SUB output(N,xcum(),x2cum(),ntrial)
    PRINT "# steps","<x>","<x^2>","<x^2> - <x>^2"
    PRINT
    FOR istep = 1 to N
        LET xbar = xcum(istep)/ntrial
        LET x2bar = x2cum(istep)/ntrial
        LET variance = x2bar - xbar*xbar
        PRINT istep,xbar,x2bar,variance
    NEXT istep
END SUB
```

Problem 7.6   Random walks in one dimension

a.   In *Program random_walk* the steps are of unit length so that $\ell = 1$. Use *Program random_walk* to estimate the number of trials needed to obtain $<x^2(N)>$ for $N = 10$ steps with an accuracy of approximately 5%. Compare your result to the exact answer (7.11). Approximately how many trials do you need to obtain the same relative accuracy for $N = 40$?

b.   Is $<x(N)>$ exactly zero in your simulations? Explain the difference between the analytical result and the results of your simulations.

c.   Modify *Program random_walk* so that $P(x, N)$, the probability that the displacement of the walker from the origin is x after N steps, is computed. The following statements might be helpful:

```
DIM prob(-64 to 64)        ! x is positive and negative
prob(x) = prob(x) + 1      ! call after N steps
```

Compute $P(x, N)$ for $N = 10$ and $N = 40$ using at least 1000 trials and plot $P(x, N)$ after every 100 trials. Remember to normalize your results. Does the qualitative form of $P(x, N)$ change as the number of trials increases? What is the approximate width of $P(x, N)$ and the value of $P(x, N)$ at its maximum for each value of N?

d.   How do your results for $<x(N)>$ and $<\Delta x^2(N)>$ change for $p \neq q$? Choose $p = 0.7$ and determine the N dependence of $<x(N)>$ and $<\Delta x^2(N)>$.

e.   Is $P(x, N)$ a continuous function of x? Can you fit the envelope of $P(x, N)$ to a continuous function such as

$$\sqrt{\frac{2}{\pi <\Delta x^2(N)>}}\, e^{-(x-<x(N)>)^2/(2<\Delta x^2(N)>)}. \tag{7.14}$$

Compare your computed values for $P(x, N)$ found in part (c) to the form (7.14) using $<x(N)>$ and $<\Delta x^2(N)>$ as input. Note that the form (7.14) is not the standard form of the Gaussian probability density $p(x)$ for a continuous function. The probability density $p(x)$ is given by

$$p(x) = \frac{1}{\sqrt{2\pi\sigma^2}}\, e^{-(x-<x>)^2/2\sigma^2}. \tag{7.15}$$

The probability density $p(x)$ is related to the probability $P(x, N)$ by the relation $p(x)\, dx = P(x, N)$. The origin of the factor of two difference between (7.14) and (7.15) is that x in the discrete case assumes only integer values separated by $\Delta x = 2$.

*f.   Determine $<x^2(N)>$ for $N = 1$ to $N = 5$ by enumerating all the possible walks. For $N = 1$, there are two possible walks: one step to the right and one step to the left. In both cases $x^2 = 1$ and hence $<x^2(1)> = 1$ (for $p = \frac{1}{2}$). For $N = 2$ there are four possible walks with the same probability: (i) two steps to the right, (ii) two steps to the left, (iii) first step to the right and second step to the left, and (iv) first step to the left and second step to the right. The value of $x^2(2)$ for these walks is 4, 4, 0, and 0 respectively, and hence $<x^2(2)> = (4 + 4 + 0 + 0)/4 = 2$. Write a program that enumerates all the possible walks of a given N and x and compute the various averages exactly.

Problem 7.7   Random walks in two dimensions

a.   Consider a two-dimensional random walk for which each step is either up, down, right, or left with equal probability. Assume $\ell = 1$. Generalize *Program* `random_walk` and determine how the mean square displacement $<R^2> = <x^2 + y^2>$ depends on N.

b.   Modify your program so that the walker can move in one of six directions, where each direction is separated by 60 degrees. Is $<R^2>$ still proportional to N?

One reason random walks are very useful in simulating many physical processes and modeling many differential equations of physical interest is that their behavior is closely related to the solutions of the *diffusion* equation. The one-dimensional diffusion equation can be written as

$$\frac{\partial P(x, t)}{\partial t} = D \frac{\partial^2 P(x, t)}{\partial x^2}, \tag{7.16}$$

where D is the self-diffusion coefficient and $P(x, t)\, dx$ is the probability of a particle being in the interval between x and $x + dx$ at time t. In a typical application $P(x, t)$ might represent the concentration of ink molecules diffusing in a fluid. In three dimensions the second derivative $\partial^2/\partial x^2$ is replaced by the Laplacian ∇^2.

In Appendix 7A we show that the solution to the diffusion equation with the boundary condition $P(x = \pm\infty, t) = 0$ yields

$$<x(N)> = 0 \tag{7.17}$$

and

$$<x^2(N)> = 2Dt. \tag{7.18}$$

If we compare the form of (7.11) with (7.18), we see that the random walk and the diffusion equation give the same time dependence if we identify t with $N\Delta t$ and $2D$ with $\ell^2/\Delta t$.

7.4   PROBLEMS IN PROBABILITY

Because most of the questions in this introductory chapter on random processes can be answered by analytical methods, why bother to simulate these processes? One reason is that it is simpler to introduce new methods in a familiar context. Another reason is that if we change the nature of the random processes slightly, it often happens that it is difficult or impossible to obtain the answers by familiar methods. Still another reason is that writing a program and doing a simulation can aid your intuitive understanding of the subtle concept of probability. Probability is an elusive concept in part because it cannot be measured at one time. To reinforce the importance of thinking about how to solve a problem on a computer, we suggest some problems in probability in the following. Does thinking about these problems in this way help lead you to a pencil and paper solution?

Problem 7.8   The three boxes: stick or switch?

Suppose that there are three identical boxes, each with a lid. When you leave the room, a friend places a $10 bill in one of the boxes and closes the lid of each box. The friend knows the location of the $10 bill, but you do not. You then reenter the room and guess which one of the boxes has the $10 bill. As soon as you do, your friend opens the lid of another box that is empty. So if you have chosen an empty box, your friend will open the lid of the other empty box. If you have chosen the right box, your friend will open at random the lid of one of the two empty boxes. You now have the opportunity to stay with your original choice, or to switch to the other unopened box. Suppose that you play this game many times and that each time you guess correctly, you keep the money. To maximize your winnings, should you maintain your initial choice or should you switch? Which strategy is better? Write a program to simulate this game and output the probability of winning for switching and for not switching. It is likely that before you finish your program, the correct strategy will become clear. To make your program more useful, consider four or five boxes.

Problem 7.9   Conditional probability

Suppose that many people in a community are tested at random for HIV. The accuracy of the test is 87% and the incidence of the disease in the general population, independent of any test, is 1%. If a person tests positive for HIV, what is the probability that this person really has HIV? Write a program to compute the probability. (The answer can be found by using Bayes' theorem (cf. Bernardo and Smith). The answer is much less than 87%.)

Problem 7.10   The roll of the dice

a.   Write a program to compute the probability of obtaining at least one double six in twenty-four throws of a pair of die.

b.   A player rolls two dice. If the sum of the two dice is 7 or 11, the player wins immediately. If the sum is 2, 3, or 12, the player loses immediately. If the game is neither won nor lost on the first throw, the initial number is either 4, 5, 6, 8, 9, or 10. The player rolls the dice again until she either wins by repeating her initial number or she loses by rolling a 7. Write a program to determine the probability that the player wins this game (a variation of the game of craps).

c. Suppose that two gamblers each begin with $100 in capital and on each throw of a coin, one gambler must win $1 and the other must lose $1. How long can they play on the average until the capital of the loser is exhausted? How long can they play if they each begin with $1000? Neither gambler is allowed to go into debt.

Problem 7.11 The Boys of Summer

Luck plays a large role in the outcome of any baseball season. The American League West standings for 1989 are given in Table 7.1. Suppose that the teams remain unchanged and their probability of winning a particular game was the same as in 1989. Do a simulation to determine the probability that Oakland would lead the division for another season. For simplicity, assume that the teams play only each other.

Much of the present day motivation for the development of probability comes from science rather than from gambling. The next problem has much to do with statistical physics even though this application is not apparent.

Problem 7.12 Money exchange

Consider a two-dimensional plane that has been subdivided into cells, e.g., a checkerboard. There can be an indefinite number of coins stacked on each cell. For simplicity, we initially assign one coin to each cell. The game proceeds as follows. Select two cells at random. If there is at least one coin on the first cell, move one coin to the second cell. If the first cell is empty, then do nothing. After many coin exchanges, what does a typical state look like? Are the coins uniformly distributed as in the initial state or are many cells empty? Does the system approach equilibrium? Write a Monte Carlo program to simulate this game and show the state of the cells visually. Consider a system with at least 16×16 cells. Plot the histogram $H(n)$ versus n, where $H(n)$ is the number of cells with n coins. Do your results change if you consider bigger systems or begin with more coins on each cell?

Team	Won	Lost	Percentage
Oakland	99	63	0.611
Kansas City	92	70	0.568
California	91	71	0.562
Texas	83	79	0.512
Minnesota	80	82	0.494
Seattle	73	89	0.451
Chicago	69	92	0.429

Table 7.1 The American League West standings for 1989.

Problem 7.13   Distribution of cooking times

An industrious physics student finds a job at a local fast food restaurant to help him pay his way through college. His task is to cook 20 hamburgers on a grill at any one time. When a hamburger is cooked, he is supposed to replace it with an uncooked hamburger. However, our physics major does not pay attention to whether the hamburger is cooked or not. His method is to choose a hamburger at random and replace it by a uncooked one and not bother to check whether the hamburger that he removes from the grill is cooked or not. What is the distribution of cooking times of the hamburgers that he removes? To simplify the problem, assume that he replaces a random hamburger at regular intervals of one minute and that there is an indefinite supply of uncooked hamburgers. Does the qualitative nature of the distribution change if he cooks 40 hamburgers at any one time?

7.5   METHOD OF LEAST SQUARES

There is a long way to go from obtaining data to determining the relation that best describes it. For example, in Problem 7.5 we did a simulation of $N(t)$, the number of unstable nuclei at time t. Given the finite accuracy of our data, how do we know if our simulation results are consistent with the exponential relation between N and t? In Problem 7.6 we computed the mean square displacement $<x^2(N)>$ as a function of N for a simple random walk. Are our simulation results consistent with the theoretical prediction that $<x^2(N)>$ is proportional to N (for $p = \frac{1}{2}$)? The approach that we have been using is to plot the measured values of $<x^2(N)>$ as a function of N and to rely on our eye to help us draw the curve that best fits the data points. Of course, this graphical approach works best when the curve is a straight line, i.e., when the relation is linear. The advantages of this approach are that it is straightforward and allows us to see what we are doing. For example, if a data point is far from the curve, or if there is a gap in the data, we will notice it easily. If the true analytical relation is not linear, it is likely that we will notice that the data points do not fit a simple straight line, but instead show curvature. If we blindly let a computer fit the data to a straight line, we might not notice that the fit is not very good unless we already have had much experience fitting data by hand. Finally, the visceral experience of using a transparent ruler and fitting the data gives us some feeling for the nature of the data that might otherwise be missed. It usually is a good idea to plot some data in this way even though a computer can do it much faster.

Although the graphical approach is simple, it does not yield precise fits and we need to use analytical methods also. The most common method for finding the best straight line fit to a series of measured points is called *linear regression* or *least squares*. Suppose we have n pairs of measurements $(x_1, y_1), (x_2, y_2), \ldots, (x_n, y_n)$ and that the errors are entirely in the values of y. For convenience, we also assume that the uncertainties in y all have the same magnitude. Our goal is to obtain the best fit to the linear function

$$y = mx + b. \tag{7.19}$$

The problem is to calculate the values of the parameters m and b for the best straight line through the n data points. The difference

$$d_i = y_i - mx_i - b \tag{7.20}$$

is a measure of the discrepancy in y_i. It is reasonable to assume that the best set of values of m and b are those that minimize the quantity

$$S = \sum_{i=1}^{n}(y_i - mx_i - b)^2. \tag{7.21}$$

Why should we minimize the sum of the squared differences between the experimental values, y_i, and the analytical values, $mx_i + b$, and not some other function of the differences? The justification is based on the assumption that if we did many simulations, then the values of d_i would be distributed according to the Gaussian distribution (see Problems 7.6 and 12.9). Based on this assumption, it can be shown that the values of m and b that minimize S yield a set of values of $mx_i + b$ that are the most probable set of measurements that we would find based on the available information.

To minimize S, we take the derivative of S with respect to b and m:

$$\frac{\partial S}{\partial m} = -2 \sum_{i=1}^{n} x_i(y_i - mx_i - b) = 0, \tag{7.22a}$$

$$\frac{\partial S}{\partial b} = -2 \sum_{i=1}^{n}(y_i - mx_i - b) = 0. \tag{7.22b}$$

From (7.22) we obtain two simultaneous equations:

$$m \sum_{i=1}^{n} x_i^2 + b \sum_{i=1}^{n} x_i = \sum_{i=1}^{n} x_i y_i \tag{7.23a}$$

$$m \sum_{i=1}^{n} x_i + bn = \sum_{i=1}^{n} y_i. \tag{7.23b}$$

It is convenient to define the average quantities

$$\bar{x} = \frac{1}{n} \sum_{i=1}^{N} x_i \tag{7.24a}$$

$$\bar{y} = \frac{1}{n} \sum_{i=1}^{N} y_i \tag{7.24b}$$

$$\overline{xy} = \frac{1}{n} \sum_{i=1}^{N} x_i y_i, \tag{7.24c}$$

and rewrite (7.23a) and (7.23b) as

$$m\overline{x^2} + b\overline{x} = \overline{xy}, \tag{7.25a}$$
$$m\overline{x} + b = \overline{y}. \tag{7.25b}$$

The solution of (7.25a) and (7.25b) can be expressed as

$$m = \frac{\overline{xy} - \overline{x}\,\overline{y}}{(\Delta x)^2} \tag{7.26a}$$
$$b = \overline{y} - m\,\overline{x}. \tag{7.26b}$$

where

$$(\Delta x)^2 = \overline{x^2} - \overline{x}^2 \tag{7.26c}$$

Equations (7.26a)–(7.26c) determine the slope m and the intercept b of the best straight line through the n data points.

As an example, consider the data shown in Table 7.2 for a one-dimensional random walk. To make the example more interesting, suppose that the walker takes steps of length 1 or 2 with equal probability. The direction of the step is random and $p = \frac{1}{2}$. As in Section 7.3, we assume that the mean square displacement $<x^2(N)>$ obeys the general relation

$$<x^2(N)> = aN^{2\nu} \tag{7.27}$$

with an unknown exponent ν. First we convert the nonlinear relation (7.27) to a linear relation by taking the logarithm of both sides:

$$\ln <x^2(N)> = \ln a + 2\nu \ln N. \tag{7.28}$$

If we take the logarithm of the data in Table 7.2 and use (7.26), we find that $m = 1.02$ and $b = 0.83$. Hence, we conclude from our limited data and the relation $2\nu = m$ that $\nu \approx 0.51$, a numerical result that is consistent with the expected result $\nu = 1/2$. The values of $y = \ln <x^2(N)>$ and $x = \ln N$ and the least squares fit are shown in Fig. 7.2. Note that the original fitting problem is nonlinear, that is, $<x^2(N)>$ depends on N^ν

N	$<x^2(N)>$
8	19.43
16	37.65
32	76.98
64	160.38

Table 7.2 Computed values of the mean square displacement $<x^2(N)>$ as a function of the total number of steps N. The results $<x^2(N)>$ are averaged over 1000 trials. The one-dimensional random walker takes steps of length 1 or 2 with equal probability, and the direction of the step is random with $p = \frac{1}{2}$.

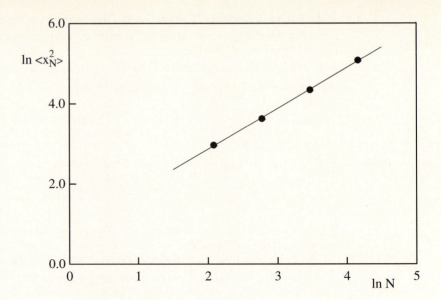

Fig. 7.2   Plot of $\ln <x^2(N)>$ versus $\ln N$ for the data listed in Table 7.2. The straight line $y = 1.02x + 0.83$ through the points is found by minimizing the sum (7.21).

rather than N. Often a problem that looks nonlinear can be turned into a linear problem by a change of variables.

The least squares fitting procedure also allows us to estimate the uncertainty or the most probable error in m and b by analyzing the measurements themselves. The result of this analysis is that the most probable error in m and b, σ_m and σ_b respectively, is given by

$$\sigma_m = \frac{1}{\sqrt{n}} \frac{\sigma_y}{\Delta x} \tag{7.29}$$

$$\sigma_b = \frac{1}{\sqrt{n}} \frac{\left(\overline{x^2}\right)^{1/2}}{\Delta x} \sigma_y, \tag{7.30}$$

where

$$\sigma_y^2 = \frac{1}{n-2} \sum_{i=1}^{n} d_i^2, \tag{7.31}$$

and d_i is given by (7.20). Because there are n data points, we might have guessed that n rather than $n - 2$ would be present in the denominator of (7.31). The reason for the factor of $n - 2$ is related to the fact that to determine σ_y, we first need to calculate *two* quantities m and b, leaving only $n - 2$ independent degrees of freedom. To see that the $n - 2$ factor is reasonable, consider the special case of $n = 2$. In this case we can find a line that passes exactly through the two data points, but we cannot deduce anything about the reliability of the set of measurements because the fit always is exact. If we use (7.31), we see that both the numerator and denominator would be zero, and hence $\sigma_y = 0/0$, i.e., σ_y is undetermined. If a factor of n appeared in (7.31) instead, we would conclude that $\sigma_y = 0/2 = 0$, an absurd conclusion. Usually $n >> 1$, and the difference between n and $n - 2$ is negligible.

For our example, $\sigma_y = 0.03$, $\sigma_b = 0.07$, and $\sigma_m = 0.02$. The uncertainties δm and δv are related by $2\delta v = \delta m$. Because we can associate δm with σ_m, we conclude that our best estimate for v is $v = 0.51 \pm 0.01$.

If the values of y_i have different uncertainties σ_i, then the data points are weighted by the quantity $w_i = 1/\sigma_i^2$. In this case it is reasonable to minimize the quantity

$$\chi^2 = \sum_{i=1}^{n} w_i (y_i - mx_i - b)^2. \tag{7.32}$$

The resulting expressions (7.26a) and (7.26b) for m and b are unchanged if we generalize the definition of the averages to be

$$\overline{f} = \frac{1}{n\overline{w}} \sum_{i=1}^{n} w_i f_i, \tag{7.33}$$

where

$$\overline{w} = \frac{1}{n} \sum_{i=1}^{n} w_i \tag{7.34}$$

In Chapter 11 we discuss how to estimate the most probable errors in $<x^2(N)>$.

Problem 7.14   Example of least squares fit

a. Write a program to find the least squares fit for a set of data. As a check on your program, compute the most probable values of m and b for the data shown in Table 7.2.

b. Modify `Program random_walk` so that steps of length 1 and 2 are taken with equal probability. Use at least 10 000 trials and do a least squares fit to $<x^2(N)>$ as done in the text. Is your most probable estimate for v closer to $v = 1/2$?

For the simple random walk problems considered in this chapter, the relation $<x^2(N)> = aN^v$ holds for all N. However, in many random walk problems (see Chapter 12), a power law relation between $<x^2(N)>$ and N holds only asymptotically for large N, and hence we should use only the larger values of N to estimate the slope. We also need to give equal weight to all intervals of the independent variable N. In the above example, we used $N = 8, 16, 32$, and 64, so that the values of $\ln N$ are equally spaced.

7.6   A SIMPLE VARIATIONAL MONTE CARLO METHOD

Many problems in physics can be formulated in terms of a variational principle. In the following, we consider examples of variational principles from geometrical optics and classical mechanics. We then discuss how Monte Carlo methods can be applied to obtain estimates for the maximum or minimum. A more sophisticated application of Monte Carlo methods to a variational problem in quantum mechanics is discussed in Chapter 18.

Our everyday experience of light leads naturally to the concept of light rays. This description of light propagation, called *geometrical* or *ray optics*, is applicable when the wavelength of light is small compared to the linear dimensions of any obstacles or openings. The propagation of light rays can be formulated in terms of a principle due to Fermat:

> *A ray of light follows the path between two points (consistent with any constraints) that requires the least amount of time.*

Fermat's principle of least time can be adopted as the basis of geometrical optics. For example, Fermat's principle implies that light travels from a point A to a point B in a straight line in a homogeneous medium. Because the speed of light is constant along any path within the medium, the path of shortest time is the path of shortest distance, i.e., a straight line from A to B. What happens if we impose the constraint that the light must strike a mirror before reaching B?

The speed of light in a medium can be expressed in terms of c, the speed of light in a vacuum, and the index of refraction n of the medium:

$$v = \frac{c}{n}. \tag{7.35}$$

Suppose that a light ray in a medium with index of refraction n_1 passes through a second medium with index of refraction n_2. The two media are separated by a plane surface. We now show how we can use Fermat's principle and a Monte Carlo method to find the path of the light. The analytical solution to this problem using Fermat's principle is found in many texts (cf. Feynman et al.).

Our strategy, as implemented in `Program fermat`, is to begin with an arbitrary path and to make changes in the path at random. These changes are accepted only if they reduce the travel time of the light. Some of the features of `Program fermat` include:

1. Light propagates from left to right through N media.

2. The width of each region is unity and the index of refraction is uniform in each region. The index i increases from left to right. There are $N - 1$ boundaries separating the N media with index of refraction `n(i)` and speed `v(i)`. We have chosen units such that the speed of light in a vacuum equals unity.

3. Because the light propagates in a straight line in each medium, the path of the light is given by the coordinates `y(i)` at each boundary.

4. The coordinates of the light source and the detector are at $(1, y(1))$ and $(N, y(N))$ respectively, where y(1) and y(N) are fixed.

5. The initial path is the connection of the set of random points at the boundary of each region.

6. The path of the light is found by choosing the boundary i at random and generating a trial value of y(i) that differs from its previous value by a random number between $-\delta$ to δ. If the trial value of y(i) yields a shorter travel time, this value becomes the new value for y(i).

7. The path is redrawn whenever it is changed.

```
PROGRAM fermat
! Monte Carlo method for finding minimum optical path
DIM y(101),v(101)
CALL initial(y(),v(),N)
CALL change_path(y(),v(),N)
END

SUB initial(y(),v(),N)
    RANDOMIZE
    LET N = 10                         ! number of different media (even)
    ! speed of light in vacuum equal to unity
    LET v1 = 1.0
    LET index = 1.5                    ! index of refraction of medium #2
    LET v2 = 1/index
    FOR i = 1 to N/2
        LET v(i) = v1                  ! speed of light in left half
    NEXT i
    FOR i = N/2 + 1 to N
        LET v(i) = v2                  ! speed of light in right half
    NEXT i
    LET y(1) = 2                       ! fixed source
    LET y(N) = 8                       ! fixed detector
    SET WINDOW 0.5,N+1,y(1)-0.5,y(N)+0.5
    BOX LINES 1,N,y(1),y(N)
    PLOT LINES: N/2,y(1);N/2,y(N)         ! boundary
    ! choose initial path at boundary between endpoints
    FOR i = 2 to N-1
        LET y(i) = (y(N) - y(1))*rnd + y(1)
    NEXT i
    SET COLOR "red"
    FOR i = 1 to N-1
        PLOT LINES: i,y(i);i+1,y(i+1)  ! initial path
    NEXT i
    SET COLOR "red/white"
END SUB
```

```
SUB change_path(y(),v(),N)
    LET delta = 0.5                    ! maximum change in y
    DO
        ! choose random x not including x = 1 or N
        LET x = int((N-2)*rnd) + 2
        LET ytrial = y(x) + (2*rnd - 1)*delta     ! new y position
        LET dy2 = (y(x) - y(x+1))^2      ! vertical distance squared
        LET dist = sqr(1 + dy2)        ! horizontal distance is unity
        LET t_original = dist/v(x+1)
        LET dy2 = (y(x) - y(x-1))^2
        LET dist = sqr(1 + dy2)
        LET t_original = t_original + dist/v(x)
        LET dy2 = (ytrial - y(x+1))^2
        LET dist = sqr(1 + dy2)
        LET t_trial = dist/v(x+1)
        LET dy2 = (ytrial - y(x-1))^2
        LET dist = sqr(1 + dy2)
        LET t_trial = t_trial + dist/v(x)
        IF t_trial < t_original then      ! new position reduces time
            SET COLOR "white"
            PLOT LINES: x-1,y(x-1);x,y(x);x+1,y(x+1)
            SET COLOR "red"
            LET y(x) = ytrial
            PLOT LINES: x-1,y(x-1);x,y(x);x+1,y(x+1)
        END IF
    LOOP until key input
END SUB
```

Problem 7.15   The law of refraction

a.   Use *Program fermat* to determine the angle of incidence θ_1 and the angle of refraction θ_2 between two media with different indices of refraction. The angles θ_1 and θ_2 are measured from the normal to the boundary. Set $N = 10$ and let the first medium be air ($n_1 \approx 1$) and the second medium be glass ($n_2 \approx 1.5$). Describe the path of the light after a number of trial paths are attempted. Add some statements to the program to determine θ_1 and θ_2, the vertical position of the intersection of the light at the boundary between the two media, and the total time for the light to go from $(1, y(1))$ and $(10, y(10))$.

b.   Modify the program so that the first medium represents glass ($n_1 \approx 1.5$) and the second medium represents water ($n_2 \approx 1.33$). Verify that your results in (a) and (b) are consistent with Snell's law, $n_2 \sin \theta_2 = n_1 \sin \theta_1$.

Problem 7.16   Inhomogeneous media

a.   The earth's atmosphere is thin at the top and dense near the earth's surface. We can model this inhomogeneous medium by dividing the atmosphere into equal width segments each of which is homogeneous. Take the index of refraction

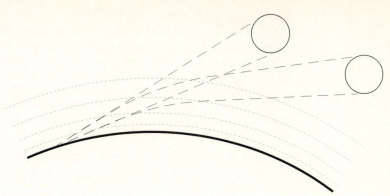

Fig. 7.3 Near the horizon, the apparent (exaggerated) position of the sun is higher than the true position of the sun. Note that the light rays from the true sun are curved due to refraction.

of region i to be $n(i) = 1 + (i-1)*dn$. Run *Program fermat* with $N = 10$ and $dn = 0.1$, and find the path of least time. Use your results to explain why when we see the sun set, the sun already is below the horizon (see Fig. 7.3).

*b. Use *Program fermat* to find the appropriate distribution of $n(i)$ for a fiber optic cable. In this case the ith region corresponds to a cross sectional slab through the cable. Although a real cable is three-dimensional, we consider a two-dimensional cable for simplicity. Imagine the cable to be a flat, long ribbon of width equal to N. The middle region is the center of the cable and the $i = 1$ and $i = N$ regions are at the edge of the cable. We want the cable to have the property that if a ray of light starts from one side of the cable and ends at the other, the slope dy/dx of the path should be near zero at the edges so that light does not escape from the cable.

Fermat's principle is an example of an extremum (maxima or minima) principle. An extremum means that a small change ϵ in an independent variable leads to a change in a function (more precisely, a function of functions) that is proportional to ϵ^2 or a higher power of ϵ. An important extremum principle in classical mechanics is based on the action S:

$$S = \int_{t_{\text{initial}}}^{t_{\text{final}}} L\, dt \qquad (7.36)$$

The Lagrangian L in (7.36) is the kinetic energy minus the potential energy. The extremum principle for the action is known as *the principle of least action*. If we were to take the extremum of (7.36), we would find that the path for which S is an extremum satisfies the differential equation of motion equivalent to Newton's second law (for conservative forces). One reason for the importance of the principle of least action is that quantum mechanics can be formulated in terms of an integral over the action. This way of doing quantum mechanics is called the path integral formulation (see Section 18.8).

Our main motivation here is to gain more experience with extremum problems. To use (7.36) to find the motion of a single particle in one dimension, we fix the position at the initial and final times, $x(t_{initial})$ and $x(t_{final})$, and then choose the velocities and positions for other times $t_{initial} < t < t_{final}$ so as to minimize the action. One way to implement this procedure numerically is to convert the integral in (7.36) to a sum:

$$S \approx \sum_{i=1}^{N-1} L(t_i)\, \Delta t, \qquad\qquad (7.37)$$

where $t_i = t_{initial} + i\,\Delta t$. (The approximation used to obtain (7.37) in known as the rectangular approximation and is discussed in Chapter 11.) For a single particle in one dimension, we can write

$$L_i \approx \frac{m}{2(\Delta t)^2}(x_{i+1} - x_i)^2 - u(x_i), \qquad\qquad (7.38)$$

where m is the mass of the particle and $u(x_i)$ is the potential energy of the particle at x_i. The velocity has been approximated as the difference in position divided by the change in time Δt.

*Problem 7.17   Principle of least action

a.   Write a program to minimize the action S given in (7.36) for the motion of a single particle in one dimension. Use the approximate form of the Lagrangian given in (7.38). One way to write the program is to modify *Program fermat* so that the vertical coordinate for the light ray becomes the position of the particle, and the horizontal region number i of width Δx becomes the discrete time interval number of duration Δt. The quantity to be minimized is different, but otherwise the algorithm is similar. It is possible to extend the principle of least action to more dimensions or particles, but it is necessary to begin with a path close to the optimum one to obtain a good approximation to the optimum path in a reasonable time.

b.   Verify your program for the case of free fall for which the potential energy is $u(x) = mgx$. Choose $x(t = 0) = 2\,\text{m}$ and $x(t = 10\,\text{s}) = 8\,\text{m}$, and begin with $N = 20$. Allow the maximum change in the position to be 5 m. Make sure that the window coordinates are sufficiently large to see the paths. What do you expect the shape of the plot of x versus t to be?

c.   Consider the harmonic potential $u(x) = \frac{1}{2}kx^2$. What shape do you expect the path $x(t)$ to be? Increase N to approximately 50.

Appendix 7A   RANDOM WALKS AND THE DIFFUSION EQUATION

To gain some insight into the relation between random walks and the diffusion equation, we show that the latter implies that $<x(t)>$ is zero and $<x^2(t)>$ is proportional to t.

We rewrite the diffusion equation (7.16) here for convenience:

$$\frac{\partial P(x,t)}{\partial t} = D\frac{\partial^2 P(x,t)}{\partial x^2},$$

(7.39)

To derive the t dependence of $<x(t)>$ and $<x^2(t)>$ from (7.39), we write the average of any function of x as

$$<f(x,t)> = \int_{-\infty}^{\infty} f(x)P(x,t)\,dx.$$

(7.40)

The average displacement is given by

$$<x(t)> = \int_{-\infty}^{\infty} xP(x,t)\,dx.$$

(7.41)

To do the integral on the right hand side of (7.41), we multiply both sides of (7.39) by x and formally integrate over x:

$$\int_{-\infty}^{\infty} x\frac{\partial P(x,t)}{\partial t}\,dx = D\int_{-\infty}^{\infty} x\frac{\partial^2 P(x,t)}{\partial x^2}\,dx.$$

(7.42)

The left-hand side can be expressed as:

$$\int_{-\infty}^{\infty} x\frac{\partial P(x,t)}{\partial t}\,dx = \frac{\partial}{\partial t}\int_{-\infty}^{\infty} xP(x,t)\,dx = \frac{\partial}{\partial t}<x>.$$

(7.43)

The right-hand side of (7.42) can be written in the desired form by doing an integration by parts:

$$D\int_{-\infty}^{\infty} x\frac{\partial^2 P(x,t)}{\partial x^2}\,dx = Dx\frac{\partial P(x,t)}{\partial x}\Big|_{x=-\infty}^{x=\infty} - D\int_{-\infty}^{\infty} \frac{\partial P(x,t)}{\partial x}\,dx.$$

(7.44)

The first term on the right hand side of (7.44) is zero because $P(x=\pm\infty, t)=0$ and all the spatial derivatives of P at $x=\pm\infty$ are zero. The second term also is zero because it integrates to $D[P(x=\infty,t) - P(x=-\infty,t)]$. Hence, we find that

$$\frac{\partial}{\partial t}<x> = 0,$$

(7.45)

or $<x>$ is a constant, independent of time. Because $x=0$ at $t=0$, we conclude that $<x>=0$ for all t. To calculate $<x^2(t)>$, two integrations by parts are necessary, and we find that

$$\frac{\partial}{\partial t}<x^2(t)> = 2D,$$

(7.46)

or

$$<x^2(t)> = 2Dt.$$

(7.47)

We see that the random walk and the diffusion equation have the same time dependence. In d-dimensional space, $2D$ is replaced by $2dD$.

References and Suggestions for Further Reading

William R. Bennett, *Scientific and Engineering Problem-Solving with the Computer*, Prentice Hall (1976). Many random processes including the spread of disease are considered in Chapter 6.

J. M. Bernardo and A. F. M. Smith, *Bayesian Theory*, John Wiley & Sons (1994). Bayes Theorem is given concisely on page 2.

Philip R. Bevington and D. Keith Robinson, *Data Reduction and Error Analysis for the Physical Sciences*, second edition, McGraw-Hill (1992).

William S. Cleveland and Robert McGill, "Graphical perception and graphical methods for analyzing scientific data," *Science* **229**, 828 (1985). There is more to analyzing data than least squares fits.

A. K. Dewdney, "Computer Recreations (Five easy pieces for a do loop and random-number generator)," *Sci. Amer.* **252**(#4), 20 (1985).

Robert M. Eisberg, *Applied Mathematical Physics with Programmable Pocket Calculators*, McGraw-Hill (1976). Chapter 7 discusses entropy and the arrow of time.

Richard P. Feynman, Robert B. Leighton, and Matthew Sands, The Feynman Lectures on Physics, Addison-Wesley (1963). See Vol. 1, Chapter 26 for a discussion of the principle of least time and Vol. 2, Chapter 19, for a discussion of the principle of least action.

Peter R. Keller and Mary M. Keller, *Visual Cues*, IEEE Press (1993). A well illustrated book on data visualization techniques.

William H. Press, Saul A. Teukolsky, William T. Vetterling, and Brian P. Flannery, *Numerical Recipes*, second edition, Cambridge University Press (1992). See Chapter 15 for a general discussion of the modeling of data including general linear least squares and nonlinear fits.

F. Reif, *Statistical and Thermal Physics*, Berkeley Physics, Vol. 5, McGraw-Hill (1965). Chapter 2 introduces random walks.

Charles Ruhla, *The Physics of Chance*, Oxford University Press (1992). A delightful book on probability in many contexts.

G. L. Squires, *Practical Physics*, third edition, Cambridge University Press (1985). An excellent text on the design of experiments and the analysis of data.

John R. Taylor, *An Introduction to Error Analysis*, University Science Books, Oxford University Press (1982).

Edward R. Tuffe, *The Visual Display of Quantitative Information*, Graphics Press (1983).

Charles A. Whitney, *Random Processes in Physical Systems,* John Wiley and Sons (1990). An excellent introduction to random processes with many applications to astronomy.

Robert S. Wolff and Larry Yaeger, *Visualization of Natural Phenomena*, Springer-Verlag (1993). A CD-ROM disk is included with the book. An extensive list of references also is given.

Hugh D. Young, *Statistical Treatment of Experimental Data*, McGraw-Hill (1962).

CHAPTER

8

The Dynamics of Many Particle Systems

We simulate the dynamical behavior of many particle systems and observe their qualitative features. Some of the basic ideas of equilibrium statistical mechanics and kinetic theory are introduced.

8.1   INTRODUCTION

Gases, liquids, and solids are examples of systems that contain many mutually interacting particles. Given our knowledge of the laws of physics at the microscopic level, how can we understand the observed behavior of these systems and more complex systems such as polymers and proteins? As an example, consider two cups of water prepared under similar conditions. Each cup contains approximately 10^{24} molecules which, to a good approximation, move according to the laws of classical physics. Although the intermolecular forces produce a complicated trajectory for each molecule, the observable properties of the water in each cup are indistinguishable and are easy to describe. For example, the temperature of the water in each cup is independent of time even though the positions and velocities of the individual molecules are changing continually.

One way to understand the behavior of a many particle system is to begin from the known intermolecular interactions and do a computer simulation of its dynamics. This approach, known as the *molecular dynamics* method, has been applied to systems of several hundred to a million particles and has given us much insight into the behavior of gases, liquids, and solids.

A knowledge of the trajectories of 10^4 or even 10^{24} particles is not helpful unless we know the right questions to ask. What are the useful parameters needed to describe these systems? What are the essential characteristics and regularities exhibited by many particle systems? From our study of chaotic systems, we might suspect that the only meaningful quantities we can compute are averages over the trajectories, rather than the trajectories themselves. Questions such as these are addressed by statistical mechanics and many of the ideas of statistical mechanics are discussed in this chapter. However, the only background needed for this chapter is a knowledge of Newton's laws of motion.

8.2   THE INTERMOLECULAR POTENTIAL

The first step is to specify the model system we wish to simulate. For simplicity, we assume that the dynamics can be treated classically and that the molecules are spherical and chemically inert. We also assume that the force between any pair of molecules depends only on the distance between them. In this case the total potential energy U is a sum of two-particle interactions:

$$U = u(r_{12}) + u(r_{13}) + \cdots + u(r_{23}) + \cdots = \sum_{i=1}^{N-1} \sum_{j=i+1}^{N} u(r_{ij}), \qquad (8.1)$$

where $u(r_{ij})$ depends only on the magnitude of the distance $\mathbf{r}_{ij}$ between particles i and j. The pairwise interaction form (8.1) is appropriate for simple liquids such as liquid argon.

In principle, the form of $u(r)$ for electrically neutral molecules can be constructed by a first principles quantum mechanical calculation. Such a calculation is very difficult, and it usually is sufficient to choose a simple phenomenological form for $u(r)$. The most important features of $u(r)$ for simple liquids are a strong repulsion for small r and a weak attraction at large r. The repulsion for small r is a consequence of the Pauli exclusion principle. That is, the electron clouds of two molecules must distort to avoid overlap, causing some of the electrons to be in different quantum states. The net effect is an increase in kinetic energy and an effective repulsive force between the electrons, known as *core repulsion*. The dominant weak attraction at larger r is due to the mutual polarization of each molecule; the resultant attractive force is called the *van der Waals* force.

One of the most common phenomenological forms of $u(r)$ is the Lennard-Jones potential:

$$u(r) = 4\epsilon \left[\left(\frac{\sigma}{r}\right)^{12} - \left(\frac{\sigma}{r}\right)^{6} \right]. \tag{8.2}$$

A plot of the Lennard-Jones potential is shown in Fig. 8.1. The r^{-12} form of the repulsive part of the interaction has been chosen for convenience only. The Lennard-Jones potential is parameterized by a length σ and an energy ϵ. Note that $u(r) = 0$ at $r = \sigma$, and that $u(r)$ is essentially zero for $r > 3\sigma$. The parameter ϵ is the depth of the potential at the minimum of $u(r)$; the minimum occurs at a separation $r = 2^{1/6}\sigma$. The parameters ϵ and σ of the Lennard-Jones potential which give good agreement with the experimental properties of liquid argon are $\epsilon = 1.65 \times 10^{-21}$ J and $\sigma = 3.4$ Å.

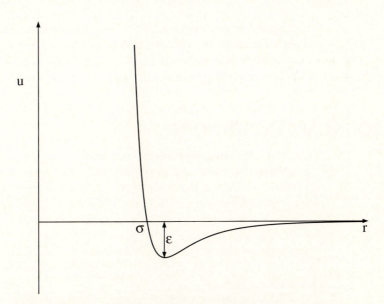

Fig. 8.1   Plot of the Lennard-Jones potential $u(r)$. Note that the potential is characterized by a length σ and an energy ϵ.

Problem 8.1   Qualitative properties of the Lennard-Jones interaction

Write a short program or use a graphics package to plot the Lennard-Jones potential (8.2) and the magnitude of the corresponding force:

$$\mathbf{f}(r) = -\nabla u(r) = \frac{24\epsilon}{r}\left[2\left(\frac{\sigma}{r}\right)^{12} - \left(\frac{\sigma}{r}\right)^{6}\right]\hat{\mathbf{r}}. \tag{8.3}$$

What is the value of $u(r)$ for $r = 0.8\sigma$? How much does u increase if r is decreased to $r = 0.72\sigma$, a decrease of 10%? What is the value of u at $r = 2.5\sigma$? At what value of r does the force equal zero?

8.3   THE NUMERICAL ALGORITHM

Now that we have specified the interaction between the particles, we need to introduce a numerical integration method for computing the trajectory of each particle. As might be expected, we need to use at least a second-order algorithm to maintain conservation of energy for the times of interest in molecular dynamics simulations. We adopt the commonly used algorithm:

$$x_{n+1} = x_n + v_n\Delta t + \frac{1}{2}a_n(\Delta t)^2 \tag{8.4a}$$

$$v_{n+1} = v_n + \frac{1}{2}(a_{n+1} + a_n)\Delta t. \tag{8.4b}$$

To simplify the notation, we have written the algorithm for only one component of the particle's motion. The new position is used to find the new acceleration a_{n+1} which is used together with a_n to obtain the new velocity v_{n+1}. The algorithm represented by (8.4) is a convenient form of the *Verlet algorithm* (see Appendix 5A).

8.4   BOUNDARY CONDITIONS

A useful simulation must incorporate all the relevant features of the physical system of interest. The ultimate goal of our simulations is to understand the behavior of bulk systems—systems of the order of $N \sim 10^{23} - 10^{25}$ particles. In bulk systems the fraction of particles near the walls of the container is negligibly small. However, the number of particles that can be studied in a molecular dynamics simulation is typically 10^3 – 10^5, although as many as 10^6 particles or more, can be studied on present-day supercomputers. For these small systems the fraction of particles near the walls of the container is significant, and hence the behavior of such a system would be dominated by surface effects.

The most common way of minimizing surface effects and to simulate more closely the properties of a bulk system is to use what are known as *periodic boundary conditions*. First consider a one-dimensional "box" of N particles that are constrained to move on a line of length L. The ends of the line serve as imaginary walls. The usual

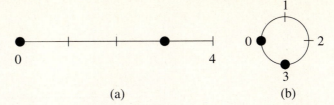

Fig. 8.2   (a) Two particles at $x = 0$ and $x = 3$ on a line of
length $L = 4$; the distance between the particles is 3.
(b) The application of periodic boundary conditions
for short range interactions is equivalent to thinking
of the line as forming a circle of circumference L.
In this case the minimum distance between the two
particles is 1.

application of periodic boundary conditions is equivalent to considering the line to be a
circle (see Fig. 8.2). The distance between the particles is measured along the arc, and
hence the maximum separation between any two particles is $L/2$.

The computer code for periodic boundary conditions is straightforward. If a particle leaves the box by crossing a boundary, we add or subtract L to the coordinate. One
simple way is to use an IF statement after the particles have been moved:

```
IF x > L then
    LET x = x - L
ELSE IF x < 0 then
    LET x = x + L
END IF
```

To compute the minimum distance dx between particles 1 and 2 at x(1) and x(2)
respectively, we can write

```
LET dx = x(1) - x(2)
IF dx > 0.5*L then
    LET dx = dx - L
ELSE IF dx < -0.5*L then
    LET dx = dx + L
END IF
```

The generalization of this application of periodic boundary conditions to two dimensions is straightforward if we imagine a box with opposite edges joined so that the box
becomes the surface of a torus (the shape of a doughnut and a bagel).

We now discuss the motivation for this choice of boundary conditions. Imagine
a set of N particles in a two-dimensional cell. The use of periodic boundary conditions implies that this central cell is duplicated an infinite number of times to fill two-

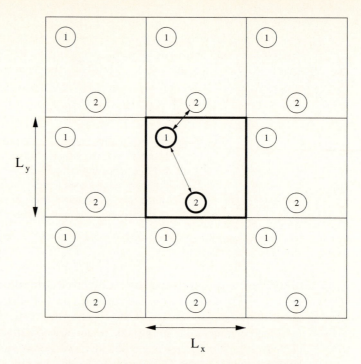

Fig. 8.3 Example of the minimum image approximation in two dimensions. The minimum image distance convention implies that the separation between particles 1 and 2 is given by the shorter of the two distances shown.

dimensional space. Each image cell contains the original particles in the same relative positions as the central cell. Fig. 8.3 shows the first several image cells for $N = 2$ particles. Periodic boundary conditions yield an infinite system, although the motion of particles in the image cells is identical to the motion of the particles in the central cell. These boundary conditions also imply that every point in the cell is equivalent and that there is no surface. The shape of the central cell must be such that the cell fills space under successive translations.

As a particle moves in the original cell, its periodic images move in the image cells. Hence only the motion of the particles in the central cell needs to be followed. When a particle enters or leaves the central cell, the move is accompanied by an image of that particle leaving or entering a neighboring cell through the opposite face.

The total force on a given particle i is due to the force from every other particle j within the central cell and from the periodic images of particle j. That is, if particle i interacts with particle j in the central cell, then particle i interacts with *all* the periodic replicas of particle j. Hence in general, there are an infinite number of contributions to the force on any given particle. For long range interactions such as the Coulomb potential, these contributions have to be included using special methods. However, for short range interactions, we may reduce the number of contributions by adopting the *minimum image* or nearest image approximation. This approximation implies that particle i in the central cell interacts only with the nearest image of particle j; the interaction is set equal to zero if the distance of the image from particle i is greater than $L/2$. An example of the minimum image condition is shown in Fig. 8.3. Note that the minimum image approximation implies that the calculation of the total force on all N particles due to pairwise interactions involves a maximum of $N(N-1)/2$ contributions.

8.5   UNITS

To reduce the possibility of roundoff error, it is useful to choose units so that the computed quantities are neither too small nor too large. Because the values of the distance and the energy associated with typical liquids are very small in SI units, we choose the Lennard-Jones parameters σ and ϵ to be the units of distance and energy, respectively. (The values of σ and ϵ for argon are given in Table 8.1.) We also choose the unit of mass to be the mass of one atom, m. We can express all other quantities in terms of σ, ϵ, and m. For example, we measure velocities in units of $(\epsilon/m)^{1/2}$, and the time in units of $\sigma(m/\epsilon)^{1/2}$. If we take $m = 6.69 \times 10^{-26}$ kg, the mass of an argon atom, then the unit of time is 2.17×10^{-12} s. The units of some of the physical quantities of interest are shown in Table 8.1. All program variables are in reduced units, e.g., the time in our molecular dynamics program is expressed in units of $\sigma(m/\epsilon)^{1/2}$. As an example, suppose that we run our molecular dynamics program for 2000 time steps with a time step $\Delta t = 0.01$. The total time of our run is $2000 \times 0.01 = 20$ in reduced units or 4.34×10^{-11} s for argon (see Table 8.1). The total time of a typical molecular dynamics simulation is in the range of $10 - 10^4$ in reduced units, corresponding to a duration of approximately $10^{-11} - 10^{-9}$ s.

quantity	unit	value for argon
length	σ	3.4×10^{-10} m
energy	ϵ	1.65×10^{-21} J
mass	m	6.69×10^{-26} kg
time	$\sigma(m/\epsilon)^{1/2}$	2.17×10^{-12} s
velocity	$(\epsilon/m)^{1/2}$	1.57×10^2 m/s
force	ϵ/σ	4.85×10^{-12} N
pressure	ϵ/σ^2	1.43×10^{-2} N $\cdot$ m^{-1}
temperature	ϵ/k	120 K

Table 8.1   The system of units used in the molecular dynamics simulations of particles interacting via the Lennard-Jones potential. The numerical values of σ, ϵ, and m are for argon. The quantity k is Boltzmann's constant and has the value $k = 1.38 \times 10^{-23}$ J/K. The unit of pressure is for a two-dimensional system.

8.6   A MOLECULAR DYNAMICS PROGRAM

In the following, we develop a molecular dynamics simulation of a two-dimensional system of particles interacting via the Lennard-Jones potential. We choose two rather than three dimensions because it is easier to visualize the results and the calculations are not as time consuming. The structure of *Program md* is given in the following:

```
PROGRAM md
PUBLIC x(36),y(36),vx(36),vy(36),ax(36),ay(36)
PUBLIC N,Lx,Ly,dt,dt2
LIBRARY "csgraphics"
CALL initial(t,ke,kecum,pecum,vcum,area)
CALL set_up_windows(#1,#2)
CALL accel(pe,virial)
LET E = ke + pe                      ! total energy
LET ncum = 0                         ! number of times data accumulated
LET flag$ = ""
DO
    CALL show_positions(flag$,#2)
    CALL Verlet(t,ke,pe,virial)
    CALL show_output(t,ke,pe,virial,kecum,vcum,ncum,area,#1)
LOOP until flag$ = "stop"
CALL save_config
END
```

The *x*- and *y*-components of the positions, velocities, and accelerations are represented by arrays and are declared as public variables (cf. Appendix 3C) because they are used in almost all of the subroutines. These arrays are dimensioned in a PUBLIC statement and are declared in a DECLARE PUBLIC statement in each subroutine in which they are used. An array that is declared in a PUBLIC statement is not dimensioned in a DIM statement. The nature of the passed variables and the subroutines are discussed in the following.

Because the system is deterministic, the nature of the motion is determined by the initial conditions. An appropriate choice of the initial conditions is more difficult than might first appear. For example, how do we choose the initial configuration (a set of positions and velocities) to correspond to a fluid at a desired temperature? We postpone a discussion of such questions until Section 8.7, and instead we use initial conditions that have been computed previously.

The initial conditions can be incorporated into the program by either reading a data file or storing the information within a program using DATA and READ statements. The following program illustrates the use of the DATA and READ statements:

```
PROGRAM example_data
DIM x(6)
DATA 4.48,3.06,0.20,2.08,3.88,3.36
FOR i = 1 to 6
    READ x(i)                        ! reads input from DATA statement
NEXT i
END
```

The location of the DATA statements is unimportant, but the data must be stored in the order in which it will be read by the READ statements.

The following version of SUB initial uses the DATA and READ statements to store an initial configuration of $N = 16$ particles in a central cell of linear dimension $Lx = Ly = 6$.

```
SUB initial(t,ke,kecum,pecum,vcum,area)
    DECLARE PUBLIC x(),y(),vx(),vy()
    DECLARE PUBLIC N,Lx,Ly,dt,dt2
    DECLARE DEF pbc
    LET dt = 0.01
    LET dt2 = dt*dt
    LET response$ = ""
    DO
        INPUT prompt "read data statements (d) or file (f)? ": start$
        IF (start$ = "d") or (start$ = "D") then
            LET response$ = "ok"
            LET N = 16
            LET Lx = 6
            LET Ly = 6
            DATA 1.09,0.98,-0.33,0.78,3.12,5.25,0.12,-1.19
            DATA 0.08,2.38,-0.08,-0.10,0.54,4.08,-1.94,-0.56
            DATA 2.52,4.39,0.75,0.34,3.03,2.94,1.70,-1.08
            DATA 4.25,3.01,0.84,0.47,0.89,3.11,-1.04,0.06
            DATA 2.76,0.31,1.64,1.36,3.14,1.91,0.38,-1.24
            DATA 0.23,5.71,-1.58,0.55,1.91,2.46,-1.55,-0.16
            DATA 4.77,0.96,-0.23,-0.83,5.10,4.63,-0.31,0.65
            DATA 4.97,5.88,1.18,1.48,3.90,0.20,0.46,-0.51
            FOR i = 1 to N
                READ x(i),y(i),vx(i),vy(i)
            NEXT i
        ELSE IF (start$ = "f") or (start$ = "F") then
            LET response$ = "ok"
            INPUT prompt "file name = ": file$
            OPEN #1: name file$, access input
            INPUT #1: N
            INPUT #1: Lx
            INPUT #1: Ly
            INPUT #1: heading$
            FOR i = 1 to N
                INPUT #1: x(i),y(i)
            NEXT i
            INPUT #1: heading$
            FOR i = 1 to N
                INPUT #1: vx(i),vy(i)
            NEXT i
            CLOSE #1
        ELSE
            PRINT
```

```
                    PRINT "d or f are the only acceptable responses."
                END IF
            LOOP until response$ = "ok"
            CLEAR
            LET ke = 0                            ! kinetic energy
            FOR i = 1 to N
                LET ke = ke + vx(i)*vx(i) + vy(i)*vy(i)
            NEXT i
            LET ke = 0.5*ke
            LET area = Lx*Ly
            LET t = 0                             ! time
            ! initialize sums
            LET kecum = 0
            LET pecum = 0
            LET vcum = 0
        END SUB
```

It is good programming practice to have an appropriate response for any input. For this reason SUB initial checks that an appropriate response is given to the INPUT statement.

SUB accel finds the total force on each particle and uses Newton's third law to reduce the number of calculations by a factor of two. (In reduced units, the mass of a particle is unity and hence acceleration and force are equivalent.) FUNCTION separation ensures that the separation between particles is no greater than Lx/2 in the x direction and Ly/2 in the y direction. The force on a given particle is assumed to be due to all the other $N - 1$ particles. Because of the short range nature of the Lennard-Jones potential, we could truncate the force at a distance $r = r_c \approx 2.5\sigma$ and ignore the forces from particles whose distance is greater than r_c. The meaning of the quantity virial in SUB accel is discussed in Section 8.7.

```
        SUB accel(pe,virial)
            DECLARE PUBLIC x(),y(),ax(),ay()
            DECLARE PUBLIC N,Lx,Ly
            DECLARE DEF separation
            FOR i = 1 to N
                LET ax(i) = 0
                LET ay(i) = 0
            NEXT i
            LET pe = 0
            LET virial = 0
            FOR i = 1 to N - 1                 ! compute total force on particle i
                FOR j = i + 1 to N             ! due to particles j > i
                    LET dx = separation(x(i) - x(j),Lx)
                    LET dy = separation(y(i) - y(j),Ly)
                    ! acceleration = force because mass = 1 in reduced units
                    CALL force(dx,dy,fxij,fyij,pot)
                    LET ax(i) = ax(i) + fxij
                    LET ay(i) = ay(i) + fyij
```

```
                        LET ax(j) = ax(j) - fxij    ! Newton's third law
                        LET ay(j) = ay(j) - fyij
                        LET pe = pe + pot
                        LET virial = virial + dx*ax(i) + dy*ay(i)
                    NEXT j
                NEXT i
            END SUB


            FUNCTION separation(ds,L)
                IF ds > 0.5*L then
                    LET separation = ds - L
                ELSE IF ds < -0.5*L then
                    LET separation = ds + L
                ELSE
                    LET separation = ds
                END IF
            END DEF


            SUB force(dx,dy,fx,fy,pot)
                LET r2 = dx*dx + dy*dy
                LET rm2 = 1/r2
                LET rm6 = rm2*rm2*rm2
                LET f_over_r = 24*rm6*(2*rm6 - 1)*rm2
                LET fx = f_over_r*dx
                LET fy = f_over_r*dy
                LET pot = 4*(rm6*rm6 - rm6)
            END SUB
```

The Verlet algorithm for the numerical solution of Newton's equations of motion is implemented in SUB Verlet. Note that the velocity is partially updated using the old acceleration. Then SUB accel is called to determine the acceleration using the new position and the velocity is updated again. SUB Verlet calls FUNCTION pbc which implements the periodic boundary conditions using IF statements.

```
            SUB Verlet(t,ke,pe,virial)
                DECLARE PUBLIC x(),y(),vx(),vy(),ax(),ay()
                DECLARE PUBLIC N,Lx,Ly,dt,dt2
                DECLARE DEF pbc
                FOR i = 1 to N
                    LET xnew = x(i) + vx(i)*dt + 0.5*ax(i)*dt2
                    LET ynew = y(i) + vy(i)*dt + 0.5*ay(i)*dt2
                    ! partially update velocity using old acceleration
                    LET vx(i) = vx(i) + 0.5*ax(i)*dt
                    LET vy(i) = vy(i) + 0.5*ay(i)*dt
                    LET x(i) = pbc(xnew,Lx)
                    LET y(i) = pbc(ynew,Ly)
                NEXT i
                CALL accel(pe,virial)              ! new acceleration
```

```
            LET ke = 0
            FOR i = 1 to N
                ! complete the update of the velocity using new acceleration
                LET vx(i) = vx(i) + 0.5*ax(i)*dt
                LET vy(i) = vy(i) + 0.5*ay(i)*dt
                LET ke = ke + vx(i)*vx(i) + vy(i)*vy(i)
            NEXT i
            LET ke = 0.5*ke
            LET t = t + dt
        END SUB


        FUNCTION pbc(pos,L)
            IF pos < 0 then
                LET pbc = pos + L
            ELSE IF pos > L then
                LET pbc = pos - L
            ELSE
                LET pbc = pos
            END IF
        END DEF
```

In SUB set_up_windows we divide the screen into two windows so that the trajectories are shown in a separate window. One advantage of the use of multiple windows is that the CLEAR statement erases the current window only, rather than the entire screen.

```
        SUB set_up_windows(#1,#2)
            DECLARE PUBLIC Lx,Ly
            OPEN #1: screen 0,1,0.90,1.0  ! numerical output
            OPEN #2: screen 0.02,1,0.02,0.90   ! particle trajectories
            CALL compute_aspect_ratio(Lx,xwin,ywin)
            SET WINDOW 0,xwin,0,ywin
            BOX LINES 0,Lx,0,Ly
            CALL headings(#1)
        END SUB


        SUB headings(#1)
            WINDOW #1
            SET CURSOR 1,1
            PRINT using "##########": "time steps";
            PRINT,
            PRINT using "###.##": "time";
            PRINT,
            PRINT using "---#.###": "energy";
            PRINT,
            PRINT using "##.##": "<T>";
            PRINT,
            PRINT using "##.##": "<P>"
        END SUB
```

The kinetic energy and the potential energy are computed at each time step and their values are accumulated in SUB show_output. Although it is inefficient to compute these quantities at every time step, we do so for simplicity.

```
SUB show_output(t,ke,pe,virial,kecum,vcum,ncum,area,#1)
    WINDOW #1
    DECLARE PUBLIC N,Lx,Ly
    SET CURSOR 2,1
    SET COLOR "black/white"
    LET ncum = ncum + 1
    PRINT using "######": ncum;
    PRINT,
    PRINT using "###.##": t;         ! time
    PRINT,
    LET E = ke + pe                  ! total energy
    PRINT using "---#.###": E;
    PRINT,
    LET kecum = kecum + ke
    LET vcum = vcum + virial
    LET mean_ke = kecum/ncum         ! still need to divide by N
    LET p = mean_ke + (0.5*vcum)/ncum  ! mean pressure * area
    LET p = p/area
    PRINT using "##.##": mean_ke/N;      ! mean kinetic temperature
    PRINT,
    PRINT using "##.##": p;
END SUB
```

The trajectories of the individual particles are displayed in SUB show_positions by representing the position of a particle at each time step by a point. The trajectories eventually fill the screen, which can be cleared by pressing the letter 'r' (refresh). If the letter 'n' is pressed, the particle positions are not shown on the screen and the program runs faster. Pressing 'r' restarts the screen updates. The program can be stopped by pressing the letter 's' (stop).

```
SUB show_positions(flag$,#2)
    DECLARE PUBLIC x(),y(),N,Lx,Ly
    IF key input then
       GET KEY k
       IF k = ord("r") then
          WINDOW #2
          CLEAR
          SET COLOR "black"
          BOX LINES 0,Lx,0,Ly
          LET flag$ = ""
       ELSEIF k = ord("s") then
          LET flag$ = "stop"
       ELSEIF k = ord("n") then
          LET flag$ = "no_show"
       END IF
    END IF
```

```
            IF flag$ <> "no_show" then
                WINDOW #2
                SET COLOR "red"
                FOR i = 1 to N
                    PLOT x(i),y(i)
                NEXT i
            END IF
        END SUB
```

Frequently, it is convenient to start a new run from the last configuration of a previous run. The final configuration is saved to a file in the following:

```
SUB save_config
    DECLARE PUBLIC x(),y(),vx(),vy()
    DECLARE PUBLIC N,Lx,Ly
    INPUT prompt "file name of saved configuration = ": file$
    OPEN #1: name file$,access output,create new
    PRINT #1: N
    PRINT #1: Lx
    PRINT #1: Ly
    PRINT #1: "x(i)","y(i)"
    ! comma added between outputs on the same line so that file
    ! can be read by True BASIC
    FOR i = 1 to N
        PRINT #1, using "----.####, ----.#####": x(i),y(i)
    NEXT i
    PRINT #1: "vx(i)","vy(i)"
    FOR i = 1 to N
        PRINT #1, using "----.####, ----.#####": vx(i),vy(i)
    NEXT i
    CLOSE #1
END SUB
```

In Problem 8.2 we use this preliminary version of *Program md* to simulate the approach of a system to equilibrium.

Problem 8.2   Approach to equilibrium

a. The initial configuration incorporated into the DATA statements in *Program md* corresponds to $N = 16$ particles interacting via the Lennard-Jones potential in a square cell of linear dimension $L = 6$. Check that the x and y coordinates of all the particles lies between 0 and 6. Set $\Delta t = 0.01$ and run the program to make sure that it is working properly. The total energy should be approximately conserved and the trajectories of all sixteen particles should be seen on the screen.

b. Suppose that at $t = 0$, the constraint that $0 \le x \le 6$ is removed and the particles can move freely in a rectangular cell with Lx = 12 and Ly = 6. Incorporate this change into your program and observe the trajectories of the particles. Does the system become more or less random as time increases?

c.   Compute $n(t)$, the number of particles in the left half of the cell, and plot $n(t)$ as a function of t. What is the qualitative behavior of $n(t)$? Also compute the time average of $n(t)$, and plot it as a function of t. What is the mean number of particles on the left half after the system has reached equilibrium? Compare your qualitative results with the results you found in Problem 7.1. Would the approach to equilibrium be better defined if you did a molecular dynamics simulation with $N = 64$ particles?

Problem 8.3   Sensitivity to initial conditions

a.   Consider the following initial condition corresponding to $N = 11$ particles moving in the same direction with the same velocity (see Fig. 8.4). Choose $Lx = Ly = 10$ and $\Delta t = 0.01$.

```
FOR i = 1 to N
    LET x(i) = Lx/2
    LET y(i) = (i - 0.5)*Ly/N
    LET vx(i) = 1
    LET vy(i) = 0
NEXT i
```

Does the system eventually reach equilibrium? Why or why not?

b.   Change the velocity of particle 6 so that $vx(6) = 0.99$ and $vy(6) = 0.01$. Is the behavior of the system qualitatively different than in part (a)? Does the system eventually reach equilibrium? Are the trajectories of the particles sensitive to

Fig. 8.4   Example of a special initial condition; the arrows represent the magnitude and the direction of each particle's velocity.

the initial conditions? Explain why almost all initial states lead to the same qualitative behavior.

* **c.** Modify *Program md* so that the program runs for a predetermined time interval. Use the initial condition included in the DATA statements, and save the positions and velocities of all the particles at $t = 0.5$ in a file. Then consider the time reversed process, i.e., the motion that would occur if the direction of time were reversed. This reversal is equivalent to letting $\mathbf{v} \rightarrow -\mathbf{v}$ for all particles. Do the particles return to their original positions? What happens if you reverse the velocities at a later time? What happens if you choose a smaller value of Δt?

* **d.** Use the initial condition included in the DATA statements, but let Lx = 12. Save the positions and velocities of all the particles in a file at $t = 2.5$. Then consider the time-reversed process. Do the particles return to their original state? What happens if you reverse the velocities at $t = 5$ rather than at $t = 2.5$?

* **e.** What can you conclude about the chaotic nature of the trajectories from your results? Are the computed trajectories the same as the "true" trajectories?

The trajectories generated in Problems 8.2 and 8.3 show the system in its greatest possible detail. From this *microscopic* viewpoint, the trajectories appear rather complex. The system can be described more simply by specifying its *macroscopic state*. For example, in Problem 8.2 we described the approach of the system to equilibrium by specifying n, the number of particles in the left half of the cell. Your observations of the macroscopic variable $n(t)$ should be consistent with the general properties of many body systems:

1. After the removal of an internal constraint, an isolated system changes in time from a "less random" to a "more random" state.

2. The equilibrium macroscopic state is characterized by relatively small fluctuations about a mean that is independent of time. A many particle system whose macroscopic state is independent of time is said to be in *equilibrium*. (The relative fluctuations become smaller as the number of particles becomes larger.)

In Problem 8.2(b) and (c) we found that the particles filled the box, and hence we were able to define a direction of time. Of course, this direction would be better defined if we considered more particles. Note that there is nothing intrinsic in Newton's laws of motion that gives time a preferred direction.

Before we consider other macroscopic variables, we need to monitor the total energy and verify our claim that the Verlet algorithm maintains conservation of energy with a reasonable choice of Δt. We also introduce a check for momentum conservation.

Problem 8.4   Tests of the Verlet algorithm

a. One essential check of a molecular dynamics program is that the total energy be conserved to the desired accuracy. Use *Program md* and determine the value of Δt necessary for the total energy to be conserved to a given accuracy over a time interval of $t = 2$. One criterion is to compute $\Delta E_{max}(t)$, the maximum of the difference $|E(t) - E(0)|$, over the time interval t, where $E(0)$ is the initial total energy, and $E(t)$ is the total energy at time t. Verify that $\Delta E_{max}(t)$ decreases when Δt is made smaller for fixed t. If your program is working properly, then $\Delta E_{max}(t)$ should decrease as approximately $(\Delta t)^2$.

b.   Because of the use of periodic boundary conditions, all points in the central cell are equivalent and the system is translationally invariant. Explain why *Program md* should conserve the total linear momentum. Floating point error and the truncation error associated with a finite difference algorithm can cause the total linear momentum to drift. Programming errors also might be detected by checking for the conservation of momentum. Hence, it is a good idea to monitor the total linear momentum at regular intervals and reset the total momentum equal to zero if necessary. SUB check_momentum, listed in the following, should be called in SUB initial after the initial configuration is obtained and in the main loop of the program at regular intervals, e.g., every $100 - 1000$ time steps. How well does *Program md* conserve the total linear momentum for $\Delta t = 0.01$?

```
SUB check_momentum
    DECLARE PUBLIC vx(),vy(),N
    LET vxsum = 0
    LET vysum = 0
    ! compute total center of mass velocity (momentum)
    FOR i = 1 to N
        LET vxsum = vxsum + vx(i)
        LET vysum = vysum + vy(i)
    NEXT i
    LET vxcm = vxsum/N
    LET vycm = vysum/N
    FOR i = 1 to N
        LET vx(i) = vx(i) - vxcm
        LET vy(i) = vy(i) - vycm
    NEXT i
END SUB
```

* c.   Another way of monitoring how well the program is conserving the total energy is to analyze the time series $E(t)$ using a least squares fit of $E(t)$ to a straight line. The slope of the line can be interpreted as the drift and the root mean square deviation from the straight line can be interpreted as the noise (σ_y in the notation of Section 7.5). How does the drift and the noise depend on Δt for a fixed time interval t? Most research applications conserve the energy to within 1 part in 10^3 or 10^4 or better over the duration of the run.

* d.   Consider one of the higher-order algorithms discussed in Appendix 5A for the solution of Newton's equations of motion. Can you choose a larger value of Δt to achieve the same degree of energy conservation that you found using the Verlet algorithm? Does the total energy fluctuate and/or eventually drift?

8.7 THERMODYNAMIC QUANTITIES

In the following we discuss how some of the macroscopic quantities of interest such as the temperature and the pressure can be related to time averages over the phase space trajectories of the particles. We then use our molecular dynamics program to explore the qualitative properties of gases and liquids.

The kinetic definition of the temperature follows from the equipartition theorem: each quadratic term in the energy of a classical system in equilibrium at temperature T has a mean value equal to $\frac{1}{2}kT$. Hence, we can define the temperature $T(t)$ at time t by the relation

$$N\frac{d}{2}kT(t) = \mathrm{K}(t),\tag{8.5}$$

or

$$kT(t) = \frac{2}{d}\frac{\mathrm{K}(t)}{N} = \frac{1}{dN}\sum_{i=1}^{N} m_i \mathbf{v}_i(t) \cdot \mathbf{v}_i(t).\tag{8.6}$$

K represents the total kinetic energy of the system, d is the spatial dimension, and the sum is over the N particles in the system. The mean temperature can be expressed as the time average of $T(t)$ over many configurations of the particles. For two dimensions ($d = 2$) and our choice of units, we write the mean temperature T as

$$T \equiv \overline{T} = \frac{1}{2N}\sum_{i=1}^{N} m_i \overline{\mathbf{v}_i(t) \cdot \mathbf{v}_i(t)}. \qquad \text{(two dimensions)}\tag{8.7}$$

The relation (8.7) is an example of the relation of a macroscopic quantity, the mean temperature, to a time average over the trajectories of the particles.

Note that the relation (8.6) holds only if the momentum of the center of mass of the system is zero — we do not want the motion of the center of mass to change the temperature. In a laboratory system the walls of the container ensure that the center of mass motion is zero (if the mean momentum of the walls is zero). In our simulation, we impose the constraint that the center of mass momentum (in d directions) be zero. Consequently, the system has $dN - d$ independent velocity components rather than dN components, and we should replace (8.6) by

$$kT(t) = \frac{1}{dN - d}\sum_{i=1}^{N} m_i \mathbf{v}_i(t) \cdot \mathbf{v}_i(t).\tag{8.8}$$

The presence of the factor $d(N - 1)$ rather than dN in (8.8) is an example of a *finite size* correction which becomes unimportant for large N. We shall ignore this correction in the following.

Another macroscopic quantity of interest is the mean pressure of the system. The pressure is related to the force per unit area acting normal to an imaginary surface in the system. By Newton's second law, this force can be related to the momentum that crosses the surface per unit time. In general, this momentum flux has two contributions. The easiest contribution to understand is the one carried by the particles due to their motion. This contribution, equal to the pressure of an ideal gas, is derived in many texts (cf. Chapter 7 of Reif) using simple kinetic theory arguments and is given by $P_{\text{ideal}} = NkT/V$.

The other contribution to the momentum flux arises from the momentum transferred across the surface due to the forces between particles on different sides of the surface. The form of this contribution to the dynamical pressure is difficult to derive if periodic boundary conditions are used (cf. Haile). The instantaneous pressure at time t including both contributions to the momentum flux is given by

$$P(t) = \frac{N}{V}kT(t) + \frac{1}{dV}\sum_{i<j}\mathbf{r}_{ij}(t)\cdot\mathbf{F}_{ij}(t),\tag{8.9}$$

where $\mathbf{r}_{ij} = \mathbf{r}_i - \mathbf{r}_j$, and $\mathbf{F}_{ij}$ is the force on particle i due to particle j.

The mean pressure $P \equiv \overline{P(t)}$ is found by computing a time average of the right-hand side of (8.9). The computed quantity of interest is not P, but the quantity

$$\frac{PV}{NkT} - 1 = \frac{1}{dNkT}\sum_{i<j}\overline{\mathbf{r}_{ij}\cdot\mathbf{F}_{ij}}.\tag{8.10}$$

In *Program md*, the right-hand side of (8.10), known as the *virial*, is computed in SUB accel and accumulated in the variable vcum. This quantity represents the correction to the ideal gas equation of state due to interactions between the particles.

The relation of information at the microscopic level to macroscopic quantities such as the temperature and pressure is one of the fundamental elements of statistical mechanics. In brief, molecular dynamics allows us to compute various time averages of the phase space trajectory over finite time intervals. The main practical question we must consider is whether our time intervals are sufficiently long to allow the system to explore phase space and yield meaningful averages. In equilibrium statistical mechanics, a time average is replaced by an *ensemble* average over all possible configurations. The quasi-ergodic hypothesis asserts the equivalence of these two averages if the same quantities are held fixed. In statistical mechanics, the ensemble of systems at fixed E, V, and N is called the microcanonical ensemble. Averages in this ensemble correspond to the time averages we find in molecular dynamics which are at fixed E, V and N. (Molecular dynamics also imposes an additional, but unimportant, constraint on the center of mass motion.) Ensemble averages are explored using Monte Carlo methods in Chapters 16 and 17.

The goal of the following problems is to explore some of the qualitative features of gases, liquids, and solids. Because we consider only small systems and relatively short run times, our results will only be suggestive.

Problem 8.5   Qualitative properties of a liquid and a gas

a. Use the initial configuration stored in the DATA statements in *Program* md for this problem. (If you do not want to type in the initial conditions, see Problem 8.11 for a discussion of how to generate your own.) For this initial condition $N = 16$ and Lx = Ly = 6. What is the reduced density? What is the initial energy of the system? Choose $\Delta t = 0.01$ and run the simulation for at least 500 time steps or $t = 5$. Compare your estimate for P with the value for an ideal gas.

b. Modify your program so that the instantaneous values of the temperature and pressure are not accumulated until the system has reached equilibrium. What is your criterion for equilibrium? One criterion is to compute the average values of T and P over finite time intervals and check that these averages do not drift with time.

c. One way of starting a simulation is to use the positions saved from an earlier run. The simplest way of obtaining an initial condition corresponding to a different density, but with the same value of N, is to rescale the positions of the particles and the linear dimensions of the cell. The following code shows one way to do so.

```
LET rscale = 0.95
FOR i = 1 to N
    LET x(i) = rscale*x(i)
    LET y(i) = rscale*y(i)
NEXT i
LET Lx = rscale*Lx
LET Ly = rscale*Ly
LET area = Lx*Ly
```

Incorporate the above code into your program in a separate subroutine. How do you expect P and T to change when the system is compressed? Use the same initial configuration as in part (a) and determine the mean temperature and pressure for the density $\rho = 16/(5.7)^2 \approx 0.49$ (rescale = 0.95). Compare your result for P to the ideal gas result. What do you think would happen if you choose a value of rescale that is much smaller? Save the final configuration of your simulation in a file and use it as the initial condition for another run with $\rho = 16/(0.95 \times 5.7)^2 \approx 0.55$. Determine T and P for this higher density and save the final configuration.

Problem 8.6   Distribution of speeds and velocities

a. Write a subroutine to compute the equilibrium probability $P(v)\Delta v$ that a particle has a speed between v and $v + \Delta v$. To do so, estimate the value of the maximum speed vmax that you need to bin. Choose bins of width dv = vmax/nbin, where nbin is the total number of bins. The following code illustrates one way of putting the speeds into their proper bins:

```
LET v2 = vx(i)*vx(i) + vy(i)*vy(i)
LET v = sqr(v2)
LET ibin = truncate(v/dv,0) + 1
IF ibin > nbin then LET ibin = nbin
LET prob(ibin) = prob(ibin) + 1
```

The array element `prob(ibin)` records the number of times the speed of a particle corresponds to `ibin`. Although it is inefficient, determine `prob(ibin)` after every time step for at least 100 time steps. Normalize `prob(ibin)` by dividing by the number of particles and by the number of time steps. Use the initial configuration stored in the DATA statements in *Program md* for this problem. Choose nbin = 50.

b. Plot the probability density $P(v)$ versus v. What is the qualitative form of $P(v)$? What is the most probable value of v? What is the approximate width of $P(v)$? Compare your measured result to the theoretical form (in two dimensions)

$$P(v)\,dv = Ae^{-mv^2/2kT}\,v\,dv. \tag{8.11}$$

The form (8.11) of the distribution of speeds is known as the Maxwell-Boltzmann distribution.

c. Determine the probability density for the x and y components of the velocity. Make sure that you distinguish between positive and negative values. What is the most probable value for the x and y velocity components? What are their average values? Plot the probability densities $P(v_x)$ versus v_x and $P(v_y)$ versus v_y. Better results can be found by plotting the average $[P(v_x = w) + P(v_y = w)]/2$ versus w. What is the qualitative form of $P(\mathbf{v})$? Is it the same as the probability density for the speed?

Another useful thermal quantity is the *heat capacity* at constant volume defined by the relation $C_V = (\partial E/\partial T)_V$. C_V is an example of a linear response function, that is, the response of the temperature to energy in the form of heat added to the system.

One way to obtain C_V is to determine $T(E)$, the temperature as a function of E. (Remember that T is obtained as a function of E in the microcanonical ensemble.) The heat capacity is given by $\Delta E/\Delta T$ for two runs that have slightly different temperatures. This method is straightforward, but requires that simulations at different energies be done before the derivative can be estimated. An alternative way of estimating C_V from the fluctuations of the kinetic energy is discussed in Problem 8.7.

Problem 8.7 Energy dependence of the temperature and pressure

a. We have seen in Problem 8.5 that the total energy is determined by the initial conditions, and the temperature is a derived quantity found only after the system has reached thermal equilibrium. For this reason it is difficult to study the system at a particular temperature. The mean temperature can be changed to the desired temperature by rescaling the velocities of the system, but we have to be careful not to increase the velocities too quickly. Why? Choose the initial configuration of the system to be an equilibrium configuration from an earlier simulation for Lx = Ly = 6 and $N = 16$ and determine $T(E)$, the energy dependence of the mean temperature, in the range $T = 1.0$ to $T = 1.2$. Rescale the velocities by the desired factor. Is the equilibrium temperature increased to the desired value? If not, you need to rescale the velocities again. In general, the desired temperature is reached by a series of velocity rescalings over a sufficiently long time such that the system remains close to equilibrium during the rescaling.

b. Use your data for $T(E)$ found in part (a) to plot the total energy E as a function of T. Is T a monotonically increasing function of E? Estimate the contribution to C_V from the potential energy. What percentage of the contribution to the heat capacity is due to the potential energy? Why are accurate determinations of C_V difficult to achieve?

* c. In our molecular dynamics simulations, the total energy is fixed, but the kinetic and potential energies can fluctuate. Another way of determining C_V is to relate it to the fluctuations of the kinetic energy. (In Chapter 17 we find that C_V is related to the fluctuations of the total energy in the constant T, V, N ensemble.) It can be shown that (cf. Ray and Graben)

$$\overline{T^2} - \overline{T}^2 = \frac{d}{2}N(k\overline{T})^2\left[1 - \frac{dNk}{2C_V}\right]. \tag{8.12}$$

Note that the relation (8.12) reduces to the ideal gas result if $\overline{T^2} = \overline{T}^2$. Write a subroutine to determine C_V from (8.12) and compare your results with the determination of C_V in part (b). What are the advantages and disadvantages of determining C_V from the fluctuations of T compared to the method used in part (b)?

How can we generate a typical initial condition for a system of particles at high density? What happens if we simply place the particles at random in the central cell? The problem is that if the system is dense, some particles will be very close to one another and exert a very large repulsive force F on one another. Because the condition $(F/m)(\Delta t)^2 << \sigma$ must be satisfied for a finite difference method to be applicable, the random placement of particles is not practical. However, it is possible to use such an initial condition if a fictitious drag force proportional to the square of the velocity is introduced to equilibrate the system. The effect of such a force is to damp the velocity of those particles whose velocities become too large due to the large forces exerted on them.

In general, the best way of obtaining a configuration with the desired density and number of particles is to place the particles on the sites of a regular lattice. If the goal is to equilibrate the system at fluid densities, the symmetry of the lattice is not important. The initial velocities can be chosen at random according to the Maxwell-Boltzmann distribution you found in Problem 8.6c. However, as you have found, the velocities equilibrate very quickly, and we may simply choose each component of $\mathbf{v}_i$ with uniform probability such that the desired temperature is obtained. Because the velocities are chosen at random, we need to use SUB check_momentum to ensure that the initial total momentum in the x and y direction is zero. After the system has come into partial equilibrium, you can increase the velocities of all the particles to reduce the time it takes to "melt" the system and reach the desired fluid state.

To simulate a solid we need to choose the shape of the central cell to be consistent with the symmetry of the solid phase of the system. This choice is necessary even though we have used periodic boundary conditions to minimize surface effects. If the cell does not correspond to the correct crystal structure, the particles cannot form a perfect crystal, and some of the particles will wander around in an endless search for their "correct" position. Consequently, a simulation of a small number of particles at high density and low temperature would lead to spurious results.

We know that the equilibrium structure of a crystalline solid at $T = 0$ is the configuration of lowest energy. In Problem 8.8 we compute the energy of a Lennard-Jones solid for both the square and triangular lattice structures (see Fig. 8.5).

Problem 8.8   Ground state energy of two-dimensional lattices

The nature of the triangular lattice can be seen from Fig. 8.5. Each particle has six nearest neighbors. Although it is possible to choose the central cell of the triangular lattice to be a rhombus, it is more convenient to choose the cell to be rectangular. We take the linear dimensions of the cell to be L_x and $L_y = \sqrt{3}L_x/2$ respectively. For simplicity, we assume that $\sqrt{N}$ is an integer so that the lattice spacings in the horizontal and vertical directions are $a_x = L_x/\sqrt{N}$ and $a_y = L_y/\sqrt{N}$, respectively. The lattice sites in each row are displaced by $\frac{1}{2}a_x$ from the preceding row. The following code generates a triangular lattice.

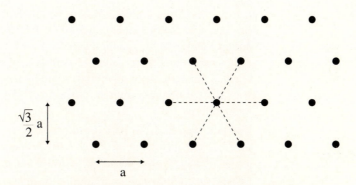

Fig. 8.5   Each particle has six nearest neighbors in a triangular lattice.

```
LET nx = sqr(N)
LET ny = nx
LET ax = Lx/nx
LET ay = Ly/ny
LET i = 0
FOR col = 1 to nx
    FOR row = 1 to ny
        LET i = i + 1
        LET x(i) = (col - 0.5)*ax
        LET y(i) = (row - 0.5)*ay
    NEXT row
NEXT col
```

Write a program to compute the potential energy per particle of a system of N particles interacting via the Lennard-Jones potential. Consider both the triangular and square lattices, and choose the linear dimension of the square lattice to be $L = \sqrt{L_x L_y}$, so that both lattices have the same density. Choose $N = 36$ and determine the energy for $L_x = 5$ and $L_x = 7$. What is the density of the system for each case? Do your results for E/N depend on the size of the lattice? Which lattice symmetry has a lower energy? Explain your results in terms of the ability of the triangular lattice to pack the particles closer together.

Problem 8.9   The solid state and melting

a.   Choose $N = 16$, $L_x = 4$, and $L_y = \sqrt{3}L_x/2$, and place the particles on a triangular lattice. Give each particle zero initial velocity. What is the total energy of the system? Do a simulation and measure the temperature and pressure as a function of time. Does the system remain a solid?

b.   Give the particles a random velocity in the interval $[-0.5, +0.5]$. What is the total energy? Equilibrate the system and determine the mean temperature and pressure. Describe the trajectories of the particles. Are the particles localized? Is the system a solid? Save an equilibrium configuration for use in part (c).

c.   Choose the initial configuration to be an equilibrium configuration from part (b), but increase the kinetic energy by a factor of two. What is the new total energy? Describe the qualitative behavior of the motion of the particles. What is the equilibrium temperature and pressure of the system? After equilibrium is reached, increase the temperature again by rescaling the velocities in the same way. Repeat this rescaling and measure $P(T)$ and $E(T)$ for several different temperatures.

d.   Use your results from part (c) to plot $E(T) - E(0)$ and $P(T)$ as a function of T. Is the difference $E(T) - E(0)$ proportional to T? What is the mean potential energy for a harmonic solid? What is its heat capacity?

e.   Choose an equilibrium configuration from part (b) and decrease the density by rescaling L_x, L_y and the particle coordinates by a factor of 1.1. What is the nature of the trajectories? Decrease the density of the system until the system melts. What is your qualitative criterion for melting?

It is possible for a system to not be in equilibrium even though its properties do not change over a long time interval. For example, if we rapidly lower the temperature of a liquid below its freezing temperature, it is likely that the resulting state is not an equilibrium crystal, but rather a supercooled liquid that will eventually nucleate to a crystal. In general, we must carefully prepare our system so as to minimize the probability that the system becomes trapped in a *metastable state*. On the other hand, there is much interesting physics and many applications to material science in the kinetics of nonequilibrium systems as they evolve toward equilibrium.

Problem 8.10   Metastability

a.   We can create a metastable state by placing the particles in a square cell whose shape is not consistent with the lowest energy state corresponding to a triangular lattice. Modify SUB initial so that the particle positions form a square lattice. If the initial velocities are set to zero, what happens when you run the program? Choose $N = 64$ and $L_x = L_y = 9$.

b.   We can show that the system in part (a) is in a metastable state by giving the particles a small random initial velocity with vmax $= 0.5$. Does the square lattice immediately change? When do you begin to see local structure resembling a triangular lattice?

c.   If time permits, repeat part (b) with vmax $= 0.1$.

8.8   RADIAL DISTRIBUTION FUNCTION

We can gain more insight into the structure of a many body system by looking at how the positions of the particles are correlated with one another due to their interactions. The *radial distribution function* $g(r)$ is the most common measure of this correlation and is defined as follows. Suppose that N particles are contained in a region of volume V with number density $\rho = N/V$. (In two and one dimensions, we replace V by the area and length respectively.) Choose one of the particles to be the origin. Then the mean number of other particles in the shell between $\mathbf{r}$ and $\mathbf{r} + d\mathbf{r}$ is given by $\rho g(\mathbf{r}) \, d\mathbf{r}$, where the volume element $d\mathbf{r} = 4\pi r^2 dr \ (d = 3)$, $2\pi r dr \ (d = 2)$, or $2 \, dr \ (d = 1)$. If the interparticle interaction is spherically symmetric and the system is a gas or a liquid, then $g(\mathbf{r})$ depends only on the separation $r = |\mathbf{r}|$. The normalization condition for $g(r)$ is

$$\rho \int g(r) \, d\mathbf{r} = N - 1 \approx N. \tag{8.13}$$

Equation (8.13) implies that if we choose one particle as the origin and count all the other particles in the system, we obtain $N - 1$ particles. For an ideal gas, there are no correlations between the particles, and $g(r) = 1$ for all r. For the Lennard-Jones interaction, we expect that $g(r) \to 0$ as $r \to 0$, because the particles cannot penetrate one another. We also expect that $g(r) \to 1$ as $r \to \infty$, because the effect of one particle on another decreases as their separation increases.

The radial distribution function can be measured indirectly by elastic radiation scattering experiments, especially by the scattering of X-rays. Several thermodynamic properties also can be obtained from $g(r)$. Because $\rho g(r)$ can be interpreted as the local density about a given particle, the potential energy of interaction between this particle and all other particles between r and $r + dr$ is $u(r)\rho g(r)\,d\mathbf{r}$, if we assume that only two-body interactions are present. The total potential energy is found by integrating over all values of r and multiplying by $N/2$. The factor of N is included because any of the N particles could be chosen as the particle at the origin, and the factor of 1/2 is included so that each pair interaction is counted only once. The result is that the mean potential energy per particle can be expressed as

$$\frac{U}{N} = \frac{\rho}{2} \int g(r)u(r)\,d\mathbf{r}. \tag{8.14}$$

It also can be shown that the relation (8.10) for the mean pressure can be rewritten in terms of $g(r)$ so that the equation of state can be expressed as

$$\frac{PV}{NkT} = 1 - \frac{\rho}{2dkT} \int g(r)\,r\,\frac{du(r)}{dr}\,d\mathbf{r}. \tag{8.15}$$

To determine $g(r)$ for a particular configuration of particles, we first compute $n(r, \Delta r)$, the number of particles in a spherical (circular) shell of radius r and small, but nonzero width Δr, with the center of the shell centered about each particle. A subroutine for computing $n(r)$ is given in the following:

```
SUB compute_g(ncorrel)
    DECLARE PUBLIC x(),y()
    DECLARE PUBLIC N,Lx,Ly
    DECLARE PUBLIC gcum(),nbin,dr
    DECLARE DEF separation
    ! accumulate data for n(r)
    FOR i = 1 to N - 1
        FOR j = i + 1 to N
            LET dx = separation(x(i) - x(j),Lx)
            LET dy = separation(y(i) - y(j),Ly)
            LET r2 = dx*dx + dy*dy
            LET r = sqr(r2)
            LET ibin = truncate(r/dr,0) + 1
            IF ibin <= nbin then
                LET gcum(ibin) = gcum(ibin) + 1
            END IF
        NEXT j
    NEXT i
    LET ncorrel = ncorrel + 1      ! # times n(r) computed
END SUB
```

The results for $n(r)$ for different configurations are accumulated in the array gcum; the latter array is normalized in SUB normalize_g listed below. The use of periodic boundary conditions in SUB compute_g implies that the maximum separation between any two particles in the x and y direction is Lx/2 and Ly/2 respectively. Hence for a square cell, we can determine $g(r)$ only for $r \leq \frac{1}{2}L$.

To obtain $g(r)$ from $n(r)$, we note that for a given particle i, we consider only those particles whose label j is greater than i (see SUB compute_g). Hence, there are a total of $\frac{1}{2}N(N-1)$ separations that are considered. In two dimensions we compute $n(r, \Delta r)$ for a circular shell whose area is $2\pi r \Delta r$. These considerations imply that $g(r)$ is related to $n(r)$ by

$$\rho g(r) = \frac{\overline{n(r, \Delta r)}}{\frac{1}{2}N \, 2\pi r \Delta r}. \qquad \text{(two dimensions)} \qquad (8.16)$$

Note the factor of $N/2$ in the denominator of (8.16). The following subroutine normalizes the array gcum and yields $g(r)$:

```
SUB normalize_g(ncorrel)
    DECLARE PUBLIC N,Lx,Ly
    DECLARE PUBLIC gcum(),dr
    LET density = N/(Lx*Ly)
    LET rmax = min(Lx/2,Ly/2)
    LET normalization = density*ncorrel*0.5*N
    LET bin = 1
    LET r = 0
    OPEN #2: name "gdata", access output,create new
    DO while r <= rmax
        LET area_shell = pi*((r + dr)^2 - r^2)
        LET g = gcum(bin)/(normalization*area_shell)
        PRINT r+dr/2,g
        PRINT #2: r+dr/2,g
        LET bin = bin + 1
        LET r = r + dr
    LOOP
    CLOSE #2
END SUB
```

The shell thickness Δr needs to be sufficiently small so that the important features of $g(r)$ are found, but large enough so that each bin has a reasonable number of contributions. The value of Δr can be specified in SUB initial; a reasonable compromise choice for its magnitude is dr = 0.025.

Problem 8.11   The structure of g(r) for a dense liquid and a solid

a.   Incorporate SUB compute_g and SUB normalize_g into your molecular dynamics program and determine $g(r)$ for some of the same densities and temperatures that you have considered in previous problems. What are the qualitative features of $g(r)$?

b.   Compute $g(r)$ for a system of $N = 64$ particles that are fixed on a triangular lattice with $L_x = 8$ and $L_y = \sqrt{3}L_x/2$. What is the density of the system?

What is the nearest neighbor distance between sites? At what value of r does the first maximum of $g(r)$ occur? What is the next nearest distance between sites? At what value of r does the second maximum of $g(r)$ occur? Does your calculated $g(r)$ have any other relative maxima? If so, relate these maxima to the structure of the triangular lattice.

*c. Use your molecular dynamics program to compute $g(r)$ for a dense fluid ($\rho >$ 0.6, $T \approx 1.0$) using at least $N = 32$ particles. How many relative maxima can you observe? In what ways do they change as the density is increased? How does the behavior of $g(r)$ for a dense liquid compare to that of a dilute gas and a solid?

8.9   HARD DISKS

How can we understand the temperature and density dependence of the equation of state and the structure of a dense liquid? One way to gain more insight is to modify the interaction and see how the properties of the system change. In particular, we would like to understand the relative role of the repulsive and attractive parts of the interaction. For this reason, we consider an idealized system of hard disks for which the interaction $u(r)$ is purely repulsive:

$$u(r) = \begin{cases} +\infty, & r < \sigma \\ 0, & r \geq \sigma . \end{cases} \tag{8.17}$$

The length σ is the diameter of the hard disks (see Fig. 8.6). In three dimensions the interaction (8.17) describes the interaction of hard spheres (billiard balls); in one dimension (8.17) describes the interaction of hard rods.

Because the interaction $u(r)$ between hard disks is a discontinuous function of r, the dynamics of hard disks is qualitatively different than it is for a continuous interaction such as the Lennard-Jones potential. For hard disks, the particles move in straight

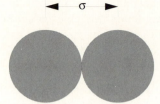

Fig. 8.6   The closest distance between two hard disks is σ. The disks exert no force on one another unless they touch.

lines at constant speed between collisions and change their velocities instantaneously when a collision occurs. Hence the problem becomes finding the next collision and computing the change in the velocities of the colliding pair. We will see that the dynamics can be computed exactly in principle and is limited only by computer roundoff errors.

The dynamics of a system of hard disks can be treated as a sequence of two-body elastic collisions. The idea is to consider all pairs of particles i and j and to find the collision time t_{ij} for their next collision ignoring the presence of all other particles. In many cases, the particles will be going away from each other and the collision time is infinite. From the collection of collision times for all pairs of particles, we find the minimum collision time. We then move all particles forward in time until the collision occurs and calculate the postcollision velocities of the colliding pair.

We first determine the particle velocities after a collision. Consider a collision between particles 1 and 2. Let $\mathbf{v}_1$ and $\mathbf{v}_2$ be their velocities before the collision and $\mathbf{v}_1'$ and $\mathbf{v}_2'$ be their velocities after the collision. Because the particles have equal mass, it follows from conservation of energy and linear momentum that

$$v_1'^2 + v_2'^2 = v_1^2 + v_2^2 \tag{8.18}$$

$$\mathbf{v}_1' + \mathbf{v}_2' = \mathbf{v}_1 + \mathbf{v}_2. \tag{8.19}$$

From (8.19) we have

$$\Delta \mathbf{v}_1 = \mathbf{v}_1' - \mathbf{v}_1 = -(\mathbf{v}_2' - \mathbf{v}_2) = -\Delta \mathbf{v}_2. \tag{8.20}$$

When two hard disks collide, the force is exerted along the line connecting their centers, $\mathbf{r}_{12} = \mathbf{r}_1 - \mathbf{r}_2$. Hence, the components of the velocities parallel to $\mathbf{r}_{12}$ are exchanged, and the perpendicular components of the velocities are unchanged. It is convenient to write the velocity of particles 1 and 2 as a vector sum of its components parallel and perpendicular to the unit vector $\hat{\mathbf{r}}_{12} = \mathbf{r}_{12}/|\mathbf{r}_{12}|$. We write the velocity of particle 1 as:

$$\mathbf{v}_1 = \mathbf{v}_{1,\|} + \mathbf{v}_{1,\perp}, \tag{8.21}$$

where $\mathbf{v}_{1,\|} = (\mathbf{v}_1 \cdot \hat{\mathbf{r}}_{12}) \hat{\mathbf{r}}_{12}$,

$$\mathbf{v}_{1,\|}' = \mathbf{v}_{2,\|} \qquad \mathbf{v}_{2,\|}' = \mathbf{v}_{1,\|} \tag{8.22a}$$

and

$$\mathbf{v}_{1,\perp}' = \mathbf{v}_{1,\perp} \qquad \mathbf{v}_{2,\perp}' = \mathbf{v}_{2,\perp}. \tag{8.22b}$$

Hence, we can write $\mathbf{v}_1'$ as

$$\begin{aligned}
\mathbf{v}_1' &= \mathbf{v}_{1,\|}' + \mathbf{v}_{1,\perp}' \\
&= \mathbf{v}_{2,\|} + \mathbf{v}_{1,\perp} \\
&= \mathbf{v}_{2,\|} - \mathbf{v}_{1,\|} + \mathbf{v}_{1,\|} + \mathbf{v}_{1,\perp} \\
&= \left[(\mathbf{v}_2 - \mathbf{v}_1) \cdot \hat{\mathbf{r}}_{12} \right] \hat{\mathbf{r}}_{12} + \mathbf{v}_1.
\end{aligned} \tag{8.23}$$

The change in the velocity of particle 1 at a collision is given by

$$\Delta \mathbf{v}_1 = \mathbf{v}_1' - \mathbf{v}_1 = -\left[(\mathbf{v}_1 - \mathbf{v}_2) \cdot \hat{\mathbf{r}}_{12}\right]\hat{\mathbf{r}}_{12} \tag{8.24}$$

or

$$\Delta \mathbf{v}_1 = -\Delta \mathbf{v}_2 = \left(\frac{\mathbf{r}_{12}\, b_{12}}{\sigma^2}\right)_{\text{contact}}, \tag{8.25}$$

where $b_{12} = \mathbf{v}_{12} \cdot \mathbf{r}_{12}$, $\mathbf{v}_{12} = \mathbf{v}_1 - \mathbf{v}_2$, and we have used the fact that $|\mathbf{r}_{12}| = \sigma$ at contact.

Problem 8.12   Velocity distribution of hard rods

Use (8.18) and (8.19) to show that $v_1' = v_2$ and $v_2' = v_1$ in one dimension, i.e., two colliding hard rods of equal mass exchange velocities. If you start a system of hard rods with velocities chosen from a uniform random distribution, will the velocity distribution approach the equilibrium Maxwell-Boltzmann distribution?

The most time consuming part of a hard disk dynamics program is computing the collision times of all pairs of particles. We now consider the criteria for a collision to occur. Consider disks 1 and 2 at positions $\mathbf{r}_1$ and $\mathbf{r}_2$ at $t = 0$. If they collide at a time t_{12} later, their centers will be separated by a distance σ:

$$|\mathbf{r}_1(t_{12}) - \mathbf{r}_2(t_{12})| = \sigma \tag{8.26}$$

During the time t_{12}, the disks move with constant velocities. Hence we have

$$\mathbf{r}_1(t_{12}) = \mathbf{r}_1(0) + \mathbf{v}_1(0)\, t_{12} \quad \text{and} \quad \mathbf{r}_2(t_{12}) = \mathbf{r}_2(0) + \mathbf{v}_2(0)\, t_{12}. \tag{8.27}$$

If we substitute (8.27) into (8.26), we find

$$[\mathbf{r}_{12} + \mathbf{v}_{12}t_{12}]^2 = \sigma^2$$

or

$$t_{12} = \frac{-\mathbf{v}_{12} \cdot \mathbf{r}_{12} \pm \sqrt{(\mathbf{v}_{12} \cdot \mathbf{r}_{12})^2 - v_{12}^2(r_{12}^2 - \sigma^2)}}{v_{12}^2}. \tag{8.28}$$

Because $t_{12} > 0$ for a collision to occur, we see from (8.28) that the condition

$$\mathbf{v}_{12} \cdot \mathbf{r}_{12} < 0 \tag{8.29}$$

must be satisfied. That is if $\mathbf{v}_{12} \cdot \mathbf{r}_{12} > 0$, the particles are moving away from each other and there is no possibility of a collision.

If the condition (8.29) is satisfied, then the discriminant in (8.28) must satisfy the condition

$$(\mathbf{v}_{12} \cdot \mathbf{r}_{12})^2 - v_{12}^2(r_{12}^2 - \sigma^2) \geq 0. \tag{8.30}$$

If the condition (8.30) is satisfied, then the quadratic in (8.28) has two roots. The smaller root corresponds to the physically significant collision because the disks are impenetrable. Hence, the physically significant solution for the time of a collision t_{ij} for particles i and j is given by

$$t_{ij} = \frac{-b_{ij} - \left[b_{ij}{}^2 - v_{ij}{}^2 \left(r_{ij}{}^2 - \sigma^2\right)\right]^{\frac{1}{2}}}{v_{ij}{}^2}. \tag{8.31}$$

Problem 8.13 Calculation of collision times

Use (8.31) and SUB check_collision listed in *Program hd* to write a program that determines the collision times (if any) of the following pairs of particles. It would be a good idea to draw the trajectories to confirm your results. Consider the three cases: $\mathbf{r}_1 = (2, 1)$, $\mathbf{v}_1 = (-1, -2)$, $\mathbf{r}_2 = (1, 3)$, $\mathbf{v}_2 = (1, 1)$; $\mathbf{r}_1 = (4, 3)$, $\mathbf{v}_1 = (2, -3)$, $\mathbf{r}_2 = (3, 1)$, $\mathbf{v}_2 = (-1, -1)$; and $\mathbf{r}_1 = (4, 2)$, $\mathbf{v}_1 = (-2, \frac{1}{2})$, $\mathbf{r}_2 = (3, 1)$, $\mathbf{v}_2 = (-1, 1)$. As usual, choose units so that $\sigma = 1$.

The main thermodynamic quantity of interest for hard disks is the mean pressure P. Because the forces act only when two disks are in contact, we have to modify the form of (8.10). We write $\mathbf{F}_{ij}(t) = \mathbf{I}_{ij}\,\delta(t - t_c)$, where t_c is the time at which the collision occurs. This form of $\mathbf{F}_{ij}$ implies the force is nonzero only when there is a collision between i and j. (The delta function $\delta(t)$ is infinite for $t = 0$ and is zero otherwise; $\delta(t)$ is defined by its use in an integral as shown in (8.32).) This form of the force yields

$$\int_0^t \mathbf{I}_{ij}\,\delta(t' - t_c)\,dt' = \mathbf{I}_{ij} = m\,\Delta\mathbf{v}_{ij}, \tag{8.32}$$

where we have used Newton's second law and assumed that a single collision has occurred during the time interval t. The quantity $\Delta\mathbf{v}_{ij} = \mathbf{v}_i' - \mathbf{v}_i - (\mathbf{v}_j' - \mathbf{v}_j)$. If we explicitly include the time average to account for all collisions during a time interval t, we can write (8.10) as

$$\frac{PV}{NkT} - 1 = \frac{1}{dNkT}\frac{1}{t}\sum_{ij}\int_0^t \mathbf{r}_{ij}\cdot\mathbf{I}_{ij}\,\delta(t' - t_c)\,dt'$$

$$= \frac{1}{dNkT}\frac{1}{t}\sum_{c_{ij}} m\,\Delta\mathbf{v}_{ij}\cdot\mathbf{r}_{ij} \tag{8.33}$$

The sum in (8.33) is over all collisions c_{ij} between disks i and j in the time interval t; $\mathbf{r}_{ij}$ is the vector between the centers of the disks at the time of a collision; the magnitude of $\mathbf{r}_{ij}$ in (8.33) is σ.

Our hard disk dynamics program implements the following steps. Given the initial condition in SUB initial, we find the collision times and the collision partners for all pairs of particles i and j. We then

1. locate the minimum collision time t_{min};

2. advance all particles using a straight line trajectory until the collision occurs, that is, displace particle i by $\mathbf{v}_i\,t_{min}$ and update the collision time;

3. compute the postcollision velocities of the colliding pair i and j;

4. calculate any quantities of interest and accumulate data;

5. update the collision partners of the colliding pair i and j and any other particles that were to collide with either i or j if i and j had not collided first;

6. repeat steps 1–5 indefinitely.

This steps are implemented in *Program hd* which is listed in the following:

```
PROGRAM hd
! dynamics of system of hard disks
! program based in part on Fortran program of Allen and Tildesley
PUBLIC x(100),y(100),vx(100),vy(100)
PUBLIC collision_time(100),partner(100)
PUBLIC N,Lx,Ly,t,timebig
LIBRARY "csgraphics"
CALL initial(vsum,rho,area)
CALL set_up_windows(ncolor,#1,#2)
CALL kinetic_energy(ke,#1)
CALL show_positions(flag$,#2)
LET temperature = ke/N
LET flag$ = ""
LET collisions = 0                    ! number of collisions
DO
    CALL minimum_collision_time(i,j,tij)
    ! move particles forward and reduce collision times by tij
    CALL move(tij)
    LET t = t + tij
    LET collisions = collisions + 1
    CALL show_positions(flag$,#2)
    CALL contact(i,j,virial)          ! compute collision dynamics
    LET vsum = vsum + virial
    CALL show_output(t,collisions,temperature,vsum,rho,area,#1)
    ! reset collision list for relevant particles
    CALL reset_list(i,j)
LOOP until flag$ = "stop"
CALL save_config(#2)
END
```

The colliding pair and the next collision time are found in SUB minimum_collision_time, and all particles are moved forward in SUB move until contact occurs. The collision dynamics of the colliding pair is computed in SUB contact, where the contribution to the pressure virial also is found. In SUB reset_list we update the collision partners of the colliding pair (i and j) and any other particles that were to collide with i and j if i and j had not collided first.

In SUB initial we initialize various variables and most importantly, call SUB uplist and SUB check_collision to compute the collision time for each particle assuming that no other particles are present. Note that the ith element in the array collision_time contains the minimum collision time for particle i with all particles j

such that $j > i$. The array partner(i) stores the particle label of the collision partner corresponding to this time with partner(i) always greater than i. The collision time for each particle is initially set to an arbitrarily large value, timebig, to account for the fact that at any given time, some particles have no collision partners.

```
SUB initial(vsum,rho,area)
    DECLARE PUBLIC x(),y(),vx(),vy()
    DECLARE PUBLIC N,Lx,Ly,t
    DECLARE PUBLIC collision_time(),partner(),timebig
    LET t = 0
    INPUT prompt "read file (f) or lattice start (l) = ": start$
    IF start$ = "f" or start$ = "F" then
        INPUT prompt "file name = ": file$
        OPEN #1: name file$, access input
        INPUT #1: N
        INPUT #1: Lx
        INPUT #1: Ly
        INPUT #1: heading$
        FOR i = 1 to N
            INPUT #1: x(i),y(i)
        NEXT i
        INPUT #1: heading$
        FOR i = 1 to N
            INPUT #1: vx(i),vy(i)
        NEXT i
        CLOSE #1
    ELSE IF start$ = "l" or start$ = "L" then
        RANDOMIZE
        INPUT prompt "N = ": N       ! choose N so that sqr(N) an integer
        INPUT prompt "Lx = ": Lx
        INPUT prompt "Ly = ": Ly
        INPUT prompt "vmax = ": vmax
        LET nx = sqr(N)
        LET ny = nx
        IF nx >= Lx or nx >= Ly then
            PRINT "box too small"
            STOP
        END IF
        LET ax = Lx/nx                 ! "lattice" spacing
        LET ay = Ly/ny
        LET i = 0
        FOR col = 1 to nx
            FOR row = 1 to ny
                LET i = i + 1
                LET x(i) = (col - 0.5)*ax
                LET y(i) = (row - 0.5)*ay
                ! choose random positions and velocities
                LET vx(i) = (2*rnd - 1)*vmax
```

```
                                    LET vy(i) = (2*rnd - 1)*vmax
                        NEXT row
                    NEXT col
                END IF
                CLEAR
                CALL check_overlap                ! check if two disks overlap
                CALL check_momentum
                LET area = Lx*Ly
                LET rho = N/area
                LET timebig = 1.0e10
                LET vsum = 0                       ! virial sum
                FOR i = 1 to N
                    LET partner(i) = N
                NEXT i
                LET collision_time(N) = timebig
                ! set up initial collision lists
                FOR i = 1 to N
                    CALL uplist(i)
                NEXT i
            END SUB


            SUB uplist(i)
                DECLARE PUBLIC N,collision_time(),timebig
                ! look for collisions with particles j > i
                IF i = N then EXIT SUB
                LET collision_time(i) = timebig
                FOR j = i + 1 to N
                    CALL check_collision(i,j)
                NEXT j
            END SUB
```

SUB check_collision uses the relations (8.29) and (8.31) to determine whether particles i and j will collide and if so, the time tij until their collision. We consider not only the nearest periodic image of j, but also the images of j in the adjoining cells, and determine the minimum collision time using all of these images. As shown in Fig. 8.7, it is possible for i to collide with an image of j that is not the image closest to i. For dense systems the probability that j is in a more distant cell is very small and these distant cells are ignored in the program.

```
            SUB check_collision(i,j)
                DECLARE PUBLIC x(),y(),vx(),vy()
                DECLARE PUBLIC Lx,Ly,collision_time(),partner()
                ! consider collisions between i and periodic images of j
                FOR xcell = -1 to 1
                    FOR ycell = -1 to 1
                        LET dx = x(i) - x(j) + xcell*Lx
                        LET dy = y(i) - y(j) + ycell*Ly
                        LET dvx = vx(i) - vx(j)
```

```
            LET dvy = vy(i) - vy(j)
            LET bij = dx*dvx + dy*dvy
            IF bij < 0 then
               LET r2 = dx*dx + dy*dy
               LET v2 = dvx*dvx + dvy*dvy
               LET discr = bij*bij - v2*(r2 - 1)
               IF discr > 0 then
                  LET tij = (-bij - sqr(discr))/v2
                  IF tij < collision_time(i) then
                     LET collision_time(i) = tij
                     LET partner(i) = j
                  END IF
               END IF
            END IF
         NEXT ycell
      NEXT xcell
END SUB
```

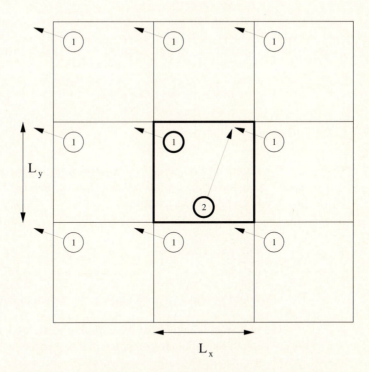

Fig. 8.7 The positions and velocities of disks 1 and 2 are such that disk 1 collides with an image of disk 2 that is not the image closest to disk 1. The periodic images of disk 2 are not shown.

The minimum collision time `tij` is found in SUB `minimum_collision_time` and all particles are moved forward by this time in SUB `move`.

```
SUB minimum_collision_time(i,j,tij)
    DECLARE PUBLIC N,collision_time(),partner(),timebig
    ! locate minimum collision time
    LET tij = timebig
    FOR k = 1 to N
        IF collision_time(k) < tij then
            LET tij = collision_time(k)
            LET i = k
        END IF
    NEXT k
    LET j = partner(i)
END SUB


SUB move(tij)
    DECLARE PUBLIC x(),y(),vx(),vy()
    DECLARE PUBLIC N,collision_time()
    DECLARE PUBLIC Lx,Ly
    DECLARE DEF pbc
    FOR k = 1 to N
        LET collision_time(k) = collision_time(k) - tij
        LET x(k) = x(k) + vx(k)*tij
        LET y(k) = y(k) + vy(k)*tij
        LET x(k) = pbc(x(k),Lx)
        LET y(k) = pbc(y(k),Ly)
    NEXT k
END SUB
```

The function `pbc` allows for the possibility that a disk has moved further than the linear dimension of the central cell between a collision. We have written it as a separate function to emphasize its purpose.

```
DEF pbc(pos,L)
    LET pbc = mod(pos,L)
END DEF
```

The function `separation` is identical to the function listed in *Program md* and is not listed here.

The collision dynamics for the colliding particles `i` and `j` and the contribution of the collision to the virial are computed in SUB `contact`:

```
SUB contact(i,j,virial)
    DECLARE PUBLIC x(),y(),vx(),vy(),Lx,Ly
    DECLARE DEF separation
    ! compute collision dynamics for particles i and j at contact
```

```
                LET dx = separation(x(i) - x(j),Lx)
                LET dy = separation(y(i) - y(j),Ly)
                LET dvx = vx(i) - vx(j)
                LET dvy = vy(i) - vy(j)
                LET factor = dx*dvx + dy*dvy
                LET delvx = - factor*dx
                LET delvy = - factor*dy
                LET vx(i) = vx(i) + delvx
                LET vx(j) = vx(j) - delvx
                LET vy(i) = vy(i) + delvy
                LET vy(j) = vy(j) - delvy
                LET virial = delvx*dx + delvy*dy
        END SUB
```

In SUB `reset_list` we find the new collision partners of particles i and j and those particles that were due to collide with i and j. Note that this update procedure must be done for particles whose labels are greater than i and j (SUB uplist) and for particles whose labels are less than i and j (SUB downlist).

```
    SUB reset_list(i,j)
        DECLARE PUBLIC N,partner()
        ! reset collision list for relevant particles
        FOR k = 1 to N
            LET test = partner(k)
            IF k = i or test = i or k = j or test = j then
                CALL uplist(k)
            END IF
        NEXT k
        CALL downlist(i)
        CALL downlist(j)
    END SUB


    SUB downlist(j)
        DECLARE PUBLIC collision_time()
        ! look for collisions with particles i < j
        IF j = 1 then EXIT SUB
        FOR i = 1 to j - 1
            CALL check_collision(i,j)
        NEXT i
    END SUB
```

SUB `check_momentum` and SUB `set_up_windows` are identical to the subroutines listed in *Program md* and are not listed here. SUB `save_config` is similar to the one listed in *Program md* except that the disks are moved so that none of the disks are in contact before the configuration is saved. Because the main loop of the program is not completed until two disks are in contact, this move is made to remove the possibility that two disks might appear to overlap due to floating point error.

```
SUB save_config(#2)
    DECLARE PUBLIC x(),y(),vx(),vy()
    DECLARE PUBLIC N,Lx,Ly
    WINDOW #2
    SET COLOR "black"
    ! move particles away from collision for final configuration
    CALL minimum_collision_time(i,j,tij)
    CALL move(0.5*tij)
    SET CURSOR 1,1
    INPUT prompt "name of saved configuration = ": file$
    OPEN #1: name file$, access output, create new
    PRINT #1: N
    PRINT #1: Lx
    PRINT #1: Ly
    PRINT #1: "x(i)","y(i)"
    FOR i = 1 to N
        PRINT #1, using "----.#####, ----.#####": x(i),y(i)
    NEXT i
    PRINT #1: "vx(i)","vy(i)"
    FOR i = 1 to N
        PRINT #1, using "----.#####, ----.#####": vx(i),vy(i)
    NEXT i
    CLOSE #1
END SUB
```

As discussed in Problem 8.14, an important check on the calculated trajectories of a hard disk system is that no two disks overlap. *SUB check_overlap* tests for this condition.

```
SUB check_overlap
    DECLARE PUBLIC x(),y()
    DECLARE PUBLIC N,Lx,Ly
    DECLARE DEF separation
    LET tol = 1.0e-4
    FOR i = 1 to N - 1
        FOR j = i + 1 to N
            LET dx = separation(x(i) - x(j),Lx)
            LET dy = separation(y(i) - y(j),Ly)
            LET r2 = dx*dx + dy*dy
            IF r2 < 1 then
                LET r = sqr(r2)
                IF (1 - r) > tol then
                    PRINT "particles ";i;" and ";j;"overlap"
                    STOP
                END IF
            END IF
        NEXT j
    NEXT i
END SUB
```

The remaining output subroutines are similar to those in *Program md*, but are listed below for completeness.

```
SUB headings(#1)
    WINDOW #1
    SET CURSOR 1,1
    PRINT using "##########": "collisions";
    PRINT,
    PRINT using "##########": "time";
    PRINT,
    PRINT using "########.##": "<P>";
    PRINT,
    PRINT using "####.##": "T";
    PRINT,
END SUB


SUB show_output(t,collisions,temperature,vsum,rho,area,#1)
    WINDOW #1
    SET CURSOR 2,1
    SET COLOR "black/white"
    PRINT using "######": collisions;
    PRINT,
    PRINT using "####.##": t;
    PRINT,
    LET mean_virial = vsum/(2*t)
    LET mean_pressure = rho*temperature + mean_virial/area
    PRINT using "####.##": mean_pressure;
END SUB


SUB show_positions(flag$,#2)
    DECLARE PUBLIC x(),y(),N,Lx,Ly
    IF key input then
       GET KEY k
       IF k = ord("r") then
          WINDOW #2
          CLEAR
          SET COLOR "black"
          BOX LINES 0,Lx,0,Ly
          LET flag$ = ""
       ELSEIF k = ord("s") then
          LET flag$ = "stop"
       ELSEIF k = ord("n") then
             LET flag$ = "no_show"
       END IF
    END IF
    IF flag$ <> "no_show" then
       SET COLOR "red"
       WINDOW #2
```

```
            FOR i = 1 to N
                PLOT x(i),y(i)
            NEXT i
        END IF
    END SUB
```

SUB show_disks can be substituted for SUB show_positions to represent the positions of the hard disks as circles rather than as points.

```
SUB show_disks(ncolor,flag$,#2)
    DECLARE PUBLIC x(),y(),N,Lx,Ly
    WINDOW #2
    IF key input then
       GET KEY k
       IF k = ord("r") then
          CLEAR
          BOX LINES 0,Lx,0,Ly
          LET flag$ = ""
       ELSEIF k = ord("s") then
          LET flag$ = "stop"
       ELSEIF k = ord("n") then
          LET flag$ = "no_show"
       END IF
    END IF
    IF flag$ <> "no_show" then
       LET ncolor = mod(ncolor,6) + 1
       IF ncolor = 1 then
          CLEAR
          BOX LINES 0,Lx,0,Ly
       END IF
       SET COLOR MIX(ncolor) rnd,rnd,rnd
       SET COLOR ncolor
       FOR i = 1 to N
           BOX CIRCLE x(i)-0.5,x(i)+0.5,y(i)-0.5,y(i)+0.5
       NEXT i
    END IF
END SUB
```

Finally, we list SUB kinetic_energy which is needed to compute the kinetic temperature:

```
SUB kinetic_energy(ke,#1)
    DECLARE PUBLIC vx(),vy()
    DECLARE PUBLIC N
    WINDOW #1
    LET ke = 0
    FOR i = 1 to N
        LET ke = ke + vx(i)*vx(i) + vy(i)*vy(i)
    NEXT i
    LET ke = 0.5*ke
    SET CURSOR 2,1
```

```
        PRINT,,,
        PRINT using "##.###": ke/N;
        PRINT,
    END SUB
```

Problem 8.14   Initial tests of `Program hd`

a. Because even a small error in computing the trajectories of the disks will
    eventually lead to their overlap and hence to a fatal error, it is necessary to test
    Program hd carefully. For simplicity, start from a lattice configuration. The
    most important test of the program is to monitor the computed positions of the
    hard disks at regular intervals for overlaps. If the distance between the centers
    of any two hard disks is less then unity (distances are measured in units of σ),
    there must be a serious error in the program. To check for the overlap of hard
    disks, include SUB overlap in the main loop of *Program hd* while you are
    testing the program.

b. The temperature for a system of hard disks is constant and can be defined as in
    (8.7). Why does the temperature not fluctuate as it does for a system of parti-
    cles interacting with a continuous potential? The constancy of the temperature
    can be used as another check on your program. What is the effect of increasing
    all the velocities by a factor of two? What is the natural unit of time? Explain
    why the state of the system is determined only by the density and not by the
    temperature.

c. Use *Program hd* to generate equilibrium configurations of a system of $N = 16$
    disks in a square cell of linear dimension $L = 6$. Suppose that at $t = 0$, the
    constraint that $0 \leq x \leq 6$ is removed, and the disks are allowed to move in a
    rectangular cell with Lx = 12 and Ly = 6. Does the system become more or
    less random? What is the qualitative nature of the time dependence of $n(t)$,
    the number of disks on the left half of the cell?

d. Modify your program so that averages are not computed until the system is
    in equilibrium. For simplicity, use the initial positions in the DATA statements
    in *Program md* as the initial condition. Compute the virial (8.33) and make a
    rough estimate of the error in your determination of the mean pressure due to
    statistical fluctuations.

e. Modify your program so that you can compute the velocity and speed distri-
    butions and verify that the computed distributions have the desired forms.

Problem 8.15   Static properties of hard disks

a. As we have seen in Section 8.7, a very time consuming part of the simulation
    is equilibrating the system from an arbitrary initial configuration. One way
    to obtain a set of initial positions is to add the hard disks sequentially with
    random positions and reject an additional hard disk if it overlaps any disks
    already present. Although this method is very inefficient at high densities, try it
    so that you will have a better idea of how difficult it is to obtain a high density

configuration in this way. A much better method is to place the disks on the sites of a lattice.

b. We first consider the dependence of the mean pressure P on the density ρ. Is P a monotonically increasing function of ρ? Is a system of hard disks always a fluid or is there a fluid to solid transition at higher densities? We will not be able to find definitive answers to these questions for $N = 16$. However, many simulations in the 1960's and 70's were done for systems of $N = 108$ hard disks and the largest simulations were for several hundred particles.

c. Compute the radial distribution function $g(r)$ for the same densities as you considered for the Lennard-Jones interaction. Compare the qualitative behavior of $g(r)$ for the two interactions. On the basis of your results, which part of the Lennard-Jones interaction plays the dominant role in determining the structure of a dense Lennard-Jones liquid?

* d. The largest number of hard disks that can be placed into a fixed volume defines the maximum density. What is the maximum density if the disks are placed on a square lattice? What is the maximum density if the disks are placed on a triangular lattice? Suppose that the initial condition is chosen to be a square lattice with $N = 100$ and $L = 11$ so that each particle has four nearest neighbors initially. What is the qualitative nature of the system after several hundred collisions have occurred? Do most particles still have four nearest neighbors or are there regions where most particles have six neighbors?

In Problem 8.16 we consider two physical quantities associated with the *dynamics* of a system of hard disks, namely the mean free time and the mean free path, quantities that are discussed in texts on kinetic theory (cf. Reif).

Problem 8.16   Mean free path and collision time

a. *Program* hd provides the information needed to determine the mean free time t_c, i.e., the average time a particle travels between collisions. For example, suppose we know that 40 collisions occurred in a time $t = 2.5$ for a system of $N = 16$ disks. Because two particles are involved in each collision, there was an average of 80/16 collisions per particle. Hence $t_c = 2.5/(80/16) = 0.5$. Write a subroutine to compute t_c and determine t_c as a function of ρ.

* b. Write a subroutine to determine the distribution of times between collisions. What is the qualitative form of the distribution? How does the width of this distribution depend on ρ?

* c. The mean free path ℓ is the mean distance a particle travels between collisions. Is ℓ simply related to t_c by the relation $\ell = \bar{v} t_c$, where $\bar{v} = \sqrt{\overline{v^2}}$? Write a subroutine to compute the the mean free path of the particles. Note that the displacement of particle i during the time t is $v_i t$, where v_i is the speed of particle i.

8.10 DYNAMICAL PROPERTIES

The mean free time and the mean free path are well defined for hard disks for which the meaning of a collision is clear. As discussed in texts on kinetic theory (cf. Reif), both quantities are related to the transport properties of a dilute gas. However, the concept of a collision is not well-defined for systems with a continuous interaction such as the Lennard-Jones potential. In the following, we take a more general approach to the dynamics of a many body system and discuss how the transport of particles in a system near equilibrium is related to the *equilibrium* properties of the system.

Suppose that we tag a certain fraction of the particles in our system (label them blue), and let $n(\mathbf{r}, t)$ be the mean number density of the tagged particles, that is, $n(\mathbf{r}, t)d\mathbf{r}$ is the mean number of particles with positions in the region $d\mathbf{r}$ about $\mathbf{r}$ at time t. In equilibrium, we expect that the tagged particles would be distributed uniformly and hence $n(\mathbf{r}, t)$ would be independent of $\mathbf{r}$ and t. Suppose however, that we start with a nonuniform distribution of tagged particles and ask how $n(\mathbf{r}, t)$ changes with $\mathbf{r}$ and t to make the density of tagged particles more uniform. We expect that $\mathbf{J}$, the flux of tagged particles, is proportional to the density gradient of the tagged particles. In one dimension we write

$$J_x = -D\frac{\partial n(x, t)}{\partial x}. \tag{8.34a}$$

More generally, we have that

$$\mathbf{J} = -D\nabla n(\mathbf{r}, t). \tag{8.34b}$$

The empirical relation (8.34) is known as Fick's law and expresses the fact that the flow of tagged particles acts to equalize the density. The quantity D in (8.34) is known as the *self-diffusion coefficient*. If we combine (8.34) with the statement that the number of tagged particles is conserved:

$$\frac{\partial n(\mathbf{r}, t)}{\partial t} + \nabla \cdot \mathbf{J}(\mathbf{r}, t) = 0, \tag{8.35}$$

we obtain the diffusion equation:

$$\frac{\partial n(\mathbf{r}, t)}{\partial t} = D\nabla^2 n(\mathbf{r}, t). \tag{8.36}$$

As the above discussion implies, we usually think of the transport of particles (and the transport of energy and momentum and other quantities) in the context of nonequilibrium situations. However, as we have already seen in our simulations, the particles in an equilibrium system are in continuous motion and hence continuously create local density fluctuations. We expect that these spontaneous density fluctuations behave in the same way as the density fluctuations that are created by weak external perturbations. Hence, we also expect (8.36) to apply to a system in equilibrium.

Consider the trajectory of a particular particle, e.g., particle 1, in equilibrium. At some arbitrarily chosen time $t = 0$, its position is $\mathbf{r}_1(0)$. At a later time t, its displacement is $\mathbf{r}_1(t) - \mathbf{r}_1(0)$. If there were no net force on the particle during this time interval, then $r_1(t) - r_1(0)$ would increase linearly with t. However, a particle in a fluid undergoes many collisions and on the average its net displacement would be zero. A more interesting quantity is the mean square displacement defined as

$$R_1^2(t) = \overline{[r_1(t) - r_1(0)]^2}. \tag{8.37}$$

The average in (8.37) is over all possible choices of the time origin. If the system is in equilibrium, the choice of $t = 0$ is arbitrary, and $\overline{R_1^2(t)}$ depends only on the time difference t. We have seen in Appendix 7A that the diffusion equation (8.36) implies that the t dependence of $\overline{R^2(t)}$ is given by

$$\overline{R^2(t)} = 2dDt, \qquad (t \rightarrow \infty) \tag{8.38}$$

where d is the spatial dimension. We have omitted the subscript because the average behavior of all the particles is the same.

The relation (8.38) relates the macroscopic transport coefficient D to a microscopic quantity, $\overline{R^2(t)}$, and gives us a straightforward way of computing D for an equilibrium system. The easiest way of computing $\overline{R^2(t)}$ is to write the position of a particle at regular time intervals into a file. We later can use a separate program to read the data file and compute $\overline{R^2(t)}$. Of course, we would find much better results if we average over all particles.

To understand the procedure for computing $\overline{R^2(t)}$, we consider a simple example. Suppose that the position of a particle in a one-dimensional system is given by $x(t = 0) = 1.65$, $x(t = 1) = 1.62$, $x(t = 2) = 1.84$, and $x(t = 3) = 2.22$. If we average over all possible time origins, we obtain

$$\overline{R^2(t = 1)} = \frac{1}{3}\left[\left(x(1) - x(0)\right)^2 + \left(x(2) - x(1)\right)^2 + \left(x(3) - x(2)\right)^2\right]$$

$$= \frac{1}{3}\left[0.0009 + 0.0484 + 0.1444\right] = 0.0646$$

$$\overline{R^2(t = 2)} = \frac{1}{2}\left[\left(x(2) - x(0)\right)^2 + \left(x(3) - x(1)\right)^2\right]$$

$$= \frac{1}{2}\left[0.0361 + 0.36\right] = 0.1981$$

$$\overline{R^2(t = 3)} = \left(x(3) - x(0)\right)^2 = 0.3249$$

Note that there are fewer combinations of the positions as the time difference increases. *Program R2*, listed in the following, reads a data file consisting of the x and y coordinates of a particle and computes $\overline{R^2(t)}$ by averaging over all possible choices of the time origin.

```
PROGRAM R2
! compute mean square displacement of one particle
! by averaging over all possible origins
DIM x(1000),y(1000),R2cum(20),norm(20)
CALL initial(Lx,Ly,R2cum(),ndiff)
CALL read_data(x(),y(),ndata)
```

```
CALL displacements(x(),y(),Lx,Ly,R2cum(),norm(),ndata,ndiff)
CALL normalization(ndiff,R2cum(),norm())
END


SUB initial(Lx,Ly,R2cum(),ndiff)
    LET Lx = 5
    LET Ly = 5
    LET ndiff = 10                      ! maximum time difference
    FOR idiff = 1 to ndiff
        LET R2cum(idiff) = 0
    NEXT idiff
END SUB


SUB read_data(x(),y(),ndata)
    ! read file for position of particle at regular intervals
    OPEN #1: name "xy.dat",access input
    LET t = 0
    DO while more #1
       LET t = t + 1
       INPUT #1: x(t),y(t)
    LOOP
    CLOSE #1
    LET ndata = t                       ! # of data points
END SUB


SUB displacements(x(),y(),Lx,Ly,R2cum(),norm(),ndata,ndiff)
    DECLARE DEF separation            ! function same as in Program md
    FOR idiff = 1 to ndiff
        FOR i = 1 to ndata - idiff
            LET dx = separation(x(i+idiff) - x(i),Lx)
            LET dy = separation(y(i+idiff) - y(i),Lx)
            LET R2cum(idiff) = R2cum(idiff) + dx*dx + dy*dy
            LET norm(idiff) = norm(idiff) + 1
        NEXT i
    NEXT idiff
END SUB


SUB normalization(ndiff,R2cum(),norm())
    PRINT "time difference","   <R2>"
    PRINT
    FOR idiff = 1 to ndiff
        IF R2cum(idiff) > 0 then
            LET R2bar = R2cum(idiff)/norm(idiff)
            PRINT idiff,R2bar
        END IF
    NEXT idiff
END SUB
```

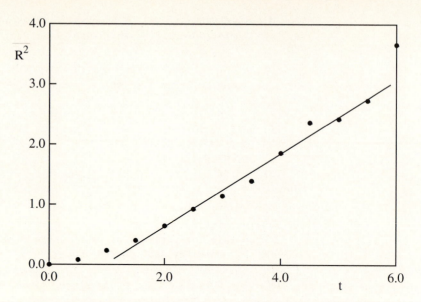

Fig. 8.8   The time dependence of the mean square displacement $\overline{R^2(t)}$ for one particle in a two-dimensional Lennard-Jones system with $N = 16$, $L = 5$, and $E = 5.8115$. The position of a particle was saved at intervals of 0.5. Much better results can be obtained by averaging over all particles and over a longer run. The least squares fit was made between $t = 1.5$ and $t = 5.5$. As expected, this fit does not pass through the origin. The slope of the fit is 0.61.

We show our results for $\overline{R^2(t)}$ for a system of Lennard-Jones particles in Fig. 8.8. Note that $\overline{R^2(t)}$ increases approximately linearly with t with a slope of roughly 0.61. From (8.38) the corresponding self-diffusion coefficient is $D = 0.61/4 \approx 0.15$. In Problem 8.17 we use *Program R2* to compute the self-diffusion coefficient. An alternative way of computing D is discussed in Project 8.1.

Problem 8.17   The self-diffusion coefficient

a.   Use *Program md* or *Program hd* and visually follow the motion of a particular particle by "tagging" it, e.g., by drawing its path with a different color. Describe its motion qualitatively.

b.   Modify *Program md* or *Program hd* so that the coordinates of a particular particle are saved at regular intervals. The desired time interval needs to be determined empirically. If we save the coordinates too often, the data file will be too large, and we will waste time writing the data to a file. If we do not save the positions often enough, we lose information. Because the time step Δt must be small compared to any interesting time scale, we know that the time interval for saving the positions must be at least a factor of ten greater than Δt. A good first guess is to choose the time interval to be the order of 10–50 time steps. The easiest procedure for hard disks is to save the positions at intervals measured in terms of the number of collisions. If we average over a sufficient number of collisions, we can find the relation between the elapsed time and the number of collisions.

c. Compute $\overline{R^2(t)}$ for conditions that correspond to a dense fluid. Does $\overline{R^2(t)}$ increase as t^2 as for a free particle or more slowly? Does $\overline{R^2(t)}$ increase linearly with t for longer times? What is the maximum value of $\overline{R^2(t)}$ that you find? Why does the use of periodic boundary conditions imply that there is an upper bound to the maximum time difference t we can consider when computing $\overline{R^2(t)}$?

d. Use the relation (8.38) to estimate the magnitude of D from the slope of $\overline{R^2(t)}$. Obtain D for several different temperatures and densities. (A careful study of $\overline{R^2(t)}$ for much larger systems and much longer times would show that $\overline{R^2(t)}$ is not proportional to t in two dimensions. Instead $\overline{R(t)^2}$ has a term proportional to $t \log t$, which dominates the linear t term if t is sufficiently large. However, we will not be able to observe the effects of this logarithmic term, and we can interpret our results for $\overline{R^2(t)}$ in terms of an "effective" diffusion coefficient. No such problem exists for three dimensions.)

e. Compute $\overline{R^2(t)}$ for an equilibrium configuration corresponding to a harmonic solid. What is the qualitative behavior of $\overline{R^2(t)}$?

f. Compute $\overline{R^2(t)}$ for an equilibrium configuration corresponding to a dilute gas. Is $\overline{R^2(t)}$ proportional to t for small times? Can we consider the particles to diffuse over short time intervals?

Another physically important single particle property is the *velocity autocorrelation function* $\psi(t)$. Suppose that particle i has velocity $\mathbf{v}_i$ at time t_1. If there were no net force on particle i, its velocity would remain constant. However, the interactions with other particles in the fluid will change the particle's velocity, and we expect that after several collisions, its velocity will not be strongly correlated with its velocity at an earlier time. We define $\psi(t)$ as

$$\psi(t) = \frac{1}{v_0^2} \overline{\mathbf{v}_i(t_2) \cdot \mathbf{v}_i(t_1)}, \tag{8.39}$$

where $v_0^2 = \overline{\mathbf{v}_i(0) \cdot \mathbf{v}_i(0)} = dkT/m$ and $t = t_2 - t_1$. As in our discussion of the mean square displacement, the average in (8.39) is over all possible time origins. We have defined $\psi(t)$ such that $\psi(t=0) = 1$. For large time differences $t_2 - t_1$, we expect $\mathbf{v}_i(t_2)$ to be independent of $\mathbf{v}_i(t_1)$, and hence $\psi(t) \to 0$ for $t \to \infty$. (Note that we have implicitly assumed that $\overline{\mathbf{v}_i(t)} = 0$.) It can be shown that the self-diffusion coefficient defined by (8.38) can be related to an integral of $\psi(t)$:

$$D = v_0^2 \int_0^\infty \psi(t)\, dt. \tag{8.40}$$

Other transport coefficients such as the shear viscosity and the thermal conductivity can also be expressed as a time integral over a corresponding autocorrelation function. The qualitative properties of the velocity autocorrelation function are explored in Problem 8.18.

*Problem 8.18   The velocity autocorrelation function

a. Modify your molecular dynamics program so that the velocity of a particular particle is saved to a file at regular time intervals. Then modify *Program R2* so that you can compute $\psi(t)$. The following code might be useful.

```
FOR idiff = 1 to ndiff
    FOR i = 1 to nmax - idiff
        LET psi(idiff) = psi(idiff) + vx(i + idiff)*vx(i)
        LET psi(idiff) = psi(idiff) + vy(i + idiff)*vy(i)
        LET norm(idiff) = norm(idiff) + 1
    NEXT i
NEXT idiff
```

Compute $\psi(t)$ for the same equilibrium configurations as in Problem 8.17c. Plot $\psi(t)$ versus t and describe its qualitative behavior. Estimate D from the relation (8.40). (To estimate the integral of $\psi(t)$, add your results for $\psi(t)$ at the different values of t and multiply the sum by the time difference between successive values of t.) How does your result for D compare to the determination using (8.38)?

b. Assume that $\psi(t)$ satisfies the form $\psi(t) = e^{-t/t_r}$ for all t. Substitute this form for $\psi(t)$ into (8.40) and determine the relationship between D and the *relaxation time* t_r. Plot the natural logarithm of $\psi(t)$ versus t and estimate t_r from the linear behavior of $\ln \psi(t)$. (At very long times, $\psi(t)$ exhibits slower than exponential decay. (This "long-time tail" is due to hydrodynamic effects.) Use your derived relationship between D and t_r to find D. Compare your estimates for D found from the slope of $\overline{R^2(t)}$, the relation (8.40), and the estimate of t_r. Are these estimates consistent?

c. Increase the density by 50% and compute $\psi(t)$. What is the qualitative behavior of $\psi(t)$? What is the implication of the fact that $\psi(t)$ becomes negative after a relatively short time?

d. Compute $\psi(t)$ for an equilibrium solid. Plot $\psi(t)$ versus t and describe its qualitative behavior. Explain your results in terms of the oscillatory motion of the particles about their lattice sites.

e. Contrast the behavior of the mean square displacement, the velocity autocorrelation function, and the radial distribution function in the solid and fluid phases and explain how these quantities can be used to indicate the nature of the phase.

f. Modify your program so that $\overline{R^2(t)}$ and $\psi(t)$ are averaged over all particles.

8.11 EXTENSIONS

The primary goals of this chapter have been to introduce the method of molecular dynamics and some of the concepts of statistical mechanics and kinetic theory. Although we found that simulations of systems as small as sixteen particles show some of the qualitative properties of macroscopic systems, we would need to simulate larger systems to make quantitative conclusions.

How do we know if the size of our system is sufficiently large to yield quantitative results that are independent of N? The straightforward answer is to repeat the simulation for larger N. Fortunately, most simulations of equilibrium systems with simple interactions require only several hundred to several thousand particles for reliable results. How do we know if our runs are long enough to give statistically meaningful averages? The simple answer is to run longer and see if the averages change significantly.

In general, the most time consuming parts of a molecular dynamics simulation are generating an appropriate initial configuration and doing the bookkeeping necessary for the force and energy calculations. If the force is sufficiently short range, there are a number of ways to reduce the equilibration time. For example, suppose we want to simulate a system of 864 particles in three dimensions. We first can simulate a system of 108 particles and allow the small system to come to equilibrium at the desired temperature. After equilibrium has been established, the small system can be replicated twice in each direction to generate the desired system of 864 particles. All of the velocities are reassigned at random using the Maxwell-Boltzmann distribution. Equilibration of the new system usually is established quickly.

The computer time required for our simple molecular dynamics program is order N^2 for each time step. The reason for this quadratic dependence on N is that the energy and force calculations require sums over all $\frac{1}{2}N(N-1)$ pairs of particles. If the interactions are short range, the time required for these sums can be reduced to approximately order N. The idea is to take advantage of the fact that at any given time, most pairs of particles are separated by a distance much greater than the effective range r_c of the interparticle interaction ($r_c \approx 2.5\,\sigma$ for the Lennard-Jones potential). Hence the calculation of the force and the energy requires the consideration of only those pairs of particles whose separation is less than r_c. Because testing whether each pair satisfies this criterion is an order N^2 calculation, we have to limit the number of pairs tested. One method is to divide the box into small cells and to compute the distance between particles that are in the same cell or in nearby cells. Another method is to maintain a list for each particle of its neighbors whose separation is less than a distance r_l, where r_l is chosen to be slightly greater than r_c so that the neighbor list can be used for several time steps before it is updated again. Both the cell method and the neighbor list method do not become efficient until N is approximately a few hundred.

So far we have discussed molecular dynamics simulations at fixed energy, volume, and number of particles. In laboratory systems we usually keep the temperature rather than the energy fixed, and frequently consider systems at fixed pressure rather than at fixed volume. It is possible to do molecular dynamics simulations at constant temperature and/or pressure. It also is possible to do simulations in which the shape of the cell is determined by the dynamics rather than imposed by the program. Such a simulation is essential for the study of solid-to-solid transitions where the major change is the shape of the crystal.

In addition to these technical advances, there is much more to learn about the properties of the system by doing averages over the trajectories. For example, how are transport properties such as the viscosity and the thermal conductivity related to the trajectories? We also have not discussed one of the most fundamental properties of a many body system, namely, its entropy. In brief, not all macroscopic properties of a many body system can be simply defined as a time average over some function of the phase space coordinates of the particles. The entropy is an example of such a quantity (but see Ma). However, changes in the entropy can be computed by using thermodynamic integration or a test particle method (see references).

There is another fundamental limitation of molecular dynamics, namely the *multiple time scale problem*. We know that we must choose the time step Δt to be smaller than any physical time scale in the system. For a solid, the smallest time scale is the period of the oscillatory motion of individual particles about their equilibrium positions. If we want to know how the solid responds if we add an interstitial particle or create a vacancy, we would have to run for millions of small time steps for the vacancy to move several interparticle distances. Although this particular problem can be overcome by using a faster computer, there are many problems for which no imaginable supercomputer would be sufficient. One of the biggest challenges of present interest is the *protein folding problem*. The biological function of a protein is determined by its three-dimensional structure which is encoded by the sequence of amino acids in the protein. At present, we know little about how the protein forms its three-dimensional structure. Such formidable computational challenges remind us that we cannot simply put a problem on a computer and let the computer tell us the answer. In particular, molecular dynamics methods need to be complemented by other simulation methods, especially Monte Carlo methods (see Chapters 16 and 17).

Now that we are familiar with the method of molecular dynamics, we briefly discuss its historical role in aiding our present understanding of simple equilibrium liquids. Simulations of systems of hard disks and hard spheres have shown that the structure of these systems does not differ significantly from the structure of systems with more complicated interactions. Given this insight, our present theories of liquids are based on the use of the hard sphere (disk) system as a reference system; the differences between the hard sphere interaction and the more complicated interaction of interest are treated as a perturbation about this reference system.

The emphasis in current applications of molecular dynamics is shifting from the studies of simple equilibrium fluids to studies of more complex fluids and studies of nonequilibrium systems. For example, how does a solid form when the temperature of a liquid is lowered quickly? How does a crack propagate in a brittle solid? What is the nature of the glass transition? Molecular dynamics and related methods will play an important role in aiding our understanding of these and many other problems.

8.12 PROJECTS

Many of the pioneering applications of molecular dynamics were done on relatively small systems. It is interesting to peruse the research literature of the past three decades and to see how much physical insight was obtained from these simulations. Many research-level problems can be generated by first reproducing previously published work and then extending the work to larger systems or longer run times to obtain better statistics. An interesting project based on recent research is suggested in the following. Some related projects are discussed in Section 17.11.

Project 8.1   Single particle fluctuation metric

As we discussed briefly in Section 8.7, the quasi-ergodic hypothesis assumes that time averages and ensemble averages are identical for a system in thermodynamic equilibrium. The assumption is that if we run a molecular dynamics simulation for a sufficiently long time, then the dynamical trajectory will fill the accessible phase space.

One way to confirm the quasi-ergodic hypothesis is to compute an ensemble average by simulating many independent copies of the system of interest using different initial configurations. Another way is to simulate a very large system and compare the behavior of different parts. A more direct and computationally efficient measure of the ergodicity has been proposed by Thirumalai and Mountain. This measure is called the fluctuation metric and is based on a comparison of the time averaged quantity $\overline{f_i(t)}$ of f_i for particle i to its average for all other particles. Effectively, we take the ensemble average of f over all particles, rather than over different parts of the system. If the system is ergodic, then all particles see the same average environment, and the time average $\overline{f_i(t)}$ for each particle will be the same if t is sufficiently long. Note that $\overline{f_i(t)}$ is the average of the quantity f_i over the time interval t and not the value of f_i at time t. The time average of f_i is defined as

$$\overline{f_i(t)} = \frac{1}{t} \int_0^t f(t')\, dt', \tag{8.41}$$

and the average of $\overline{f_i(t)}$ over all particles is given by

$$<f(t)> = \frac{1}{N} \sum_{i=1}^{N} \overline{f_i(t)}. \tag{8.42}$$

One of the physical quantities of interest is the energy of a particle, e_i, defined as

$$e_i = \frac{p_i^2}{2m_i} + \frac{1}{2} \sum_{i \neq j} u(r_{ij}). \tag{8.43}$$

The factor of $1/2$ is included in the potential energy term in (8.43) because the interaction energy is shared between pairs of particles. The above considerations lead us to define the energy fluctuation metric, $\Omega_e(t)$, as

$$\Omega_e(t) = \frac{1}{N} \sum_{i=1}^{N} \left[\overline{e_i(t)} - <e(t)>\right]^2. \tag{8.44}$$

a. Compute $\Omega_e(t)$ for a system of Lennard-Jones particles at a relatively high temperature. Determine $e_i(t)$ at time intervals of 0.5 or less and average Ω_e over as many time origins as possible. If the system is displaying behavior that is expected of an ergodic system over the time interval t, it can be shown that $\Omega_e(t)$ decreases as $1/t$. Do you find $1/t$ behavior for relatively short times? Nonergodic behavior might be found by rapidly reducing the kinetic energy (a temperature quench) and obtaining an amorphous solid or glass rather than a crystalline solid. However, it would be necessary to consider three-dimensional rather than two-dimensional systems because the latter nucleate to a crystalline solid very quickly.

b. Another quantity of interest is the velocity fluctuation metric Ω_v:

$$\Omega_v(t) = \frac{1}{dN} \sum_{i=1}^{N} \left[\overline{\mathbf{v}_i(t)} - <\mathbf{v}(t)>\right]^2. \tag{8.45}$$

The factor of $1/d$ in (8.45) is included because the velocity is a vector with d components. If we choose the total momentum of the system to be zero, then $<\mathbf{v}(t)> = 0$, and we can write (8.45) as

$$\Omega_v(t) = \frac{1}{dN} \sum_{i=1}^{N} \overline{\mathbf{v}_i(t)} \cdot \overline{\mathbf{v}_i(t)}. \tag{8.46}$$

We now show that the t dependence of $\Omega_v(t)$ is not a good indicator of ergodicity, but can be used to determine the diffusion coefficient D. We write

$$\overline{\mathbf{v}_i(t)} = \frac{1}{t} \int_0^t \mathbf{v}_i(t')\, dt' = \frac{1}{t}\left[\mathbf{r}_i(t) - \mathbf{r}_i(0)\right]. \tag{8.47}$$

If we substitute (8.47) into (8.46), we can express the velocity fluctuation metric in terms of the mean square displacement:

$$\Omega_v(t) = \frac{1}{dNt^2} \sum_{i=1}^{N} [\mathbf{r}_i(t) - \mathbf{r}_i(0)]^2 = \frac{<R^2(t)>}{d\,t^2}. \tag{8.48}$$

The average in (8.48) is over all particles. If the particles are diffusing during the time interval t, then $<R^2(t)> = 2d\,Dt$, and

$$\Omega_v(t) = 2D/t. \tag{8.49}$$

From (8.49) we see that $\Omega_v(t)$ goes to zero as $1/t$ as claimed in part (a). However, if the particles are localized (as in a crystalline solid and a glass), then $<R^2>$ is bounded for all t, and $\Omega_v(t) \sim 1/t^2$. Because a crystalline solid is ergodic and a glass is not, the velocity fluctuation metric is not a good measure of the lack of ergodicity. Use the t dependence of $\Omega_v(t)$ in (8.49) to determine D for the same configurations as in Problem 8.17. Note that the determination of D from Ω_v does not require a correction for the use of periodic boundary conditions.

References and Suggestions for Further Reading

Farid F. Abraham, "Computational statistical mechanics: methodology, applications and super-computing," *Adv. Phys.* **35**, 1 (1986). The author discusses both molecular dynamics and Monte Carlo techniques.

M. P. Allen and D. J. Tildesley, *Computer Simulation of Liquids*, Clarendon Press (1987). See Chapter 7 for a discussion of the test particle method.

R. P. Bonomo and F. Riggi, "The evolution of the speed distribution for a two-dimensional ideal gas: A computer simulation," *Am. J. Phys.* **52**, 54 (1984). The authors consider a system of hard disks and show that the system always evolves toward the Maxwell-Boltzmann distribution.

J. P. Boon and S. Yip, *Molecular Hydrodynamics*, Dover (1991). Their discussion of transport properties is an excellent supplement to our brief discussion.

Giovanni Ciccotti and William G. Hoover, editors, *Molecular-Dynamics Simulation of Statistical-Mechanics Systems*, North-Holland (1986).

Giovanni Ciccotti, Daan Frenkel, and Ian R. McDonald, editors, *Simulation of Liquids and Solids*, North-Holland (1987). A collection of reprints on the simulation of many body systems. Of particular interest are B. J. Alder and T. E. Wainwright, "Phase transition in elastic disks," *Phys. Rev.* **127**, 359 (1962) and earlier papers by the same authors; A. Rahman, "Correlations in the motion of atoms in liquid argon," *Phys. Rev.* **136**, A405 (1964), the first application of molecular dynamics to systems with continuous potentials; and Loup Verlet, "Computer 'experiments' on classical fluids. I. Thermodynamical properties of Lennard-Jones molecules," *Phys. Rev.* **159**, 98 (1967).

J. M. Haile, *Molecular Dynamics Simulation*, John Wiley & Sons (1992). A derivation of the mean pressure using periodic boundary conditions is given in Appendix B.

Jean Pierre Hansen and Ian R. McDonald, *Theory of Simple Liquids*, second edition, Academic Press (1986). An excellent reference that derives most of the theoretical results used in this chapter.

R. M. Hockney and J. W. Eastwood, *Computer Simulation Using Particles*, Adam Hilger (1988).

W. G. Hoover, *Molecular Dynamics*, Springer-Verlag (1986) and W. G. Hoover, *Computational Statistical Mechanics*, Elsevier (1991).

J. Kushick and B. J. Berne, "Molecular dynamics methods: continuous potentials" in *Statistical Mechanics Part B: Time-Dependent Processes*, Bruce J. Berne, editor, Plenum Press (1977). Also see the article by Jerome J. Erpenbeck and William Wood on "Molecular dynamics techniques for hard-core systems" in the same volume.

Shang-keng Ma, "Calculation of entropy from data of motion," *J. Stat. Phys.* **26**, 221, (1981). See also Chapter 25 of Ma's graduate level text, *Statistical Mechanics*, World Scientific (1985). Ma discusses a novel approach for computing the entropy directly from the trajectories. Note that the coincidence rate in Ma's approach is related to the recurrence time for a finite system to return to an arbitrarily small neighborhood of almost any given initial state.

S. Ranganathan, G. S. Dubey, and K. N. Pathak, "Molecular-dynamics study of two-dimensional Lennard-Jones fluids," *Phys. Rev.* A **45**, 5793 (1992). Two-dimensional systems are of interest because they are simpler theoretically and computationally and are related to single layer films.

Dennis Rapaport, *The Art of Molecular Dynamics Simulation*, Cambridge University Press (1995).

John R. Ray and H. W. Graben, "Direct calculation of fluctuation formulae in the microcanonical ensemble," *Mol. Phys.* **43**, 1293 (1981).

F. Reif, *Fundamentals of Statistical and Thermal Physics*, McGraw-Hill (1965.) An intermediate level text on statistical physics with a more thorough discussion of kinetic theory than found in most undergraduate texts. *Statistical Physics*, Vol. 5 of the Berkeley Physics Course, McGraw-Hill (1965) by the same author was one of the first texts to use computer simulations to illustrate the approach of macroscopic systems to equilibrium.

Marco Ronchetti and Gianni Jacucci, editors, *Simulation Approach to Solids*, Kluwer Academic Publishers (1990). Another collection of reprints.

R. M. Sperandeo Mineo and R. Madonia, "The equation of state of a hard-particle system: a model experiment on a microcomputer," *Eur. J. Phys.* **7**, 124 (1986).

D. Thirumalai and Raymond D. Mountain, "Ergodic convergence properties of supercooled liquids and glasses," *Phys. Rev.* A **42**, 4574 (1990).

James H. Williams and Glenn Joyce, "Equilibrium properties of a one-dimensional kinetic system," *J. Chem. Phys.* **59**, 741 (1973). Simulations in one dimension are even easier than in two.

9

Normal Modes and Waves

We discuss the physics of wave phenomena and the motivation and use of Fourier transforms.

9.1   COUPLED OSCILLATORS AND NORMAL MODES

Terms such as period, amplitude, and frequency are used to describe both waves and oscillatory motion. To understand the relation between the latter two phenomena, consider a flexible rope that is under tension with one end fixed. If we flip the free end, a pulse propagates along the rope with a speed that depends on the tension and on the inertial properties of the rope. At the *macroscopic* level, we observe a transverse wave that moves along the length of the rope. In contrast, at the *microscopic* level we see discrete particles undergoing oscillatory motion in a direction perpendicular to the motion of the wave.

One goal of this chapter is to use simulations to understand the relation between the microscopic dynamics of a simple mechanical model and the macroscopic wave motion that the model can support. For simplicity, we consider a one-dimensional chain of L particles each of mass M. The particles are coupled by massless springs with force constant K. The equilibrium separation between the particles is a. We denote the displacement of particle j from its equilibrium position at time t by $u_j(t)$ (see Fig. 9.1). For many purposes the most realistic boundary conditions are to attach particles $j = 1$ and $j = L$ to springs which are attached to fixed walls. We denote the walls by $j = 0$ and $j = L + 1$, and require that $u_0(t) = u_{L+1}(t) = 0$.

The force on an individual particle is determined by the compression or extension of the adjacent springs. The equation of motion of particle j is given by

$$M\frac{d^2u_j(t)}{dt^2} = -K\big[u_j(t) - u_{j+1}(t)\big] - K\big[u_j(t) - u_{j-1}(t)\big]$$
$$= -K\big[2u_j(t) - u_{j+1}(t) - u_{j-1}(t)\big]. \tag{9.1}$$

As expected, the motion of particle j is coupled to its two nearest neighbors. The equations of motion (9.1) describe *longitudinal* oscillations, i.e., motion along the length of the system. It is straightforward to show that identical equations hold for the *transverse* oscillations of L identical mass points equally spaced on a stretched massless string (cf. French).

The equations of motion (9.1) are linear, that is, only terms proportional to the displacements appear. It is straightforward to obtain analytical solutions of (9.1). Although the analytical solution will help us interpret the numerical solutions in terms of normal modes, it is not necessary to understand the analytical solution in detail to understand the numerical solutions.

To find the normal modes, we look for solutions for which the displacement of each particle is proportional to $\sin\omega t$ or $\cos\omega t$. We write

$$u_j(t) = u_j\cos\omega t, \tag{9.2}$$

where u_j is the amplitude of vibration of the jth particle. If we substitute the form (9.2) into (9.1), we obtain

$$-\omega^2 u_j = -\frac{K}{M}\big[2u_j - u_{j+1} - u_{j-1}\big]. \tag{9.3}$$

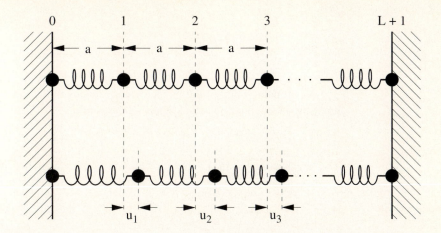

Fig. 9.1   A one-dimensional chain of L particles of mass M coupled by massless springs with force constant K. The first and last particles (0 and $L + 1$) are attached to fixed walls. The top chain shows the oscillators in equilibrium. The bottom chain shows the oscillators displaced from equilibrium.

We next assume that u_j depends sinusoidally on the distance ja:

$$u_j = C \sin qja. \tag{9.4}$$

The magnitude of the constant C will be determined later. If we substitute the form (9.4) into (9.3), we find the following condition for ω:

$$-\omega^2 \sin qja = -\frac{K}{M}\big[2 \sin qja - \sin q(j-1)a - \sin q(j+1)a\big]. \tag{9.5}$$

We write $\sin q(j \pm 1)a = \sin qja \cos qa \pm \cos qja \sin qa$ and find that (9.4) is a solution if

$$\omega^2 = 2\frac{K}{M}\big(1 - \cos qa\big). \tag{9.6}$$

We need to find the values of the wavenumber q that satisfy the boundary conditions $u_0 = 0$ and $u_{L+1} = 0$. The former condition is automatically satisfied by assuming a sine instead of a cosine solution in (9.4). The latter boundary condition implies that

$$q = q_n = \frac{\pi n}{a(L+1)} \qquad n = 1, \ldots, L \qquad \text{(fixed boundary conditions)} \tag{9.7}$$

What are the corresponding possible values of the wavelength λ? The latter is related to q by $q = 2\pi/\lambda$. The corresponding values of the angular frequencies are given by

$$\omega_n{}^2 = 2\frac{K}{M}\big[1 - \cos q_n a\big] = 4\frac{K}{M}\sin^2\frac{q_n a}{2} \tag{9.8}$$

or

$$\omega_n = 2\sqrt{\frac{K}{M}} \sin \frac{q_n a}{2}. \tag{9.9}$$

The relation (9.9) between ω_n and q_n is known as a dispersion relation.

A particular value of the integer n corresponds to the nth *normal mode*. We write the (time-independent) normal mode solutions as

$$u_{j,n} = C \sin q_n j a. \tag{9.10}$$

The linear nature of the equation of motion (9.1) implies that the time dependence of the displacement of the jth particle can be written as a superposition of normal modes:

$$u_j(t) = C \sum_{n=1}^{L} \left(A_n \cos \omega_n t + B_n \sin \omega_n t \right) \sin q_n j a \tag{9.11}$$

The coefficients A_n and B_n are determined by the initial conditions:

$$u_j(t = 0) = C \sum_{n=1}^{L} A_n \sin q_n j a \tag{9.12a}$$

and

$$v_j(t = 0) = C \sum_{n=1}^{L} \omega_n B_n \sin q_n j a. \tag{9.12b}$$

To solve (9.12) for A_n and B_n, we note that the normal mode solutions, $u_{j,n}$, are *orthogonal*, that is, they satisfy the condition

$$\sum_{j=1}^{L} u_{j,n} u_{j,m} \propto \delta_{n,m}. \tag{9.13}$$

The Kronecker δ symbol $\delta_{n,m} = 1$ if $n = m$ and is zero otherwise. It is convenient to normalize the $u_{j,n}$ so that they are *orthonormal*, i.e.,

$$\sum_{j=1}^{L} u_{j,n} u_{j,m} = \delta_{n,m}. \tag{9.14}$$

It is easy to show that the choice, $C = 1/\sqrt{(L+1)/2}$, in (9.4) and (9.10) insures that (9.14) is satisfied.

We now use the orthonormality condition to determine the coefficients A_n and B_n. If we multiply both sides of (9.12) by $C \sin q_m j a$, sum over j, and use the orthogonality condition (9.14), we obtain

$$A_n = C \sum_{j=1}^{L} u_j(0) \sin q_n j a \tag{9.15}$$

and

$$B_n = C \sum_{j=1}^{L} (v_j(0)/\omega_n) \sin q_n ja. \tag{9.16}$$

For example, if the initial displacement of every particle is zero, and the initial velocity of every particle is zero except for $v_1(0) = 1$, we find $A_n = 0$ for all n, and

$$B_n = \frac{C}{\omega_n} \sin q_n a. \tag{9.17}$$

The corresponding solution for $u_j(t)$ is

$$u_j(t) = \frac{2}{L+1} \sum_{n=1}^{L} \frac{1}{\omega_n} \cos \omega_n t \sin q_n a \sin q_n ja. \tag{9.18}$$

What is the solution if the particles start in a normal mode, i.e., $u_j(t=0) \propto \sin q_2 ja$?

The analytical solution (9.11) together with the initial conditions represent the complete solution of the displacement of the particles. If we wish, we can use a computer to compute the sum in (9.11) and plot the time dependence of the displacements $u_j(t)$. There are many interesting extensions that are amenable to analytical solutions. What is the effect of changing the boundary conditions? What happens if the spring constants are not all equal, but are chosen from a probability distribution? What happens if we vary the masses of the particles? For these cases we can follow a similar approach and look for the eigenvalues ω_n and eigenvectors $u_{j,n}$ of the matrix equation

$$\mathbf{T} \mathbf{u} = \omega^2 \mathbf{u}. \tag{9.19}$$

The matrix elements $T_{i,j}$ are zero except for

$$T_{i,i} = \frac{1}{M_i} \big[K_{i,i+1} + K_{i,i-1} \big] \tag{9.20a}$$

$$T_{i,i+1} = -\frac{K_{i,i+1}}{M_i} \tag{9.20b}$$

and

$$T_{i,i-1} = -\frac{K_{i,i-1}}{M_i}, \tag{9.20c}$$

where $K_{i,j}$ is the spring constant between particles i and j. The solution of matrix equations is a well studied problem in linear programming, and a commercial subroutine package such as IMSL or a symbolic programming language such as Maple, MatLab, or Mathematica can be used to obtain the solutions.

For our purposes it is easier to find the numerical solution of the equations of motion (9.1) directly because we also are interested in the effects of nonlinear forces between the particles, a case for which the matrix approach is inapplicable. In *Program oscillators* we use the Euler-Richardson algorithm to simulate the dynamics of L

linearly coupled oscillators. The particle displacements are displayed as transverse oscillations using techniques similar to Program animation in Chapter 5. Note that we have used the MAT instruction to assign the array usave to u (see SUB initial). This instruction is equivalent to assigning every element of the array usave the corresponding value of the array u.

```
PROGRAM oscillators
! simulate coupled linear oscillators in one dimension
DIM u(0 to 21),v(0 to 21),usave(0 to 21)
CALL initial(L,u(),v(),t,dt,usave(),mass$,erase$)
DO
    CALL update(L,u(),v(),ke,t,dt,usave())
    CALL animate(L,u(),ke,t,usave(),mass$,erase$)
LOOP until key input
END

SUB initial(L,u(),v(),t,dt,usave(),mass$,erase$)
    DATA 0,0.5,0,0,0,0,0,0,0,0,0
    DATA 0,0,0,0,0,0,0,0,0,0,0
    LET t = 0
    LET dt = 0.025
    LET L = 10                          ! number of particles
    SET WINDOW -1,L+1,-4,4
    SET COLOR "red"
    BOX AREA 0.9,1.1,-0.1,0.1
    BOX KEEP 0.9,1.1,-0.1,0.1 in mass$
    SET COLOR "background"
    BOX AREA -0.1,0.1,-0.1,0.1
    BOX KEEP -0.1,0.1,-0.1,0.1 in erase$
    PLOT line -2,0;L+2,0
    FOR j = 1 to L
        READ u(j)                       ! initial displacements
        BOX SHOW mass$ at j-0.1,u(j)-0.1
    NEXT j
    FOR j = 1 to L
        READ v(j)                       ! initial velocities
    NEXT j
    LET u(0) = 0                        ! fixed wall boundary conditions
    LET u(L+1) = 0
    MAT usave = u           ! note use of matrix assignment instruction
END SUB

SUB update(L,u(),v(),ke,t,dt,usave())
    ! Euler-Richardson algorithm
    DIM a(20),amid(20),umid(0 to 21),vmid(20)
    LET ke = 0
    ! K/M equal to unity
    FOR j = 1 to L
        LET usave(j) = u(j)
```

```
                    LET a(j) = -2*u(j) + u(j+1) + u(j-1)
                    LET umid(j) = u(j) + 0.5*v(j)*dt
                    LET vmid(j) = v(j) + 0.5*a(j)*dt
                NEXT j
                LET umid(0) = 0
                LET umid(L+1) = 0
                FOR j = 1 to L
                    LET amid(j) = -2*umid(j) + umid(j+1) + umid(j-1)
                    LET u(j) = u(j) + vmid(j)*dt
                    LET v(j) = v(j) + amid(j)*dt
                    LET ke = ke + v(j)*v(j)
                NEXT j
                LET t = t + dt
            END SUB

            SUB animate(L,u(),ke,t,usave(),mass$,erase$)
                LET pe = (u(1) - u(0))^2        ! interaction with left spring
                FOR j = 1 to L
                    ! compute potential energy
                    LET pe = pe + (u(j+1) - u(j))^2
                    ! transverse oscillation
                    BOX SHOW erase$ at j-0.1,usave(j)-0.1
                    BOX SHOW mass$ at j-0.1,u(j)-0.1
                NEXT j
                PLOT line -2,0;L+2,0
                SET CURSOR 1,1
                SET COLOR "black"
                PRINT using "t = ###.##": t
                LET E = 0.5*(ke + pe)
                PRINT using "E = #.####": E
            END SUB
```

Problem 9.1   Motion of coupled oscillators

a. Run *Program oscillators* for $L = 2$ and choose the initial values of $u(1)$
    and $u(2)$ so that the system is in one of its two normal modes, e.g., $u(1) =$
    $u(2) = 0.5$. Set the initial velocities equal to zero. Note that the program sets
    the ratio $K/M = 1$. Describe the displacement of the particles. Is the motion
    of each particle periodic in time? To answer this question, add a subroutine
    that plots the displacement of each particle versus the time. Then consider the
    other normal mode, e.g., $u(1) = 0.5$, $u(2) = -0.5$. What is the period in this
    case? Does the system remain in a normal mode indefinitely? Finally, choose
    the initial particle displacements equal to random values between -0.5 and
    $+0.5$. Is the motion of each particle periodic in this case?

b. Consider the same questions as in part (a), but with $L = 4$ and $L = 10$. Con-
    sider the $n = 2$ mode for $L = 4$ and the $n = 3$ and $n = 8$ modes for $L = 10$.
    (See (9.10) for the form of the normal mode solutions.) Also consider random
    initial displacements.

c.   *Program oscillators* depicts the oscillations as transverse because they are easier to visualize. Modify the program to represent longitudinal oscillations instead. Define the density as the number of particles within a certain range of x. For example, set $L = 20$ and describe how the average density varies as a function of the time within the region defined by $8 < x < 12$. Use the initial condition $u_j = \cos(3j\pi/(L+1))$ corresponding to the third normal mode. Repeat for another normal mode.

d.   Write a program to verify that the normal mode solutions (9.10) are orthonormal. Then compare the analytical results and the numerical results for $L = 10$ using the initial conditions listed in the DATA statements in *Program oscillators*. How much faster is it to calculate the analytical solution? What is the maximum deviation between the analytical and numerical solution of $u_j(t)$? How well is the total energy conserved in *Program oscillators*? How does the maximum deviation and the conservation of the total energy change when the time step Δt is reduced?

Problem 9.2   Motion of coupled oscillators with external forces

a.   Modify *Program oscillators* so that an external force F_e is exerted on the first particle,

$$F_e/m = 0.5 \cos \omega_e t, \qquad\qquad\qquad (9.21)$$

where ω_e is the angular frequency of the external force. (Note that ω_e is an angular frequency, but as is common practice, we frequently refer to ω_e as a frequency.) Let the initial displacements and velocities of all L particles be zero. Choose $L = 3$ and then $L = 10$ and consider the response of the system to an external force for $\omega = 0.5$ to 4.0 in steps of 0.5. Record $A(\omega)$, the maximum amplitude of any particle, for each value of ω. Explain how this system can be used as a high frequency filter.

b.   Choose ω_e to be one of the normal mode frequencies. Does the maximum amplitude remain constant or does it increase with time? How can you use the response of the system to an external force to determine the normal mode frequencies? Discuss your results in terms of the power input, $F_e v_1$?

c.   In addition to the external force exerted on the first particle, add a damping force equal to $-\gamma v_i$ to all the oscillators. Choose the damping constant $\gamma = 0.05$. How do you expect the system to behave? How does the maximum amplitude depend on ω_e? Are the normal mode frequencies changed when $\gamma \neq 0$?

Problem 9.3   Different boundary conditions

a.   Modify *Program oscillators* so that periodic boundary conditions are used, i.e., $u(L + 1) = u(1)$ and $u(0) = u(L)$. Choose $L = 10$, and the initial condition corresponding to the normal mode (9.10) with $n = 2$. Does this initial condition yield a normal mode solution for periodic boundary conditions? It might be easier to answer this question by plotting u(i) versus time for two or more particles. For fixed boundary conditions there are $L + 1$ springs, but for periodic boundary conditions there are L springs. Why? Choose the initial condition corresponding to the $n = 2$ normal mode, but replace $L + 1$ by L in (9.7). Does this initial condition correspond to a normal mode? Now try $n = 3$, and other values of n. Which values of n give normal modes? Only sine functions can be normal modes for fixed boundary conditions (see (9.4)). Can there be normal modes with cosine functions if we use periodic boundary conditions?

b. Modify *Program oscillators* so that free boundary conditions are used, that is, u(L + 1) = u(L) and u(0) = u(1). Choose $L = 10$. Use the initial condition corresponding to the $n = 3$ normal mode found using fixed boundary conditions. Does this condition correspond to a normal mode for free boundary conditions? Is $n = 2$ a normal mode for free boundary conditions? Are the normal modes purely sinusoidal?

c. Choose free boundary conditions and $L \geq 10$. Let the initial condition be a pulse of the form, u(1) = 0.2, u(2) = 0.6, u(3) = 1.0, u(4) = 0.6, u(5) = 0.2, and all other u(j) = 0. After the pulse reaches the right end, what is the phase of the reflected pulse, i.e., are the displacements in the reflected pulse in the same direction as the incoming pulse (a phase shift of zero degrees) or in the opposite direction (a phase shift of 180 degrees)? What happens for fixed boundary conditions? Choose L to be as large as possible so that it is easy to distinguish the incident and reflected waves.

*d. Set $L = 20$ and let the spring constants on the right half of the system be four times greater than the spring constants on the left half. Use fixed boundary conditions. Set up a pulse on the left side. Is there a reflected pulse at the boundary between the two types of springs? If so, what is its relative phase? Compare the amplitude of the reflected and transmitted pulses. Consider the same questions with a pulse that is initially on the right side.

9.2 FOURIER TRANSFORMS

In Section 9.1, we showed that the displacement of a single particle can be written as a linear combination of normal modes, that is, a linear superposition of sinusoidal terms. In general, an arbitrary periodic function $f(t)$ of period T can be expressed as a Fourier series of sines and cosines:

$$f(t) = \frac{1}{2}a_0 + \sum_{k=1}^{\infty}\left(a_k \cos \omega_k t + b_k \sin \omega_k t\right), \tag{9.22}$$

where

$$\omega_k = k\omega_0 \quad \text{and} \quad \omega_0 = \frac{2\pi}{T}. \tag{9.23}$$

The quantity ω_0 is the fundamental frequency. The sine and cosine terms in (9.22) for $k = 2, 3, \ldots$ represent the second, third, $\ldots$, and higher order harmonics. The *Fourier coefficients* a_k and b_k are given by

$$a_k = \frac{2}{T}\int_0^T f(t)\cos\omega_k t\, dt \tag{9.24a}$$

$$b_k = \frac{2}{T}\int_0^T f(t)\sin\omega_k t\, dt. \tag{9.24b}$$

The constant term $\frac{1}{2}a_0$ in (9.22) is the average value of $f(t)$. The expressions (9.24) for the coefficients follow from the orthogonality conditions:

$$\frac{2}{T}\int_0^T \sin\omega_k t\,\sin\omega_{k'}t\, dt = \delta_{k,k'} \tag{9.25a}$$

$$\frac{2}{T}\int_0^T \cos\omega_k t\,\cos\omega_{k'}t\, dt = \delta_{k,k'}. \tag{9.25b}$$

$$\frac{2}{T}\int_0^T \sin\omega_k t\,\cos\omega_{k'}t\, dt = 0. \tag{9.25c}$$

In general, an infinite number of terms is needed to represent an arbitrary periodic function exactly. In practice, a good approximation usually can be obtained by including a relatively small number of terms. Unlike a power series, which can approximate a function only near a particular point, a Fourier series can approximate a function at all points. *Program* synthesize, listed in the following, plots the sum (9.22) for various values of N, the number of terms in the series. One purpose of the program is to help us visualize how well a finite sum of harmonic terms can represent an arbitrary periodic function.

```
PROGRAM synthesize
CALL plotf(0,0.5,0.5,1)
CALL plotf(0,0.5,0,0.5)
CALL plotf(0.5,1,0.5,1)
CALL plotf(0.5,1,0,0.5)
END

SUB plotf(xmin,xmax,ymin,ymax)
    OPEN #1: screen xmin,xmax,ymin,ymax
    SET WINDOW -4,4,-2,2
    PLOT LINES: -pi,0;pi,0
    PLOT LINES: 0,-1.5;0,1.5
    INPUT prompt "number of modes = ": N
    SET COLOR "red"
    CALL fourier(N)
    CLOSE #1
END SUB
```

```
SUB fourier(N)
    ! compute Fourier series and plot function
    DIM a(0 to 1000),b(1000)
    CALL coefficients(N,a(),b())
    LET nplot = 100
    LET t = -pi
    LET dt = pi/100
    DO while t <= pi
       LET f = a(0)/2
       FOR k = 1 to N
            IF a(k) <> 0 then LET f = f + a(k)*cos(k*t)
            IF b(k) <> 0 then LET f = f + b(k)*sin(k*t)
       NEXT k
       PLOT LINES: t,f;
       LET t = t + dt
    LOOP
END SUB

SUB coefficients(N,a(),b())
    ! generate Fourier coefficients for special case
    LET a(0) = 0
    FOR k = 1 to N
        LET a(k) = 0
        IF mod(k,2) <> 0 then
            LET b(k) = 2/(k*pi)
        ELSE
            LET b(k) = 0
        END IF
    NEXT k
END SUB
```

Problem 9.4   Fourier synthesis

a.   The process of constructing a function by adding together a fundamental fre-
     quency and harmonics of various amplitudes is called *Fourier synthesis*. Use
     Program synthesize to visualize how a sum of harmonic functions can rep-
     resent an arbitrary periodic function. Consider the series

$$f(t) = \frac{2}{\pi}(\sin t + \frac{1}{3}\sin 3t + \frac{1}{5}\sin 5t + \cdots). \qquad (9.26)$$

     Describe the nature of the plot when only the first three terms in (9.26) are
     retained. Increase the number of terms until you are satisfied that (9.26) repre-
     sents the function sufficiently accurately. What function is represented by the
     infinite series?

b.   Modify *Program synthesize* so that you can "zoom in" on the visual repre-
     sentation of $f(t)$ for different intervals of t. Consider the series (9.26) with at

least 32 terms. For what values of t does the finite sum most faithfully represent the exact function? For what values of t does it not? Why is it necessary to include a large number of terms to represent $f(t)$ where it has sharp edges? The small oscillations that increase in amplitude as a sharp edge is approached are known as the Gibbs phenomenon.

c. Use *Program synthesize* to determine the function that is represented by the Fourier series with coefficients $a_k = 0$ and $b_k = (2/k\pi)(-1)^{k-1}$ for $k = 1, 2, 3, \ldots$ Approximately how many terms in the series are required?

So far we have considered how a sum of sines and cosines can approximate a known periodic function. More typically, we measure a time series consisting of N data points, $f(t_i)$, where $t_i = 0, \Delta, 2\Delta, \ldots (N-1)\Delta$. We assume that the data repeats itself with a period T given by $T = N\Delta$. (The time interval Δ between the measurements should not be confused with the finite time step Δt used in the numerical solution of a differential equation.) Our goal is to determine the Fourier coefficients a_k and b_k because, as we will see, these coefficients contain important physical information.

If we know only a finite number of terms in a time series, it is possible to find only a finite set of Fourier coefficients. For a given value of Δ, what is the largest frequency component we can extract? In the following, we give a plausibility argument that suggests that the maximum frequency we can analyze is

$$\omega_c = \frac{\pi}{\Delta}. \qquad \text{(Nyquist critical frequency)} \qquad (9.27)$$

One way to understand this result is to imagine that $f(t)$ is a sine wave. If $f(t_i)$ has the same value for all t_i, the period is equal to either Δ or Δ/n, where n is an integer. The largest frequency component we can determine in this case is $\omega = 2\pi n/\Delta$, an arbitrarily large quantity. Hence, a constant data set does not impose any limitations on the maximum frequency. Now suppose that $f(t_i)$ has one value for even i and another value for odd i. In this case we know that the period is 2Δ, and hence the maximum possible frequency of this function is $\omega = 2\pi/(2\Delta) = \pi/\Delta$. More variations in $f(t_i)$ would correspond to lower frequencies, and hence we conclude that the highest frequency is π/Δ.

One consequence of (9.27) is that there are $\omega_c/\omega_0 + 1$ independent coefficients for a_k (including a_0), and ω_c/ω_0 independent coefficients for b_k, a total of $N+1$ independent coefficients. (Recall that $\omega_c/\omega_0 = N/2$, where $\omega_0 = 2\pi/T$ and $T = N\Delta$.) However, because $\sin \omega_c t = 0$ for all values of t that are multiples of Δ, we have that $b_{N/2} = 0$ from (9.24b). Consequently, there are $N/2 - 1$ values for b_k, and hence a total of N Fourier coefficients that can be computed. This conclusion is reasonable because the number of meaningful Fourier coefficients should be the same as the number of data points.

Program analyze computes the Fourier coefficients a_k and b_k of a function $f(t)$ defined between $t = 0$ and $t = T$ at intervals of Δ, and plots a_k and b_k versus k. To compute the coefficients we do the integrals in (9.24) numerically using the simple rectangular approximation (see Section 11.1):

$$a_k \approx \frac{2\Delta}{T} \sum_{i=0}^{N-1} f(t_i) \cos \omega_k t_i \qquad (9.28a)$$

$$b_k \approx \frac{2\Delta}{T} \sum_{i=0}^{N-1} f(t_i) \sin \omega_k t_i, \qquad (9.28b)$$

where the ratio $2\Delta/T = 2/N$.

```
PROGRAM analyze
! determine the Fourier coefficients a_k and b_k
CALL parameters(N,nmax,delta,period)
CALL screen(nmax,period,#1,#2)
CALL coefficents(N,nmax,delta,period,#1,#2)
END

SUB parameters(N,nmax,delta,period)
    INPUT prompt "number of data points N (even) = ": N
    INPUT prompt "sampling time dt = ": delta
    LET period = N*delta          ! assumed period
    ! maximum value of mode corresponding to Nyquist frequency
    LET nmax = N/2
END SUB

SUB screen(nmax,period,#1,#2)
    LET ymax = 2
    LET ticksize = ymax/50
    OPEN #1: screen 0,1,0.5,1
    PRINT "      a_k";
    PRINT "        ";
    PRINT using "frequency interval = #.#####": 2*pi/period
    SET WINDOW -1,nmax+1,-ymax,ymax
    CALL plotaxis(nmax,ticksize)
    SET COLOR "red"
    OPEN #2: screen 0,1,0,0.5
    PRINT "      b_k"
    SET WINDOW -1,nmax+1,-ymax,ymax
    CALL plotaxis(nmax,ticksize)
    SET COLOR "red"
END SUB

SUB plotaxis(nmax,ticksize)
    PLOT LINES: 0,0;nmax,0
    FOR k = 1 to nmax
        PLOT LINES: k,-ticksize;k,ticksize
    NEXT k
END SUB
```

```
SUB coefficents(N,nmax,delta,period,#1,#2)
    DECLARE DEF f
    FOR k = 0 to nmax
        LET ak = 0
        LET bk = 0
        LET wk = 2*pi*k/period
        ! rectangular approximation
        FOR i = 0 to N - 1
            LET t = i*delta
            LET ak = ak + f(t)*cos(wk*t)
            LET bk = bk + f(t)*sin(wk*t)
        NEXT i
        LET ak = 2*ak/N
        LET bk = 2*bk/N
        WINDOW #1
        PLOT LINES: k,0;k,ak
        WINDOW #2
        PLOT LINES: k,0;k,bk
    NEXT k
END SUB

FUNCTION f(t)
    LET w0 = 0.1*pi
    LET f = sin(w0*t)                    ! simple example
END DEF
```

In Problem 9.5 we compute the Fourier coefficients for several known functions. We will see that if $f(t)$ is a sum of sinusoidal functions with different periods, it is essential that the period T in *Program analyze* be an integer multiple of the periods of all the functions in the sum. If T does not satisfy this condition, then the results for some of the Fourier coefficients will be spurious. In practice, the solution to this problem is to vary the sampling rate and the total time over which the signal $f(t)$ is sampled. Fortunately, the results for the power spectrum (see below) are less ambiguous than the values for the Fourier coefficients themselves.

Problem 9.5   Fourier analysis

a. Use *Program analyze* with $f(t) = \sin \pi t/10$. Determine the Fourier coefficients by doing the integrals in (9.24) analytically before running the program. Choose the number of data points to be $N = 200$ and the sampling time $\Delta = 0.1$. Which Fourier components are nonzero? Repeat your analysis for $N = 400$, $\Delta = 0.1$; $N = 200$, $\Delta = 0.05$; $N = 205$, $\Delta = 0.1$; and $N = 500$, $\Delta = 0.1$, and other combinations of N and Δ. Explain your results by comparing the period of $f(t)$ with $N\Delta$, the assumed period. If the combination of N and Δ are not chosen properly, do you find any spurious results for the coefficients?

b. Consider the functions $f_1(t) = \sin \pi t / 10 + \sin \pi t / 5$, $f_2(t) = \sin \pi t / 10 + \cos \pi t / 5$, and $f_3(t) = \sin \pi t / 10 + \frac{1}{2} \cos \pi t / 5$, and answer the same questions as in part (a). What combinations of N and Δ give reasonable results for each function?

c. Consider a function that is not periodic, but falls to zero for large $\pm t$. For example, try $f(t) = t^4 e^{-t^2}$ and $f(t) = t^3 e^{-t^2}$. Interpret the difference between the Fourier coefficients of these two functions.

Fourier analysis can be simplified by using exponential notation and combining the sine and cosine functions in one expression. We express $f(t)$ as

$$f(t) = \sum_{k=-\infty}^{\infty} c_k e^{i\omega_k t}, \tag{9.29}$$

and use (9.22) to express the complex coefficients c_k in terms of a_k and b_k:

$$c_k = \frac{1}{2}(a_k - ib_k) \tag{9.30a}$$

$$c_0 = \frac{1}{2}a_0 \tag{9.30b}$$

$$c_{-k} = \frac{1}{2}(a_k + ib_k). \tag{9.30c}$$

The coefficients c_k can be expressed in terms of $f(t)$ by using (9.30) and (9.24) and the fact that $e^{\pm i\omega_k t} = \cos \omega_k t \pm i \sin \omega_k t$. The result is

$$c_k = \frac{1}{T} \int_0^T f(t) e^{-i\omega_k t} \, dt. \tag{9.31}$$

As in (9.28), we can approximate the integral in (9.31) using the rectangular approximation. We write

$$g(\omega_k) \equiv c_k \frac{T}{\Delta} \approx \sum_{j=0}^{N-1} f(j\Delta) e^{-i\omega_k j\Delta} = \sum_{j=0}^{N-1} f(j\Delta) e^{-i2\pi kj/N}. \tag{9.32}$$

If we multiply (9.32) by $e^{i2\pi kj'/N}$, sum over k, and use the orthogonality condition

$$\sum_{k=0}^{N-1} e^{i2\pi kj/N} e^{-i2\pi kj'/N} = N\delta_{j,j'}, \tag{9.33}$$

we obtain the inverse Fourier transform

$$f(j\Delta) = \frac{1}{N} \sum_{k=0}^{N-1} g(\omega_k) e^{i2\pi kj/N} = \frac{1}{N} \sum_{k=0}^{N-1} g(\omega_k) e^{i\omega_k t_j}. \tag{9.34}$$

The frequencies ω_k for $k > N/2$ in the summations in (9.34) are greater than the Nyquist frequency ω_c. However, from (9.32), we see that $g(\omega_k) = g(\omega_k - \omega_N)$. Hence, we can interpret all frequencies for $k > N/2$ as negative frequencies equal to $(\frac{1}{2}N - k)\omega_0$. If $f(t)$ is real, then $g(-\omega_k) = g(\omega_k)$. The occurrence of negative frequency components is a consequence of the use of the exponential functions rather than a sum of sines and cosines.

The importance of a particular frequency component within a signal is measured by the power $P(\omega)$ associated with that frequency. To obtain this power, we use the discrete form of Parseval's theorem which can be written as

$$\sum_{j=0}^{N-1} |f(t_j)|^2 = \frac{1}{N} \sum_{k=0}^{N-1} |g(\omega_k)|^2. \tag{9.35}$$

In most measurements the function $f(t)$ corresponds to an amplitude, and the power or intensity is proportional to the square of this amplitude or for complex functions, the modulus squared. Note that the left-hand sum in (9.35) (and hence the right-hand side) is proportional to N, and hence we need to divide both sides by N to obtain a quantity independent of N. The power in the frequency component ω_k is proportional to

$$P(\omega_k) = \frac{1}{N^2}\big[|g(\omega_k)|^2 + |g(-\omega_k)|^2\big] = \frac{2}{N^2}|g(\omega_k)|^2. \qquad (0 < \omega_k < \omega_c) \tag{9.36a}$$

The last equality follows if $f(t)$ is real. Because the Fourier coefficients for $\omega = \omega_c$ and $\omega = -\omega_c$ are identical, we write for this case:

$$P(\omega_c) = \frac{1}{N^2}|g(\omega_c)|^2. \tag{9.36b}$$

Similarly, there is only one term with zero frequency, and hence $P(0)$ is given by

$$P(0) = \frac{1}{N^2}|g(0)|^2. \tag{9.36c}$$

The *power spectrum* $P(\omega)$ defined in (9.36) is proportional to the power associated with a particular frequency component embedded in the quantity of interest.

What happens to the power associated with frequencies greater than the Nyquist frequency? To answer this question, consider two choices of the Nyquist frequency, ω_c^a and $\omega_c^b > \omega_c^a$, and the corresponding sampling times, $\Delta^b < \Delta^a$. The calculation with $\Delta = \Delta^b$ represents the more accurate calculation because the sampling time is smaller. Suppose that this calculation of the spectrum yields the result that $P(\omega > \omega_c^a) > 0$. What happens if we compute the power spectrum using $\Delta = \Delta^a$? The power associated with $\omega > \omega_c^a$ must be "folded" back into the $\omega < \omega_c^a$ frequency components. For example, the frequency component at $\omega + \omega_c^a$ is added to the true value at $\omega - \omega_c^a$ to produce an incorrect value at $\omega - \omega_c^a$ in the computed power spectrum. This phenomenon is called *aliasing* and leads to spurious results. Aliasing occurs in calculations of $P(\omega)$ if the latter does not vanish above the Nyquist frequency. To avoid aliasing, it is necessary to sample more frequently, or to remove the high frequency components from the signal before computing the Fourier transform.

The power spectrum can be computed by a simple modification of *Program ana-lyze*. The procedure is order N^2, because there are N integrals for the N Fourier components, each of which is divided into N intervals. However, many of the calculations are redundant, and it is possible to organize the calculation so that the computational time is order $N \log N$. Such an algorithm is called a *fast Fourier transform* (FFT) and is discussed in Appendix 9A. It is a good idea to use the FFT for many of the following problems.

Problem 9.6   Examples of power spectra

a.   Create a data set corresponding to $f(t) = 0.3 \cos(2\pi t / T) + r$, where r is a random number between 0 and 1. Plot $f(t)$ versus t in intervals of $\Delta = 4T/N$ for $N = 128$ values. Can you visually detect any periodicity? Then compute the power spectrum using the same sampling interval $\Delta = 4T/N$. Does the behavior of the power spectrum indicate that there are any special frequencies?

b.   Simulate a one-dimensional random walk, and compute $x^2(t)$, where $x(t)$ is the distance from the origin of the walk after t steps. Compute the power spectrum for a walk of $t = 256$. In this case $\Delta = 1$, the time between steps. Do you observe any special frequencies? Remember to average over several samples.

c.   Let f_n be the nth number of a random number sequence so that the time $t = n$ with $\Delta = 1$. Compute the power spectrum of the random number generator. Do you detect any periodicities? If so, is the random number generator acceptable?

Problem 9.7   Power spectrum of coupled oscillators

a.   Modify *Program oscillators* so that the power spectrum of one of the L particles is computed at the end of the simulation. Set $\Delta = 0.1$ so that the Nyquist frequency is $\omega_c = \pi / \Delta \approx 31.4$. Choose the time of the simulation equal to $T = 25.6$ and let $K/M = 1$. Plot the power spectrum $P(\omega)$ at frequency intervals equal to $\Delta \omega = \omega_0 = 2\pi / T$. First choose $L = 2$ and choose the initial conditions so that the system is in a normal mode. What do you expect the power spectrum to look like? What do you find? Then choose $L = 10$ and choose initial conditions corresponding to various normal modes.

b.   Repeat part (a) for $L = 2$ and $L = 10$ with the initial particle displacements equal to random values between -0.5 and 0.5. Can you detect all the normal modes in the power spectrum? Repeat for a different set of random initial displacements.

c.   Repeat part (a) for initial displacements corresponding to the sum of two normal modes.

d.   Recompute the power spectrum for $L = 10$ with $T = 6.4$. Is this time long enough? How can you tell?

*Problem 9.8   Quasiperiodic power spectra

a.   Write a program to compute the power spectrum of the circle map (6.54). Begin by exploring the power spectrum for $K = 0$. Plot $\ln P(\omega)$ versus ω, where $P(\omega)$ is proportional to the modulus squared of the Fourier transform of x_n. Begin with $N = 256$ iterations. How does the power spectra differ for rational and irrational values of the parameter Ω? How are the locations of the peaks in the power spectra related to the value of Ω?

b.   Set $K = 1/2$ and compute the power spectra for $0 < \Omega < 1$. Does the power spectra differ from the spectra found in part (a)?

c.   Set $K = 1$ and compute the power spectra for $0 < \Omega < 1$. How does the power spectra compare to those found in parts (a) and (b)?

In Problem 9.7 we found that the peaks in the power spectrum yield information about the normal mode frequencies. In Problem 9.9 and 9.10 we compute the power spectra for a system of coupled oscillators where disorder is present. Disorder can be generated by having random masses and/or random spring constants. We will see that one effect of disorder is that the normal modes are no longer simple sinusoidal functions. Instead, some of the modes are localized, meaning that only some of the particles move significantly while the others remain essentially at rest. This effect is known as *Anderson localization*. Typically, we find that modes above a certain frequency are *localized*, and those below this threshold frequency are *extended*. This threshold frequency is well defined for large systems. In one dimension with a finite disorder (e.g., a finite density of defects) all states are localized in the limit of an infinite chain.

Problem 9.9   Localization with a single defect

a.   Modify *Program oscillators* so that the mass of one oscillator is equal to one fourth that of the others. Set $L = 20$ and use fixed boundary conditions. Compute the power spectrum over a time $T = 51.2$ using random initial displacements between -0.5 and 0.5 and zero initial velocities. Sample the data at intervals of $\Delta = 0.1$. The normal mode frequencies correspond to the well defined peaks in $P(\omega)$. Consider at least three different sets of random initial displacements to insure that you find all the normal mode frequencies.

b.   Apply an external force $F_e = 0.3 \sin \omega_e t$ to each particle. (The steady state behavior occurs sooner if we apply an external force to each particle instead of just one particle.) Because the external force pumps energy into the system, it is necessary to add a damping force to prevent the oscillator displacements from becoming too large. Add a damping force equal to $-\gamma v_i$ to all the oscillators with $\gamma = 0.1$. Choose random initial displacements and zero initial velocities and use the frequencies found in part (a) as the driving frequencies ω_e. Describe the motion of the particles. Is the system driven to a normal mode? Take a "snapshot" of the particle displacements after the system has run for a sufficiently long time so that the patterns repeat themselves. Are the particle displacements simple sinusoidal functions of position? Sketch the approximate normal mode patterns for each normal mode frequency. Which of the modes appear localized and which modes appear to be extended? What is the approximate cutoff frequency that separates the localized from the extended modes?

Problem 9.10   Localization in a disordered chain of oscillators

a. Modify *Program oscillators* so that the spring constants can be varied by the user. Set $L = 10$ and use fixed wall boundary conditions. Consider the following set of 11 spring constants:

    DATA 0.704,0.388,0.707,0.525,0.754,0.721
    DATA 0.006,0.479,0.470,0.574,0.904

To help you determine all the normal modes, we provide two of the normal mode frequencies: $\omega \approx 0.28$ and 1.15. Find the power spectrum using the procedure outlined in Problem 9.9a.

b. Apply an external force $F_e = 0.3 \sin \omega_e t$ to each particle, and find the normal modes as outlined in Problem 9.9b.

c. Repeat parts (a) and (b) for another set of random spring constants. If you have sufficient computer resources, consider $L = 40$. Discuss the nature of the localized modes in terms of the specific values of the spring constants. For example, is the edge of a localized mode at a spring that has a relatively large or small spring constant?

d. Repeat parts (a) and (b) for uniform spring constants, but random masses between 0.5 and 1.5. Is there a qualitative difference between the two types of disorder?

In 1955 Fermi, Pasta, and Ulam used the Maniac I computer at Los Alamos to study a chain of oscillators. Their surprising discovery might have been the first time a qualitatively new result, instead of a more precise number, was found from a computer simulation. To understand their results (known as the FPU problem), we need to discuss an idea from statistical mechanics that was mentioned briefly in Project 8.1. Some of the ideas of statistical mechanics are introduced in greater depth in later chapters.

A fundamental assumption of statistical mechanics is that an isolated system of particles is ergodic, that is, the system will evolve through all configurations consistent with the conservation of energy. Clearly, a set of linearly coupled oscillators is not ergodic, because if the system is initially in a normal mode, it stays in that normal mode forever. Before 1955 it was believed that if the interaction between the particles is slightly nonlinear (and the number of particles is sufficiently large), the system would be ergodic and evolve through the different normal modes of the linear system. In Problem 9.11 we will find, as did Fermi, Pasta, and Ulam, that the behavior of the system is much more complicated.

Problem 9.11   Nonlinear oscillators

a. Modify *Program oscillators* so that cubic forces between the particles are added to the linear spring forces, i.e., let the force on particle i due to particle j be

$$F_{ij} = -(u_i - u_j) - \alpha(u_i - u_j)^3, \tag{9.37}$$

where α is the amplitude of the nonlinear term. Choose the masses of the particles to be unity. Consider $L = 10$ and choose initial displacements corresponding to a normal mode of the linear ($\alpha = 0$) system. Compute the power spectrum over a time interval of 51.2 with $\Delta = 0.1$ for $\alpha = 0, 0.1, 0.2$, and 0.3. For what value of α does the system become ergodic, i.e., the heights of all the normal mode peaks are approximately the same?

b. Repeat part (a) for the case where the displacements of the particles are initially random. Make sure the same set of random displacements are used for each value of α.

* c. We now know that the number of oscillators is not as important as the magnitude of the nonlinear interaction. Repeat parts (a) and (b) for $L = 20$ and 40 and discuss the effect of increasing the number of particles.

* Problem 9.12   Spatial Fourier transforms

a. So far we have considered Fourier transforms in time and frequency. Spatial Fourier transforms are of interest in many contexts. The main difference is that spatial transforms usually involve positive and negative values of x, whereas we have considered only nonnegative values of t. Modify *Program analyze* so that it computes the real and imaginary parts of the Fourier transform $\phi(k)$ of a complex function $\psi(x)$, where both x and k can have negative values. That is, instead of doing the integral (9.31) from 0 to T, integrate from $-L/2$ to $L/2$, where $\psi(x + L) = \psi(x)$.

b. Compute the Fourier transform of the Gaussian function $\psi(x) = Ae^{-bx^2}$. Plot $\psi(x)$ and $\phi(k)$ for at least three values of b. Does $\phi(k)$ appear to be a Gaussian? Choose a reasonable criterion for the half-width of $\psi(x)$ and measure its value. Use the same criterion to measure the half-width of $\phi(k)$. How do these widths depend on b? How does the width of $\phi(k)$ change as the width of $\psi(x)$ increases?

c. Repeat part (b) with the function $\psi(x) = Ae^{-bx^2}e^{ik_0x}$ for various values of k_0.

9.3   WAVE MOTION

Our simulations of coupled oscillators have shown that the microscopic motion of the individual oscillators leads to macroscopic wave phenomena. To understand the transition between microscopic and macroscopic phenomena, we reconsider the oscillations of a linear chain of L particles with equal spring constants K and equal masses M. As we found in Section 9.1, the equations of motion of the particles can be written as (see (9.1))

$$\frac{d^2 u_j(t)}{dt^2} = -\frac{K}{M}\big[2u_j(t) - u_{j+1}(t) - u_{j-1}(t)\big]. \qquad (i = 1, \cdots L). \qquad (9.38)$$

We consider the limits $L \to \infty$ and $a \to 0$ with the length of the chain La fixed. We will find that the discrete equations of motion (9.38) can be replaced by the continuous *wave equation*

$$\frac{\partial^2 u(x, t)}{\partial t^2} = c^2 \frac{\partial^2 u(x, t)}{\partial x^2}, \qquad (9.39)$$

where c has the dimension of velocity.

We can obtain the wave equation (9.39) as follows. First we replace $u_j(t)$, where j is a discrete variable, by the function $u(x, t)$, where x is a *continuous* variable, and rewrite (9.38) in the form

$$\frac{\partial^2 u(x, t)}{\partial t^2} = \frac{Ka^2}{M}\frac{1}{a^2}\big[u(x + a, t) - 2u(x, t) + u(x - a, t)\big]. \qquad (9.40)$$

We have written the time derivative as a partial derivative because the function u depends on two variables. If we use the Taylor series expansion

$$u(x \pm a) = u(x) \pm a\frac{\partial u}{\partial x} + \frac{a^2}{2}\frac{\partial^2 u}{\partial x^2} + \dots, \qquad (9.41)$$

it is easy to show that as $a \to 0$, the quantity

$$\frac{1}{a^2}\big[u(x + a, t) - 2u(x, t) + u(x - a, t)\big] \to \frac{\partial^2 u(x, t)}{\partial x^2}. \qquad (9.42)$$

The wave equation (9.39) is obtained by substituting (9.42) into (9.40) with $c^2 = Ka^2/M$. If we introduce the linear mass density $\mu = M/a$ and the tension $T = Ka$, we can express c in terms of μ and T and obtain the familiar result $c^2 = T/\mu$.

It is straightforward to show that any function of the form $f(x \pm ct)$ is a solution to (9.39). Among these many solutions to the wave equation are the familiar forms:

$$u(x, t) = A\cos\frac{2\pi}{\lambda}(x \pm ct) \qquad (9.43a)$$

$$u(x, t) = A\sin\frac{2\pi}{\lambda}(x \pm ct). \qquad (9.43b)$$

Because the wave equation is linear, and hence satisfies a superposition principle, we can understand the behavior of a wave of arbitrary shape by representing its shape as a sum of sinusoidal waves.

One way to solve the wave equation numerically is to retrace our steps back to the discrete equations (9.38) to find a discrete form of the wave equation that is convenient for numerical calculations. This procedure of converting a continuum equation to a physically motivated discrete form frequently leads to useful numerical algorithms. From (9.42) we see how to approximate the second derivative by a finite difference. If we replace a by Δx and take Δt to be the time step, we can rewrite (9.38) by

$$\frac{1}{(\Delta t)^2}\big[u(x, t + \Delta t) - 2u(x, t) + u(x, t - \Delta t)\big] =$$

$$\frac{c^2}{(\Delta x)^2}\big[u(x + \Delta x, t) - 2u(x, t) + u(x - \Delta x, t)\big]. \tag{9.44}$$

The quantity Δx is the spatial interval. The result of solving (9.44) for $u(x, t + \Delta t)$ is

$$u(x, t + \Delta t) = 2\big[1 - b\big]u(x, t)$$
$$+ b\big[u(x + \Delta x, t) + u(x + \Delta x, t)\big] - u(x, t - \Delta t), \tag{9.45}$$

where $b \equiv (c\Delta t/\Delta x)^2$. Equation (9.45) expresses the displacements at time $t + \Delta t$ in terms of the displacements at the current time t and at the previous time $t - \Delta t$.

Problem 9.13   Solution of the discrete wave equation

a.   Write a program to compute the numerical solutions of the discrete wave equation (9.45). Three spatial arrays corresponding to $u(x)$ at times $t + \Delta t$, t, and $t - \Delta t$ are needed, where Δt is the time step. We denote the displacement $u(j\Delta x)$ by the array element u(j), where Δx is the size of the spatial grid. Use periodic boundary conditions so that u(0) = u(L) and u(L + 1) = u(1), where L is the total number of spatial intervals. Draw lines between the displacements at neighboring values of x. Note that the initial conditions require the specification of the array u at $t = 0$ and at $t = -\Delta t$. Let the waveform at $t = 0$ and $t = -\Delta t$ be $u(x, t = 0) = \exp(-(x - 10)^2)$ and $u(x, t = -\Delta t) = \exp(-(x - 10 + c\Delta t)^2)$, respectively. What is the direction of motion implied by these initial conditions?

b.   Our first task is to determine the optimum value of the parameter b. Let $\Delta x = 1$ and $L \geq 100$, and try the following combinations of c and Δt: $c = 1$, $\Delta t = 0.1$; $c = 1$, $\Delta t = 0.5$; $c = 1$, $\Delta t = 1$; $c = 1$, $\Delta t = 1.5$; $c = 2$, $\Delta t = 0.5$; and $c = 2$, $\Delta t = 1$. Verify that the value $b = (c\Delta t)^2 = 1$ leads to the best results, that is, for this value of b, the initial form of the wave is preserved.

c.   It is possible to show that the discrete form of the wave equation with $b = 1$ is exact up to numerical roundoff error (cf. DeVries). Hence, we can replace (9.45) by the simpler algorithm

$$u(x, t + \Delta t) = u(x + \Delta x, t) + u(x - \Delta x, t) - u(x, t - \Delta t). \tag{9.46}$$

That is, the solutions of (9.46) are equivalent to the solutions of the original partial differential equation (9.39). Try several different initial waveforms, and show that if the displacements have the form $f(x \pm ct)$, then the waveform maintains its shape with time. For the remaining problems we use (9.46) corresponding to $b = 1$. Unless otherwise specified, choose $c = 1$, $\Delta x = \Delta t = 1$, and $L \geq 100$ in the following problems.

Problem 9.14   Velocity of waves

a.   Use the waveform given in Problem 9.13a and measure the speed of the wave by determining the distance traveled on the screen in a given amount of time. Add tick marks to the x axis. Because we have set $\Delta x = \Delta t = 1$ and $b = 1$, the speed $c = 1$. (A way of incorporating different values of c is discussed in Problem 9.15d.)

b.   Replace the waveform considered in part (a) by a sinusoidal wave that fits exactly, i.e., choose $u(x, t) = \sin(qx - \omega t)$ such that $\sin q(L + 1) = 0$. Measure the period T of the wave by measuring the time it takes for successive maxima to pass a given point. What is the wavelength λ of your wave? Does it depends on the value of q? The frequency of the wave is given by $f = 1/T$. Verify that $\lambda f = c$.

Problem 9.15   Reflection of waves

a.   Consider a wave of the form $u(x, t) = e^{-(x-10-ct)^2}$. Use fixed boundary conditions so that $u(0) = u(L + 1) = 0$. What happens to the reflected wave?

b.   Modify your program so that free boundary conditions are incorporated: $u(0) = u(1)$ and $u(L + 1) = u(L)$. Compare the phase of the reflected wave to your result from part (a).

c.   Modify your program so that a "sluggish" boundary condition, e.g., $u(0) = \frac{1}{2}u(1)$ and $u(L + 1) = \frac{1}{2}u(L)$, is used. What do you expect the reflected wave to look like? What do you find from your numerical solution?

d.   What happens to a pulse at the boundary between two media? Set $c = 1$ and $\Delta t = 1$ on the left side of your grid and $c = 2$ and $\Delta t = 0.5$ on the right side. These choices of c and Δt imply that $b = 1$ on both sides, but that the right side is updated twice as often as the left side. What happens to a pulse that begins on the left side and moves to the right? Is there both a reflected and transmitted wave at the boundary between the two media? What is their relative phase? Find a relation between the amplitude of the incident pulse and the amplitudes of the reflected and transmitted pulses. Repeat for a pulse starting from the right side.

Problem 9.16   Superposition of waves

a.   Consider the propagation of the wave determined by $u(x, t = 0) = \sin(4\pi x/L)$. What must $u(x, -\Delta t)$ be such that the wave moves in the positive x direction? Test your answer by doing the simulation. Use periodic boundary conditions. Repeat for a wave moving in the negative x direction.

b. Simulate two waves moving in opposite directions each with the same spatial dependence given by $u(x, 0) = \sin 4\pi x/L$. Describe the resultant wave pattern. Repeat the simulation for $u(x, 0) = \sin 8\pi x/L$.

c. Assume that $u(x, 0) = \sin q_1 x + \sin q_2 x$, with $\omega_1 = cq_1 = 10\pi/L$, $\omega_2 = cq_2 = 12\pi/L$, and $c = 1$. Describe the qualitative form of $u(x, t)$ for fixed t. What is the distance between modulations of the amplitude? Estimate the wavelength associated with the fine ripples of the amplitude. Estimate the wavelength of the *envelope* of the wave. Find a simple relationship for these two wavelengths in terms of the wavelengths of the two sinusoidal terms. This phenomena is known as *beats*.

d. Consider the motion of the wave described by $u(x, 0) = e^{-(x-10)^2} + e^{-(x-90)^2}$; the two Gaussian pulses move in opposite directions. What happens to the two pulses as they travel through each other? Do they maintain their shape? While they are going through each other, is the displacement $u(x, t)$ given by the sum of the displacements of the individual pulses?

Problem 9.17   Standing waves

a. In Problem 9.16c we considered a *standing wave*, the continuum analog of a normal mode of a system of coupled oscillators. As is the case for normal modes, each point of the wave has the same time dependence. For fixed boundary conditions, the displacement is given by $u(x, t) = \sin qx \cos \omega t$, where $\omega = cq$ and q is chosen so that $\sin qL = 0$. Choose an initial condition corresponding to a standing wave for $L = 100$. Describe the motion of the particles, and compare it with your observations of standing waves on a rope.

b. Establish a standing wave by displacing one end of a system periodically. The other end is fixed. Let $u(x, 0) = u(x, -\Delta t) = 0$, and $u(x = 0, t) = A \sin \omega t$ with $A = 0.1$.

We have seen that the wave equation can support pulses that propagate indefinitely without distortion. In addition, because the wave equation is linear, the sum of any two solutions also is a solution, and the principle of superposition is satisfied. As a consequence, we know that two pulses can pass through each other unchanged. We also have seen that similar phenomena exist in the discrete system of linearly coupled oscillators. What happens if we create a pulse in a system of nonlinear oscillators? As an introduction to nonlinear wave phenomena, we consider a system of L coupled oscillators with the potential energy of interaction given by

$$V = \frac{1}{2} \sum_{j=1}^{L} \left(e^{-(u_j - u_{j-1})} - 1 \right)^2. \tag{9.47}$$

This form of the interaction is known as the Morse potential. All parameters in the potential (such as the overall strength of the potential) have been set to unity. The force on the jth particle is

$$F_j = -\frac{\partial V}{\partial u_j} = Q_j(1 - Q_j) - Q_{j+1}(1 - Q_{j+1}), \tag{9.48a}$$

where

$$Q_j \equiv e^{-(u_j - u_{j-1})}. \tag{9.48b}$$

In linear systems it is possible to set up a pulse of any shape and maintain the shape of the pulse indefinitely. In a nonlinear system there also exist solutions that maintain their shape, but we will find in Problem 9.18 that not all pulse shapes do so. The pulses that maintain their shape are called *solitons*.

Problem 9.18   Solitons

a.   Modify *Program oscillators* so that the force on particle j is given by (9.48). Use periodic boundary conditions. Choose $L \geq 60$ and an initial Gaussian pulse of the form $u(x, t) = 0.5\, e^{-(x-10)^2}$. You should find that the initial pulse splits into two pulses plus some noise. Describe the motion of the pulses (solitons). Do they maintain their shape, or is this shape modified as they move? Describe the motion of the particles far from the pulse. Are they stationary?

b.   Save the displacements of the particles when the peak of one of the solitons is located near the center of your screen. Is it possible to fit the shape of the soliton to a Gaussian? Continue the simulation, and after one of the solitons is relatively isolated, set u(j) = 0 for all j far from this soliton. Does the soliton maintain its shape?

c.   Repeat part (b) with a pulse given by $u(x, 0) = 0$ everywhere except for $u(20, 0) = u(21, 0) = 1$. Do the resulting solitons have the same shape as in part (b)?

d.   Begin with the same Gaussian pulse as in part (a), and run until the two solitons are well separated. Then change at random the values of u(j) for particles in the larger soliton by about 5%, and continue the simulation. Is the soliton destroyed? Increase this perturbation until the soliton is no longer discernible.

e.   Begin with a single Gaussian pulse as in part (a). The two resultant solitons will eventually "collide." Do the solitons maintain their shape after the collision? The principle of superposition implies that the displacement of the particles is given by the sum of the displacements due to each pulse. Does the principle of superposition hold for solitons?

f.   Compute the speeds, amplitudes, and width of the solitons produced from a single Gaussian pulse. Take the amplitude of a solition to be the largest value of its displacement and the half-width to correspond to the value of x at which the displacement is half its maximum value. Repeat these calculations for solitons of different amplitudes by choosing the initial amplitude of the Gaussian pulse to be 0.1, 0.3, 0.5, 0.7, and 0.9. Plot the soliton speed and width versus the corresponding soliton amplitude.

g. Change the boundary conditions to free boundary conditions and describe the behavior of the soliton as it reaches a boundary. Compare this behavior with that of a pulse in a system of linear oscillators.

h. Begin with an initial sinusoidal disturbance that would be a normal mode for a linear system. Does the sinusoidal mode maintain its shape? Compare the behavior of the nonlinear and linear systems.

9.4 INTERFERENCE AND DIFFRACTION

Interference is one of the most fundamental characteristics of all wave phenomena. The term *interference* is used when relatively few sources of waves separately derived from the same source are brought together. The term *diffraction* has a similar meaning and is commonly used if there are many sources. Because it is relatively easy to observe interference and diffraction phenomena with light, we discuss these phenomena in this context.

The classic example of interference is Young's double slit experiment (see Fig. 9.2). Imagine two narrow parallel slits separated by a distance a and illuminated by a light source that emits light of only one frequency (monochromatic light). If the light source is placed on the line bisecting the two slits and the slit opening is very narrow, the two slits become coherent light sources with equal phases. We first assume that the slits act as point sources, e.g., pinholes. A screen that displays the intensity of the light from the two sources is placed a distance L away. What do we see on the screen?

The electric field at position $\mathbf{r}$ associated with the light emitted from a monochromatic point source at $\mathbf{r}_1$ has the form

$$E(\mathbf{r}, t) = \frac{A}{|\mathbf{r} - \mathbf{r}_1|} \cos(q|\mathbf{r} - \mathbf{r}_1| - \omega t), \tag{9.49}$$

where $|\mathbf{r} - \mathbf{r}_1|$ is the distance between the source and the point of observation. The superposition principle implies that the total electric field at $\mathbf{r}$ from N point sources at $\mathbf{r}_i$ is

$$E(\mathbf{r}, t) = \sum_{i=1}^{N} \frac{A}{|\mathbf{r} - \mathbf{r}_i|} \cos(q|\mathbf{r} - \mathbf{r}_i| - \omega t). \tag{9.50}$$

Equation 9.50 assumes that the amplitude of each source is the same. The observed intensity is proportional to the time-averaged value of $|E|^2$.

In Problem 9.19 we discuss writing a program to determine the intensity of light that is observed on a screen due to an arrangement of point sources. The wavelength of the light sources, the positions of the sources $\mathbf{r}_i$, and the observation points on

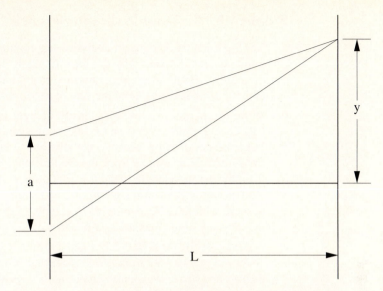

Fig. 9.2 Young's double slit experiment. The figure defines the quantities a, L, and y that are used in Problem 9.19.

the screen need to be specified. The program sums the fields due to all the sources for a given observation point, and computes $|E|^2$. The part of the program that is not straightforward is the calculation of the time average of $|E|^2$. One way of obtaining the time average is to compute the integral

$$\overline{E^2} = \frac{1}{T} \int_0^T |E|^2 \, dt, \tag{9.51}$$

where $T = 1/f$ is the period, and f is the frequency of the light sources. We now show that such a calculation is not necessary if the sources are much closer to each other than they are to the screen. In this case (the *far field condition*), we can ignore the slow r dependence of $|\mathbf{r} - \mathbf{r}_i|^{-1}$ and write the field in the form

$$E(\mathbf{r}) = E_0 \cos(\phi - \omega t). \tag{9.52}$$

We have absorbed the factors of $|\mathbf{r} - \mathbf{r}_i|^{-1}$ into E_0. The phase ϕ is a function of the source positions and $\mathbf{r}$. The form of (9.52) allows us to write

$$\frac{1}{T} \int_0^T \cos^2(\phi - \omega t) \, dt = \frac{1}{M} \sum_{m=1}^M \cos^2(\phi - \frac{2\pi m}{M}) = \frac{1}{2}. \qquad (M > 2) \tag{9.53}$$

The result (9.53) is independent of ϕ and allows us to perform the time average by using the summation in (9.53) with $M = 3$.

In the following problems we discuss a variety of geometries. The main part of your program that must be changed is the specification of the source positions.

Problem 9.19   Double slit interference

a.   Verify (9.52) by finding an analytical expression for E from two and three
point sources. Explain why the form (9.52) is valid for an arbitrary number
of sources. Then write a program to verify the result (9.53) for $M = 3, 4, 5$,
and 10. What is the value of the sum in (9.53) for $M = 2$?

b.   Write a program to plot the intensity of light on a screen due to two slits.
Calculate E using (9.50) and do the time average using the relation (9.53).
Let a be the distance between the slits and y be the vertical position along
the screen as measured from the central maximum. Set $L = 200$ mm, $a =
0.1$ mm, the wavelength of light $\lambda = 5000$ Å (1 Å $= 10^{-7}$ mm), and consider
-5.0 mm $\leq y \leq 5.0$ mm (see Fig. 9.2). Describe the interference pattern you
observe with $M = 3$. Identify the locations of the intensity maxima, and plot
the intensity of the maxima as a function of y.

c.   Repeat part (b) for $L = 0.5$ mm and 1.0 mm $\leq y \leq 1.0$ mm. Do your results
change as you vary M from 3 to 5? Note that in this case L is not much greater
than a, and hence we cannot ignore the r dependence of $|\mathbf{r} - \mathbf{r}_i|^{-1}$ in (9.4).
How large must M be so that your results are approximately independent of
M?

Problem 9.20   Diffraction grating

High resolution optical spectroscopy is done with multiple slits. In its simplest
form, a diffraction grating consists of N parallel slits made, for example, by ruling
with a diamond stylus on aluminum plated glass. Compute the intensity of light
for $N = 3, 4, 5$, and 10 slits with $\lambda = 5000$ Å, slit separation $a = 0.01$ mm, screen
distance $L = 200$ mm, and -15 mm $\leq y \leq 15$ mm. How does the intensity of the
peaks and their separation vary with N?

   In our analysis of the double slit and the diffraction grating, we assumed that each
slit was a pinhole that emits spherical waves. In practice, real slits are much wider
than the wavelength of visible light. In Problem 9.21 we consider the pattern of light
produced when a plane wave is incident on an aperture such as a single slit. To do so,
we use Huygens' principle and replace the slit by many coherent sources of spherical
waves. This equivalence is not exact, but is applicable when the aperture width is large
compared to the wavelength.

Problem 9.21   Single slit diffraction

a.   Compute the time averaged intensity of light diffracted from a single slit of
width 0.02 mm by replacing the slit by $N = 20$ point sources spaced 0.001 mm
apart. Choose $\lambda = 5000$ Å, $L = 200$ mm, and consider -30 mm $\leq y \leq 30$ mm.
What is the width of the central peak? How does the width of the central peak
compare to the width of the slit? Do your results change if N is increased?

b. Determine the position of the first minimum of the diffraction pattern as a function of wavelength, slit width, and distance to the screen.

c. Compute the intensity pattern for $L = 1$ mm and 50 mm. Is the far field condition satisfied in this case? How do the patterns differ?

Problem 9.22   A more realistic double slit simulation

Reconsider the intensity distribution for double slit interference using slits of finite width. Modify your program to simulate two "thick" slits by replacing each slit by 20 point sources spaced 0.001 mm apart. The centers of the thick slits are $a = 0.1$ mm apart. How does the intensity pattern change?

*Problem 9.23   Diffraction pattern due to a rectangular aperture

We can use a similar approach to determine the diffraction pattern due to an aperture of finite width and height. The simplest approach is to divide the aperture into little squares and to consider each square as a source of spherical waves. Similarly we can divide the screen or photographic plate into small regions or cells and calculate the time averaged intensity at the center of each cell. The calculations are straightforward, but time consuming, because of the necessity of evaluating the cosine function many times. The less straightforward part of the problem is deciding how to plot the different values of the calculated intensity on the screen. One way is to plot "points" at random locations in each cell so that the number of points is proportional to the computed intensity at the center of the cell. Suggested parameters are $\lambda = 5000$ Å and $L = 200$ mm for an aperture of dimensions 1mm × 3 mm.

Appendix 9A   FAST FOURIER TRANSFORM

The fast Fourier transform (FFT) has been discovered independently by many workers in a variety of contexts, and there are a number of variations on the basic algorithm. In the following, we describe a version due to Danielson and Lanczos. The goal is to compute the Fourier transform $g(\omega_k)$ given the data set $f(j\Delta) \equiv f_j$ of (9.32). For convenience we rewrite the relation:

$$g_k \equiv g(\omega_k) = \sum_{j=0}^{N-1} f(j\Delta)\, e^{-i2\pi kj/N},$$
(9.54)

and introduce the complex number W given by

$$W = e^{-i2\pi/N}.$$
(9.55)

The following algorithm works with any complex data set if we require that N is a power of two. Real data sets can be transformed by setting the array elements corresponding to the imaginary part equal to 0.

To understand the FFT algorithm, we consider the case $N = 8$, and rewrite (9.54) as

$$
\begin{aligned}
g_k &= \sum_{j=0,2,4,8} f(j\Delta)\, e^{-i2\pi kj/N} + \sum_{j=1,3,5,7} f(j\Delta)\, e^{-i2\pi kj/N} \\
&= \sum_{j=0,1,2,3} f(2j\Delta)e^{-i2\pi k2j/N} + \sum_{j=0,1,2,3} f((2j+1)\Delta)e^{-i2\pi k(2j+1)/N} \\
&= \sum_{j=0,1,2,3} f(2j\Delta)e^{-i2\pi kj/(N/2)} + W^k \sum_{j=0,1,2,3} f((2j+1)\Delta)\, e^{-i2\pi kj/(N/2)} \\
&= g_k^{\mathrm{e}} + W^k g_k^{\mathrm{o}},
\end{aligned}
\tag{9.56}
$$

where $W^k = e^{-i2\pi k/N}$. The quantity g^{e} is the Fourier transform of length $N/2$ formed from the even components of the original $f(j\Delta)$; g^{o} is the Fourier transform of length $N/2$ formed from the odd components.

Of course, we do not have to stop here, and we can continue this decomposition if N is a power of two. That is, we can decompose g^{e} into its $N/4$ even and $N/4$ odd components, g^{ee} and g^{eo}, and decompose g^{o} into its $N/4$ even and $N/4$ odd components, g^{oe} and g^{oo}. We find

$$
g_k = g_k^{\mathrm{ee}} + W^{2k} g_k^{\mathrm{eo}} + W^k g_k^{\mathrm{oe}} + W^{3k} g_k^{\mathrm{oo}}.
\tag{9.57}
$$

One more decomposition leads to

$$
\begin{aligned}
g_k = {} & g_k^{\mathrm{eee}} + W^{4k} g_k^{\mathrm{eeo}} + W^{2k} g_k^{\mathrm{eoe}} + W^{6k} g_k^{\mathrm{eoo}} \\
& + W^k g_k^{\mathrm{oee}} + W^{5k} g_k^{\mathrm{oeo}} + W^{3k} g_k^{\mathrm{ooe}} + W^{7k} g_k^{\mathrm{ooo}}.
\end{aligned}
\tag{9.58}
$$

At this stage each of the Fourier transforms in (9.58) use only one data point. We see from (9.54) with $N = 1$, that the value of each of these Fourier transforms, $g_k^{\mathrm{eee}}, g_k^{\mathrm{eeo}}, \cdots$, is equal to the value of f at the corresponding data point. Note that for $N = 8$, we have performed $3 = \log_2 N$ decompositions. In general, we would perform $\log_2 N$ decompositions.

There are two steps to the FFT. First, we reorder the components so that they appear in the order given in (9.58). This step makes the subsequent calculations easier to organize. To see how to do the reordering, we rewrite (9.58) using the values of f:

$$
\begin{aligned}
g_k = {} & f(0) + W^{4k} f(4\Delta) + W^{2k} f(2\Delta) + W^{6k} f(6\Delta) \\
& + W^k f(\Delta) + W^{5k} f(5\Delta) + W^{3k} f(3\Delta) + W^{7k} f(7\Delta).
\end{aligned}
\tag{9.59}
$$

We use a trick to obtain the ordering in (9.59) from the original order $f(0\Delta)$, $f(1\Delta)$, $\cdots$, $f(7\Delta)$. Part of the trick is to refer to each g in (9.58) by a string of 'e' and 'o' characters. We assign 0 to 'e' and 1 to 'o' so that each string represents the binary representation of a number. If we reverse the order of the representation, i.e., set 110 to 011, we obtain the value of f we want. For example, the fifth term in (9.58) contains g^{oee} corresponding to the binary number 100. The reverse of this number is 001 which equals 1 in decimal notation, and hence the fifth term in (9.59) contains the function

$f(1\Delta)$. Convince yourself that this bit reversal procedure works for the other seven terms.

The first step in the FFT algorithm is to use this bit reversal procedure on the original array representing the data. In the next step this array is replaced by its Fourier transform. If you want to save your original data, it is necessary to first copy the data to another array before passing the array to a FFT subroutine. The Danielson-Lanczos algorithm involves three loops. The outer loop runs over $\log_2 N$ steps. For each of these steps, N calculations are performed in the two inner loops. As can be seen in SUB FFT, in each pass through the innermost loop each element of the array g is updated once by the quantity temp formed from a power of W multiplied by the current value of an appropriate element of g. The power of W used in temp is changed after each pass through the innermost loop. The power of the FFT algorithm is that we do not separately multiply each $f(j\Delta)$ by the appropriate power of W. Instead, we first take pairs of $f(j\Delta)$ and multiply them by an appropriate power of W to create new values for the array g. Then we repeat this process for pairs of the new array elements (each array element now contains four of the $f(j\Delta)$). We repeat this process until each array element contains a sum of all N values of $f(j\Delta)$ with the correct powers of W multiplying each term to form the Fourier transform.

```
SUB FFT(g(,),p)
    ! fast Fourier transform of complex input data set g
    ! transform returned in g
    DIM Wp(2),factor(2),temp(2)
    LET N = 2^p                          ! number of data points
    LET N2 = N/2
    ! rearrange input data according to bit reversal
    LET j = 1
    FOR i = 1 to N-1
        ! g(i,1) is real part of f((i-1)*del_t)
        ! g(i,2) is imaginary part of f((i-1)*del_t)
        ! set g(i,2) = 0 if data real
        IF i < j then                ! swap values
            LET temp(1) = g(j,1)
            LET temp(2) = g(j,2)
            LET g(j,1) = g(i,1)
            LET g(j,2) = g(i,2)
            LET g(i,1) = temp(1)
            LET g(i,2) = temp(2)
        END IF
        LET k = N2
        DO while k < j
            LET j = j-k
            LET k = k/2
        LOOP
        LET j = j + k
    NEXT i
```

```
! begin Danielson-Lanczos algorithm
LET jmax = 1
FOR L = 1 to p
    LET del_i = 2*jmax
    LET Wp(1) = 1                    !  Wp initialized at W^0
    LET Wp(2) = 0
    LET angle = pi/jmax
    LET factor(1) = cos(angle)       ! ratio of new to old W^p
    LET factor(2) = -sin(angle)
    FOR j = 1 to jmax
        FOR i = j to N step del_i
            ! calculate transforms of length 2^L
            LET ip = i + jmax
            LET temp(1) = g(ip,1)*Wp(1) - g(ip,2)*Wp(2)
            LET temp(2) = g(ip,1)*Wp(2) + g(ip,2)*Wp(1)
            LET g(ip,1) = g(i,1) - temp(1)
            LET g(ip,2) = g(i,2) - temp(2)
            LET g(i,1) = g(i,1) + temp(1)
            LET g(i,2) = g(i,2) + temp(2)
        NEXT i
        ! find new W^p
        LET temp(1) = Wp(1)*factor(1) - Wp(2)*factor(2)
        LET temp(2) = Wp(1)*factor(2) + Wp(2)*factor(1)
        MAT Wp = temp
    NEXT j
    LET jmax = del_i
NEXT L
END SUB
```

Exercise 9.1   Testing the FFT algorithm

a.   Test sub FFT for $N = 8$ by going through the code by hand and showing that the subroutine reproduces (9.59).

b.   Write a subroutine to do the Fourier transform in the conventional manner based on (9.54). Make sure that your subroutine has the same arguments as SUB FFT, that is, write a subroutine to convert $f(j\Delta)$ to the two-dimensional array g. If the data is real, let $g(i, 2) = 0$. The subroutine should consist of two nested loops, one over k and one over j. Print the Fourier transform of random real values of $f(j\Delta)$ for $N = 8$, using both SUB FFT and the direct computation of the Fourier transform. Compare the two sets of data to insure that there are no errors in SUB FFT. Repeat for a random collection of complex data points.

c.   Compute the CPU time as a function of N for $N = 16, 64, 256$, and 1024 for the FFT algorithm and the direct computation. You can use the time function in True BASIC before and after each call to the subroutines. Verify that the dependence on N is what you expect.

d. Modify SUB FFT to compute the inverse Fourier transform defined by (9.34). The inverse Fourier transform of a Fourier transformed data set should be the original data set.

References and Suggestions for Further Reading

E. Oran Brigham, *The Fast Fourier Transform*, Prentice Hall (1988). A classic text on Fourier transform methods.

David C. Champeney, *Fourier Transforms and Their Physical Applications*, Academic Press (1973).

James B. Cole, Rudolph A. Krutar, Susan K. Numrich, and Dennis B. Creamer, "Finite-difference time-domain simulations of wave propagation and scattering as a research and educational tool," *Computers in Physics* **9**, 235 (1995).

Frank S. Crawford, *Waves*, Berkeley Physics Course, Vol. 3, McGraw-Hill (1968). A delightful book on waves of all types. The home experiments are highly recommended. One observation of wave phenomena equals many computer demonstrations.

Paul DeVries, *A First Course in Computational Physics*, John Wiley & Sons (1994). Part of our discussion of the wave equation is based on Chapter 7. There also are good sections on the numerical solution of other partial differential equations, Fourier transforms, and the FFT.

N. A. Dodd, "Computer simulation of diffraction patterns," *Phys. Educ.* **18**, 294 (1983).

P. G. Drazin and R. S. Johnson, *Solitons: an Introduction*, Cambridge (1989). This book focuses on analytical solutions to the Korteweg-de Vries equation which has soliton solutions.

Richard P. Feynman, Robert B. Leighton, and Matthew Sands, *The Feynman Lectures on Physics*, Vol. 1, Addison-Wesley (1963). Chapters relevant to wave phenomena include Chapters 28–30 and Chapter 33.

A. P. French, *Vibrations and Waves*, W. W. Norton & Co. (1971). An introductory level text that emphasizes mechanical systems.

Eugene Hecht, *Optics*, second edition, Addison-Wesley & Sons (1987). An intermediate level optics text that emphasizes wave concepts.

Akira Hirose and Karl E. Lonngren, *Introduction to Wave Phenomena*, John Wiley & Sons (1985). An intermediate level text that treats the general properties of waves in various contexts.

Amy Kolan, Barry Cipra, and Bill Titus, "Exploring localization in nonperiodic systems," *Computers in Physics* **9**, to be published (1995). An elementary discussion of how to solve the problem of a chain of coupled oscillators with disorder using transfer matrices.

William H. Press, Saul A. Teukolsky, William T. Vetterling, and Brian P. Flannery, *Numerical Recipes*, second edition, Cambridge University Press (1992). See Chapter 12 for a discussion of the fast Fourier transform.

Timothy J. Rolfe, Stuart A. Rice, and John Dancz, "A numerical study of large amplitude motion on a chain of coupled nonlinear oscillators," *J. Chem. Phys.* **70**, 26 (1979). Problem 9.18 is based on this paper.

Garrison Sposito, *An Introduction to Classical Dynamics*, John Wiley & Sons (1976). A good discussion of the coupled harmonic oscillator problem is given in Chapter 6.

William J. Thompson, *Computing for Scientists and Engineers*, John Wiley & Sons (1992). See Chapters 9 and 10 for a discussion of Fourier transform methods.

Michael L. Williams and Humphrey J. Maris, "Numerical study of phonon localization in disordered systems," *Phys. Rev.* B **31**, 4508 (1985). The authors consider the normal modes of a two-dimensional system of coupled oscillators with random masses. The idea of using mechanical resonance to extract the normal modes is the basis of a new numerical method for finding the eigenmodes of large lattices. See Kousuke Yukubo, Tsuneyoshi Nakayama, and Humphrey J. Maris, "Analysis of a new method for finding eigenmodes of very large lattice systems," *J. Phys. Soc. Japan* **60**, 3249 (1991).

C H A P T E R

10
Electrodynamics

We compute the electric fields due to static and moving charges, describe methods for computing the electric potential in boundary value problems, and solve Maxwell's equations numerically.

10.1 STATIC CHARGES

Suppose we want to know the electric field $\mathbf{E}(\mathbf{r})$ at the point $\mathbf{r}$ due to N point charges $q_1, q_2, \ldots, q_N$ at fixed positions $r_1, r_2, \ldots, r_N$. We know that $\mathbf{E}(\mathbf{r})$ satisfies a superposition principle and is given by

$$\mathbf{E}(\mathbf{r}) = K \sum_{i}^{N} \frac{q_i}{|\mathbf{r} - \mathbf{r}_i|^3} (\mathbf{r} - \mathbf{r}_i), \qquad (10.1)$$

where $\mathbf{r}_i$ is the fixed location of the ith charge and K is a constant that depends on the choice of units. One of the difficulties associated with electrodynamics is the competing systems of units. In the SI (or rationalized MKS) system of units, the charge is measured in coulombs (C) and the constant K is given by

$$K = \frac{1}{4\pi \epsilon_0} \approx 9.0 \times 10^9 \, \text{N} \cdot \text{m}^2/\text{C}^2. \qquad \text{(SI units)} \qquad (10.2)$$

The constant ϵ_0 is known as the electrical permittivity of free space. This choice of units is not convenient for computer programs because $K \gg 1$. Another popular system of units is the Gaussian (cgs) system for which the constant K is absorbed into the unit of charge so that $K = 1$. Charge is in "electrostatic units" or esu. One feature of Gaussian units is that the electric and magnetic fields have the same units. For example, the (Lorentz) force on a particle of charge q and velocity $\mathbf{v}$ in an electric field $\mathbf{E}$ and a magnetic field $\mathbf{B}$ has the form

$$\mathbf{F} = q(\mathbf{E} + \frac{\mathbf{v}}{c} \times \mathbf{B}). \qquad \text{(Gaussian units)} \qquad (10.3)$$

These virtues of the Gaussian system of units lead us to adopt this system for this chapter even though SI units are used in introductory texts.

The usual way of visualizing the electric field is to draw *electric field lines*. The properties of these lines are as follows:

1. An electric field line is a directed line whose tangent at every position is parallel to the electric field at that position.

2. The lines are smooth and continuous except at singularities such as point charges. (It makes no sense to talk about the electric field *at* a point charge.)

3. The density of lines at any point in space is proportional to the magnitude of the field at that point. This property implies that the total number of electric field lines from a point charge is proportional to the magnitude of that charge. The value of the proportionality constant is chosen to provide the clearest pictorial representation of the field. The drawing of field lines is art plus science.

Program fieldline, which is listed in the following, draws electric field lines in two dimensions starting at positive charges if the net charge $q_{net} \geq 0$ or at negative charges if $q_{net} < 0$. The program implements the following algorithm:

1. Begin at a point (x, y) near a charge and compute the components E_x and E_y of the electric field vector $\mathbf{E}$ using (10.1).

2. Draw a small line segment of size $\Delta s = |\Delta \mathbf{s}|$ tangent to $\mathbf{E}$ at that point. If $q_{net} \geq 0$, then $\Delta \mathbf{s} > 0$, otherwise $\Delta \mathbf{s} < 0$. The components of the line segment are given by

$$\Delta x = \Delta s \frac{E_x}{|\mathbf{E}|} \text{ and } \Delta y = \Delta s \frac{E_y}{|\mathbf{E}|}. \tag{10.4}$$

The program uses a small value for Δs if the field line is close to the charges or if the field magnitude is large. To speed up the program, a large value of Δs is used when a field line moves off the screen and the field has a small magnitude.

3. Repeat the process beginning at the new point $(x + \Delta x, y + \Delta y)$. Continue until the field line approaches another charge.

4. Repeat steps (1)–(3) for equally spaced starting positions on a circle around the charge. The spacing is inversely proportional to the magnitude of the charge.

5. Repeat steps (1)–(4) for each charge of the same sign.

Program fieldline draws the correct density of lines in most cases. In some cases a field line will go far away from the charges, but eventually return to a negative charge. Because such a field line might take too much time to draw, the user can hit any key to stop drawing this field line. There are some other situations where the algorithm breaks down. For example, if a field line is directed along the line connecting two charges that are equal in magnitude and opposite in sign, and begins by going away from both charges, then the field line will never return. Note the use of the GET POINT statement which allows the user to point to a position in the current window and click on the mouse to select it.

```
PROGRAM fieldline
! draw electric field lines in two dimensions
LIBRARY "csgraphics"
DIM x(10),y(10),q(10)
CALL screen(a,pos$,neg$)
CALL charges(N,x(),y(),q(),qtotal,a,pos$,neg$)
! draw field lines
CALL draw_lines(N,x(),y(),q(),qtotal,a)
END

SUB screen(a,pos$,neg$)
    LET L = 10
    CALL compute_aspect_ratio(L,xwin,ywin)
    SET WINDOW -xwin,xwin,-ywin,ywin
    LET a = 0.2                     ! "radius" of visual image of charges
    SET COLOR "blue"
    BOX CIRCLE -a,a,-a,a
    FLOOD 0,0
    BOX KEEP -a,a,-a,a in pos$
    CLEAR
    SET COLOR "red"
    BOX CIRCLE -a,a,-a,a
```

```
            FLOOD 0,0
            BOX KEEP -a,a,-a,a in neg$
            CLEAR
        END SUB

        SUB charges(N,x(),y(),q(),qtotal,a,pos$,neg$)
            ! input charge values and location
            LET N = 0                    ! # of point charges
            SET COLOR "black"
            DO
                CALL draw_charges(N,x(),y(),q(),a,pos$,neg$)
                INPUT prompt "charge (0 to exit input mode) = ": charge
                IF charge <> 0 then
                    PRINT "place mouse at charge location and click."
                    LET N = N + 1
                    GET POINT x(N),y(N)      ! location of charge
                    LET q(N) = charge
                    LET qtotal = qtotal + charge
                END IF
                CLEAR
            LOOP until charge = 0
            ! redraw charges
            CALL draw_charges(N,x(),y(),q(),a,pos$,neg$)
        END SUB

        SUB draw_charges(N,x(),y(),q(),a,pos$,neg$)
            FOR i = 1 to N
                IF q(i) > 0 then
                    BOX SHOW pos$ at x(i)-a,y(i)-a
                ELSE
                    BOX SHOW neg$ at x(i)-a,y(i)-a
                END IF
            NEXT i
        END SUB

        SUB draw_lines(N,x(),y(),q(),qtotal,a)
            ! number of lines per unit charge leaving positive charges
            SET COLOR "black"
            LET lpc = 8                  ! lines per charge
            LET Emin = 0.01              ! if E < Emin stop drawing fieldline
            LET sign = sgn(qtotal)
            IF sign = 0 then LET sign = 1
            LET ds_small = 0.01*sign
            LET ds_big = sign
            FOR i = 1 to N               ! loop over all charges
                IF q(i)*sign > 0 then    ! start fields at positive charges
                    LET dtheta = 2*pi/(lpc*abs(q(i)))
                    FOR theta = 0 to 2*pi step dtheta
                        LET stop_plot$ = "no"
```

```
                    LET xline = x(i) + a*cos(theta)
                    LET yline = y(i) + a*sin(theta)
                    DO
                        LET Ex = 0
                        LET Ey = 0
                        FOR j = 1 to N
                            ! x-distance from point to charge j
                            LET dx = xline - x(j)
                            ! y-distance from point to charge j
                            LET dy = yline - y(j)
                            LET r = sqr(dx*dx + dy*dy)
                            IF r > 0.9*a then
                                LET E0 = q(j)/(r*r*r)
                                LET Ex = Ex + E0*dx
                                LET Ey = Ey + E0*dy
                            ELSE
                                ! field line reached another charge
                                LET stop_plot$ = "yes"
                            END IF
                        NEXT j
                        LET E = sqr(Ex*Ex + Ey*Ey)
                        IF E > Emin or r < 20 then
                            ! new position on fieldline
                            LET xline = xline + ds_small*Ex/E
                            LET yline = yline + ds_small*Ey/E
                        ELSE
                            ! new position on fieldline
                            LET xline = xline + ds_big*Ex/E
                            LET yline = yline + ds_big*Ey/E
                        END IF
                        PLOT xline,yline;
                        IF key input then
                            ! user can stop drawing field line
                            GET KEY key
                            LET stop_plot$ = "yes"
                        END IF
                    LOOP until stop_plot$ = "yes"
                    PLOT              ! turn beam off
                NEXT theta
            END IF
        NEXT i
    END SUB
```

Problem 10.1   Electric field lines from point charges

a. *Program* `fieldline` is written so that the user inputs the value of the charge
    from the keyboard and its position using the mouse. Enter some simple charge
    configurations and check that the program is working properly. Then let

q(1) = 1, q(2) = −4, and q(3) = 3, and place the three charges at the vertices of an approximate equilateral triangle. Are the units of charge and distance relevant? Remember that the number of field lines entering a negative charge might not be correct. Verify that the field lines never connect charges of the same sign. Why do field lines never cross?

b. Modify SUB charge so that values of q(i), x(i), and y(i) can be assigned without entering them from the keyboard. In this way we can assign their values more precisely. Draw the field lines for an electric dipole.

c. Draw the field lines for the electric quadrupole with q(1) = 1, x(1) = 1, y(1) = 1, q(2) = −1, x(2) = −1, y(2) = 1, q(3) = 1, x(3) = −1, y(3) = −1, and q(4) = −1, x(4) = 1, and y(4) = −1.

d. A continuous charge distribution can be approximated by a large number of closely spaced point charges. Draw the electric field lines due to a row of ten equally spaced unit charges located between −2.5 and +2.5 on the x axis. How does the electric field distribution compare to the distribution due to a single point charge?

e. Repeat part (d) with two rows of equally spaced positive charges on the lines y = 0 and y = 1, respectively. Then consider one row of positive charges and one row of negative charges.

f. Another way of representing the electric field is to divide space into a discrete grid and to draw arrows in the direction of **E** at the vertices of the grid. The magnitude of the arrow can be chosen to be proportional to the magnitude of the electric field. Another possibility is to use color or gray scale to represent the magnitude. Which representation do you think conveys more information?

Problem 10.2   Field lines due to infinite line of charge

a. *Program fieldline* plots field lines in two dimensions. Sometimes this restriction can lead to spurious results (see Freeman). Consider four identical charges placed at the corners of a square. Use *Program fieldline* to plot the field lines. What is wrong with the results? What should happen to the field lines near the center of the square?

b. The two-dimensional analog of a point charge is an infinite line of charge perpendicular to the plane. The electric field due to an infinite line of charge is proportional to the linear charge density and inversely proportional to the distance (instead of the distance squared) from the line of charge to a point in the plane. Modify the calculation of the electric field in *Program fieldline* so that field lines from infinite lines of charge are drawn. Use your program to draw the field lines due to four identical infinite lines of charge located at the corners of a square, and compare the field lines with your results in part (a).

c. Use your modified program from part (b) to draw the field lines for the two-dimensional analogs of the distributions considered in Problem 10.1. Compare the results for two and three dimensions, and discuss any qualitative differences.

Problem 10.3   Motion of a charged particle in an electric field

a. Write a program to compute the motion of a particle of mass m and charge q in the presence of the electric field created by a fixed distribution of point charges. Use the Euler-Richardson algorithm to update the position and velocity of the particle. The acceleration of the charge is given by $q\mathbf{E}/m$, where $\mathbf{E}$ is the electric field due to the fixed point charges. (We ignore the effects of radiation due to accelerating charges.) Incorporate the relevant subroutines from *Program fieldline* to visualize the electric field.

b. Assume that $\mathbf{E}$ is due to a charge $q(1) = 1.5$ fixed at the origin. Simulate the motion of a charged particle of mass $m = 0.1$ and charge $q = 1$ initially at $x = 1$, $y = 0$. Consider the following initial conditions for its velocity: (i) $v_x = 0$, $v_y = 0$; (ii) $v_x = 1$, $v_y = 0$; (iii) $v_x = 0$, $v_y = 1$; and (iv) $v_x = -1$, $v_y = 0$. Draw electric field lines beginning at the initial values of (x, y) and a line representing the particle's trajectory. Why does the trajectory of the particle not always follow a field line?

c. Assume that the electric field is due to two fixed point charges: $q(1) = 1$ at $x(1) = 2$, $y(1) = 0$ and $q(2) = -1$ at $x(2) = -2$, $y(2) = 0$. Place a charged particle of unit mass and unit positive charge at the point $x = 0.05$, $y = 0$. What do you expect the motion of this charge to be? Do the simulation and determine the qualitative nature of the motion.

*d. Consider the motion of a charged particle in the vicinity of the electric dipole defined in part (c). Choose the initial position to be five times the separation of the charges in the dipole. Do you find any bound orbits? Can you find any closed orbits or do all orbits show some precession?

We know that it is often easier to analyze the behavior of a system using energy rather than force concepts. We define the electric potential $V(\mathbf{r})$ by the relation

$$V(\mathbf{r}_2) - V(\mathbf{r}_1) = -\int_{\mathbf{r}_1}^{\mathbf{r}_2} \mathbf{E} \cdot d\mathbf{r} \tag{10.5}$$

or

$$\mathbf{E}(\mathbf{r}) = -\nabla V(\mathbf{r}). \tag{10.6}$$

Only differences in the potential between two points have physical significance. The gradient operator ∇ is given in Cartesian coordinates by

$$\nabla = \frac{\partial}{\partial x}\hat{\mathbf{x}} + \frac{\partial}{\partial y}\hat{\mathbf{y}} + \frac{\partial}{\partial z}\hat{\mathbf{z}}, \tag{10.7}$$

where the vectors $\hat{\mathbf{x}}$, $\hat{\mathbf{y}}$, and $\hat{\mathbf{z}}$ are unit vectors along the x, y, and z axes respectively. If V depends only on the magnitude of $\mathbf{r}$, then (10.6) becomes $E(r) = -dV(r)/dr$. Recall that $V(r)$ for a point charge q relative to a zero potential at infinity is given by

$$V(r) = \frac{q}{r}. \qquad \text{(Gaussian units)} \tag{10.8}$$

The surface on which the electric potential has an equal value everywhere is called an *equipotential surface* (curve in two dimensions). Because **E** is in the direction in which the electric potential decreases most rapidly, the electric field lines are orthogonal to the equipotential surfaces at any point. We can use this relation between the electric field lines and the equipotential lines to modify *Program* `fieldline` so that it draws the latter. Because the components of the line segment Δ**s** parallel to the electric field line are given by $\Delta x = \Delta s(E_x/E)$ and $\Delta y = \Delta s(E_y/E)$, the components of the line segment perpendicular to **E**, and hence parallel to the equipotential line, are given by $\Delta x = -\Delta s(E_y/E)$ and $\Delta y = \Delta s(E_x/E)$. It is unimportant whether the minus sign is assigned to the x or y component, because the only difference would be the direction that the equipotential lines are drawn.

Problem 10.4   Equipotential lines

a.   Modify *Program* `fieldline` to draw some of the equipotential lines for the charge distributions considered in Problem 10.1. One way to do so is to add a subroutine that uses a mouse click to determine the initial position of an equipotential line. The following code determines this position from a mouse click.

```
DO
    GET MOUSE x0,y0,s
LOOP until s <> 0
```

The variable s has the value 0 until the mouse is clicked. After a mouse click, the subroutine should draw the equipotential line starting from x0,y0 and ending when the line closes on itself.

b.   What would a higher density of equipotential lines mean if we drew lines such that each adjacent line differed from a neighboring one by a fixed potential difference?

c.   Explain why equipotential surfaces never cross.

Problem 10.5   The electric potential due to a finite sheet of charge

Consider a uniformly charged dielectric plate of total charge Q and linear dimension L centered at $(0, 0, 0)$ in the x-y plane. In the limit $L \to \infty$ with the charge density $\sigma = Q/L^2$ a constant, we know that the electric field is normal to the sheet and is given by $E_n = 2\pi\sigma$ (Gaussian units). What is the electric field due to a finite sheet of charge? A simple method is to divide the plate into a grid of p cells on a side such that each cell is small enough to be approximated by a point charge of magnitude $q = Q/p^2$. Because the potential is a scalar quantity, it is easier to compute the total potential rather than the total electric field from the $N = p^2$ point charges. Use the relation (10.8) for the potential from a point charge and write a program to compute $V(z)$ and hence $E_z = -\partial V(z)/\partial z$ for points along the z-axis and perpendicular to the sheet. Take $L = 1$, $Q = 1$, and $p = 10$ for your initial calculations. Increase p until your results for $V(z)$ do not change significantly. Plot $V(z)$ and E_z as a function of z and compare their z-dependence to their infinite sheet counterparts.

*Problem 10.6 Electrostatic shielding

We know that the (static) electric field is zero inside a conductor, all excess charges reside on the surface of the conductor, and the surface charge density is greatest at the points of greatest curvature. Although these properties are plausible, it is instructive to do a simulation to see how these properties follow from Coulomb's law. For simplicity, consider the conductor to be two-dimensional so that the potential energy is proportional to $\ln r$ rather than $1/r$ (see Problem 10.2). It also is convenient to choose the surface of the conductor to be an ellipse.

a. If we are interested only in the final distribution of the charges and not in the dynamics of the system, we can use a Monte Carlo method. Our goal is to find the minimum energy configuration beginning with the N charges randomly placed within the ellipse. The method is to choose a charge i at random, and make a trial change in the position of the charge. The trial position should be no more than d_{max} from the old position and still within the ellipse. The parameter d_{max} should be chosen to be approximately $b/10$, where b is the semiminor axis of the ellipse. Compute the change in the total potential energy given by (in arbitrary units)

$$\Delta U = -\sum_j [\ln r_{ij}^{(new)} - \ln r_{ij}^{(old)}].$$ (10.9)

The sum is over all charges in the system not including i. If $\Delta U > 0$, then reject the trial move, otherwise accept it. Repeat this procedure many times until very few trial moves are accepted. Write a program to implement this Monte Carlo algorithm. Run the simulation for $N \geq 20$ charges inside a circle and then repeat the simulation for an ellipse. How are the charges distributed in the (approximately) minimum energy distribution? Which parts of the ellipse have a higher charge density?

b. Repeat part (a) for a two-dimensional conductor, but assume that the potential energy $U \sim 1/r$. Do the charges move to the surface? Is it sufficient that the interaction be repulsive?

c. Repeat part (a) with the added condition that there is a fixed positive charge of magnitude $N/2$ located outside the ellipse. How does this fixed charge effect the charge distribution? Are the excess free charges still at the surface? Try different positions for the fixed charge.

d. Repeat parts (a) and (b) for $N = 50$ charges located within an ellipsoid in three dimensions.

10.2 NUMERICAL SOLUTIONS OF LAPLACE'S EQUATION

In Section 10.1 we found the electric fields and potentials due to a fixed distribution of charges. Suppose that we do not know the positions of the charges and instead know only the potential on a set of boundaries surrounding a charge-free region. This information is sufficient to determine the potential $V(\mathbf{r})$ at any point within the charge-free region.

The direct method of solving for $V(x, y, z)$ is based on Laplace's equation which can be expressed in Cartesian coordinates as

$$\nabla^2 V(x, y, z) \equiv \frac{\partial^2 V}{\partial x^2} + \frac{\partial^2 V}{\partial y^2} + \frac{\partial^2 V}{\partial z^2} = 0. \tag{10.10}$$

The problem is to find the function $V(x, y, z)$ that satisfies (10.10) and the specified boundary conditions. This type of problem is an example of a *boundary value* problem. Because analytical methods for regions of arbitrary shape do not exist, the only general approach is to use numerical methods.

Laplace's equation is not a new law of physics, but can be derived directly from (10.6) and the relation $\nabla \cdot \mathbf{E} = 0$ or indirectly from Coulomb's law in regions of space where there is no charge. For simplicity, we consider only two-dimensional boundary value problems for $V(x, y)$. We use a finite difference method and divide space into a discrete grid of points. In the following, we show that in the absence of a charge at (x, y), the discrete form of Laplace's equation satisfies the relation

$$\begin{aligned} V(x, y) \approx \frac{1}{4}[&V(x + \Delta x, y) + V(x - \Delta x, y) \\ &+ V(x, y + \Delta y) + V(x, y - \Delta y)]. \end{aligned} \qquad \text{(two dimensions)} \tag{10.11}$$

That is, $V(x, y)$ is the average of the potential at the four nearest neighbor points. This remarkable property of $V(x, y)$ can be derived by approximating the partial derivatives in (10.10) by finite differences (see Problem 10.7b).

In Problem 10.7a we verify (10.11) by calculating the potential due to a point charge at a point in space selected by the user and at the four nearest neighbors. As the form of (10.11) implies, the average of the potential at the four neighboring points should equal the potential at the center point. We assume the form (10.8) for the potential $V(r)$ due to a point charge, a form that satisfies Laplace's equation for $r \neq 0$. The program is listed in the following.

```
PROGRAM verify
! verify the discrete form of Laplace's equation
! for a point charge on a square lattice
LIBRARY "csgraphics"
CALL initial(L,dx,dy,#1,#2)
CALL draw_grid(L,dx,dy,#3)
DO
    CALL potential(L,dx,dy,stop_plot$,#1,#2,#3)
LOOP until stop_plot$ = "yes"
END
```

```
SUB initial(L,dx,dy,#1,#2)
    INPUT prompt "lattice dimension = ": L
    INPUT prompt "grid spacing in x direction = ": dx
    INPUT prompt "grid spacing in y direction = ": dy
    OPEN #1: screen 0,0.25,0.8,1
    SET WINDOW -1.2,2,-1,2
    OPEN #2: screen 0,0.25,0,0.6
END SUB

SUB draw_grid(L,dx,dy,#1)
    OPEN #1: screen 0.25,1,0,1
    CALL compute_aspect_ratio(L,xwin,ywin)
    SET WINDOW -xwin,xwin,-ywin,ywin
    LET a = 0.2*dx                    ! "radius" of visual image of charge
    SET COLOR "blue"
    BOX CIRCLE -a,a,-a,a
    FLOOD 0,0
    SET COLOR "black"
    FOR y = -L to L step dy
        FOR x = -L to L step dx
            PLOT POINTS: x,y
        NEXT x
    NEXT y
    BOX LINES -L,L,-L,L
END SUB

SUB potential(L,dx,dy,stop_plot$,#1,#2,#3)
    WINDOW #3
    SET CURSOR 1,20
    PRINT "click on mouse to choose site"
    LET s = 0
    DO
        GET MOUSE xs,ys,s
    LOOP until s = 2
    WINDOW #1
    CLEAR
    LET stop_plot$ = ""
    IF abs(xs) <= L and abs(ys) <= L then
        CALL showpotential(xs,ys,0,0,V0)
        CALL showpotential(xs+dx,ys,1,0,V1)
        CALL showpotential(xs-dx,ys,-1,0,V2)
        CALL showpotential(xs,ys+dy,0,1,V3)
        CALL showpotential(xs,ys-dy,0,-1,V4)
        WINDOW #2
        SET CURSOR 1,1
        PRINT " average potential"
        PRINT " of four neighbors:"
        PRINT truncate(0.25*(V1+V2+V3+V4),4)
    ELSE
```

```
        LET stop_plot$ = "yes"
      END IF
END SUB

SUB showpotential(x,y,xp,yp,V)
      LET V = 1/sqr(x*x + y*y)      ! potential of a point charge
      LET V = truncate(V,4)
      PLOT TEXT, AT xp,yp: str$(V)
END SUB
```

Problem 10.7   Verification of the difference equation for the potential

a.   Choose reasonable values for the grid spacings Δx and Δy and consider a point that is not too close to the source charge. Compare the computed potential at a point to the average of the potential at its four nearest neighbor points. Do similar measurements for other points. Does the relative agreement with (10.11) depend on the distance of the point to the source charge? Choose smaller values of Δx and Δy and determine if results are in better agreement with (10.11). Does it matter whether Δx and Δy have the same value?

b.   Derive the finite difference equation (10.11) for $V(x, y)$ using the second-order Taylor expansion:

$$V(x + \Delta x, y) = V(x, y) + \Delta x \frac{\partial V(x, y)}{\partial x} + \frac{1}{2}(\Delta x)^2 \frac{\partial^2 V(x, y)}{\partial x^2} + \cdots$$

$$V(x, y + \Delta y) = V(x, y) + \Delta y \frac{\partial V(x, y)}{\partial y} + \frac{1}{2}(\Delta y)^2 \frac{\partial^2 V(x, y)}{\partial y^2} + \cdots$$

The effect of including higher derivatives is discussed by MacDonald (see references).

Now that we have found that (10.11), a finite difference form of Laplace's equation, is consistent with Coulomb's law, we adopt (10.11) as the basis for computing the potential for systems for which we cannot calculate the potential directly. In particular, we consider problems where the potential is specified on a closed surface that divides space into interior and exterior regions in which the potential is independently determined. For simplicity, we consider only two-dimensional geometries. The approach, known as the *relaxation method*, is based on the following algorithm:

1. Divide the region of interest into a rectangular grid of points spanning the region. The region is enclosed by a surface (curve in two dimensions) with specified values of the potential along the curve.

2. Assign to a boundary point the potential of the boundary nearest the point.

3. Assign all interior points an arbitrary potential (preferably a reasonable guess).

4. Compute new values for the potential *V* for each interior point. Each new value is obtained by finding the average of the previous values of the potential at the four nearest neighbor points.

5. Repeat step (4) using the values of *V* obtained in the previous iteration. This iterative process is continued until the potential at each interior point is computed to the desired accuracy.

In Problems 10.8–10.10 we use this method and its variants to compute the potential for various geometries.

```
PROGRAM laplace
! implementation of Jacobi relaxation method to solve
! Laplace's equation in rectangular geometry
DIM V(0:100,0:100)
CALL assign(V(,),nx,ny,min_change)
CALL iterate(V(,),nx,ny,min_change,iterations)
END

SUB assign(V(,),nx,ny,min_change)
    ! linear dimension of rectangular region
    LET nx = 9       ! number of interior points in x-direction
    LET ny = 9       ! number of interior points in y-direction
    LET V0 = 10                ! boundary potential of rectangle
    INPUT prompt "percentage change = ": min_change
    LET min_change = min_change/100
    ! fix potential on boundary of rectangle
    FOR x = 0 to nx + 1
        LET V(x,0) = V0
        LET V(x,ny+1) = V0
    NEXT x
    FOR y = 0 to ny + 1
        LET V(0,y) = V0
        LET V(nx+1,y) = V0
    NEXT y
    ! guess initial values of potentials of interior sites
    FOR y = 1 to ny
        FOR x = 1 to nx
            LET V(x,y) = 0.9*V0
        NEXT x
    NEXT y
    CALL show_output(V(,),nx,ny,0)
END SUB

SUB iterate(V(,),nx,ny,min_change,iterations)
    DIM Vave(100,100)
    LET iterations = 0
    DO
        ! maximum difference of iterated potential at any site
```

```
            LET change = 0
            FOR y = 1 to ny
                FOR x = 1 to nx
                    ! average of potential of neighboring cells
                    LET Vave(x,y) = V(x+1,y) + V(x-1,y)
                    LET Vave(x,y) = Vave(x,y) + V(x,y+1) + V(x,y-1)
                    LET Vave(x,y) = 0.25*Vave(x,y)
                    ! compute percentage change in potential
                    IF Vave(x,y) <> 0 then
                        LET dV = abs((V(x,y) - Vave(x,y))/Vave(x,y))
                        IF dV > change then LET change = dV
                    END IF
                NEXT x
            NEXT y
            FOR y = 1 to ny         ! update potential at each site
                FOR x = 1 to nx
                    LET V(x,y) = Vave(x,y)
                NEXT x
            NEXT y
            LET iterations = iterations + 1
            CALL show_output(V(,),nx,ny,iterations)
        LOOP until change <= min_change
    END SUB

    SUB show_output(V(,),nx,ny,iterations)        ! print potential
        PRINT
        PRINT "iteration ="; iterations
        FOR y = 0 to ny + 1
            FOR x = nx + 1 to 0 step - 1
                PRINT using "###.##": V(x,y);
            NEXT x
            PRINT
        NEXT y
    END SUB
```

Problem 10.8   Numerical solution of the potential within a rectangular region

a. Use *Program laplace* to determine the potential $V(x, y)$ in a square region
with linear dimension $L = 9$. The boundary of the square is at a potential
$V = 10$. Let the number of interior grid points in the horizontal and vertical
directions be nx = ny = 9, respectively. Before you run the program, guess the
exact form of $V(x, y)$ and set the initial values of the interior potential close to
the exact answer. How many iterations are necessary to achieve 1% accuracy?
Increase both Δx and Δy by a factor of two, and determine the number of
iterations that are now necessary to achieve 1% accuracy.

b. Consider the same geometry as in part (a), but set the initial potential at the interior points equal to zero except for the center point whose potential is set equal to four. Does the potential distribution evolve to the same values as in part (a)? What is the effect of a poor initial guess? Are the final results independent of your initial guess?

c. Modify SUB assign in *Program laplace* so that the value of the potential at the four sides is 5, 10, 5, and 10, respectively (see Fig. 10.1). Sketch the equipotential surfaces. What happens if the potential is 10 on three sides and 0 on the fourth? Start with a reasonable guess for the initial values of the potential at the interior points and iterate until 1% accuracy is obtained.

*d. Consider the same initial choice of the potential as in part (b) and focus your attention on the potential at the points near the center of the square. If there were four random walkers at the central point corresponding to an initial potential of four, how many walkers would there be at the nearest neighbor points after the first iteration? Follow the distribution of the "walkers" as a function of the number of iterations and verify that the nature of the relaxation of the potential to its correct distribution is closely related to *diffusion* (see Chapters 7 and 12). It would be helpful to increase the number of points in the grid and the initial value of the potential at the central point to see the nature of the relaxation more clearly.

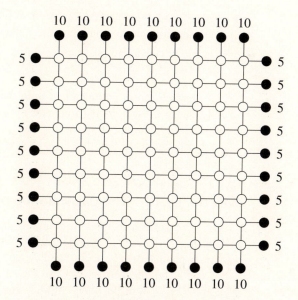

Fig. 10.1 Potential distribution considered in Problem 10.8c. The number of interior points in each direction is nine.

In Problem 10.8, we implemented a simple version of the relaxation method known as the Jacobi method. In particular, the new potential of each point is found based on the values of the potentials at the neighboring points at the previous iteration. After the entire lattice was visited, the potential at each point was updated *simultaneously*. The difficulty with this relaxation method is that it converges very slowly. The use of more general relaxation methods is discussed in many texts (cf. Koonin and Meredith or Press et al.). In Problem 10.9 we consider a method known as Gauss-Seidel relaxation.

Problem 10.9   Gauss-Seidel relaxation

a.   Modify the program that you used in Problem 10.8 so that the potential at each point is updated sequentially. That is, after the average potential of the nearest neighbor points of point i is computed, update the potential at i immediately. In this way the new potential of the next point is computed using the most recently computed values of its nearest neighbor potentials. Are your results better, worse, or about the same as the simple relaxation method?

b.   Imagine coloring the alternate points of a grid red and black, so that the grid resembles a checkerboard. Modify the program so that all the red points are updated first, and then all the black points are updated. This ordering is repeated for each iteration. Do your results converge any more quickly than in part (a)?

*c.   The slow convergence of the relaxation methods we have explored is due to the fact that it takes a long time for a change in the potential at one point to effect changes further away. We can improve the Gauss-Seidel method by using an overrelaxation method which updates the new potential as follows:

```
LET V(x,y) = w*Vave(x,y) + (1-w)*V(x,y)
```

The overrelaxation parameter w is in the range $1 < w < 2$. The effect of w is to cause the potential to change by a greater amount than in the simple relaxation procedure. Explore the dependence of the rate of convergence on w. A relaxation method that does increase the rate of convergence is explored in Project 10.1.

Problem 10.10   The capacitance of concentric squares

a.   Use a relaxation method to compute the potential distribution between the two concentric square cylinders shown in Fig. 10.2. The potential of the outer square conductor is $V_{out} = 10$ and the potential of the inner square conductor is $V_{in} = 5$. The linear dimensions of the exterior and interior squares are $L_{out} = 25$ and $L_{in} = 5$, respectively. Modify your program so that the potential of the interior square is fixed. Sketch the equipotential surfaces.

b.   A system of two conductors with charge Q and $-Q$ respectively has a capacitance C that is defined as the ratio of Q to the potential difference ΔV between

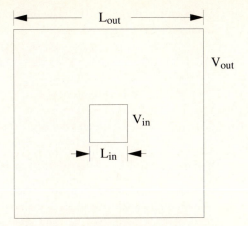

Fig. 10.2 The geometry of the two concentric squares considered in Problem 10.10.

the two conductors. Determine the capacitance per unit length of the concentric cylinders considered in part (a). In this case $\Delta V = 5$. The charge Q can be determined from the fact that near a conducting surface, the surface charge density σ is given by $\sigma = E_n/4\pi$, where E_n is the magnitude of the electric field normal to the surface. E_n can be approximated by the relation $-\delta V/\delta r$, where δV is the potential difference between a boundary point and an adjacent interior point a distance δr away. Use the result of part (a) to compute δV for each point adjacent to the two square surfaces. Use this information to determine E_n for the two surfaces and the charge per unit length on each conductor. Are the charges equal and opposite in sign? Compare your numerical result to the capacitance per unit length, $1/2 \ln r_{out}/r_{in}$, of a system of two concentric circular cylinders of radii r_{out} and r_{in}. Assume that the circumference of each cylinder equals the perimeter of the corresponding square, i.e., $2\pi r_{out} = 4L_{out}$ and $2\pi r_{in} = 4L_{in}$.

c. Move the inner square 1 cm off center and repeat the calculations of parts (a) and (b). How do the potential surfaces change? Is there any qualitative difference if we set the inner conductor potential equal to -5 statvolt instead of $+5$ statvolt?

Laplace's equation holds only in charge-free regions. If there is a charge density $\rho(x, y, z)$ in the region, we need to use *Poisson's* equation which can be written as

$$\nabla^2 V(\mathbf{r}) = \frac{\partial^2 V}{\partial x^2} + \frac{\partial^2 V}{\partial y^2} + \frac{\partial^2 V}{\partial z^2} = -4\pi\rho(\mathbf{r}), \tag{10.12}$$

where $\rho(\mathbf{r})$ is the charge density. The difference form of Poisson's equation is given in two dimensions by

$$V(x, y) \approx \frac{1}{4}\big[V(x + \Delta x, y) + V(x - \Delta x, y) + V(x, y + \Delta y) + V(x, y - \Delta y)\big]$$
$$+ \frac{1}{4}\Delta x \Delta y \, 4\pi\rho(x, y). \tag{10.13}$$

Note that the product $\rho(x, y)\Delta x \Delta y$ is the total charge in the cell centered at (x, y).

Problem 10.11   Numerical solution of Poisson's equation

a.   Consider a square of linear dimension $L = 25$ whose boundary is fixed at a potential equal to $V = 10$. Assume that each interior cell has a uniform charge density ρ such that the total charge is $Q = 1$. Use a modification of *Program laplace* to compute the potential distribution for this case. Compare the equipotential surfaces obtained for this case to that found in Problem 10.10.

b.   Find the potential distribution if the charge distribution of part (a) is restricted to a 5×5 square at the center.

c.   Find the potential distribution if the charge distribution of part (a) is restricted to a 1×1 square at the center. How does the potential compare to that of a point charge without the boundary?

10.3   RANDOM WALK SOLUTION OF LAPLACE'S EQUATION

In Section 10.2 we found that the solution to Laplace's equation in two dimensions at the point (x, y) is given by

$$V(x, y) = \frac{1}{4}\sum_{i=1}^{4} V(i), \tag{10.14}$$

where $V(i)$ is the value of the potential at the ith neighbor. A generalization of this result is that the potential at any point equals the average of the potential on a circle (or sphere in three dimensions) centered about that point.

The relation (10.14) can be given a probabilistic interpretation in terms of random walks (see Problem 10.8d). Suppose that many random walkers are at the point (x, y) and each walker "jumps" to one of its four neighbors (on a square grid) with equal probability $p = 1/4$. From (10.14) we see that the average potential found by the walkers after jumping one step is the potential at (x, y). This relation generalizes to walkers that visit a point on a closed surface with fixed potential. The random walk algorithm for computing the solution to Laplace's equation can be stated as:

1.   Begin at a point (x, y) where the value of the potential is desired, and take a step in a random direction.

2.   Continue taking steps until the walker reaches the surface. Record $V_b(i)$, the potential at the boundary point i. A typical walk is shown in Fig. 10.3.

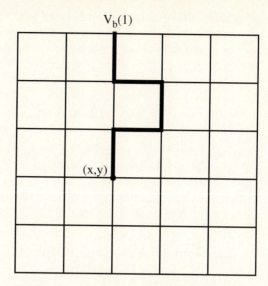

Fig. 10.3   A random walk on a 6 × 6 grid
starting at the point $(x, y) = (3, 3)$
and ending at the boundary point
$V_b(3, 6)$ where the potential is
recorded.

3. Repeat steps (1) and (2) n times and sum the potential found at the surface each
   time.

4. The value of the potential at the point (x, y) is estimated by

$$V(x, y) = \frac{1}{n} \sum_{i=1}^{n} V_b(i) \qquad\qquad (10.15)$$

   where n is the total number of random walkers.

Problem 10.12   Random walk solution of Laplace's equation

a. Consider the square region shown in Fig. 10.1 and compare the results of
    the random walk method with the results of the relaxation method (see Prob-
    lem 10.8c). Try $n = 100$ and $n = 1000$ walkers, and choose a point near the
    center of the square.

b. Repeat part (a) for other points within the square. Do you need more or less
    walkers when the potential near the surface is desired? How quickly do your
    answers converge as a function of n?

   The disadvantage of the random walk method is that it requires many walkers
to obtain a good estimate of the potential at each point. However, if the potential is
needed at only a small number of points, then the random walk method might be more
appropriate than the relaxation method which requires the potential to be computed at

all points within the region. Another case where the random walk method is appropriate is when the geometry of the boundary is fixed, but the potential in the interior for a variety of different boundary potentials is needed. In this case the quantity of interest is $G(x, y, x_b, y_b)$, the number of times that a walker from the point (x, y) lands at the boundary (x_b, y_b). The random walk algorithm is equivalent to the relation

$$V(x, y) = \frac{1}{n} \sum_b G(x, y, x_b, y_b) V(x_b, y_b), \tag{10.16}$$

where the sum is over all points on the boundary. We can use the same function G for different distributions of the potential on a given boundary. G is an example of a Green's function, a function that you will encounter in advanced treatments of electrodynamics and quantum mechanics (cf. Section 18.7). Of course, if we change the geometry of the boundary, we have to recompute the function G.

Problem 10.13 Green's function solution of Laplace's equation

a. Compute the Green's function $G(x, y, x_b, y_b)$ for the same geometry considered in Problem 10.12. Use at least 200 walkers at each interior point to estimate G. Because of the symmetry of the geometry, you can determine some of the values of G from other values without doing an additional calculation. Store your results for G in a file.

b. Use your results for G found in part (a) to determine the potential at each interior point when the boundary potential is the same as in part (a) except for five boundary points which are held at $V = 20$. Find the locations of the five boundary points that maximize the potential at the interior point located at $(3, 5)$. Repeat the calculation to maximize the potential at $(5, 3)$. Use trial and error guided by your physical intuition.

The random walk algorithm can help us gain additional insight into the nature of Laplace's equation. Suppose that you have a boundary similar to the one shown in Fig. 10.4. The potentials on the left and right boundaries are V_L and V_R, respectively. If the neck between the two sides is narrow, it is clear that a random walker starting on the left side has a low probability of reaching the other side. Hence, we can conclude that the potential in the interior of the left side is approximately V_L except very near the neck.

Poisson's equation also can be solved using the random walk method. In this case, the potential is given by

$$V(x, y) = \frac{1}{n} \sum_\alpha V(\alpha) + \frac{\pi \Delta x \Delta y}{n} \sum_{i,\alpha} \rho(x_{i,\alpha}, y_{i,\alpha}), \tag{10.17}$$

where α labels the walker, and i labels the point visited by the walker. That is, each time a walker is at a point i, we add the charge density at that point to the second sum in (10.17).

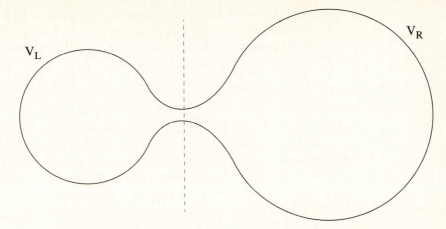

Fig. 10.4 Two regions of space connected by a narrow neck. The boundary of the left region has a potential V_L, and the boundary of the right region has a potential V_R.

*10.4 FIELDS DUE TO MOVING CHARGES

The fact that accelerating charges radiate electromagnetic waves is one of the most important results in the history of physics. In this section we discuss a numerical algorithm for computing the electric and magnetic fields due to the motion of charged particles. The algorithm is very general, but requires some care in its application.

To understand the algorithm, we need a few results that can be found conveniently in Feynman's lectures. We begin with the fact that the scalar potential at the observation point $\mathbf{R}$ due to a stationary particle of charge q is

$$V(\mathbf{R}) = \frac{q}{|\mathbf{R} - \mathbf{r}|}, \tag{10.18}$$

where $\mathbf{r}$ is the position of the charged particle. The electric field is given by

$$\mathbf{E}(\mathbf{R}) = -\frac{\partial V(\mathbf{R})}{\partial \mathbf{R}}, \tag{10.19}$$

where $\partial V(\mathbf{R})/\partial \mathbf{R}$ is the gradient with respect to the coordinates of the observation point. (Note that our notation for the observation point differs from that used in other sections in this chapter.) How do the relations (10.18) and (10.19) change when the particle is moving? We might guess that because it takes a finite time for the disturbance due to a charge to reach the point of observation, we should modify (10.18) by writing

$$V(\mathbf{R}) \overset{?}{=} \frac{q}{r_{\text{ret}}}, \tag{10.20}$$

where

$$r_{\text{ret}} = |\mathbf{R} - \mathbf{r}(t_{\text{ret}})|. \tag{10.21}$$

The quantity r_{ret} is the separation of the charged particle from the observation point $\mathbf{R}$ at the retarded time t_{ret}. The latter is the time at which the particle was at $\mathbf{r}(t_{\text{ret}})$ such that a disturbance starting at $\mathbf{r}(t_{\text{ret}})$ and traveling at the speed of light would reach $\mathbf{R}$ at time t; t_{ret} is given by the implicit equation

$$t_{\text{ret}} = t - \frac{r_{\text{ret}}(t_{\text{ret}})}{c}, \tag{10.22}$$

where t is the observation time and c is the speed of light.

Although the above reasoning is plausible, the relation (10.20) is not quite correct (cf. Feynman et al. for a derivation of the correct result). We need to take into account that the potential due to the charge is a maximum if the particle is moving toward the observation point and a minimum if it is moving away. The correct result can be written as

$$V(\mathbf{R}, t) = \frac{q}{r_{\text{ret}}\left(1 - \hat{\mathbf{r}}_{\text{ret}} \cdot \mathbf{v}_{\text{ret}}/c\right)}, \tag{10.23}$$

where

$$\mathbf{v}_{\text{ret}} = \frac{d\mathbf{r}(t)}{dt}\Big|_{t=t_{\text{ret}}}. \tag{10.24}$$

To find the electric field of a moving charge, we recall that the electric field is related to the time rate of change of the magnetic flux. Hence, we expect that the total electric field at the observation point $\mathbf{R}$ has a contribution due to the magnetic field created by the motion of the charge. We know that the magnetic field due to a moving charge is given by

$$\mathbf{B} = \frac{1}{c}\frac{q\mathbf{v} \times \mathbf{r}}{r^3}. \tag{10.25}$$

If we define the vector potential $\mathbf{A}$ as

$$\mathbf{A} = \frac{q}{r}\frac{\mathbf{v}}{c}, \tag{10.26}$$

we can express $\mathbf{B}$ in terms of $\mathbf{A}$ as

$$\mathbf{B} = \nabla \times \mathbf{A}. \tag{10.27}$$

As we did for the scalar potential V, we argue that the correct formula for $\mathbf{A}$ is

$$\mathbf{A}(\mathbf{R}, t) = q\frac{\mathbf{v}_{\text{ret}}/c}{r_{\text{ret}}\left(1 - \hat{\mathbf{r}}_{\text{ret}} \cdot \mathbf{v}_{\text{ret}}/c\right)}. \tag{10.28}$$

Equations (10.23) and (10.28) are known as the Liénard-Wiechert potentials.

The contribution to the electric field $\mathbf{E}$ from V and $\mathbf{A}$ is given by

$$\mathbf{E} = -\nabla V - \frac{1}{c}\frac{\partial \mathbf{A}}{\partial t}. \tag{10.29}$$

The derivatives in (10.29) are with respect to the observation coordinates. The difficulty associated with calculating these derivatives is that the potentials depend on t_{ret} which in turn depends on $\mathbf{R}$, $\mathbf{r}$, and t. The result can be expressed as

$$\mathbf{E}(\mathbf{R}, t) = \frac{q r_{ret}}{\left(\mathbf{r}_{ret} \cdot \mathbf{u}_{ret}\right)^3} \left[\mathbf{u}_{ret}(c^2 - v_{ret}^2) + \mathbf{r}_{ret} \times \left(\mathbf{u}_{ret} \times \mathbf{a}_{ret}\right)\right], \tag{10.30}$$

where

$$\mathbf{u}_{ret} \equiv c\hat{\mathbf{r}}_{ret} - \mathbf{v}_{ret}. \tag{10.31}$$

The acceleration of the particle $\mathbf{a}_{ret} = d\mathbf{v}(t)/dt|_{t=t_{ret}}$. We also can show using (10.27) that the magnetic field $\mathbf{B}$ is given by

$$\mathbf{B} = \hat{\mathbf{r}}_{ret} \times \mathbf{E}. \tag{10.32}$$

The above discussion is not rigorous, but we can accept (10.30) and (10.32) in the same spirit as we accept Coulomb's law and the Biot-Savart law. All of classical electrodynamics can be reduced to (10.30) and (10.32), if we assume that the sources of all fields are charges, and all electric currents are due to the motion of charged particles. Note that (10.30) and (10.32) are consistent with the special theory of relativity and reduce to known results in the limit of stationary charges and steady currents.

Although the equations (10.30) and (10.32) are deceptively simple (we do not even have to solve any differential equations), it is difficult to calculate the fields analytically even if the position of a charged particle is an analytic function of time. The difficulty is that we must find the retarded time t_{ret} from (10.22) for each observation position $\mathbf{R}$ and time t. For example, consider a charged particle whose motion is sinusoidal, i.e., $x(t_{ret}) = A \cos \omega t_{ret}$. To calculate the fields at the position $\mathbf{R} = (X, Y, Z)$ at time t, we need to solve the following transcendental equation for t_{ret}:

$$t_{ret} = t - \frac{r_{ret}}{c} = t - \frac{1}{c}\sqrt{(X - A\cos^2 \omega t_{ret})^2 + Y^2 + Z^2}. \tag{10.33}$$

The solution of (10.33) can be expressed as a root finding problem for which we need to find the zero of the function $f(t_{ret})$:

$$f(t_{ret}) = t - t_{ret} - \frac{r_{ret}}{c}. \tag{10.34}$$

One way to find the root is to use a bracketing method. The idea is to first find a value t_a such that $f(t_a) > 0$, and another value t_b such that $f(t_b) < 0$. Because $f(t_{ret})$ is continuous, there is a value of t_{ret} in the interval $t_a < t_{ret} < t_b$ such that $f(t_{ret}) = 0$. There are various ways of guessing the solution and reducing the size of the interval. In the *bisection method* (see Section 6.6), we use the midpoint $t_m = (t_a + t_b)/2$ as the guess, and compute $f(t_m)$. If $f(t_m) > 0$, find the next midpoint with $t_a = t_m$; otherwise, find the next midpoint with $t_b = t_m$. Continue until $f(t_m) = 0$ to the desired accuracy.

In the *method of false position*, we find the intersection of the line connecting the points $(t_a, f(t_a))$ and $(t_b, f(t_b))$ with the t_{ret} axis. The value of t_{ret} at the intersection, t_1, gives the first estimate for the zero of (10.33). If $f(t_1) > 0$, repeat the calculation with $t_a = t_1$, else repeat the calculation with $t_b = t_1$. This procedure is continued until the desired level of accuracy is achieved.

The bisection method and the method of false position are guaranteed to find a root. *Newton's method*, which uses the slope $df(t_{ret})/dt_{ret}$ to estimate the solution, is much faster, but does not always give the correct answer and must be used with care. In this method the guess for t_{ret} at the nth iteration is given by

$$t_n = t_{n-1} - \frac{f(t_{n-1})}{df(t_{ret})/dt_{ret}|_{t_{ret}=t_{n-1}}}.$$

(10.35)

We use Newton's method unless the interval between two guesses is increasing. In that case we use the bisection method to decrease the interval. *Program radiation* uses this strategy to calculate the electric field due to an oscillating charge. We choose units such that the velocity is measured in terms of the speed of light c.

```
PROGRAM radiation
! compute electric fields due to accelerating charge and
! plot electric field lines
LIBRARY "csgraphics"
DIM snapshot$(100)
CALL initial(L,nsnap,radius,lpc,dt,dtheta)
FOR isnap = 1 to nsnap
    CALL draw_field(L,isnap,radius,dt,dtheta,snapshot$())
NEXT isnap
CALL animate(L,nsnap,snapshot$())
END

SUB initial(L,nsnap,a,lpc,dt,dtheta)
    LET L = 20
    CALL compute_aspect_ratio(2*L,xwin,ywin)
    SET WINDOW -xwin,xwin,-ywin,ywin
    LET a = 0.3                      ! "radius" of visual image of charge
    LET lpc = 10                     ! number of field lines
    LET dt = 0.4                     ! time between plots
    LET dtheta = 2*pi/lpc
    LET nsnap = 15                   ! number of snapshots
END SUB

SUB draw_field(L,isnap,a,dt,dtheta,snapshot$())
    ! adopted from Program fieldline
    DIM E(3),R(3),r_ret(3),v_ret(3),a_ret(3)
    DECLARE DEF dotproduct
    LET ds = 1
    LET t = isnap*dt                 ! observation time
    LET R(1) = 0
    LET R(2) = 0
    CALL motion(R(),t,r_ret(),v_ret(),a_ret())
    ! find charge position at time t
```

```
                    LET x = R(1) - r_ret(1)
                    LET y = R(2) - r_ret(2)
                    BOX CIRCLE x-a,x+a,y-a,y+a
                    FOR theta = 0 to 2*pi step dtheta
                        LET xline = x + a*cos(theta)
                        LET yline = y + a*sin(theta)
                        PLOT xline,yline;
                        DO
                            LET R(1) = xline
                            LET R(2) = yline
                            CALL field(R(),t,E())  ! compute fields
                            LET Emag = sqr(dotproduct(E(),E()))
                            ! new position on field line
                            LET xline = xline + ds*E(1)/Emag
                            LET yline = yline + ds*E(2)/Emag
                            PLOT xline,yline;
                        LOOP until abs(xline) > L or abs(yline) > L
                        PLOT                        ! turn beam off
                    NEXT theta
                BOX LINES -L,L,-L,L
                BOX KEEP -L,L,-L,L in snapshot$(isnap)
                CLEAR
            END SUB

            SUB motion(R(),t_ret,r_ret(),v_ret(),a_ret())
                ! compute motion of source
                LET r_ret(1) = R(1) - 0.2*cos(t_ret)     ! harmonic motion
                LET r_ret(2) = R(2)
                LET r_ret(3) = R(3)
                LET v_ret(1) = -0.2*sin(t_ret)       ! particle velocity
                LET v_ret(2) = 0
                LET v_ret(3) = 0
                LET a_ret(1) = -0.2*cos(t_ret)       ! particle acceleration
                LET a_ret(2) = 0
                LET a_ret(3) = 0
            END SUB

            SUB field(R(),t,E())
                ! compute electric field vector
                DECLARE DEF dotproduct
                DIM r_ret(3),v_ret(3),a_ret(3),u_ret(3),uxa(3),w_ret(3)
                CALL retarded_time(R(),t,t_ret)      ! find retarded time
                ! dynamical variables of moving charge
                CALL motion(R(),t_ret,r_ret(),v_ret(),a_ret())
                LET v2 = dotproduct(v_ret(),v_ret())
                LET dist_ret = sqr(dotproduct(r_ret(),r_ret()))
                FOR i = 1 to 3
                    LET u_ret(i) = r_ret(i)/dist_ret - v_ret(i)
                NEXT i
```

```
              CALL crossproduct(u_ret(),a_ret(),uxa())
              CALL crossproduct(r_ret(),uxa(),w_ret())
              LET ru = dotproduct(r_ret(),u_ret())
              LET E0 = dist_ret/ru^3
              FOR i = 1 to 3
                  LET E(i) = E0*(u_ret(i)*(1 - v2) + w_ret(i))
              NEXT i
         END SUB


         SUB retarded_time(R(),t,t_ret)
              LET tb = t                      ! upper guess for retarded time
              CALL fanddf(fb,dfdt,R(),t,tb)
              LET ta = -1/1.6                 ! lower guess for retarded time
              ! insure that f(ta) > 0 and f(tb) < 0
              DO
                  LET ta = ta*1.6
                  CALL fanddf(fa,dfdt,R(),t,ta)
              LOOP until fa > 0
              LET t_ret = 0.5*(ta + tb)
              CALL fanddf(f,dfdt,R(),t,t_ret)
              CALL zero_of_f(f,dfdt,R(),t,t_ret,ta,tb)
         END SUB


         SUB fanddf(f,dfdt,R(),t,t_ret)
              ! calculate f and df/dt_r
              DIM r_ret(3),v_ret(3),a_ret(3)
              DECLARE DEF dotproduct
              CALL motion(R(),t_ret,r_ret(),v_ret(),a_ret())
              LET dist_ret = sqr(dotproduct(r_ret(),r_ret()))
              LET f = t - t_ret - dist_ret
              ! derivative evaluated at retarded time
              LET dfdt = -1 + dotproduct(r_ret(),v_ret())/dist_ret
         END SUB


         SUB zero_of_f(f,dfdt,R(),t,t_ret,ta,tb)
              ! do no more than 100 iterations to find the value of the reduced
              ! time t_ret such that f = t - t_ret - dist_ret = 0
              LET eps = 1e-6
              LET dt_r = tb - ta
              FOR j = 1 to 100
                  ! Newton difference between successive t_ret estimates
                  LET dt = f/dfdt
                  LET sign = ((t_ret - ta) - dt)*((t_ret - tb) - dt)
                  LET dt_old = dt_r
                  IF sign >= 0 or abs(2*dt) > abs(dt_old) then
                      ! use bisection method if next Newton iteration is not
                      ! between ta and tb, or if dt is not less than half
                      ! of the old value of dt_r
                      LET dt_r = 0.5*(ta - tb)
```

```
                        LET t_ret = tb + dt_r
                    ELSE                            ! use Newton's method
                        LET dt_r = dt
                        LET t_ret = t_ret - dt_r
                    END IF
                    IF abs(dt_r) < eps then EXIT SUB      ! convergence test
                    CALL fanddf(f,dfdt,R(),t,t_ret)
                    IF f < 0 then
                        LET tb = t_ret
                    ELSE
                        LET ta = t_ret
                    END IF
                NEXT j
                PRINT "too many iterations"
                STOP
            END SUB

            SUB animate(L,nsnap,snapshot$())
                FOR isnap = 1 to nsnap           ! show motion picture
                    BOX SHOW snapshot$(isnap) at -L,-L
                    PAUSE 1
                NEXT isnap
            END SUB

            SUB crossproduct(a(),b(),c())
                LET c(1) = a(2)*b(3) - a(3)*b(2)
                LET c(2) = a(3)*b(1) - a(1)*b(3)
                LET c(3) = a(1)*b(2) - a(2)*b(1)
            END SUB

            DEF dotproduct(a(),b())
                LET dotproduct = a(1)*b(1) + a(2)*b(2) + a(3)*b(3)
            END DEF
```

Problem 10.14   Field lines from an accelerating charge

a. Read *Program radiation* carefully to understand the correspondence be-
    tween the program and the calculation discussed in the text of the electric field
    lines due to an accelerating point charge. Describe qualitatively the nature of
    the electric field lines from an oscillating point charge.

b. Use *Program radiation* to calculate **E** due to a positively charged particle
    oscillating about the origin according to $x(t') = 0.2 \cos t'$. The program draws
    field lines in the x-y plane starting from a small circle surrounding the origin.
    Let the observation time be $t = 1$ and stop drawing each field line when $|x| >
    20$ or $|y| > 20$. How do the field lines differ from those of a static charge at the
    origin?

c. What happens to the field lines as you increase the observation time t?

d. One way of visualizing how the field lines change with time is by drawing the field lines at successive times, capturing the screen image each time. In True BASIC we can use the BOX KEEP statement and then show the images sequentially using the BOX SHOW statement. What new information does this mode of presentation convey?

e. Repeat the above observations for a charge moving with uniform circular motion about the origin.

Problem 10.15   Spatial dependence of radiating fields

a. As waves propagate from an accelerating point source, the total power that passes through a spherical surface of radius R remains constant. Because the surface area is proportional to R^2, the power per unit area or intensity is proportional to $1/R^2$. Also, because the intensity is proportional to E^2, we expect that $E \propto 1/R$ far from the source. Modify *Program radiation* to verify this result for a charge that is oscillating along the x-axis according to $x(t') = 0.2 \cos t'$. Plot $|E|$ as a function of the observation time t for a fixed position such as $\mathbf{R} = (10, 10, 0)$. The field should oscillate in time. Find the amplitude of this oscillation. Next double the distance of the observation point from the origin. How does the amplitude depend on R?

b. Repeat part (a) for several directions and distances. Generate a polar diagram showing the amplitude as a function of angle in the x-y plane. Is the radiation greatest along the line in which the charge oscillates?

Problem 10.16   Field lines from a charge moving at constant velocity

a. Use *Program radiation* to calculate $\mathbf{E}$ due to a charged particle moving at constant velocity toward the origin, i.e., $x(t_{\text{ret}}) = 1 - 2t_{\text{ret}}$. Take a snapshot at $t = 0.5$ and compare the field lines with those you expect from a stationary charge.

b. Modify *SUB motion* so that $x(t_{\text{ret}}) = 1 - 2t_{\text{ret}}$ for $t_{\text{ret}} < 0.5$ and $x(t_{\text{ret}}) = 0$ for $t_{\text{ret}} > 0.5$. Describe the field lines for $t > 0.5$. Does the particle accelerate at any time? Is there any radiation?

Problem 10.17   Frequency dependence of an oscillating charge

a. The radiated power at any point in space is proportional to E^2. Plot $|E|$ versus time at a fixed observation point (e.g., $X = 10$, $Y = Z = 0$), and calculate the frequency dependence of the amplitude of $|E|$ due to a charge oscillating at the frequency ω. It is shown in standard textbooks that the power associated with radiation from an oscillating dipole is proportional to ω^4. How does the

ω-dependence that you measured compare to that for dipole radiation? Repeat for a much bigger value of R, and explain any differences.

b. Repeat part (a) for a charge moving in a circle. Are there any qualitative differences.

*10.5 MAXWELL'S EQUATIONS

In Section 10.4 we found that accelerating charges produce electric and magnetic fields which depend on position and time. We now investigate the direct relation between changes in **E** and **B** given by the differential form of Maxwell's equations:

$$\frac{\partial \mathbf{B}}{\partial t} = -\frac{1}{c} \nabla \times \mathbf{E} \tag{10.36}$$

$$\frac{\partial \mathbf{E}}{\partial t} = c \nabla \times \mathbf{B} - 4\pi \mathbf{j}, \tag{10.37}$$

where **j** is the electric current density. We can regard (10.36) and (10.37) as the basis of electrodynamics. In addition to (10.36) and (10.37), we need the relation between **j** and the charge density ρ that expresses the conservation of charge:

$$\frac{\partial \rho}{\partial t} = -\nabla \cdot \mathbf{j}. \tag{10.38}$$

A complete description of electrodynamics requires (10.36), (10.37), and (10.38) and the initial values of all currents and fields.

For completeness, we obtain the Maxwell's equations that involve $\nabla \cdot \mathbf{B}$ and $\nabla \cdot \mathbf{E}$ by taking the divergence of (10.36) and (10.37), substituting (10.38) for $\nabla \cdot \mathbf{j}$, and then integrating over time. If the initial fields are zero, we obtain (using the relation $\nabla \cdot (\nabla \times \mathbf{A}) = 0$ for any vector **A**):

$$\nabla \cdot \mathbf{E} = 4\pi \rho \tag{10.39}$$
$$\nabla \cdot \mathbf{B} = 0. \tag{10.40}$$

If we introduce the electric and magnetic potentials, it is possible to convert the first-order equations (10.36) and (10.37) to second-order differential equations. However, the familiar first-order equations are better suited for numerical analysis. To solve (10.36) and (10.37) numerically, we need to interpret the curl and divergence of a vector. As its name implies, the curl of a vector measures how much the vector twists around a point. A coordinate free definition of the curl of an arbitrary vector **W** is

$$(\nabla \times \mathbf{W}) \cdot \hat{\mathbf{S}} = \lim_{S \to 0} \frac{1}{S} \oint_C \mathbf{W} \cdot d\mathbf{l}, \tag{10.41}$$

where S is the area of any surface bordered by the closed curve C, and $\hat{\mathbf{S}}$ is a unit vector normal to the surface S.

Equation (10.41) gives the component of $\nabla \times \mathbf{W}$ in the direction of $\hat{\mathbf{S}}$ and suggests a way of computing the curl numerically. We divide space into cubes of linear dimension Δl. The rectangular components of **W** can be defined either on the edges or on

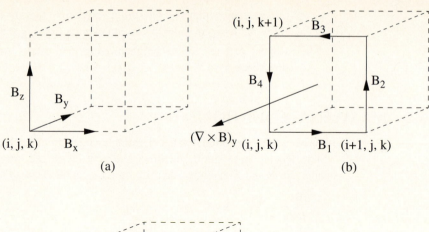

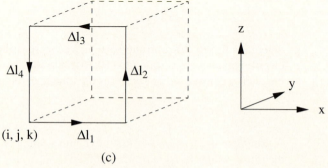

Fig. 10.5 Calculation of the curl of **B** defined on the edges of a cube. (a) The edge vector **B** associated with cube (i, j, k). (b) The components B_i along the edges of the front face of the cube. $B_1 = B_x(i, j, k)$, $B_2 = B_z(i + 1, j, k)$, $B_3 = -B_x(i, j, k + 1)$, and $B_4 = -B_z(i, j, k)$. (c) The vector components $\Delta\mathbf{l}_i$ on the edges of the front face. (The y-component of $\nabla \times \mathbf{B}$ defined on the face points in the negative y direction.)

the faces of the cubes. We compute the curl using both definitions. We first consider a vector **B** that is defined on the edges of the cubes so that the curl of **B** is defined on the faces. (We use the notation **B** because we will find that it is convenient to define the magnetic field in this way.) Associated with each cube is one edge vector and one face vector. We label the cube by the coordinates corresponding to its lower left front corner (see Fig. 10.5a). The three components of **B** associated with this cube are shown in Fig. 10.5a. The other edges of the cube are associated with B vectors defined at neighboring cubes.

The discrete version of (10.41) for the component of $\nabla \times \mathbf{B}$ defined on the front face of the cube (i, j, k) is

$$(\nabla \times \mathbf{B}) \cdot \hat{\mathbf{S}} = \frac{1}{(\Delta l)^2} \sum_{i=1}^{4} B_i \Delta l_i, \tag{10.42}$$

where $S = (\Delta l)^2$, and B_i and l_i are shown in Figs. 10.5b and 10.5c, respectively. Note that two of the components of **B** are associated with neighboring cubes.

The components of a vector also can be defined on the faces of the cubes. We call this vector **E** because it will be convenient to define the electric field in this way. In Fig. 10.6a we show the components of **E** associated with the cube (i, j, k). Because **E** is normal to a cube face, the components of $\nabla \times \mathbf{E}$ lie on the edges. The components E_i and l_i are shown in Figs. 10.6b and 10.6c respectively. The form of the discrete version of $\nabla \times \mathbf{E}$ is similar to (10.42) with B_i replaced by E_i, where E_i and l_i are shown in Figs. 10.6b and 10.6c respectively. The z-component of $\nabla \times \mathbf{E}$ is along the left edge of the front face.

A coordinate free definition of the divergence of the vector field **W** is

$$\nabla \cdot \mathbf{W} = \lim_{V \to 0} \frac{1}{V} \oint_S \mathbf{W} \cdot d\mathbf{S}, \tag{10.43}$$

where V is the volume enclosed by the closed surface **S**. The divergence measures the average flow of the vector through a closed surface. An example of the discrete version of (10.43) is given in (10.44).

We now discuss where to define the quantities $\rho, \mathbf{j}, \mathbf{E},$ and **B** on the grid. It is natural to define the charge density ρ at the center of a cube. From the continuity equation (10.38), we see that this definition leads us to define **j** at the faces of the cube. Hence, each face of a cube has a number associated with it corresponding to the current density flowing parallel to the outward normal to that face. Given the definition of **j** on the grid, we see from (10.37) that the electric field **E** and **j** should be defined at the same places, and hence we define the electric field on the faces of the cubes. Because **E** is defined on the faces, it is natural to define the magnetic field **B** on the edges of the cubes. Our definitions of the vectors **j**, **E**, and **B** on the grid are now complete.

We label the faces of cube c by the symbol f_c. If we use the simplest finite difference method with a discrete time step Δt and discrete spatial interval $\Delta x = \Delta y = \Delta z \equiv \Delta l$, we can write the continuity equation as:

$$\left[\rho(c, t + \tfrac{1}{2}\Delta) - \rho(c, t - \tfrac{1}{2}\Delta t) \right] = -\frac{\Delta t}{\Delta l} \sum_{f_c=1}^{6} j(f_c, t). \tag{10.44}$$

The factor of $1/\Delta l$ comes from the area of a face $(\Delta l)^2$ used in the surface integral in (10.43) divided by the volume $(\Delta l)^3$ of a cube. In the same spirit, the discretization of (10.37) can be written as:

$$E(f, t + \tfrac{1}{2}\Delta t) - E(f, t - \tfrac{1}{2}\Delta t) = \Delta t \left[\nabla \times \mathbf{B} - 4\pi j(f, t) \right]. \tag{10.45}$$

Note that **E** in (10.45) and ρ in (10.44) are defined at different times than **j**. As usual, we choose units such that $c = 1$.

We next need to define a square around which we can discretize the curl. If **E** is defined on the faces, it is natural to use the square that is the border of the faces. As

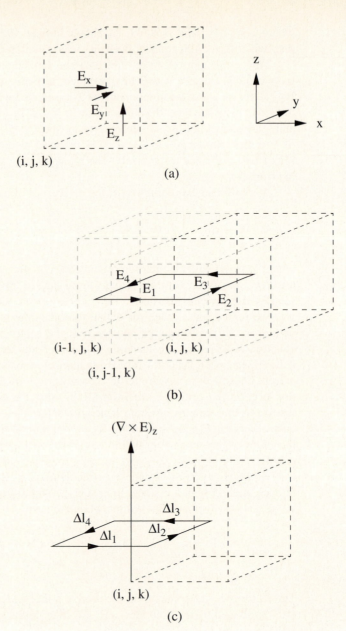

Fig. 10.6 Calculation of the curl of the vector **E** defined on the faces of a cube. (a) The face vector **E** associated with the cube (i, j, k). The components associated with the left, front, and bottom faces are $E_x(i, j, k)$, $E_y(i, j, k)$, $E_z(i, j, k)$ respectively. (b) The components E_i on the faces that share the front left edge of the cube (i, j, k). $E_1 = E_x(i, j - 1, k)$, $E_2 = E_y(i, j, k)$, $E_3 = -E_x(i, j, k)$, and $E_4 = -E_y(i - 1, j, k)$. The cubes associated with E_1 and E_4 also are shown. (c) The vector components Δl_i on the faces that share the left front edge of the cube. (The z-component of the curl of **E** defined on the left edge points in the positive z direction.)

we have discussed, this choice implies that we should define the magnetic field on the edges of the cubes. We write (10.45) as:

$$E(f, t + \frac{1}{2}\Delta t) - E(f, t - \frac{1}{2}\Delta t) = \Delta t \Big[\frac{1}{\Delta l} \sum_{e_f=1}^{4} B(e_f, t) - 4\pi j(f, t) \Big], \quad (10.46)$$

where the sum is over e_f, the four edges of the face f (see Fig. 10.6b). Note that B is defined at the same time as j. In a similar way we can write the discrete form of (10.36) as:

$$B(e, t + \Delta t) - B(e, t) = -\frac{\Delta t}{\Delta l} \sum_{f_e=1}^{4} E(f_e, t + \frac{1}{2}\Delta t), \quad (10.47)$$

where the sum is over f_e, the four faces that share the same edge e (see Fig. 10.6b).

We now have a well defined algorithm for computing the spatial dependence of the electric and magnetic field, the charge density, and the current density as a function of time. This algorithm was developed by Yee, an electrical engineer, in 1966, and independently by Visscher, a physicist, in 1988 who also showed that all of the integral relations and other theorems that are satisfied by the continuum fields are also satisfied for the discrete fields.

Perhaps the most difficult part of the method is specifying the initial conditions since we cannot simply place a charge somewhere. The reason is that the initial fields appropriate for this charge would not be present. Indeed, our rules for updating the fields and the charge densities reflect the fact that the electric and magnetic fields do not appear instantaneously at all positions in space when a charge appears, but instead evolve from the initial appearance of a charge. Of course, charges do not appear out of nowhere, but appear by disassociating the charges from neutral objects. Conceptually, the simplest initial condition corresponds to two charges of opposite sign moving oppositely to each other. This condition corresponds to an initial current on one face. From this current, a charge density and electric field appears using (10.44) and (10.46), respectively, and a magnetic field appears using (10.47).

Because we cannot compute the fields for an infinite lattice, we need to specify the boundary conditions. The easiest method is to use fixed boundary conditions such that the fields vanish at the edges of the lattice. If the lattice is sufficiently large, fixed boundary conditions are a reasonable approximation. However, fixed boundary conditions usually lead to nonphysical reflections off the edges, and a variety of approaches have been used including boundary conditions equivalent to a conducting medium that gradually absorbs the fields. In some cases physically motivated boundary conditions can be employed. For example, in simulations of microwave cavity resonators (see Problem 10.19), the appropriate boundary conditions are that the tangential component of **E** and the normal component of **B** vanish at the boundary.

As we have noted, **E** and ρ are defined at different times than **B** and **j**. This "half-step" approach leads to well behaved equations that are stable over a range of parameters. An analysis of the stability requirement for the Yee-Visscher algorithm shows that the time step Δt must be smaller than the spatial grid Δl by:

$$c\Delta t \leq \frac{\Delta l}{\sqrt{3}}. \qquad \text{(stability requirement)} \qquad (10.48)$$

Your understanding of the Yee-Visscher algorithm for finding solutions to Maxwell's equations will be enhanced by carefully reading the program listing for *Program maxwell* given in the following. The program uses a special True BASIC graphics subroutine PICTURE arrow that is called by a DRAW statement. Such a subroutine has arguments just like any other subroutine. Its utility is that it creates a graphical image that can then be rotated or shifted in space.

```
PROGRAM maxwell
! implementation of Yee-Visscher algorithm
LIBRARY "csgraphics"
PUBLIC E(0 to 21,0 to 21,0 to 21,3),B(0 to 21,0 to 21,0 to 21,3)
PUBLIC j(0 to 21,0 to 21,0 to 21,3)
PUBLIC n(3),dt,dl
PUBLIC mid,fpi,dl_1
CALL initial(escale,bscale,jscale)
CALL screen(#1,#2,#3)
DO
  CALL current(t)
  CALL newE
  CALL newB
  CALL plotfields(escale,bscale,jscale,#1,#2,#3)
  DO
  LOOP until key input
  GET KEY k
LOOP until k = ord("s")
END

SUB initial(escale,bscale,jscale)
    DECLARE PUBLIC E(,,,),B(,,,)
    DECLARE PUBLIC n(),dt
    DECLARE PUBLIC mid,fpi,dl_1
    LET dt = 0.03
    LET n(1) = 8
    LET n(2) = 8
    LET n(3) = 8
    LET mid = n(1)/2
    LET dl = 0.1
    LET dl_1 = 1/dl
    LET fpi = 4*pi
    LET escale = dl/(4*pi*dt)
    LET bscale = escale*dl/dt
    LET jscale = 1
    FOR x = 1 to n(1)
```

```
                  FOR y = 1 to n(2)
                    FOR z = 1 to n(3)
                      FOR comp = 1 to 3
                        LET E(x,y,z,comp) = 0
                        LET B(x,y,z,comp) = 0
                      NEXT comp
                    NEXT z
                  NEXT y
              NEXT x
          END SUB

          SUB current(t)
              DECLARE PUBLIC j(,,,),n(),mid
              ! steady current loop in x-y plane turned on at t = 0 and left on
              LET j(mid,mid,mid,2) = 1
              LET j(mid,mid,mid,1) = -1
              LET j(mid-1,mid,mid,2) = -1
              LET j(mid,mid-1,mid,1) = 1
          END SUB

          SUB newE
              ! E defined at the faces
              DECLARE PUBLIC E(,,,),B(,,,),j(,,,)
              DECLARE PUBLIC n(),dt,dl_1,fpi
              FOR x = 1 to n(1)
                FOR y = 1 to n(2)
                  FOR z = 1 to n(3)
                    LET curlBx = B(x,y,z,2)+B(x,y+1,z,3)-B(x,y,z+1,2)-B(x,y,z,3)
                    LET curlBx = curlBx*dl_1
                    LET E(x,y,z,1) = E(x,y,z,1) + dt*(curlBx - fpi*j(x,y,z,1))
                    LET curlBy = B(x,y,z,3)-B(x,y,z,1)+B(x,y,z+1,1)-B(x+1,y,z,3)
                    LET curlBy = curlBy*dl_1
                    LET E(x,y,z,2) = E(x,y,z,2) + dt*(curlBy - fpi*j(x,y,z,2))
                    LET curlBz = B(x,y,z,1)+B(x+1,y,z,2)-B(x,y+1,z,1)-B(x,y,z,2)
                    LET curlBz = curlBz*dl_1
                    LET E(x,y,z,3) = E(x,y,z,3) + dt*(curlBz - fpi*j(x,y,z,3))
                  NEXT z
                NEXT y
              NEXT x
          END SUB

          SUB newB
              ! B defined at the edges
              DECLARE PUBLIC E(,,,),B(,,,)
              DECLARE PUBLIC n(),dt,dl_1
              FOR x = 1 to n(1)
                FOR y = 1 to n(2)
                  FOR z = 1 to n(3)
                    LET curlEx = E(x,y,z,3)-E(x,y,z,2)-E(x,y-1,z,3)+E(x,y,z-1,2)
```

```
              LET curlEx = curlEx*dl_1
              LET B(x,y,z,1) = B(x,y,z,1) - curlEx*dt
              LET curlEy = E(x,y,z,1)-E(x,y,z,3)-E(x,y,z-1,1)+E(x-1,y,z,3)
              LET curlEy = curlEy*dl_1
              LET B(x,y,z,2) = B(x,y,z,2) - curlEy*dt
              LET curlEz = E(x,y,z,2)-E(x,y,z,1)-E(x-1,y,z,2)+E(x,y-1,z,1)
              LET curlEz = curlEz*dl_1
              LET B(x,y,z,3) = B(x,y,z,3) - curlEz*dt
          NEXT z
        NEXT y
     NEXT x
END SUB

SUB screen(#1,#2,#3)
    DECLARE PUBLIC n()
    LET L = n(1)
    CALL compute_aspect_ratio(L+1,xwin,ywin)
    SET BACKGROUND COLOR "black"
    OPEN #1: screen 0,.5,0,.5
    SET WINDOW 0,xwin,0,ywin
    SET COLOR "white"
    OPEN #2: screen 0.5,1,0.5,1
    SET WINDOW 0,xwin,0,ywin
    SET COLOR "white"
    OPEN #3: screen 0.5,1,0,0.5
    SET WINDOW 0,xwin,0,ywin
    SET COLOR "white"
    OPEN #4: screen 0,0.5,0.5,1
    SET COLOR "white"
    SET CURSOR 1,1
    PRINT "Type s to stop"
    PRINT
    PRINT "Type any other key for next time step"
END SUB

SUB plotfields(escale,bscale,jscale,#1,#2,#3)
    DECLARE PUBLIC E(,,,),B(,,,),j(,,,)
    DECLARE PUBLIC n(),dt,mid
    WINDOW #1
    CLEAR
    PRINT "E(x,y)"
    FOR x = 1 to n(1)
      FOR y = 1 to n(2)
        CALL plotarrow(E(x,y,mid,1),x,y,escale,0,0.5,0,pi)
        CALL plotarrow(E(x,y,mid,2),x,y,escale,0.5,0,pi/2,3*pi/2)
      NEXT y
    NEXT x
    WINDOW #2
    CLEAR
```

```
            PRINT "B(x,z)"
            FOR x = 1 to n(1)
              FOR z = 1 to n(3)
                CALL plotarrow(B(x,mid,z,1),x,z,bscale,0.5,0,0,pi)
                CALL plotarrow(B(x,mid,z,3),x,z,bscale,0,0.5,pi/2,3*pi/2)
              NEXT z
            NEXT x
            WINDOW #3
            CLEAR
            PRINT "j(x,y)"
            FOR x = 1 to n(1)
              FOR y = 1 to n(2)
                CALL plotarrow(j(x,y,mid,1),x,y,jscale,0,0.5,0,pi)
                CALL plotarrow(j(x,y,mid,2),x,y,jscale,0.5,0,pi/2,3*pi/2)
              NEXT y
            NEXT x
        END SUB

        SUB plotarrow(V,x,y,scale,shiftx,shifty,angle1,angle2)
            IF V > 0 then
                DRAW arrow(V/scale) with rotate(angle1)*shift(x+shiftx,y+shifty)
            ELSE IF V < 0 then
                DRAW arrow(-V/scale) with rotate(angle2)*shift(x+shiftx,y+shifty)
            END IF
        END SUB

        PICTURE arrow(x)
            SET COLOR "yellow"
            PLOT LINES: -0.25*x,0;0.25*x,0;0.12*x,0.12*x
            PLOT LINES: 0.25*x,0;0.12*x,-0.12*x
            SET COLOR "white"
        END PICTURE
```

Problem 10.18   Fields from a current loop

a. *Program maxwell* shows the electric field in the x-y plane and the magnetic field in the x-z plane in separate windows. The fields are represented by arrows, whose length is proportional to the field magnitude at each position where the field is defined. A steady current loop in the middle of the x-y plane is turned on at $t = 0$ and left on for all time (see SUB current). Before running the program, predict what you expect to see. Compare your expectations with the results of the simulation. Use $\Delta t = 0.03$, $\Delta l = 0.1$, and take the number of cubes in each direction to be $n(1) = n(2) = n(3) = 8$.

b. Verify the stability requirement (10.48), by running your program with $\Delta t = 0.1$ and $\Delta l = 0.1$. Then try $\Delta t = 0.05$ and $\Delta l = \Delta t \sqrt{3}$. What happens to the results in part (a) if the stability requirement is not satisfied?

c. Modify the current density in part (a) so that **j** is nonzero only for one time step. What happens to the electric and magnetic field vectors?

* d. The amplitude of the fields far from the current loop should be characteristic of radiation fields for which the amplitude falls off as $1/r$, where r is the distance from the current loop to the observation point. Try to detect this dependence (if you have sufficient patience or computer resources).

Problem 10.19 Microwave cavity resonators

a. Cavity resonators are a practical way of storing energy in the form of oscillating electric and magnetic fields without losing as much energy as would be dissipated in a resonant LC circuit. Consider a cubical resonator of linear dimension L whose walls are made of a perfectly conducting material. The tangential components of **E** and the normal component of **B** vanish at the walls. Standing microwaves can be set up in the box of the form (cf. Reitz et al.)

$$E_x = E_{x0} \cos k_x x \sin k_y y \sin k_z z \, e^{i\omega t} \tag{10.49a}$$

$$E_y = E_{y0} \cos k_y y \sin k_x x \sin k_z z \, e^{i\omega t} \tag{10.49b}$$

$$E_z = E_{z0} \cos k_z z \sin k_x x \sin k_y y \, e^{i\omega t}. \tag{10.49c}$$

The wave vector $\mathbf{k} = (k_x, k_y, k_z) = (m_x \pi / L, m_y \pi / L, m_z \pi / L)$, where m_x, m_y, and m_z are integers. A particular mode is labeled by the integers (m_x, m_y, m_z). The initial electric field is perpendicular to **k**, and $\omega = ck$. Implement the boundary conditions at $(x = 0, y = 0, z = 0)$ and $(x = L, y = L, z = L)$. Set $\Delta t = 0.05$, $\Delta l = 0.1$, and $L = 1$. At $t = 0$, set $\mathbf{B} = 0$, $\mathbf{j} = 0$ (there are no currents within the cavity), and use (10.49) with $(m_x, m_y, m_z) = (0, 1, 1)$, and $E_{x0} = 1$. Plot the field components at specific positions as a function of t and find the resonant frequency ω. Compare your computed value of ω with the analytical result. Do the magnetic fields change with time? Are they perpendicular to **k** and **E**?

b. Repeat part (a) for two other modes.

c. Repeat part (a) with a uniform random noise added to the initial field at all positions. Assume the amplitude of the noise is δ and describe the resulting fields for $\delta = 0.1$. Are they similar to those without noise? What happens for $\delta = 0.5$? More quantitative results can be found by computing the power spectrum $|E(\omega)|^2$ for the electric field at a few positions. What is the order of magnitude of δ for which the maximum of $|E(\omega)|^2$ at the standing wave frequency is swamped by the noise?

d. Change the shape of the container slightly by removing a 0.1×0.1 cubical box from each of the corners of the original resonator. Do the standing wave frequencies change? Determine the standing wave frequency by adding noise to the initial fields and looking at the power spectrum. How do the standing wave patterns change?

e. Change the shape of the container slightly by adding a 0.1×0.1 cubical box at the center of one of the faces of the original resonator. Do the standing wave frequencies change? How do the standing wave patterns change?

f. Cut a 0.2×0.2 square hole in a face in the y-z plane, and double the computational region in the x direction. Begin with a $(0, 1, 1)$ standing wave, and observe how the fields "leak" out of the hole.

Problem 10.20 Billiard microwave cavity resonators

a. Repeat part (a) of Problem 10.19 for $L_x = L_y = 2$, $L_z = 0.2$, $\Delta l = 0.1$, and $\Delta t = 0.05$. Indicate the magnitude of the electric field in the $L_z = 0.1$ plane by a color code. Choose an initial normal mode field distribution and describe the pattern that you obtain. Then repeat your calculation for a random initial field distribution.

b. Place an approximately circular conductor in the middle of the cavity of radius $r = 0.4$. Describe the patterns that you see. Such a geometry leads to chaotic trajectories for particles moving within such a cavity (see Project 6.6). Is there any evidence of chaotic behavior in the field pattern?

c. Repeat part (b) with the circular conductor placed off center.

10.6 PROJECT

Much of the difficulty in understanding electromagnetic phenomena is visualizing its three-dimensional character. Although True BASIC has an excellent three-dimensional graphics toolkit, we have not used it here because of the difficulty of translating its statements to other programming languages. Many interesting problems can be posed based on the simple, but nontrivial question of how the electromagnetic fields can best be represented visually in various contexts.

Many of the techniques used in this chapter, e.g., the random walk method and the relaxation method for solving Laplace's equation, have applications in other fields, especially problems in fluid flow and transport. Similarly, the multigrid method, discussed below, has far reaching applications.

Project 10.1 Multigrid method

In general, the relaxation method for solving Laplace's equation is very slow even using overrelaxation. The reason is that the local updates of the relaxation method cannot quickly take into account effects at very large length scales. The *multigrid method* greatly improves performance by using relaxation at many length scales. The important idea is to use a relaxation method to find the values of the potential on coarser and coarser grids, and then use the coarse grid values to determine the fine grid values. The fine grid relaxation updates take into account effects at short length scales. If we define the initial grid by a lattice spacing $b = 1$, then the coarser grids are characterized by $b = 2^n$, where n is the grid level. We need to decide how to use the fine grid values of the potential to assign values to a coarser grid, and then how to use a coarse grid to assign values to a finer grid. The first step is called prolongation and the second step is called restriction. There is some flexibility on how to do these two operations. We discuss one approach.

We define the points of the coarse grid as every other point of the fine grid. That is, if the set $\{i, j\}$ represents the positions of the points of the fine grid, then $\{2i, 2j\}$ represents the positions of the coarse grid points. The fine grid points that are at the same position as a coarse grid point are assigned the value of the potential of the corresponding coarse grid point. The fine grid points that have two coarse grid points as nearest neighbors are assigned the average value of these two coarse grid points. The other fine grid points have four coarse grid points as next nearest neighbors and are assigned the average value of these four coarse grid points. This prescription defines the restriction of the coarse grid to the fine grid.

In the full weighting prolongation method, each coarse grid point receives one fourth of the potential of the fine grid point at the same position, one eighth of the potential for the four nearest neighbor points of the fine grid, and one sixteenth of the potential for the four next nearest neighbor points of the fine grid. Note that the sum of these fractions, $1/4 + 4(1/8) + 4(1/16)$, adds up to unity. An alternative procedure, known as half weighting, ignores the next nearest neighbors and uses one half of the potential of the fine grid point at the same position as the coarse grid point.

a. Write a program that implements the multigrid method using Gauss-Seidel relaxation on a checkerboard lattice (see Problem 10.9b). In its simplest form the program should allow the user to intervene and decide whether to go to a finer or coarser grid, or to remain at the same level for the next relaxation step. Also have the program print the potential at each point of the current level after each relaxation step. Test your program on a 4×4 grid whose boundary points are all equal to unity, and whose initial internal points are set to zero. Make sure that the boundary points of the coarse grids also are set to unity.

b. The exact solution for part (a) gives a potential of unity at each point. How many relaxation steps does it take to reach unity within 0.1% at every point by simply using the 4×4 grid? How many steps does it take if you use one coarse grid and continue until the coarse grid values are within 0.1% of unity? Is it necessary to carry out any fine grid relaxation steps to reach the desired accuracy on the fine grid? Next start with the coarsest scale, which is just one point. How many relaxation steps does it take now?

c. Repeat part (b), but change the boundary so that one side of the boundary is held at a potential of 0.5. Experiment with different sequences of prolongation, restriction, and relaxation.

d. Assume that the boundary points alternate between zero and unity, and repeat part (b). Does the multigrid method work? Should one go up and down in levels many times instead of staying at the coarsest level and then going down to the finest level?

References and Suggestions for Further Reading

Forman S. Acton, *Numerical Methods That Work*, Harper & Row (1970); corrected edition, Mathematical Association of America (1990). Chapter 18 discusses solutions to Laplace's equation using the relaxation method and alternative approaches.

Charles K. Birdsall and A. Bruce Langdon, *Plasma Physics via Computer Simulation*, McGraw-Hill (1985).

D. H. Choi and W. J. R. Hoefer, "The finite-difference-time-domain method and its application to eigenvalue problems," *IEEE Trans. Microwave Theory and Techniques* **34**, 1464 (1986). The authors use Yee's algorithm to model microwave cavity resonators.

David M. Cook, *The Theory of the Electromagnetic Field*, Prentice Hall (1975). One of the first books to introduce numerical methods in the context of electromagnetism.

Robert Ehrlich, Jaroslaw Tuszynski, Lyle Roelofs, and Ronald Stoner, *Electricity and Magnetism Simulations: The Consortium for Upper-Level Physics Software*, John Wiley (1995).

Richard P. Feynman, Robert B. Leighton, and Matthew Sands, *The Feynman Lectures on Physics*, Vol. 2, Addison-Wesley (1963).

T. E. Freeman, "One-, two- or three-dimensional fields?," *Am. J. Phys.* **63**, 273 (1995).

Robert M. Eisberg and Lawrence S. Lerner, *Physics*, Vol. 2, McGraw-Hill (1981). An introductory level text that uses numerical methods to find the electric field lines and solutions to Laplace's equation.

R. L. Gibbs, Charles W. Beason, and James D. Beason, "Solutions to boundary value problems of the potential type by random walk method," *Am. J. Phys.* **43**, 782 (1975).

R. H. Good, "Dipole radiation: Simulation using a microcomputer," *Am. J. Phys.* **52**, 1150 (1984). The author reports on a graphical simulation of dipole radiation.

R. W. Hockney and J. W. Eastwood, *Computer Simulation Using Particles*, McGraw-Hill (1981).

Steven E. Koonin and Dawn C. Meredith, *Computational Physics*, Addison-Wesley (1990). See Chapter 6 for a discussion of the numerical solution of elliptic partial differential equations of which Laplace's and Poisson's equations are examples.

William M. MacDonald, "Discretization and truncation errors in a numerical solution of Laplace's equation," *Amer. J. Phys.* **62**, 169 (1994).

William H. Press and Saul A. Teukolsky, "Multigrid Methods for Boundary Value Problems I," *Computers in Physics* **5**(5), 514 (1991).

Edward M. Purcell, *Electricity and Magnetism*, second edition, Berkeley Physics Course, Vol. 2, McGraw-Hill (1985). A well known text that discusses the relaxation method.

John R. Reitz, Frederick J. Milford, and Robert W. Christy, *Foundations of Electromagnetic Theory*, third edition, Addison-Wesley (1979). This text discusses microwave cavity resonators.

Matthew N. O. Sadiku, *Numerical Techniques in Electromagnetics*, CRC Press (1992).

A. Taflove and M. E. Brodwin, "Numerical solution of steady state electromagnetic scattering problems using the time dependent Maxwell equations," *IEEE Trans. Microwave Theory and Techniques* **23**, 623 (1975). The authors derive the stability conditions for the Yee algorithm.

P. B. Visscher, *Fields and Electrodynamics*, John Wiley & Sons (1988). An intermediate level text that incorporates computer simulations and analysis into its development.

P. J. Walker and I. D. Johnston, "Computer model clarifies spontaneous charge distribution in conductors," *Computers in Physics* **9**, 42 (1995).

Gregg Williams, "An introduction to relaxation methods," *Byte* **12**(1), 111 (1987). The author discusses the application of relaxation methods to the solution of the two-dimensional Poisson's equation.

K. S. Yee, *IEEE Trans. Antennas and Propagation* **14**, 302 (1966). Yee uses the discretized Maxwell's equations to model the scattering of electromagnetic waves off a perfectly conducting rectangular obstacle.

CHAPTER

11
Numerical Integration and Monte Carlo Methods

Simple classical and Monte Carlo methods are illustrated in the context of the numerical evaluation of definite integrals.

11.1   NUMERICAL INTEGRATION METHODS IN ONE DIMENSION

Monte Carlo methods were introduced in Chapter 7 in the context of systems that are intrinsically random. In this chapter we will find that we can use sequences of random numbers to estimate definite integrals, a problem that seemingly has nothing to do with randomness. To place the Monte Carlo numerical integration methods in perspective, we first discuss the common classical methods of determining the numerical value of definite integrals. We will see that the classical methods, although usually preferable in low dimensions, are impractical for multidimensional integrals and that Monte Carlo methods are essential for the evaluation of the latter if the number of dimensions is sufficiently high.

Consider a one-dimensional definite integral of the form

$$F = \int_a^b f(x)\, dx. \tag{11.1}$$

For some choices of the integrand $f(x)$, the integration in (11.1) can be done analytically, found in reference books, or evaluated as an asymptotic series. However, there are many common functions whose integrals are intractable and must be evaluated numerically.

The classical methods of numerical integration are based on the geometrical interpretation of the integral (11.1) as the area under the curve of the function $f(x)$ from $x = a$ to $x = b$ (see Fig. 11.1). The x-axis is divided into n equal intervals of width Δx, where Δx is given by

$$\Delta x = \frac{b - a}{n}, \tag{11.2a}$$

and

$$x_n = x_0 + n\,\Delta x. \tag{11.2b}$$

In the above, $x_0 = a$ and $x_n = b$.

The simplest estimate of the area under the curve $f(x)$ is the sum of rectangles shown in Fig. 11.2. In the usual *rectangular* approximation, $f(x)$ is evaluated at the *beginning* of the interval, and the estimate F_n of the integral is given by

$$F_n = \sum_{i=0}^{n-1} f(x_i)\Delta x. \qquad \text{(rectangular approximation)} \tag{11.3}$$

In the *trapezoidal* approximation the integral is estimated by computing the area under a trapezoid with one side equal to $f(x)$ at the beginning of the interval and the other side equal to $f(x)$ at the end of the interval. This approximation is equivalent to replacing the function by a straight line connecting the values of $f(x)$ at the beginning and the end of each interval. Because the approximate area under the curve from x_i to x_{i+1} is given by $\frac{1}{2}[f(x_{i+1}) + f(x_i)]\Delta x$, the total area F_n is given by

$$F_n = \left[\frac{1}{2}f(x_0) + \sum_{i=1}^{n-1} f(x_i) + \frac{1}{2}f(x_n)\right]\Delta x. \quad \text{(trapezoidal approximation)} \tag{11.4}$$

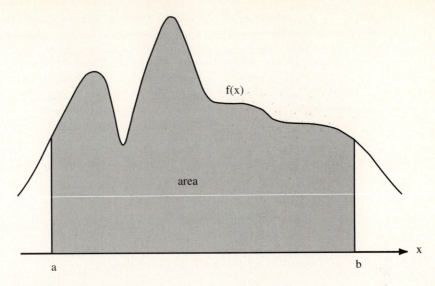

Fig. 11.1   The integral F equals the area under the curve $f(x)$.

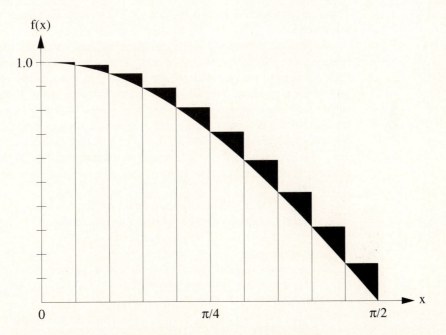

Fig. 11.2   The rectangular approximation for $f(x) = \cos x$ for $0 \le x \le \pi/2$. The error in the rectangular approximation is shaded. Numerical values of the errors for various values of the number of intervals n are given in Table 11.1.

A generally more accurate method is to use a quadratic or parabolic interpolation procedure through adjacent triplets of points. For example, the equation of the second-order polynomial that passes through the points (x_0, y_0), (x_1, y_1), and (x_2, y_2) can be written as

$$y(x) = y_0 \frac{(x - x_1)(x - x_2)}{(x_0 - x_1)(x_0 - x_2)} + y_1 \frac{(x - x_0)(x - x_2)}{(x_1 - x_0)(x_1 - x_2)}$$
$$+ y_2 \frac{(x - x_0)(x - x_1)}{(x_2 - x_0)(x_2 - x_1)}. \tag{11.5}$$

What is the value of $y(x)$ at $x = x_1$? The area under the parabola $y(x)$ between x_0 and x_2 can be found by simple integration and is given by

$$F_0 = \frac{1}{3} (y_0 + 4y_1 + y_2) \, \Delta x, \tag{11.6}$$

where $\Delta x = x_1 - x_0 = x_2 - x_1$. The total area under all the parabolic segments yields the parabolic approximation for the total area:

$$F_n = \frac{1}{3} \big[f(x_0) + 4f(x_1) + 2f(x_2) + 4f(x_3) + \ldots$$
$$+ 2f(x_{n-2}) + 4f(x_{n-1}) + f(x_n) \big] \Delta x. \qquad \text{(Simpson's rule)} \tag{11.7}$$

This approximation is known as *Simpson's rule*. Note that Simpson's rule requires that n be even.

In practice, Simpson's rule is adequate for functions $f(x)$ that are reasonably well behaved, i.e., functions that can be adequately represented by a polynomial. If $f(x)$ is such a function, we can evaluate the area for a given number of intervals n and then double the number of intervals and evaluate the area again. If the two evaluations are sufficiently close to one another, we may stop. Otherwise, we again double n until we achieve the desired accuracy. Of course, this strategy will fail if $f(x)$ is not well-behaved. An example of a poorly-behaved function is $f(x) = x^{-1/3}$ at $x = 0$, where $f(x)$ diverges. Another example where this strategy might fail is a case where a limit of integration is equal to $\pm\infty$. In many cases we can eliminate the problem by a change of variables.

A program that applies the rectangular approximation to the integral of $f(x)$ and successively doubles the number of intervals is given below.

```
PROGRAM integ
! compute integral of f(x) from x = a to x = b
! using rectangular approximation
CALL initial(a,b,sum,dx,delta,n)
DO
    CALL sumf(a,b,sum,dx,delta)
    CALL output(sum,dx,n)
    CALL double(dx,delta,n)
LOOP until key input
GET KEY k
END
```

```
SUB initial(a,b,sum,dx,delta,n)
    DECLARE DEF f
    INPUT prompt "lower limit a = ": a
    INPUT prompt "upper limit b = ": b
    LET n = 2                          ! initial number of intervals
    LET dx = (b - a)/n
    LET delta = dx                     ! delta = dx only for n = 2
    LET sum = f(a)    ! -> (f(a) + f(b))/2 for trapezoidal rule
    PRINT
    PRINT "    n","rectangular approximation"
    PRINT
END SUB

SUB sumf(a,b,sum,dx,delta)
    DECLARE DEF f
    LET x = a + dx                     ! dx is integration interval
    DO while x < b
       LET sum = sum + f(x)
       LET x = x + delta
    LOOP
END SUB

SUB double(dx,delta,n)
    ! double number of intervals
    LET delta = dx
    LET n = 2*n
    LET dx = 0.5*dx                    ! dx does not equal delta for n > 2
END SUB

SUB output(sum,dx,n)
    PRINT using "#######": n;
    PRINT,
    PRINT using "####.#######": sum*dx
END SUB

DEF f(x) = cos(x)
```

Let us consider the accuracy of the rectangular approximation for the integral of $f(x) = \cos x$ from $x = 0$ to $x = \pi/2$ by comparing the numerical results shown in Table 11.1 with the exact answer of unity. We see that the error decreases as n^{-1}. This observed n dependence of the error is consistent with the analytical derivation of the n dependence of the error obtained in Appendix 11A. We explore the n dependence of the error associated with other numerical integration methods in Problems 11.1 and 11.2.

Problem 11.1   The rectangular and midpoint approximations

a. *Program integ* implements the rectangular approximation and doubles the number of intervals until the desired accuracy is reached. To understand the nature of the program, assume that $a = 0$ and $b = 1$. For $n = 2$, SUB initial

n	F_n	Δ_n
2	1.34076	0.34076
4	1.18347	0.18347
8	1.09496	0.09496
16	1.04828	0.04828
32	1.02434	0.02434
64	1.01222	0.01222
128	1.00612	0.00612
256	1.00306	0.00306
512	1.00153	0.00153
1024	1.00077	0.00077

Table 11.1   Rectangular approximation estimates of the integral of $\cos x$ from $x = 0$ to $x = \pi/2$ as a function of n, the number of intervals. The error Δ_n is the difference between the rectangular approximation and the exact result of unity. Note that Δ_n decreases approximately as n^{-1}, i.e., if n is increased by a factor of 2, Δ_n decreases by a factor 2.

assigns the values sum $= f(0)$, dx $= 0.5$, and delta $= 0.5$. The first call to SUB sumf yields sum $=$ f(0) $+$ f(0.5). For $n = 4$, we have delta $= 0.5$ and dx $= 0.25$, and the second call to SUB sumf yields sum $=$ f(0) $+$ f(0.25) $+$ f(0.5) $+$ f(0.75). Note that for $n \geq 4$, the value of delta lags behind the value of dx. Explain how *Program integ* works by doing a hand calculation of a particular choice of $f(x)$. Then test *Program integ* by using several functions that you can integrate exactly.

b.   Use the rectangular approximation to determine numerical estimates for the definite integrals of $f(x) = 2x + 3x^2 + 4x^3$ and $f(x) = e^{-x}$ for $0 \leq x \leq 1$. What is the approximate n dependence of the error in each case?

c.   A straightforward modification of the rectangular approximation is to evaluate $f(x)$ at the *midpoint* of each interval. Make the necessary modifications of *Program integ* and estimate the integral of $f(x) = \cos x$ in the interval $0 \leq x \leq \pi/2$. How does the magnitude of the error compare with the results shown in Table 11.1? What is the approximate dependence of the error on n?

d.   Use the midpoint approximation to estimate the definite integrals considered in part (b). What is the approximate n dependence of the error in each case?

The trapezoidal approximation can be obtained by replacing the initial value of the sum in *Program integ* by

```
LET sum = 0.5*(f(a) + f(b)) + f(a+dx)
```

Simpson's rule can be obtained using the results of the trapezoidal approximation. If the quantity trapezoid is the trapezoidal approximation to the integral, then the following additional code in SUB output will generate the parabolic approximation (Simpson's rule) to the integral:

```
LET simpson = (4*sum*dx - trapezoid)/3
LET trapezoid = sum*dx
```

The quantity trapezoid used to compute simpson is computed in the previous call to SUB output and must be saved between subroutine calls. Note also that in the computation of simpson, the quantity 4*sum*dx uses a value of dx that is one-half of the value used in the previous subroutine call to compute the quantity trapezoid.

Problem 11.2   The trapezoidal approximation and Simpson's rule

a. Modify *Program* integ so that that the trapezoidal approximation and Simpson's rule are computed and printed simultaneously.

b. Use both approximations to estimate the integrals of $f(x) = 2x + 3x^2 + 4x^3$ and $f(x) = e^{-x}$ for $0 \le x \le 1$. What is the approximate n dependence of the error in each case? Which approximation yields the best results for the same computation time?

c. Use Simpson's rule to estimate the integral of $f(x) = (2\pi)^{-1/2} e^{-x^2}$ for $-1 \le x \le 1$. Do you obtain the same estimate of the integral by choosing the interval [0, 1] and then multiplying by two? Why or why not?

d. Evaluate the integral of the function $f(x) = 4\sqrt{1 - x^2}$ for $-1 \le x \le 1$. What value of n is needed for four decimal accuracy? The reason for the slow convergence can be understood by reading Appendix 11A.

* e. So far, our strategy for numerically estimating the value of definite integrals has been to choose one or more of the classical integration formulae and to compute F_n and F_{2n} for reasonable values of n. If the difference $|F_{2n} - F_n|$ is too large, then we double n until the desired accuracy is reached. The success of this strategy is based on the implicit assumption that the sequence $F_n, F_{2n}, \cdots$ converges to the true integral F. Is there a way of extrapolating this sequence to the limit? Let us explore this idea by using the trapezoidal approximation. Because the error for this approximation decreases approximately as n^{-2}, we can write $F = F_n + Cn^{-2}$, and plot F_n as a function of n^{-2} to obtain the extrapolated result F. Apply this procedure to the integrals considered in some of the above problems and compare your results to those found from the trapezoidal approximation and Simpson's rule alone. A more sophisticated application of this idea is known as *Romberg* integration (cf. Press et al.).

11.2   SIMPLE MONTE CARLO EVALUATION OF INTEGRALS

We now explore a totally different method of estimating integrals. Let us introduce this method by asking, "Can we use a pile of stones to measure the area of a pond whose shape is irregular?" Suppose the pond is in the middle of a field of known area A. One way to estimate the area of the pond is to throw the stones so that they land at random within the boundary of the field and count the number of splashes that occur when a stone lands in a pond. The area of the pond is approximately the area of the field times the fraction of stones that make a splash. This simple procedure is an example of a *Monte Carlo* method.

More explicitly, imagine a rectangle of height H, width $(b - a)$, and area $A = H(b - a)$ such that the function $f(x)$ is within the boundaries of the rectangle (see Fig. 11.3). Compute n pairs of random numbers x_i and y_i with $a \leq x_i \leq b$ and $0 \leq y_i \leq H$. The fraction of points x_i, y_i that satisfy the condition $y_i \leq f(x_i)$ is an estimate of the ratio of the integral of $f(x)$ to the area of the rectangle. Hence, the estimate F_n in the *hit or miss* method is given by

$$F_n = A \frac{n_s}{n}, \qquad \text{(hit or miss method)} \qquad (11.8)$$

where n_s is the number of "splashes" or points below the curve, and n is the total number of points. Note that n in (11.8) should not be confused with the number of intervals used in the numerical methods discussed in Section 11.1.

Another Monte Carlo integration method is based on the mean-value theorem of calculus, which states that the definite integral (11.1) is determined by the average value of the integrand $f(x)$ in the range $a \leq x \leq b$. To determine this average, we choose the x_i at random instead of at regular intervals and *sample* the value of $f(x)$. For the one-dimensional integral (11.1), the estimate F_n of the integral in the *sample mean* method is given by

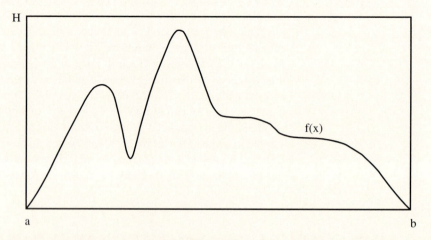

Fig. 11.3   The function $f(x)$ is in the domain determined by the rectangle of height H and width $(b - a)$.

$$F_n = (b - a) <f> = (b - a)\frac{1}{n}\sum_{i=1}^{n} f(x_i), \qquad \text{(sample mean method)} \qquad (11.9)$$

where the x_i are random numbers distributed uniformly in the interval $a \le x_i \le b$, and n is the number of *trials*. Note that the forms of (11.3) and (11.9) are identical except that the n points are chosen with equal spacing in (11.3) and with random spacing in (11.9). We will find that for low dimensional integrals (11.3) is more accurate, but for higher dimensional integrals (11.9) does better.

Problem 11.3 Monte Carlo integration in one dimension

a. Write a program that implements the hit or miss Monte Carlo method. Find the estimate F_n of the integral of $f(x) = 4\sqrt{1 - x^2}$ in the interval $0 \le x \le 1$ as a function of n. Choose $a = 0, b = 1, H = 1$, and compute the mean value of the function $\sqrt{1 - x^2}$. Multiply the estimate by 4 to determine F_n. Calculate the difference between F_n and the exact result of π. This difference is a measure of the error associated with the Monte Carlo estimate. Make a log-log plot of the error as a function of n. What is the approximate functional dependence of the error on n for large n, e.g., $n \ge 10^4$?

b. Estimate the same integral using the sample mean Monte Carlo method (11.9) and compute the error as a function of the number of trials n for $n \ge 10^4$. How many trials are needed to determine F_n to two decimal places? What is the approximate functional dependence of the error on n for large n?

c. Determine the computational time per trial using the two Monte Carlo methods. Which Monte Carlo method is preferred in this case?

*11.3 NUMERICAL INTEGRATION OF MULTIDIMENSIONAL INTEGRALS

Many problems in physics involve averaging over many variables. For example, suppose we know the position and velocity dependence of the total energy of ten interacting particles. In three dimensions each particle has three velocity components and three position components. Hence the total energy is a function of 60 variables, and a calculation of the average energy per particle involves computing a $d = 60$ dimensional integral. (More accurately, the total energy is a function of $60 - 6 = 54$ variables if we use center of mass and relative coordinates.) If we divide each coordinate into p intervals, there would be p^{60} points to sum. Clearly, standard numerical methods such as Simpson's rule would be impractical for this example.

A discussion of the n dependence of the error associated with the standard numerical methods for d-dimensional integrals is given in Appendix 11A. We show that if the error decreases as n^{-a} for $d = 1$, then the error decreases as $n^{-a/d}$ in d dimensions. In contrast, we find (see Section 11.4) that the error for Monte Carlo integration decreases as $n^{-1/2}$ independently of the dimension of the integral. Hence for higher dimensions, Monte Carlo methods are essential.

To illustrate the general method for evaluating multidimensional integrals, we consider the two-dimensional integral

$$F = \int_R f(x, y)\, dx dy,\tag{11.10}$$

where R denotes the region of integration. The extension to higher dimensions is straightforward, but tedious. Form a rectangle that encloses the region R, and divide this rectangle into squares of length h. Assume that the rectangle runs from x_a to x_b in the x direction and from y_a to y_b in the y direction. The total number of squares is $n_x n_y$, where $n_x = (x_b - x_a)/h$ and $n_y = (y_b - y_a)/h$. If we use the midpoint approximation, the integral F is estimated by

$$F \approx \sum_{i=1}^{n_x} \sum_{j=1}^{n_y} f(x_i, y_j) H(x_i, y_j)\, h^2,\tag{11.11}$$

where $x_i = x_a + (i - \frac{1}{2})h$, $y_j = y_a + (j - \frac{1}{2})h$, and the function $H(x, y)$ equals unity if (x, y) is in R and is zero otherwise.

A simple Monte Carlo method for evaluating a two-dimensional integral uses the same rectangular region as in the above, but the n points (x_i, y_i) are chosen at random within the rectangle. The estimate for the integral is then

$$F_n = \frac{A}{n} \sum_{i=1}^{n} f(x_i, y_i) H(x_i, y_i),\tag{11.12}$$

where A is the area of the rectangle. Note that if $f(x, y) = 1$ everywhere, then (11.12) is equivalent to the hit or miss method of calculating the area of the region R. In general, (11.12) represents the area of the region R multiplied by the average value of $f(x, y)$ in R. In Section 11.7 we discuss a more efficient Monte Carlo method for evaluating definite integrals.

Problem 11.4   Two-dimensional numerical integration

a. Write a program to implement the midpoint approximation in two dimensions and integrate the function $f(x, y) = x^2 + 6xy + y^2$ over the region defined by the condition $x^2 + y^2 \le 1$. Use $h = 0.1$, 0.05, 0.025, and if time permits 0.0125. Print the number of squares, n, and the estimate for the integral.

b. Repeat part (a) using a Monte Carlo method and the same number of points n. For each value of n repeat the calculation several times to obtain a crude estimate of the random error.

* Problem 11.5   Volume of a hypersphere

a. The interior of a d-dimensional hypersphere of unit radius is defined by the condition $x_1^2 + x_2^2 + \ldots x_d^2 \le 1$. Write a program that finds the volume of a hypersphere using the midpoint approximation. If you are clever, you can write a program that does any dimension using recursive subroutines. Test your program for $d = 2$ and $d = 3$, and then find the volume for $d = 4$ and $d = 5$. Begin with $h = 0.2$, and decrease h until your results do not change by more than 1%, or until you run out of patience or resources.

b. Repeat part (a) using a Monte Carlo technique. For each value of n, repeat the calculation several times to obtain a rough estimate of the random error. Is a program valid for any d easier to write in this case than in part (a)?

11.4 MONTE CARLO ERROR ANALYSIS

Both the classical numerical integration methods and the Monte Carlo methods yield approximate answers whose accuracy depends on the number of intervals or on the number of trials respectively. So far, we have used our knowledge of the exact value of various integrals to determine that the error in the Monte Carlo method approaches zero as approximately $n^{-1/2}$ for large n, where n is the number of trials. In the following, we will find how to estimate the error when the exact answer is unknown. Our main result is that the n dependence of the error is independent of the nature of the integrand and, most importantly, independent of the number of dimensions.

We found in Problem 11.2 that the error using Simpson's rule for a one-dimensional integral is proportional to n^{-4}, where n is the number of intervals. Because the computational time is roughly proportional to n in both the classical and Monte Carlo methods, we conclude that for low dimensions the classical numerical methods such as Simpson's rule are preferable to Monte Carlo methods unless the domain of integration is very complicated. However, the error in the conventional numerical methods increases with dimension (see Appendix 11A), and Monte Carlo methods are essential for higher dimensional integrals.

Because the appropriate measure of the error in Monte Carlo calculations is subtle, we first find the error for an explicit example. Consider the Monte Carlo evaluation of the integral of $f(x) = 4\sqrt{1 - x^2}$ in the interval [0, 1] (see Problem 11.3). Our calculated result for a particular sequence of $n = 10^4$ random numbers using the sample mean method is $F_n = 3.1489$. How does this result for F_n compare with your result found in Problem 11.3 for the same value of n? By comparing F_n to the exact result of $F = \pi \approx 3.1416$, we find that the error associated with $n = 10^4$ trials is approximately 0.0073.

How can we estimate the error if the exact result is unknown? How can we know if $n = 10^4$ trials is sufficient to achieve the desired accuracy? Of course, we cannot answer these questions definitively because if the actual error in F_n were known, we could correct F_n by the required amount and obtain F. The best we can do is to calculate the *probability* that the true value F is within a certain range centered on F_n.

If the integrand were a constant, then the error would be zero, that is, F_n would equal F for any n. Why? This limiting behavior suggests that a possible measure of the error is the *variance* σ^2 defined by

$$\sigma^2 = <f^2> - <f>^2,$$

(11.13)

where

$$<f> = \frac{1}{n}\sum_{i=1}^{n} f(x_i),$$ (11.14a)

and

$$<f^2> = \frac{1}{n}\sum_{i=1}^{n} f(x_i)^2.$$ (11.14b)

From the definition of the *standard deviation* σ, we see that if f is independent of x, σ is zero. For our example and the same sequence of random numbers used to obtain $F_n = 3.1489$, we obtain $\sigma_n = 0.8850$. This value of σ is two orders of magnitude larger than the actual error, and we conclude that σ cannot be a direct measure of the error. Instead, σ is a measure of how much the function $f(x)$ varies in the interval of interest.

Another clue to finding an appropriate measure of the error can be found by increasing n and seeing how the actual error decreases as n increases. In Table 11.2 we see that as n goes from 10^2 to 10^4, the actual error decreases by a factor of 10, that is, as $\sim 1/n^{\frac{1}{2}}$. However, we also see that σ_n is roughly constant and is much larger than the actual error.

One way to obtain an estimate for the error is to make additional runs of n trials each. Each run of n trials yields a mean value or a single *measurement* that we denote as M_α. In general, these measurements are not equal because each measurement uses a different sequence of random numbers. Table 11.3 shows the results of ten separate measurements of $n = 10^4$ trials each. We see that the actual error varies from measurement to measurement. Qualitatively, the magnitude of the differences between the measurements is similar to the actual errors, and hence these differences are a measure of the error associated with a single measurement. To obtain a quantitative measure of this error, we determine the differences of these measurements using the *standard deviation of the means* σ_m which is defined as

$$\sigma_m{}^2 = <M^2> - <M>^2,$$ (11.15)

where

$$<M> = \frac{1}{m}\sum_{\alpha=1}^{m} M_\alpha,$$ (11.16a)

n	F_n	actual error	σ_n
10^2	3.0692	0.0724	0.8550
10^3	3.1704	0.0288	0.8790
10^4	3.1489	0.0073	0.8850

Table 11.2　Examples of Monte Carlo measurements of the mean value of $f(x) = 4\sqrt{1 - x^2}$ in the interval [0, 1]. The actual error is given by the difference $|F_n - \pi|$. The standard deviation σ_n is found using (11.13).

run α	M_α	actual error
1	3.14892	0.00735
2	3.13255	0.00904
3	3.14042	0.00117
4	3.14600	0.00441
5	3.15257	0.01098
6	3.13972	0.00187
7	3.13107	0.01052
8	3.13585	0.00574
9	3.13442	0.00717
10	3.14047	0.00112

Table 11.3 Examples of Monte Carlo measurements of the mean value of $f(x) = 4\sqrt{1-x^2}$ in the interval [0, 1]. A total of 10 measurements of $n = 10^4$ trials each were made. The mean value M_α and the actual error $|M_\alpha - \pi|$ are shown.

and

$$<M^2> = \frac{1}{m} \sum_{\alpha=1}^{m} M_\alpha{}^2. \tag{11.16b}$$

From the values of M_α in Table 11.3 and the relation (11.15), we find that $\sigma_m = 0.0068$. This value of σ_m is consistent with the results for the actual errors shown in Table 11.3 which we see vary from 0.00112 to 0.01098. Hence we conclude that σ_m, the standard deviation of the means, is a measure of the error for a single measurement. The more precise interpretation of σ_m is that a single measurement has a 68% chance of being within σ_m of the "true" mean. Hence the probable error associated with our first measurement of F_n with $n = 10^4$ is 3.149 ± 0.007.

Although σ_m gives an estimate of the probable error, our method of obtaining σ_m by making additional measurements is impractical. In Appendix 11B we derive the relation

$$\sigma_m = \frac{\sigma}{\sqrt{n-1}} \tag{11.17a}$$

$$\approx \frac{\sigma}{\sqrt{n}}. \tag{11.17b}$$

The reason for the expression $1/\sqrt{n-1}$ rather than $1/\sqrt{n}$ in (11.17a) is similar to the reason for the expression $1/\sqrt{n-2}$ in the error estimates of the least squares fits (see (7.31)). The idea is that to compute σ, we need to use the n measurements to compute the mean, $\overline{f(x)}$, and, loosely speaking, we have only $n-1$ independent measurements left for calculating σ. Because we almost always make a large number of measurements, we will use the relation (11.17b) and consider only this limit in Appendix 11B. Note that (11.17) implies that the most probable error decreases with the square root of the number of trials. For our example we find that the most probable error of our initial measurement is approximately $0.8850/100 \approx 0.009$, an estimate consistent with the known error of 0.007 and with our estimated value of $\sigma_m \approx 0.007$.

One way to verify the relation (11.17) is to divide the initial measurement of n trials into s subsets. This procedure does not require additional measurements. We denote the mean value of $f(x_i)$ in the kth subset by S_k. As an example, we divide the 10^4 trials of the first measurement into $s = 10$ subsets of $n/s = 10^3$ trials each. The results for S_k are shown in Table 11.4. As expected, the mean values of $f(x)$ for each subset k are not equal. A reasonable candidate for a measure of the error is the standard deviation of the means of each subset. We denote this quantity as σ_s where

$$\sigma_s{}^2 = \, <S^2> - <S>^2, \tag{11.18}$$

where the averages are over the subsets. From Table 11.4 we obtain $\sigma_s = 0.025$, a result that is approximately three times larger than our estimate of 0.007 for σ_m. Moreover, we need to obtain an error estimate that is independent of how we subdivide the data. This quantity is not σ_s, but the ratio $\sigma_s/\sqrt{s}$, which for our example is approximately $0.025/3.16 \approx 0.008$. This value is consistent with both σ_m and the ratio $\sigma/\sqrt{n}$. We conclude that we can interpret the n trials either as a single measurement or as a collection of s measurements. In the former interpretation, the probable error is given by the standard deviation of the n trials divided by the square root of the number of trials. In the same spirit, the latter interpretation implies that the probable error is given by the standard deviation of the s measurements of the subsets divided by the square root of the number of measurements.

Note that we can make the error as small as we wish by either increasing the number of trials or by increasing the efficiency of the individual trials and thereby reducing the standard deviation σ. Several *reduction of variance* methods are introduced in Sections. 11.7 and 11.8.

subset k	S_k
1	3.14326
2	3.15633
3	3.10940
4	3.15337
5	3.15352
6	3.11506
7	3.17989
8	3.12398
9	3.17565
10	3.17878

Table 11.4   The values of S_k for $f(x) = 4\sqrt{1-x^2}$ for $0 \le x \le 1$ is shown for 10 subsets of 10^3 trials each. The average value of $f(x)$ over the 10 subsets is 3.14892, in agreement with the result for F_n for the first measurement shown in Table 11.3.

Problem 11.6 Estimate of the Monte Carlo error

a. Estimate the integral of $f(x) = e^{-x}$ in the interval $0 \leq x \leq 1$ using the sample mean Monte Carlo method with $n = 10^2$, $n = 10^3$, and $n = 10^4$. Compute the standard deviation σ as defined by (11.13). Does your estimate of σ change significantly as n is increased? Determine the exact answer analytically and estimate the n dependence of the error. How does your estimated error compare with the error estimate obtained from the relation (11.17)

b. Generate nineteen additional measurements of the integral each with $n = 10^3$ trials. Compute σ_m, the standard deviation of the twenty measurements. Is the magnitude of σ_m consistent with your estimate of the error obtained in part (a)? Will your estimate of σ_m change significantly if more measurements are made?

c. Divide your first measurement of $n = 10^3$ trials into $s = 20$ subsets of 50 trials each. Compute the standard deviation of the subsets σ_s. Is the magnitude $\sigma_s/\sqrt{s}$ consistent with your previous error estimates?

d. Divide your first measurement into $s = 10$ subsets of 100 trials each and again compute the standard deviation of the subsets. How does the value of σ_s compare to what you found in part (c)? What is the value of $\sigma_s/\sqrt{s}$ in this case? How does the standard deviation of the subsets compare using the two different divisions of the data?

e. Estimate the integral

$$\int_0^1 e^{-x^2}\, dx \tag{11.19}$$

to two decimal places using $\sigma_n/\sqrt{n}$ as an estimate of the probable error.

*Problem 11.7 Importance of randomness

We will learn in Chapter 12 that the random number generator included with many programming languages is based on the linear congruential method. In this method each term in the sequence can be found from the preceding one by the relation

$$x_{n+1} = (ax_n + c) \bmod m, \tag{11.20}$$

where x_0 is the seed, and a, c, and m are nonnegative integers. The random numbers r in the unit interval $0 \leq r < 1$ are given by $r_n = x_n/m$. The notation $y = x \bmod m$ means that if x exceeds m, then the *modulus* m is subtracted from x as many times as necessary until $0 \leq y < m$. Eventually, the sequence of numbers generated by (11.20) will repeat itself, yielding a *period* for the random number generator. To examine the effect of a poor random number generator, we choose values of x_0, m, a, and c such that (11.20) has poor statistical properties, e.g., a short period. What is the period for $x_0 = 1$, $a = 5$, $c = 0$, and $m = 32$? Estimate the integral in Problem 11.6a by making a single measurement of $n = 10^3$ trials using the "random number" generator (11.20) with the above values of x_0, a, c, and m.

Analyze your measurement in the same way as before, i.e., calculate the mean, the mean of each of the twenty subsets, and the standard deviation of the means of the subsets. Then divide your data into ten subsets and calculate the same quantities. Are the standard deviations of the subsets related as before? If not, why?

11.5 NONUNIFORM PROBABILITY DISTRIBUTIONS

In the previous two sections we learned how uniformly distributed random numbers can be used to estimate definite integrals. In general, it is desirable to sample the integrand $f(x)$ more often in regions of x where the magnitude of $f(x)$ is large or rapidly varying. Because such importance sampling methods require nonuniform probability distributions, we now consider several methods for generating random numbers that are not distributed uniformly. In the following, we will denote r as a member of a uniform random number sequence in the unit interval $0 \leq r < 1$.

Suppose that two discrete events occur with probabilities p_1 and p_2 such that $p_1 + p_2 = 1$. How can we choose the two events with the correct probabilities using a uniform probability distribution? For this simple case, it is obvious that we choose event 1 if $r < p_1$; otherwise, we choose event 2. If there are three events with probabilities p_1, p_2, and p_3, then if $r < p_1$ we choose event 1; else if $r < p_1 + p_2$, we choose event 2; else we choose event 3. We can visualize these choices by dividing a line segment of unit length into three pieces whose lengths are as shown in Fig. 11.4. A random point r on the line segment will land in the ith segment with a probability equal to p_i.

Now consider n discrete events. How do we determine which event, i, to choose given r? The generalization of the procedure we have followed for $n = 2$ and 3 is to find the value of i that satisfies the condition

$$\sum_{j=0}^{i-1} p_j \leq r \leq \sum_{j=0}^{i} p_j, \tag{11.21}$$

where we have defined $p_0 \equiv 0$. Check that (11.21) reduces to the correct procedure for $n = 2$ and $n = 3$.

Now let us consider a continuous nonuniform probability distribution. One way to generate such a distribution is to take the limit of (11.21) and associate p_i with $p(x)\,dx$, where we have introduced the *probability density* $p(x)$ such that $p(x)\,dx$ is

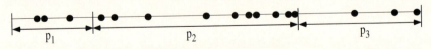

Fig. 11.4   The unit interval is divided into three segments of lengths $p_1 = 0.2$, $p_2 = 0.5$, and $p_3 = 0.3$. Sixteen random numbers are represented by the filled circles uniformly distributed on the unit interval. The fraction of circles within each segment is approximately equal to the value of p_i for that segment.

the probability that the event x is in the interval between x and $x + dx$. The probability density $p(x)$ is normalized such that

$$\int_{-\infty}^{+\infty} p(x)\, dx = 1. \tag{11.22}$$

In the continuum limit the two sums in (11.21) become the same integral and the inequalities become equalities. Hence we can write

$$P(x) \equiv \int_{-\infty}^{x} p(x')\, dx' = r. \tag{11.23}$$

From (11.23) we see that the uniform random number r corresponds to the *cumulative probability distribution* function $P(x)$, which is the probability of choosing a value less than or equal to x. The function $P(x)$ should not be confused with the probability density $p(x)$ or the probability $p(x)\, dx$. In many applications the meaningful range of values of x is positive. In that case, we have $p(x) = 0$ for $x < 0$.

The relation (11.23) leads to the *inverse transform* method for generating random numbers distributed according to the function $p(x)$. This method involves generating a random number r and solving (11.23) for the corresponding value of x. As an example of the method, we use (11.23) to generate a random number sequence according to the uniform probability distribution on the interval $a \le x \le b$. The desired probability density $p(x)$ is

$$p(x) = \begin{cases} (1/(b-a), & a \le x \le b \\ 0, & \text{otherwise.} \end{cases} \tag{11.24}$$

The cumulative probability distribution function $P(x)$ for $a \le x \le b$ can be found by substituting (11.24) into (11.23) and performing the integral. The result is

$$P(x) = \frac{x-a}{b-a}. \tag{11.25}$$

If we substitute the form (11.25) for $P(x)$ into (11.23) and solve for x, we find the desired relation

$$x = a + (b-a)r. \tag{11.26}$$

The variable x given by (11.26) is distributed according to the probability distribution $p(x)$ given by (11.24). Of course, the relation (11.26) is rather obvious, and we already have used (11.26) in our Monte Carlo programs.

Let us now apply the inverse transform method to the probability density function

$$p(x) = \begin{cases} (1/\lambda)\, e^{-x/\lambda}, & \text{if } 0 \le x \le \infty \\ 0, & x < 0. \end{cases} \tag{11.27}$$

In Section 11.6 we will use this probability density to find the distance between scattering events of a particle whose mean free path is λ. If we substitute (11.27) into (11.23) and do the integration, we find

$$r = P(x) = 1 - e^{-x/\lambda}. \tag{11.28}$$

The solution of (11.28) for x yields $x = -\lambda \ln(1 - r)$. Because $1 - r$ is distributed in the same way as r, we can write

$$x = -\lambda \ln r. \tag{11.29}$$

The variable x found from (11.29) is distributed according to the probability density $p(x)$ given by (11.27). On many computers the computation of the natural logarithm in (11.29) is relatively slow, and hence the inverse transform method might not necessarily be the most efficient method to use.

From the above examples, we see that two conditions must be satisfied in order to apply the inverse transform method. Specifically, the form of $p(x)$ must allow the integral in (11.23) to be performed analytically or numerically, and it must be feasible to invert the relation $P(x) = r$ for x. The Gaussian probability density

$$p(x) = \frac{1}{(2\pi\sigma^2)^{1/2}} e^{-x^2/2\sigma^2} \tag{11.30}$$

is an example of a probability density for which the cumulative distribution $P(x)$ cannot be obtained analytically. However, we can generate the two-dimensional probability $p(x, y)\,dx\,dy$ given by

$$p(x, y)\,dx\,dy = \frac{1}{2\pi\sigma^2} e^{-(x^2+y^2)/2\sigma^2}\,dx\,dy. \tag{11.31}$$

First, we make a change of variables to polar coordinates:

$$r = (x^2 + y^2)^{1/2} \qquad \theta = \tan^{-1}\frac{y}{x}. \tag{11.32}$$

Let $\rho = r^2/2$, and write the two-dimensional probability as

$$p(\rho, \theta)\,d\rho d\theta = \frac{1}{2\pi} e^{-\rho} d\rho\, d\theta, \tag{11.33}$$

where we have set $\sigma = 1$. If we generate ρ according to the exponential distribution (11.27) and generate θ uniformly in the interval $0 \le \theta < 2\pi$, then the variables

$$x = (2\rho)^{1/2}\cos\theta \quad \text{and} \quad y = (2\rho)^{1/2}\sin\theta \qquad \text{(Box-Muller method)} \tag{11.34}$$

will each be generated according to (11.30) with zero mean and $\sigma = 1$. (Note that the two-dimensional density (11.31) is the product of two independent one-dimensional Gaussian distributions.) This way of generating a Gaussian distribution is known as the *Box-Muller* method. We discuss other methods for generating the Gaussian distribution in Problem 11.12 and Appendix 11C.

Problem 11.8 Nonuniform probability densities

a. Write a program to simulate the simultaneous rolling of two dice. In this case the events are discrete and occur with nonuniform probability. You might wish to revisit Problem 7.10 and simulate the game of craps.

b. Write a program to verify that the sequence of random numbers $\{x_i\}$ generated by (11.29) is distributed according to the exponential distribution (11.27).

c. Generate random variables according to the probability density function

$$p(x) = \begin{cases} 2(1-x), & \text{if } 0 \le x \le 1; \\ 0, & \text{otherwise}. \end{cases} \tag{11.35}$$

d. Verify that the variables x and y in (11.34) are distributed according to the Gaussian distribution. What is the mean value and the standard deviation of x and of y?

e. How can you use the relations (11.34) to generate a Gaussian distribution with arbitrary mean and standard deviation?

*11.6 NEUTRON TRANSPORT

We now consider the application of a nonuniform probability distribution to the simulation of the transmission of neutrons through bulk matter, one of the original applications of a Monte Carlo method. Suppose that a neutron is incident on a plate of thickness t. We assume that the plate is infinite in the x and y directions and that the z axis is normal to the plate. At any point within the plate, the neutron can either be captured with probability p_c or scattered with probability p_s. These probabilities are proportional to the capture cross section and scattering cross section, respectively. If the neutron is scattered, we need to find its new direction as specified by the polar angle θ (see Fig. 11.5). Because we are not interested in how far the neutron moves in the x or y direction, the value of the azimuthal angle ϕ is irrelevant.

If the neutrons are scattered equally in all directions, then the probability $p(\theta, \phi)\, d\theta d\phi$ equals $d\Omega/4\pi$, where $d\Omega$ is an infinitesimal solid angle and 4π is the total solid angle. Because $d\Omega = \sin\theta\, d\theta d\phi$, we have

$$p(\theta, \phi) = \frac{\sin\theta}{4\pi}. \tag{11.36}$$

We can find the probability density for θ and ϕ separately by integrating over the other angle. For example,

$$p(\theta) = \int_0^{2\pi} p(\theta, \phi)\, d\phi = \frac{1}{2}\sin\theta, \tag{11.37}$$

and

$$p(\phi) = \int_0^{\pi} p(\theta, \phi)\, d\theta = \frac{1}{2\pi}. \tag{11.38}$$

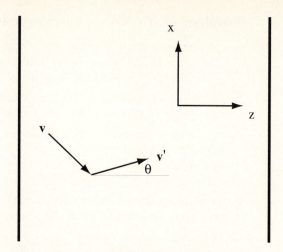

Fig. 11.5   The definition of the scattering angle θ.
The velocity before scattering is **v** and
the velocity after scattering is **v'**. The
scattering angle θ is independent of **v**
and is defined relative to the z axis.

Because the point probability $p(\theta, \phi)$ is the product of the probabilities $p(\theta)$ and $p(\phi)$,
θ and ϕ are independent variables. Although we do not need to generate a random angle
ϕ, we note that since $p(\phi)$ is a constant, ϕ can be found from the relation

$$\phi = 2\pi r. \tag{11.39}$$

To find θ according to the distribution (11.37), we substitute (11.37) in (11.23) and
obtain

$$r = \frac{1}{2} \int_0^\theta \sin x \, dx \tag{11.40}$$

If we do the integration in (11.40), we find

$$\cos \theta = 1 - 2r. \tag{11.41}$$

Note that (11.39) implies that ϕ is uniformly distributed between 0 and 2π and (11.41)
implies that $\cos \theta$ is uniformly distributed between -1 and $+1$.

We could invert the cosine in (11.41) to solve for θ. However, to find the z component of the path of the neutron through the plate, we need to multiply $\cos \theta$ by the
path length ℓ, and hence we need $\cos \theta$ rather than θ. The path length, which is the distance traveled between subsequent scattering events, is obtained from the exponential
probability density, $p(\ell) \propto e^{-\ell/\lambda}$ (see (11.27)). From (11.29), we have

$$\ell = -\lambda \ln r, \tag{11.42}$$

where λ is the mean free path.

Now we have all the needed ingredients for calculating the probabilities for a neutron to pass through the plate, be reflected off the plate, or be captured and absorbed in the plate. The input parameters are the thickness of the plate t, the capture and scattering probabilities p_c and p_s, and the mean free path λ. We begin with $z = 0$, and implement the following steps:

1. Determine if the neutron is captured or scattered. If it is captured, then add one to the number of captured neutrons, and go to step 5.

2. If the neutron is scattered, compute $\cos \theta$ from (11.41) and ℓ from (11.42). Change the z coordinate of the neutron by $\ell \cos \theta$.

3. If $z < 0$, add one to the number of reflected neutrons. If $z > t$, add one to the number of transmitted neutrons. In either case, skip to step 5 below.

4. Repeat steps 1–3 until the fate of the neutron has been determined.

5. Repeat steps 1–4 with additional incident neutrons until sufficient data has been obtained.

Problem 11.9   Elastic neutron scattering

a.   Write a program to implement the above algorithm for neutron scattering through a plate. Assume $t = 1$ and $p_c = p_s/2$. Find the transmission, reflection, and absorption probabilities for the mean free path λ equal to 0.01, 0.05, 0.1, 1, and 10. Begin with 100 incident neutrons, and increase this number until satisfactory statistics are obtained. Give a qualitative explanation of your results.

b.   Choose $t = 1$, $p_c = p_s$, and $\lambda = 0.05$, and compare your results with the analogous case considered in part (a).

c.   Repeat part (b) with $t = 2$ and $\lambda = 0.1$. Do the various probabilities depend on λ and t separately or only on their ratio? Answer this question before doing the simulation.

d.   Draw some typical paths of the neutrons. From the nature of these paths, explain the results in parts (a)–(c). For example, how does the number of scattering events change as the absorption probability changes?

Problem 11.10   Inelastic neutron scattering

a.   In Problem 11.9 we assumed elastic scattering, i.e., no energy is lost during scattering. Here we assume that some of the neutron energy E is lost and that the mean free path is proportional to the speed and hence to $\sqrt{E}$. Modify your program so that a neutron loses a fraction f of its energy at each scattering event, and assume that $\lambda = \sqrt{E}$. Consider $f = 0.05, 0.1$, and 0.5, and compare your results with those found in Problem 11.9a.

b.   Make a histogram for the path lengths between scattering events and plot the path length distribution function for $f = 0.1, 0.5$, and 0 (elastic scattering).

The above procedure for simulating neutron scattering and absorption is more computer intensive than necessary. Instead of considering a single neutron at a time, we can consider a collection of neutrons at each position. Then instead of determining whether one neutron is captured or scattered, we determine the fraction that is captured and the fraction that is scattered. For example, at the first scattering site, a fraction p_c of the neutrons are captured and a fraction p_s are scattered. We accumulate the fraction p_c for the captured neutrons. We also assume that all the scattered neutrons move in the same direction with the same path length, both of which are generated at random as before. At the next scattering site there are p_s^2 scattered neutrons and $p_s p_c$ captured neutrons. At the end of m steps, the fraction of neutrons remaining is $w = p_s^m$ and the total fraction of captured neutrons is $p_c + p_c p_s + p_c p_s^2 + \ldots + p_c p_s^{m-1}$. If the new position at the mth step is at $z < 0$, we add w to the sum for the reflected neutrons; if $z > t$, we add w to the neutrons transmitted. When the neutrons are reflected or absorbed, we start over again at $z = 0$ with another collection of neutrons.

Problem 11.11   Improved neutron scattering method

Apply the improved Monte Carlo method to neutron transmission through a plate. Repeat the simulations suggested in Problem 11.9a and compare your new and previous results. Also compare the computational times for the two approaches to obtain comparable statistics.

The power of the Monte Carlo method becomes apparent when the geometry of the material is complicated or when the material is spatially nonuniform so that the cross sections vary from point to point. A difficult problem of current interest is the absorption of various forms of radiation in the human body.

Problem 11.12   Transmission through layered materials

Consider two plates with the same thickness $t = 1$ that are stacked on top of one another with no space between them. For one plate, $p_c = p_s$, and for the other, $p_c = 2p_s$, that is, the top plate is a better absorber. Assume that $\lambda = 1$ in both plates. Find the transmission, reflection, and absorption probabilities for elastic scattering. Does it matter which plate receives the incident neutrons?

11.7   IMPORTANCE SAMPLING

Because the analysis of Section 11.4 showed that the error estimate associated with a Monte Carlo estimate is proportional to σ, we wish to introduce *importance sampling* techniques that reduce σ and improve the efficiency of each sample. In the context of numerical integration, we introduce a positive function $p(x)$ such that

$$\int_a^b p(x)\,dx = 1, \tag{11.43}$$

and rewrite the integral (11.1) as

$$F = \int_a^b \left[\frac{f(x)}{p(x)} \right] p(x)\, dx. \tag{11.44}$$

We can evaluate the integral (11.44) by sampling according to the probability distribution $p(x)$ and constructing the sum

$$F_n = \frac{1}{n} \sum_{i=1}^n \frac{f(x_i)}{p(x_i)}. \tag{11.45}$$

The sum (11.45) reduces to (11.9) for the uniform case $p(x) = 1/(b-a)$.

   We wish to choose a form for $p(x)$ that minimizes the variance of the integrand $f(x)/p(x)$. Because we cannot evaluate σ analytically in general, we determine σ a *posteriori* and choose a form of $p(x)$ that mimics $f(x)$ as much as possible, particularly where $f(x)$ is large. If we are able to determine an appropriate form of $p(x)$, the integrand $f(x)/p(x)$ will be slowly varying and hence the variance will be reduced. As an example, we again consider the integral (see Problem 11.6e)

$$F = \int_0^1 e^{-x^2}\, dx. \tag{11.46}$$

The estimate of F with $p(x) = 1$ for $0 \leq x \leq 1$ is shown in the first column of Table 11.5. A reasonable choice of a weight function is $p(x) = Ae^{-x}$, where A is chosen such that $p(x)$ is normalized on the unit interval. Note that this choice of $p(x)$ is positive definite and is qualitatively similar to $f(x)$. The results are shown in the second column of Table 11.5. We see that although the computation time per trial for the nonuniform case is larger, the smaller value of σ makes the use of the nonuniform probability distribution more efficient.

	$p(x) = 1$	$p(x) = Ae^{-x}$
n (trials)	4×10^5	8×10^3
F_n	0.7471	0.7469
σ	0.2010	0.0550
$\sigma/\sqrt{n}$	3×10^{-4}	6×10^{-4}
Total CPU time (s)	35	1.35
CPU time per trial (s)	10^{-4}	2×10^{-4}

Table 11.5   Comparison of the Monte Carlo estimates of the integral (11.46) using the uniform probability density $p(x) = 1$ and the nonuniform probability density $p(x) = Ae^{-x}$. The normalization constant A is chosen such that $p(x)$ is normalized on the unit interval. The value of the integral to four decimal places is 0.7468. The estimates F_n, variance σ, and the probable error $\sigma/n^{1/2}$ are shown. The CPU time (seconds) is shown for comparison only and was found on a Macintosh Quadra 840AV running True BASIC.

Problem 11.13   Importance sampling

a.   Choose the importance function $p(x) = e^{-x}$ and evaluate the integral

$$\int_0^3 x^{3/2} e^{-x} \, dx. \tag{11.47}$$

b.   Choose $p(x) = e^{-ax}$ and estimate the integral

$$\int_0^\pi \frac{1}{x^2 + \cos^2 x} \, dx. \tag{11.48}$$

Determine the value of a that minimizes the variance of the integral.

11.8   METROPOLIS MONTE CARLO METHOD

Another way of generating an arbitrary nonuniform probability distribution was introduced by Metropolis, Rosenbluth, Rosenbluth, Teller and Teller in 1953. The *Metropolis* method is a special case of an importance sampling procedure in which certain possible sampling attempts are rejected (see Appendix 11C). The Metropolis method is useful for computing averages of the form

$$<f> = \frac{\int p(x) f(x) \, dx}{\int p(x) \, dx}, \tag{11.49}$$

where $p(x)$ is an arbitrary probability distribution that need not be normalized. In Chapter 17 we will discuss the application of the Metropolis method to problems in statistical mechanics.

For simplicity, we introduce the Metropolis method in the context of estimating one-dimensional definite integrals. Suppose we wish to use importance sampling to generate random variables according to an arbitrary probability density $p(x)$. The Metropolis method produces a random walk of points $\{x_i\}$ whose asymptotic probability distribution approaches $p(x)$ after a large number of steps. The random walk is defined by specifying a *transition probability* $T(x_i \rightarrow x_j)$ from one value x_i to another value x_j such that the distribution of points $x_0, x_1, x_2, \ldots$ converges to $p(x)$. It can be shown that it is sufficient (but not necessary) to satisfy the "detailed balance" condition

$$p(x_i)T(x_i \rightarrow x_j) = p(x_j)T(x_j \rightarrow x_i). \tag{11.50}$$

The relation (11.50) does not specify $T(x_i \rightarrow x_j)$ uniquely. A simple choice of $T(x_i \rightarrow x_j)$ that is consistent with (11.50) is

$$T(x_i \rightarrow x_j) = \min\left[1, \frac{p(x_j)}{p(x_i)}\right]. \tag{11.51}$$

If the "walker" is at position x_i and we wish to generate x_{i+1}, we can implement this choice of $T(x_i \rightarrow x_j)$ by the following steps:

1. Choose a trial position $x_{\text{trial}} = x_i + \delta_i$, where δ_i is a random number in the interval $[-\delta, \delta]$.

2. Calculate $w = p(x_{\text{trial}})/p(x_i)$.

3. If $w \geq 1$, accept the change and let $x_{i+1} = x_{\text{trial}}$.

4. If $w < 1$, generate a random number r.

5. If $r \leq w$, accept the change and let $x_{i+1} = x_{\text{trial}}$.

6. If the trial change is not accepted, then let $x_{i+1} = x_i$.

It is necessary to sample many points of the random walk before the asymptotic probability distribution $p(x)$ is attained. How do we choose the maximum "step size" δ? If δ is too large, only a small percentage of trial steps will be accepted and the sampling of $p(x)$ will be inefficient. On the other hand, if δ is too small, a large percentage of trial steps will be accepted, but again the sampling of $p(x)$ will be inefficient. A rough criterion for the magnitude of δ is that approximately one third to one half of the trial steps should be accepted. We also wish to choose the value of x_0 such that the distribution $\{x_i\}$ will approach the asymptotic distribution as quickly as possible. An obvious choice is to begin the random walk at a value of x at which $p(x)$ is a maximum.

A subroutine that implements the Metropolis algorithm is given below. Can you make the subroutine more efficient?

```
SUB Metropolis(x,delta,naccept)
    DECLARE DEF p
    LET xtrial = x + delta*(2*rnd - 1)
    LET w = p(xtrial)/p(x)
    IF rnd <= w then
        LET x = xtrial
        LET naccept = naccept + 1   ! number of acceptances
    END IF
END SUB
```

Problem 11.14   The Gaussian distribution

a.   Write a program using SUB Metropolis to generate the Gaussian distribution, $p(x) = Ae^{-x^2/2\sigma^2}$. Is the value of the normalization constant A relevant? Determine the qualitative dependence of the acceptance ratio and the equilibration time on the maximum step size δ. One possible criterion for equilibrium is that $\langle x^2 \rangle \approx \sigma^2$. For $\sigma = 1$, what is a reasonable choice for δ? How many trials are needed to reach equilibrium for your choice of δ? Choose $x_0 = 0$.

b.   Modify your program so that it plots the asymptotic probability distribution generated by the Metropolis algorithm.

*c.   Calculate the autocorrelation function $C(j)$ defined by

$$C(j) = \frac{\langle x_{i+j}x_i \rangle - \langle x_i \rangle^2}{\langle x_i^2 \rangle - \langle x_i \rangle^2}, \tag{11.52}$$

where $<...>$ indicates an average over the random walk. What is the value of $C(j = 0)$? What would be the value of $C(j \neq 0)$ if x_i were completely random? Calculate $C(j)$ for different values of j and determine the value of j for which $C(j)$ is essentially zero.

Problem 11.15   Application of the Metropolis method

a.   Although the Metropolis method is not the most efficient method in this case, write a program to estimate the average

$$<x> = \frac{\int_0^\infty x e^{-x}\, dx}{\int_0^\infty e^{-x}\, dx},$$ (11.53)

with $p(x) = Ae^{-x}$ for $x \geq 0$ and $p(x) = 0$ for $x < 0$. Incorporate into the program a computation of the histogram $H(x)$ showing the fraction of points in the random walk in the region x to $x + \Delta x$, with $\Delta x = 0.2$. Begin with $n = 1000$ and maximum step size $\delta = 1$. Allow the system to equilibrate for 200 steps before computing averages. Is the integrand sampled uniformly? If not, what is the approximate region of x where the integrand is sampled more often?

b.   Calculate analytically the exact value of $<x>$. How do your Monte Carlo results compare with the exact value for $n = 100$ and $n = 1000$ with $\delta = 0.1, 1$, and 10? Estimate the standard error of the mean. Does this error give a reasonable estimate of the error? If not, why?

c.   In part (b) you should have found that the estimated error is much smaller than the actual error. The reason is that the $\{x_i\}$ are not statistically independent. The Metropolis algorithm produces a random walk whose points are correlated with each other over short times (measured in the number of Monte Carlo steps). The correlation of the points decays exponentially with time. If τ is the characteristic time for this decay, then only points separated by approximately 2 to 3τ can be considered statistically independent. Rerun your program with the data grouped into 20 sets of 50 points each and 10 sets of 100 points each. If the sets of 50 points each are statistically independent (i.e., if τ is significantly smaller than 50), then your estimate of the error for the two groupings should be approximately the same.

Appendix 11A   ERROR ESTIMATES FOR NUMERICAL INTEGRATION

We derive the dependence of the truncation error estimates on the number of intervals for the numerical integration methods considered in Sections. 11.1 and 11.3. These estimates are based on the assumed adequacy of the Taylor series expansion of the integrand $f(x)$:

$$f(x) = f(x_i) + f'(x_i)(x - x_i) + \frac{1}{2} f''(x_i)(x - x_i)^2 + \ldots,$$ (11.54)

and the integration of (11.1) in the interval $x_i \leq x \leq x_{i+1}$:

$$\int_{x_i}^{x_{i+1}} f(x)\,dx = f(x_i)\Delta x + \frac{1}{2}f'(x_i)(\Delta x)^2 + \frac{1}{6}f''(x_i)(\Delta x)^3 + \dots \qquad (11.55)$$

We first estimate the error associated with the rectangular approximation with $f(x)$ evaluated at the left side of each interval. The error Δ_i in the interval $[x_i, x_{i+1}]$ is the difference between (11.55) and the estimate $f(x_i)\Delta x$ is

$$\Delta_i = \left[\int_{x_i}^{x_{i+1}} f(x)\,dx \right] - f(x_i)\Delta x \approx \frac{1}{2}f'(x_i)(\Delta x)^2. \qquad (11.56)$$

We see that to leading order in Δx, the error in each interval is order $(\Delta x)^2$. Because there are a total of n intervals and $\Delta x = (b-a)/n$, the total error associated with the rectangular approximation is $n\Delta_i \sim n(\Delta x)^2 \sim n^{-1}$.

The estimated error associated with the trapezoidal approximation can be found in the same way. The error in the interval $[x_i, x_{i+1}]$ is the difference between the exact integral and the estimate, $\frac{1}{2}[f(x_i) + f(x_{i+1})]\Delta x$.

$$\Delta_i = \left[\int_{x_i}^{x_{i+1}} f(x)\,dx \right] - \frac{1}{2}[f(x_i) + f(x_{i+1})]\Delta x. \qquad (11.57)$$

If we use (11.55) to estimate the integral and (11.54) to estimate $f(x_{i+1})$ in (11.57), we find that the term proportional to f' cancels and that the error associated with one interval is order $(\Delta x)^3$. Hence, the total error in the interval $[a, b]$ associated with the trapezoidal approximation is order n^{-2}.

Because Simpson's rule is based on fitting $f(x)$ in the interval $[x_{i-1}, x_{i+1}]$ to a parabola, error terms proportional to f'' cancel. We might expect that error terms of order $f'''(x_i)(\Delta x)^4$ contribute, but these terms cancel by virtue of their symmetry. Hence the $(\Delta x)^4$ term of the Taylor expansion of $f(x)$ is adequately represented by Simpson's rule. If we retain the $(\Delta x)^4$ term in the Taylor series of $f(x)$, we find that the error in the interval $[x_i, x_{i+1}]$ is of order $f''''(x_i)(\Delta x)^5$ and that the total error in the interval $[a, b]$ associated with Simpson's rule is $O(n^{-4})$.

The error estimates can be extended to two dimensions in a similar manner. The two-dimensional integral of $f(x, y)$ is the volume under the surface determined by $f(x, y)$. In the "rectangular" approximation, the integral is written as a sum of the volumes of parallelograms with cross sectional area $\Delta x \Delta y$ and a height determined by $f(x, y)$ at one corner. To determine the error, we expand $f(x, y)$ in a Taylor series

$$f(x, y) = f(x_i, y_i) + \frac{\partial f(x_i, y_i)}{\partial x}(x - x_i) + \frac{\partial f(x_i, y_i)}{\partial y}(y - y_i) + \dots, \qquad (11.58)$$

and write the error as

$$\Delta_i = \left[\int\!\!\int f(x, y)\,dx\,dy \right] - f(x_i, y_i)\Delta x \Delta y. \qquad (11.59)$$

If we substitute (11.58) into (11.59) and integrate each term, we find that the term proportional to f cancels and the integral of $(x - x_i)\,dx$ yields $\frac{1}{2}(\Delta x)^2$. The integral of this term with respect to dy gives another factor of Δy. The integral of the term proportional to $(y - y_i)$ yields a similar contribution. Because Δy also is order Δx, the error associated with the intervals $[x_i, x_{i+1}]$ and $[y_i, y_{i+1}]$ is to leading order in Δx:

$$\Delta_i \approx \frac{1}{2}[f_x'(x_i, y_i) + f_y'(x_i, y_i)](\Delta x)^3. \tag{11.60}$$

We see that the error associated with one parallelogram is order $(\Delta x)^3$. Because there are n parallelograms, the total error is order $n(\Delta x)^3$. However in two dimensions, $n = A/(\Delta x)^2$, and hence the total error is order $n^{-1/2}$. In contrast, the total error in one dimension is order n^{-1}, as we saw earlier.

The corresponding error estimates for the two-dimensional generalizations of the trapezoidal approximation and Simpson's rule are order n^{-1} and n^{-2} respectively. In general, if the error goes as order n^{-a} in one dimension, then the error in d dimensions goes as $n^{-a/d}$. In contrast, Monte Carlo errors vary as order $n^{-1/2}$ independent of d. Hence for large enough d, Monte Carlo integration methods will lead to smaller errors for the same choice of n.

Appendix 11B   THE STANDARD DEVIATION OF THE MEAN

In Section 11.4 we gave empirical reasons for the claim that the error associated with a single measurement consisting of n trials equals $\sigma/\sqrt{n}$, where σ is the standard deviation in a single measurement. We now present an analytical derivation of this relation. The quantity of experimental interest is denoted as x. Consider m sets of measurements each with n trials for a total of mn trials. We use the index α to denote a particular measurement and the index i to designate the ith trial within a measurement. We denote $x_{\alpha,i}$ as trial i in the measurement α. The value of a measurement is given by

$$M_\alpha = \frac{1}{n} \sum_{i=1}^{n} x_{\alpha,i}. \tag{11.61}$$

The mean $\overline{M}$ of the *total mn* individual trials is given by

$$\overline{M} = \frac{1}{m} \sum_{\alpha=1}^{m} M_\alpha = \frac{1}{nm} \sum_{\alpha=1}^{m} \sum_{i=1}^{n} x_{\alpha,i}. \tag{11.62}$$

The difference between measurement α and the mean of all the measurements is given by

$$e_\alpha = M_\alpha - \overline{M}. \tag{11.63}$$

We can write the variance of the means as

$$\sigma_m{}^2 = \frac{1}{m} \sum_{\alpha=1}^{m} e_\alpha{}^2. \tag{11.64}$$

We now wish to relate σ_m to the variance of the individual trials. The discrepancy $d_{\alpha,i}$ between an individual sample $x_{\alpha,i}$ and the mean is given by

$$d_{\alpha,i} = x_{\alpha,i} - \overline{M}. \tag{11.65}$$

Hence, the variance σ^2 of the nm individual trials is

$$\sigma^2 = \frac{1}{mn} \sum_{\alpha=1}^{m} \sum_{i=1}^{n} d_{\alpha,i}{}^2. \tag{11.66}$$

We write

$$e_\alpha = M_\alpha - \overline{M} = \frac{1}{n} \sum_{i=1}^{n} \left(x_{\alpha,i} - \overline{M} \right)$$

$$= \frac{1}{n} \sum_{i=1}^{n} d_{\alpha,i}. \tag{11.67}$$

If we substitute (11.67) into (11.64), we find

$$\sigma_m{}^2 = \frac{1}{m} \sum_{\alpha=1}^{n} \left(\frac{1}{n} \sum_{i=1}^{n} d_{\alpha,i} \right) \left(\frac{1}{n} \sum_{j=1}^{n} d_{\alpha,j} \right). \tag{11.68}$$

The sum in (11.68) over trials i and j in set α contains two kinds of terms—those with $i = j$ and those with $i \neq j$. We expect that $d_{\alpha,i}$ and $d_{\alpha,j}$ are independent and equally positive or negative on the average. Hence in the limit of a large number of measurements, we expect that only the terms with $i = j$ in (11.68) will survive, and we write

$$\sigma_m{}^2 = \frac{1}{mn^2} \sum_{\alpha=1}^{n} \sum_{i=1}^{n} d_{\alpha,i}{}^2. \tag{11.69}$$

If we combine (11.69) with (11.66), we arrive at the desired result

$$\sigma_m{}^2 = \frac{\sigma^2}{n}. \tag{11.70}$$

Appendix 11C THE ACCEPTANCE-REJECTION METHOD

Although the inverse transform method discussed in Section 11.5 can in principle be used to generate any desired probability distribution, in practice the method is limited to functions for which the equation, $r = P(x)$, can be solved analytically for x or by simple numerical approximation. Another method for generating nonuniform probability distributions is the *acceptance-rejection* method due to von Neumann.

Suppose that $p(x)$ is a (normalized) probability density function that we wish to generate. For simplicity, we assume $p(x)$ is nonzero in the unit interval. Consider a positive definite *comparison function* $w(x)$ such that $w(x) > p(x)$ in the entire range of interest. A simple although not generally optimum choice of w is a constant greater than the maximum value of $p(x)$. Because the area under the curve $p(x)$ in the range x to $x + \Delta x$ is the probability of generating x in that range, we can follow a procedure similar to that used in the hit or miss method. Generate two numbers at random to define the location of a point in two dimensions which is distributed uniformly in the area under the comparison function $w(x)$. If this point is outside the area under $p(x)$, the point is rejected; if it lies inside the area, we accept it. This procedure implies that the accepted points are uniform in the area under the curve $p(x)$ and that their x values are distributed according to $p(x)$.

One procedure for generating a uniform random point (x, y) under the comparison function $w(x)$ is as follows.

1. Choose a form of $w(x)$. One choice would be to choose $w(x)$ such that the values of x distributed according to $w(x)$ can be generated by the inverse transform method. Let the total area under the curve $w(x)$ be equal to A.

2. Generate a uniform random number in the interval $[0, A]$ and use it to obtain a corresponding value of x distributed according to $w(x)$.

3. For the value of x generated in step (2), generate a uniform random number y in the interval $[0, w(x)]$. The point (x, y) is uniformly distributed in the area under the comparison function $w(x)$. If $y \leq p(x)$, then accept x as a random number distributed according to $p(x)$.

Steps (2) and (3) are repeated many times.

Note that the acceptance-rejection method is efficient only if the comparison function $w(x)$ is close to $p(x)$ over the entire range of interest.

References and Suggestions for Further Reading

Forman S. Acton, *Numerical Methods That Work*, Harper & Row (1970); corrected edition, Mathematical Association of America (1990). A delightful book on numerical methods.

Steven E. Koonin and Dawn C. Meredith, *Computational Physics*, Addison-Wesley (1990). Chapter 8 covers much of the same material on Monte Carlo methods as discussed in this chapter.

Malvin H. Kalos and Paula A. Whitlock, *Monte Carlo Methods, Vol. 1: Basics,* John Wiley & Sons (1986). The authors are well known experts on Monte Carlo methods.

William H. Press, Saul A. Teukolsky, William T. Vetterling, and Brian P. Flannery, *Numerical Recipes*, second edition, Cambridge University Press (1992).

Reuven Y. Rubinstein, *Simulation and the Monte Carlo Method*, John Wiley & Sons (1981). An advanced, but clearly written treatment of Monte Carlo methods.

I. M. Sobol, *The Monte Carlo Method*, Mir Publishing (1975). A very readable short text with excellent sections on nonuniform probability densities and the neutron transport problem.

C H A P T E R
12
Random Walks

We explore applications of random walks to a variety of systems.

12.1 INTRODUCTION

We introduced the random walk problem in Section 7.3 in the context of the motion of drunken sailors and the one-dimensional motion of particles. Of course, random walks are not restricted to one dimension nor are the applications limited to the wanderings of inebriates. We already know (see Section 11.8) that we can use random walk methods to estimate a definite integral. In this chapter we introduce some of the more popular random walk models and discuss several applications. In succeeding chapters we discuss other applications of random walk methods to problems that in many cases have no obvious connection to random walks.

For convenience, we start with the most elementary version of a random walk model, a walk on a one-dimensional translationally invariant lattice with steps occurring at uniform time intervals. In Section 7.3 we showed that the variance or dispersion

$$<\Delta x^2(N)> = <x^2(N)> - <x(N)>^2 \qquad (12.1)$$

depends linearly on N, where N is the number of steps. Many applications of random walk models make use of asymptotic results for large N. For example, in many cases $<\Delta x^2(N)>$ satisfies a power law for sufficiently large N, that is,

$$<\Delta x^2(N)> \sim N^{2\nu}. \qquad (N \gg 1) \qquad (12.2)$$

In this context the symbol $\sim$ is interpreted as "asymptotically equal to" and the relation (12.2) is an example of an asymptotic *scaling law*. For the simple one-dimensional random walk model we know that the relation (12.2) is valid for all N and that $\nu = \frac{1}{2}$. In many of the random walk problems introduced in this chapter, we will determine if such a power law dependence exists for large N.

Although we introduced the simple walk model in one dimension in Section 7.3, we did not consider all its properties. Some additional properties of this random walk are considered in Problem 12.1.

Problem 12.1   Discrete time random walks in one dimension

a. Suppose that the probability of moving to the right is $p = 0.7$. Compute $<x(N)>$ and $<x^2(N)>$ for $N = 4, 8, 16$, and 32. What is the interpretation of $<x(N)>$ in this case? What is the qualitative dependence of $<\Delta x^2(N)>$ on N? Does $<x^2(N)>$ depend simply on N?

b. Use the error analysis discussed in Section 11.4 to estimate the number of trials needed to obtain $<\Delta x^2(N)>$ to 1% accuracy for $N = 8$ and $N = 32$.

c. An interesting property of random walks is the mean number $<D(N)>$ of *distinct* lattice sites visited during the course of an N step walk. Do a Monte Carlo simulation of $<D(N)>$ and determine its N dependence.

We can equally well consider either a large number of successive walks as in Problem 12.1 or a large number of similar (noninteracting) walkers moving at the same time. In Problem 12.2 we consider the motion of many random walkers moving independently of one another on a two-dimensional lattice.

Problem 12.2 A random walk in two dimensions

Consider a collection of walkers initially at the origin of a square lattice. At each time step, each of the walkers moves at random with equal probability in one of the four possible directions. The following program implements this algorithm and shows the sites that have been visited.

```
PROGRAM random_walk2
! random walk in two dimensions
DIM x(500),y(500)
LIBRARY "csgraphics"
CALL initial(x(),y(),N,nwalkers,walker$)
FOR step = 1 to N
    SET CURSOR 1,1
    PRINT step
    CALL move(x(),y(),N,nwalkers,walker$)
NEXT step
END


SUB initial(x(),y(),N,nwalkers,walker$)
    RANDOMIZE
    INPUT prompt "number of walkers = ": nwalkers
    INPUT prompt "total number of steps = ": N
    CLEAR
    CALL compute_aspect_ratio(N,xwin,ywin)
    SET WINDOW -xwin,xwin,-ywin,ywin
    ! place N particles at origin
    SET COLOR "red"
    LET r = 0.5
    BOX AREA -r,r,-r,r
    BOX KEEP -r,r,-r,r in walker$
    SET COLOR "black"
    FOR i = 1 to nwalkers
        LET x(i) = 0
        LET y(i) = 0
    NEXT i
END SUB


SUB move(x(),y(),N,nwalkers,walker$)
    FOR i = 1 to nwalkers
        CALL choice(x(),y(),i)
        BOX SHOW walker$ at x(i)-0.5,y(i)-0.5
    NEXT i
END SUB
```

```
SUB choice(x(),y(),i)
    LET p = rnd
    IF p <= 0.25 then
        LET x(i) = x(i) + 1
    ELSEIF p <= 0.5 then
        LET x(i) = x(i) - 1
    ELSEIF p <= 0.75 then
        LET y(i) = y(i) - 1
    ELSE
        LET y(i) = y(i) + 1
    END IF
END SUB
```

a. Run *Program random_walk2* with the number of walkers nwalkers ≥ 200 and the number of steps taken by each walker N ≥ 500. If each walker represents a bee, describe the qualitative nature of the shape of the swarm of bees. Describe the qualitative nature of the surface of the swarm as a function of N. Is the surface jagged or smooth?

b. Compute the quantities $<x(N)>$, $<y(N)>$, $<\Delta x^2(N)>$, and $<\Delta y^2(N)>$ as a function of N. The average is over the walkers. Also compute the net mean square displacement $<\Delta R^2(N)>$ given by

$$<\Delta R^2(N)> \; = \; <x^2(N)> + <y^2(N)> - <x(N)>^2 - <y(N)>^2. \quad (12.3)$$

What is the dependence of each quantity on N?

Problem 12.3   Random walks on two and three-dimensional lattices

a. Enumerate all the random walks on a square lattice for $N = 4$ and obtain exact results for $<x(N)>$, $<y(N)>$ and $<\Delta R^2(N)>$ (see (12.3)). Verify your program by comparing your Monte Carlo and exact enumeration results. Consider the case where all four directions are equally probable.

b. Do a Monte Carlo simulation to estimate $<\Delta R^2(N)>$ for $N = 8, 16, 32$, and 64 using a reasonable number of trials for each value of N. Assume that $<\Delta R^2(N)>$ has the asymptotic N dependence:

$$<\Delta R^2(N)> \; \sim \; N^{2\nu}, \qquad\qquad (N \gg 1) \qquad\qquad\qquad (12.4)$$

and estimate the exponent ν from a log-log plot of $<\Delta R^2(N)>$ versus N. If $\nu \approx \frac{1}{2}$, estimate the magnitude of the self-diffusion coefficient D given by

$$<R^2(N)> \; \sim \; 2d\,DN. \qquad\qquad\qquad\qquad\qquad\qquad (12.5)$$

The form (12.5) is similar to (8.38) with the time t in (8.38) replaced by the number of steps N.

c. Compute the quantities $<x(N)>$, $<y(N)>$, $<R^2(N)> = <x^2(N) + y^2(N)>$, and $<\Delta R^2(N)>$ for the same values of N as in part (b), with the step probabilities 0.4, 0.2, 0.2, 0.2 corresponding to right, left, up, and down, respectively. This choice of probabilities corresponds to a biased random walk with a drift to the right. What is the interpretation of $<x(N)>$ in this case? What is the dependence of $<\Delta R^2(N)>$ on N? Does $<R^2(N)>$ depend simply on N?

*d. Consider a random walk that starts at a site that is a distance $y = h$ above a horizontal line (see Fig. 12.1). If the probability of a step down is greater than the probability of a step up, we expect that the walker will eventually reach a site on the horizontal line. This walk is a simple model of the fall of a rain drop in the presence of a random swirling breeze. Do a Monte Carlo simulation to determine the mean time τ for the walker to reach any site on the line $x = 0$ and find the functional dependence of τ on h. Is it possible to define a velocity in the vertical direction? Because the walker does not always move vertically, it suffers a net displacement Δx in the horizontal direction. How does $<\Delta x^2>$ depend on h and τ? Reasonable values for the step probabilities are 0.1, 0.6, 0.15, 0.15, corresponding to up, down, right, and left, respectively.

*e. Do a Monte Carlo simulation of $<R^2(N)>$ on the triangular lattice (see Fig. 8.5) and estimate ν. Can you conclude that ν is independent of the symmetry of the lattice? Does D depend on the symmetry of the lattice? If so, give a qualitative explanation for this dependence.

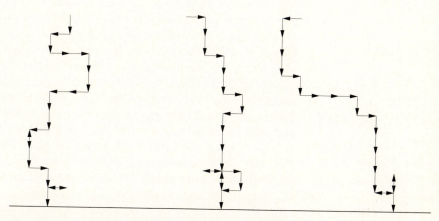

Fig. 12.1 Examples of the random path of a raindrop to the ground. The step probabilities are given in Problem 12.3d.

12.2 MODIFIED RANDOM WALKS

So far we have considered random walks on one- and two-dimensional lattices where the walker has no "memory" of the previous step. What happens if the walkers remember the nature of their previous steps? What happens if there are multiple random walkers, with the condition that no double occupancy is allowed? We explore these and other variations of the simple random walk in this section. All these variations have applications to physical systems, and we suggest that you think of additional phenomena that require these and other modifications of the random walk.

Problem 12.4   A persistent random walk

a. In a "persistent" random walk, the *transition* or "jump" probability depends on the previous transition. Consider a walk on a one-dimensional lattice, and suppose that step $N - 1$ has been made. Then step N is made in the same direction with probability α; a step in the opposite direction occurs with probability $1 - \alpha$. Write a program to do a Monte Carlo simulation of the persistent random walk in one dimension. Compute $<x(N)>$, $<x^2(N)>$, $<\Delta x^2(N)>$, and $P(x, N)$. Note that it is necessary to specify both the initial position and an initial direction of the walker. What is the $\alpha = \frac{1}{2}$ limit of the persistent random walk?

b. Consider the cases $\alpha = 0.25$ and $\alpha = 0.75$ and determine $<\Delta x^2(N)>$ for $N = 8, 64, 256$, and 512. Estimate the value of ν from a log-log plot of $<\Delta x^2(N)>$ versus N for large N. Does ν depend on α? If $\nu \approx \frac{1}{2}$, determine the self-diffusion coefficient D for $\alpha = 0.25$ and 0.75. Give a physical argument why $D(\alpha \neq 0.5)$ is greater (smaller) than $D(\alpha = 0.5)$.

c. A persistent random walk can be considered as an example of a *multistate* walk in which the state of the walk is defined by the last transition. In the above example, the walker is in one of two states; at each step the probabilities of remaining in the same state or switching states are α and $1 - \alpha$ respectively. One of the earliest applications of a two state random walk was to the study of diffusion in a chromatographic column. Suppose that a molecule in a chromatographic column can be either in a mobile phase (constant velocity v) or in a trapped phase (zero velocity). Instead of each step changing the position by ± 1, the position at each step changes by $+v$ or 0. A quantity of experimental interest is the probability $P(x, N)$ that a molecule has traveled a distance x in N steps. Choose $v = 1$ and $\alpha = 0.75$ and compute the qualitative behavior of $P(x, N)$. Explain why the molecule cannot diffuse in either state, but that it is still possible to define an effective diffusion coefficient for the molecule.

d. You might have expected that the persistent random walk yields a nonzero value for $<x(N)>$. Verify that $<x(N)> = 0$, and explain why this result is exact. How does the persistent random walk differ from the biased random walk for which $p \neq q$?

The fall of a raindrop considered in Problem 12.3d is an example of a *restricted random walk*, i.e., a walk in the presence of a boundary. (Another example of a restricted random walk was considered in Problem 7.10c.) In the following problem, we discuss in a more general context the effects of various types of restrictions or boundaries on random walks. Another example of a restricted random walk is given in Problem 12.11.

Problem 12.5   Restricted random walks

a. Consider a one-dimensional lattice with "trap" sites at $x = 0$ and $x = a$ ($a > 0$). A walker begins at site x_0 ($0 < x_0 < a$) and takes unit steps to the left and right with equal probability. When the walker arrives at a trap, it vanishes. Do a Monte Carlo simulation and verify that the mean number of steps τ for the particle to be trapped (*the first passage time*) is given by

$$\tau = (2D)^{-1} x_0 (a - x_0). \tag{12.6}$$

D is the self-diffusion coefficient in the absence of the traps, and the average is over all possible walks.

b. Random walk models in the presence of traps have had an important role in condensed matter science. For example, consider the following idealized model of energy transport in solids. The solid is represented as a lattice with two types of sites: hosts and traps. An incident photon is absorbed at a host site and excites the host molecule or atom. The excitation energy or *exciton* is transferred at random to one of the host's nearest neighbors and the original excited molecule returns to its ground state. In this way the exciton wanders through the lattice until it reaches a trap site. The exciton is then trapped and a chemical reaction occurs.

A simple version of this energy transport model is given by a one-dimensional lattice with traps placed on a periodic sublattice. Because the traps are placed at regular intervals, we can replace the random walk on an infinite lattice by a random walk on a ring. Consider a ring of N host or non-trapping sites and one trap site. If a walker has an equal probability of starting from any host site and an equal probability of a step to each nearest neighbor site, what is the N dependence of the mean survival time τ (the mean number of steps taken before a trap site is reached)? Use the results of part (a) rather than doing another simulation.

c. Consider a one-dimensional lattice with reflecting sites at $x = -a$ and $x = a$. For example, if a walker reaches the reflecting site at $x = a$, it is reflected at the next step to $x = a - 1$. At $t = 0$, the walker starts at $x = 0$ and steps with equal probability to nearest neighbor sites. Write a Monte Carlo program to determine $P(x, N)$, the probability that the walker is at site x after N steps. Compare the form of $P(x, N)$ with and without the presence of the reflecting "walls." Can you distinguish the two probability distributions if N is the order of a? At what value of N can you first distinguish the two distributions?

Although all of the above problems involved random walks on a lattice, it was not necessary to store the positions of the lattice sites or the path of the walker. In the following problem, we consider a random walk model that requires us to store the positions of a "gas" of random walkers.

*Problem 12.6   Individual particle diffusion in a lattice gas

Consider a square lattice with a nonzero density ρ of particles. Each particle moves at random to *empty* nearest neighbor sites. Double occupancy of sites is excluded; otherwise the particles are noninteracting. Such a model is an example of a *lattice gas*. The physical motivation of this model arises from solid state physics where the diffusing particles are thermal vacancies whose density depends on the temperature. The main physical quantity of interest is the self-diffusion coefficient D of an individual (tagged) particle. The model can be summarized by the following algorithm:

1. Occupy at random the $L \times L$ sites of a square lattice with N particles subject to the conditions that no double occupancy is allowed, and the density $\rho = N/L^2$ has the desired value. (Remember that $\rho < 1$.) Tag each particle, i.e., distinguish it from the others, and record its initial position in an array.

2. At each step choose a particle and one of its nearest neighbor sites at random. If the neighbor site is empty, the particle is moved to this site; otherwise the particle remains in its present position. The measure of "time" in this context is arbitrary. The usual definition is that during one unit of time or one *Monte Carlo step per particle*, each particle attempts one jump *on the average*. That is, the time is advanced by $1/N$ each time a particle is chosen even if the particle does not move.

The diffusion coefficient D is obtained as the limit $t \to \infty$ of $D(t)$, where $D(t)$ is given by

$$D(t) = \frac{1}{2dt} <\Delta R(t)^2>, \qquad (12.7)$$

and $<\Delta R(t)^2>$ is the net mean square displacement per tagged particle after t units of time. An example of a program that implements this algorithm is given in the following.

```
PROGRAM lattice_gas
! simulation of particle diffusion in a lattice gas
DIM site(40,40),occup$(-1:0)
DIM x(1200),y(1200),x0(1200),y0(1200)
LIBRARY "csgraphics"
CALL initial(L,L_2,N,tmax,#2)
CALL lattice(site(,),x(),y(),x0(),y0(),L,N,occup$(),#1)
FOR t = 1 to tmax
    CALL move(site(,),x(),y(),x0(),y0(),L,N,occup$(),#1)
```

```
                    CALL data(x(),y(),x0(),y0(),L,L_2,N,t,#2)
          NEXT t
          END

          SUB initial(L,L_2,N,tmax,#2)
              RANDOMIZE
              LET L = 40                    ! linear dimension of lattice
              LET L_2 = 0.5*L
              LET N = 500                   ! number of particles
              LET tmax = 50                 ! # Monte Carlo steps per particle
              SET BACKGROUND COLOR "black"
              OPEN #2: screen 0.52,1.0,0,1.0
              SET WINDOW 0,tmax + 1,0,L_2 + 1
              SET COLOR "white"
              PRINT "Plot of <R^2(t)> versus t"
              BOX LINES 0,tmax,0,L_2
          END SUB

          SUB lattice(site(,),x(),y(),x0(),y0(),L,N,occup$(),#1)
              OPEN #1: screen 0,0.5,0.2,1
              LET r = 0.5                   ! size of box
              CALL compute_aspect_ratio(L+r,xwin,ywin)
              SET WINDOW -0.1*xwin,xwin,-0.1*ywin,ywin
              SET COLOR "blue"
              BOX CIRCLE 2,3,2,3
              FLOOD 2.5,2.5
              BOX KEEP 2,3,2,3 in occup$(-1)
              CLEAR
              SET COLOR "black"
              BOX CIRCLE 2,3,2,3
              FLOOD 2.5,2.5
              BOX KEEP 2,3,2,3 in occup$(0)
              CLEAR
              SET COLOR "white"
              PRINT "density = "; N/(L*L)
              SET COLOR "red"
              BOX LINES 1-r,L+r,1-r,L+r
              FOR iy = 1 to L               ! draw lattice sites
                  FOR ix = 1 to L
                      LET site(ix,iy) = 0
                      PLOT POINTS: ix,iy
                  NEXT ix
              NEXT iy
              LET i = 0
              DO
                  LET xadd = int(L*rnd) + 1
                  LET yadd = int(L*rnd) + 1
                  IF site(xadd,yadd) = 0 then
                      LET i = i + 1         ! number of particles added
```

```
                            LET site(xadd,yadd) = -1      ! site occupied
                            BOX SHOW occup$(-1) at xadd - 0.5,yadd - 0.5
                            LET x(i) = xadd
                            LET y(i) = yadd
                            LET x0(i) = x(i)    ! x-coordinate at t = 0
                            LET y0(i) = y(i)
                        END IF
                    LOOP until i = N
                END SUB

                SUB move(site(,),x(),y(),x0(),y0(),L,N,occup$(),#1)
                    WINDOW #1
                    FOR particle = 1 to N
                        LET i = int(N*rnd) + 1
                        LET xtrial = x(i)
                        LET ytrial = y(i)
                        CALL direction(xtrial,ytrial,L)
                        IF site(xtrial,ytrial) >= 0 then    ! unoccupied site
                            LET site(x(i),y(i)) = 0
                            BOX SHOW occup$(0) at x(i) - 0.5,y(i) - 0.5
                            PLOT POINTS: x(i),y(i)
                            LET x(i) = xtrial
                            LET y(i) = ytrial
                            LET site(x(i),y(i)) = -1    ! new site occupied
                            BOX SHOW occup$(-1) at x(i) - 0.5,y(i) - 0.5
                        END IF
                    NEXT particle
                    SET COLOR "red"
                    BOX LINES 0.5,L+0.5,0.5,L+0.5
                END SUB

                SUB direction(xtemp,ytemp,L)
                    ! choose random direction and use periodic boundary conditions
                    LET dir = int(4*rnd) + 1
                    SELECT CASE dir
                    CASE 1
                        LET xtemp = xtemp + 1
                        IF xtemp > L then LET xtemp = 1
                    CASE 2
                        LET xtemp = xtemp - 1
                        IF xtemp < 1 then LET xtemp = L
                    CASE 3
                        LET ytemp = ytemp + 1
                        IF ytemp > L then LET ytemp = 1
                    CASE 4
                        LET ytemp = ytemp - 1
                        IF ytemp < 1 then LET ytemp = L
                    END SELECT
                END SUB
```

```
SUB data(x(),y(),x0(),y0(),L,L_2,N,t,#2)
    LET R2 = 0
    FOR i = 1 to N
        LET dx = x(i) - x0(i)
        LET dy = y(i) - y0(i)
        CALL separation(dx,dy,L,L_2)
        LET R2 = R2 + dx*dx + dy*dy
    NEXT i
    LET R2bar = R2/N
    WINDOW #2
    PLOT t,R2bar
END SUB


SUB separation(dx,dy,L,L_2)
    ! use periodic boundary conditions to determine separation
    IF abs(dx) > L_2 then LET dx = dx - sgn(dx)*L
    IF abs(dy) > L_2 then LET dy = dy - sgn(dy)*L
END SUB
```

a. Do a Monte Carlo simulation to determine D on a square lattice for $\rho = 0.1, 0.2, 0.3, 0.5$, and 0.7. Choose $L \geq 40$. Although D is defined as the limit $t \to \infty$ of (12.7), for this model $D(t)$ fluctuates with t after a short equilibration time and no improvement in accuracy is achieved by increasing t. Better statistics for D can be obtained by averaging D over as many tagged particles as possible and hence by considering a lattice with L as large as possible. The accuracy of D also can be increased by averaging $<R(t)^2>$ over different initial starting times. Why is it necessary to limit the number of Monte Carlo steps so that $<R(t)^2>$ is less than $(L/2)^2$? Verify that deviations of $D(t)$ from its mean value are proportional to the inverse square root of the total number of particles that enter into the average in (12.7).

b. Why is D a monotonically decreasing function of the density ρ? To gain some insight into this dependence, determine the dependence on ρ of the probability that if a particle jumps to a vacancy at time t, it returns to its previous position at time $t + 1$. Is there a qualitative relation between the density dependence of D and this probability?

c. Consider a one-dimensional lattice model for which particles move at random, but double occupancy of sites is excluded. This restriction implies that particles cannot pass by each other. Compute $<\Delta x^2>$ as a function of t. Do the particles diffuse, i.e., is $<\Delta x^2>$ proportional to t? If not, what is the t dependence of $<\Delta x^2>$?

Problem 12.7   Random walk on a continuum

One of the first continuum models of a random walk was proposed by Rayleigh in 1919. The model is known as the freely jointed chain in polymer physics. In this model the length a of each step is a random variable with probability density $p(a)$, and the direction of each step is uniformly random. For simplicity, we first consider a walker in two dimensions with $p(a)$ chosen so that each step has unit length. At each step the walker takes a step of unit length at a random angle. Write a Monte Carlo program to compute $P(r, N)\,dr$, the probability that the displacement of the walker is in the range r to $r + dr$ after N steps, where r is the distance from the origin. Verify that for sufficiently large N, $P(r, N)$ can be approximated by a Gaussian. Is a Gaussian a good approximation for small N? Is it necessary to do a Monte Carlo simulation to confirm that $<R(N)^2> \sim N$, or can you give a simple argument for this dependence based on the form of $P(r, N)$?

Problem 12.8   Random walks with steps of variable length

a.   Consider a random walk in one dimension with jumps of all lengths allowed. The probability density that the length of a single step is a is denoted by $p(a)$. If the form of $p(a)$ is given by $p(a) = e^{-a}$, what is the form of $P(x, N)$? Suggestions: Use the inverse transform method discussed in Section 11.5 to generate step lengths according to the probability density $p(a)$. Consider a walk of N steps and determine the net displacement x. Generate many such walks and determine $P(x, N)$. Plot $P(x, N)$ versus x and confirm that the form of $P(x, N)$ is consistent with a Gaussian distribution. Is this random walk equivalent to a diffusion process for sufficiently large N?

b.   Assume that the probability density $p(a)$ is given by $p(a) = C/a^2$ for $a \geq 1$. Determine the normalization constant C using the condition $C \int_1^\infty a^{-2}\, da = 1$. Does the second moment of $p(a)$ exist? Do a Monte Carlo simulation as in part (a) and verify that the form of $P(x, N)$ is given by

$$P(x, N) \sim \frac{bN}{x^2 + b^2 N^2}, \tag{12.8}$$

What is the magnitude of the constant b? Does the variance $<x^2> - <x>^2$ of $P(x, N)$ exist? Is this random walk equivalent to a diffusion process?

Problem 12.9   The central limit theorem

Consider a continuous random variable x with probability density $f(x)$. That is, $f(x)\Delta x$ is the probability that x has a value between x and $x + \Delta x$. The mth *moment* of $f(x)$ is defined as

$$<x^m> = \int x^m f(x)\, dx. \tag{12.9}$$

The mean value $<x>$ is given by (12.9) with $m = 1$. The variance σ_x^2 of $f(x)$ is defined as

$$\sigma_x^2 = <x^2> - <x>^2. \tag{12.10}$$

Consider the sum y_n corresponding to the average of n values of x:

$$y_n = \frac{1}{n}(x_1 + x_2 + \ldots + x_n).$$ (12.11)

We adopt the notation $y = y_n$. Suppose that we make many measurements of y. We know that the values of y are not identical, but are distributed according to a probability density $P(y)$ different from $f(x)$, where $P(y)\Delta y$ is the probability that the measured value of y is in the range y to $y + \Delta y$. The main quantities of interest are the mean $<y>$, the variance $\sigma_y^2 = <y^2> - <y>^2$, and $P(y)$ itself.

a. Suppose that $f(x)$ is uniform in the interval $[-1, 1]$. Calculate $<x>$ and σ_x analytically. Use a Monte Carlo method to make a sufficient number of measurements of y to determine $P(y)$, $<y>$, and σ_y with reasonable accuracy. For example, choose $n = 1000$ and make 100 measurements of y. Verify that σ_y is approximately equal to $\sigma_x/\sqrt{n}$. Plot $P(y)$ versus y and discuss its qualitative form. Does the form of $P(y)$ change significantly if n is increased? Does the form of $P(y)$ change if the number of measurements of y is increased?

b. To test the generality of the results of part (a), consider the exponential probability density

$$f(x) = \begin{cases} e^{-x}, & \text{if } x \geq 0; \\ 0, & \text{if } x < 0. \end{cases}$$ (12.12)

Calculate $<x>$ and σ_x analytically. Modify your Monte Carlo program and estimate $<y>$, σ_y, and $P(y)$. Is σ_y related to σ_x as in part (a)? Plot $P(y)$ and discuss its qualitative form and its dependence on n and on the number of measurements of y.

c. Let y be the Monte Carlo estimate of the integral (see Problem 11.3a)

$$4 \int_0^1 dx \sqrt{1 - x^2}.$$ (12.13)

In this case y is found by sampling the integrand $f(x) = 4\sqrt{1 - x^2}$ n times. Choose $n \geq 1000$ and make at least 100 measurements of y. Show that the values of y are distributed according to a Gaussian distribution. How is the variance of $P(y)$ related to the variance of $f(x)$?

d. Consider the Lorentzian probability density

$$f(x) = \frac{1}{\pi} \frac{1}{x^2 + 1}.$$ (12.14)

Calculate the mean value $<x>$. Does the second moment and hence the variance of $f(x)$ exist? Do a Monte Carlo calculation of $<y>$, σ_y, and $P(y)$. Plot $P(y)$ as a function of y and discuss its qualitative form. What is the dependence of $P(y)$ on the number of trials?

Problem 12.9 illustrates the *central limit theorem* which states that the probability distribution of a variable y is a Gaussian centered at $<y>$ with a standard deviation $1/\sqrt{n}$ times the standard deviation of $f(x)$. The requirements are that $f(x)$ has finite first and second moments, that the measurements of y are statistically independent, and that n is large. Use the central limit theorem to explain your results in Problem 12.9 and in Problem 12.8a. What is the relation of the present calculation of $P(y)$ to the calculations of the probability distribution in the random walk models that we already have considered?

Problem 12.10   Generation of the Gaussian distribution

Consider the sum

$$y = \sum_{i=1}^{12} r_i, \tag{12.15}$$

where r_i is a uniform random number in the unit interval. Make many "measurements" of y and show that the probability distribution of y approximates the Gaussian distribution with mean value 6 and variance 1. Discuss how to use this result to generate a Gaussian distribution with arbitrary mean and variance. This way of generating a Gaussian distribution is particularly useful when a "quick and dirty" approximation is appropriate.

Many of the problems we have considered have revealed the slow convergence of Monte Carlo simulations and the difficulty of obtaining quantitative results for asymptotic quantities. We conclude this section with a cautionary note and consider a "simple" problem for which straightforward Monte Carlo methods give misleading asymptotic results.

*Problem 12.11   Random walk on lattices containing traps

a. We have considered the mean survival time of a one-dimensional random walker in the presence of a periodic distribution of traps (see Problem 12.5b). Now suppose that the trap sites are distributed at random on a one-dimensional lattice with density ρ. If a walker is placed at random at any nontrap site, determine its mean survival time τ, the mean number of steps before a trap site is reached. Assume that the walker has an equal probability of moving to nearest neighbor sites at each step and use periodic boundary conditions. This problem is more difficult than it might first appear, and there are a number of pitfalls. The major complication is that it is necessary to perform *three* averages: the distribution of traps, the origin of the walker, and the different walks for a given trap distribution and origin. Choose reasonable values for the number of trials associated with each average and do a Monte Carlo simulation to estimate the mean survival time τ. If τ exhibits a power law dependence, e.g., $\tau \approx \tau_0 \rho^{-z}$, estimate the exponent z.

b. A seemingly straightforward extension of part (a) is to estimate the probability of survival $P(N)$ of an N step random walk. Choose $\rho = 0.5$ and do a Monte

Carlo simulation of $P(N)$ for N as large as possible. (Published results are for $N = 2 \times 10^3$ on lattices with $L = 50\,000$ sites and $50\,000$ trials.) Assume that the asymptotic form of $P(N)$ is given by

$$P(N) \sim e^{-bN^{\alpha}}, \tag{12.16}$$

where b is a constant that depends on ρ. Are your results consistent with this form? Is it possible to make a meaningful estimate of the exponent α?

c. The object of part (b) is to convince you that it is not possible to use Monte Carlo methods directly to obtain the correct asymptotic behavior of $P(N)$. The difficulty is that we are trying to estimate $P(N)$ in the asymptotic region where $P(N)$ is very small, and the inherent fluctuations of the Monte Carlo method prevent us from obtaining meaningful results. It has been proved using analytical methods, that the asymptotic N dependence of $P(N)$ has the form (12.16), but with $\alpha = 1/3$. Are your Monte Carlo results consistent with this value of α?

d. A method that reduces the number of required averages and hence reduces the fluctuations is to determine exactly, for a given distribution of trap sites, the probability that the walker is at site i after N steps. The method is illustrated in Fig. 12.2. The first line represents a given configuration of traps distributed randomly on a one-dimensional lattice. One walker is placed at each regular site; trap sites are assigned the value 0. Because each walker moves with probability $\frac{1}{2}$ to each neighbor, the number of walkers $W_i(N + 1)$ on site i at step $N + 1$ is given by

$$W_i(N + 1) = \frac{1}{2}\left[W_{i+1}(N) + W_{i-1}(N)\right]. \tag{12.17}$$

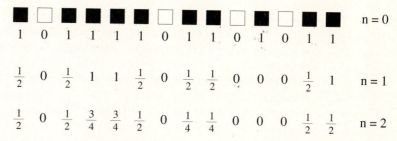

Fig. 12.2 Example of the exact enumeration of walks on a given configuration of traps. The filled and empty squares denote regular and trap sites respectively. At step $N = 0$, a walker is placed at each regular site. The numbers at each site i represent the number of walkers W_i. Periodic boundary conditions are used. The initial number of walkers in this example is $m_0 = 10$. The mean survival probability at step $N = 1$ and $N = 2$ is found to be 0.6 and 0.475 respectively.

(Compare the relation (12.17) to (12.31) and to the relation that you found in Problem 7.6f.) The survival probability $P(N)$ after N steps for a given configuration of traps is given exactly by

$$P(N) = \frac{1}{m_0} \sum_i W_i(N), \tag{12.18}$$

where m_0 is the initial number of walkers and the sum is over all sites in the lattice. Explain the relation (12.18), and write a program that computes $P(N)$ using (12.18) and (12.17). Then obtain the average $<P(N)>$ over several configurations of traps. Choose $\rho = 0.5$ and determine $P(N)$ for $N = 32, 64, 128, 512,$ and 1024. Choose periodic boundary conditions and as large a lattice as possible. How well can you estimate the exponent α? For comparison, Havlin et al. consider a lattice of $L = 50\ 000$ and values of N up to 10^7.

12.3 APPLICATIONS TO POLYMERS

Random walk models play an important role in polymer physics (cf. de Gennes). A polymer consists of N repeat units (monomers) with N very large ($N \sim 10^3 - 10^5$). For example, polyethylene can be represented as $\cdots -CH_2-CH_2-CH_2- \cdots$. The detailed structure of the polymer is important for many practical applications. For example, if we wish to improve the fabrication of rubber, a good understanding of the local motions of the monomers in the rubber chain is essential. However, if we are interested in the *global* properties of the polymer, the details of the chain structure often can be ignored.

Let us consider a familiar example of a polymer chain in a good solvent: a noodle in warm water. A short time after we place the noodle in the water, the noodle becomes flexible, and it neither collapse into a little ball or becomes fully stretched. Instead, it adopts a random structure as shown schematically in Fig. 12.3. If we do not add too many noodles, we can say that the noodles behave as a dilute solution of polymer chains in a good solvent. The dilute nature of the solution implies that we can ignore entanglement effects of the noodles and consider each noodle individually. The presence of a good solvent implies that the polymers can move freely and adopt many different configurations.

A fundamental geometrical property that characterizes a polymer in solution is the mean square end-to-end distance $<R^2(N)>$, where N is the number of monomers. It is known that for a dilute solution of polymer chains in a good solvent, the asymptotic dependence of $<R^2(N)>$ is given by (12.4) with the exponent $\nu \approx 0.592$ in three dimensions. The result for ν in two dimensions is known to be exactly $\nu = 3/4$ for the model of polymers to be discussed below. The proportionality constant in (12.4) depends on the structure of the monomers and on the solvent. In contrast, the exponent ν is independent of these details.

We now discuss a random walk model that incorporates the global features of linear polymers in solution. We already have introduced a model of a polymer chain consisting of straight line segments of the same size joined together at random angles

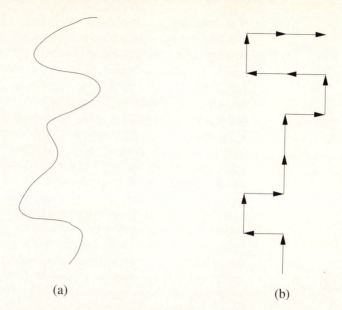

Fig. 12.3   (a) Schematic illustration of a linear polymer in a
good solvent. (b) Example of the corresponding
walk on a square lattice.

(see Problem 12.7). A further idealization is to place the polymer chain on a lattice
(see Fig. 12.3b). If we ignore the interactions of the monomers, we know that this
simple random walk model gives $\nu = \frac{1}{2}$ independent of the dimension and symmetry
of the lattice. Because this result for ν does not agree with experiment, we know we are
overlooking an important physical feature of polymers. Let us consider a more realistic
model of a chain that accounts for the most important physical feature of a polymer—
two monomers cannot occupy the same spatial position. This constraint, known as the
excluded volume condition, implies that the walk cannot be adequately described by
a purely random walk. A well known lattice model for a flexible polymer chain that
incorporates this constraint is known as the *self-avoiding* walk (SAW). Consider the
set of all N step walks starting from the origin subject to the global constraint that
no lattice site can be visited more than once in each walk; this constraint accounts for
the excluded volume condition. The calculation of the properties of the self-avoiding
walk is formidable. Computer enumeration and Monte Carlo simulation have played an
important role in our current understanding. We consider Monte Carlo simulations of
the self-avoiding walk in two dimensions in Problem 12.12.

Problem 12.12   The two-dimensional self-avoiding walk

a.   Consider the self-avoiding walk on the square lattice. Choose an arbitrary site
as the origin and assume that the first step is "up." The walks generated by the

three other possible initial steps only differ by a rotation of the whole lattice, and do not have to be considered explicitly. The second step can be in three possible directions because of the constraint that the walk cannot return to the origin. To obtain unbiased results, we generate a random number to choose one of the three directions. Successive steps are generated in the same way. Unfortunately, the walk will not usually continue indefinitely. The difficulty is that to obtain unbiased results, we must generate a random number (e.g., 1, 2, or 3) as usual, even though one or more of the steps might lead to a self-intersection. If the next step does lead to a self-intersection, the walk must be terminated to keep the statistics correct. A new walk is started again at the origin. An example for $N = 3$ is shown in Fig. 12.4a. The next step leads to a self-intersection and violates the constraint. Write a program that implements this straightforward algorithm and record the fraction $f(N)$ of successful attempts of constructing polymer chains with N total monomers. It is convenient to represent the lattice as a two-dimensional array so that you can record the sites that already have been visited. What is the qualitative dependence of $f(N)$ on N? What is the maximum value of N that you can reasonably consider? Determine the mean square end-to-end distance $<R^2(N)>$ for these values of N.

b.   The disadvantage of the straightforward sampling method in part (a) is that it becomes very inefficient for long chains, i.e., the fraction of successful attempts decreases exponentially fast. To overcome this attrition, several "enrichment" techniques have been developed. We first discuss a relatively simple procedure proposed by Rosenbluth and Rosenbluth in which each walk of N steps is associated with a weighting function $W(N)$. Because the first step to the north is always possible, we have $W(1) = 1$. In order that all allowed configurations of a given N are counted equally, the weights $W(N)$ for $N > 1$ are determined according to the following possibilities:

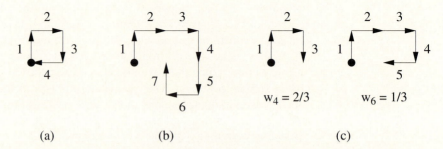

(a)   (b)   (c)

Fig. 12.4   Examples of self-avoiding walks on the square lattice. The origin is denoted by a filled circle. (a) A $N = 3$ walk. The fourth step shown is forbidden. (b) A $N = 7$ walk that leads to a self-intersection at the next step; the weight of the $N = 8$ walk is zero. (c) Examples of the weights of walks in the enrichment method.

1. All three possible steps violate the self-intersection constraint (see Fig. 12.4b). The walk is terminated with a weight $W(N) = 0$, and a new walk is generated at the origin.
2. All three steps are possible and $W(N) = W(N-1)$.
3. Only m steps are possible with $1 \leq m < 3$ (see Fig. 12.4c). In this case $W(N) = (m/3)W(N-1)$, and a random number is generated to choose one of the m possible steps.

The correct value of $<R^2(N)>$ is obtained by weighting $R_i^2(N)$, the value of $R^2(N)$ found in the ith trial, by the value of $W_i(N)$, the weight found for the particular walk. Hence we write

$$<R^2(N)> = \frac{\sum_i W_i(N) R_i^2(N)}{\sum_i W_i(N)}, \tag{12.19}$$

where the sum is over all trials. Incorporate the Rosenbluth method into your Monte Carlo program, and calculate $<R^2(N)>$ for $N = 4, 8, 16$, and 32. Estimate the exponent ν from a log-log plot of $<R^2(N)>$ versus N. Can you distinguish your estimate for ν from its random walk value $\nu = \frac{1}{2}$?

*Problem 12.13 Application of the reptation method

One of the more efficient enrichment algorithms is the "reptation" method (see Wall and Mandel). For simplicity, consider a model polymer chain in which all bond angles are $\pm 90°$. As an example of this model, the five independent $N = 5$ polymer chains are shown in Fig. 12.5. (Other chains differ only by a rotation or a reflection.) The reptation method can be stated as follows:

1. Choose a chain at random and remove the tail link.
2. Attempt to add a link to the head of the chain. There is a maximum of two directions in which the new head link can be added.
3. If the attempt violates the self-intersection constraint, return to the original chain and interchange the head and tail. Include the chain in the statistical sample.

The above steps are repeated many times to obtain a statistical average of $R^2(N)$.

As an example of the reptation method, suppose we choose chain a of Fig. 12.5. A new link can be added in two directions (see Fig. 12.6a), so that on the average we find, $a \to \frac{1}{2}c + \frac{1}{2}d$. In contrast, a link can be added to chain b in only one direction, and we obtain $b \to \frac{1}{2}e + \frac{1}{2}b$, where the tail and head of chain b have been interchanged (see Fig. 12.6b). Confirm that $c \to \frac{1}{2}e + \frac{1}{2}a$, $d \to \frac{1}{2}c + \frac{1}{2}d$, and $e \to \frac{1}{2}a + \frac{1}{2}b$, and that all five chains are equally probable. That is, the transformations in the reptation method preserve the proper statistical weights of the chains without attrition. There is just one problem: unless we begin with a double ended "cul-de-sac" configuration such as shown in Fig. 12.7, we will never obtain such a configuration using the above transformation. Hence, the reptation method introduces a small statistical bias, and the calculated mean end-to-end distance will be slightly larger than if all configurations were considered. However, the probability

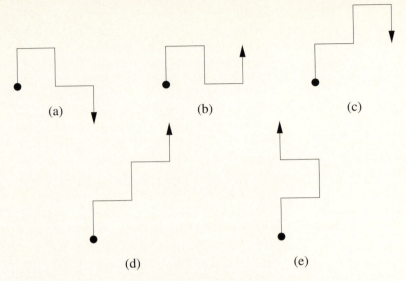

Fig. 12.5   The five independent possible walks of $N = 5$ steps on a square
lattice with $\pm 90°$ bond angles. The tail and head of each walk are
denoted by a circle and arrow respectively.

of such trapped configurations is very small, and the bias can be neglected for most
purposes.

Adopt the $\pm 90°$ bond angle restriction and calculate by hand the exact value
of $<R^2(N)>$ for $N = 5$. Then write a Monte Carlo program that implements the
reptation method. Generate one walk of $N = 5$ and use the reptation method to
generate a statistical sample of chains. As a check on your Monte Carlo program,
compute $<R^2(N)>$ for $N = 5$ and compare your result with the exact result. Then
extend your Monte Carlo computations of $<R^2(N)>$ to larger N. Modify the
reptation model so that the bond angle also can be 180°. This modification leads
to a maximum of three directions for a new bond. Compare the results of the two
models.

Problem 12.14   The dynamics of polymers in a dilute solution

In principle, the dynamics of a polymer chain undergoing collisions with solvent
molecules can be simulated by using a molecular dynamics method. However, in
practice only relatively small chains can be simulated in this way. An alternative
approach is to use a Monte Carlo model that simplifies the effect of the random
collisions of the solvent molecules with the atoms of the chain. Most of these
models (cf. Verdier and Stockmayer) consider the chain to be composed of beads
connected by bonds and restrict the motion of the beads to a lattice. For simplicity,
we assume that the bond angles can be either $\pm 90°$ or 180°. Begin with an allowed

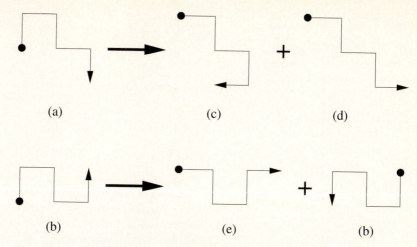

(a)   (c)   (d)

(b)   (e)   (b)

Fig. 12.6 The possible transformations of chains a and b. One of the two possible transformations of chain b violates the self-intersection restriction and the head and tail are interchanged.

configuration of N beads ($N + 1$ bonds). A possible starting configuration can be generated by taking successive steps in the positive y direction and positive x directions. The dynamics of the model is summarized by the following algorithm.

1. Select at random a bead (occupied site) on the polymer chain. If the bead is not an end site, then the bead can move to a nearest neighbor site of another bead if this site is empty and if the new angle between adjacent bonds is either $\pm 90°$ or $180°$. For example, bead 4 in Fig. 12.8 can move to position $4'$ while bead 3 cannot move if selected. That is, a selected bead can move to a diagonally

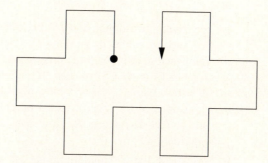

Fig. 12.7 Example of a double cul-de-sac configuration for the usual self-avoiding walk that cannot be obtained by the reptation method.

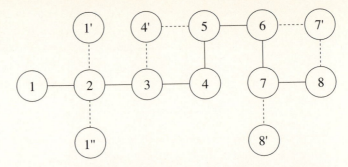

Fig. 12.8   Examples of possible moves of the simple polymer
dynamics model considered in Problem 12.14. For
this configuration beads 2, 3, 5, and 6 cannot move,
while beads 1, 4, 7, and 8 can move to the positions
shown if they are selected. Only one bead can move
at a time. This figure is adopted from the article by
Verdier and Stockmayer.

opposite unoccupied site only if the two bonds to which it is attached are
mutually perpendicular.

2. If the selected bead is an end site, move it to one of two (maximum) possible
   unoccupied sites so that the bond to which it is connected changes its orientation by $\pm 90°$ (see Fig. 12.8).

3. If the selected bead cannot move, retain the previous configuration.

   Some of the physical quantities of interest are the mean square end-to-end
distance $<R^2(N)>$, and the mean square displacement of the center of mass of
the chain $<\Delta R_{cm}^2(N)>$. The unit of time is the number of Monte Carlo steps per
bead (during which all beads have one chance on the average to move to a different
site).

a.   Consider a two-dimensional lattice and compute $<R^2(N)>$ and $<\Delta R_{cm}^2(N)>$
for various values of N. How do these quantities depend on N? (The first
published results for three dimensions were limited to 32 Monte Carlo steps
per bead for $N = 8$, 16, and 32 and only 8 Monte Carlo steps per bead for
$N = 64$.) Estimate the accuracy of your calculation by calculating the standard
deviations of the means. Also compute the probability $P(R, N)dR$ that the
end-to-end distance is R. How does this probability compare to a Gaussian
distribution?

* **b.**   We know that two configurations are strongly correlated if they differ by only
the position of one bead. Hence, it would be a waste of computer time to measure the end-to-end distance and the position of the center of mass after every
single move. Ideally, we wish to compute these quantities for configurations
that are approximately statistically independent. Because we do not know *a*

priori the mean number of Monte Carlo steps per bead needed to obtain configurations that are statistically independent, we need to estimate this time in our preliminary calculations. A simple way to estimate this time is to plot the time average of R^2 as a function of the time. If you start with a configuration that is not typical, then you will notice that the time average of R^2 approaches a value that no longer changes with time except for small fluctuations. Can the difference between the time average of R^2 and its equilibrium value be approximated by an exponential of the form e^{-t/τ_r}? The time τ_r is known as the *relaxation time*. If τ_r depends on N, try to quantify this relationship.

*c. The relaxation time τ_r is usually the same order of magnitude as the correlation time τ_c, where τ_c is the time needed to obtain statistically independent configurations. This time can be obtained by computing the equilibrium averaged autocorrelation function for a chain of fixed N:

$$C(t) = \frac{<R^2(t'+t)R^2(t')> - <R^2>^2}{<R^4> - <R^2>^2}. \tag{12.20}$$

Note that $C(t)$ has the same form as the velocity correlation function (8.39) and the autocorrelation function that is introduced in (12.40). $C(t)$ is defined so that $C(t=0) = 1$ and $C(t) = 0$ if the configurations are not correlated. Because the configurations will become uncorrelated if the time t between the configurations is sufficiently long, we have that $C(t) \to 0$ for $t >> 1$. In general, we expect that $C(t) \sim e^{-t/\tau_c}$, that is, $C(t)$ decays exponentially with a decay or correlation time τ_c, that is, $C(t)$. Estimate τ_c from a plot of $\ln C(t)$ versus t. Another way of estimating τ_c is from the integral $\int_0^\infty dt\, C(t)$. (Because we determine $C(t)$ at discrete values of t, this integral is actually a sum.) How do your two estimates of τ_c compare? A more detailed discussion of the estimation of correlation times can be found in Section 17.4.

Another type of random walk that is less constrained than the self-avoiding random walk is the "true" self-avoiding walk (TSAW). The TSAW describes the path of a random walker that avoids visiting a lattice site with a probability that is a function of the number of times the site has been visited already. This constraint leads to a reduced excluded volume interaction in comparison to the usual self-avoiding walk.

Problem 12.15 The true self-avoiding walk in one dimension

In one dimension the true self-avoiding walk corresponds to a walker who can jump to one of two nearest neighbors with a probability that depends on the number of times these neighbors already have been visited. Suppose that the walker is at site i at step t. The probability that at time $t+1$, the walker will jump to site $i+1$ is given by

$$P_{i+1} = \frac{e^{-gn_{i+1}}}{e^{-gn_{i+1}} + e^{-gn_{i-1}}}, \tag{12.21}$$

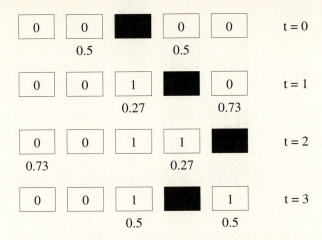

Fig. 12.9 Example of the time evolution of the true self-avoiding walk with $g = 1$. The shaded site represents the location of the walker at time t. The number of visits to each site are given within each site and the probability of a step to a nearest neighbor site is given below it. Note the use of periodic boundary conditions.

where $n_{i\pm1}$ is the number of times that the walker has already visited site $i \pm 1$. The probability of a jump to site $i - 1$ is $P_{i-1} = 1 - P_{i+1}$. The parameter g $(g > 0)$ is a measure of the "desire" of the path to avoid itself. The first few steps of a typical true self-avoiding walk walk are shown in Fig. 12.9. The main quantity of interest is the exponent ν. We know that $g = 0$ corresponds to the usual random walk with $\nu = \frac{1}{2}$ and that the limit $g \to \infty$ corresponds to the self-avoiding walk. What is the value of ν for a self-avoiding walk in one dimension? Is the value of ν for any finite value of g different than these two limiting cases? We explore these questions in the following.

Write a program to do a Monte Carlo simulation of the true self-avoiding walk in one dimension. Use an array to record the number of visits to every site. At each step calculate the probability P of a jump to the right. Generate a random number r and compare it to P. If $r \leq P$, then move the walker to the right; otherwise move the walker to the left. Compute $<\Delta x^2(N)>$, where x is the distance of the walker from the origin, as a function of the number of steps N. Make a log-log plot of $<\Delta x^2(N)>$ versus N and estimate ν. Can you distinguish ν from its random walk and self-avoiding walk values? Reasonable choices of parameters are $g = 0.1$ and $N \sim 10^3$. Averages over 10^3 trials give qualitative results. For comparison, published results (see Bernasconi and Pietronero) are for $N = 10^4$ and for 10^3 trials; extended results for $g = 2$ are given for $N = 2 \times 10^5$ and 10^4 trials.

12.4   DIFFUSION CONTROLLED CHEMICAL REACTIONS

Imagine a system containing particles of a single species A. The particles diffuse, and when two particles collide, a "reaction" occurs such that either one particle is annihilated or the two combine to form an inert species which is no longer involved in the reaction. In the latter case we can represent the chemical reaction as

$$A + A \rightarrow 0. \qquad \text{(inert)} \qquad (12.22)$$

If we ignore the spatial fluctuations of the density of species A, we can describe the kinetics by a simple rate equation:

$$\frac{dA(t)}{dt} = -kA^2(t), \qquad (12.23)$$

where A is the concentration of A particles at time t and k is the rate constant. (In the chemical kinetics literature it is traditional to use the term concentration rather than the number density.) For simplicity, we assume that all reactants are entered into the system at $t = 0$ and that no reactants are added later (a closed system). It is easy to show that the solution of the first-order differential equation (12.23) is

$$A(t) = \frac{1}{kt + 1/A(0)}, \qquad (12.24)$$

and $A(t) \sim t^{-1}$ in the limit of long times.

Another interesting case is the bimolecular reaction

$$A + B \rightarrow 0. \qquad (12.25)$$

If we neglect fluctuations in the concentration as before (this neglect yields the mean-field approximation), we can write the corresponding rate equation as

$$\frac{dA(t)}{dt} = \frac{dB(t)}{dt} = -kA(t)B(t). \qquad (12.26)$$

We also have that

$$A(t) - B(t) = \text{constant}, \qquad (12.27)$$

because each reaction leaves the difference between the concentration of A and B particles unchanged. For the special case of equal initial concentrations, the solution of (12.26) with (12.27) is the same as (12.24). What is the solution for the case $A(0) \neq B(0)$?

The above derivation of the time dependence of A for the kinetics of the one and two species annihilation process is straightforward, but is based on the neglect of spatial fluctuations. In the following two problems, we simulate the kinetics of these processes and test this assumption.

Problem 12.16 Monte Carlo simulation of diffusion controlled chemical reactions in one dimension

a. Assume that N particles do a random walk on a one-dimensional lattice of length L with periodic boundary conditions. Every particle moves once in one unit of time. Because we are interested in the long time behavior of the system when the concentration $A = N/L$ of particles is small, it is efficient to maintain an array of the particle positions, posx, in addition to an array for the occupancy of the lattice sites. That is, site(posx(i)) = i. For example, if particle 5 is located at site 12, then posx(5) = 12 and site(12) = 5. We also need a second array site_new to maintain the new positions of the walkers as they are moved one at a time. If two walkers land on the same position, k, then set site_new(k) = 0, and set the value of posx for these two walkers to 0. After all the walkers have moved, let MAT site= site_new, and remove all the reacting particles in posx that have values equal to 0. This operation can be accomplished by replacing any reacting particle in posx by the last particle in the array. Begin with all sites occupied, $A(t = 0) = 1$. Make a log-log plot of the quantity $1/A(t) - 1$ versus the time t. The times should be separated by exponential intervals so that your data is equally spaced on a logarithmic plot. For example, you might include data with times equal to 2^p, with $p = 1, 2, 3, \ldots$ Does your log-log plot yield a straight line in the limit of long times? If so, calculate its slope. Is the mean-field approximation valid in one dimension? You can obtain crude results for small lattices of order $L = 100$ and times of order $t = 10^2$. To obtain results to within ten percent, you need lattices of order $L = 10^4$ and times of order $t = 2^{13}$.

b. More insight into the origin of the time dependence of $A(t)$ can be gained from the behavior of the quantity $P(r, t)$, the probability distribution of the nearest neighbor distances at time t. The nearest neighbor distance of a particle is defined as the minimum distance between the particle and all other particles. ($P(r)$ is not the same as the radial distribution function $g(r)$ considered in Section 8.8.) The distribution of these distances changes dramatically as the reaction proceeds, and this change can give information about the reaction mechanism. Place particles at random on a one-dimensional lattice and verify that the most probable nearest neighbor distance is $r = 1$ (one lattice constant) for all concentrations. (This result is true in any dimension.) Then verify that the distribution of nearest neighbor distances on a $d = 1$ lattice is given by

$$P(r, t = 0) = 2A\,e^{-2A(r-1)}. \tag{12.28}$$

Is the form (12.28) properly normalized? Start with $A(t = 0) = 0.1$ and find $P(r, t)$ for $t = 10$, 100, and 1000. Average over all particles. How does $P(r, t)$ change as the reaction proceeds? Does it retain the same form as the concentration decreases?

* **c.**   Compute the quantity $D(t)$, the number of *distinct* sites visited by an individual walker. How does the time dependence of $D(t)$ compare to the computed time dependence of $1/A(t) - 1$?

* **d.**   Write a program to simulate the $A + B = 0$ reaction. For simplicity, assume that multiple occupancy of the same site is not allowed, e.g., an A particle cannot jump to a site already occupied by an A particle. The easiest procedure is to allow a walker to choose one of its nearest neighbor sites at random, but to not move the walker if the chosen site is already occupied by a particle of the same type. If the site is occupied by a walker of another type, then the pair of reacting particles is annihilated. Keep separate arrays for the A and B particles, with the value of the array denoting the label of the particle as before. An easy way to distinguish A and B walkers is to make the array element site(k) positive if the site is occupied by an A particle and negative if the site is occupied by a B particle. Start with equal concentrations of A and B particles and occupy the sites at random. Some of the interesting questions are similar to those that we posed in parts (a)–(c). Color code the particles and observe what happens to the relative positions of the particles.

*Problem 12.17   Reaction diffusion in two dimensions

a.   Do a similar simulation as in Problem 12.16 on a two-dimensional lattice for the reaction $A + A \rightarrow 0$. In this case it is necessary to have one array for each dimension, e.g., posx and posy. Set $A(t = 0) = 1$, and choose $L = 50$. Show the configuration of walkers after each Monte Carlo step per walker. Describe the geometry of the clusters of particles as the diffusion process proceeds. Are the particles uniformly distributed throughout the lattice for all times? Calculate $A(t)$ and compare your results for $1/A(t) - 1/A(0)$ to the t-dependence of $D(t)$, the number of distinct lattice sites that are visited in time t. ($D(t) \sim t/\log t$ for two dimensions.) How well do the slopes compare? Do a similar simulation with $A(t = 0) = 0.01$. What slope do you obtain in this case? What can you conclude about the initial density dependence? Is the mean-field approximation valid in this case? If time permits, do a similar simulation in three dimensions.

b.   Begin with A and B type random walkers initially segregated on the left and right halves (in the x direction) of a square lattice. The process $A + B \rightarrow C$ exhibits a reaction front where the production of particles of type C is nonzero. Some of the quantities of interest are the time dependence of the mean position $\overline{x}(t)$ and the width $w(t)$ of the reaction front. The rules of this process are the same as in part (a) except that a particle of type C is added to a site when a reaction occurs. A particular site can be occupied by one particle of type A or type B as well as any number of particles of type C. If $n(x, t)$ is the number of particles of type C at a distance x from the initial boundary of the reactants, then $\overline{x(t)}$ and $w(t)$ can be written as

$$\overline{x(t)} = \frac{\sum_x x\, n(x,t)}{\sum_x n(x,t)} \tag{12.29}$$

$$w^2(t) = \frac{\sum_x \left[x - \overline{x(t)}\right]^2 n(x,t)}{\sum_x n(x,t)}. \tag{12.30}$$

Choose lattice sizes of order 100×100, and average over at least 10 trials. The fluctuations in $\overline{x(t)}$ and $w(t)$ can be reduced by averaging $n(x,t)$ over the order of 100 time units centered about t. More details can be found in Jiang and Ebner.

12.5 THE CONTINUUM LIMIT

In Chapter 7 we showed that the solution of the diffusion equation is equivalent to the long time behavior of a simple random walk on a lattice. In the following, we show directly that the continuum limit of the one-dimensional random walk model is a diffusion equation.

If there is an equal probability of taking a step to the right or left, the random walk can be written in terms of the simple "master" equation

$$P(i, N) = \frac{1}{2} P(i+1, N-1) + \frac{1}{2} P(i-1, n-1), \tag{12.31}$$

where $P(i, N)$ is the probability that the walker is at site i after N steps. To obtain a differential equation for the probability density $P(x,t)$, we identify $t = N\tau$, $x = ia$, and $P(i, N) = aP(x,t)$, where τ is the time between steps and a is the lattice spacing. This notation allows us to rewrite (12.31) in the equivalent form

$$P(x, t) = \frac{1}{2} P(x+a, t-\tau) + \frac{1}{2} P(x-a, t-\tau). \tag{12.32}$$

We rewrite (12.32) by subtracting $P(x, t-\tau)$ from both sides of (12.32) and dividing by τ:

$$\frac{1}{\tau}\left[P(x,t) - P(x, t-\tau)\right] = \frac{a^2}{2\tau}\left[P(x+a, t-\tau) - 2P(x, t-\tau) \right. \tag{12.33}$$
$$\left. + P(x-a, t-\tau)\right]a^{-2}.$$

If we expand $P(x, t-\tau)$ and $P(x \pm a, t-\tau)$ in a Taylor series and take the limit $a \to 0$ and $\tau \to 0$ with the ratio $D \equiv a^2/2\tau$ finite, we obtain the diffusion equation

$$\frac{\partial P(x,t)}{\partial t} = D \frac{\partial^2 P(x,t)}{\partial x^2}. \tag{12.34a}$$

The generalization of (12.34a) to three dimensions is

$$\frac{\partial P(x, y, z, t)}{\partial t} = D \nabla^2 P(x, y, z, t). \tag{12.34b}$$

where $\nabla^2 = \partial^2/\partial x^2 + \partial^2/\partial y^2 + \partial^2/\partial x^2$ is the Laplacian operator. Equation (12.34) is known as the *diffusion* equation and is frequently used to describe the dynamics of fluid molecules.

The numerical solution of the prototypical *parabolic* partial differential equation (12.34) is a nontrivial problem in numerical analysis (cf. Press et al. or Koonin and Meredith.) An indirect method of analysis of (12.34) is to use a Monte Carlo method, that is, replace (12.34) by a corresponding random walk on a lattice with discrete time steps. Because the asymptotic behavior of the partial differential equation and the random walk model are equivalent, this approach uses the Monte Carlo technique as a method of *numerical analysis*. In contrast, if our goal is to understand a random walk lattice model directly, the Monte Carlo technique is a *simulation* method. The difference between simulation and numerical analysis is sometimes only in the eyes of the beholder.

Problem 12.18   Biased random walk

Show that the form of the differential equation satisfied by $P(x, t)$ corresponding to a random walk with a drift, i.e., a walk for $p \neq q$, is

$$\frac{\partial P(x, t)}{\partial t} = D \nabla^2 P(x, y, z, t) - v \frac{\partial P(x, t)}{\partial x} \tag{12.35}$$

How is v related to p and q?

12.6   RANDOM NUMBER SEQUENCES

So far we have used the random number generator supplied with True BASIC to generate the desired "random" numbers for our Monte Carlo applications. In principle, we might have generated these numbers from a random physical process, such as the decay of radioactive nuclei or the thermal noise from a semiconductor device. In practice, random number sequences are generated from a physical process only for specific purposes such as a lottery. Although we can store the outcome of a random physical process so that the random number sequence would be both truly random and reproducible, such a method would be inconvenient and inefficient in part because we often require very long sequences. Hence, in practice we use a digital computer, a deterministic machine, to generate sequences of random numbers. These sequences cannot be truly random and are sometimes referred to as *pseudorandom*. However, such a distinction is not important if the sequence satisfies all our criteria for randomness.

Most random number generators yield a sequence in which each number is used to find the succeeding one according to a well defined algorithm. The most important features of a desirable random number generator are that its sequence satisfies the known statistical tests for randomness, the probability distribution is uniform, the sequence has a long period, the method is efficient, the sequence is reproducible, and the algorithm is machine independent.

The most widely used random number generator is based on the *linear congruential* method. That is, given the *seed* x_0, each number in the sequence is determined by the one-dimensional map

$$x_n = (ax_{n-1} + c) \bmod m, \tag{12.36}$$

where a, c, and m are integers. (The notation $y = z \bmod m$ means that m is subtracted from z until $0 \leq y < m$.) The map (12.36) is characterized by three parameters, the *multiplier* a, the *increment* c, and the *modulus* m. Because m is the largest integer generated by (12.36), the maximum possible *period* is m. In general, the period depends on all three parameters. For example, if $a = 3$, $c = 4$, $m = 32$, and $x_0 = 1$, the sequence generated by (12.36) is 1, 7, 25, 15, 17, 23, 9, 31, 1, 7, 25, ... and the period is 8 rather than the maximum possible value of 32. If we are careful to choose a, c, and m such that the maximum period is obtained, then all possible integers between 0 and $m - 1$ will occur in the sequence. Because we usually wish to have random numbers r in the unit interval $0 \leq r < 1$ rather than random integers, random number generators usually return the ratio x_n/m which is always less than unity. One advantage of the linear congruential method is that it is very fast. As the above example illustrates, m, a, and c must be chosen carefully to achieve optimum results. Several rules have been developed (see Knuth) to obtain the longest period. Some of the properties of the linear congruential method are explored in Problem 12.19.

Another popular random number generator is the *generalized feedback shift register* (GFSR) method which uses bit manipulation. Every integer is represented as a series of 1's and 0's called bits. These bits can be shuffled by using the bitwise *exclusive or* operator $\oplus$ defined by $a \oplus b = 1$ if the bits $a \neq b$; $a \oplus b = 0$ if $a = b$. The nth member of the sequence is given by

$$x_n = x_{n-p} \oplus x_{n-q}, \tag{12.37}$$

where $p > q$, and p, q and x_n are integers. The first p random numbers must be supplied by another random number generator. As an example of how the exclusive or operator works, suppose that $n = 6$, $p = 5$, $q = 3$, $x_3 = 11$, and $x_1 = 6$. Then $x_6 = x_1 \oplus x_3 = 0110 \oplus 1011 = 1101 = 2^3 + 2^2 + 2^0 = 8 + 4 + 1 = 13$. Not all values of p and q lead to good results. Some common pairs are $(p, q) = (31, 3)$, $(250, 103)$, and $(521, 168)$. In Fortran the exclusive or operation on the integers m and n is given by the intrinsic function IEOR(m,n); the equivalent operation in C is given by m^n. True BASIC does not provide intrinsic bitwise operations.

The above two examples of random number generators illustrate their general nature. That is, numbers in the sequence are used to find the succeeding ones according to a well defined algorithm, and the sequence is determined by the *seed*, the first number of the sequence (or the first p members of the sequence for the GFSR method). In general, the maximum possible period is related to the size of the computer word, e.g., 16, 32, or 64 bits, that is used. We also note that the choice of the constants and the proper initialization of the sequence is very important and that the algorithm must be implemented with care.

There is no necessary and sufficient test for the randomness of a finite sequence of numbers; the most that can be said about any finite sequence of numbers is that it is "apparently" random. Because no single statistical test is a reliable indicator, we need to consider several tests. Some of the best known tests are discussed in Problem 12.19. Many of these tests can be stated in terms of random walks.

Problem 12.19   Statistical tests of randomness

a. The most obvious requirement for a random number generator is that its period be much greater than the number of random numbers needed in a specific calculation. One way to visualize the period of the random number generator is to use it to generate a plot of the displacement x of a random walker as a function of the number of steps N. When the period of the random number is reached, the plot will begin to repeat itself. Generate such a plot using (12.36) for the two choices of parameters $a = 899$, $c = 0$, and $m = 32768$, and $a = 16807$, $c = 0$, $m = 2^{31} - 1$ with $x_0 = 12$. What are the periods of the corresponding random number generators? Obtain similar plots using different values for the parameters a, c, and m. Why is the seed value $x_0 = 0$ forbidden for this choice of c? Do some combinations of these parameters give longer periods than others? What is the period of the random number generator supplied with True BASIC or the random number generator that you are using?

b. A random number sequence should contain numbers distributed in the unit interval with equal probability. The simplest test of uniformity is to divide this interval into M equal size subintervals or bins and place each member of the sequence into one of the bins. For example, consider the first $N = 10^4$ numbers generated by (12.36) with $a = 106$, $c = 1283$, and $m = 6075$ (see Press et al.). Place each number into one of $M = 100$ bins. Is the number of entries in each bin approximately equal? What happens if you increase N?

c. Is the distribution of numbers in the bins of part (b) consistent with the laws of statistics? The most common test of this consistency is the *chi-square* or χ^2 test. Let y_i be the observed number in bin i and E_i be the expected value. For the example in part (b) with $N = 10^4$ and $M = 100$, we have $E_i = 100$. The chi-square statistic is

$$\chi^2 = \sum_{i=1}^{M} \frac{(y_i - E_i)^2}{E_i}.$$

(12.38)

The magnitude of the number χ^2 is a measure of the agreement between the observed and expected distributions. In general, the individual terms in the sum (12.38) are expected to be order 1, and because there are M terms in the sum, we expect $\chi^2 \leq M$. As an example, we did five independent runs of a random number generator with $N = 10^4$ and $M = 100$, and found

$\chi^2 \approx 92, 124, 85, 91$, and 99. These values of χ^2 are consistent with this expectation. Although we usually want χ^2 to be as small as possible, we should be suspicious if $\chi^2 \approx 0$, because such a small value suggests that N is a multiple of the period of the generator and that each value in the sequence appears an equal number of times.

A more quantitative measure of our confidence that the discrepancy $(y_i - E_i)$ is distributed according to the Gaussian distribution is given by the chi-square probability function $P(x, v)$ defined as:

$$P(x, v) = \frac{1}{2^{v/2}\Gamma(v/2)} \int_0^x t^{(v-2)/2} e^{-t/2} dt. \tag{12.39}$$

The Gamma function $\Gamma(z)$ in (12.39) is given by $\Gamma(z) = \int_0^\infty t^{z-1} e^{-t} dt$; the familiar relation $\Gamma(z+1) = z!$ holds if z is a positive integer. The quantity v in (12.39) is the number of degrees of freedom. In our case $v = M - 1$, because we have imposed the constraint that $\sum_{i=1}^M E_i = N$. The function $Q(x, v) = 1 - P(x, v)$ is the probability that the measured value of χ^2 is greater than x. For our example we can solve for x in the equation $Q(x, v) = q$ with $v = 99$ for various values of q or find the solution in a statistical table. For $v = 99$, we find that $x \approx 139$ for $q = 0.005$, $x \approx 123$ for $q = 0.05$, $x \approx 111$ for $q = 0.2$, and $x \approx 98$ for $q = 0.5$. Our above results for χ^2 show that $\chi^2 > 123$ for one run out of five (20%). Because 123 is the value of x at the 5% level, we expect to see $\chi^2 \geq 123$ in only one out of twenty runs. Hence, our confidence level is less than 95%. Instead, we can assume an approximately 80% confidence level in our random number generator because the value of x for this confidence level is 111. We might be able to increase our confidence level by doing more runs. Suppose that we make twenty runs and we still find only one measurement of χ^2 greater than 123. In this case our confidence level would rise to 95%. Determine χ^2 for twenty independent runs using the values of a, c, and m given in parts (a) and (b). Estimate your level of confidence in these random number generators. Additional entries from a chi-square statistical table would be useful.

d. Although a random number sequence might contain numbers that are distributed in the unit interval with equal probability, consecutive numbers might not appear in a perfectly uniform way, but have a tendency to be clumped or correlated in some other way. One test of this correlation is to fill a square lattice of L^2 sites at random. Consider an array $n(x, y)$ that is initially empty, where $1 \leq x_i, y_i \leq L$. A point is selected randomly by choosing its two coordinates x_i and y_i from two consecutive numbers in the sequence. If the site is empty, it is filled and $n(x_i, y_i) = 1$; otherwise it is not changed. This procedure is repeated tL^2 times, where t is the number of Monte Carlo steps per site. Because this process is analogous to the decay of radioactive nuclei, we expect that the fraction of empty lattice sites should decay as e^{-t}. Determine the fraction of unfilled sites using the random number generator that you have been using for $L = 10, 15$, and 20. Are your results consistent with the expected fraction? Repeat the same test using (12.36) with $a = 65549$, $c = 0$, and $m = 231$. The existence of triplet correlations can be determined by a similar test on a simple cubic lattice by choosing the three coordinates x_i, y_i, and z_i from three consecutive random numbers.

e.   Sometimes a picture is worth a thousand numbers. Another way of checking for correlations is to plot x_{i+k} versus x_i. If there are any obvious patterns in the plot, then there is something wrong with the generator. Use the generator (12.36) with $a = 16807$, $c = 0$, and $m = 2^{31} - 1$. Can you detect any structure in the plotted points for $k = 1$ to $k = 5$? Also test the random number generator that you have been using. Do you see any evidence of lattice structure, e.g., equidistant parallel lines? Is the logistic map $x_{n+1} = 4x_n(1 - x_n)$ a suitable random number generator?

f.   Another measure of short term correlations is the autocorrelation function

$$C(k) = \frac{<x_{i+k}x_i> - <x_i>^2}{<x_i x_i> - <x_i><x_i>}, \qquad (12.40)$$

where x_i is the ith term in the sequence. We have used the fact that $<x_{i+k}> = <x_i>$, that is, the choice of the origin of the sequence is irrelevant. The quantity $<x_{i+k}x_i>$ is found for a particular choice of k by forming all the possible products of $x_{i+k}x_i$ and dividing by the number of products. If x_{i+k} and x_i are not correlated, then $<x_{i+k}x_i> = <x_{i+k}><x_i>$ and $C(k) = 0$. Is $C(k)$ identically zero for any finite sequence? Compute $C(k)$ for $a = 106$, $c = 1283$, and $m = 6075$.

g.   Another test based on the properties of random walks has been proposed recently (see Vattulainen et al.). Assume that a walker begins at the origin of the x-y plane and generate n_w walks of N steps. Count the number of walks in each quadrant q_i of the x-y plane, and use the χ^2 test (12.38) with $y_i \to q_i$, $M = 4$, and $E_i = n_w/4$. If $\chi^2 > 7.815$ (a 5% probability if the random number generator is perfect), we say that the run fails. The random number generator fails if two out of three independent runs fail. The probability of a perfect generator failing two out of three runs would be approximately $3 \times 0.95 \times (0.05)^2 \approx 0.007$. Test several random number generators.

*Problem 12.20   An "improved" random number generator

One way to reduce sequential correlation and to lengthen the period is to mix or *shuffle* two different random number generators. The following procedure illustrates the approach for two random number generators that we denote as RAN1 and RAN2.

1.   Make a list of 256 random numbers using RAN1. (The number 256 is arbitrary, but should be less than the period of RAN1.)

2.   Choose a random number x from this list by using RAN2 to generate a random index between 1 and 256.

3. Replace the number chosen in step 2 by a new random number generated by RAN1.

Consider two random number generators with relatively short periods and strong sequential correlation and show that the above shuffling scheme improves the quality of the random numbers.

At least some of the statistical tests given in Problem 12.19 should be done whenever serious calculations are contemplated. However, even if a random number generator passes all these tests, there still can be problems in rare cases. Typically, these problems arise when a small number of events have a large weight. In these cases a very small bias in the random number generator might lead to a systematic error in the final results, and two generators, which appear equally good as determined by various statistical tests, might give statistically different results when used in a specific application. For this reason, it is important that the random number generator that is used be reported along with the actual results. Confidence in the results also can be increased by repeating the calculation with another random number generator.

Because all random number generators are based on a deterministic algorithm, it always is possible to construct a test generator for which a particular algorithm will fail. The success of a random number generator in passing various statistical tests is necessary and improves our overall confidence in its statistical properties, but it is not a sufficient condition for their use in all applications. In Project 17.2 we discuss an application of Monte Carlo methods to the Ising model for which some commonly used random number generators give incorrect results.

12.7 PROJECTS

Almost all of the problems in this chapter can be done using more efficient programs, greater number of trials, and larger systems. More applications of random walks and random number sequences are discussed in subsequent chapters. Many more ideas for projects can be gained from the references.

Project 12.1   Application of the pivot algorithm

The algorithms that we have discussed for generating self-avoiding random walks are all based on making *local* deformations of the walk (chain) for a given value of N, the number of bonds. As discussed in Problem 12.14, the time τ_c between statistically independent configurations is nonzero. The problem is that τ_c increases with N as some power, e.g., $\tau_c \sim N^3$. This power law dependence of τ_c on N is called *critical slowing down* and implies that it becomes increasingly more time consuming to generate long walks. We now discuss an example of a *global* algorithm that reduces the dependence of τ_c on N. Another example of a global algorithm that reduces critical slowing down is discussed in Project 17.1.

    a.   Consider the walk shown in Fig. 12.10a. Select a site at random and one of the four possible directions. The shorter portion of the walk is rotated (pivoted) to

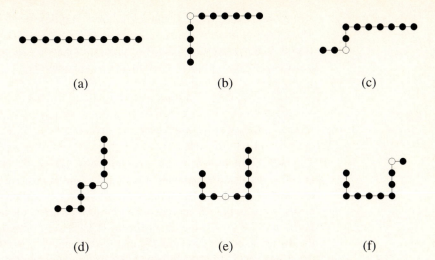

Fig. 12.10 Examples of the first several changes generated by the pivot algorithm for a self-avoiding walk of $N = 10$ steps (11 sites). The open circle denotes the pivot point. This figure is adopted from the article by MacDonald et al.

this new direction by treating the walk as a rigid structure. The new walk is accepted only if the new walk is self-avoiding; otherwise the old walk is retained. (The shorter portion of the walk is chosen to save computer time.) Some typical moves are shown in Fig. 12.10. Note that if an end point is chosen, the previous walk is retained. Write a program to implement this algorithm and compute the dependence of the mean square end-to-end distance $<R^2(N)>$ on N. Consider values of N in the range $10 \leq N \leq 80$. A discussion of the results and the implementation of the algorithm can be found in MacDonald et al. and Madras and Sokal, respectively.

b. Compute the correlation time τ_c for different values of N using the approach discussed in Problem 12.14c.

Project 12.2 A simple reaction diffusion model

In Problem 12.17 we saw that simple patterns can develop as a result of random behavior. The phenomenon of pattern formation is of much interest in a variety of contexts ranging from the large scale structure of the universe to the roll patterns seen in convection (e.g., smoke rings). In the following, we explore the patterns that can develop in a simple reaction diffusion model based on the reactions, $A + 2B \rightarrow 3B$, and $B \rightarrow C$, where C is inert. Such a reaction is called *autocatalytic*. In Problem 12.17 we considered chemical reactions in a closed system where the reactions can proceed to equilibrium. In contrast, open systems allow a continuous supply of fresh reactants and a removal of products. These two processes allow steady states to be realized and oscillatory conditions to be maintained indefinitely. The model of interest also assumes that A is added at a constant rate and that both A and B are removed by the feed process. Pearson (see references) models these processes by two coupled reaction diffusion equations:

$$\frac{\partial A}{\partial t} = D_A \nabla^2 A - AB^2 + f(1 - A) \qquad\qquad (12.41a)$$

$$\frac{\partial B}{\partial t} = D_B \nabla^2 B + AB^2 - (f + k)B \qquad\qquad (12.41b)$$

The AB^2 term represents the reaction $A + 2B \rightarrow 3B$. This term is negative in (12.41a) because the reactant A decreases, and is positive in (12.41b) because the reactant B increases. The term $+f$ represents the constant addition of A, and the terms $-fA$ and $-fB$ represent the removal process; the term $-kB$ represents the reaction $B \rightarrow C$. All the quantities in (12.41) are dimensionless. We assume that the diffusion coefficients are $D_A = 2 \times 10^{-5}$ and $D_B = 10^{-5}$, and the behavior of the system is determined by the values of the rate constant k and the feed rate f.

a.   We first consider the behavior of the reaction kinetics that results when the diffusion terms in (12.41) are neglected. It is clear from (12.41) that there is a trivial steady state solution $A = 1$, $B = 0$. Are there other solutions, and if so, are they stable? The steady state solutions can easily be found by solving (12.41) with $\partial A/\partial t = \partial B/\partial t = 0$. To determine the stability, we can add a perturbation and determine whether the perturbation grows or not. However, without the diffusion terms, it is more straightforward to solve (12.41) numerically using a simple Euler algorithm. Choose a time step equal to unity, and let $A = 0.1$ and $B = 0.5$ at $t = 0$. Determine the steady state values for $0 < f \leq 0.3$ and $0 < k \leq 0.07$ in increments of $\Delta f = 0.02$ and $\Delta k = 0.005$. Record the steady state values of A and B. Then repeat this exercise for the initial values $A = 0.5$ and $B = 0.1$. You should find that for some values of f and k, only one steady state solution can be obtained for the two initial conditions, and for other initial values of A and B there are two steady state solutions. Try other initial conditions. If you obtain a new solution, change the initial A or B slightly to see if your new solution is stable. On an f versus k plot indicate where there are two solutions and where there are one. In this way you can determine the approximate phase diagram for this process.

b.   There is a small region in f-k space where one of the steady state solutions becomes unstable and periodic solutions occur (the mechanism is known as a Hopf bifurcation). Try $f = 0.009$, $k = 0.03$, and set $A = 0.1$ and $B = 0.5$ at $t = 0$. Plot the values of A and B versus the time t. Are they periodic? Try other values of f and k and estimate where the periodic solutions occur.

c.   Numerical solutions of the full equation with diffusion (12.41) can be found by making a finite difference approximation to the spatial derivatives as in (3.19) and using a simple Euler algorithm for the time integration. Adopt periodic boundary conditions. Although it is straightforward to write a program to do the numerical integration, an exploration of the dynamics of this sys-

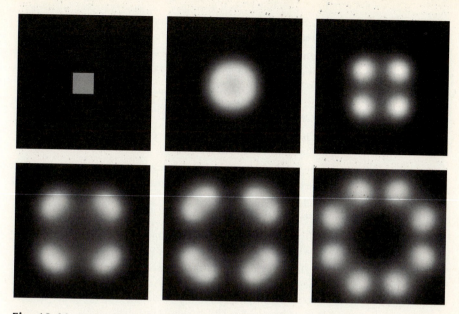

Fig. 12.11 Evolution of the pattern starting from the initial conditions suggested in
Project 12.2c.

tem requires a supercomputer. However, we can find some preliminary results
with a small system and a coarse grid. Consider a 0.5×0.5 system with a
spatial mesh of 128×128 grid points on a square lattice. Choose $f = 0.18$,
$k = 0.057$, and $\Delta t = 0.1$. Let the entire system be in the initial trivial state
($A = 1$, $B = 0$) except for a 20×20 grid located at the center of the system
where the sites are $A = 1/2$, $B = 1/4$ with a $\pm 1\%$ random noise. The effect of
the noise is to break the square symmetry. Let the system evolve for approximately 80,000 time steps and look at the patterns that develop. Color code
the grid according to the concentration of A, with red representing $A = 1$ and
blue representing $A \approx 0.2$ and with several intermediate colors. Very interesting patterns have been found by Pearson.

References and Suggestions for Further Reading

Panos Argyrakis, "Simulation of diffusion-controlled chemical reactions," *Computers in Physics*
6, 525 (1992).

J. Bernasconi and L. Pietronero, "True self-avoiding walk in one dimension," *Phys. Rev.* B **29**,
5196 (1984). The authors present results for the exponent ν accurate to 1%.

Neal Madras and Alan D. Sokal, "The pivot algorithm: a highly efficient Monte Carlo method
for the self-avoiding walk," *J. Stat. Phys.* **50**, 109 (1988). The pivot algorithm was invented
in 1969 by Lai. See also, "Monte Carlo methods for the self-avoiding walk," in *Monte*

Carlo and Molecular Dynamics Simulations in Polymer Science, Kurt Binder, editor, Oxford University Press (1995).

S. Chandrasekhar, "Stochastic problems in physics and astronomy," *Rev. Mod. Phys.* **15**, 1 (1943). This article is reprinted in M. Wax, *Selected Papers on Noise and Stochastic Processes*, Dover (1954).

Mohamed Daoud, "Polymers," Chapter 6 in Armin Bunde and Shlomo Havlin, editors, *Fractals in Science*, Springer-Verlag (1994).

Robert Ehrlich, *Physics and Computers*, Houghton Mifflin (1973). See Chapter 4 for a discussion of the linear congruential method.

R. Everaers, I. S. Graham, and M. J. Zuckermann, "End-to-end distance and asymptotic behavior of self-avoiding walks in two and three dimensions," *J. Phys.* A **28**, 1271 (1995).

Pierre-Giles de Gennes, *Scaling Concepts in Polymer Physics,* Cornell University Press (1979). An important but difficult text.

Shlomo Havlin and Daniel Ben-Avraham, "Diffusion in disordered media," *Adv. Phys.* **36**, 695 (1987). Section 7 of this review article discusses trapping and diffusion-limited reactions.

Shlomo Havlin, George H. Weiss, James E. Kiefer, and Menachem Dishon, "Exact enumeration of random walks with traps," *J. Phys. A: Math. Gen.* **17**, L347, (1984). The authors discuss a method based on exact enumeration for calculating the survival probability of random walkers on a lattice with randomly distributed traps.

Z. Jiang and C. Ebner, "Simulation study of reaction fronts," *Phys. Rev.* A **42**, 7483 (1990).

Steven E. Koonin and Dawn C. Meredith, *Computational Physics*, Addison-Wesley (1990).

Donald E. Knuth, *Seminumerical Algorithms*, second edition, Vol. 2 of *The Art of Computer Programming,* Addison-Wesley (1981). The standard reference on random number generators.

Bruce MacDonald, Naeem Jan, D. L. Hunter, and M. O. Steinitz, "Polymer conformations through 'wiggling'," *J. Phys.* A **18**, 2627 (1985). A discussion of the pivot algorithm discussed in Project 12.1.

Elliott W. Montroll and Michael F. Shlesinger, "On the wonderful world of random walks," in *Nonequilibrium Phenomena II: From Stochastics to Hydrodynamics,* J. L. Lebowitz and E. W. Montroll, editors, North-Holland Press (1984). The first part of this delightful review article chronicles the history of the random walk.

John E. Pearson, "Complex patterns in a simple fluid," *Science* **261**, 189 (1993). See also P. Gray and S. K. Scott, "Sustained oscillations and other exotic patterns of behavior in isothermal reactions," *J. Phys. Chem.* **89**, 22 (1985).

William H. Press, Saul A. Teukolsky, William T. Vetterling, and Brian P. Flannery, *Numerical Recipes*, second edition, Cambridge University Press (1992).

Sidney Redner and Francois Leyvraz, "Kinetics and spatial organization of competitive reactions," Chapter 7 in Armin Bunde and Shlomo Havlin, editors, *Fractals in Science*, Springer-Verlag (1994).

F. Reif, *Fundamentals of Statistical and Thermal Physics*, McGraw-Hill (1965). This well known text on statistical physics has a good discussion on random walks (Chapter 1) and diffusion (Chapter 12).

Marshall N. Rosenbluth and Arianna W. Rosenbluth, "Monte Carlo calculation of the average extension of molecular chains," *J. Chem. Phys.* **23**, 356 (1955). One of the first Monte Carlo calculations for the self-avoiding walk.

John R. Taylor, *An Introduction to Error Analysis*, University Science Books, Oxford University Press (1982). See Chapter 12 for a discussion of the χ^2 test.

I. Vattulainen, T. Ala-Nissila, and K. Kankaala, "Physical tests for random numbers in simulations," *Phys. Rev. Lett.* **73**, 2513 (1994).

Peter H. Verdier and W. H. Stockmayer, "Monte Carlo calculations on the dynamics of polymers in dilute solution," *J. Chem. Phys.* **36**, 227 (1962).

Frederick T. Wall and Frederic Mandel, "Macromolecular dimensions obtained by an efficient Monte Carlo method without sample attrition," *J. Chem. Phys.* **63**, 4592 (1975). An exposition of the "reptation" method.

George H. Weiss, "A primer of random walkology," Chapter 5 in Armin Bunde and Shlomo Havlin, editors, *Fractals in Science*, Springer-Verlag (1994).

George H. Weiss and Shlomo Havlin, "Trapping of random walks on the line," *J. Stat. Phys.* **37**, 17 (1984). The authors discuss an analytical approach to the asymptotic behavior of one-dimensional random walkers with randomly placed traps.

George H. Weiss and Robert J. Rubin, "Random walks: theory and selected applications," *Adv. Chem Phys.* **52**, 363 (1983). In spite of its research orientation, much of this review article can be understood by the well motivated student.

Charles A. Whitney, *Random Processes in Physical Systems*, John Wiley and Sons (1990). A good introduction to random processes including random walks.

Charles A. Whitney, "Generating and testing pseudorandom numbers," *Byte*, pp. 128 (October, 1984). An accessible and informative article.

13
Percolation

We introduce several concepts associated with critical phenomena in the context of percolation.

13.1 INTRODUCTION

Our discussion of percolation and geometrical phase transitions requires little background in physics, e.g., no classical or quantum mechanics and little statistical physics. All that is required is some understanding of geometry and probability. Much of the appeal of percolation models is their game-like aspects and their intuitive simplicity. Moreover, these models serve as an excellent introduction to discrete computer models and the importance of graphical analysis. On the other hand, if you have a background in physics, this chapter will be more meaningful and can serve as an introduction to phase transitions and to important ideas such as scaling relations, critical exponents, and the renormalization group.

Although the term "percolation" might be familiar to you in the context of the brewing of coffee, our use of the term has little to do with it. Instead, we consider another example from the kitchen and imagine a large metal sheet on which we randomly place drops of cookie dough. Assume that each drop of cookie dough can spread while the cookies are baking in an oven. If two cookies touch, they coalesce to form one cookie. If we are not careful, we might find a very large cookie that spans from one edge of the sheet to the opposite edge (see Fig. 13.1). If such a spanning cookie exists, we say that there has been a *percolation transition*. If such a cookie does not exist, the cookie sheet is below the percolation threshold. (There is no percolation transition for ground coffee, because the water dissolves some of the ground coffee beans and flows, regardless of the density of the ground coffee.)

Let us make the cookie example more abstract to make the concept of percolation more clear. We represent the cookie sheet by a lattice where each site can be in one of two states, occupied (by a cookie) or empty. Each site is occupied independently of its neighbors with probability p. This model of percolation is called *site* percolation. The occupied sites either are isolated or form groups of nearest neighbors. We define a *cluster* as a group of occupied nearest neighbor lattice sites (see Fig. 13.2).

An easy way to study percolation uses the random number generator on a calculator. The procedure is to generate a random number r in the unit interval $0 < r \leq 1$ for each site in the lattice. A site is occupied if its random number satisfies the condition $r \leq p$. If p is small, we expect that only small isolated clusters will be present (see Fig. 13.3a). If p is near unity, we expect that most of the lattice will be occupied, and the occupied sites will form a large cluster that extends from one end of the lattice to the other (see Fig. 13.3c). Such a cluster is said to be a *spanning cluster*. Because there is no spanning cluster for small p and there is a spanning cluster for p near unity, there must be an intermediate value of p at which a spanning cluster first exists (see Fig. 13.3b). We shall see that in the limit of an infinite lattice, there exists a well defined threshold probability p_c such that:

> For $p < p_c$, no spanning cluster exists and all clusters are finite.
> For $p \geq p_c$, one spanning cluster exists.

We emphasize that the defining characteristic of percolation is *connectedness*. Because the connectedness exhibits a qualitative change at a well defined value of a continuous parameter, we shall see that the transition from a state with no spanning cluster to a state with one spanning cluster is a type of *phase transition*.

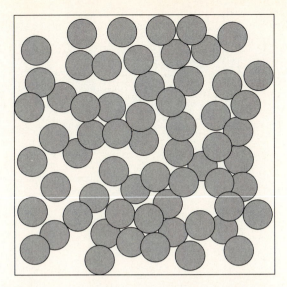

Fig. 13.1 Cookies (circles) placed at random on a large sheet. Note that there is a path of overlapping circles that connects the bottom and top edges of the cookie sheet. If such a path exists, we say that the cookies "percolate" the lattice or that there is a "spanning path." See Problem 13.1d for a discussion of the algorithm used to generate this configuration.

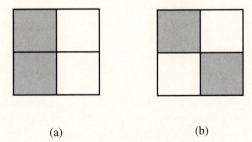

(a)   (b)

Fig. 13.2 Example of a site percolation cluster on a square lattice of linear dimension $L = 2$. The two nearest neighbor occupied sites (shaded) in (a) are part of the same cluster; the two occupied sites in (b) are not nearest neighbor sites and do not belong to the same cluster.

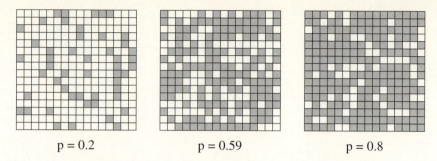

p = 0.2   p = 0.59   p = 0.8

Fig. 13.3   Examples of site percolation clusters on a square lattice of linear
dimension $L = 16$ for $p = 0.2, 0.59$, and 0.8. On the average, the
fraction of occupied sites (shaded squares) is equal to p. Note that in
this example, there exists a cluster that "spans" the lattice horizontally
and vertically for $p = 0.59$.

Our real interest is not in large cookies or in abstract models, but in the applications
of percolation. An example of the application of percolation is to the electrical conductivity of composite systems made of a mixture of metallic and insulating materials. An
easy way to make such a system is to place a mixture of small plastic and metallic
spheres of equal size into a container (see Fitzpatrick et al.). Care must be taken to pack
the spheres at random. If the metallic domains constitute a small fraction of the volume
of the system, electricity cannot be conducted and the composite system is an insulator.
However, if the metallic domains comprise a sufficiently large fraction of the container,
electricity can flow from one domain to another and the composite system is a conductor. The description of the conduction of electricity through composite materials can be
made more precise by introducing the parameter ϕ, the volume fraction of the container
that consists of metallic spheres. The transition between the two types of behavior (non-
conducting and conducting) occurs abruptly as ϕ (the analog of p) is increased and is
associated with the nonexistence or existence of a *connected path* of metallic spheres.
More realistic composite systems are discussed in Zallen's book.

Percolation phenomena also can be observed with a piece of chicken wire or wire
mesh. Watson and Leath measured the electrical conductivity of a large piece of uniform steel-wire screen mesh as a function of the fraction of the nodes that are present.
The coordinates of the nodes to be removed were given by a random number generator.
The measured electrical conductivity is a rapidly decreasing function of the fraction
of nodes p still present and vanishes below a critical threshold. A related conductivity
measurement on a sheet of conducting paper with random "holes" has been performed
(see Mehr et al.).

The applications of percolation phenomena go beyond metal-insulator transitions
and the conductivity of chicken wire, and include the spread of disease in a population,
the behavior of magnets diluted by nonmagnetic impurities, the flow of oil through
porous rock, and the characterization of gels. We concentrate on understanding several
simple models of percolation that have an intuitive appeal of their own. Many of the
applications of percolation phenomena are discussed in the references.

13.2   THE PERCOLATION THRESHOLD

Because it is not convenient to generate many percolation configurations using a calculator, we develop a simple program to do so. Consider a square lattice of linear dimension L and unit lattice spacing, and associate a random number between zero and one with each site in the lattice. A site is occupied if its random number is less than p. *Program* site, listed below, generates site percolation configurations and shows the occupied sites as filled circles of diameter unity. The program uses the FLOOD statement in True BASIC so that if the user clicks on an occupied site, all the sites that are connected to it are changed to the same color and the clusters can be identified visually. A click of the mouse outside the lattice increases p by the desired amount. The array rsite stores the random number associated with each lattice site.

```
PROGRAM site
! draw site percolation configurations
DIM rsite(64,64)
LIBRARY "csgraphics"
! define lattice and screen parameters
CALL initial(rsite(,),L,p,delta_p,occup$)
! occupy sites for given probability p
CALL configuration(rsite(,),L,p,delta_p,occup$)
END

SUB initial(rsite(,),L,p,delta_p,occup$)
    RANDOMIZE
    INPUT prompt "linear dimension of square lattice = ": L
    INPUT prompt "initial probability p = ": p
    ! desired increase in p upon mouse click
    INPUT prompt "delta p = ": delta_p
    CALL compute_aspect_ratio(L,xwin,ywin)
    SET WINDOW -0.1*xwin,xwin,-0.1*ywin,ywin
    ! represent occupied site as blue circle with unit diameter
    SET COLOR "blue"
    BOX CIRCLE 2,3,2,3
    FLOOD 2.5,2.5
    BOX KEEP 2,3,2,3 in occup$
    CLEAR
    SET COLOR "black"
    LET r = 0.5
    BOX LINES 1-r,L+r,1-r,L+r
    FOR y = 1 to L                    ! draw lattice sites
        FOR x = 1 to L
            ! assign random number to each lattice site
            LET rsite(x,y) = rnd
            PLOT POINTS: x,y
        NEXT x
    NEXT y
END SUB
```

```
SUB configuration(rsite(,),L,p,delta_p,occup$)
    DO while p <= 1
        SET CURSOR 1,1
        PRINT "p =";p
        FOR y = 1 to L
            LET yplot = y - 0.5
            FOR x = 1 to L
                IF rsite(x,y) < p then
                    LET xplot = x - 0.5
                    BOX SHOW occup$ at xplot,yplot
                END IF
            NEXT x
        NEXT y
        BOX LINES 0.5,L+0.5,0.5,L+0.5    ! redraw lines
        DO
            DO
                GET MOUSE xs,ys,state
            LOOP until state = 2
            ! click on occupied site to see cluster
            ! click mouse to right of lattice to increase p by delta_p
            IF (xs < L and xs > 0) and ys <= L then
                SET COLOR "red"
                FLOOD xs,ys
            END IF
        LOOP until xs > L
        LET p = p + delta_p
        SET COLOR "black"
    LOOP
END SUB
```

The percolation threshold p_c is defined as the probability p at which a spanning cluster first appears in an infinite lattice. However, for the finite lattices of linear dimension L that we can simulate on a computer, there is a nonzero probability of a spanning cluster connecting one side of the lattice to the opposite side for any value of p. For small p, this probability is order p^L (see Fig. 13.4). This probability goes to zero as L becomes large, and hence for small p and sufficiently large L, only finite clusters exist. For a finite lattice, the definition of spanning is arbitrary. For example, we can define a connected path as one that (i) spans the lattice either horizontally or vertically; (ii) spans the lattice in a fixed direction, e.g., vertically; or (iii) spans the lattice both horizontally and vertically. In addition, the criteria for defining $p_c(L)$ for a finite lattice are somewhat arbitrary. One possibility is to define $p_c(L)$ as the average value of p at which a spanning cluster first appears. Another possibility is to define $p_c(L)$ as the value of p for which half of the configurations generated at random span the lattice. These criteria should lead to the same extrapolated value for p_c in the limit $L \to \infty$. In Problem 13.1 we will find an estimated value for $p_c(L)$ that is accurate to about 10%. A more sophisticated analysis discussed in Project 13.2c allows us to extrapolate our results for $p_c(L)$ to $L \to \infty$.

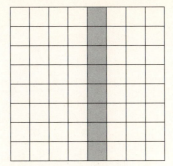

Fig. 13.4   An example of a spanning cluster of probability p^L on a $L = 8$ lattice. How many other ways are there of realizing a spanning cluster of L sites?

Problem 13.1   Site percolation on the square lattice

a.   Use *Program* site to generate random site configurations on a square lattice. Estimate $p_c(L)$ by finding the average value of p at which a spanning cluster is first attained. Choose $L = 4$ and begin at a value of p for which a spanning cluster is unlikely to be present. Then increase p in increments of delta_p $= 0.01$ until you find a spanning cluster. Record the value of p at which spanning first occurs for each spanning criteria. Repeat this process for a total of ten configurations and find the average value of $p_c(L)$. (Remember that each configuration corresponds to a different set of random numbers.) Are your results for $p_c(L)$ using the three spanning criteria consistent with your expectations?

b.   Repeat part (a) for $L = 16$ and 32. Is $p_c(L)$ better defined for larger L, that is, are the values of $p_c(L)$ spread over a smaller range of values? How quickly can you visually determine the existence of a spanning cluster? Describe your visual "algorithm" for determining if a spanning cluster exists.

The value of p_c depends on the symmetry of the lattice and on its dimension. In addition to the square lattice, the most common two-dimensional lattice is the triangular lattice. As discussed in Chapter 8, the essential difference between the square and triangular lattices is in the number of nearest neighbors.

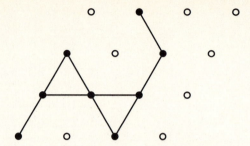

Fig. 13.5 Example of a spanning cluster on a $L = 4$ triangular lattice. The bonds between the occupied sites are drawn to clarify the symmetry of the lattice.

*Problem 13.2   Site percolation on the triangular lattice

Modify `Program site` to simulate random site percolation on a triangular lattice. Assume that a connected path connects the top and bottom sides of the lattice (see Fig. 13.5). Do you expect p_c for the triangular lattice to be smaller or larger than the value of p_c for the square lattice? Estimate $p_c(L)$ for $L = 4$, 16, and 32. Are your results for p_c consistent with your expectations?

In *bond* percolation each lattice site is occupied, and only a fraction of the sites have connections or bonds between them and their nearest neighbor sites (see Fig. 13.6). Each bond either is occupied with probability p or not occupied with probability $1 - p$. A cluster is a group of sites connected by occupied bonds. The wire mesh described in Section 13.1 is an example of bond percolation if we imagine cutting the bonds between the nodes rather than removing the nodes themselves. An application of bond percolation to the description of gelation is discussed in Problem 13.3.

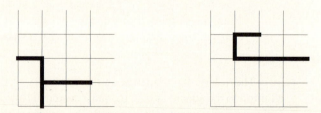

Fig. 13.6   Two examples of bond clusters. The occupied bonds are shown as bold lines.

* Problem 13.3 Bond percolation on a square lattice

Suppose that all the lattice sites of a square lattice are occupied by monomers, each with functionality four, i.e., each monomer can react to form a maximum of four bonds. This model is equivalent to bond percolation on a square lattice. Also assume that the presence or absence of a bond between a given pair of monomers is random and is characterized by a probability p. For small p, the system consists of only finite polymers (groups of monomers) and the system is in the *sol* phase. For some threshold value p_c, there will be a single polymer that is infinite in spatial extent. We say that for $p \geq p_c$, the system is in the *gel* phase. How does a bowl of jello, an example of a gel phase, differ from a bowl of broth? Write a program to simulate bond percolation on a square lattice and determine the bond percolation threshold. Are your results consistent with the exact result, $p_c = 1/2$?

We also can consider *continuum* percolation models. For example, we can place disks at random into a two-dimensional box. Two disks are in the same cluster if they touch or overlap. A typical continuum (off-lattice) percolation configuration is depicted in Fig. 13.7. One quantity of interest is the quantity ϕ, the fraction of the area (volume

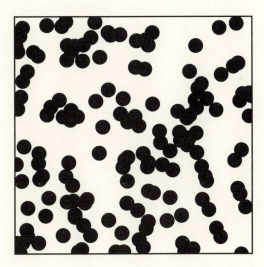

Fig. 13.7 A model of continuum (off-lattice) percolation realized by placing disks of unit diameter at random into a square box of linear dimension L. If we concentrate on the voids between the disks rather than the disks, then this model of continuum percolation is known as the Swiss cheese model.

in three dimensions) in the system that is covered by disks. In the limit of an infinite size box, it can be shown that

$$\phi = 1 - e^{-\rho \pi r^2}, \tag{13.1}$$

where ρ is the number of disks per unit area, and r is the radius of a disk (see Xia and Thorpe). Equation (13.1) is significantly inaccurate for small boxes because disks located near the edge of the box might have a significant fraction of their area located outside of the box. *Program site* can be modified to simulate continuum percolation. Instead of placing the disks on regular lattice sites, place them at random within a square box of area L^2. The relevant parameter is the density ρ, the number of disks per unit area, instead of the probability p. Because the disks overlap, it is convenient to replace the BOX SHOW statement in *Program site* with

```
BOX SHOW occup$ at x(i)-0.5,y(i)-0.5 using "or"
```

where the arrays x(i) and y(i) are used to store the disk positions of disk i. It also is a good idea to set the background color to red (not black or white).

Problem 13.4 Continuum percolation

a. For site percolation, we can define ϕ as the area fraction covered by the disks that are placed on the sites as in *Program site*. Convince yourself that $\phi_c = (\pi/4)p_c$ (for disks of unit diameter and unit lattice spacing). It is easy to do a Monte Carlo calculation of the area covered by the disks to confirm this result. (Choose points at random in the box and calculate the fraction of points within any disk.)

b. Modify *Program site* to simulate continuum percolation as discussed in the text. Estimate the value of the percolation threshold ρ_c. Given this value of ρ_c, use a Monte Carlo method to estimate the corresponding area fraction ϕ_c, and compare the value of ϕ_c for site and continuum percolation. Explain why you might expect ϕ_c to be bigger for continuum percolation than for site percolation. Compare your direct Monte Carlo estimate of ϕ_c with the indirect value of ϕ_c obtained from (13.1) using the value of ρ_c. Explain any discrepancy.

c. Consider the simple model of the cookie problem discussed in Section 13.1. Write a program that places disks at random into a square box and chooses their diameter randomly between 0 and 1. Estimate the value of ρ_c at which a spanning cluster first appears. How is the value of ρ_c changed from your estimate found in part (b)? Is your value for ϕ_c more or less than what was found in part (b)?

*d. A more realistic model of the cookie problem is to place disks with unit diameter at random into a square box with the constraint that the disks do not overlap. Continue to add disks until the probability of placing an additional disk becomes less than 1%, i.e., when one hundred successive attempts at adding a disk are not successful. Then increase the diameters of all the disks at a constant rate (in analogy to the baking of the cookies) until a spanning cluster is attained. How does ϕ_c for this model compare with ϕ_c found in part (c)?

*e. A continuum model that is applicable to random porous media is known as the *Swiss cheese* model. In this model the relevant quantity (the cheese) is the space between the disks. For the Swiss cheese model in two dimensions, the cheese area fraction at the percolation threshold, $\tilde{\phi}_c$, is given by $\tilde{\phi}_c = 1 - \phi_c$, where ϕ_c is the disk area fraction at the threshold of the disks. Do you think such a relation holds in three dimensions (see Project 13.3)? Imagine that the disks are conductors and that the cheese is an insulator and let $\sigma(\phi)$ denote the conductivity of this system. Alternatively, we can imagine that the cheese is a conductor and the disks are insulators and define a conductivity $\sigma(\tilde{\phi})$. Do you think that $\sigma(\phi) = \sigma(\tilde{\phi})$ when $\phi = \tilde{\phi}$? This question is investigated in Project 13.3.

Our discussion of percolation has emphasized the existence of the percolation threshold p_c and the appearance of a spanning path or cluster for $p \geq p_c$. Another quantity that characterizes percolation is $P_\infty(p)$, the probability that an occupied site belongs to the spanning cluster. P_∞ is defined as

$$P_\infty = \frac{\text{number of sites in the spanning cluster}}{\text{total number of occupied sites}}. \tag{13.2}$$

As an example, $P_\infty(p = 0.6) = 36/47$ for the single configuration shown in Fig. 13.3c. A realistic calculation of P_∞ involves an average over many configurations for a given value of p. For an infinite lattice, $P_\infty(p) = 0$ for $p < p_c$ and $P_\infty(p) = 1$ for $p = 1$. Between p_c and 1, $P_\infty(p)$ increases monotonically.

More information can be obtained from the *mean cluster size distribution $n_s(p)$* defined by

$$n_s(p) = \frac{\text{average number of clusters of size } s}{\text{total number of lattice sites}}. \tag{13.3}$$

For $p \geq p_c$, the spanning cluster is excluded from n_s. (For historical reasons, the *size* of a cluster refers to the *number* of sites in the cluster rather than to its spatial extent.) As an example, we see from Fig. 13.3a that $n_s(p = 0.2) = 5/64$, $1/64$, and $2/64$ for $s = 1$, 2, and 3 respectively, and is zero otherwise. Because $N \sum_s s n_s$ is the total number of occupied sites (N is the total number of lattice sites), and $N s n_s$ is the number of occupied sites in clusters of size s, the quantity

$$w_s = \frac{s n_s}{\sum_s s n_s} \tag{13.4}$$

is the probability that an occupied site chosen at random is part of an s-site cluster. Hence, the *mean cluster size S* is given by

$$S = \sum_s s w_s = \frac{\sum_s s^2 n_s}{\sum_s s n_s}. \tag{13.5}$$

The sum in (13.5) is over the finite clusters only. As an example, the mean cluster size corresponding to the eight clusters in Fig. 13.3a is $S = 27/13$.

Problem 13.5   Qualitative behavior of n_s(p), S(p), and P_∞(p)

a.   Visually determine the cluster size distribution $n_s(p)$ for a square lattice with $L = 16$ and $p = 0.4$, $p = p_c$, and $p = 0.8$. Take $p_c = 0.5927$. One way to help you identify the clusters is to modify `Program site` so that you can use the FLOOD statement to show different clusters in different colors. Consider at least five configurations for each value of p and average $n_s(p)$ over the configurations. Because the lattice is finite, more consistent results can be obtained by discarding those configurations that have a spanning cluster for $p < p_c$ and those that do not have a spanning cluster for $p \geq p_c$. For each value of p, plot n_s as a function of s and describe the observed s-dependence. Does n_s decrease more rapidly with s for $p = p_c$ or for $p \neq p_c$?

b.   Use the same configurations considered in part (a) to compute the mean cluster size S as a function of p. Remember that for $p > p_c$, the spanning cluster is excluded.

c.   Similarly, compute $P_\infty(p)$ for $L = 16$, and for various values of $p \geq p_c$. Plot $P(p)$ as a function of p and discuss its qualitative behavior.

d.   Verify that $\sum_s s n_s(p) = p$ for $p < p_c$ and explain this relation. How is this relation modified for $p \geq p_c$?

13.3   CLUSTER LABELING

Your visual algorithm for determining the existence of a connected path probably is very sophisticated. Although using the FLOOD command in True BASIC helps us to automate the process for a single configuration, we need to average over many configurations to obtain quantitative results. Hence, we need to develop an algorithm that finds the clusters. In the following, we will find that this task is not easy. The difficulty is that the assignment of a site to a cluster is a *global* rather than a *local* property of the site.

We consider the multiple cluster labeling method of Hoshen and Kopelman. The algorithm can best be described by an example. Consider the configuration shown in Fig. 13.8. We define an array `site` to store the occupancy of the sites; an occupied site initially is assigned the value -1 and an unoccupied site is assigned the value 0. We assign cluster labels to sites beginning at the lower left corner and continue from left to right. Because `site(1, 1)` is occupied, we assign to it cluster label 1. The next site is

7	7		9	3	3	
7			8	3	3	
	6	3			3	3
6	6	5	5	3	3	3
		5		4	3	3
1	1			4	3	3
1		2	2		3	

(a)

7	7		3	3	3	
7			3	3	3	
	3	3			3	3
3	3	3	3	3	3	3
		3		3	3	3
1	1			3	3	3
1		2	2		3	

(b)

Fig. 13.8 A percolation configuration on a square lattice with $L = 7$. Site coordinates are measured from the origin at the lower left corner $(1, 1)$. Part (a) shows the improper cluster labels initially assigned by *Program cluster*, a modified implementation of the Hoshen-Kopelman algorithm. Part (b) shows the proper cluster labels.

empty, and hence is not labeled. The next occupied site in the first row is $\text{site}(3, 1)$. Because its left neighbor is unoccupied, we assign to it the next available cluster label, label 2. The assignment of cluster labels to the remainder of the row is straightforward, and we proceed to $\text{site}(1, 2)$ of the second row. Because this site is occupied and its nearest neighbor in the preceding row is labeled 1, we assign label 1 to $\text{site}(1, 2)$. We continue from left to right along the second row checking the occupancy of each site. If a site is occupied, we check the occupancy of its nearest neighbors in the previous row and column. If neither neighbor is occupied, we assign the next available cluster label. If only one nearest neighbor site is occupied, the site is assigned the label of its occupied neighbor. For example, $\text{site}(2, 2)$ is assigned label 1 since its occupied neighbor, $\text{site}(1, 2)$ has label 1.

The difficulty arises when we come to an occupied site at which two clusters coalesce and cluster labels need to be reassigned. This case first occurs at $\text{site}(6, 2)$ — its two neighbors in the previous row and column have labels 3 and 4, respectively. We define the *proper* cluster label assignment at $\text{site}(6, 2)$ as the smaller of labels 3 and 4. Hence $\text{site}(6, 2)$ is assigned cluster label 3 and label 4 should be reassigned to label 3. It is inefficient to continually relabel the clusters, because there likely will be further reassignments. Hence, we delay the reassignment of cluster labels until the entire lattice is surveyed, and instead, keep track of the connections of the labels through a *label tree*. We introduce an array np that distinguishes proper and improper labels and provides their connections. Let us return to the configuration shown in Fig. 13.8 to explain the use of this array. Before we came to $\text{site}(6, 2)$, labels 1 through 4 were proper labels and we set

$$np(1) = 1, \qquad np(2) = 2, \qquad np(3) = 3, \qquad np(4) = 4.$$

At site$(6, 2)$ where labels 3 and 4 are linked, we set $np(4) = 3$. This reassignment of $np(4)$ tells us that label 4 is improper, and the numerical value of $np(4)$ tells us that label 4 is linked to label 3. Note that the argument i of np(i) always is greater than or equal to the value of np(i).

This procedure is still not complete. What should we do when we come to a site with two previously labeled neighbors one or both of which are improper? For example, consider site$(5, 4)$ which has two occupied neighbors with labels 5 and 4. We might be tempted to assign site$(5, 4)$ the label 4 and set $np(5) = 4$. However, instead of assigning to a site the minimum label of its two neighbors, we should assign to it the minimum of the *proper* labels of the two neighboring sites. In addition, if the two neighboring sites have different proper labels, then we should set np of the maximum proper label equal to the minimum proper label. In this example, we have $np(5) = 3$.

The above version of the Hoshen-Kopelman cluster algorithm is implemented in the following subroutines. The arrays site and np are declared in a main program. After *SUB* assign is called, site contains the proper labels for each occupied site. The array site is given an extra empty column on the left $(x = 0)$ and an extra empty row on the bottom $(y = 0)$ so that the first column and row do not have to be treated differently. The following declaration in the main program for the arrays would be appropriate.

```
DIM site(0:64,0:64),np(0:1000)
```

A more efficient and different version of the Hoshen-Kopelman algorithm written in Fortran has been given by Stauffer and Aharony (see references). Although the Hoshen-Kopelman algorithm is the most efficient cluster identification method for two-dimensional systems, it is not obvious that this approach is the most efficient in higher dimensions. Can you think of another method for identifying the clusters?

```
SUB assign(site(,),np(),L)
    ! assign cluster numbers to occupied sites
    DECLARE DEF proper
    MAT np = 0
    LET ncluster = 0                    ! cluster number
    FOR y = 1 to L
        FOR x = 1 to L
            IF site(x,y) < 0 then        ! site occupied
                LET down = y - 1          ! square lattice
                LET left = x - 1
                IF site(x,down) + site(left,y) = 0 then
                    LET ncluster = ncluster + 1     ! new cluster
                    LET site(x,y) = ncluster
                    LET np(ncluster) = ncluster     ! proper label
                ELSE
                    CALL neighbor(site(,),np(),x,y)
                END IF
            END IF
        NEXT x
```

```
                NEXT y
                ! assign proper labels to cluster array
                FOR y = 1 to L
                    FOR x = 1 to L
                        LET site(x,y) = proper(np(),site(x,y))
                    NEXT x
                NEXT y
            END SUB


        SUB neighbor(site(,),np(),x,y)      ! determine occupancy of neighbors
            LET down = y - 1
            LET left = x - 1
            IF site(x,down)*site(left,y) > 0 then
                ! both neighbors occupied
                CALL label_min(site(,),np(),x,y,left,down)
            ELSE IF site(x,down) > 0 then       ! down neighbor occupied
                LET site(x,y) = site(x,down)
            ELSE                                    ! left neighbor occupied
                LET site(x,y) = site(left,y)
            END IF
        END SUB


        SUB label_min(site(,),np(),x,y,left,down)
            ! both neighbors occupied, determine minimum cluster number
            DECLARE DEF proper
            IF site(left,y) = site(x,down) then
                ! both neighbors have same cluster label
                LET site(x,y) = site(left,y)
            ELSE
                ! determine minimum cluster label
                LET cl_left = proper(np(),site(left,y))
                LET cl_down = proper(np(),site(x,down))
                LET nmax = max(cl_left,cl_down)
                LET nmin = min(cl_left,cl_down)
                LET site(x,y) = nmin
                IF nmin <> nmax then
                    LET np(nmax) = nmin      ! set improper label nmax = nmin
                END IF
            END IF
        END SUB


        FUNCTION proper(np(),label)
            ! recursive function
            IF np(label) = label then
                LET proper = label
            ELSE
                LET proper = proper(np(),np(label))
            END IF
        END DEF
```

Problem 13.6   Test of the cluster labeling algorithm

Incorporate SUB assign and its associated subroutines into *Program site*. You will need to add array site (in addition to array rsite) to the original subroutines of *Program site*. Use the following subroutine to show the improper and proper cluster labels and to help you check that the cluster labeling is being done correctly. Choose a particular configuration and explain how the Hoshen-Kopelman algorithm is implemented.

```
SUB show_label(site(,),np(),L,proper$)
    DECLARE DEF proper
    FOR y = 1 to L
        LET yplot = y - 0.5
        FOR x = 1 to L
            LET xplot = x - 0.5
            LET label = site(x,y)
            IF label > 0 then
                IF proper$ = "yes" then LET label = proper(np(),label)
                BOX CLEAR xplot-0.5,xplot+0.5,yplot-0.5,yplot+0.5
                BOX LINES xplot-0.5,xplot+0.5,yplot-0.5,yplot+0.5
                PLOT TEXT, at xplot-0.25,yplot-0.25: using$("##",label)
            END IF
        NEXT x
    NEXT y
END SUB
```

After the clusters are labeled, we can obtain a number of geometrical quantities of interest. Vertical spanning can be checked by seeing if there is a nonzero label in the bottom row equal to a nonzero label in the top row. The following subroutine checks for the existence of a vertical spanning cluster.

```
SUB span(site(,),np(),vspan,L)
    DECLARE DEF proper
    ! check for existence of vertical spanning cluster
    LET vspan = 0
    FOR bottom = 1 to L
        IF site(bottom,1) > 0 then
            LET nbot = site(bottom,1)
            FOR top = 1 to L
                IF site(top,L) > 0 then
                    LET ntop = site(top,L)
                    IF ntop = nbot then
                        LET vspan = ntop
                        EXIT SUB
                    END IF
                END IF
            NEXT top
        END IF
    NEXT bottom
END SUB
```

Note that if vspan <> 0, there is a vertically spanning cluster with label vspan. Horizontal spanning can be determined by comparing the labels of the left and right columns.

The cluster distribution n_s can be found by computing the number of sites m(i) in cluster i, counting the number of clusters of size s, and normalizing the results by dividing by L^2, the number of sites. The probability of an occupied site belonging to the spanning cluster P_∞, can be determined by finding the label of the spanning cluster, vspan, and taking the ratio of m(vspan) to the total number of occupied sites. The mean cluster size S of the nonspanning clusters can be computed from the relation

$$S = \frac{\sum_i m^2(i)}{\sum_i m(i)}. \tag{13.6}$$

Convince yourself that (13.6) is equivalent to the definition (13.5) of S. Note that the spanning cluster is not included in the sums in (13.6).

In Problem 13.7 we apply the Hoshen-Kopelman cluster algorithm to a more systematic study of site percolation. In Section 13.4 we use a finite size scaling analysis to estimate the critical exponents.

*Problem 13.7   Applications of the cluster labeling algorithm

a.   Compute $F(p)\, dp$, the probability of *first* spanning an $L \times L$ square lattice for p in the range p to $p + dp$. Write a subroutine to find when the spanning first occurs as p is increased. Do a minimum of 100 configurations for each value of L and plot $F(p)$ as a function of p. Consider $L = 4, 16$, and 32. How does the shape of $F(p)$ change with increasing L? At what value of p is $F(p)\, dp \approx 0.5$ for the various spanning criteria and for each value of L? Call this value $p_c(L)$. How strongly does $p_c(L)$ depend on L for a given spanning criterion? How strongly does $p_c(L)$ depend on the spanning criterion for fixed L?

b.   Compute P_∞ for $p = p_c$, $p = 0.65$, $p = 0.75$, and $p = 0.9$ for $L = 4, 16$, and 32. Do a minimum of 100 configurations for each value of p. Use either the estimated value of $p_c(L)$ determined in part (a) or the known value $p_c = 0.5927$ (to four decimal places). What is the qualitative p-dependence of P_∞? Is $P_\infty(p = p_c)$ an increasing or decreasing function of L? (Discard those configurations that do not have a spanning cluster.)

c.   Write a subroutine to compute $n_s(p)$. Consider $p = p_c$ and $p = p_c \pm 0.1$ for $L = 4, 16$, and 32 and average over at least ten configurations. Why is n_s a decreasing function of s? Does n_s decrease more quickly for $p = p_c$ or for $p \neq p_c$?

d.   Write a subroutine to compute the mean cluster size S for $p = p_c$ and $p = p_c \pm 0.1$ for $L = 4, 16$, and 32. Average over at least ten configurations. What is the qualitative p-dependence of $S(p)$? How does $S(p = p_c)$ depend on L? For $p < p_c$, discard the configurations that contain a spanning cluster and for $p > p_c$ discard the configurations that do not have a spanning cluster.

It is convenient to associate a characteristic linear dimension or *connectedness length* $\xi(p)$ with the clusters. One way to do so is to define the radius of gyration R_s of a single cluster of s particles as

$$R_s^2 = \frac{1}{s}\sum_{i=1}^{s}(\mathbf{r}_i - \bar{\mathbf{r}})^2,$$ (13.7)

where

$$\bar{\mathbf{r}} = \frac{1}{s}\sum_{i=1}^{s}\mathbf{r}_i,$$ (13.8)

and $\mathbf{r}_i$ is the position of the ith site in the same cluster. The quantity $\bar{\mathbf{r}}$ is the familiar definition of the center of mass of the cluster. From (13.7), we see that R_s is the root mean square radius of the cluster measured from its center of mass. The connectedness length ξ can be defined as an average over the radii of gyration of all the finite clusters. The choice of the appropriate average is neither unique nor obvious. To find an expression for ξ, consider a site on a cluster of s sites. The site is connected to $s-1$ other sites and the average square distance to these sites is R_s^2 (see Problem 13.8a). The probability that a site belongs to a cluster of site s is $w_s = sn_s$. These considerations suggest that one definition of ξ is

$$\xi^2 = \frac{\sum_s(s-1)w_s R_s^2}{\sum_s(s-1)w_s}.$$

To simplify the expression for ξ, we write s instead of $s-1$ and let $w_s = sn_s$:

$$\xi^2 = \frac{\sum_s s^2 n_s R_s^2}{\sum_s s^2 n_s}$$ (13.9)

As before, the sum in (13.9) is over the nonspanning clusters only. The definition (13.9) and a simpler definition of $\xi(p)$ are explored in Problem 13.8.

* Problem 13.8   The connectedness length

a.   An alternative way of defining the radius of gyration of a cluster of s sites is as follows:

$$R_s^2 = \frac{1}{2s^2}\sum_{i,j}(\mathbf{r}_i - \mathbf{r}_j)^2.$$ (13.10)

The sum (13.10) is over all pairs of particles in the cluster. What is the physical interpretation of (13.10)? Show that the form (13.10) is equivalent to (13.7). Which expression for R_s^2 is easier to compute?

b.   Write a subroutine to compute R_s for a nonspanning cluster of size s. Choose $L=64$ and $p=0.57$ and compute ξ using the definition (13.9). Average over at least five configurations. Does the largest nonspanning cluster make the dominant contribution to the sum?

    c.   Compute the p-dependence of either $\xi(p)$ using (13.9) or associate ξ with the radius of gyration of the largest nonspanning cluster. Choose $L = 64$ and consider at least 50 configurations for each value of p. Consider values of p in steps of 0.01 in the interval $[p_c - 0.05, p_c - 0.01]$ and $[p_c + 0.01, p_c + 0.05]$ with $p_c = 0.5927$. For $p < p_c$ discard those configurations that contain a spanning cluster, and for $p > p_c$ discard those configurations that do not have a spanning cluster. Plot $\xi(p)$ and discuss its qualitative dependence on p. Is $\xi(p)$ a monotonically increasing or decreasing function of p for $p < p_c$ and $p > p_c$?

13.4   CRITICAL EXPONENTS AND FINITE SIZE SCALING

We are familiar with different phases of matter from our everyday experience. The most familiar example is water which can exist as a vapor, liquid, or solid. It is well known that water changes from one phase to another at a well defined temperature and pressure, e.g., the transition from ice to liquid water occurs at $0\,°C$ at atmospheric pressure. Such a change of phase is an example of a *thermodynamic phase transition*. Most substances also exhibit a *critical point*; that is, beyond a particular temperature and pressure, it is no longer possible to distinguish between the liquid and gaseous phases and the phase boundary terminates.

Another example of a critical point occurs in magnetic systems at the Curie temperature T_c and zero magnetic field. We know that at low temperatures some substances exhibit ferromagnetism, a spontaneous magnetization in the absence of an external magnetic field. If we raise the temperature of a ferromagnet, the spontaneous magnetization decreases and vanishes continuously at a critical temperature T_c. For $T > T_c$, the system is a paramagnet. In Chapter 17 we use Monte Carlo methods to investigate the behavior of a magnetic system near the magnetic critical point.

In the following, we will find that the properties of the *geometrical* phase transition in the percolation problem are qualitatively similar to the properties of thermodynamic phase transitions. Hence, a discussion of the percolation phase transition also can serve as an introduction to thermodynamic phase transitions. We will see that in the vicinity of a phase transition, the qualitative behavior of the system is governed by the appearance of long-range correlations.

We have seen that the essential physics near the percolation threshold is associated with the existence of large but finite clusters. For example, for $p \neq p_c$, we found in Problem 13.7c that n_s decays rapidly with s. However for $p = p_c$, the s-dependence of n_s is qualitatively different, and n_s decreases much more slowly. This different behavior of n_s at $p = p_c$ is due to the presence of clusters of all length scales, e.g., the "infinite" spanning cluster and the finite clusters of all sizes. We also found (see Problem 13.8) that $\xi(p)$ is finite, and an increasing function of p for $p < p_c$ and a decreasing function of p for $p > p_c$ (see Fig. 13.9). Moreover, we know that $\xi(p = p_c)$ is approximately equal to L and hence diverges as $L \to \infty$. This qualitative behavior of $\xi(p)$ is consistent with our physical picture of the clusters, that is, as p approaches p_c, the probability that two occupied sites are in the same cluster increases. These

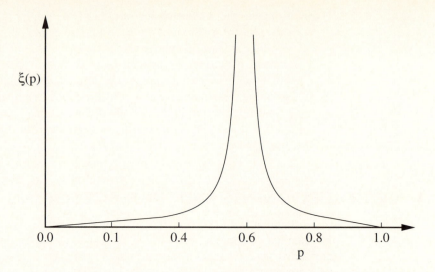

Fig. 13.9   Qualitative p-dependence of the connectedness length $\xi(p)$. The divergent
behavior of $\xi(p)$ in the critical region is characterized by the exponent ν
(see (13.11)).

qualitative considerations lead us to conjecture that in the limit $L \to \infty$, $\xi(p)$ grows
rapidly in the *critical region, $|p - p_c| \ll 1$*. We can describe the divergence of $\xi(p)$
more quantitatively by introducing the *critical exponent ν*, defined by the relation

$$\xi(p) \sim |p - p_c|^{-\nu}. \tag{13.11}$$

Of course, there is no *a priori* reason why the divergence of $\xi(p)$ can be characterized
by a simple power law. Note that ν is assumed to be the same above and below p_c.

   How do the other quantities that we have considered behave in the critical region in
the limit $L \to \infty$? According to the definition (13.2) of P_∞, $P_\infty = 0$ for $p < p_c$ and is
an increasing function of p for $p > p_c$. We conjecture that in the critical region, the
increase of P_∞ with increasing p is characterized by an exponent β defined by the
relation

$$P_\infty(p) \sim (p - p_c)^{\beta}. \tag{13.12}$$

Note that P_∞ is assumed to approach zero continuously from above p_c. We say that
the percolation transition is a *continuous* phase transition. In the language of critical
phenomena, P_∞ is an example of an *order parameter* of the system. An order parameter
is a quantity that measures the "order" of a system, and is nonzero in the ordered phase
and zero in the disordered phase. In the percolation context, we consider the phase
with a spanning cluster to be ordered and the phase without a spanning cluster to be
disordered.

The mean number of sites in the finite clusters, $S(p)$, also diverges in the critical region. Its critical behavior is written as

$$S(p) \sim |p - p_c|^{-\gamma}, \tag{13.13}$$

which defines the critical exponent γ. The common critical exponents for percolation are summarized in Table 13.1. For comparison, the analogous critical exponents of a magnetic critical point also are shown.

Because we can simulate only finite lattices, a direct fit of the measured quantities ξ, P_∞, and $S(p)$ to their assumed critical behavior for an infinite lattice would not yield good estimates for the corresponding exponents ν, β, and γ (see Problem 13.9a). The problem is that if p is close to p_c, the extent of the largest cluster becomes comparable to L, and the nature of the cluster distribution is affected by the finite size of the system. In contrast, for p far from p_c, $\xi(p)$ is small in comparison to L and the measured values of ξ, and hence the values of other physical quantities, are not appreciably affected by the finite size of the lattice. Hence for $p \ll p_c$ and $p \gg p_c$, the properties of the system are indistinguishable from the corresponding properties of a truly macroscopic system ($L \to \infty$). However, if p is close to p_c, $\xi(p)$ is comparable to L and the behavior of the system differs from that of an infinite system. In particular, a finite lattice cannot exhibit a true phase transition characterized by divergent physical quantities. Instead, ξ and S reach a finite maximum at $p = p_c(L)$.

The effects of the finite size of the system can be made more quantitative by the following argument. Consider for example, the critical behavior (13.12) of P_∞. As long as ξ is much less than L, the power law behavior given by (13.12) is expected to hold. However, if ξ is comparable to L, ξ cannot change appreciably and (13.12) is no longer applicable. This qualitative change in the behavior of P_∞ and other physical quantities occurs for

$$\xi(p) \sim L \sim |p - p_c|^{-\nu}. \tag{13.14}$$

Quantity	Functional form	Exponent	$d = 2$	$d = 3$
Percolation				
order parameter	$P_\infty \sim (p - p_c)^\beta$	β	5/36	0.4
mean size of finite clusters	$S(p) \sim |p - p_c|^{-\gamma}$	γ	43/18	1.8
connectedness length	$\xi(p) \sim |p - p_c|^{-\nu}$	ν	4/3	0.9
cluster numbers	$n_s \sim s^{-\tau} \quad p = p_c$	τ	187/91	2.2
Ising model				
order parameter	$M(T) \sim (T_c - T)^\beta$	β	1/8	0.32
susceptibility	$\chi(T) \sim |T - T_c|^{-\gamma}$	γ	7/4	1.24
correlation length	$\xi(T) \sim |T - T_c|^{-\nu}$	ν	1	0.63

Table 13.1 Several of the critical exponents for the percolation and magnetism phase transitions in $d = 2$ and $d = 3$ dimensions. Ratios of integers correspond to known exact results. The critical exponents for the Ising model are discussed in Chapter 17.

We invert (13.14) and write

$$|p - p_c| \sim L^{-1/\nu}. \tag{13.15}$$

The difference $|p - p_c|$ in (13.15) is the "distance" from the critical point at which "saturation" or finite size effects occur. Hence if ξ and L are approximately the same size, we can replace (13.12) by the relation

$$P_\infty(p = p_c) \sim L^{-\beta/\nu} \qquad (L \to \infty) \tag{13.16}$$

for the value of P_∞ at $p = p_c$ for a finite lattice. The relation (13.16) between P_∞ and L at $p = p_c$ is consistent with the fact that a phase transition, i.e., a singularity, is defined only for infinite systems.

One implication of (13.16) is that we can use it to determine the critical exponents. This method of analysis is known as *finite size scaling*. Suppose that we generate percolation configurations at $p = p_c$ for different values of L and analyze P_∞ as a function of L. If our values of L are sufficiently large, we can use the asymptotic relation (13.16) to estimate the ratio β/ν. A similar analysis can be used for $S(p)$ and other quantities of interest. We use this method in Problem 13.9.

Problem 13.9 Finite size scaling analysis of critical exponents

a. Compute P_∞ at $p = p_c$ for at least 100 configurations. Consider $L = 10, 20, 40,$ and 60. Include in your average only those configurations that have a spanning cluster. Best results are obtained using the value of p_c for the infinite square lattice, $p_c \approx 0.5927$. Plot $\ln P_\infty$ versus $\ln L$, and estimate the ratio β/ν.

b. Use finite size scaling arguments to determine the dependence of the mean cluster size S on L at $p = p_c$. Average S over the same configurations as considered in part (b). Remember that S is the mean number of sites in the nonspanning clusters.

* c. Analyze your data for the p-dependence of $S(p)$ obtained in Problem 13.5b for $L = 16$ and estimate the value of γ according to the assumed behavior given in (13.13). How does your estimate for γ compare with the answer that you obtained in part (b)?

* d. Find the mass (number of particles) M in the spanning cluster at $p = p_c$ as a function of L. Use the same configurations as in part (a). Determine an exponent from a plot of $\ln M$ versus $\ln L$. This exponent is called the fractal dimension of the cluster and is discussed in Chapter 14.

We found in Section 13.2 that the numerical value of the percolation threshold p_c depends on the symmetry and dimension of the lattice, e.g., $p_c \approx 0.5927$ for the square lattice and $p_c = 1/2$ for the triangular lattice. A remarkable feature of the power law dependencies summarized in Table 13.1 is that the values of the critical exponents do not depend on the symmetry of the lattice and are independent of the existence of the lattice itself, e.g., they are identical for the continuum percolation model discussed in Problem 13.4. Moreover, it is not necessary to distinguish between the exponents for site and bond percolation. In the vocabulary of critical phenomena, we say that site, bond, and continuum percolation all belong to the same *universality class* and that their critical exponents are identical.

Another important idea in critical phenomena is the existence of relations between the critical exponents. An example of such a *scaling law* is

$$2\beta + \gamma = \nu d,$$ (13.17)

where d is the spatial dimension of the lattice. The scaling law (13.17) indicates that the universality class depends on the spatial dimension. A more detailed discussion of finite size scaling and the scaling laws can be found in Chapter 17 and in the references.

13.5 THE RENORMALIZATION GROUP

In Section 13.4, we studied the properties of various quantities on different length scales to determine the values of the critical exponents. The idea of examining physical quantities near the critical point on different length scales can be extended beyond finite size scaling and is the basis of the *renormalization group* method, probably the most important new method developed in theoretical physics during the past twenty-five years. Kenneth Wilson was honored in 1981 with the Nobel prize in physics for his contributions to the development of the renormalization group method. Although the method was first applied to thermodynamic phase transitions, it is simpler to introduce the method in the context of the percolation transition. We will find that the renormalization group method yields the critical exponents directly, and in combination with Monte Carlo methods, it is frequently more powerful than Monte Carlo methods alone.

To introduce the method, consider a photograph of a percolation configuration generated at $p = p_0 < p_c$. If we view the photograph (or screen) from further and further distances, what would we see? Convince yourself that when you are far away from the photograph, you cannot distinguish occupied sites that are adjacent to each other and you cannot observe single site clusters. In addition, branches emanating from larger clusters and narrow bridges connecting large "blobs" are lost in your distant view of the photograph. Hence for $p_0 < p_c$, the distant photograph looks like a percolation configuration generated at a value of $p = p_1$ less than p_0. In addition, the connectedness length $\xi(p_1)$ of the remaining clusters is smaller than $\xi(p_0)$. If we move even further away from the photograph, the new clusters look even smaller with a value of $p = p_2$ less than p_1. Eventually we will not be able to distinguish any clusters and the photograph will appear as if it were at the trivial *fixed point* $p = 0$.

What would we observe as we go away from the photograph for $p_0 > p_c$? We can use the same reasoning to deduce that we would see only small regions of unoccupied sites. As we move further away from the photograph, these spaces become less discernible and the configuration looks as though a larger percentage of the lattice were occupied. Hence, the photograph will look like a configuration generated at a value of

$p = p_1$ greater than p_0 with $\xi(p_1) < \xi(p_0)$. As we move further and further away from the photograph, it will eventually appear to be at the other trivial fixed point $p = 1$.

What would we observe at $p_0 = p_c$? We know that at the percolation threshold, all length scales are present and it does not matter which length scale we use to observe the system. Hence, the photograph appears the same (although smaller overall) regardless of the distance at which we observe it. In this sense, p_c is a special, *nontrivial* fixed point.

We now consider a way of using a computer to change the configurations in a way that is similar to moving away from the photograph. Consider a square lattice that is partitioned into *cells* or *blocks* that cover the lattice (see Fig. 13.10). If we view the lattice from the perspective in which the sites in a cell merge to become a new supersite or renormalized site, then the new lattice has the same symmetry as the original lattice. However, the replacement of cells by the new sites has changed the length scale — all distances are now smaller by a factor of b, where b is the linear dimension of the cell. Hence, the effect of a "renormalization" is to replace each group of sites with a single renormalized site and to rescale the connectedness length for the renormalized lattice by a factor of b.

How can we decide whether the renormalized site is occupied or not? Because we want to preserve the main features of the original lattice and hence its connectedness (and its symmetry), we assume that a renormalized site is occupied if the original group of sites spans the cell. We adopt the vertical spanning criterion for convenience. The effect of performing a renormalization transformation on typical percolation configurations for p above and below p_c is illustrated in Fig. 13.11 and Fig. 13.12 respectively. In both cases, the effect of the successive transformations is to move the system away from p_c. We see that for $p = 0.7$, the effect of the transformations is to drive the system toward $p = 1$. For $p = 0.5$, the trend is to drive the system toward $p = 0$. As we discuss in the following, we can associate p_c with an unstable fixed point of the renormalization transformation. Of course, because we began with a finite lattice, we cannot continue the renormalization transformation indefinitely.

Program rg implements a visual interpretation of the renormalization group. The program divides the screen into four windows with the original lattice in the first win-

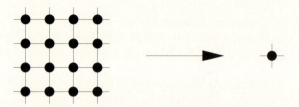

Fig. 13.10 An example of a $b = 4$ cell used on the square lattice. The cell contains b^2 sites which are rescaled to a single supersite after a renormalization group transformation.

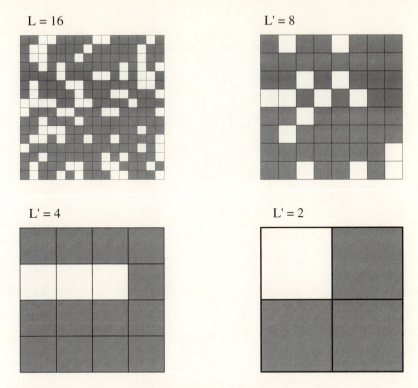

Fig. 13.11 A percolation configuration generated at $p = 0.7$. The original configuration has been renormalized three times by transforming cells of four sites into one new supersite. What would be the effect of an additional transformation?

dow and three renormalized lattices in windows 2 through 4. In *Program site* we represented an occupied site at lattice point x,y as a filled circle of unit diameter centered about the point (x, y). In contrast, *Program rg* represents an occupied site at x,y as a filled box whose lower left corner is at $x - 1, y - 1$.

```
PROGRAM rg
DIM r(32,32)
LIBRARY "csgraphics"
CALL initial(r(,),L,b,#1,#2,#3,#4)
CALL configuration(r(,),L,b,#1,#2,#3,#4)
END

SUB initial(r(,),L,b,#1,#2,#3,#4)
    RANDOMIZE
    LET L = 32
    ! program must be modified for b <> 2
    LET b = 2
```

L = 16   L' = 8

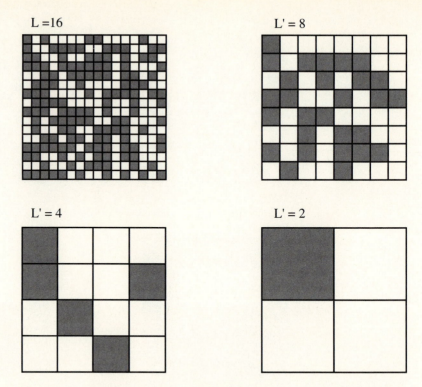

L' = 4   L' = 2

Fig. 13.12   A percolation configuration generated at $p = 0.5$. The original
                  configuration has been renormalized three times by transforming
                  blocks of four sites into one new site. What would be the effect of
                  an additional transformation?

```
! assign random number to each site
FOR y = 1 to L
    FOR x = 1 to L
        LET r(x,y) = rnd
    NEXT x
NEXT y
OPEN #1: screen 0,0.5,0.5,1
CALL lattice(L)                        ! draw original lattice in window #1
OPEN #2: screen 0.5,1,0.5,1
CALL lattice(L/b)
OPEN #3: screen 0,0.5,0,0.5
CALL lattice(L/(b*b))
OPEN #4: screen 0.5,1,0,0.5
CALL lattice(L/(b*b*b))
END SUB

SUB lattice(Ln)
    CALL compute_aspect_ratio(Ln,xwin,ywin)
```

```
                SET WINDOW -0.07*xwin,1.07*xwin,-0.07*ywin,1.07*ywin
                BOX LINES 0,Ln,0,Ln
                SET CURSOR 1,1
                PRINT "L ="; Ln
                FOR y = 1 to Ln
                    FOR x = 1 to Ln
                        PLOT POINTS: x - 0.5,y - 0.5
                    NEXT x
                NEXT y
        END SUB

        SUB configuration(r(,),L,b,#1,#2,#3,#4)
            DIM s(32,32),s1(16,16),s2(8,8),s3(4,4)
            DO while p <=1
                WINDOW #1
                SET COLOR "black"
                SET CURSOR 1,14
                PRINT "        "                ! erase previous value of p
                SET CURSOR 1,10
                INPUT prompt "p = ": p
                SET COLOR "red"
                FOR y = 1 to L
                    FOR x = 1 to L
                        IF r(x,y) <= p then
                            BOX AREA x - 1, x, y - 1,y
                            LET s(x,y) = 1
                        END IF
                    NEXT x
                NEXT y
                CALL block(#2,L/b,s(,),s1(,))
                CALL block(#3,L/(b*b),s1(,),s2(,))
                CALL block(#4,L/(b*b*b),s2(,),s3(,))
            LOOP
        END SUB

        SUB block(#9,Ln,w(,),wr(,))
            WINDOW #9
            SET COLOR "red"
            FOR y = 1 to Ln
                LET yc = 2*y - 1
                FOR x = 1 to Ln
                    LET xc = 2*x - 1
                    ! cell spans vertically -> renormalized site occupied
                    IF w(xc,yc)*w(xc,yc+1)=1 or w(xc+1,yc)*w(xc+1,yc+1)=1 then
                        LET wr(x,y) = 1
                        BOX AREA x - 1, x, y - 1,y
                    END IF
                NEXT x
            NEXT y
        END SUB
```

Problem 13.10   Visual renormalization group

Use *Program rg* with $L = 32$ to estimate the value of the percolation threshold. For example, confirm that for small p, e.g., $p = 0.4$, the renormalized lattice almost always renormalizes to a nonspanning cluster. What happens for $p = 0.8$? How can you use the properties of the renormalized lattices to estimate p_c?

Although a visual implementation of the renormalization group allows us to estimate p_c, it does not allow us to estimate the critical exponents. In the following, we present a renormalization group method that allows us to obtain p_c and the critical exponent ν associated with the connectedness length. This analysis follows closely the method presented by Reynolds et al. (see references).

The implementation of a renormalization group method consists of two parts: (i) an average over the underlying variables together with a specification of the variables that determine the state of the renormalized configuration, and (ii) a parameterization of the renormalized configuration in terms of the original parameters and possibly others. We adopt the same average as before, i.e., we replace the b^d sites within a cell of linear dimension b by a single site that represents whether or not the original lattice sites span the cell. The second step is to determine which parameters specify the new configuration after the averaging. We make the simple approximation that each cell is independent of all the other cells and is characterized only by the probability p' that the cell is occupied. The renormalization transformation between p and p' reflects the fact that the basic physics of percolation is connectedness, because we define a cell to be occupied only if it contains a set of sites that span the cell. If the sites are occupied with probability p, then the cells are occupied with probability p', where p' is given by a *renormalization transformation* or a *recursion relation* of the form

$$p' = R(p). \tag{13.18}$$

The quantity $R(p)$ is the total probability that the sites form a spanning path.

An example will make the formal relation (13.18) more clear. In Fig. 13.13, we show the seven vertically spanning site configurations for a $b = 2$ cell. The probability p' that the renormalized site is occupied is given by the sum of the probabilities of all spanning configurations:

$$p' = R(p) = p^4 + 4p^3(1 - p) + 2p^2(1 - p)^2. \tag{13.19}$$

In general, the probability p' of the occupied renormalized sites is different than the occupation probability p of the original sites. For example, suppose that we begin with $p = p_0 = 0.5$. After a single renormalization transformation, the value of p' obtained from (13.19) is $p_1 = p' = R(p_0 = 0.5) = 0.44$. If we perform a second renormalization transformation, we have $p_2 = R(p_1) = 0.35$. It is easy to see that further transformations drive the system to the fixed point $p = 0$. Similarly, if we begin with $p = p_0 = 0.7$, we find that successive transformations drive the system to the fixed point $p = 1$. This behavior is qualitatively similar to what we observed in the visual renormalization group.

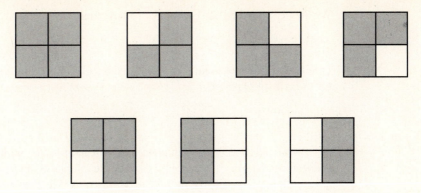

Fig. 13.13 The seven (vertically) spanning configurations on a $b = 2$ cell.

To find the nontrivial fixed point associated with the critical threshold p_c, we need to find the special value of p such that

$$p^* = R(p^*). \tag{13.20}$$

For the recursion relation (13.19), we find that the solution of the fourth degree equation for p^* yields the two trivial fixed points, $p^* = 0$ and $p^* = 1$, and the nontrivial fixed point $p^* = 0.61804$ which we associate with p_c. This calculated value of p^* for $b = 2$ should be compared with the estimate $p_c = 0.5927$.

To calculate the critical exponent ν, we recall that all lengths are reduced on the renormalized lattice by a factor of b in comparison to the lengths in the original system. Hence the connectedness length transforms as

$$\xi' = \xi / b. \tag{13.21}$$

Because $\xi(p) = \text{const}|p - p_c|^{-\nu}$ for $p \sim p_c$, we have

$$|p' - p^*|^{-\nu} = b^{-1}|p - p^*|^{-\nu}, \tag{13.22}$$

where we have identified p_c with p^*. To find the relation between p' and p near p_c, we expand the renormalization transformation (13.18) in a Taylor series about p^* and obtain to first order in $(p - p^*)$:

$$p' - p^* = R(p) - R(p^*) \approx \lambda\,(p - p^*), \tag{13.23}$$

where

$$\lambda = \frac{dR(p = p^*)}{dp}. \tag{13.24}$$

We need to do a little algebra to obtain an explicit expression for ν. We first raise both sides of (13.23) to the νth power and write

$$|p' - p^*|^\nu = \lambda^\nu (p - p^*)^\nu. \tag{13.25}$$

We then compare (13.25) and (13.22) and obtain

$$b = \lambda^{\nu}. \tag{13.26}$$

Finally, we take the logarithm of both sides of (13.26) and obtain the desired relation for the critical exponent ν:

$$\nu = \frac{\log b}{\log \lambda}. \tag{13.27}$$

As an example, let us calculate λ for a square lattice with $b = 2$. We write (13.19) in the form $R(p) = -p^4 + 2p^2$. The derivative of $R(p)$ with respect to p yields $\lambda = 4p(1 - p^2) = 1.5279$ at $p = p^* = 0.61804$. We then use the relation (13.27) to obtain

$$\nu = \frac{\log 2}{\log 1.5279} = 1.635\ldots \tag{13.28}$$

A comparison of (13.28) with the exact result $\nu = 4/3$ (see Table 13.1) in two dimensions shows remarkable agreement for such a simple calculation. (What would we be able to conclude if we were to measure $\xi(p)$ directly on a 2×2 lattice?) However, the accuracy of our calculation of ν is not known. What is the nature of our approximations? Our major assumption has been that the occupancy of each cell is independent of all other cells. This assumption is correct for the original sites, but after one renormalization, we lose some of the original connecting paths and gain connecting paths that are not present in the original lattice. An example of this interface problem is shown in Fig. 13.14. Because this surface effect becomes less probable with increasing cell size, one way to improve the renormalization group calculation is to consider larger cells. We consider a $b = 3$ calculation in Problem 13.11d. In Project 13.2 we combine the renormalization group method with a Monte Carlo approach to treat still larger cells.

Problem 13.11   Renormalization group method for small cells

a. Enumerate the spanning configurations for a $b = 2$ cell assuming that a cell is occupied if a spanning path exists in either the vertical or the horizontal directions. Obtain the recursion relation and solve for the fixed point p^*. Although you could use a root finding algorithm to solve for p^*, it is easy to use trial and error to find the value of p such that $R(p) - p$ is zero. Or you can plot the

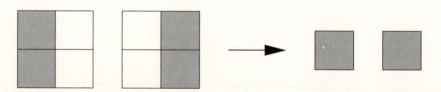

Fig. 13.14   Example of the interface problem between cells. The two cells are not connected at the original site level but are connected at the cell level.

function $R(p) - p$ versus p and find the value of p at which $R(p) - p$ crosses the horizontal axis. How do p^* and ν compare to their values using the vertical spanning criterion?

b. Repeat the simple renormalization group calculation in (a) using the criterion that a cell is occupied only if a spanning path exists in both directions.

* c. The association of p_c with p^* is not the only possible one. Two alternatives involve the derivative $R'(p) = dR/dp$. For example, we could let $p_c = \overline{p} = \int_0^1 p R'(p) \, dp$. Alternatively, we could choose $p_c = p_{max}$, where p_{max} is the value of p at which $R'(p)$ has its maximum value. Compute p_c using these two alternative definitions and the various spanning criteria. In the limit of large cells, all three definitions should lead to the same values of p_c.

* d. Enumerate the possible spanning configurations of a $b = 3$ cell, assuming that a cell is occupied if a cluster spans the cell vertically. Determine the probability of each configuration, and verify that the renormalization transformation $R(p) = p^9 + 9p^8q + 36p^7q^2 + 67p^6q^3 + 59p^5q^4 + 22p^4q^5 + 3p^3q^6$. It is possible to do the exact enumeration by hand. Solve the recursion relation (13.20) for p^*. Use this value of p^* to find the slope λ and the exponent ν. Then assume a cell is occupied if a cluster spans the cell both vertically and horizontally and obtain $R(p)$. Determine $p^*(b = 3)$ and $\nu(b = 3)$ for the two spanning criteria. Are your results for p^* and ν closer to their known values than for $b = 2$ for the same spanning criteria?

Problem 13.12 Renormalization group method for triangular lattice

a. There are some difficulties with the above renormalization group method in the infinite cell limit, if a cell is said to span when there is a path in one fixed direction (see Ziff). This problem is absent for the triangular lattice. For this symmetry a cell can be formed by grouping three sites that form a triangle into one renormalized site. The only reasonable spanning criterion is that the cell spans if any two sites are occupied. Verify that $R(p) = p^3 + 3p^2(1 - p)$ and find $p_c = p^*$. Hoes does p^* compare to the exact result $p_c = 1/2$?

b. Calculate the critical exponent ν and compare its value with the exact result. Explain why b is given by $b^2 = 3$. Give a qualitative argument why the renormalization group argument might work better for small cells on a triangular lattice than on a square lattice.

It is possible to improve our renormalization group results for p_c and ν by enumerating the spanning clusters for larger b. However, because the 2^{b^2} possible configurations for a $b \times b$ cell increase rapidly with b, exact enumeration is not practical for $b > 7$, and we must use Monte Carlo methods if we wish to proceed further. Two Monte Carlo approaches are discussed in Project 13.2. The combination of methods, Monte Carlo and renormalization group (MCRG), provides a powerful tool for obtaining information on phase transitions and other properties of materials.

13.6   PROJECTS

We have seen that the percolation problem illustrates many of the important ideas in critical phenomena. In later chapters we apply similar ideas and approaches to thermal systems. The following projects require larger systems and more computer resources than the problems in this chapter, but they are not much more difficult conceptually. More ideas for projects can be obtained from the references.

Project 13.1   Cell-to-cell renormalization group method

In Section 13.5 we discussed the cell-to-site renormalization group transformation for a system of cells of linear dimension b. An alternative transformation is to go from cells of linear dimension b_1 to cells of linear dimension b_2. For such a "cell-to-cell" transformation, the rescaling length b_1/b_2 can be made close to unity. Many errors in a cell-to-cell renormalization group transformation cancel, resulting in a transformation that is more accurate in the limit in which the change in length scale is infinitesimal. We can use the fact that the connectedness lengths of the two systems are related by $\xi(p_2) = (b_1/b_2)^{-1}\xi(p_1)$ to derive the relation

$$\nu = \frac{\ln b_1/b_2}{\ln \lambda_1/\lambda_2},\tag{13.29}$$

where $\lambda_i = dR(p^*, b_i)/dp$ is evaluated at the solution to the fixed point equation, $R(b_2, p^*) = R(b_1, p^*)$. Note that (13.29) reduces to (13.27) for $b_2 = 1$. Use the results you found in Problem 13.11d for one of the spanning criteria to estimate ν from a $b_1 = 3$ to $b_2 = 2$ transformation. Then consider larger values of b_2 and b_1.

Project 13.2   Monte Carlo renormalization group

a.   One way to estimate $R(p)$, the total probability of all the spanning clusters, can be understood by writing $R(p)$ in the form

$$R(p) = \sum_{n=1}^{N} \binom{N}{n} p^n q^{(N-n)} S(n),\tag{13.30}$$

where $N = b^2$. The binomial coefficient $\binom{N}{n} = N!/((N-n)!\,n!)$ represents the number of possible configurations of n occupied sites and $N - n$ empty sites. The quantity $S(n)$ is the probability that a random configuration of n occupied sites spans the cell. A comparison of (13.19) and (13.30) shows that for $b = 2$ and the vertical spanning criterion, $S(1) = 0$, $S(2) = 2/6$, $S(3) = 1$, and $S(4) = 1$. What are the values of $S(n)$ for $b = 3$?

   We can estimate the probability $S(n)$ by straightforward Monte Carlo methods. One way to sample $S(n)$ is to add a particle at random to an unoccupied site and check if a spanning path exists. If a spanning path does not exist, add another particle at random to a previously unoccupied site. If a spanning path exists after s particles are added, then let $S(n) = S(n) + 1$ for $n \geq s$ and generate a new configuration. After a reasonable number of configurations, the results for $S(n)$ can be normalized. Of course, this procedure can be made more efficient by checking for a spanning cluster only after the total number of particles added is near $s \sim p^*N$ and by checking for spanning after adding several particles.

Write a Monte Carlo program to sample $S(n)$. (Hint: store the location of the unoccupied sites in a separate array.) To check your program, first sample $S(n)$ for $b = 2$ and $b = 3$ and compare your results to the exact results for $S(n)$. Consider larger values of b and determine $S(n)$ for $b = 5, 8, 16$, and 32. For $b \geq 16$, the total probability $R(p)$ can be found by using (13.30) and the Gaussian approximation for the probability of a configuration of n occupied sites:

$$P_N(n) = \binom{N}{n} p^n q^{(N-n)} \approx (2\pi Npq)^{-\frac{1}{2}} e^{-(n-pN)^2/2Npq}. \tag{13.31}$$

Because $P_N(n)$ is sharply peaked for large b, it is necessary to sample $S(n)$ only near $n = p^*N$. This method has been investigated by Hu (see references).

b. In part (a) the number of particles rather than the occupation probability p was varied. Another Monte Carlo procedure is to vary p and sample $F(p)\,dp$, the probability of *first* spanning a $b \times b$ cell in the range p to $p + dp$. Because the renormalization group transformation defines p' as the *total* probability of spanning at p, p' can be interpreted as the *cumulative distribution function* and is related to $F(p)$ by

$$p' = R(p) = \int_0^p F(p)\,dp. \tag{13.32}$$

The sampling of $F(p)$ for finite width bins Δp implies that the integral in (13.32) reduces to a sum. Because $\lambda = dR(p = p^*)/dp$, we have $\lambda = F(p^*)$. The simplest way to estimate λ is by setting $\lambda = F(p_{\max})$, where $p_{\max}$ is the value of p for which $F(p)$ is a maximum. Determine $p_c(b)$ and $\nu(b)$ for $b = 5, 8, 16$, and 32. How do your results compare with those found in part (a)? Which method yields smaller error estimates for p_c and ν?

c. It is possible to extrapolate the results for the successive estimates $p_c(b)$ and $\nu(b)$ to the limit $b \to \infty$. Finite size scaling arguments (cf. Stauffer and Aharony) suggest that

$$\nu(b) \approx \nu - c_1/\ln b \tag{13.33a}$$

and

$$p^*(b) \approx p_c - a_1 b^{-1/\nu} \tag{13.33b}$$

for b sufficiently large. The relation (13.33a) suggests that the sequence $\nu(b)$ should be plotted as a function of $1/\ln b$ and the extrapolated result should be a straight line with an intercept of ν. The quantities a_1 and c_1 in (13.33) are fitting parameters. The relation (13.33b) suggests that we should plot $p_c(b)$ versus $b^{-1/\nu}$ using the value of ν found from (13.33a). How sensitively does your result for p_c depend on the assumed value of ν? It is necessary to consider cells on the order of $b = 500$, and to do a more sophisticated analysis of $\nu(b)$ and $p^*(b)$, to obtain extrapolated values that agree to four places with the exact value $\nu = 4/3$ and the estimate $p_c = 0.5927$.

Project 13.3   Percolation in three dimensions

a. The value of p_c for site percolation on the simple cubic lattice is approximately 0.311. Write a program using the Hoshen-Kopelman cluster labeling method to verify this value. Compute ϕ_c, the volume fraction occupied at p_c, if a sphere with a diameter equal to the lattice spacing is placed on each occupied site.

b. Consider continuum percolation in three dimensions where spheres of unit diameter are placed at random in a cubical box of linear dimension L. Two spheres that overlap are in the same cluster. As each sphere is added to the box, determine if the sphere overlaps with any other sphere in the box. If it does not, then the sphere constitutes a new cluster. If it does, then the sphere adopts the cluster label of the sphere with which it overlaps. If it overlaps with spheres of different cluster labels, then it is necessary to either relabel the clusters or to use the Hoshen-Kopelman algorithm to determine the proper label and generate a label tree. The volume fraction occupied by the spheres is given by

$$\phi = 1 - e^{-\rho 4\pi r^3/3}, \tag{13.34}$$

where ρ is the number density of the spheres, and r is their radius. Write a program to simulate continuum percolation in three dimensions and find the percolation threshold ρ_c. Use the Monte Carlo procedure discussed in Problem 13.4 to estimate ϕ_c and compare its value with the value obtained using (13.34). How does ϕ_c for continuum percolation compare with the value of ϕ_c found for site percolation in part (a)? Which do you expect to be larger and why?

* c. In the Swiss cheese model in three dimensions, we are concerned with the percolation of the space between the spheres. This model is appropriate for porous rock with the spheres representing solid material and the space between the spheres representing the pores. Superimpose a regular grid with lattice spacing equal to $0.1r$ on the system, where r is the radius of the spheres. If a point on the grid is not within any sphere, it is "occupied." The use of the grid allows us to determine the connectivity between different regions of the pore space. Use your cluster labeling routine from part (a) to label the clusters, and determine $\tilde{\phi}_c$, the volume fraction occupied by the pores at threshold. You might be surprised to find that $\tilde{\phi}_c$ is relatively small. If time permits, use a finer grid and repeat the calculation to improve the accuracy of your results.

*d. Use finite size scaling to estimate the critical percolation exponents for the three models presented in parts (a)–(c). Are they the same within the accuracy of your calculation?

Project 13.4   Conductivity in a random resistor network

a. An important critical exponent for percolation is the conductivity exponent t defined by

$$\sigma \sim (p - p_c)^t, \tag{13.35}$$

where σ is the conductance (or inverse resistance) per unit length in two dimensions. Consider bond percolation on a square lattice where each occupied bond between two neighboring sites is a resistor of unit resistance. Unoccupied bonds have infinite resistance. Because the total current into any node must equal zero by Kirchhoff's law, the voltage at any site (node) is equal to the average of the voltages of all nearest neighbor sites connected by resistors (occupied bonds). Since this relation for the voltage is the same as the algorithm for solving Laplace's equation on a lattice, the voltage at each site can be computed using a relaxation method discussed in Chapter 10. To compute the conductivity for a given $L \times L$ resistor network, we fix the voltage $V = 0$ at sites for which $x = 0$ and fix $V = 1$ at sites for which $x = L + 1$. In the y direction we use periodic boundary conditions. We then compute the voltage at all sites using the relaxation method. The current through each resistor connected to a site at $x = 0$ is simply $I = \Delta V/R = (V - 0)/1 = V$. The conductivity is the sum of the currents through all the resistors connected to $x = 0$ divided by L. In a similar way, the conductivity can be computed from the resistors attached to the $x = L + 1$ boundary. Write a program to implement the relaxation method for the conductivity of a random resistor network on a square lattice. An indirect, but easier way of computing the conductivity, is considered in Problem 14.8.

b. The bond percolation threshold on a square lattice is $p_c = 0.5$. Use your program to compute the conductivity for a $L = 30$ square lattice. Average over at least ten spanning configurations for $p = 0.51, 0.52$, and 0.53. Note that you can eliminate all bonds that are not part of the spanning cluster and all occupied bonds connected to only one other occupied bond. Why? If possible, consider more values of p. Estimate the critical exponent t defined in (13.35).

c. Fix p at $p = p_c = 1/2$ and use finite size scaling to estimate the conductivity exponent t.

*d. Use larger lattices and the multigrid method (see Project 10.1) to improve your results. If you have sufficient computing resources, compute t for a simple cubic lattice for which $p_c \approx 0.247$. (In two dimensions t is the same for lattice and continuum percolation. However, in three dimensions t can be different.)

References and Suggestions for Further Reading

Joan Adler, "Series expansions," *Computers in Physics* **8**, 287 (1994). The critical exponents and the value of p_c also can be determined by doing exact enumeration.

I. Balberg, "Recent developments in continuum percolation," *Phil. Mag.* **56**, 991 (1987). An earlier paper on continuum percolation is by Edward T. Gawlinski and H. Eugene Stanley "Continuum percolation in two dimensions: Monte Carlo tests of scaling and universality for non-interacting discs," *J. Phys. A: Math. Gen.* **14**, L291 (1981). These workers divide the system into cells and use the Poisson distribution to place the appropriate number of disks in each cell.

Armin Bunde and Shlomo Havlin, editors, *Fractals and Disordered Systems*, Springer-Verlag (1991). Chapter 2 by the editors is on percolation.

C. Domb, E. Stoll, and T. Schneider, "Percolation clusters," *Contemp. Phys.* **21**, 577 (1980). This review paper discusses the nature of the percolation transition using illustrations from a film of a Monte Carlo simulation of a percolation process.

J. W. Essam, "Percolation theory," *Reports on Progress in Physics* **53**, 833 (1980). A mathematically oriented review paper.

Jens Feder, *Fractals*, Plenum Press (1988). See Chapter 7 on percolation. We discuss the fractal properties of the spanning cluster at the percolation threshold in Chapter 14.

J. P. Fitzpatrick, R. B. Malt, and F. Spaepen, "Percolation theory of the conductivity of random close-packed mixtures of hard spheres," *Phys. Lett.* A **47**, 207 (1974). The authors describe a demonstration experiment done in a first year physics course at Harvard.

J. Hoshen and R. Kopelman, "Percolation and cluster distribution. I. Cluster multiple labeling technique and critical concentration algorithm," *Phys. Rev.* B **14**, 3438 (1976). The original paper on an efficient cluster labeling algorithm.

Chin-Kun Hu, Chi-Ning Chen, and F. Y. Wu, "Histogram Monte Carlo position-space renormalization proup: applications to site percolation," preprint. The authors use a histogram Monte Carlo method which is similar to the method discussed in Project 13.2. A similar Monte Carlo method also was used by M. Ahsan Khan, Harvey Gould, and J. Chalupa, "Monte Carlo renormalization group study of bootstrap percolation," *J. Phys.* C **18**, L223 (1985).

Ramit Mehr, Tal Grossman, N. Kristianpoller, and Yuval Gefen, "Simple percolation experiment in two dimensions," *Am. J. Phys.* **54**, 271 (1986). A simple experiment for an undergraduate physics laboratory is proposed.

D. C. Rapaport, "Percolation on large lattices," *Phil. Mag.* **56**, 1027 (1987). See also D. C. Rapaport, "Cluster size distribution at criticality," *J. Stat. Phys.* **66**, 679 (1992). The author generates many independent sublattices and then combines them to study percolation on a $640\,000^2$ lattice.

Peter J. Reynolds, H. Eugene Stanley, and W. Klein, "Large-cell Monte Carlo renormalization group for percolation," *Phys. Rev.* B**21**, 1223 (1980). An especially clearly written research paper. Our discussion on the renormalization group in Section 13.5 is based upon this paper.

Muhammad Sahimi, *Applications of Percolation Theory*, Taylor & Francis (1994). The author is a chemical engineer, and the emphasis is on modeling various phenomena in disordered media.

Muhammad Sahimi and Hossein Rassamdana, "On position-space renormalization group approach to percolation," *J. Stat. Phys.* **78**, 1157 (1995).

Dietrich Stauffer and Amnon Aharony, *Introduction to Percolation Theory*, second edition, Taylor & Francis (1994). A delightful book by two of the leading workers in the field. An efficient Fortran implementation of the Hoshen-Kopelman algorithm is given in Appendix A.3.

D. Stauffer, "Percolation clusters as teaching aid for Monte Carlo simulation and critical exponents," *Am. J. Phys.* **45**, 1001 (1977); D. Stauffer, "Scaling theory of percolation clusters," *Physics Reports* **54**, 1 (1979).

B. P. Watson and P. L. Leath, "Conductivity in the two-dimensional-site percolation problem," *Phys. Rev.* B**9**, 4893 (1974). A research paper on the conductivity of chicken wire.

Kenneth G. Wilson, "Problems in physics with many scales of length," *Sci. Am.* **241**, 158 (1979). An accessible article on the renormalization group method and its applications in particle and condensed matter physics. See also K. G. Wilson, "The renormalization group and critical phenomena," *Rev. Mod. Phys.* **55**, 583 (1983). The latter article is the text of Wilson's lecture on the occasion of the presentation of the 1982 Nobel Prize in Physics. In this lecture he claims that he "... found it very helpful to demand that a correctly formulated field theory be soluble by computer, the same way an ordinary differential equation can be solved on a computer ..."

W. Xia and M. F. Thorpe, "Percolation properties of random ellipses," *Phys. Rev.* A **38**, 2650 (1988). The authors consider continuum percolation of elliptical shapes, and show that $\phi = e^{-A\rho}$, where A is the area of the object, and ρ is the number density.

Richard Zallen, *The Physics of Amorphous Solids,* Wiley-Interscience (1983). Chapter 4 discusses many of the applications of percolation concepts to realistic systems.

Robert M. Ziff, "Spanning probability in 2D percolation," *Phys. Rev. Lett.* **69**, 2670 (1992). The author finds $p_c = 0.592\,7460 \pm 0.000\,0005$ for a square lattice.

C H A P T E R
14
Fractals

We introduce the concept of fractal dimension and discuss several processes that generate fractal objects.

14.1   THE FRACTAL DIMENSION

One of the more interesting geometrical properties of objects is their shape. As an example, we show in Fig. 14.1 a spanning cluster generated at the percolation threshold. Although the visual description of such a cluster is subjective, such a cluster can be described as ramified, airy, tenuous, and stringy, and would not be described as compact or space-filling.

In recent years a new *fractal* geometry has been developed by Mandelbrot and others to describe the characteristics of ramified objects. One quantitative measure of the structure of these objects is their *fractal dimension D*. To define D, we first review some simple ideas of dimension in ordinary Euclidean geometry. Consider a circular or spherical object of mass M and radius R. If the radius of the object is increased from R to $2R$, the mass of the object is increased by a factor of 2^2 if the object is circular, or by 2^3 if the object is spherical. We can express this relation between mass and length as

$$M(R) \sim R^D, \qquad \text{(mass dimension)} \qquad (14.1)$$

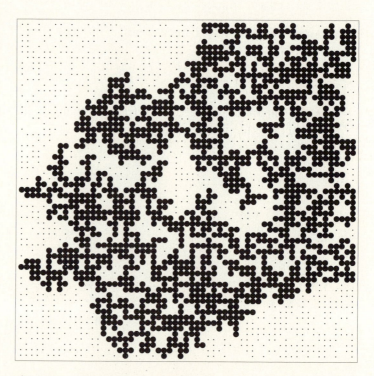

Fig. 14.1   Example of a percolation cluster generated at $p = 0.5927$ on a $L = 61$ square lattice. Occupied sites that are not part of the spanning cluster are shown as points; unoccupied sites are not shown.

where D is the dimension of the object. Equation (14.1) implies that if the linear dimensions of an object are increased by a factor of b while preserving its shape, then the mass of the object is increased by b^D. This mass-length scaling relation is closely related to our intuitive understanding of dimension. Note that if the dimension of the object, D, and the dimension of the Euclidean space in which the object is embedded, d, are identical, then the mass density $\rho = M/R^d$ scales as

$$\rho(R) \propto M(R)/R^d \sim R^0. \tag{14.2}$$

An example of a two-dimensional object is shown in Fig. 14.2. An object whose mass-length relation satisfies (14.1) with $D = d$ is said to be *compact*.

Equation (14.1) can be used to define the fractal dimension. We denote objects as fractals if they satisfy (14.1) with a value of D different from the spatial dimension d. Note that if an object satisfies (14.1) with $D < d$, its density is not the same for all R, but scales as

$$\rho(R) \propto M/R^d \sim R^{D-d}. \tag{14.3}$$

Because $D < d$, a fractal object becomes less dense at larger length scales. The scale dependence of the density is a quantitative measure of the ramified or stringy nature of fractal objects. That is, one characteristic of fractal objects is that they have holes of all sizes.

The percolation cluster shown in Fig. 14.1 is an example of a *random* or statistical fractal because the mass-length relation (14.1) is satisfied only on the average, i.e., only if the quantity $M(R)$ is averaged over many different origins in a given cluster and over many clusters.

In physical systems, the relation (14.1) does not extend over all length scales, but is bounded by both upper and lower cut-off lengths. For example, a lower cut-off length

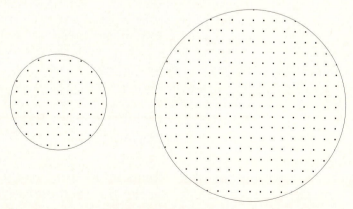

Fig. 14.2 The number of dots per unit area in each circle is uniform. How does the total number of dots (mass) vary with the radius of the circle?

is provided by a microscopic distance such as a lattice spacing or the mean distance between the constituents of the object. In computer simulations a maximum length usually is provided by the finite system size. The presence of these cut-offs complicates the determinations of the fractal dimension.

Another important characteristic of fractal objects is that they look the same over a range of length scales. This property of self-similarity or scale invariance means that if we take part of a fractal object and magnify it by the same magnification factor in all directions, the magnified picture is indistinguishable from the original.

In Problem 14.1 we compute the fractal dimension of percolation clusters using straightforward Monte Carlo methods. A renormalization group method for estimating the fractal dimension is considered in Problem 14.2. Remember that data extending over several decades is required to obtain convincing evidence for a power law relationship between M and R and to determine accurate estimates for the fractal dimension. Hence, conclusions based on the limited simulations posed in the problems need to be interpreted with caution.

Problem 14.1   The fractal dimension of percolation clusters

a.   Generate a site percolation configuration on a square lattice with $L = 61$ at $p = p_c \approx 0.5927$. Why might it be necessary to generate a number of configurations before a spanning cluster is obtained? Obtain a feel for the ramified nature of the spanning cluster by printing a configuration and marking the positions of the sites in the spanning cluster as in Fig. 14.1. Does the spanning cluster have many dangling ends?

b.   Choose a point on the spanning cluster and count the number of points in the spanning cluster $M(b)$ within a square of area b^2 centered about that point. Then double b and count the number of points within the larger box. Repeat this procedure until you can estimate the b-dependence of the number of points. Can you repeat this procedure indefinitely? Use the b-dependence of $M(b)$ to estimate D according to the definition, $M(b) \sim b^D$ (see (14.1)). Choose another point in the cluster and repeat this procedure. Are your results similar? A better estimate for D can be found by averaging $M(b)$ over several origins in each spanning cluster and averaging over many spanning clusters.

c.   If you have not done Problem 13.9d, compute D by determining the mean size (mass) M of the spanning cluster at $p = p_c$ as a function of the linear dimension L of the lattice. Consider $L = 11, 21, 41$, and 61 and estimate D from a log-log plot of M versus L.

*Problem 14.2   Renormalization group calculation
of the fractal dimension

Compute $<M^2>$, the mean square number of occupied sites in the spanning cluster at $p = p_c$, and the quantity $<M'^2>$, the mean square number of occupied sites in the spanning cluster on the renormalized lattice of linear dimension $L' = L/b$.

Because $<M^2> \sim R^{2D}$ and $<M'^2> \sim (R/b)^{2D}$, we can obtain D from the relation $b^{2D} = <M^2>/<M'^2>$. Choose the length rescaling factor to be $b = 2$ and adopt the same blocking procedure as used in Section 13.5. An average over ten spanning clusters for $L = 16$ and $p = 0.5927$ is sufficient for qualitative results.

In Problems 14.1 and 14.2 we were interested only in the properties of the spanning clusters. For this reason, our algorithm for generating percolation configurations is inefficient because it generates many clusters. There is a more efficient way of generating single percolation clusters due independently to Hammersley, Leath, and Alexandrowicz. This algorithm, commonly known as the Leath algorithm, is equivalent to the following steps (see Fig. 14.3):

1. Occupy a single seed on the lattice. The nearest neighbors (four on the square lattice) of the seed represent the *perimeter* sites.

2. For each perimeter site, generate a random number r in the unit interval. If $r \leq p$, the site is occupied and added to the cluster; otherwise the site is not occupied. In order that sites be unoccupied with probability $1 - p$, these sites are not tested again.

3. For each site that is occupied, determine if there are any new perimeter sites, i.e., untested neighbors. Add the new perimeter sites to the perimeter list.

4. Continue steps 2 and 3 until there are no untested perimeter sites to test for occupancy.

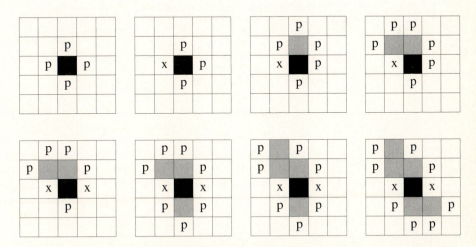

Fig. 14.3   An example of the growth of a percolation cluster. Sites are occupied with probability $p = 0.5927$. Occupied sites are represented by a shaded square, perimeter sites are labeled by 'p,' and tested unoccupied sites are labeled by 'x.' Because the seed site is occupied but not tested, we have represented it differently than the other occupied sites. The perimeter sites are chosen at random.

Program `perc_cluster` implements this algorithm and computes the number of occupied sites within a radius r of the seed particle. The seed site is placed at the center of a square lattice. Two one-dimensional arrays, `perx` and `pery`, store the x and y positions of the perimeter sites. The status of a site is stored in the array `site` with $site(x, y) = 1$ an occupied site, $site(x, y) = 2$ a perimeter site, and $site(x, y) = -1$ a site that has already been tested and not occupied, and $site(x, y) = 0$ an untested and unvisited site. To avoid checking for the boundaries of the lattice, we add an extra row and column at the boundary and set these sites equal to -1.

```
PROGRAM perc_cluster
! cluster generated by Hammersley, Leath, and Alexandrowicz algorithm
DIM xs(10000),ys(10000),status$(-1:2)
LIBRARY "csgraphics"
CALL initial(p,L,status$())
CALL initialize_arrays(xs(),ys())
CALL grow(p,L,N,xs(),ys(),status$())
CALL mass_dist(L,N,xs(),ys())
END

SUB initial(p,L,status$())
    RANDOMIZE
    INPUT prompt "value of L (odd) = ": L
    INPUT prompt "site occupation probability = ": p
    CALL compute_aspect_ratio(L,xwin,ywin)
    SET WINDOW 0,xwin,0,ywin
    SET COLOR "red"
    BOX CIRCLE 0,1,0,1
    FLOOD 0.5,0.5
    BOX KEEP 0,1,0,1 in status$(1)      ! occupied site
    CLEAR
    SET COLOR "blue"
    BOX CIRCLE 0,1,0,1
    FLOOD 0.5,0.5
    BOX KEEP 0,1,0,1 in status$(2)      ! perimeter site
    CLEAR
    SET COLOR "black"
    BOX CIRCLE 0,1,0,1
    FLOOD 0.5,0.5
    BOX KEEP 0,1,0,1 in status$(-1)     ! tested, unoccupied site
    CLEAR
    BOX LINES 0.5,L+0.5,0.5,L+0.5
    FOR y = 1 to L
        FOR x = 1 to L
            PLOT POINTS: x,y
        NEXT x
    NEXT y
    BOX SHOW status$(1) at 1,L+2
    PLOT TEXT, AT 1,L+2: "  occupied site"
    BOX SHOW status$(2) at 30,L+2
```

```
              PLOT TEXT, AT 30,L+2: "  perimeter site"
              BOX SHOW status$(-1) at 60,L+2
              PLOT TEXT, AT 60,L+2: "  tested site"
          END SUB

          SUB initialize_arrays(xs(),ys())
              MAT xs = 0
              MAT ys = 0
          END SUB

          SUB grow(p,L,N,xs(),ys(),status$())
              ! generate single percolation cluster
              DIM perx(25000),pery(25000),site(0:131,0:131)
              DIM nx(4),ny(4)              ! set up direction vectors for lattice
              DATA 1,0,-1,0,0,1,0,-1
              ! set up boundary sites
              FOR i = 1 to L
                  LET site(0,i) = -1
                  LET site(L+1,i) = -1
                  LET site(i,L+1) = -1
                  LET site(i,0) = -1
              NEXT i
              ! seed at center of lattice
              LET xseed = int(L/2) + 1
              LET yseed = xseed
              LET site(xseed,yseed) = 1     ! seed site
              LET xs(1) = xseed
              LET ys(1) = yseed
              LET N = 1                     ! number of sites in the cluster
              BOX SHOW status$(1) at xseed-0.5,yseed-0.5
              FOR i = 1 to 4
                  ! nx,ny direction vectors for new perimeter sites
                  READ nx(i),ny(i)
                  ! perx,pery, positions of perimeter sites of seed
                  LET perx(i) = xseed + nx(i)
                  LET pery(i) = yseed + ny(i)
                  ! perimeter sites labeled by 2
                  LET site(perx(i),pery(i)) = 2  ! site placed on perimeter list
                  BOX SHOW status$(2) at perx(i)-0.5,pery(i)-0.5
              NEXT i
              LET nper = 4                  ! initial number of perimeter sites
              DO
                  ! randomly choose perimeter site
                  LET iper = int(rnd*nper) + 1
                  LET x = perx(iper)      ! coordinate of a perimeter site
                  LET y = pery(iper)
                  ! relabel remaining perimeter sites so that
                  ! last perimeter site in array replaces newly chosen site
                  LET perx(iper) = perx(nper)
                  LET pery(iper) = pery(nper)
```

```
            LET nper = nper - 1
            IF rnd < p then          ! site occupied
                LET site(x,y) = 1
                LET N = N + 1
                LET xs(N) = x         ! save position of occupied site
                LET ys(N) = y
                BOX SHOW status$(1) at x-0.5,y-0.5
                FOR nn = 1 to 4       ! find new perimeter sites
                    LET xnew = x + nx(nn)
                    LET ynew = y + ny(nn)
                    IF site(xnew,ynew) = 0 then
                        LET nper = nper + 1
                        LET perx(nper) = xnew
                        LET pery(nper) = ynew
                        ! place site on perimeter list
                        LET site(xnew,ynew) = 2
                        BOX SHOW status$(2) at xnew-0.5,ynew-0.5
                    END IF
                NEXT nn
            ELSE                     ! rnd >= p so site is not occupied
                LET site(x,y) = -1
                BOX SHOW status$(-1) at x-0.5,y-0.5
            END IF
        LOOP until nper < 1        ! all perimeter sites tested
        BOX LINES 0.5,L+0.5,0.5,L+0.5       ! redraw box
END SUB

SUB mass_dist(L,N,xs(),ys())
    DIM mass(10000)
    PRINT "press any key or click mouse to see data"
    DO
        GET MOUSE xm,ym,s
    LOOP until key input or s <> 0
    FOR i = 1 to N
        LET xcm = xcm + xs(i)       ! compute center of mass
        LET ycm = ycm + ys(i)
    NEXT i
    LET xcm = xcm/N
    LET ycm = ycm/N
    FOR i = 1 to N
        LET dx = xs(i) - xcm
        LET dy = ys(i) - ycm
        LET r = int(sqr(dx*dx + dy*dy))
        ! distance from center of mass
        ! mass(r) = number of sites at distance r from center of mass
        IF r > 1 then LET mass(r) = mass(r) + 1
    NEXT i
    LET rprint = 2
    CLEAR
    PRINT " r "," m "," ln(r) "," ln(m) "
```

```
      FOR r = 2 to L/2
          LET masstotal = masstotal + mass(r)
          IF r = rprint then
              PRINT r, masstotal,log(r),log(masstotal)
              LET rprint = 2*rprint  ! use logarithmic scale for r
          END IF
      NEXT r
  END SUB
```

We use the growth algorithm in Problem 14.3 to generate a spanning cluster at the percolation threshold. The fractal dimension is determined by counting the number of sites M in the cluster within a distance r of the center of mass of the cluster. The center of mass is defined by

$$\mathbf{r}_{cm} = \frac{1}{N} \sum_i \mathbf{r}_i, \tag{14.4}$$

where N is the total number of particles in the cluster. A typical plot of $\ln M$ versus $\ln r$ is shown in Fig. 14.4. Because the cluster cannot grow past the edge of the lattice, we do not include data for $r \approx L$.

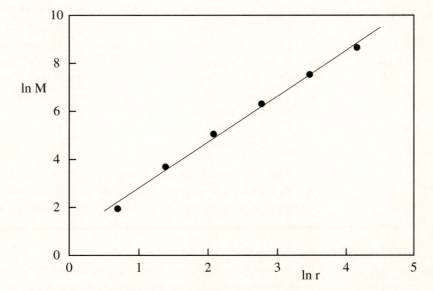

Fig. 14.4 Plot of $\ln M$ versus $\ln r$ for a single spanning percolation cluster generated at $p = 0.5927$ on a $L = 129$ square lattice. The straight line is a linear least squares fit to the data. The slope of this line is 1.91 and is an estimate of the fractal dimension D. The exact value of D for a percolation cluster is $D = 91/48 \approx 1.896$.

Problem 14.3 Single cluster growth and the fractal dimension

a. Explain how the Leath algorithm generates single clusters in a way that is equivalent to the multiple clusters that are generated by visiting all sites. More precisely, the Leath algorithm generates percolation clusters with a distribution of cluster sizes, sn_s. The additional factor of s is due to the fact that each site of the cluster has an equal chance of being the seed of the cluster, and hence the same cluster can be generated in s ways. See Project 14.3 for a discussion of the scaling form of n_s.

b. Use `Program perc_cluster` to grow percolation clusters using the Leath algorithm. Consider a spanning cluster to be one that connects the top and bottom rows of the lattice. Can you grow a spanning cluster for $p = 0.4$ or does the growth usually stop after a few sites are occupied? Choose $L \geq 31$.

c. Choose $p = 0.5927$ and $L \geq 31$ and generate several pictures of spanning clusters. Do all your trials generate a spanning cluster? Explain. Determine the number of occupied sites $M(r)$ within a distance r of the center of mass of the cluster. Determine M for several values of r and average $M(r)$ over at least ten spanning clusters. Estimate D from the log-log plot of M versus r (see Fig. 14.4). If time permits, generate percolation clusters on larger lattices.

d. Grow as large a spanning cluster as you can and look at it on different length scales. One way to do so is to divide the screen into four windows, each of which magnifies a part of the cluster shown in the previous window. Does the part of the cluster shown in each window look approximately self-similar?

e. Generate clusters at $p = 0.65$, a value of p greater than p_c, for $L = 61$. Make a log-log plot of $M(r)$ versus r. Is the slope approximately equal to the value of D found in part (b)? Does the slope increase or decrease for larger r? Repeat for $p = 0.80$. Is a spanning cluster generated at $p > p_c$ a fractal?

f. The fractal dimension of percolation clusters is not an independent exponent, but satisfies the scaling law

$$D = d - \beta/\nu, \tag{14.5}$$

where β and ν are defined in Table 13.1. The relation (14.5) can be understood by a finite-size scaling argument which we now summarize. The number of sites in the spanning cluster on a lattice of linear dimension L is given by

$$M(L) \sim P_\infty(L)L^d, \tag{14.6}$$

where P_∞ is the probability that an occupied site belongs to the spanning cluster and L^d is the total number of sites in the lattice. In the limit of an infinite lattice and p near p_c, we know that $P_\infty(p) \sim (p - p_c)^\beta$ and $\xi(p) \sim (p - p_c)^{-\nu}$ independent of L. Hence for $L \sim \xi$, we have that $P_\infty(L) \sim L^{-\beta/\nu}$ (see (13.12)), and we can write

$$M(L) \sim L^{-\beta/\nu}L^d \sim L^D. \tag{14.7}$$

The relation (14.5) follows. Use the exact values of β and ν from Table 13.1 to find the exact value of D for $d = 2$. Is your estimate for D consistent with this value?

* **g.**   Estimate the fractal dimension for percolation clusters on a simple cubic lattice. Take $p_c = 0.3117$.

14.2   REGULAR FRACTALS

As we have seen, one characteristic of random fractal objects is that they look the same on a range of length scales. To gain a better understanding of the meaning of self-similarity, consider the following example of a *regular* fractal, a mathematical object that is self-similar on *all* length scales. Begin with a line one unit long (see Fig. 14.5a). Remove the middle third of the line and replace it by two lines of length 1/3 each so that the curve has a triangular bump in it and the total length of the curve is 4/3 (see Fig. 14.5b). In the next stage, each of the segments of length 1/3 is divided into lines of length 1/9 and the procedure is repeated as shown in Fig. 14.5c. What is the length of the curve shown in Fig. 14.5c?

The three stages shown in Fig. 14.5 can be extended an infinite number of times. The resulting curve is infinitely long, containing an infinite number of infinitesimally small segments. Such a curve is known as the triadic Koch curve. A True BASIC program that uses a recursive procedure (see Section 6.3) to draw this curve is given

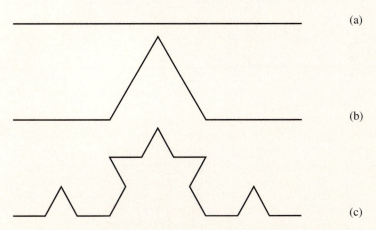

Fig. 14.5   The first three stages (a)–(c) of the generation of a self-similar Koch curve. At each stage the displacement of the middle third of each segment is in the direction that increases the area under the curve. The curves were generated using *Program koch*. The Koch curve is an example of a continuous curve for which there is no tangent defined at any of its points. The Koch curve is self-similar on each length scale.

in the following. Note that SUB draw calls itself. Use *Program koch* to generate the curves shown in Fig. 14.5.

```
PROGRAM Koch
! generate triadic Koch curve using recursion
CALL initial(x1,y1,x2,y2,n)
! draw Koch curve for different number of iterations
DO
    LET k = 0
    CALL draw(x1,y1,x2,y2,n)
    DO                                  ! pause until any key is hit
        GET KEY k
    LOOP UNTIL k <> 0
    LET n = n + 1                       ! number of stages of generation
    CLEAR
LOOP UNTIL k = ord("s")
END

SUB initial(x1,y1,x2,y2,n)
    LET n = 0                           ! number of iterations
    LET x1 = 0                          ! coordinates at left end of line
    LET y1 = 0                          ! arbitrary units
    LET x2 = 10                         ! coordinates at right end of line
    LET y2 = 0
    SET WINDOW x1-1,x2+1,-1,6
    ASK MAX COLOR mc
    IF mc > 2 then SET COLOR "blue"
END SUB

SUB draw(x1,y1,x2,y2,n)
    IF n > 0 then
        LET dx = (x2 - x1)/3
        LET dy = (y2 - y1)/3
        LET x1n = x1 + dx
        LET y1n = y1 + dy
        LET x2n = x1 + 2*dx
        LET y2n = y1 + 2*dy
        ! rotate line segment (dx,dy) by 60 degs and add to (x1n,y1n)
        LET xmid = 0.5*dx - 0.866*dy + x1n
        LET ymid = 0.5*dy + 0.866*dx + y1n
        CALL draw(x1,y1,x1n,y1n,n-1)
        CALL draw(x1n,y1n,xmid,ymid,n-1)
        CALL draw(xmid,ymid,x2n,y2n,n-1)
        CALL draw(x2n,y2n,x2,y2,n-1)
    ELSE
        PLOT LINES: x1,y1;x2,y2
    END IF
END SUB
```

d = 1   d = 2

Fig. 14.6   Examples of one-dimensional and
two-dimensional objects.

How can we determine the fractal dimension of the Koch and similar mathematical objects? In Section 14.5 we will see that there are several generalizations of the
Euclidean dimension that lead naturally to a definition of the fractal dimension. Here
we consider a definition based on counting boxes. Consider a one-dimensional curve
of unit length that has been divided into N equal segments of length ℓ so that $N = 1/\ell$
(see Fig. 14.6). As ℓ is decreased, N increases linearly—the expected result for a one-
dimensional curve. Similarly if we divide a two-dimensional square of unit area into
N equal subsquares of length ℓ, we have $N = 1/\ell^2$, the expected result for a two-
dimensional object (see Fig. 14.6). In general, we can write that $N = 1/\ell^D$, where D
is the fractal dimension of the object. If we take the logarithm of both sides of this
relation, we can express the fractal dimension as

$$D = \frac{\log N}{\log(1/\ell)}. \qquad \text{(box dimension)} \qquad (14.8)$$

Now let us apply these ideas to the Koch curve. We found that each time the length
ℓ of our measuring unit is reduced by a factor of 3, the number of segments is increased
by a factor of 4. Hence, we have $N = 4$ and $\ell = 1/3$, and the fractal dimension of the
triadic Koch curve is given by

$$D = \frac{\log 4}{\log 3} \approx 1.2619. \qquad \text{(triadic Koch curve)} \qquad (14.9)$$

From (14.9) we see that the Koch curve has a dimension between that of a line and a
plane. Is this statement consistent with your visual interpretation of the degree to which
the triadic Koch curve fills space?

Problem 14.4   The recursive generation of regular fractals

a.   The concept of recursive programming as illustrated in *Program koch* is probably one of the most difficult programming concepts you will encounter. Explain the nature of recursion and the way it is implemented in *Program koch*.

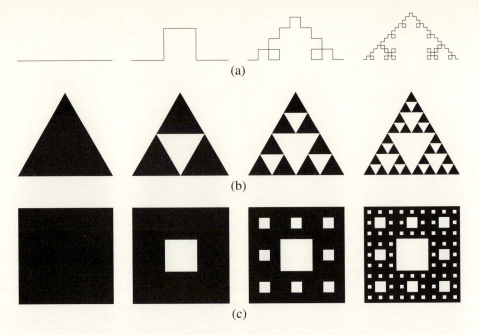

Fig. 14.7   (a) The first few iterations of the quadric Koch curve; (b) The first few
iterations of the Sierpiński gasket; (c) The first few iterations of the Sierpiński
carpet.

b. Regular fractals can be generated from a pattern that is used in a self-
replicating manner. Write a program to generate the quadric Koch curve shown
in Fig. 14.7a. What is its fractal dimension?

c. What is the fractal dimension of the Sierpiński gasket shown in Fig. 14.7b?
Write a program that generates the next several iterations.

d. What is the fractal dimension of the Sierpiński carpet shown in Fig. 14.7c?
How does the fractal dimension of the Sierpiński carpet compare to the fractal
dimension of a percolation cluster? Are the two fractals visually similar?

14.3   FRACTAL GROWTH PROCESSES

Many systems occurring in nature exhibit fractal geometry. Fractals have been used to
describe the irregular shapes of such varied objects as coastlines, clouds, coral reefs,
and the human lung. Why are fractal structures so common? How do fractal structures
form? In this section we discuss several simple "growth" models that generate structures which show a remarkable similarity to forms observed in nature. The first two
models are already familiar to us and exemplify the flexibility and general utility of
kinetic growth models.

Epidemic model. In the context of the spread of disease, we usually want to know the conditions for an epidemic. A simple lattice model of the spread of a disease can be formulated as follows. Suppose that an occupied site corresponds to an infected person. Initially there is a single infected site and the four nearest neighbor perimeter sites (on the square lattice) are susceptible. At the next time step, we visit the four susceptible sites and occupy (infect) each site with probability p. If a susceptible site is not occupied, we say that the site is immune and we do not test it again. We then find the new susceptible sites and continue until either the disease is controlled or reaches the boundary of the lattice. Convince yourself that this growth model of a disease generates a cluster of infected sites that is identical to a percolation cluster at probability p. The only difference is that we have introduced a discrete time step into the model. Some of the properties of this model are explored in Problem 14.5.

Problem 14.5 A simple epidemic model

a. Explain why the simple epidemic model discussed in the text generates the same clusters as the Leath algorithm if the probability that a susceptible site becomes infected is p. What is the minimum value of p necessary for an epidemic to occur? Recall that in one time step, all susceptible sites are visited *simultaneously* and infected with probability p. Determine how N, the number of infected sites, depends on the time t (the number of time steps) for various values of p. A straightforward way to proceed is to modify `Program perc_cluster` so that all perimeter sites are visited and occupied with probability p before new perimeter sites are found. In Chapter 15 we will learn that this model is an example of a cellular automaton.

b. The susceptible (or growth) sites S are the only sites from which the disease can spread. Verify that for p near p_c^+, S increases as $f(p)N^{\delta_s}$ with $f(p) \propto (p - p_c)^y$. Estimate the numerical values of the exponents δ_s and y. How does S depend on the time? Does δ_s have a different value at p_c for clusters that grow indefinitely? Choose $L \geq 61$ and average over at least 10 realizations.

c. A similar growth exponent can be defined for $p < p_c$. In this case the number of susceptible sites does not increase without bound, and S usually first increases and then goes to zero. Determine the maximum number $S_{\max}$ of susceptible sites and show that $S_{\max} \propto (p_c - p)^{-x}$ near p_c. Estimate the numerical value of the exponent x.

Eden model. An even simpler example of a growth model was proposed by Eden in 1958 to simulate the growth of cell colonies. Although we will find that the resultant mass distribution is not a fractal, the description of the Eden growth algorithm illustrates the general nature of the fractal growth models we discuss.

The algorithm can be summarized as follows. Place a seed site at the origin, e.g., the center of the lattice. The unoccupied nearest neighbors of the occupied sites are denoted as *growth* or perimeter sites. In the simplest version of the model, a growth site is chosen at random and occupied. The newly occupied site is removed from the list of growth sites and the new growth sites are added to the list. This growth process is

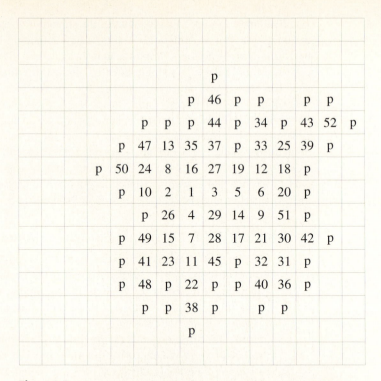

Fig. 14.8 An example of a cluster grown on the square lattice according to the Eden model. The numbers on the sites denote the order in which these sites were occupied and the growth sites are denoted by the letter p.

repeated many times until a large cluster of occupied sites are formed (see Fig. 14.8). The basic difference between this model and the previous one is that all tested sites are occupied. In other words, no sites are ever "immune." Some of the properties of Eden clusters are investigated in Problem 14.6.

Problem 14.6   Eden model

a. Modify *Program* `perc_cluster` so that clusters are generated on a square lattice according to the Eden model. A straightforward modification is to occupy perimeter sites with probability $p = 1$ until the cluster reaches the edge of the lattice. What would happen if we were to occupy perimeter sites indefinitely? Follow the procedure of Problem 14.3 and determine the number of occupied sites $M(r)$ within a distance r of the seed site. Assume that $M(r) \sim r^D$ for sufficiently large r, and estimate D from the slope of a log-log plot of M versus r. A typical log-log plot is shown in Fig. 14.9 for $L = 61$. Can you conclude from your data that Eden clusters are compact?

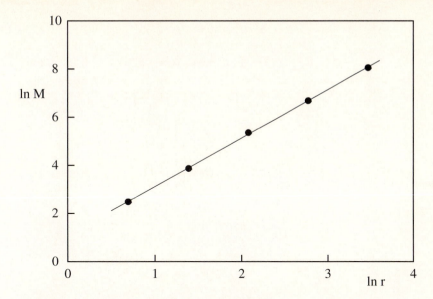

Fig. 14.9 Plot of ln M versus ln r for a single Eden cluster generated on a
$L = 61$ square lattice. A least squares fit to the data from $r = 2$ to
$r = 32$ yields a slope of approximately 2.01.

    b.   Modify your program so that only the perimeter or growth sites are shown. Where are the majority of the perimeter sites relative to the center of the cluster? Grow as big a cluster as your time and patience permits.

    Invasion percolation. A dynamical process known as *invasion percolation* can be used to model the shape of the oil-water interface which occurs when water is forced into a porous medium containing oil. The idea is to use the water to recover as much oil as possible. In this process a water cluster grows into the oil through the path of least resistance. Consider a lattice of size $L \times 2L$, with the water (the invader) initially occupying the left edge (see Fig. 14.10). The resistance to the invader is given by uniformly distributed random numbers between 0 and 1 which are assigned to each site in the lattice and held fixed throughout the invasion. Sites that are nearest neighbors of the invader sites are the perimeter sites. At each time step, the perimeter site with the lowest random number is occupied by the invader and the oil (the defender) is displaced. The invading cluster grows until a path forms which connects the left and right edges of the lattice. Note that after this path forms, there is no need for the water to occupy any additional sites. To minimize boundary effects, periodic boundary conditions are used for the top and bottom edges and all quantities are measured only over the central $L \times L$ region of the lattice.

    Program invasion implements the invasion percolation algorithm. The two-dimensional array element site(i,j) initially stores a random number for the site at (i,j). If the site at (i,j) is occupied, then site(i,j) is increased by 1. If the site

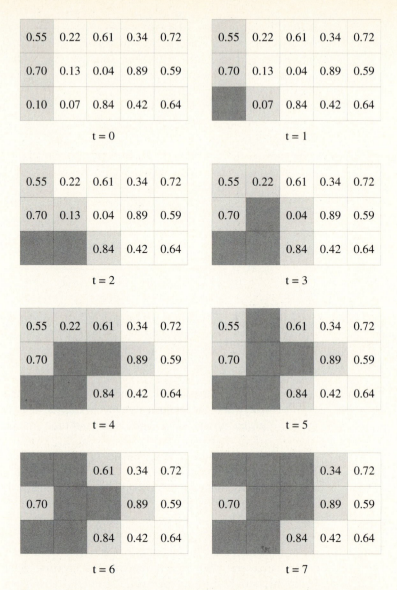

Fig. 14.10   Example of a cluster formed by invasion percolation. The lattice at $t = 0$ shows the random numbers that have been assigned to the sites. The darkly shaded sites are occupied by the invader that occupies the perimeter site (lightly shaded) with the smallest random number.

at (i,j) is a perimeter site, then site(i,j) is increased by 2, and is inserted into its proper ordered position in the perimeter lists perx and pery. The perimeter lists are ordered with the site with the largest random number at the beginning of the list.

Two searching routines are provided in *Program invasion* for determining the position of a new perimeter site in the perimeter lists. In a *linear search* we go through the list in order until the random number associated with the new perimeter site is between two random numbers in the list. In a *binary search* we divide the list in two, and determine the half in which the new random number belongs. Then we divide this half into half again and so on until the correct position is found. A comparison of the linear and binary search methods is investigated in Problem 14.7d. The binary search is the default method used in *Program invasion*.

The main quantities of interest are the fraction of sites occupied by the invader, and the probability $P(r)\, dr$ that a site with a random number between r and $r + dr$ is occupied. The properties of the invasion percolation model are explored in Problem 14.7.

```
PROGRAM invasion
! generate invasion percolation cluster
DIM site(0 to 200,0 to 100),perx(5000),pery(5000)
LIBRARY "csgraphics"
CALL initial(Lx,Ly,water$,#1)
CALL assign(site(,),perx(),pery(),nper,Lx,Ly,water$,#1)
CALL invade(site(,),perx(),pery(),nper,Lx,Ly,water$,#1)
CALL average(site(,),Lx,Ly,#1)
END

SUB initial(Lx,Ly,water$,#1)
    RANDOMIZE
    INPUT prompt "length in y direction = ": Ly
    LET Lx = 2*Ly
    OPEN #1: screen 0.01,0.65,0.2,0.99
    CALL compute_aspect_ratio(Lx,xwin,ywin)
    SET WINDOW 0,xwin,0,ywin
    SET COLOR "blue"
    BOX AREA 0,1,0,1
    FLOOD 0.5,0.5
    BOX KEEP 0,1,0,1 in water$
    CLEAR
    SET COLOR "black"
    BOX LINES 0.5,Lx+0.5,0.5,Ly+0.5
    FOR y = 1 to Ly
        FOR x = 1 to Lx
            PLOT POINTS: x,y
        NEXT x
    NEXT y
END SUB
```

```
SUB assign(site(,),perx(),pery(),nper,Lx,Ly,water$,#1)
    FOR y = 1 to Ly
        LET site(1,y) = 1            ! occupy first column
        BOX SHOW water$ at 0.5,y-0.5
    NEXT y
    ! assign random numbers to remaining sites
    FOR y = 1 to Ly
        FOR x = 2 to Lx
            LET site(x,y) = rnd
        NEXT x
    NEXT y
    ! sites in second column are initial perimeter sites
    ! site(x,y) greater than 2 if perimeter site
    LET x = 2
    LET nper = 0
    FOR y = 1 to Ly
        LET site(x,y) = 2 + site(x,y)
        LET nper = nper + 1          ! number of perimeter sites
        ! sort perimeter sites
        ! order perimeter list
        CALL insert(site(,),perx(),pery(),nper,x,y)
    NEXT y
END SUB

SUB insert(site(,),perx(),pery(),nper,x,y)
    ! call linear or binary sort
    CALL binary_search(site(,),perx(),pery(),nper,x,y,ninsert)
    ! move sites with smaller random numbers to next higher array index
    FOR ilist = nper to ninsert + 1 step - 1
        LET perx(ilist) = perx(ilist-1)
        LET pery(ilist) = pery(ilist-1)
    NEXT ilist
    LET perx(ninsert) = x            ! new site inserted in list
    LET pery(ninsert) = y
END SUB

SUB binary_search(site(,),perx(),pery(),nper,x,y,ninsert)
    ! divide list in half and determine half in which random # belongs
    ! continue this division until position of number is determined
    LET nfirst = 1                   ! beginning of list
    LET nlast = nper - 1             ! end of list
    IF nlast < 1 then LET nlast = 1
    LET nmid = int((nfirst + nlast)/2)       ! middle of list
    ! determine which half of list new number is located
    DO
        IF nlast - nfirst <= 1 then      ! exact position equal to nlast
            LET ninsert = nlast
            EXIT SUB
        END IF
```

```
                    LET xmid = perx(nmid)
                    LET ymid = pery(nmid)
                    IF site(x,y) > site(xmid,ymid) then
                        LET nlast = nmid          ! search upper half
                    ELSE
                        LET nfirst = nmid         ! search lower half
                    END IF
                    LET nmid = int((nfirst + nlast)/2)
                LOOP
            END SUB

            SUB linear_search(site(,),perx(),pery(),nper,x,y,ninsert)
                IF nper = 1 then
                    LET ninsert = 1
                ELSE
                    FOR iper = 1 to nper - 1
                        LET xperim = perx(iper)
                        LET yperim = pery(iper)
                        IF site(x,y) > site(xperim,yperim) then
                            LET ninsert = iper   ! insert new site
                            EXIT SUB
                        END IF
                    NEXT iper
                END IF
                LET ninsert = nper
            END SUB

            SUB invade(site(,),perx(),pery(),nper,Lx,Ly,water$,#1)
                ! nx and ny are components of vectors pointing to nearest neighbors
                DIM nx(4),ny(4)
                DATA 1,0,-1,0,0,1,0,-1
                FOR i = 1 to 4
                    READ nx(i),ny(i)
                NEXT i
                DO
                    LET x = perx(nper)
                    LET y = pery(nper)
                    LET nper = nper - 1
                    ! mark site occupied and no longer perimeter site
                    LET site(x,y) = site(x,y) - 1
                    BOX SHOW water$ at x-0.5,y-0.5
                    FOR i = 1 to 4                ! find new perimeter sites
                        LET xnew = x + nx(i)
                        LET ynew = y + ny(i)
                        IF ynew > Ly then          ! periodic boundary conditions in y
                            LET ynew = 1
                        ELSE IF ynew < 1 then
                            LET ynew = Ly
                        END IF
```

```
              IF site(xnew,ynew) < 1 then        ! new perimeter site
                  LET site(xnew,ynew) = site(xnew,ynew) + 2
                  LET nper = nper + 1
                  CALL insert(site(,),perx(),pery(),nper,xnew,ynew)
              END IF
          NEXT i
      LOOP until x >= Lx                    ! until cluster reaches right boundary
  END SUB

  SUB average(site(,),Lx,Ly,#1)
      ! compute probability density P(r)
      DIM P(0 to 20),nr(0 to 20)
      LET Lmin = Lx/3
      LET Lmax = 2*Lmin
      LET n = (Lmax - Lmin + 1)*Ly  ! # sites in middle half of lattice
      LET dr = 0.05
      LET nbin = 1/dr
      FOR x = Lmin to Lmax
          FOR y = 1 to Ly
              LET ibin = nbin*(mod(site(x,y),1))
              LET nr(ibin) = nr(ibin) + 1
              IF site(x,y) >= 1 and site(x,y) < 2 then
                  LET occupied = occupied + 1  ! total # of occupied sites
                  LET P(ibin) = P(ibin) + 1
              END IF
          NEXT y
      NEXT x
      WINDOW #1
      PRINT "# occupied sites ="; occupied
      OPEN #2: screen 0.66,1.0,0.01,0.99
      PRINT " r","  P(r)"
      PRINT
      FOR ibin = 0 to nbin
          LET r = dr*ibin
          IF nr(ibin) > 0 then PRINT r,P(ibin)/nr(ibin)
      NEXT ibin
      PRINT
  END SUB
```

Problem 14.7   Invasion percolation

a. Use *Program invasion* to generate an invasion percolation cluster on a 20×40 lattice and describe the qualitative nature of the cluster.

b. Modify *Program invasion* so that $M(L)$, the number of sites occupied by the invader in the central $L \times L$ region of the $L \times 2L$ lattice at the time that the invader first reaches the right edge, is averaged over at least twenty trials. Assume that $M(L) \sim L^D$ and estimate D from a plot of $\ln M$ versus $\ln L$. Compare your estimate for D with the fractal dimension of ordinary percolation. (The published results for $M(L)$ by Wilkinson and Willemsen are for 2000 realizations each for L in the range 20 to 100.)

c. Determine the probability $P(r) dr$ that a site with a random number between r and $r + dr$ is occupied. It is sufficient to choose $dr = 0.05$. Plot $P(r)$ versus r for $L = 20$ and also for larger values of L up to $L = 50$. Can you define a critical value of r near which $P(r)$ changes rapidly? How does this critical value of r compare to the value of p_c for ordinary site percolation on the square lattice? On the basis of your numerical estimate for the exponent D found in part (b) and the qualitative behavior of $P(r)$, make an hypothesis about the relation between the nature of the geometrical properties of the invasion percolation cluster and the spanning percolation cluster at $p = p_c$.

d. Explain the nature of the two searching subroutines given in *Program inva-sion*. Which method yields the fastest results on a 30×60 lattice? Verify that the CPU time for a linear and binary search is proportional to n and $\log n$ respectively, where n is the number of items in the list to be searched. Hence, for sufficiently large n, a binary search usually is preferred.

* e. Modify your program so that the invasion percolation clusters are grown from a seed at the origin. Grow a cluster until it either occupies a given fraction of the lattice or it reaches a boundary of the lattice. Estimate the fractal dimension as you did for the spanning percolation clusters in Problem 14.3 and compare your two estimates. On the basis of this estimate and your results from part (b) and (c), can you conclude that the spanning cluster is a fractal? Note that this process of occupying the minimum number of sites to obtain a spanning cluster is an example of a self-organized critical phenomenon (see Chapter 15).

Diffusion in disordered media. In Chapters 7 and 12 we considered random walks on perfect lattices and on simple continuum systems. We found that the mean-square displacement of a random walker, $<R^2(t)>$, is proportional to the time t for sufficiently large t. (For a simple random walk this relation holds for all t.) Now let us suppose that the random walker is restricted to a disordered lattice, e.g., the occupied sites of a percolation cluster. What is the asymptotic t-dependence of $<R^2(t)>$ in this case? This simple model of a random walk on a percolation cluster is known as the "ant in the labyrinth" problem.

There are at least two reasons for our interest in random walks on disordered lattices. Just as a random walk on a lattice is a simple model of diffusion, a random walk on a disordered lattice is a simple example of the general problem of diffusion and transport in disordered media. Because most materials of interest are noncrystalline and disordered, there are many physical phenomena that can be related to the motion of an ant in the labyrinth. Another reason for the interest in diffusion in disordered media is that the diffusion coefficient is proportional to the electrical conductivity of the medium. This relation between the conductivity and the diffusion coefficient is known as the Einstein relation (cf. Reif). We can understand this relation as follows. Consider for example, a system of electrons. Classically, we can follow the individual motion of

the electrons and determine their mean square displacement. In the absence of external forces we can measure the self-diffusion coefficient D. In the presence of a "small"
electric field, we can measure the electron's mean velocity in the direction of the field
and deduce the electron's *mobility* μ, the ratio of the mean velocity to the applied force.
Einstein's contribution was to show that μ is proportional to D, that is, the linear response of the system is related to an equilibrium quantity. Because the mean velocity of
the electrons is proportional to the electron current and the applied force is proportional
to the voltage, the mobility and the electrical conductivity are proportional. Hence, we
conclude that the conductivity is proportional to the self-diffusion coefficient.

In the usual formulation of the ant in the labyrinth problem we place a walker
(the ant) at random on one of the occupied sites of a percolation cluster generated with
probability p. At each time step, the ant tosses a coin with four possible outcomes (on a
square lattice). If the outcome corresponds to a step to an occupied site, the ant moves;
otherwise it remains in its present position. Either way, the time t is increased by one
unit. The main quantity of interest is $R^2(t)$, the square of the distance between the ant's
position at $t = 0$ and its position at time t. We can generate many walks with different
initial positions on the same cluster as well as over many percolation clusters to obtain
the ant's mean square displacement $<R^2(t)>$. How does $<R^2(t)>$ depend on p and t?
How do the laws of diffusion change on a fractal lattice (e.g., the percolation cluster at
$p = p_c$)? We consider these questions in Problem 14.8.

Problem 14.8   The ant in the labyrinth

a.   For $p = 1$, the ants walk on a perfect lattice, and hence, $<R^2(t)> \propto t$. Suppose that an ant does a random walk on a two-dimensional percolation cluster
with $p > p_c$. Assume that $<R^2(t)> \sim 4D_s(p)t$ for $p > p_c$. We have denoted
the diffusion coefficient by D_s to remind ourselves that we are considering
random walks on spanning clusters only and are not considering walks on
the finite clusters that also exist for $p > p_c$. Generate a percolation cluster at
$p = 0.7$ using the growth algorithm considered in Problem 14.3. Choose the
initial position of the ant to be the seed site and modify your program to observe the motion of the ant on the screen. Where does the ant spend much
of its time? If the ant diffuses, what can you say qualitatively about the ratio
$D_s(p)/D(p = 1)$?

b.   Compute $<R^2(t)>$ for $p = 0.4$ and confirm that for $p < p_c$, the clusters are
finite, $<R^2(t)>$ is bounded, and diffusion is impossible.

c.   As in part (a) compute the mean square displacement for $p = 1.0, 0.8, 0.7$,
0.65, and 0.62 with $L = 61$. If time permits, average over several clusters.
Make a log-log plot of $<R^2(t)>$ versus t. What is the qualitative t-dependence
of $<R^2(t)>$ for relatively short times? Decide whether $<R^2(t)>$ is proportional to t for longer times. (Remember that the maximum value of $<R^2>$
is bounded by the finite size of the lattice.) If $<R^2(t)> \sim t$, estimate $D_s(p)$.
Plot the ratio $D_s(p)/D(p = 1)$ as a function of p and discuss its qualitative
behavior.

d.   Because there is no diffusion for $p < p_c$, we might expect that D_s vanishes as $p \to p_c$, that is, $D_s(p) \sim (p - p_c)^{\mu_s}$ for $p \geq p_c$. Extend your calculations of part (c) to larger L and more values of p near p_c and estimate the dynamical exponent μ_s.

e.   At $p = p_c$, we might expect a different type of t-dependence of $<R^2(t)>$ to be observed, e.g., $<R^2(t)> \sim t^{2/z}$ for large t. Do you expect the exponent z to be greater or less than two? Do a Monte Carlo simulation of $<R^2(t)>$ at $p = p_c$ and estimate z. Choose $L \geq 61$ and average over several spanning clusters.

f.   The chicken wire measurements by Watson and Leath (see Section 13.1) found that the dc electrical conductivity σ vanishes near the percolation threshold as $\sigma \sim (p - p_c)^{\mu}$, with $\mu \approx 1.38 \pm 0.12$. More precise estimates give $\mu \approx 1.30$. The difficulty of doing a direct Monte Carlo calculation of σ was considered in Project 13.4. From the Einstein relation we know that the electrical conductivity and the self-diffusion coefficient behave in the same way. However, we measured the self-diffusion coefficient D_s by always placing the ant on a spanning cluster rather than on *any* cluster. In contrast, the conductivity is measured for the entire system including all finite clusters. Hence, the self-diffusion coefficient D that enters into the Einstein relation should be determined by placing the ant at random anywhere on the lattice, including sites that belong to the spanning cluster and sites that belong to the many finite clusters. Because only those ants that start on the spanning cluster can contribute to D, D is related to D_s by $D = P_{\infty} D_s$, where P_{∞} is the probability that the ant would land on a spanning cluster. Since P_{∞} scales as $P_{\infty} \sim (p - p_c)^{\beta}$, we have that $(p - p_c)^{\mu} \sim (p - p_c)^{\beta}(p - p_c)^{\mu_s}$ or $\mu_s = \mu - \beta$. Use your result for μ_s found in part (d) and the exact result $\beta = 5/36$ (see Table 13.1) to estimate μ and compare your result to the critical exponent for the dc electrical conductivity given above.

*g.   We have found that $D_s(p) \sim (p - p_c)^{\mu_s}$ for $p > p_c$ and $<R^2(t)> \sim t^{2/z}$ for $p = p_c$. We now give a simple scaling argument to find a relation between z and μ_s. For $p > p_c$, we know that $<R^2(t)> \sim (p - p_c)^{\mu_s} t$ in the limit of $t \gg 1$ such that the root mean square displacement is much larger than the connectedness length ξ. We also know that $<R^2(t)> \sim t^{2/z}$ for shorter time scales that satisfy the condition $<R^2(t)> \ll \xi^2$. We expect that the crossover between the two dependencies on t occurs when $<R^2> \sim \xi^2$ or when $t \sim \xi^z$. Hence we have $\xi^2 \sim (p - p_c)^{\mu_s} \xi^z$, or since $\xi \sim (p - p_c)^{-\nu}$, we have $(p - p_c)^{\nu(z-2)} \sim (p - p_c)^{\mu_s}$. If we equate powers of $(p - p_c)$, we have

$$z = 2 + \frac{\mu_s}{\nu} = 2 + \frac{\mu - \beta}{\nu}. \tag{14.10}$$

Is it easier to determine μ_s or z accurately from a Monte Carlo simulation on a finite lattice? That is, if our real interest is estimating the best value of the critical exponent μ for the conductivity, should we determine the conductivity directly or should we measure the self-diffusion coefficient at $p = p_c$ or at $p > p_c$? What is your best estimate of the conductivity exponent μ?

*h.   A better method for treating random walks on a random lattice is to use an exact enumeration approach. The essence of the exact enumeration method is that $W_{t+1}(i)$, the probability that the ant is at site i at time $t + 1$, is determined solely by the probabilities of the ant being at the neighbors of site i at time t. Store the positions of the occupied sites in an array and introduce two arrays corresponding to $W_{t+1}(i)$ and $W_t(i)$ for all sites i in the cluster. Use the probabilities $W_t(i)$ to obtain $W_{t+1}(i)$ (see Fig. 14.11). Spatial averages such as the mean square displacement can be calculated from the probability distribution function at different times. Details of the method and the results are discussed in Majid et al. These workers considered walks of 5000 steps on clusters with $\sim 10^3$ sites and averaged their results over 1000 different clusters.

Diffusion-limited aggregation (DLA). Many objects in nature grow by the random addition of subunits. Examples include snow flakes, lightning, crack formation along a geological fault, and the growth of bacterial colonies. Although it might seem unlikely that such phenomena have much in common, the behavior observed in many models that have been developed in recent years gives us clues that these and many other natural phenomena can be understood in terms of a few unifying principles. One

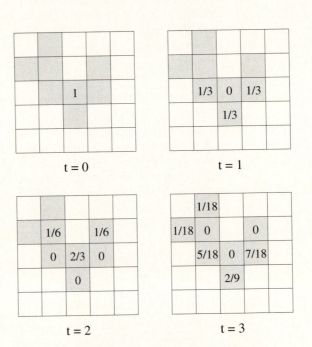

Fig. 14.11   The time evolution of the probability distribution function $W_t(i)$ for three successive time steps.

model that has provided much insight is known as *diffusion limited aggregation* or DLA. The model provides an example of how random motion can give rise to beautiful self-similar clusters.

The first step is to occupy a site with a seed particle. Next, a particle is released from the perimeter of a large circle whose center coincides with the seed. The particle undergoes a random walk, i.e., diffuses, until it reaches a perimeter site of the seed and sticks. Then another random walker is released and allowed to walk until it reaches a perimeter site of one of the two particles in the cluster and sticks. The process is repeated many times (typically on the order of several thousand to several million) until a large cluster is formed. A typical DLA cluster is shown in Fig. 14.12. Some of the properties of DLA clusters are explored in Problem 14.9.

Problem 14.9   Diffusion limited aggregation

a. Write a program to generate diffusion limited aggregation clusters on a square lattice. Let each walker begin at a random site on a circle of radius $r = R_{max} + 2$, where R_{max} is the maximum distance of any cluster particle from the origin. To save computer time, assume that a walker that reaches a distance $2R_{max}$ from the seed site is removed and a new walker is placed at random on the

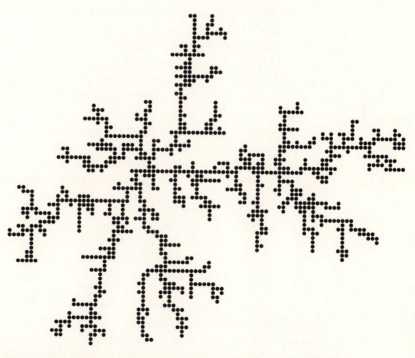

Fig. 14.12   An example of a DLA cluster of 1000 particles on a square lattice.

circle of radius $r = R_{max} + 2$. Choose a lattice of linear dimension $L \geq 31$. Color code the cluster sites according to their time of arrival, e.g., choose the first 100 sites to be blue, the next 100 sites to be yellow, etc. Which parts of the cluster grow faster? If the clusters appear to be fractals, make a visual estimate of the fractal dimension. (Experts can make a visual estimate of D to within a few percent!)

b. At $t = 0$ the four perimeter (growth) sites on the square lattice each have a probability $p_i = 1/4$ of growing, that is, of becoming part of the cluster. At $t = 1$, the cluster has mass two and six perimeter sites. Identify the perimeter sites and convince yourself that their growth probabilities are not uniform. Do a Monte Carlo simulation and verify that two perimeter sites have $p_i = 2/9$ and the other four have $p_i = 5/36$. We discuss a more direct way of determining the growth probabilities in Problem 14.10.

c. It is likely that your program generates DLA clusters inefficiently, because most of the CPU time is spent while the random walker is wandering far from the perimeter sites of the cluster. There are several ways of overcoming this problem. One way is to let the walker take bigger steps the further the walker is from the cluster. For example, if the random walker is at a distance $R > R_{max}$, a step of length greater than or equal to $R - R_{max} - 1$ may be permitted if this distance is greater than one lattice unit. If the walker is very close to the cluster, the step length is one lattice unit. Other possible modifications are discussed by Meakin (see references). Modify your program (or see *Program dla* listed below) and estimate the fractal dimension of diffusion limited clusters generated on a square lattice.

* d. Modify your program so that DLA clusters are generated on a triangular lattice. Do the clusters have the same visual appearance as on the square lattice? Estimate the fractal dimension and compare your estimate to your result for the square lattice.

* e. In Chapter 13 we found that the exponents describing the percolation transition are independent of the symmetry of the lattice, e.g., the exponents for the square and triangular lattices are the same. We might expect that the fractal dimension of DLA clusters would also show such universal behavior. However, the presence of a lattice introduces a small anisotropy that becomes apparent only when very large clusters with on the order of 10^6 sites are grown. We again are reminded of the difficulty of extrapolating from finite L to infinite L. The best estimates of D for the square and triangular lattices are $D \approx 1.5$ and $D \approx 1.7$ respectively. We consider the growth of diffusion-limited aggregation clusters in a continuum in Project 14.1.

The following program provides a reasonably efficient simulation of DLA. Walkers begin just outside a circle of radius R0 enclosing the existing cluster and centered at the seed site $(0, 0)$. If the walker moves away from the cluster, the step size for the random walker increases. If the walker wonders too far away (further than twice R0), then the walk is started over.

```
PROGRAM dla
DIM site(-100 to 100,-100 to 100)
LIBRARY "csgraphics"
CALL initial(site(,),L)
CALL grow_cluster(site(,),L)
END

SUB initial(site(,),L)
    RANDOMIZE
    INPUT prompt "L = ": L
    CALL compute_aspect_ratio(L+2,xwin,ywin)
    SET WINDOW -xwin,xwin,-ywin,ywin
    BOX LINES -L-0.5,L+0.5,-L-0.5,L+0.5
    MAT site = 0                 ! initialize sites
    FOR y = -L to L
        FOR x = -L to L
            PLOT POINTS: x,y
        NEXT x
    NEXT y
    SET COLOR "red"
END SUB

SUB grow_cluster(site(,),L)
    LET R0 = 3                 ! start walker at distance R0 from origin
    LET site(0,0) = 1          ! seed site
    LET N = 1                  ! number of particles in cluster
    BOX AREA -0.5,0.5,-0.5,0.5
    DO
        ! find random initial position of new walker
        LET theta = 2*pi*rnd
        LET x = int(R0*cos(theta))
        LET y = int(R0*sin(theta))
        CALL walk(site(,),L,x,y,R0,N)    ! random walk
    LOOP until key input
END SUB

SUB walk(site(,),L,x,y,R0,N)
    ! walk until on a perimeter site of cluster
    ! or walker strays too far from cluster
    DO
        LET onperimeter$ = "no"
        LET r = sqr(x*x + y*y)
        IF r < R0 + 1 then
            ! test if walker on perimeter
            CALL test(site(,),L,x,y,r,R0,N,onperimeter$)
        END IF
        LET step = int(r - R0) - 1       ! big step
        IF step < 1 then LET step = 1
        IF onperimeter$ = "no" then
```

```
          LET random = rnd
          IF random < 0.25 then
              LET x = x + step
          ELSE IF random < 0.5 then
              LET x = x - step
          ELSE IF random < 0.75 then
              LET y = y + step
          ELSE
              LET y = y - step
          END IF
       END IF
    LOOP until onperimeter$ = "yes" or r > 2*R0
END SUB

SUB test(site(,),L,x,y,r,R0,N,onperimeter$)
    LET sum = site(x,y+1) + site(x,y-1) + site(x+1,y) + site(x-1,y)
    IF sum > 0 then                  ! walker on perimeter site
       LET site(x,y) = 1
       LET onperimeter$ = "yes"
       IF abs(x) <= L and abs(y) <= L then
          BOX AREA x-0.5,x+0.5,y-0.5,y+0.5
          LET N = N + 1
          SET CURSOR 2,1
          PRINT using "N = ####": N
       ELSE
          STOP                       ! cluster outside box
       END IF
       IF r >= R0 then LET R0 = int(r+2)
    END IF
END SUB
```

Laplacian growth model. As we discussed in Section 10.3, we can formulate the solution of Laplace's equation in terms of a random walk. We now do the converse and formulate the DLA algorithm in terms of a solution to Laplace's equation. Consider the probability $P(\mathbf{r})$ that a random walker reaches a site $\mathbf{r}$ between the external boundary and the growing cluster without having visited the cluster or the external boundary. This probability satisfies the relation

$$p(\mathbf{r}) = \frac{1}{4} \sum_{\mathbf{a}} p(\mathbf{r} + \mathbf{a}), \qquad (14.11)$$

where the sum in (14.11) is over the four nearest neighbor sites (on a square lattice). If we set $p = 1$ on the boundary and $p = 0$ on the cluster, then (14.11) also applies to sites that are neighbors of the external boundary and the cluster. A comparison of the form of (14.11) with the form of (10.11) shows that the former is a discrete version of Laplace's equation, $\nabla^2 p = 0$. Hence $p(\mathbf{r})$ has the same behavior as the electrical potential between two electrodes connected to the outer boundary and the cluster respectively, and the growth probability at a perimeter site of the cluster is proportional to the value of the potential at that site.

*Problem 14.10   Laplacian growth models

a. Solve the discrete Laplace equation (14.11) by hand for the growth probabilities of a DLA cluster of mass 1, 2, and 3. Set $p = 1$ on the boundary and $p = 0$ on the cluster.

b. You are probably familiar with the complicated and random nature of electrical discharge patterns that occur in atmospheric lightning. Although this phenomenon, known as *dielectric breakdown*, is complicated, we will see that a simple model leads to discharge patterns that are similar to those observed experimentally. Because lightning occurs in an inhomogeneous medium with differences in the density, humidity and conductivity of air, we want to develop a model of an electrical discharge in an inhomogeneous insulator. We know that when an electrical discharge occurs, the electrical potential ϕ satisfies Laplace's equation $\nabla^2 \phi = 0$. One version of the model (see Family et al.) is specified by the following steps:

1. Consider a large boundary circle of radius R and place a charge source at the origin. Choose the potential $\phi = 0$ at the origin (occupied site) and $\phi = 1$ for sites on the circumference of the circle. The radius R should be larger than the radius of the growing pattern.
2. Use the relaxation method (see Chapter 10) to compute the values of the potential ϕ_i for (empty) sites within the circle.
3. Assign a random number r to each empty site within the boundary circle. The random number r_i at site i represents a breakdown coefficient and the random inhomogeneous nature of the insulator.
4. The perimeter sites are the nearest neighbor sites of the discharge pattern (occupied sites). Form the product $r_i \phi_i^a$ for each perimeter site i, where a is an adjustable parameter. (Because the potential for the discharge pattern is zero, ϕ_i for perimeter site i can be interpreted as the magnitude of the potential gradient at site i.)
5. The perimeter site with the maximum value of the product $r\phi^a$ breaks down, i.e., set ϕ for this site equal to zero. (We can say that the bond between the discharge pattern and the perimeter site breaks down.)
6. Use the relaxation method to recalculate the values of the potential at the remaining unoccupied sites and repeat steps (4)–(6).

Choose $a = \frac{1}{4}$ and analyze the structure of the discharge pattern. Does the pattern appear qualitatively similar to lightning? Does the pattern appear to have a fractal geometry? Estimate the fractal dimension by counting $M(b)$, the average number of sites belonging to the discharge pattern that are within a $b \times b$ box. Consider other values of a, e.g., $a = \frac{1}{6}$ and $a = \frac{1}{3}$, and show that the patterns have a fractal structure with a tunable fractal dimension. Published results (Family et al.) are for patterns generated with 800 occupied sites.

c. The usual version of the dielectric breakdown model associates a growth probability $p_i = \phi_i^a / \sum_j \phi_j^a$ with each perimeter site i, where the sum is over all perimeter sites. One of the perimeter sites is occupied with probability p_i. That is, choose a perimeter site at random and generate a random number r between

0 and 1. If $r \le p_i$, the perimeter site i is occupied. As before, the exponent a is a free parameter. Convince yourself that $a = 1$ corresponds to diffusion-limited aggregation. (The boundary condition used in the latter corresponds to a zero potential at the perimeter sites.) To what type of cluster does $a = 0$ correspond? Choose $a = 1/2$, 1, and 2 and explore the dependence of the visual appearance of the clusters on a. If time permits, estimate the fractal dimension of the clusters.

d. Consider a deterministic growth model for which *all* perimeter sites are tested for occupancy at each growth step. We adopt the same geometry and boundary conditions as in part (b) and use the relaxation method to solve Laplace's equation for ϕ_i. Then we find the perimeter site with the largest value of ϕ and set this value equal to ϕ_{max}. Only those perimeter sites for which the ratio ϕ_i/ϕ_{max} is larger than a parameter p become part of the cluster and ϕ_i is set equal to unity for these sites. After each growth step, the new perimeter sites are determined and the relaxation method is used to recalculate the values of ϕ_i at each unoccupied site. Choose $p = 0.35$ and determine the nature of the regular fractal pattern. What is the fractal dimension? Consider other values of p and determine the corresponding fractal dimension. These patterns have been termed *Laplace fractal carpets* (see Family et al.).

*Cluster-cluster aggregation**. Fractal structures commonly occur in aggregates that have been formed by the clustering of particles that are diffusing in a fluid. For example, colloids consist of particles that stick together in a liquid solvent, and aerosols are the analog in a gas. In DLA, all the particles that stick to a cluster are the same size (the growth occurs by cluster-monomer contact), and the cluster that is formed is motionless. In the following, we consider a cluster-cluster aggregation (CCA) model in which the clusters diffuse as they aggregate.

In a typical simulation we begin with a dilute collection of N particles. Each of these particles is a cluster with unit mass. The particles do random walks until one of them becomes a nearest neighbor of another particle. At that point they stick together to form a cluster of two particles. This new cluster now moves as a single random walker with a reduced diffusion coefficient. As this process continues, the clusters become larger and fewer in number. For simplicity, we assume a square lattice with periodic boundary conditions. The CCA algorithm can be summarized as follows:

1. Place N particles at random positions on the lattice. Do not allow a site to be occupied by more than one particle. Identify the ith particle with the ith cluster.

2. Check if any two clusters have particles that are nearest neighbors. If so, join these two clusters to form a single cluster.

3. Choose a cluster at random. Decide whether to move the cluster as discussed below. If so, move it randomly in one of the four possible directions. In the following, we discuss the strategy for deciding when to move a cluster.

4. Repeat steps 2 and 3 until the desired time or until there is only a single cluster.

What rule should we use to decide whether to move a cluster? One possibility is to select a cluster at random and simply move it. This possibility corresponds to all clusters having the same diffusion coefficient, regardless of the mass s of the cluster. A more realistic rule is to assume that the diffusion coefficient D_s is inversely related to the mass, for example as s^{-x} with $x \neq 0$. A common assumption in the literature is to assume $x = 1$. If we assume instead that D_s is inversely proportional to the linear dimension of the cluster, an assumption consistent with the Stokes-Einstein relation, it is reasonable to take $x = 1/d$. However because the clusters are fractals, we really should take $x = 1/D$, where D is the fractal dimension of the cluster. In Problem 14.11 we explore some of the possible forms of D_s.

To implement the cluster-cluster aggregation algorithm, we need to store the position of each particle and the cluster to which each particle belongs. In *Program cca* the position of a particle is given by its x and y coordinates and stored in the arrays x and y respectively. The array element site(x,y) equals zero if there is no particle at (x, y); otherwise the element equals the label of the cluster to which the particle at (x, y) belongs.

The labels of the clusters are found as follows. The array element first_particle(k) gives the particle label of the first particle in the kth cluster. To determine all the particles in a given cluster, we use a data structure called a *linked list*. This list is implemented using an array such that the value of an element of the array is the index for the next element in the linked list. The linked list is an example of a *circular linked list*, because the value of the last element in the linked list is the index for the first element. The array next_particle contains a series of circular linked lists, one for each cluster, such that next_particle(i) equals the particle label of another particle in the same cluster as the ith particle. If next_particle(i) = i, then the ith particle is a cluster with only one particle. To see how these arrays work, consider three particles 5, 9, and 16 which constitute cluster 4. We have first_particle(4) = 5, next_particle(5) = 9, next_particle(9) = 16, and next_particle(16) = 5.

As the clusters undergo a random walk, we need to check if any pair of particles in different clusters have become nearest neighbors. If such a situation occurs, their respective clusters have to be merged. The check for nearest neighbors is done in SUB neighbor. If for example, site(x,y) and site(x+1,y) are both nonzero and are not equal, then the two clusters associated with these sites need to be combined. To do so, we combine the two circular lists for each cluster into one circular list as is done in SUB merge. Hence if p1 and p2 are the first particles of their respective clusters, then

```
LET p1next = next_particle(p1)
LET p2next = next_particle(p2)
```

gives the second particle in each cluster. The following code merges the two clusters.

```
LET next_particle(p1) = p2next
LET next_particle(p2) = p1next
```

To complete the merger, all the entries in site(x,y) corresponding to the second cluster are relabeled with the label for the first cluster.

```
PROGRAM cca
PUBLIC site(50,50),x(1000),y(1000),L,N
LIBRARY "csgraphics"
DIM next_particle(1000),first_particle(1000)
CALL initial(first_particle(),next_particle(),ncl)
DO
    CALL move(next_particle(),first_particle(),ncl)
LOOP until ncl = 1
END

SUB initial(first_particle(),next_particle(),ncl)
    PUBLIC nx(4),ny(4)
    DECLARE PUBLIC site(,),x(),y()
    DECLARE PUBLIC L,N
    RANDOMIZE
    DATA 1,0,-1,0,0,1,0,-1
    FOR nn = 1 to 4
        READ nx(nn),ny(nn)
    NEXT nn
    INPUT prompt "L = ": L
    INPUT prompt "N = ": N
    CALL compute_aspect_ratio(L+1,xwin,ywin)
    SET WINDOW 0,xwin,0,ywin
    CLEAR
    BOX LINES 1,L+1,1,L+1
    SET COLOR "blue"
    LET ncl = 0                          ! number of clusters
    FOR i = 1 to N
        DO
            LET x(i) = int(L*rnd) + 1
            LET y(i) = int(L*rnd) + 1
        LOOP until site(x(i),y(i)) = 0
        LET ncl = ncl + 1
        LET site(x(i),y(i)) = ncl
        BOX AREA x(i),x(i)+1,y(i),y(i)+1
        LET first_particle(ncl) = i
        LET next_particle(i) = i
        LET xi = x(i)
        LET yi = y(i)
        CALL neighbor(xi,yi,next_particle(),first_particle(),ncl)
    NEXT i
END SUB

SUB neighbor(xi,yi,next_particle(),first_particle(),ncl)
    DECLARE PUBLIC site(,),x(),y()
    DECLARE PUBLIC L,nx(),ny()
    DECLARE DEF pbc
    FOR nn = 1 to 4
        LET px = pbc(xi + nx(nn),L)
        LET py = pbc(yi + ny(nn),L)
```

```
                    LET perim = site(px,py)
                    LET part = site(xi,yi)
                    IF perim <> 0 and perim <> part then
                        CALL merge(perim,part,first_particle(),next_particle(),ncl)
                    END IF
            NEXT nn
    END SUB


    SUB merge(c1,c2,first_particle(),next_particle(),ncl)
        DECLARE PUBLIC site(,), x(), y()
        LET p1 = first_particle(c1)
        LET p2 = first_particle(c2)
        LET p1next = next_particle(p1)
        LET p2next = next_particle(p2)
        LET next_particle(p1) = p2next
        LET next_particle(p2) = p1next
        DO
            LET site(x(p2next),y(p2next)) = c1
            LET p2next = next_particle(p2next)
        LOOP until p2next = p1next
        LET plast = first_particle(ncl)
        IF c2 <> ncl then
            LET p = plast
            DO
                LET site(x(p),y(p)) = c2
                LET p = next_particle(p)
            LOOP until p = plast
            LET first_particle(c2) = plast
        END IF
        LET ncl = ncl - 1
    END SUB

    SUB move(next_particle(),first_particle(),ncl)
        DECLARE PUBLIC site(,),x(),y()
        DECLARE PUBLIC L,nx(),ny()
        DECLARE DEF pbc
        LET c = int(ncl*rnd) + 1
        LET direction = int(4*rnd) + 1
        LET p1 = first_particle(c)
        LET i = p1
        LET dx = nx(direction)
        LET dy = ny(direction)
        DO
            LET site(x(i),y(i)) = 0
            SET COLOR "white"
            BOX AREA x(i),x(i)+1,y(i),y(i)+1
            LET x(i) = pbc(x(i) + dx,L)
            LET y(i) = pbc(y(i) + dy,L)
            LET i = next_particle(i)
        LOOP until i = p1
```

```
              DO
                  SET COLOR "blue"
                  BOX AREA x(i),x(i) + 1, y(i),y(i) + 1
                  LET site(x(i),y(i)) = c
                  LET i = next_particle(i)
              LOOP until i = p1
              DO
                  LET xi = x(i)
                  LET yi = y(i)
                  CALL neighbor(xi,yi,next_particle(),first_particle(),ncl)
                  LET i = next_particle(i)
              LOOP until i = p1
              SET COLOR "black"
              BOX LINES 1,L + 1,1,L + 1
        END SUB

        FUNCTION pbc(s,L)
            IF s > L then
                LET pbc = 1
            ELSE IF s < 1 then
                LET pbc = L
            ELSE
                LET pbc = s
            END IF
        END DEF
```

*Problem 14.11   Cluster-cluster aggregation

a.   *Program* cca assumes that the diffusion coefficient is independent of the cluster mass. Run *Program* cca with $L = 40$ and $N = 300$ and describe the qualitative appearance of the clusters as they form. Do they appear to be fractals? Compare their appearance to DLA clusters.

b.   Choose $L = 50$ and $N = 500$ and compute the fractal dimension of the final cluster. (Speed up the program by eliminating the visual display of the formation of the clusters.) Use the center of mass, r_{cm}, as the origin of the cluster, where $r_{cm} = (1/N)(\sum_i x_i, \sum_i y_i)$ and (x_i, y_i) is the position of the ith particle. Average your results over at least ten final clusters. Do the same for other values of L and N. Are the clusters formed by cluster-cluster aggregation more or less space filling than DLA clusters?

*c.   Assume that the diffusion coefficient of a cluster of s particles varies as $D_s \propto s^{-1/d}$, where d is the spatial dimension. Let D_{max} be the diffusion coefficient of the largest cluster. Add to your program an array that tracks the mass of each cluster. Choose a random number r between 0 and 1 and move the cluster if $r < D_s/D_{max}$. Repeat the above simulations and discuss any changes in your results. What effect does this dependence of D on s have on the motion of the clusters? In any case increase the time by one step. The time-dependence of the cluster size distribution is investigated in Project 14.4.

Surface growth. The fractal objects we have discussed so far are self-similar, that is, if we look at a small piece of the object and magnify it isotropically to the size of the original, then the original and the magnified object look similar (on the average). In the following, we introduce some simple models that generate a class of fractals that are self-similar only for scale changes in certain directions.

One of the problems in surface science is understanding the formation of rough surfaces. Suppose that we have a flat surface at time $t = 0$. Let us ask how the surface grows as a result of vapor deposition and sedimentation. For example, consider a surface which initially is a line of L occupied sites. Growth is confined to the vertical direction (see Fig. 14.13).

As before, we simply choose a perimeter site at random and occupy it. The average height of the cluster is given by

$$\overline{h} = \frac{1}{N_s} \sum_{i=1}^{N_s} h_i, \tag{14.12}$$

where h_i is the distance of the ith surface site from the substrate, and the sum is over all surface sites N_s. (The precise definition of a surface site for the Eden model is discussed in Problem 14.12.)

Each time a particle is deposited, the time t is increased by unity. Our main interest is how the "width" of the surface changes with t. We define the width of the surface by

$$w^2 = \frac{1}{N_s} \sum_{i=1}^{N_s} (h_i - \overline{h})^2. \tag{14.13}$$

In general, the surface width w, which is a measure of the surface roughness, depends on L and t. Initially w grows with time. We expect that

$$w(L, t) \sim t^\beta. \tag{14.14}$$

The exponent β describes the growth of the correlations with time along the vertical direction. Fig. 14.13 illustrates the evolution of the surface generated according to the Eden model. After a characteristic time, the length over which the fluctuations are

Fig. 14.13   Surface growth according to the Eden model. The surface site in column x is the perimeter site with the maximum value h_x in the vertical direction. The average height for this surface is 20.46 and the width is 2.33.

correlated becomes comparable to L, and the width reaches a steady state value that depends only on L. We write

$$w(L, t \gg 1) \sim L^{\alpha},$$ (14.15)

where α is known as the roughness exponent.

From (14.15) we see that in the steady state, the width of the surface in the direction perpendicular to the substrate grows as L^{α}. This steady-state behavior of the width is characteristic of a *self-affine fractal*. Such a fractal is invariant (on the average) under anisotropic scale changes, that is, different scaling relations exist along different directions. For example, if we rescale the surface by a factor b in the horizontal direction, then the surface must be rescaled by a factor of b^{α} in the direction perpendicular to the surface to preserve the similarity along the original and rescaled surfaces.

Note that on short length scales, that is, lengths shorter than the width of the interface, the surface is rough and its roughness can be characterized by the exponent α. (Imagine an ant walking on the surface.) However on length scales much larger than the width of the surface, the surface appears to be flat and, in our example, it is a one-dimensional object. The properties of the surface as given by several growth models are explored in Problem 14.12.

Problem 14.12   Growing surfaces

a. *Eden model.* In the Eden model a perimeter site is chosen at random and occupied. In this model there can be "overhangs" as shown in Fig. 14.13, and the height h_x corresponds to the maximum distance of any perimeter site in column x from the surface. Use periodic boundary conditions in the horizontal directions to determine the perimeter sites. Note that the growth rule is the same as the usual Eden model, but the growth is started from the top of a strip of length L. Choose a square lattice with $L = 100$. Describe the visual appearance of the surface as the surface grows. Is the surface well-defined visually? Where are most of the perimeter sites? We have defined the surface sites as a subset of the perimeter sites (i.e., those with maximum h for a given x). Do you think our results would be qualitatively different if we included all perimeter sites?

b. Plot the width $w(t)$ as a function of t for $L = 32, 64$, and 128 on the same graph and estimate the exponents α and β for the Eden model. What type of plot is most appropriate? Does the width initially grow as a power law? If so, estimate the exponent β. Is there a L-dependent crossover time after which the width of the surface approaches its steady state value? How can you estimate the exponent α? The best numerical estimates for β and α are consistent with the presumed exact values $\beta = 1/3$ and $\alpha = 1/2$, respectively.

*c. The dependence of $w(L, t)$ on t and L can be combined into the scaling form

$$w(L, t) \approx L^{\alpha} f(t / L^{\alpha/\beta})$$ (14.16)

where

$$f(x) \approx x^\beta \qquad \text{for } x \ll 1 \tag{14.17a}$$
$$f(x) = \text{constant} \qquad \text{for } x \gg 1 \tag{14.17b}$$

Verify the existence of the scaling form (14.16) by plotting the ratio $w(L, t)/L^\alpha$ versus $t/L^{\alpha/\beta}$ for the different values of L considered in part (b). If the scaling forms holds, the results for w for the different values of L should fall on a universal curve. Use either the estimated values of α and β that you found in part (b) or the exact results.

*d. *Random deposition.* The Eden model is not really a surface growth model, because any perimeter site can become part of the cluster. In the simplest deposition model, a column is chosen at random and a particle is deposited at the top of the column of already deposited particles. There is no horizontal correlation between neighboring columns. Do a simulation of this growth model and visually inspect the surface of the interface. Show that the heights of the columns follow a Poisson distribution (see (7.7)) and that $\overline{h} \sim t$ and $w \sim t^{1/2}$. This structure does not depend on L and hence $\alpha = 0$.

*e. *Ballistic deposition.* In this model a column is chosen at random and a particle is assumed to fall vertically until it reaches the first perimeter site that is a nearest neighbor of a site that already is part of the surface. This condition allows for growth parallel to the substrate. Only one particle falls at a time. How do the rules for this growth model differ from those of the Eden model? How does the deposit that you obtain compare to that of the Eden model? Suppose that instead of the particle falling vertically, we let it do a random walk as in DLA. Would the resultant surface be the same?

14.4 FRACTALS AND CHAOS

In Chapter 6 we explored dynamical systems that exhibited chaos under certain conditions. We found that after an initial transient, the trajectory of a dynamical system consists of a set of points in phase space called an attractor. For chaotic motion this attractor often is an object that can be described by a fractal dimension. Such attractors are called *strange attractors*.

We first consider the familiar logistic map (see (6.1)), $x_{n+1} = 4rx_n(1 - x_n)$. For most values of the control parameter $r > r_\infty = 0.892486417967\ldots$, the trajectories are chaotic. Are these trajectories fractals? We explore this question in Problem 14.13.

To calculate the fractal dimension for dynamical systems, we use the *box counting* method in which space is divided into d-dimensional boxes of length ℓ. Let $N(\ell)$ equal the number of boxes that contain a piece of the trajectory. The fractal dimension is defined by the relation

$$N(\ell) \sim \lim_{\ell \to 0} \ell^{-D}. \qquad \text{(box counting dimension)} \tag{14.18}$$

Equation (14.18) is accurate only when the number of boxes is much larger than $N(\ell)$ and the number of points on the trajectory is sufficiently large. If the trajectory moves through many dimensions, i.e., the phase space is very large, box counting becomes too memory intensive because we need an array of size $\propto \ell^{-d}$. This array becomes very large for small ℓ and large d.

A more efficient approach is to calculate the *correlation dimension*. In this approach we store in an array the position of N points on the trajectory. We compute the number of points $N_i(r)$, and the fraction of points $f_i(r) = N_i(r)/(N-1)$ within a distance r of the point i. The correlation function $C(r)$ is defined by

$$C(r) \equiv \frac{1}{N} \sum_i f_i(r), \tag{14.19}$$

and the *correlation dimension* D_c is defined by

$$C(r) \sim \lim_{r \to 0} r^{D_c}. \qquad \text{(correlation dimension)} \tag{14.20}$$

From (14.20) we see that the slope of a log-log plot of $C(r)$ versus r yields an estimate of the correlation dimension. In practice, small values of r must be discarded because we cannot sample all the points on the trajectory, and hence there is a cutoff value of r below which $C(r) = 0$. In the large r limit, $C(r)$ saturates to unity if the trajectory is localized as it is for chaotic trajectories. We expect that for intermediate values of r, there is a scaling regime where (14.20) holds.

In Problems 14.13–14.15 we consider the fractal properties of some of the dynamical systems that we considered in Chapter 6.

Problem 14.13 Fractal dimension of logistic map trajectories

a. Write a program that uses box counting to determine the fractal dimension of the attractor for the logistic map. Compute $N(\ell)$, the number of boxes of length ℓ that have been visited by the trajectory. Test your program for $r < r_\infty$. How does the number of boxes containing a piece of the trajectory change with ℓ? What does this dependence tell you about the dimension of the trajectory?

b. Compute $N(\ell)$ for $r = 0.9$ using at least five different values of ℓ, e.g., $1/\ell = 100, 300, 1000, 3000, \ldots$. Iterate the map at least 1000 times before determining $N(\ell)$. What is the fractal dimension of the attractor? Repeat for $r \approx r_\infty$, $r = 0.95$, and $r = 1$.

c. Generate points at random in the unit interval and estimate the fractal dimension using the same method as in part (b). What do you expect to find? Use your results to estimate the accuracy of the fractal dimension that you found in part (b).

d. Write a program to compute the correlation dimension for the logistic map and repeat the calculations for parts (b) and (c).

Problem 14.14 Strange attractor of the Hénon map

a. Use two-dimensional boxes of linear dimension ℓ to estimate the fractal dimension of the strange attractor of the Hénon map (see (6.24)) with $a = 1.4$ and $b = 0.3$. Iterate the map at least 100 times before computing $N(\ell)$. Does it matter what initial condition you choose?

b. Compute the correlation dimension for the same parameters used in part (a) and compare D_c with the box dimension computed in part (a).

c. Iterate the Hénon map and view the trajectory on the screen by plotting x_{n+1} versus x_n in one window and y_n versus x_n in another window. Do the two ways of viewing the trajectory look similar? Estimate the correlation dimension, where the ith data point is defined by (x_i, x_{i+1}) and the distance R_{ij} between the ith and jth data point is given by $R_{ij}^2 = (x_i - x_j)^2 + (x_{i+1} - x_{j+1})^2$.

d. Estimate the correlation dimension with the ith data point defined by x_i, and $R_{ij}^2 = (x_i - x_j)^2$. What do you expect to obtain for D_c? Repeat the calculation for the ith data point given by (x_i, x_{i+1}, x_{i+2}) and $R_{ij}^2 = (x_i - x_j)^2 + (x_{i+1} - x_{j+1})^2 + (x_{i+2} - x_{j+2})^2$. What do you find for D_c?

*Problem 14.15 Strange attractor of the Lorenz model

a. Use three-dimensional graphics or three two-dimensional plots of $x(t)$ versus $y(t)$, $x(t)$ versus $z(t)$, and $y(t)$ versus $z(t)$ to view the structure of the Lorenz attractor. Use $\sigma = 10$, $b = 8/3$, $r = 28$, and the time step $\Delta t = 0.01$. Then compute the correlation dimension for the Lorenz attractor.

b. Repeat the calculation of the correlation dimension using $x(t)$, $x(t + \tau)$, and $x(t + 2\tau)$ instead of $x(t)$, $y(t)$, and $z(t)$. Choose the delay time τ to be at least ten times greater than the time step Δt.

c. Compute the correlation dimension in the two-dimensional space of $x(t)$ and $x(t + \tau)$. Do the same calculation in four dimensions using $x(t)$, $x(t + \tau)$, $x(t + 2\tau)$, and $x(t + 3\tau)$. What can you conclude about the results for the correlation dimension using two, three, and four-dimensional spaces. What do you expect to see for $d > 4$?

Problems 14.14 and 14.15 illustrate a practical method for determining the underlying structure of systems when, for example, the data consists only of a single time series, that is, measurements of a single quantity over time. The dimension $D_c(d)$ computed by increasing the dimension of the space, d, using the delayed coordinate τ eventually saturates when d is approximately equal to the number of variables that actually determine the dynamics. Hence, if we have extensive data for a single variable, e.g., the atmospheric pressure, we can use this method to determine the number of independent variables that determine the dynamics of the pressure. This information can then be used to help create models of the atmosphere.

14.5   MANY DIMENSIONS

So far we have discussed three ways of defining the fractal dimension: the mass dimension (14.1), the box counting dimension (14.18), and the correlation dimension (14.20). These methods do not always give the same results for the fractal dimension. Indeed, there are many other dimensions that we could compute. For example, instead of just counting the boxes that contain a part of an object, we can count the number of points of the object in each box, n_i, and compute $p_i = n_i/N$, where N is the total number of points. A generalized dimension D_q can be defined as

$$D_q = \frac{1}{q-1} \lim_{\ell \to 0} \frac{\ln \sum_{i=1}^{N(\ell)} p_i^q}{\ln \ell}. \tag{14.21}$$

The sum in (14.21) is over all the boxes and involves the probabilities raised to the qth power. For $q = 0$, we have

$$D_0 = - \lim_{\ell \to 0} \frac{\ln N(\ell)}{\ln \ell}. \tag{14.22}$$

If we compare the form of (14.22) with (14.18), we can identify D_0 with the box-counting dimension. For $q = 1$, we need to take the limit of (14.21) as $q \to 1$. Let

$$u(q) = \ln \sum_i p_i{}^q, \tag{14.23}$$

and do a Taylor-series expansion of $u(q)$ about $q = 1$. We have

$$u(q) = u(1) + (q-1)\frac{du}{dq} + \dots \tag{14.24}$$

The quantity $u(1) = 0$ because $\sum_i p_i = 1$. The first derivative of $u(q)$ is given by

$$\frac{du}{dq} = \frac{\sum_i p_i{}^q \ln p_i}{\sum_i p_i{}^q} = \sum_i p_i \ln p_i, \tag{14.25}$$

where the last equality follows by setting $q = 1$. If we use the above relations, we find that D_1 is given by

$$D_1 = \lim_{\ell \to 0} \frac{\sum_i p_i \ln p_i}{\ln \ell}. \qquad \text{(information dimension)} \tag{14.26}$$

D_1 is called the *information dimension* because of the similarity of the $p \ln p$ term in the numerator of (14.25) to the information form of the entropy.

It is possible to show that D_2 as defined by (14.21) is the same as the mass dimension defined in (14.1) and the correlation dimension D_c. That is, box counting gives D_0 and correlation functions give D_2 (cf. Sander et al. 1994).

There are many objects in nature that have similar fractal dimensions, but which nevertheless differ in appearance. An example of this difference is the visual appearance in three spatial dimensions of the clusters generated by diffusion-limited aggregation and the percolation clusters generated by the Leath algorithm at the percolation threshold. (Both objects have a fractal dimension of approximately 2.5.) In some cases this difference can be accounted for by *multifractal* properties of an object. For objects called *multifractals* the various D_q are different, in contrast to *monofractals* for which the different measures are the same. Percolation clusters are an example of a monofractal, because $p_i \sim \ell^{D_0}$, the number of boxes $N(\ell) \sim \ell^{-D_0}$, and from (14.21), $D_q = D_0$ for all q. Multifractals occur when the quantities p_i are not the same throughout the object, as frequently happens for the strange attractors produced by chaotic dynamics. DLA might be an example of a multifractal, and the appropriate probabilities p_i might correspond to the probability that the next perimeter site to be occupied is at i.

14.6 PROJECTS

Although the kinetic growth models yield beautiful pictures and fractal objects, there is much we do not understand. Why do the fractal dimensions have the values that we found by various numerical experiments? Can we trust our numerical estimates of the various exponents or is it necessary to consider much larger systems to obtain their true asymptotic values? Can we find unifying features for the many kinetic growth models that presently exist? What is the relation of the various kinetic growth models to physical systems? What are the essential quantities needed to characterize the geometry of an object?

One of the reasons that growth models are difficult to understand is that typically the end product depends on the history of the growth. We say that these models are examples of "nonequilibrium behavior." This combination of simplicity, beauty, complexity, and relevance to many experimental systems suggests that the study of fractal objects will continue to involve a wide range of workers in many disciplines.

Project 14.1 Off-lattice DLA

a. In the continuum (off-lattice) version of diffusion-limited aggregation. the diffusing particles are assumed to be disks of radius a. A particle executes a random walk until its center is within a distance $2a$ of the center of a particle already attached to the DLA cluster. At each step the walker changes its position by $(r \cos\theta, r \sin\theta)$, where r is the step size, and θ is a random variable between 0 and 2π. Modify your DLA program or `Program dla` to simulate off-lattice DLA.

b. Compare the appearance of an off-lattice DLA cluster with one generated on a square lattice. It is necessary to grow very large clusters (approximately 10^6 particles) to see any real differences.

c. Use the mass dimension to estimate the fractal dimension of the off-lattice DLA cluster and compare its value with the value you found for the square lattice. Explain why the fractal dimensions might be different. $D \approx 1.71$ for off-lattice DLA in two dimensions while $D \approx 1.55$ for a square lattice (also see Problem 14.9). However, it is necessary to grow very large clusters to determine the effect of the lattice.

Project 14.2   More efficient simulation of DLA

In *Program dla* we use a variable step size for the walkers to improve the efficiency of the algorithm. However, when the walker is within the distance R0 of the seed, no optimization is used. Because there can be a great deal of empty space within this distance, we describe an additional optimization technique (see Ball and Brady). The basic idea is to use a simple geometrical object (a circle or square) centered at the walker such that none of the cluster is within the object. Then in one step move the walker to the perimeter of the object. For a circle the walker can move to any location with equal probability on the circle. For the square you need the probability of moving to various locations on the square. The major difficulty is to find the largest object that does not contain a part of the DLA cluster. To do this we consider coarse grained lattices. For example, each 2×2 group of sites on the original lattice corresponds to one site on the coarser lattice, and then each 2×2 group of sites on the coarse lattice corresponds to a site on an even coarser lattice, etc. If a site is occupied then any coarse site made from this site also is occupied.

a.   Because we have considered DLA clusters on a square lattice, we use squares centered at a walker. First we must find the probability $p(\Delta x, \Delta y, s)$ that a walker centered on a square of length $l = 2s + 1$, will be displaced by the vector $(\Delta x, \Delta y)$. This probability can be computed by simulating a random walk starting at the origin and ending at the edge of the square. These simulations are then repeated for many walkers, and then for each value of s. $p(\Delta x, \Delta y, s)$ is the fraction of walkers that reached the position $(\Delta x, \Delta y)$. Determine $p(\Delta x, \Delta y, s)$ for $s = 1$ to 16. Store your results in a file.

b.   We next need to produce an array such that for a given value of s and a random number r between 0 and 1, we can quickly find $(\Delta x, \Delta y)$. To do so create four arrays. The first array lists the probability distribution determined from p in part (a) such that the values for $s = 1$ are listed first, then the values for $s = 2$, etc. Call this array p. For example, $p(1) = p(-1, -1, 1)$, $p(2) = p(1) + p(-1, 0, 1)$, $p(3) = p(2) + p(-1, 1, 1)$, etc. Next create an array start that tells you where to start in the array p for each value of s. Then create the two arrays dx(i) and dy(i) which give the values of Δx and Δy corresponding to p(i). To see how these arrays are used, consider a walker located at (x, y), centered on a square of size $2s + 1$. First compute a random number r and find i = start(s). If $p(i) > r$, then the walker moves to (x + dx(i), y + dy(i)). If not, increment i by unity and check again. Repeat until $p(i) > r$. Write a program to create these four arrays and store them in a file.

c.   Write a subroutine to determine the maximum value of the parameter s such that a square of size $2s + 1$ centered at the position of the walker does not contain any part of the DLA cluster. Use coarse grained lattices to do this determination more efficiently.

d.   Modify *Program dla* to incorporate the subroutine from part (c) and read in the arrays from part (b). How much faster is your modified program than the original *Program* dla for clusters of size 500 and 5000 particles? What is the largest cluster you can grow on your computer in one hour?

e.   Grow as large a cluster as you can. Is there any evidence for anisotropy? For example, does the cluster tend to extend further along the axes or along any other direction?

The following two projects introduce some of the important ideas associated with scaling.

Project 14.3   Scaling properties of the percolation cluster size distribution

a.   The scaling hypothesis for n_s, the number of percolation clusters of size s, near the percolation threshold p_c is

$$n_s = s^{-\tau} f_{\pm}\left(|p - p_c|^{1/\sigma} s\right), \qquad (s \gg 1) \tag{14.27}$$

where the indices $+$ and $-$ refer to $p > p_c$ and $p < p_c$, respectively. The critical exponents τ and σ are the same above and below p_c. To understand the scaling form of n_s, first note that (14.27) implies that $n_s \sim s^{-\tau}$ at $p = p_c$ (compare with Table 13.1). We now show that the exponent σ can be related to ν and τ. Recall that the connectedness length ξ is given in terms of n_s by (see (13.9)):

$$\xi^2 = \frac{\sum_s s^2 n_s R_s^2}{\sum_s s^2 n_s}, \tag{14.28}$$

where R_s is the radius of gyration of the clusters. Close to p_c, the large clusters dominate the sum in (14.28). On length scales less than ξ, the clusters have no way of telling that the system is not at p_c and hence the large clusters are fractals with $R_s \sim s^{1/D}$. If we substitute this dependence and the scaling form (14.27) for n_s in (14.28), we obtain

$$\xi^2 \sim \frac{\sum_s s^{2-\tau+2/D} f_{\pm}\left(|p - p_c|^{1/\sigma} s\right)}{\sum_s s^{2-\tau} f_{\pm}\left(|p - p_c|^{1/\sigma} s\right)}. \tag{14.29}$$

For an infinite system the sums in (14.29) range from $s = 1$ to $s = \infty$. To calculate these sums, we transform them into integrals. For example, we can write the numerator of (14.29) as

$$\sum_{s=1}^{\infty} s^{2-\tau+2/D} f_{\pm}\left(|p - p_c|^{1/\sigma} s\right) \sim \int_1^{\infty} s^{2-\tau+2/D} f_{\pm}\left(|p - p_c|^{1/\sigma} s\right) ds$$

$$\propto (p - p_c)^{(\tau D - 3D - 2)/(D\sigma)} \int_{(p-p_c)^{1/\sigma}}^{\infty} x^{2-\tau+2/D} f_{\pm}(x) \, dx, \tag{14.30}$$

where $x = (p - p_c)^{1/\sigma} s$. The denominator can be written in a similar form. If we assume that the integrands are nonsingular, then to lowest order in $(p - p_c)$, the lower integration limit can be set equal to zero. Show that we obtain

$$\xi^2 \sim |p - p_c|^{-2/(D\sigma)}. \tag{14.31}$$

Because $\xi \sim |p - p_c|^{-\nu}$, we obtain the desired relation between ν, σ, and D:

$$\nu = 1/(D\sigma). \tag{14.32}$$

If we write the argument $x = |p - p_c|^{1/\sigma} s$ as $x = s/\xi^D$, we can express the scaling form of n_s (14.27) in a more physical form:

$$n_s \sim s^{-\tau} f_{\pm}(s/\xi^D). \tag{14.33}$$

Equation (14.33) implies that n_s depends on s only through the ratio s/ξ^D or R_s/ξ. That is, the connectedness length represents the only important length near p_c. Show that the integrands in (14.30) are nonsingular if $2 - \tau > -1$ and fill in the missing algebra leading to (14.33).

b. Let us try to verify the scaling form of n_s by generating percolation clusters of all sizes as we did in Chapter 13. We use the Leath algorithm to generate clusters of size s that are grown from a seed. Modify Program `perc_cluster` so that many clusters are generated and n_s is computed for a given input probability p. Remember that the number of clusters of size s that are grown from a seed is the product sn_s, rather than n_s itself (see Problem 14.3a). Run your program at $p = p_c \approx 0.592746$ and grow at least 100 clusters for a square lattice with $L \geq 61$. From (14.27) we see that at $p = p_c$, $n_s \sim s^{-\tau} f(0) \sim s^{-\tau}$. Hence a log-log plot of n_s versus s should give a straight line for $s \gg 1$ with a slope of $-\tau$. Estimate τ from your data. If time permits, use a bigger lattice and average over more clusters, and also estimate the accuracy of your estimate of τ.

c. Determine n_s for at least three different values of p close to p_c, e.g., 0.57, 0.58, and 0.61. Plot the product $s^{\tau} n_s$ versus the product $|p - p_c|^{\sigma} s$ using the value of τ found in part (b). Try various values of σ until your points for all values of p follow the same curve. Initially, try $\sigma = 0.4, 0.45$, and 0.5.

d. The mean cluster size $S(p)$ is related to n_s by (see (13.5)):

$$S(p) = \frac{\sum_s s^2 n_s}{\sum_s s n_s}. \tag{14.34}$$

Note that $S(p)$ can be regarded as the second moment of n_s. Use arguments similar to those used in part (a) to show that $S(p) \sim |p - p_c|^{(\tau-3)/\sigma}$ for p near p_c. Since $S(p) \sim |p - p_c|^{-\gamma}$, we have the relation

$$\gamma = (3 - \tau)/\sigma. \tag{14.35}$$

Use the values of τ and σ that you found in parts (b) and (c) to estimate γ.

Similar arguments can be made for P_∞. Every site in the lattice is either empty with probability $1 - p$, or occupied and part of the spanning cluster with probability pP_∞, or occupied but not part of the spanning cluster with probability $p(1 - P_\infty) = \sum_s sn_s$. Hence we have the exact relation

$$P_\infty(p) = 1 - \frac{1}{p}\sum_s sn_s. \tag{14.36}$$

Note that P_∞ can be regarded as the first moment of n_s. Hence using the same argument that led to (14.35), we find

$$\beta = (\tau - 2)/\sigma. \tag{14.37}$$

Compare this relation to the form of (14.35). A careful derivation of this scaling relation can be found on page 70 of Bunde and Havlin. Use (14.37) to estimate the critical exponent β.

Project 14.4   Dynamical scaling properties of the cluster size distribution in cluster-cluster aggregation

a.   The dynamical scaling assumption for the number of clusters of size s at time t for cluster-cluster aggregation is

$$n_s(t) \sim s^{-\theta} f(s/t^z), \tag{14.38}$$

where θ and z are exponents and $f(x)$ is a scaling function. The density of the particles is fixed and is related to n_s by the normalization condition:

$$\rho = \sum_s sn_s(t) \sim \int sn_s(t)\,ds. \tag{14.39}$$

Use the normalization condition (14.39) to show that $\theta = 2$. The scaling form (14.38) is expected to be applicable in the low density limit at large s and t.

b.   Show that the mean cluster size $S(t)$ given by

$$S(t) = \frac{\sum_s s^2 n_s(t)}{\sum_s sn_s(t)} \tag{14.40}$$

diverges for $t \to \infty$ as

$$S(t) \sim t^z \tag{14.41}$$

Hence the scaling form of $n_s(t)$ can be written as

$$n_s(t) \sim t^{-2} f(s/S(t)) \tag{14.42}$$

c.   Modify Program cca so that the cluster size distribution, $n_s(t)$, is computed for $t = 2^p$ with $p = 1, 2, 3 \ldots$ and $s = 1, 3, 10, 30,$ and 100. For simplicity, assume that the diffusion coefficient of the clusters is size independent. Remember that $n_s = N_s(t)/L^2$, where $N_s(t)$ is the number of clusters of s particles at time t. The unit of time can be defined in various ways. The easiest way is to increase t by unity after a cluster has been moved one lattice unit. However, this choice would lead to the time changing faster when the number of clusters decreases. A better choice is to increase the time by an amount $\Delta t = s/N$, where s is the number of sites in the selected cluster and N is the (original) number of particles in the lattice. Choose $L = 50$ and $N = 500$ and average over at least ten runs. One way to test the dynamical scaling form of $n_s(t)$ is to plot $s^2 n_s(t)$ versus s/t^z for different choices of z. Another way is to make a log-log plot of $S(t)$ versus t and extract the exponent z from the slope of the linear portion of the log-log plot.

d.   Repeat part (c) for other values of L and N and discuss the accuracy of your results.

e.   A similar type of dynamics occurs in one dimension. What do you think the fractal dimension of the final cluster would be? In one dimension the surface of a cluster is two sites regardless of its size. As a result, the large clusters do not grow as fast as they do for higher dimensions. Can $n_s(t)$ still be described by a scaling form?

References and Suggestions for Further Reading

We have considered only a few of the models that lead to self-similar patterns. Use your imagination to design your own model of real-world growth processes. You are encouraged to read the research literature and recent books on growth models.

R. C. Ball and R. M. Brady, "Large scale lattice effect in diffusion-limited aggregation," *J. Phys. A* **18**, L809 (1985). The authors discuss the optimization algorithm used in Project 14.2.

Albert-László Barabási and H. Eugene Stanley, *Fractal Concepts in Surface Growth*, Cambridge University Press (1995).

K. S. Birdi, *Fractals in Chemistry, Geochemistry, and Biophysics*, Plenum Press (1993).

Armin Bunde and Shlomo Havlin, editors, *Fractals and Disordered Systems*, Springer-Verlag (1991).

Fereydoon Family and David P. Landau, editors, *Kinetics of Aggregation and Gelation*, North-Holland (1984). A collection of research papers that give a wealth of information, pictures, and references on a variety of growth models.

Fereydoon Family, Daniel E. Platt, and Tamás Vicsek, "Deterministic growth model of pattern formation in dendritic solidification," *J. Phys. A* **20**, L1177 (1987). The authors discuss the nature of Laplace fractal carpets.

Fereydoon Family and Tamás Vicsek, editors, *Dynamics of Fractal Surfaces*, World Scientific (1991). A collection of reprints.

Fereydoon Family, Y. C. Zhang, and Tamás Vicsek, "Invasion percolation in an external field: dielectric breakdown in random media," *J. Phys. A.* **19**, L733 (1986).

Jens Feder, *Fractals*, Plenum Press (1988). This text discusses the applications as well as the mathematics of fractals.

J.-M. Garcia-Ruiz, E. Louis, P. Meakin, and L. M. Sander, editors, *Growth Patterns in Physical Sciences and Biology*, NATO ASI Series B304, Plenum (1993).

J. M. Hammersley and D. C. Handscomb, *Monte Carlo Methods*, Methuen (1964). The chapter on percolation processes discusses a growth algorithm for percolation.

Shlomo Havlin and Daniel Ben-Avraham, "Diffusion in disordered media," *Adv. Phys.* **36**, 695 (1987).

H. J. Herrmann, "Geometrical Cluster Growth Models and Kinetic Gelation," *Phys. Repts.* **136**, 154 (1986).

Robert C. Hilborn, *Chaos and Nonlinear Dynamics*, Oxford University Press (1994).

Benoit B. Mandelbrot, *The Fractal Geometry of Nature*, W. H. Freeman (1983). An influential and beautifully illustrated book on fractals.

Imtiaz Majid, Daniel Ben-Avraham, Shlomo Havlin, and H. Eugene Stanley, "Exact-enumeration approach to random walks on percolation clusters in two dimensions," *Phys. Rev. B* **30**, 1626 (1984).

P. Meakin, "The growth of rough surfaces and interfaces," *Physics Reports* **235** , 189 (1993). The author has written many seminal articles on DLA and other aggregation models.

Paul Meakin, *Fractals, Scaling and Growth Far From Equilibrium*, Cambridge University Press (1995).

L. Niemeyer, L. Pietronero, and H. J. Wiesmann, "Fractal dimension of dielectric breakdown," *Phys. Rev. Lett.* **52**, 1033 (1984).

H. O. Peitgen and P. H. Richter, *The Beauty of Fractals*, Springer-Verlag (1986).

Luciano Pietronero and Erio Tosatti, editors, *Fractals in Physics*, North-Holland (1986). A collection of research papers, many of which are accessible to the motivated reader.

Mark Przyborowski and Mark van Woerkom, "Diffusion of many interacting random walkers on a three-dimensional lattice with a personal computer," *Eur. J. Phys.* **6**, 242 (1985). This work was done while the authors were high school students in West Germany.

F. Reif, *Fundamentals of Statistical and Thermal Physics,* McGraw-Hill (1965). Einstein's relation between the diffusion and mobility is discussed in Chapter 15.

John C. Russ, *Fractal Surfaces*, Plenum Press (1994). A disk also is included.

Evelyn Sander, Leonard M. Sander, and Robert M. Ziff, "Fractals and Fractal Correlations," *Computers in Physics* **8**, 420 (1994). An introduction to fractal growth models and the calculation of their properties. One of the authors, Leonard Sander, is a co-developer of the diffusion limited aggregation model (see Problem 14.9).

H. Eugene Stanley and Nicole Ostrowsky, editors, *On Growth and Form,* Martinus Nijhoff Publishers, Netherlands (1986). A collection of research papers at approximately the same level as the Family and Landau collection. The article by Paul Meakin on DLA was referenced in the text.

Hideki Takayasu, *Fractals in the Physical Sciences*, John Wiley & Sons (1990).

David D. Thornburg, *Discovering Logo*, Addison-Wesley (1983). The book is more accurately described by its subtitle, *An Invitation to the Art and Pattern of Nature*. The nature of recursive procedures and fractals are discussed using many simple examples.

Donald L. Turcotte, *Fractals and Chaos in Geology and Geophysics*, Cambridge University Press (1992).

Tamás Vicsek, *Fractal Growth Phenomena*, second edition, World Scientific Publishing (1991). This book contains an accessible introduction to diffusion limited and cluster-cluster aggregation.

Bruce J. West, *Fractal Physiology and Chaos in Medicine*, World Scientific Publishing (1990).

David Wilkinson and Jorge F. Willemsen, "Invasion percolation: a new form of percolation theory," *J. Phys. A* **16**, 3365 (1983).

CHAPTER

15
Complexity

We introduce cellular automata models, neural networks, genetic algorithms, and explore the concepts of self-organization and complexity.

15.1 CELLULAR AUTOMATA

Part of the fascination of physics is that it allows us in many cases to reduce natural phenomena to a few simple laws. It is perhaps even more fascinating to think about how a few simple laws can produce the enormously rich behavior that we see in nature. In this chapter we will discuss several models that illustrate some of the new ideas that are emerging from the study of "complex systems."

The first class of models we discuss are known as *cellular automata*. Cellular automata were originally introduced by von Neumann and Ulam in 1948 as an idealization of biological self-reproduction, and are examples of discrete dynamical systems that can be simulated exactly on a digital computer. A cellular automaton can be thought of as a checkerboard with colored squares (the cells). Each cell changes its color at the tick of an external clock according to a rule based on the present configuration (microstate) of the cells in its neighborhood.

More formally, cellular automata are mathematical idealizations of dynamical systems in which space and time are discrete and the quantities of interest have a finite set of discrete values that are updated according to a local rule. The important characteristics of cellular automata include the following:

1. Space is discrete, and there is a regular array of sites (cells). Each site has a finite set of values.
2. Time is discrete, and the value of each site is updated in a sequence of discrete time steps.
3. The rule for the new value of a site depends only on the values of a *local* neighborhood of sites near it.
4. The variables at each site are updated *simultaneously* ("synchronously") based on the values of the variables at the previous time step.

Because the original motivation for studying cellular automata was their biological aspects, the lattice sites frequently are referred to as cells. More recently, cellular automata have been applied to a wide variety of physical systems ranging from fluids to galaxies. We will refer to sites rather then cells, except when we are explicitly discussing biological systems.

We first consider one-dimensional cellular automata with the neighborhood of a given site assumed to be the site itself and the sites immediately to the left and right of it. Each site also is assumed to have two states (a Boolean automata). An example of such a rule is illustrated in Fig. 15.1, where we see that a rule can be labeled by the binary representation of the update for each of the eight possible neighborhoods and by the base ten equivalent of the binary representation. Because any eight digit binary number specifies an one-dimensional cellular automata, there are $2^8 = 256$ possible rules.

Program ca1 takes as input the decimal representation of the rule and produces the rule matrix (array update). This array is used to update each site on the lattice using periodic boundary conditions. On a single processor computer, it is necessary to use an

t:	111	110	101	100	011	010	001	000
t + 1:	0	1	0	1	1	0	1	0

Fig. 15.1   Example of a local rule for the time evolution of a one-dimensional cellular automaton. The variable at each site can have values 0 or 1. The top row shows the $2^3 = 8$ possible combinations of three sites. The bottom row gives the value of the central site at the next time step. This rule is termed 01011010 in binary notation (see the second row), the modulo-two rule, or rule 90. Note that 90 is the base ten (decimal) equivalent of the binary number 01011010, i.e., $90 = 2^1 + 2^3 + 2^4 + 2^6$.

additional array so that the state of each site can be updated using the previous values of the sites in its local neighborhood. The state of the sites as a function of time is shown on the screen with time running downwards.

```
PROGRAM ca1
! one-dimensional Boolean cellular automata
DIM update(0 to 7),site(0 to 501)
CALL setrule(update())
CALL initial(site(),L,tmax,#2)
CALL iterate(site(),L,update(),tmax,#2)
END

SUB setrule(update())
    INPUT prompt "rule number = ": rule
    OPEN #1: screen 0,0.5,0.2,0.8
    SET BACKGROUND COLOR "black"
    SET COLOR "white"
    FOR i = 7 to 0 step -1
        LET update(i) = int(rule/2^i)  ! find binary representation
        LET rule = rule - update(i)*2^i
        LET bit2 = int(i/4)
        LET bit1 = int((i - 4*bit2)/2)
        LET bit0 = i - 4*bit2 - 2*bit1
        ! show possible neighborhoods
        PRINT using "#": bit2,bit1,bit0;
        PRINT "   ";
    NEXT i
    PRINT
    FOR i = 7 to 0 step -1
        PRINT using "##": update(i);    ! print rules
        PRINT "    ";
    NEXT i
    CLOSE #1
END SUB
```

```
SUB initial(site(),L,tmax,#2)
    RANDOMIZE
    OPEN #2: screen 0.5,1,0.1,0.9
    ASK PIXELS px,py
    SET WINDOW 1,px,py,1
    SET COLOR "yellow"
    LET L = 2*int(px/8) - 8
    LET tmax = L
    LET site(L/2) = 1              ! center site
    BOX AREA 1+2*L,2*L+4,1,4       ! each site 4 x 4 pixels
END SUB

SUB iterate(site(),L,update(),tmax,#2)
    ! update lattice
    ! need to introduce additional array, sitenew, to temporarily
    ! store values of newly updated sites
    DIM sitenew(0 to 501)
    FOR t = 1 to tmax
        FOR i = 1 to L
            LET index = 4*site(i-1) + 2*site(i) + site(i+1)
            LET sitenew(i) = update(index)
            IF sitenew(i) = 1 then BOX AREA 1+i*4,i*4+4,1+t*4,t*4+4
        NEXT i
        MAT site = sitenew
        LET site(0) = site(L)      ! periodic boundary conditions
        LET site(L+1) = site(1)
    NEXT t
END SUB
```

The properties of all 256 one-dimensional cellular automata have been cataloged (see Wolfram). We explore some of the properties of one-dimensional cellular automata in Problems 15.1 and 15.2.

Problem 15.1   One-dimensional cellular automata

a.   Use *Program ca1* and rule 90 shown in Fig. 15.1. This rule also is known as the "modulo-two" rule, because the value of a site at step $t + 1$ is the sum modulo 2 of its two neighbors at step t. Choose the initial configuration to be a single nonzero site (seed) at the midpoint of the lattice. It is sufficient to consider the time evolution for approximately twenty steps. Is the resulting pattern of nonzero sites self-similar? If so, characterize the pattern by a fractal dimension.

b.   Consider the properties of a rule for which the value of a site at step $t + 1$ is the sum modulo 2 of the values of its neighbors *plus* its own value at step t. This rule is termed rule 10010110 or rule $150 = 2^1 + 2^2 + 2^4 + 2^7$. Start with a single seed site.

c.   Choose a random initial configuration for which the independent probability for each site to have the value 1 is 50%; otherwise, the value of a site is 0.

Consider the time evolution of rule 90, rule 150, rule $18 = 2^1 + 2^4$ (00010010), rule $73 = 2^0 + 2^3 + 2^6$ (01001001), and rule 136 (10001000). How sensitive are the patterns that are formed to changes in the initial conditions? Does the nature of the patterns depend on the use or nonuse of periodic boundary conditions?

Because the dynamical behavior of many of the 256 one-dimensional Boolean cellular automata is uninteresting, we also consider one-dimensional Boolean cellular automata with larger neighborhoods. The larger neighborhood implies that there are many more possible update rules, and it is convenient to place some reasonable restrictions on the rules. First, we assume that the rules are symmetric, e.g., the neighborhood 100 produces the same value for the central site as 001. We also assume that the zero neighborhood 000 yields 0 for the central site, and that the value of the central site depends only on the sum of the values of the sites in the neighborhood, e.g., 011 produces the same value for the central site as 101 (Wolfram 1984).

A simple way of coding the rules that is consistent with these requirements is as follows. Call the size of the neighborhood z if the neighborhood includes $2z + 1$ sites. Each rule is labeled by $\sum_m a_m 2^m$, where $a_m = 1$ if the central cell is 1 when the sum of all values in the neighborhood equals m; else $a_m = 0$. As an example, take $z = 2$ and suppose that the central site equals one when two or four sites are unity. This rule is labeled by $2^2 + 2^4 = 20$.

Problem 15.2   More one-dimensional cellular automata

a. Modify *Program* `ca1` so that it incorporates the possible rules discussed in the text for a neighborhood of $2z + 1$ sites. How many possible rules are there for $z = 1$? Choose $z = 1$ and a random initial configuration, and determine if the long time behavior for each rule belongs to one of the following classes:

1. A homogeneous state where every site is either 0 or 1. An example is rule 8.
2. A pattern consisting of separate stable or periodic regions. An example is rule 4.
3. A chaotic, aperiodic pattern. An example is rule 10.
4. A set of complex, localized structures that may not live forever. There are no examples for $z = 1$.

b. Modify your program so that $z = 2$. Wolfram (1984) claims that rules 20 and 52 are the only examples of complex behavior (class 4). Describe how the behavior of these two rules differs from the behavior of the other rules. Determine the fraction of the rules belonging to the four classes.

c. Repeat part (b) for $z = 3$.

d. Assume that sites can have three values, 0, 1, and 2. Classify the behavior of the possible rules for the case $z = 1$.

The results of Problem 15.2 suggest that an important feature of cellular automata is their capability for "self-organization." In particular, the class of complex localized

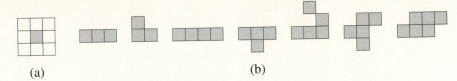

Fig. 15.2   (a) The local neighborhood of a site is given by the sum of its eight neighbors. (b) Examples of initial configurations for the Game of Life, some of which lead to interesting patterns. Live cells are shaded.

structures is distinct from regular as well as aperiodic structures. This intermediate structure is the focus of *complexity theory* whose goal is to explain complex phenomena in nature.

One-dimensional models are too limited to study the complexity of nature, and we now consider several two-dimensional models. The philosophy is the same except that the neighborhood contains more sites. *Program ca2* sets up the rule matrix and updates sites using the eight neighbor sites shown in Fig. 15.2a. There are now $2^9 = 512$ possible configurations for the eight neighbors and the center site, and 2^{512} possible rules. Clearly, we cannot go through all these rules in any systematic fashion as we did for one-dimensional cellular automata. For this reason, we will set up our rule matrix based on other considerations.

The rule matrix incorporated in *Program ca2* implements the best known two-dimensional cellular automata model: the *Game of Life*. This model, invented in 1970 by the mathematician John Conway, produces many fascinating patterns. The rules of the game are simple. For each cell determine the sum of the values of its four nearest and four next-nearest neighbors (see Fig. 15.2a). A "live" cell (value 1) remains alive only if this sum equals 2 or 3. If the sum is greater than 3, the cell will "die" (become 0) at the next time step due to overcrowding. If the sum is less than 2, the cell will die due to isolation. A dead cell will come to life only if the sum equals 3.

```
PROGRAM ca2
LIBRARY "csgraphics"
DIM update(0 to 511),cell(50,50)
CALL setrule(update(),L)
LET flag$ = ""
DO
    CALL initial(cell(,),L)
    DO
        CALL iterate(cell(,),update(),L)
        IF key input then
            GET KEY k
            IF (k = ord("s")) or (k = ord("S")) then
                LET flag$ = "stop"
            END IF
        END IF
```

```
                    LOOP UNTIL k <> 0
                 LET k = 0
              LOOP until flag$ = "stop"
              END

              SUB setrule(update(),L)
                  ! rule for Game of Life
                  FOR i = 0 to 511
                      LET update(i) = 0
                  NEXT i
                  ! three neighbors alive
                  FOR nn1 = 0 to 5
                      FOR nn2 = nn1+1 to 6
                          FOR nn3 = nn2+1 to 7
                              LET index = 2^nn1 + 2^nn2 + 2^nn3
                              LET update(index) = 1   ! center dead
                              LET update(index+256) = 1    ! center alive
                          NEXT nn3
                      NEXT nn2
                  NEXT nn1
                  ! two neighbors and center alive
                  FOR nn1 = 0 to 6
                      FOR nn2 = nn1+1 to 7
                          LET index = 256 + 2^nn1 + 2^nn2
                          LET update(index) = 1
                      NEXT nn2
                  NEXT nn1
                  SET BACKGROUND COLOR "black"
                  SET COLOR "white"
                  INPUT prompt "lattice size = ": L
                  CALL compute_aspect_ratio(L,xwin,ywin)
                  SET WINDOW -0.2*xwin,1.2*xwin,-0.2*ywin,1.2*ywin
              END SUB

              SUB initial(cell(,),L)
                  FOR i = 1 to L
                      FOR j = 1 to L
                          LET cell(i,j) = 0
                          SET COLOR "yellow"
                          BOX AREA i,i+1,j,j+1
                          SET COLOR "black"
                          BOX LINES i,i+1,j,j+1
                      NEXT j
                  NEXT i
                  SET CURSOR 1,1
                  ! click on cell to change its state or outside of lattice
                  ! to update cells
                  SET COLOR "white"
                  PRINT "click on cell to toggle or outside of lattice to continue."
```

```
            DO
                GET POINT x,y
                IF x > 1 and x < L and y > 1 and y < L then
                    LET i = truncate(x,0)
                    LET j = truncate(y,0)
                    IF cell(i,j) = 0 then
                        SET COLOR "black"
                        BOX AREA i,i+1,j,j+1
                        LET cell(i,j) = 1
                    ELSE
                        SET COLOR "yellow"
                        BOX AREA i,i+1,j,j+1
                        LET cell(i,j) = 0
                        SET COLOR "black"
                        BOX LINES i,i+1,j,j+1
                    END IF
                END IF
            LOOP until x < 1 or x > L or y < 1 or y > L
            SET CURSOR 1,1
            SET COLOR "white"
            PRINT "Hit any key for new lattice, 's' to stop";
            PRINT "                                        "
        END SUB

        SUB iterate(cell(,),update(),L)
            DIM cellnew(50,50)
            FOR i = 1 to L
                FOR j = 1 to L
                    CALL neighborhood(cell(,),i,j,sum,L)
                    LET cellnew(i,j) = update(sum)
                    IF cell(i,j) = 1 and cellnew(i,j) = 0 then
                        SET COLOR "yellow"
                        BOX AREA i,i+1,j,j+1
                        SET COLOR "black"
                        BOX LINES i,i+1,j,j+1
                    ELSE IF cell(i,j) = 0 and cellnew(i,j) = 1 then
                        SET COLOR "black"
                        BOX AREA i,i+1,j,j+1
                    END IF
                NEXT j
            NEXT i
            MAT cell = cellnew
        END SUB

        SUB neighborhood(cell(,),i,j,sum,L)
            LET ip = i + 1
            LET im = i - 1
            LET jp = j + 1
            LET jm = j - 1
```

```
            IF i = 1 then
                LET im = L
            ELSE IF i = L then
                LET ip = 1
            END IF
            IF j = 1 then
                LET jm = L
            ELSE IF j = L then
                LET jp = 1
            END IF
            LET sum = cell(i,jp) + 2*cell(i,jm) + 4*cell(im,j)
            LET sum = sum + 8*cell(ip,j)+ 16*cell(ip,jp) + 32*cell(ip,jm)
            LET sum = sum + 64*cell(im,jp) + 128*cell(im,jm) + 256*cell(i,j)
    END SUB
```

Program ca2 allows the user to use any update rule by changing SUB setrule. *Program* ca2 has not been optimized for the Game of Life and is written so that it can be easily modified for any cellular automata rule.

Problem 15.3   The Game of Life

a.   *Program* ca2 allows the user to determine the initial configuration interactively. Choose several initial configurations with a small number of live cells and investigate the different types of patterns that emerge. Some suggested initial configurations are shown in Fig. 15.2b. Does it matter whether you use fixed or periodic boundary conditions?

b.   Modify *Program* ca2 so that each cell is initially alive with a 50% probability. What types of patterns typically result after a long time? What happens for 20% live cells? What happens for 70% live cells?

* **c.**   Assume that each cell is initially alive with probability p. Given that the density of live cells at time t is $\rho(t)$, what is $\rho(t+1)$, the expected density at time $t+1$? Do the simulation and plot $\rho(t+1)$ versus $\rho(t)$. If $p = 0.5$, what is the steady-state density of live cells?

* **d.**   As we found in part (b), the system will develop structure even if each cell is randomly populated at $t = 0$. One measure of the increasing order in the system has been introduced by Schulman and Seiden and is analogous to the entropy in statistical mechanics. The idea is to divide the $L \times L$ system into boxes of linear dimension b and determine n_i, the number of live cells in the ith box. The quantity S is given by

$$S = \frac{1}{L^2} \log_2 \prod_{i=1}^{(L/b)^2} \binom{b^2}{n_i}. \tag{15.1}$$

The argument of the logarithm in (15.1) is the total number of microscopic states associated with a given sequence of n_i. Roughly speaking, S measures the extent to which live cells are correlated. Determine S as a function of time starting from a 50% random configuration. First consider $L = 50$ and $b = 2, 3, 4$, and 5. Average over at least ten separate runs. Describe how the behavior of the entropy depends on the level of "coarse graining" determined by the value of b. Does S decrease monotonically with time? Does it reach an equilibrium value? Increase L and the number of independent runs, and repeat your averages.

The Game of Life is an example of a universal computing machine. That is, we can set up an initial configuration of live cells to represent any possible program and any set of input data, run the Game of Life, and in some region of the lattice the output data will appear. The proof of this result (see Berlekamp et al.) involves showing how various configurations of cells represent the components of a computer including wires, storage, and the fundamental components of a CPU — the digital logic gates that perform and, or, and other logical and arithmetic operations.

Problem 15.4   Other two-dimensional cellular automata

a.   Consider a Boolean automaton with each site labeled by 1 ("on") and 0 ("off"). We adopt an update rule such that the value of each site at time $t + 1$ is determined by the vote of its four nearest neighbors (on a square lattice) at time t. The update rule is that a site becomes on if 2, 3, or 4 of its four neighbors are on. Consider initial configurations for which 1-sites occur with probability p and 0-sites occur with probability $1 - p$. Because the voting rule favors the growth of 1-sites, it is interesting to begin with a minority of 1-sites. Choose $p = 0.1$ and determine what happens to isolated 1-sites. How do they grow initially? For what shape (convex or concave) does a cluster of 1-sites stop growing? What happens to clusters of 1-sites such as those shown in Fig. 15.3? (If necessary, create such a configuration.) Show that for $p = 0.1$, the system eventually freezes in a pattern made of 1-site rectangular islands in a sea of 0-sites. What happens for $p = 0.14$? Can you define a "critical density" p_c at which the behavior of the system changes? Consider square lattices with linear dimension $L = 128$ and $L = 256$ (see Vichniac).

b.   Suppose that the update rule is determined by the sum of the center cell and its eight nearest and next-nearest neighbors. If this sum equals 4 or more, then the center site equals 1 at the next time step (see Vichniac). This rule also favors the growth of 1-sites and leads to a phenomenon similar to that found in part (a). Consider an initial configuration for which 1-sites occur with probability p and 0-sites occur with probability $1 - p$. Choose $L = 128$ and show that for $p = 0.2$, the system eventually freezes. What is the shape of the 1-clusters? Show that if a single 0-site is changed to a 1-site at the surface of the largest 1-cluster, the cluster of 1-sites grows. What is the eventual state of the system? What is the behavior of the system for $p = 0.3$? Is it possible to define a "critical density" p_c such that for $p \geq p_c$, the growth of the 1-clusters continues until all sites change to the 1-state? Consider larger values of L and

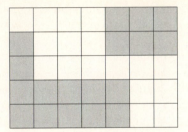

Fig. 15.3   What is the evolution
of these clusters using
the rule discussed
in Problem 15.4a.
The shaded squares
correspond to the
1-sites.

show that the value of p_c appears to be insensitive to the value of L. What is
your estimated value of p_c in the limit of an infinite lattice?

c.   There is one problem with the conclusions that you were able to reach in parts
(a) and (b) — they are incorrect! The finite lattice results are misleading and p_c
is zero in the limit of an infinite lattice. The probability of *any* configuration of
1-sites is unity for an infinite lattice. That is, somewhere in the lattice there is a
"critical cluster" that will grow indefinitely until all sites in the lattice change
to the 1-state. The moral of the story is, "Do not trust a simulation without a
theory" (a paraphrase of a quote usually attributed to Eddington).

*Problem 15.5   Ising model cellular automata

a.   One of the most frequently studied models in statistical physics is the Ising
model. In this model each cell has two states s_i that are labeled by ± 1 instead
of 0 and 1. The energy is given by

$$E = -J \sum_{i,j=\text{nn}(i)}^{N} s_i s_j. \tag{15.2}$$

The notation $j = \text{nn}(i)$ indicates that j is a nearest neighbor of i. If we think of
s_i as representing the magnetic moment or spin at cell i, then the Ising model
can be understood as a model of magnetism with J being the strength of the
interaction between nearest neighbor spins. For $J > 0$, the lowest energy state
occurs when all the spins are either all up ($s_i = +1$) or all down ($s_i = -1$). The
magnetization m per spin is given by

$$m = \frac{1}{N} \sum_{i=1}^{N} s_i. \tag{15.3}$$

Monte Carlo algorithms for the simulation of the Ising model are discussed in Chapters 16 and 17.

In the cellular automata implementation of the Ising model, the energy of the entire lattice is fixed, and a spin may flip only if the energy of the lattice does not change. Such a situation occurs if a spin has two up neighbors and two down neighbors. Hence the update rule is to flip the spin at i, i.e., $s_i \to -s_i$, if it has precisely two up neighbors; otherwise do not change s_i. But because we want to update all sites simultaneously, we have a problem. That is, if two neighboring spins have opposite signs and each has a total of two up neighbors, then flipping both spins would change the total energy. Why? The trick is to divide the lattice into two kinds of sites corresponding to the red and black squares of a checkerboard. First we update simultaneously all the red squares (hence keeping their neighbors, the black squares fixed), and then we update simultaneously all the black squares. Implement this cellular automata update of the Ising model by modifying *Program ca2* and write a subroutine to compute the mean magnetization per spin and the total energy per spin E/N. (The latter should not change with time.) The average is over the different configurations generated by the cellular automata updates. Use periodic boundary conditions.

b. Compute the mean magnetization as a function of E/N for a square lattice with $L = 20$. One way to generate an initial configuration is to let each spin be up with probability p. Allow the system to "equilibrate" before accumulating values of the magnetization. Does the mean magnetization change from being close to zero to being close to unity as E/N is lowered? If such a qualitative change occurs, we say that there is a phase transition. To improve your results, average over many different initial configurations with the same value of E/N and determine the dependence of the mean magnetization on E/N. Because the same value of p will occasionally lead to different initial energies, write a subroutine that flips spins until the desired energy is obtained.

c. Repeat part (b) for a 40×40 lattice. Do your qualitative conclusions change?

d. One difficulty with the cellular automata version of the Ising model is that it is not ergodic, that is, there are configurations with the same energy that cannot be reached for a given initial condition. In Chapter 16 we will see how to avoid this problem using Monte Carlo methods. Here we might allow the total energy E_0 to vary a little, say $\pm nJ$, where n is an integer. During a run we can periodically flip a spin at random such that the total energy is in the range $E_0 \pm nJ$. Try this method for different values of n. Do your results for the mean magnetization change significantly?

Program ca1 and *ca2* are inefficient due in part to the limitations imposed by the True BASIC language which does not have *bit* manipulation capability. The smallest element of computer memory contains a bit, which is a 0 or a 1. A *byte* is the size of memory needed to hold a single character, e.g., a letter or a digit. More precisely, a byte is eight bits. Because there are $2^8 = 256$ possible arrangements of 1's and 0's in a byte, a byte can represent the ASCII character set, including all upper and lower case letters, numerals, punctuation, and other control characters such as a line feed. A computer *word* is usually two, four, or eight bytes, and is the unit of storage that can be accessed simultaneously and moved back and forth from the central processing unit (CPU) to various storage devices. If we could manipulate bits directly, then we could represent each site in a Boolean cellular automaton by a bit and update a whole word of sites (32 sites on a 32 bit machine) simultaneously. This type of update is a simple example of parallel processing on a single processor machine.

The C language has intrinsic bit manipulation operations; most Fortran 77 and all Fortran 90 compilers also have intrinsic functions that perform bit manipulation. One of the enticements of cellular automata is that they can run very quickly on parallel processors. Special processing boards exist for personal computers and workstations that run cellular automata models faster than on a general purpose supercomputer. A major interest in cellular automata is how they can be used as a general paradigm for massively parallel computation (cf. Toffoli and Margolus).

15.2 LATTICE GAS MODELS OF FLUID FLOW

One of the most promising applications of cellular automata models is to simulations of fluid flow. Fluid flow is very difficult to simulate because the partial differential equation describing fluid flow, the Navier-Stokes equation, is nonlinear. As we have found, nonlinear equations can lead to the breakdown of standard numerical algorithms. In addition, there are typically many length scales that must be considered simultaneously. These length scales include the microscopic motion of the fluid particles, the length scales associated with fluid structures such as vortices, and the length scales of macroscopic objects such as pipes or obstacles. Because of all these considerations, simulations of fluid flow based on direct numerical solutions of the Navier-Stokes equation typically require sophisticated numerical methods (cf. Oran and Boris).

The cellular automata models of fluids are known as *lattice gas* models. These models are based on the idea that if we maintain the fundamental conservation laws and symmetries associated with fluids, then we can produce the correct physics at the macroscopic level if we average over many particles. In a lattice gas model the positions of the particles are restricted to the sites of a lattice and the velocities are restricted to a small number of velocity vectors. A microscopic model needs to include two processes, the free motion between collisions and the collisions. In the simplest models, the particles move freely to their nearest neighbor lattice site in one time step. Then the velocities of the particles at each lattice site are updated according to a collision rule that conserves mass, momentum, and kinetic energy. The free motion and the collisions for all sites are computed simultaneously.

To understand the nature of lattice gas models, it is easier to discuss a specific model. Three-dimensional models are being studied, but they are much more difficult to visualize and to understand theoretically, and we will consider only two-dimensional models. We assume a triangular lattice, because its symmetry is more closely related to that of a continuum than a square lattice. In addition, collision rules for square lattices do not typically mix the horizontal and vertical motions of the particles. All particles are assumed to have the same speed and mass. The possible velocity vectors lie only along the links connecting sites, and hence there are only six possible velocities (labeled 0 to 5 because the first bit in a computer byte is in the 0 position):

$$\mathbf{v}_0 = (1, 0) \qquad \mathbf{v}_1 = (1, -\sqrt{3})/2 \qquad \mathbf{v}_2 = -(1, \sqrt{3})/2$$
$$\mathbf{v}_3 = (-1, 0) \qquad \mathbf{v}_4 = (-1, \sqrt{3})/2 \qquad \mathbf{v}_5 = (1, \sqrt{3})/2 \; .$$

In some models a rest particle also is allowed. Each site can have at most one particle moving in a particular direction. The update process proceeds by first moving all the particles in the direction of their velocity to a neighboring site. Then at each lattice site the velocity vectors are changed according to a collision rule. Particle number and kinetic energy are easily conserved because all particles have the same speed, and we need only insure momentum conservation. There is some flexibility in the choice of rules. One set of collision rules is illustrated in Fig. 15.4. These rules are deterministic with only one possible set of velocities after a collision for each possible set of velocities before a collision.

Because True BASIC does not provide intrinsic bit manipulation functions, we do not list a program in True BASIC. Instead, we list a Fortran and C lattice gas program in Appendices A and B, respectively. You can convert either program to True BASIC by using the fact that mod(int(A/2i^N),2) is the nth bit of the integer A. We can determine if there is a particle at site x, y with velocity $\mathbf{v}_2$ by testing whether mod(int(lat(x,y)/2^N,2) is equal to 1. Or it is possible to introduce a string of eight characters, each of which can be 0 or 1 as a model for a computer byte. For example, the nth lattice site with two particles of velocity $\mathbf{v}_2$ and $\mathbf{v}_4$ can be represented by the string array lat$(n) = "001010". The string segment of astring between the pth and qth character is given by astring$[p:q]. We can determine if there is a particle with velocity $\mathbf{v}_2$ by testing whether lat$(n)[3:3] is equal to "1".

We now describe the procedure for storing information about the particles in an array of integers, lat. The eight bits in a byte are labeled from 0 to 7. In principle, we could use a four byte integer to update four sites simultaneously, but to avoid complications we will not do so. Instead each site of the lattice is represented by one element of lat. We use the first six bits from 0 to 5 of an integer to represent particles moving in the six possible directions with bit 0 corresponding to a particle moving with velocity v_0. For example, if bit 0 equals unity, we know there is a particle at this site with velocity v_0. If there are three particles with velocities v_0, v_2, and v_4 at a site, then this situation is represented by 00010101 in bit notation; the corresponding decimal equivalent is 21. From Fig. 15.4 we see that after the collision there are three particles with velocities v_1, v_3, and v_5 corresponding to 42. We can express this collision as rule(21) = 42. Similar decimal equivalents can be expressed for all the other possible collisions.

We reserve bit 6 for a possible rest particle. If we want to occupy a site with a fixed particle to represent a barrier, we use bit 7, that is, we set the value of barrier sites equal to $2^7 = 128$. What boundary condition should we use when a particle is adjacent to a barrier site and heading toward it? The simplest rule that insures that we do not lose any particles is to set the velocity $\mathbf{v}$ of such a particle equal to $-\mathbf{v}$. Other possibilities are to set the angle of incidence equal to the angle of reflection or to set the velocity to

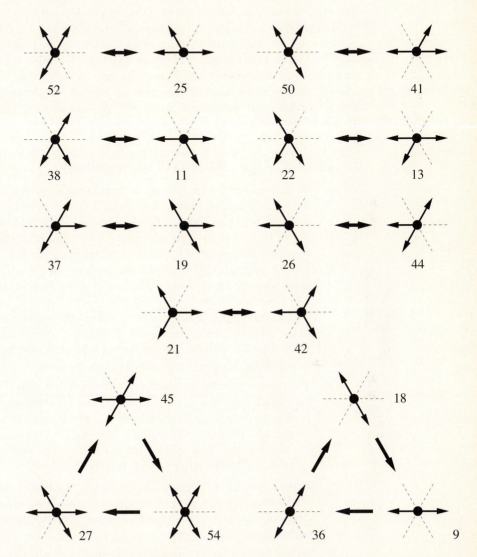

Fig. 15.4 Examples of collision rules for a lattice gas on a triangular lattice. The rule for configurations that are not shown is that the velocities do not change after a collision. The numbers represent the way that the velocities at a lattice site are encoded.

an arbitrary value. The latter case corresponds to a rough surface and is more difficult to implement because we need to insure that no site has more than one particle with the same velocity.

An important use of lattice gas models is to simulate flow in and around various geometries. The fluid velocity field can be seen to develop vortices, wakes, and other fluid structures near obstacles. Typically, particles are injected at one end and absorbed as they reach the other end. Periodic boundary conditions are used in the other direction. Very large lattices are required to obtain quantitative results, because it is necessary to average the velocity over many sites. The density of the fluid is determined by the average number of particles per site, and the pressure can be varied by changing the flux of particles that are injected at one end. We discuss some typical applications in Problems 15.6–15.8.

Problem 15.6   Approach to equilibrium

a.   To maintain zero total velocity or momentum at all times, use an initial configuration where all sites contain either no particles or six particles whose net momentum is zero. The number density $\rho = 6N/(L_x L_y)$, where N is the number of sites with six particles and the total number of sites is $L_x \times L_y$. Use *Program latgas.f* in Appendix A or *Program latgas.c* in Appendix B as a basis for your program for simulating a lattice gas. Use periodic boundary conditions in both directions. The output of the program should be a pictorial representation of the average velocity at each site, e.g., an arrow pointing in the direction of the average velocity with a size proportional to the magnitude of the average velocity. Begin with a dilute gas with $N = 10$ on a 10×10 triangular lattice. Plot the velocity vector field after each iteration of the lattice gas to check that your program is working correctly.

b.   We now consider the approach to equilibrium. Use a 30×30 lattice and place six particles at every site in a 4×4 region. Describe qualitatively what happens to the particles as a function of time. Approximately how many time steps does it take for the particles to fill the box? Do the particles appear to be at random positions with random velocities? Describe your visual algorithm for determining when equilibrium has been reached.

c.   Repeat part (b) for an initial $b \times b$ region of particles with $b = 2, 6, 8$, and 10. Estimate the equilibration time in each case. What is the qualitative dependence of the equilibration time on b? How does the equilibration time depend on the number density ρ?

d.   Repeat part (b) for an initial 4×4 region of particles, but vary the size of the box. Try $L_x = L_y = 10, 20$, and 40. Estimate the equilibration time in each case. How does the equilibration time depend on the number density ρ?

Problem 15.7   Flow past a barrier

a.   Modify your program from Problem 15.6 so that at each time step three particles are injected with velocities v_0, v_1, and v_5 at each site on the left-hand side. The particles are removed on the right-hand side. Include barrier sites in the middle of the cell representing a $b_x \times b_y$ rectangular barrier. Use the barrier boundary condition that the directions of the reflected particles are reversed, $\mathbf{v} \to -\mathbf{v}$. Use periodic boundary conditions in the vertical direction. Besides the left-hand column and the barrier sites, all other sites are initially empty. Represent the velocity field visually as described in Problem 15.6a.

b. Choose $L_x = 50$ and $L_y = 20$ with a barrier of dimensions $b_x = 5$ and $b_y = 1$. Describe the flow once a steady state velocity field begins to appear. Can you see a wake appearing behind the obstacle? Are there vortices (circular fluid flow)?

c. Repeat part (b) with different size obstacles. Are there any systematic trends?

d. Reduce the pressure by injecting particles at the left every other time step. Are there any noticeable changes in behavior from parts (b) and (c)? Reduce the pressure still further and describe any changes in the fluid flow.

* e. Increase the size of the lattice by a factor of 10 in each direction and average the velocity in each 5×5 region. Compare the flow patterns that you obtain with those obtained in parts (b)–(d).

Problem 15.8   Fluid flow in porous media

a. Modify your lattice gas program so that instead of a rectangular barrier, the barrier sites are placed at random in the lattice. Add a subroutine that sums the horizontal velocity of those particles that reach the right edge of the lattice. The current density is the average of this sum per unit height of the lattice. Compute the current density as a function of porosity, the fraction of sites not containing barriers. If time permits, average over at least ten pore configurations for each value of the porosity. Use $L_x = 50$ and $L_y = 20$.

* b. Vary the size of the lattice and use the finite size scaling procedure discussed in Section 13.4 to estimate the critical exponent μ defined by the dependence of the current density J on the porosity ϕ. That is, $J \sim (\phi - \phi_c)^\mu$. Assume that you know the value of the percolation exponent ν defined by the critical behavior of the connectedness length $\xi \sim |p - p_c|^{-\nu}$ (see Table 13.1).

The principle virtues of lattice gas models are their use of simultaneous updating, which makes them very fast on parallel computers, and their use of integer or boolean arithmetic, which might be faster than floating point arithmetic. Their major limitation is that many sites must be averaged over to obtain quantitative results. It is not yet clear whether lattice gas models are more efficient than standard simulations of the Navier-Stokes equation. The greatest promise for lattice gas models may not be with simple single component fluids, but with multicomponent fluids such as binary fluids and fluids containing bubbles.

15.3   SELF-ORGANIZED CRITICAL PHENOMENON

In nature we rarely see very large events such as a magnitude eight earthquake, an avalanche on a snow covered mountain, the sudden collapse of an empire (e.g., the Soviet Union), or the crash of the stock market. When such rare events occur, are they due to some special set of circumstances or are they a part of a more general pattern of events that would occur without any specific external intervention? The idea of *self-organized criticality* is that in many cases very large events are part of a distribution of events and do not depend on special conditions or external forces.

If s represents the magnitude of an event, such as the energy released in an earthquake or the amount of snow in an avalanche, then a system is said to be *critical* if the number of events $N(s)$ follows a power law:

$$N(s) \sim s^{-\alpha}. \qquad \text{(no characteristic scale)} \qquad (15.4)$$

One implication of the form (15.4) is that there is no characteristic scale. Systems whose correlation or distribution functions decay as power laws are said to be *scale invariant*. This terminology reflects the fact that power laws look the same on all scales. For example, the replacement $s \to bs$ in the function $N(s) = As^{-\alpha}$ yields a function $\tilde{N}(s)$ that is indistinguishable from $N(s)$, except for a change in the amplitude A by the factor $b^{-\alpha}$. If $\alpha \approx 1$, there would be one large event of size 1000 for every 1000 events of size one.

Contrast the power law dependence of $N(s)$ in (15.4) to the result of combining a large number of independently acting random events. In this case we know that the distribution of the sum is a Gaussian (see Problem 12.9) and that $N(s)$ has the form

$$N(s) \sim e^{-(s/s_0)^2}. \qquad \text{(characteristic scale)} \qquad (15.5)$$

Scale invariance does not hold for functions that decay exponentially, because the replacement $s \to bs$ in the function $e^{-(s/s_0)^2}$ changes s_0 (the characteristic scale of s) by the factor b^2. We note that for a power law distribution, there are events of all sizes, but for a Gaussian distribution, there are practically speaking no large events.

A common example of self-organized critical phenomena is an idealized sand pile. Suppose that we construct a sand pile by randomly adding one grain at a time onto a flat surface with open edges. Initially, the grains will stay more or less where they land, but after a while there will be small avalanches during which the grains move so that the local slope of the pile is not too large. Eventually, the pile reaches a statistically stationary (time-independent) state and the amount of sand added balances the sand that falls off the edge (on the average). When a single grain of sand is added to a configuration belonging to this state, a rearrangement might occur that triggers an avalanche of any size (up to the size of the system), so that the mean slope again equals the critical value. We say that the statistically stationary state is critical because there are avalanches of all sizes. The stationary state is self-organized because no external parameter (such as the temperature) needs to be tuned to force the system to this state. In contrast, the concentration of fissionable material in a nuclear chain reaction has to be carefully controlled for the nuclear chain reaction to be exactly critical.

The nature of self-organized critical phenomena can be understood by considering some simple models. We begin with a one-dimensional model based on the simplified behavior of a sand pile. Consider a lattice of L sites and let the height at each site be represented by the array element $h(i)$. One grain of sand is added to the leftmost site, $h(1) = h(1) + 1$, at each time step. During this time step all the sites are checked to see if $h(i) - h(i + 1) > 1$. All sites that satisfy this condition are marked for "toppling." Next each marked site is toppled. A simple rule is to let $h(i) = h(i) - 1$ and $h(i + 1) = h(i + 1) + 1$, i.e., the sand falls to the right. Any grains of sand that go beyond $i = L$ are lost forever. *Program sandpile* implements this simple model. We will find in Problem 15.9 that slightly more complicated rules are necessary to find nontrivial behavior.

```
PROGRAM sandpile
! simple one-dimensional sandpile model
DIM height(51),move(100),N(0:51)
CALL initial(height(),N(),L,sand$,#1,#2)
LET grain = 0
DO
    LET height(1) = height(1) + 1  ! add grain to first site
    WINDOW #1
    BOX SHOW sand$ at 1,height(1) - 1
    LET topple = 0                 ! number of grains that topple
    LET grain = grain + 1          ! number of grains added
    DO
        CALL check(height(),L,move(),unstable)
        IF unstable = 1 then
            CALL slide(height(),L,move(),sand$,#1)
            LET topple = topple + 1
        END IF
    LOOP until unstable = 0
    LET N(topple) = N(topple) + 1
    CALL show_distribution(N(),L,grain,#2)
LOOP until key input
END

SUB initial(height(),N(),L,sand$,#1,#2)
    OPEN #1: screen 0,0.7,0,1
    INPUT prompt "lattice size L = ": L      ! suggest L = 20
    FOR i = 1 to L
        LET height(i) = 0
    NEXT i
    LET height(L+1) = 0                ! boundary condition
    SET WINDOW -1,L+1,-2,40
    SET COLOR "blue"
    BOX AREA 1.1,1.9,0.1,0.9
    BOX KEEP 1,2,0,1 in sand$
    CLEAR
    SET COLOR "black"
    PLOT LINES: 1,-0.05;L+1,-0.05
```

```
            ! initialize array
            FOR topple = 0 to 20
                LET N(topple) = 0
            NEXT topple
            OPEN #2: screen 0.72,1,0.4,1
            SET WINDOW -0.1,L+1,-0.05,1.01
            PLOT LINES: 0,-0.01;L+0.2,-0.01
            PLOT LINES: -0.05,-0.01;-0.05,1
        END SUB

        SUB check(height(),L,move(),unstable)
            LET unstable = 0
            FOR i = 1 to L
                IF height(i) - height(i+1) > 1 then
                    LET move(i) = 1
                    LET unstable = 1
                ELSE
                    LET move(i) = 0
                END IF
            NEXT i
        END SUB

        SUB slide(height(),L,move(),sand$,#1)
            WINDOW #1
            FOR i = 1 to L
                IF move(i) = 1 then
                    LET height(i) = height(i) - 1     ! sand topples
                    BOX CLEAR i,i + 1,height(i),height(i) + 1
                    IF i < L then
                        ! next site receives grain
                        LET height(i+1) = height(i+1) + 1
                        BOX SHOW sand$ at i+1,height(i+1) - 1
                    END IF
                END IF
            NEXT i
        END SUB

        SUB show_distribution(N(),L,grain,#2)
            WINDOW #2
            ! erase previous graph
            SET COLOR "white"
            BOX AREA 0,L+0.2,0,1
            SET COLOR "black"
            FOR topple = 0 to L
                IF N(topple) > 0 then
                    BOX AREA topple,topple+0.2,0,N(topple)/grain
                END IF
            NEXT topple
        END SUB
```

Problem 15.9   One-dimensional sand piles

a.   Use *Program sandpile* with $L = 20$. Where is the sand added? What is the slope of the sand pile after a long time?

b.   *Program sandpile* computes the number of sites s that topple during each time step and plots the distribution of toppling sizes, $N(s)$, versus s. Modify the program so that $N(s)$ is computed only after the sand pile reaches a steady state. Is the behavior of $N(s)$ interesting for this model?

c.   Modify the rules so that a site which topples loses *two* grains of sand, one to its nearest neighbor to the right and the other to its next nearest neighbor to the right. Plot the distribution $N(s)$ versus s after steady state behavior is reached. Is there a range of values of s for which $N(s)$ shows power law behavior? If so, make the appropriate plot and estimate the critical exponent α.

d.   Introduce the variable $m(i) = h(i + 1) - h(i)$, and convince yourself that the same results are obtained by replacing $h(i)$ by $m(i)$ and using the rule $m(i) > 1$ for toppling. The variable $m(i)$ is called the local slope. Modify your program so that $m(i)$ is used instead of $h(i)$ and adopt the toppling rule used in part (c). Do your results change if you add a grain of sand at each time step at random anywhere in the lattice?

e.   Use the same rule that you considered in parts (c) and (d) and compute $N(s)$ for different values of L and systematically investigate the importance of finite size effects. Average $N(s)$ over many updates and obtain your best estimate for the critical exponent α.

In Problem 15.9 we saw that the fundamental variable is the local slope. Most sand pile models are described using this variable. In the literature many authors refer to the height when they really mean the local slope. In Problem 15.10 we explore the behavior of a two-dimensional model of a sand pile.

Problem 15.10   Two-dimensional sand pile

a.   Write a program to simulate a two-dimensional sand pile using the rule that a site topples if $m(i) > 3$. The pile is grown by choosing a site at random and adding one grain, i.e., m(i) → m(i) + 1. If site i topples (exceeds its critical value), it distributes four grains of sand to its four nearest neighbors (on a square lattice), i.e., m(i) → m(i) − 4. At the edges or corners, only three or two neighbors, respectively, are affected, and the sand that goes outside the boundary of the lattice is lost. Color code the sites according to their value of m(i). In *Program sandpile* we checked all sites for stability, but in two dimensions such a check would take too much time if the size of the lattice were sufficiently large. For this reason we suggest that you maintain two arrays posx and posy which contain the x and y coordinates of those sites that need to be checked for stability. Each time a site topples, add its neighbors' positions to these arrays. Then remove the last site from the arrays and check its stability. Continue this process of removing sites and checking their stability, and adding neighbors to the array until there are no more sites to check.

b. After the critical state has been reached, compute the number of sites s that topple in response to the addition of a single grain. Then compute the distribution of toppling sizes $N(s)$ and determine if $N(s)$ exhibits power law behavior. Begin with $L = 10$ and then consider larger values of L.

* c. Although the model in part (a) might seem reasonably realistic, it is of course over simplified. Laboratory experiments indicate that real sand piles show power law behavior if the piles are small, but larger sand piles do not (see Jaeger et al.). Modify the model to make its behavior more realistic.

Do model sand piles have anything in common with earthquakes? The Gutenberg-Richter law for $N(E)$, the number of earthquakes with energy release E, has been observed empirically and is consistent with power law behavior:

$$N(E) \sim E^{-b}, \tag{15.6}$$

with $b \approx 0.5$. The magnitude of earthquakes on the Richter scale is approximately the logarithm of the energy release. One implication of the power law dependence in (15.6) is that there is nothing special about large earthquakes. That is, if we could wait a couple of million years, we would likely observe earthquakes of size ten following the same Gutenberg-Richter law. In Problem 15.11 we explore the behavior of a model of tectonic plate motion which suggests that the Gutenberg-Richter law is a consequence of self-organized criticality.

Problem 15.11   Earthquake model

Given the long time scales between earthquakes and the complexity of the historical record, there is considerable interest in developing ways of studying earthquakes using simulations. The Burridge and Knopoff model of fault systems consists of a system of coupled masses in contact with a moving rough surface. The masses are subjected to static and dynamic frictional forces, and also are pulled by an external force corresponding to slow tectonic plate motion. The major difficulty with this model and similar ones that have been proposed is that the numerical solution of the corresponding Newton's equations of motion is computationally intensive. For this reason we first study a simple cellular automaton model that retains some of the basic physics of the Burridge and Knopoff type models. A more realistic cellular automaton model is discussed in Project 15.1.

Define a real variable $F(i, j)$ on a square lattice, where F represents the force on the block at position (i, j). The linear dimension of the lattice is L, and the number of sites N is given by $N = L^2$. The initial state of the lattice at time $t = 0$ is found by assigning small random values to $F(i, j)$. The lattice is updated according to the following rules:

1. Increase F everywhere by a small amount ΔF, e.g., choose $\Delta F = 10^{-5}$, and increase the time t by 1. This increase represents the effect of the driving force due to the slow motion of the tectonic plate.

2. Check if $F(i, j)$ is greater than F_c, the critical threshold value of the force. If not, the system is stable and step 1 is repeated. If the system is unstable, go to step 3. Choose $F_c = 4$ for convenience.

3. The release of force due to slippage of a block is represented by letting $F(i, j) = F(i, j) - F_c$. The transfer of force is represented by updating the force at the sites of the four neighbors at $(i, j \pm 1)$ and $(i \pm 1, j)$: $F \to F + 1$.

As an example, let the linear dimension of the lattice $L = 10$ so that the number of sites $N = L^2 = 100$. Do the simulation and show that the system eventually comes to a statistically stationary state, where the average value of the force at each site stops growing. Monitor the distribution of the size s, where s is the total number of sites (blocks) that are affected by an instability. Increase L to $L = 30$ and repeat your simulations.

The behavior of some other simple models is explored in various contexts in the following four problems.

Problem 15.12   Forest fire model

a.   Consider the following simple model of the spread of a forest fire. Suppose that at $t = 0$ the $L \times L$ sites of a square lattice either have a tree or are empty with probability p_t and $1 - p_t$ respectively. Those sites which have a tree, are on fire with probability f. At each time step an empty site grows a tree with probability p, a tree that has a nearest neighbor site on fire catches fire, and a site that is already on fire dies and becomes empty. Note that the changes in each site occur synchronously, and that this model is an example of a probabilistic cellular automaton. Write a program to simulate this model and color code the three types of sites. Use periodic boundary conditions.

b.   Use $L \geq 30$ and determine the values of p for which the forest maintains fires indefinitely. Note that as long as $p > 0$, new trees will always grow.

c.   Choose the value of p that you found in part (b) and compute the distribution of the number of sites s_f on fire. If the distribution is critical, determine the exponent α that characterizes this distribution. Also compute the distribution for the number of trees, s_t. Is there any relation between these two distributions?

d.   To obtain reliable results it is frequently necessary to average over many initial configurations. However, it is possible that the behavior of a system is independent of the initial configuration and averaging over many initial configurations is unnecessary. This latter possibility is called *self-averaging*. Repeat parts (b) and (c), but average your results over ten initial configurations. Is this forest fire model self-averaging?

Problem 15.13   Another forest fire model

a.   Consider a simple variation of the model discussed in Problem 15.12. At $t = 0$ each site is occupied by a tree with probability p_t; otherwise, it is empty. The system is updated in successive time steps as follows:

1.   Randomly grow new trees at time t with a small probability p from sites that are empty at time $t - 1$;
2.   A tree that is not on fire at $t - 1$ catches fire due to lightning with probability f;
3.   Trees on fire ignite neighboring trees, which in turn ignite their neighboring trees, etc. The spreading of the fire occurs instantaneously.
4.   Trees on fire at time $t - 1$ die (become empty sites) and are removed at time t (after they have set their neighbors on fire);

As in Problem 15.12, the changes in each site occur synchronously. Determine $N(s)$, the number of clusters of trees of size s that catch fire in each time step. Two trees are in the same cluster if they are nearest neighbors.

b.   Do the simulation and determine if the behavior of $N(s)$ is consistent with $N(s) \sim s^{-\alpha}$. If so, estimate the exponent α for several values of p and f.

c.   The balance between the mean rate of birth and burning of trees in the steady state suggests a value for the ratio f/p at which this model is likely to be scale invariant. If the average steady state density of trees is ρ, then at each time step the mean number of new trees appearing is $pN(1 - \rho)$, where $N = L^2$ is the total number of sites. In the same spirit, we can say that for small f, the mean number of trees destroyed by lightning is $f\rho N <s>$, where $<s>$ is the mean number of trees in a cluster. Is this reasoning consistent with the results of your simulation? If we equate these two rates, we find that $<s> \sim ((1 - \rho)/\rho)(p/f)$. Because $0 < \rho < 1$, it follows that $<s> \to \infty$ in the limit $f/p \to 0$. Given the relation $<s> = \sum_{s=1}^{\infty} sN(s)/\sum_s N(s)$ and the divergent behavior of $<s>$, why does it follow that $N(s)$ must decay more slowly than exponentially with s? This reasoning suggests that $N(s) \sim s^{-\alpha}$ with $\alpha < 2$. Is this expectation consistent with the results that you obtained in part (b)?

   In this model there are three well separated time scales, that is, the time for lightning to strike ($\propto f^{-1}$), the time for trees to grow ($\propto p^{-1}$), and the instantaneous spreading of fire through a connected cluster. This separation of time scales seems to be an essential ingredient for self-organized criticality (cf. Grinstein and Jayaprakash).

*Problem 15.14   Is "Life" critical?

Despite the simplicity of the rules of the Game of Life (see Problem 15.1), the dynamics of the game are not well understood. For example, if each cell is randomly populated at $t = 0$, how does the system develop structure and how can we characterize it? The existence of self-organized criticality in various cellular automata models has led several workers to investigate if similar phenomena exist in Life. The results are not yet definitive, and this problem will require considerable computational power and most likely some insight before it is solved. The idea is to begin with a random distribution ($p = 0.5$) of live cells and let the system evolve until only static or simple local periodic activity is found. Then a single dead cell is chosen at random and changed to a live cell. This change is analogous to adding a grain of sand to the sand pile model. The system is allowed to evolve until it again reaches a stable or periodic configuration. The quantities of interest include s, the total number of sites that are changed after the initial change, and T, the number of updates that are needed to return to a stable or periodic configuration. Then choose another dead cell at random, and repeat the above procedure. Compute $N(s)$ and $D(T)$, the distribution of s and T, respectively. If a site is changed several times after a perturbation, each change counts as part of the response. The difficult part of the program is determining whether a configuration is part of a periodic cycle. For small periods the lattice can be stored for a few times and then compared to previous configurations to check whether the state has repeated itself. It is very difficult to check for all types of periodic states, but if the system is started from a random distribution of live sites, cyclic structures with long periods are very rare. Another way of improving the efficiency is to change a dead cell to a live cell only if there is at least one live cell in its neighborhood (e.g., the cell's twenty nearest neighbors). In this way we do not waste time counting small avalanches, because if there are very few live cells in a neighborhood, then usually the system will return to a stable or periodic state very quickly.

Consider at least a 50×50 lattice with periodic boundary conditions. Are your results for $N(s)$ and $D(T)$ consistent with power law behavior? Consider progressively larger lattices and compute the mean values $<s>$ and $<T>$ in addition to $N(s)$ and $D(T)$. If $N(s)$ and $D(T)$ exhibit critical behavior, then the corresponding mean values would increase with the size of the lattice. Why? At present, some workers believe that $N(s)$ and $D(T)$ exhibit critical behavior, while others believe that this apparent behavior is an artifact resulting from the relatively small lattices that have been considered. (The largest lattices considered at present are 1024^2.) Can you reach any tentative conclusions on the basis of your results?

Problem 15.15 Model of punctuated equilibrium

a. The idea of *punctuated equilibrium* is that biological evolution occurs episodically rather than as a steady, gradual process. That is, most of the major changes in life forms occur in relatively short periods of time. Bak and Sneppen have proposed a simple model that exhibits some of the dynamical behavior expected of punctuated equilibrium. The model consists of a one-dimensional cellular automata of size L, where cell i represents the biological fitness of species i, normalized to unity. Initially, all cells receive a random fitness f_i between 0 and 1. Then the cell with the lowest fitness and its two nearest neighbors are randomly given new fitness values. This update rule is repeated indefinitely. Write a program to simulate the behavior of this model.

Use periodic boundary conditions, and show the fitness of each cell as a bar of height f_i.

b. Begin with $L = 64$ and describe what happens to the distribution of fitness values after a long time. We can crudely think of the update process as replacing a species and its neighbors by three new species. In this sense the fitness represents a barrier to creating a new species. If the barrier is low, it is easier to create a new species. Do the low fitness species die out? What is the average value of fitness of the species after the model is run for a long time (10^3, 10^4, or more time steps)? Compute the distribution of fitness values, $N(f)$, averaged over all cells and over a long time. Allow the system to come to a fluctuating steady state before computing $N(f)$. Plot $N(f)$ versus f. Is there a critical value f_c below which $N(f)$ is much less than the values above f_c? Is the update rule reasonable from a evolutionary point of view?

c. Modify your program to compute the distance x between successive fitness changes and the distribution of these distances $C(x)$. Make a log-log plot of $C(x)$ versus x. Is there any evidence of self-organized criticality?

d. Another way to visualize the results is to make a plot of the time at which a cell changed versus the position of the cell. Is the distribution of the plotted points approximately uniform? We might expect that the time of survival of a species depends exponentially on its fitness, and hence each update corresponds to an elapsed time of e^{-cf_i}, where the constant c sets the time scale and f_i is the fitness of the cell which has been changed. Choose $c = 100$ for convenience and make a similar plot with the time axis replaced by the logarithm of the time, i.e., the quantity $100 f_i$. Is this plot more meaningful?

e. Another way of visualizing the meaning of punctuated equilibrium is to plot the number of times groups of cells change as a function of time. Divide the time into units of 100 updates and compute the number of fitness changes for cells $i = 1$ to 10 as a function of time. Do you see any evidence of punctuated equilibrium?

Stuart Kauffman has devised a cellular automaton model of genetics in which each site interacts with K neighbors through a randomly selected cellular automata rule. The K neighbors of each site also are randomly chosen at the beginning of a run. For $K = 4$, there are $2^4 = 16$ neighbor configurations and $2^{16} = 65536$ possible rules. In the Kauffman model each site is initially assigned one of the 65536 possible rules (for $K = 4$). After some time the system goes either to a limit cycle (repeating itself after a finite number of states) or to a fixed point (the state does not change). The idea is to test if a change of the update rule of a single site drives the entire system out of its steady state, i.e., if a small mutation drastically changes the genetic behavior. For $K = 2$, simulations show that the genetic behavior withstands the mutation. However, for large K, e.g., $K = 10$, a single mutation leads to a very different state. Perhaps, Nature is an example of a complex system where small local changes can lead to large global changes.

15.4 NEURAL NETWORKS

Can computers think? This question has occupied philosophers, computer scientists, cognitive psychologists, and many others ever since the first computer was imagined. The assumption of workers in "strong" *artificial intelligence* is that it will be possible someday for computers to think. The reasoning is that thinking is based on symbolic manipulation of inputs from the external world. Of course, everyone agrees that we are far from having a computer think like a human. Some contend that computers will be able to only simulate the human brain and not be able to think like it. Part of the argument is that the brain is not analogous to the hardware of a computer, and the mind is not analogous to its software.

Recent developments in two classes of models, known as neural networks and genetic algorithms, tend to weaken one argument against AI. These models are based on the idea that the program can change itself based on the inputs, that is, the program statements and its data are the "mind" of the computer, and the program and its data can change depending on its own internal state and outside inputs. Indeed, we should consider the entire memory of the computer to constitute its mind. The result is that the state of the computer can change itself in ways that the programmer cannot anticipate. As we have learned, simple algorithms can lead to complex and unpredictable outcomes, and it probably comes as no surprise that the state of the computer can evolve through a sequence of states that can be very complex, i.e., neither completely random nor completely ordered. Perhaps, the mind exists at the edge of chaos where the complex behavior just begins.

Neural networks model a piece of this ultimate computer mind. The idea is to store memories so that a computer can recall them when inputs are given that are close to a particular memory. As humans we have our own algorithms for doing so. For example, if we see someone more than once, the person's face might provide input that helps us recall the person's name. In the same spirit, a neural network can be given a pattern, e.g., a string of 0's and 1's, that partially reflect a previously memorized pattern. The network then attempts to recall the memorized pattern. The significant difference between what the computer usually does to retrieve data from its memory and the memory recall of a neural network is that in the latter we consider *content addressable memory* in contrast to computer programs themselves which retrieve memory based on the address or location of the data, not on its content.

Neural network models have been motivated by how neurons in the brain might collectively store and recall memories. It is known that a neuron "fires" once it receives electrical inputs from other neurons whose strength reaches a certain threshold. An important characteristic of a neuron is that its output is a nonlinear function of the sum of its inputs. Usually, a neuron is in one of two states, a resting potential (not firing) or firing at the maximum rate. The assumption is that when memories are stored in the brain, the strengths of the connections between neurons change. Neural network models attempt to maintain the key functions of biological neurons without the specific biological substrate.

We now consider an example of a neural network due to Hopfield. The network consists of N neurons and the state of the network is defined by the potential at each neuron, V_i. The strength of the connection between the ith and jth neuron is denoted by T_{ij} and is determined by the M stored memories:

$$T_{ij} = \sum_{s=1}^{M} (2V_i^s - 1)(2V_j^s - 1). \tag{15.7}$$

V^s is the state of the sth stored memory. Given an initial state V_i^0, the dynamics of the network is simple, that is, choose a neuron i at random and change its state according to its input. Its input strength, S_i, is defined as

$$S_i = \sum_{i \neq j} T_{ij} V_j, \tag{15.8}$$

where V_j represent the current state of the jth neuron. Change the state of neuron i by setting

$$V_i = \begin{cases} 1, & \text{if } S_i > 0, \\ 0, & S_i \leq 0. \end{cases} \tag{15.9}$$

Note that the threshold value has been set equal to zero, but other values could be used as well.

Program hopfield, listed in the following, implements this model of a neural network and stores memories and recalls them based on user input.

```
PROGRAM hopfield
DIM T(50,50)
CALL memorize(T(,),N)
DO
    CALL recall(T(,),N)
LOOP
END

SUB memorize(T(,),N)
    DIM V(50)
    RANDOMIZE
    INPUT prompt "number of stored memories = ": M
    ! N corresponds to number of neurons
    INPUT prompt "size of memories = ": N
    PRINT "enter M strings of N 0's and 1's"
    FOR memory = 1 to M
        LINE INPUT s$
        CALL convert(s$,N,V())
        FOR i = 1 to N
            FOR j = 1 to N
                IF i <> j then
                    LET T(i,j) = T(i,j) + (2*V(i) - 1)*(2*V(j) - 1)
                END IF
            NEXT j
        NEXT i
    NEXT memory
    PRINT
END SUB
```

```
SUB recall(T(,),N)
    DIM V(50)
    PRINT "enter a string of N 0's and 1's"
    LINE INPUT s$
    CALL convert(s$,N,V())
    DO
        FOR k = 1 to N
            LET i = int(N*rnd) + 1        ! choose neuron at random
            LET sum = 0
            FOR j = 1 to N
                IF i <> j then LET sum = sum + T(i,j)*V(j)
            NEXT j
            IF sum > 0 then
                LET V(i) = 1           ! above threshold
            ELSE
                LET V(i) = 0           ! below threshold
            END IF
        NEXT k
        FOR i = 1 to N
            PRINT using "#": V(i);
        NEXT i
        PRINT
        PAUSE 1
    LOOP until key input
    GET KEY kk
END SUB

SUB convert(s$,N,V())
    ! convert string to array of 0's and 1's
    FOR i = 1 to N
        LET c$ = s$[i:i]
        IF c$ = "0" then
            LET V(i) = 0
        ELSE
            LET V(i) = 1
        END IF
    NEXT i
END SUB
```

Problem 15.16   Memory recall in the Hopfield model

a. Use *Program hopfield* to explore the ability of the Hopfield neural net-
    work to store and recall memories. Begin by storing $M = 2$ memories of
    $N = 20$ characters each. For example, store 11111000000000011111 and
    11001100110011001100. Then try to recall a memory using the input string
    11111110000001111111. This input is similar to the first memory. Record the
    "Hamming" distance between the final state and the closest memory, where
    the Hamming distance is the number of characters that differ between two

strings. Repeat the above procedure for a number of different values of the number of memories M and the memory length N.

b. Estimate how many memories can be stored for a given sized string before recall becomes severely reduced. Make estimates for $N = 10, 20$, and 30. What criteria did you adopt for correct recall?

c. Modify *Program hopfield* so that two-dimensional patterns can be memorized. Instead of a string of 1's and 0's, you will need a grid of $L \times L$ cells each of which can be on or off. The major change in your program is that the indices i and j for S_i and T_{ij} must now refer to a particular cell. For example, if (x, y) is the location of the ith cell, then $i = x + (y - 1)L$. T_{ij} will be represented by an $L^2 \times L^2$ array, and V_i will correspond to an array of length L^2. Also, you will need to write input and output subroutines to show the patterns. Consider a grid with $L \geq 10$ and store three patterns. Patterns could be simple geometric shapes or symbols. Then input a pattern similar to one of the stored memories and see how well the Hopfield algorithm is able to recall the correct pattern. Repeat for a number of different input patterns, and then increase the number of stored memories.

In addition to helping us understand biological neural networks, neural networks can be used to determine an optimal solution to a difficult optimization problem. In Problem 15.17 we consider the problem of finding the minimum energy of a model spin glass.

Problem 15.17   Minimum energy of an Ising spin glass

a. We can define an energy for the Hopfield model in analogy to the Ising model:

$$E = -\frac{1}{2} \sum_i \sum_{j \neq i} T_{ij} V_i V_j, \tag{15.10}$$

where we assume that $T_{ij} = T_{ji}$. Note that the form of (15.10) is very similar to (15.2). If we give the T_{ij} random values, then the model is an example of a "spin glass" with long-range interactions (see Project 17.5). Modify your program so that the T_{ij} are given random values between -1 and 1 with $T_{ii} = 0$. The program should ask for an input string of N characters. Have the program display the output string and the energy after every N attempts to change a neuron. Begin with $N = 20$.

b. Describe what happens to the energy after a long time. For different initial states, but the same set of T_{ij}, is the energy the same after the system has evolved for a long time? Explain your results in terms of the number of local energy minima.

c. What is the behavior of the states? Do you find periodic behavior and/or random behavior or do the states evolve to a state that does not change?

15.5 GENETIC ALGORITHMS

There are many people who find it difficult to accept that evolution is sufficiently powerful to generate the biological complexity seen in nature. Part of this difficulty arises from the inability of humans to intuitively grasp time scales that are much greater than their own lifetimes. Another reason is that it is very difficult to appreciate how random changes can lead to emergent complex structures. Genetic algorithms provide one way of understanding the nature of evolution. Their principal utility at present is in optimization problems, but they also are being used to model biological and social evolution. One of the important examples is due to the biologist Tom Ray (see Lewin) who used a genetic algorithm in conjunction with low level machine coding. The memory of the computer was loaded with code segments which reproduce with small changes in their code. Because each code segment requires memory and the computer's processing time, the code segments compete with one another for memory and time. In what was a surprise to many, the memory of the computer, which initially contained a few simple code segments, evolved into a complicated "ecosystem" of code segments of many different sizes and structures.

The idea of genetic algorithms is to model the process of evolution by natural selection. This process involves two steps: random changes in the genetic code during reproduction, and selection according to some fitness criteria. In biological organisms the genetic code is stored in the DNA. We will store the genetic code as a string of 0's and 1's. (In Fortran and C the genetic code can be stored as an integer variable and bit manipulation can be used.) The genetic code constitutes the *genotype*. The conversion of this string to the organism or *phenotype* depends on the problem. The selection criteria is applied to the phenotype. First we describe how change is introduced into the genotype.

Typically, nature changes the genetic code in two ways. The most obvious, but least often used method, is *mutation*. Mutation corresponds to changing a character at random in the genetic code string from 0 to 1 or from 1 to 0. The second and much more powerful method is associated with sexual reproduction. We can take two strings, remove a piece from one string and exchange it with the same length piece from the other string. For example, if string $A = 0011001010$ and string $B = 0001110001$, then exchanging the piece from position 4 to position 7 leads to two new strings $A' = 0011110010$ and $B' = 0001001001$. This type of change is called *recombination* or *crossover*. At each generation we produce changes using recombination and mutation. We then select from the enlarged population of strings (including strings from the previous generation), a new population for the next generation. Usually, a constant population size is maintained from one generation of strings to the next.

We next have to choose a selection criterion. If we want to model an actual ecosystem, we can include a physical environment and other sets of populations corresponding to different species. The fitness could depend on the interaction of the different species with one another, the interaction within each species, and the interaction with the physical environment. In addition, the behavior of the populations might change the environment from one generation to the next. For simplicity, we will introduce the idea of genetic algorithms with a single population of strings, a simple phenotype, and a simple criteria for fitness.

The phenotype we consider is a variant of the Ising spins considered in Problems 15.5 and 15.17. We consider a square lattice of linear dimension L occupied by spins that have one of the two values $s_i = \pm 1$. The energy of the system is given by

$$E = -\sum_{i,j=nn(i)} T_{ij}s_is_j,\qquad\qquad (15.11)$$

where the sum is over all pairs of spins that are nearest neighbors. Note that the energy function in (15.11) assumes that only nearest neighbor spins interact, in contrast to the energy function in (15.10) which assumes that every spin interacts with every other spin. The coupling constants T_{ij} can be either $+1$ (the ferromagnetic Ising model), -1 (the antiferromagnetic Ising model), randomly distributed (a spin glass), or have some other distribution. We adopt the energy as a measure of fitness. If we assume that $|T_{ij}| = 1$, then the minimum energy equals $-2L^2$ and the maximum energy is $2L^2$. More precisely, we choose the combination $2L^2 - E$, a quantity that is always positive, as our measure of fitness, and take the probability of selecting a particular string with energy E for the next generation to be proportional to the fitness $2L^2 - E$.

How does a genotype become "expressed" as a phenotype? The genotypes consists of a list or string of length L^2 with 1's and 0's. Lattice site (i, j) corresponds to the nth position in the string where $n = (j - 1)L + i$. If the character in the string at position n is 0, then the spin at site (i, j) equals -1. If the character is 1, then the spin equals $+1$.

We now have all the ingredients we need to apply the genetic algorithm. *Program genetic* chooses a random size and a random position for the process of recombination, and periodic boundary conditions are applied to the string for the purpose of recombination. We use the True BASIC concatenation operator, &, to piece two strings together. For example, "Magic" & "and" & "Larry" is equivalent to "MagicandLarry". Note that the selection of the population for the new generation uses the method of discrete nonuniform probability distributions discussed in Section 11.5.

```
PROGRAM genetic
DIM s$(1000),T(400,2),Eselect(1000)
CALL initial(s$(),L,L2,T(,),npop,nrecombine,nmutation,ngeneration)
FOR igeneration = 1 to ngeneration
    LET ntot = npop
    FOR iswap = 1 to nrecombine
        CALL recombine(s$(),L2,npop,ntot)
    NEXT iswap
    FOR i = 1 to nmutation
        CALL mutate(s$(),L2,npop,ntot)
    NEXT i
    CALL selection(s$(),L,L2,T(,),npop,ntot,Eselect())
NEXT igeneration
CALL showoutput(s$(),npop,Eselect())
END

SUB initial(s$(),L,L2,T(,),npop,nrecombine,nmutation,ngeneration)
    RANDOMIZE
    INPUT prompt "string (lattice) size = ": L
```

```
                LET L2 = L*L
                INPUT prompt "population number = ": npop
                ! L is linear dimension of phenotype
                INPUT prompt "number of recombinations per generation = ": nrecombine
                INPUT prompt "number of mutations per generation = ": nmutation
                INPUT prompt "number of generations = ": ngeneration
                ! create random population of genotypes
                FOR ipop = 1 to npop
                    LET s$(ipop) = ""
                    FOR i = 1 to L2
                        IF rnd > 0.5 then
                            LET s$(ipop) = s$(ipop) & "1"
                        ELSE
                            LET s$(ipop) = s$(ipop) & "0"
                        END IF
                    NEXT i
                NEXT ipop
                ! create random bond network of Tij's
                FOR i = 1 to L2
                    FOR j = 1 to 2
                        IF rnd > 0.5 then
                            LET T(i,j) = 1
                        ELSE
                            LET T(i,j) = -1
                        END IF
                    NEXT j
                NEXT i
            END SUB

            SUB recombine(s$(),L2,npop,ntot)
                ! choose two strings (genotypes) to recombine
                LET i = int(npop*rnd) + 1
                DO
                    LET j = int(npop*rnd) + 1
                LOOP until i <> j
                LET size = int(rnd*(L2/2)) + 1
                LET pos = int(rnd*L2) + 1
                LET s1$ = s$(i)
                LET s2$ = s$(j)
                IF pos + size <= L2 then
                    LET s$(ntot+1) = s1$[1:pos-1]&s2$[pos:pos+size]&s1$[pos+size+1:L2]
                    LET s$(ntot+2) = s2$[1:pos-1]&s1$[pos:pos+size]&s2$[pos+size+1:L2]
                ELSE                              ! apply periodic eboundarey conditions
                    LET pbc = pos + size - L2
                    LET s$(ntot+1) = s2$[1:pbc] & s1$[pbc+1:pos-1] & s2$[pos:L2]
                    LET s$(ntot+2) = s1$[1:pbc] & s2$[pbc+1:pos-1] & s1$[pos:L2]
                END IF
                LET ntot = ntot + 2
            END SUB
```

```
SUB mutate(s$(),L2,npop,ntot)
    LET i = int(rnd*npop) + 1
    LET pos = int(rnd*L2) + 1
    LET c$ = s$(i)[pos:pos]
    IF c$ = "1" then
       LET c$ = "0"
    ELSE
       LET c$ = "1"
    END IF
    LET s$(ntot + 1) = s$(i)[1:pos-1] & c$ & s$(i)[pos+1:L2]
    LET ntot = ntot + 1
END SUB

SUB convert(a$,L,s(,))
    ! coverts strings of 0's and 1's to 2D array of spins
    ! that is, genotype to phenotype
    FOR i = 1 to L
        FOR j = 1 to L
            LET n = (j-1)*L + i
            IF a$[n:n] = "1" then
               LET s(i,j) = 1
            ELSE
               LET s(i,j) = -1
            END IF
        NEXT j
    NEXT i
END SUB

SUB energy(L,s(,),T(,),E)
    LET E = 0
    FOR i = 1 to L
        LET ip = i + 1
        IF ip > L then LET ip = 1
        FOR j = 1 to L
            LET jp = j + 1
            IF jp > L then LET jp = 1
            LET n = (j-1)*L + i
            LET E = E - T(n,1)*s(i,j)*s(ip,j) - T(n,2)*s(i,j)*s(i,jp)
        NEXT j
    NEXT i
END SUB

SUB selection(s$(),L,L2,T(,),npop,ntot,Eselect())
    DIM s(30,30),save$(1000),Elist(0:1000)
    LET Esum = 0
    LET Elist(0) = 0
    FOR i = 1 to ntot
        CALL convert(s$(i),L,s(,))
        CALL energy(L,s(,),T(,),E)
```

```
                    LET Esum = Esum -  E + 2*L2    ! contribution to Esum > 0
                    LET Elist(i) = Esum
              NEXT i
              MAT save$ = s$
              ! select new population
              FOR ipop = 1 to npop
                    LET E = Esum*rnd
                    LET i = 0
                    DO
                        LET i = i + 1
                    LOOP until E < Elist(i)    ! choose according to energy
                    LET s$(ipop) = save$(i)
                    LET Eselect(ipop) = Elist(i-1) - Elist(i) + 2*L2
              NEXT ipop
        END SUB

        SUB showoutput(s$(),npop,Eselect())
              FOR ipop = 1 to npop
                    PRINT ipop,s$(ipop),Eselect(ipop)
              NEXT ipop
              PRINT
        END SUB
```

Problem 15.18   Ground state of Ising-like models

a. Use *Program genetic* to find the ground state of the ferromagnetic Ising model for which $T_{ij} = 1$ and the ground state energy is $E = -2L^2$. Choose $L = 4$, and consider a population of 20 strings, with 10 recombinations and 4 mutations per generation. How long does it take to find the ground state energy? You might wish to modify the program slightly so that each new generation is shown on the screen and a pause statement is added so that you can look at the new generations as they appear.

b. Find the mean number of generations needed to find the ground state for $L = 4, 6$, and 8. Repeat each run at least twice. Use a population of 100, a recombination rate of 50% and a mutation rate of 20%. Are there any general trends as L is increased? How do your results change if you double the population size? What happens if you double the recombination rate or mutation rate? Use larger lattices if you have sufficient computer resources.

c. Repeat part (b) for the antiferromagnetic model.

d. Repeat part (b) for the spin glass model where $T_{ij} = \pm 1$ randomly. In this case we do not know the ground state energy in advance. What criteria can you use to terminate a run?

One of the important features of the genetic algorithm is that the change in the genetic code is selected not in the genotype directly, but in the phenotype. Note that the way we change the strings (particularly with recombination) is not closely related to the two-dimensional lattice of spins. Indeed, we could have used some other prescription for converting our string of 0's and 1's to a configuration of spins on a two-dimensional lattice. If the phenotype is a three-dimensional lattice, we could use the same procedure for modifying the genotype, but a different prescription for converting the genetic sequence (the string of 0's and 1') to the phenotype (the three-dimensional lattice of spins). The point is that it is not necessary for the genetic coding to mimic the phenotypic expression. This point becomes distorted in the popular press when a gene is tied to a particular trait, because specific pieces of DNA rarely correspond directly to any explicitly expressed trait in the phenotype.

15.6   OVERVIEW AND PROJECTS

All of the models we have discussed in this chapter have been presented in the form of a computer algorithm rather than in terms of a differential equation. These models are an example of the development of a "computer culture" and are a reflection of the way that technology affects the way we think (cf. Vichniac). Can you discuss the models in this chapter without thinking about their computer implementation? Can you imagine understanding these models without the use of computer graphics?

We have given only a brief introduction to cellular automata and other models that are relevant to the newly developing study of complexity, and there are many models and applications that we have not discussed. These models range from attempts to understand the most fundamental processes of nature such as biological evolution and fluid flow to practical studies of the setting of cement (cf. Bentz et al.). In addition, one of the major motivations for the study of cellular automata is their relation to theories of computation and the development of new computer architectures (cf. Hillis). Some researchers believe that ultimately all models of nature can be reduced to cellular automata. One of the attractive features of these models is that the complexity of nature can be ultimately understood as the result of simple and local rules of evolution.

Project 15.1   Cellular automata spring-block model of earthquakes

Mechanical models of earthquakes based on Newton's equations of motion are difficult to simulate because of the wide range of time scales inherent in these models (cf. Carlson and Langer). For this reason various cellular automata models have been proposed (cf. Klein et al.) that approximate the dynamics on short time scales. The idea is that such approximations are not important because the time interval between earthquakes is much larger than the time of an individual slip.

Consider a set of L blocks that are constrained to move in one dimension. Each block is connected to its two nearest neighbors by harmonic springs with stiffness constant k_c and to a loader plate by a leaf spring with stiffness constant k_L. The loader plate moves at velocity v. The stress σ_j on block j at time t after the external loader plate has advanced a distance nv is given by

$$\sigma_j = k_L\big(nv - x_j(t)\big) + k_c\big(x_{j+1}(t) + x_{j-1}(t) - 2x_j(t)\big), \tag{15.12}$$

where $x_j(t)$ is the displacement of block j at time t from its initial equilibrium position. The time t is associated with local rupture and stress release, which typically is the order of a few seconds. The integer n is the number of loader plate updates and is the time associated with the transmission of the tectonic loading force, typically the order of years or decades. We assume that the equilibrium distance between blocks is unity.

A block slips only if its stress exceeds the static friction threshold in which case it advances by an amount proportional to the stress. The displacement of block j at time $t + 1$ is given by the equation of motion:

$$x_j(t + 1) = x_j(t) + \frac{\sigma_j(t)}{k_j}\theta\big(\sigma_j(t) - S\big), \tag{15.13}$$

where

$$k_j = k_L + q_j k_c, \tag{15.14}$$

and S is the static friction threshold. The value of the coordination number q in (15.14) is two except if free boundary conditions are used in which case $q = 1$ for the end blocks. The step function $\theta(x)$ is unity for $x \geq 0$ and zero for $x < 0$. Note that the equation of motion sets the residual stress on block j equal to zero. The evolution of this model is given by the following steps.

1. Choose the initial displacements of the blocks at random, i.e., let $x_j(t = 0) = (r_j - 0.5)$, where r_j is a uniform random number in the unit interval. Set $n = 0$.

2. Compute the stress σ_j on each block according to (15.12).

3. Blocks j for which $\sigma_j \geq S$ slip according to the rule, $x_j \to x_j + \sigma_j/k_j$ (see (15.13)). Increment the "slip" time by unity, $t \to t + 1$.

4. Repeat steps 2 and 3 until $\sigma_j < S$ for all blocks.

5. Increase the stress on each block by the amount $k_L v \Delta n$ and increase n by $\Delta n = 1$.

6. Repeat steps 2–5 until a sufficient number of loader plate updates have been obtained.

Note that the evolution is deterministic and that randomness is introduced only in the initial conditions. The nonlinearity in the model is in the rule for slips. In the following, we set $k_L = k_c = 1$.

a. Explain how the dynamics of the model can lead to avalanches of failed blocks. Each avalanche of failed blocks represents an earthquake. Write a program to implement the model. For your preliminary simulations, choose a system of $L = 64$ blocks. Equilibrate the system for a time long enough for each block to fail (slip) at least once. Choose $S = 1$ and $v = 0.02$. After equilibrium has been established, compute s, the number of blocks that slip during one loader update. (If a block slips more than once, count it each time it slips.) Compute the histogram $H(s)$ averaged over many loader plate updates. How would you characterize the dependence of $H(s)$ on s? Do you see any evidence of power law behavior? Does the behavior of $H(s)$ depend on the size of the system? Choose either periodic or free boundary conditions and determine if the boundary conditions influence your results.

b.   Another quantity of interest is the average slip deficit ϕ_n defined as

$$\phi_n = \frac{1}{L} \sum_j (x_j - nv). \tag{15.15}$$

The average slip deficit is a measure of the lag of the blocks with respect to the movement of the loader plate. Describe the nature of the dependence of ϕ_n on n. Is the n-dependence of ϕ_n purely random? Compute the power spectrum of ϕ_n and characterize its dependence on ω. Try different values of the parameters v, S, and k_c. The properties of this and similar earthquake models are an area of much current interest.

References and Suggestions for Further Reading

Preben Alstrøm and João Leão, "Self-organized criticality in the game of life," *Phys. Rev.* E **49**, R2507 (1994). The authors find evidence for self-organized criticality using finite size scaling, but suggest that further studies are needed.

P. Bak, "Catastrophes and self-organized criticality," *Computers in Physics* **5**(4), 430 (1991). A good introduction to self-organized critical phenomena.

Per Bak, Kan Chen, and Michael Creutz, "Self-organized criticality in the Game of Life," *Nature* **342**, 780 (1989). These workers consider lattices up to 150×150 and find evidence that the local configurations self-organize into a critical state.

Per Bak and Kim Sneppen, "Punctuated equilibrium and criticality in a simple model of evolution," *Phys. Rev. Lett.* **71**, 4083 (1993); Henrik Flyvbjerg, Kim Sneppen, and Per Bak, "Mean field theory for a simple model of evolution," *Phys. Rev. Lett.* **71**, 4087 (1993);

P. Bak, C. Tang, and K. Wiesenfeld, "Self-organized criticality," *Phys. Rev.* A **38**, 364 (1988).

Per Bak and Michael Creutz, "Fractals and self-organized criticality," in *Fractals in Science*, Armin Bunde and Shlomo Havlin, editors, Springer-Verlag (1994).

Charles Bennett and Marc S. Bourzutschky, "Life not critical?," *Nature* **350**, 468 (1991). The authors present evidence that the criticality of the Game of Life is an artifact resulting from the simulation of small lattices.

Dale P. Bentz, Peter V. Coveny, Edward J. Garboczi, Michael F. Kleyn, and Paul E. Stutzman, "Cellular automaton simulations of cement hydration and microstructural development," *Modelling Simul. Mater. Sci. Eng.* **2**, 783 (1994).

E. R. Berlekamp, J. H. Conway, and R. K. Guy, *Winning Ways for your Mathematical Plays*, *Vol. 2*, Academic Press (1982). A discussion is given of how the Game of Life simulates a universal computer.

J. M. Carlson and J. S. Langer, "Mechanical model of an earthquake fault," *Phys. Rev.* A **40**, 6470 (1989).

John W. Clark, Johann Rafelski, and Jeffrey V. Winston, "Brain without mind: Computer simulation of neural networks with modifiable neuronal interactions," *Phys. Repts.* **123**, 215 (1985).

Michael Creutz, "Deterministic Ising dynamics," *Ann. Phys.* **167**, 62 (1986). A deterministic cellular automaton rule for the Ising model is introduced.

Gary D. Doolen, Uriel Frisch, Brosl Hasslacher, Steven Orszag, and Stephen Wolfram, editors, *Lattice Gas Methods for Partial Differential Equations*, Addison-Wesley (1990). A collection of reprints and original articles by many of the leading workers in lattice gas methods.

Stephanie Forrest, editor, *Emergent Computation: Self-Organizing, Collective, and Cooperative Phenomena in Natural and artificial Computing Networks*, MIT Press (1991).

Stephen I. Gallant, *Neural Network Learning and Expert Systems*, MIT Press (1993).

M. Gardner, *Wheels, Life and other Mathematical Amusements*, W. H. Freeman (1983).

G. Grinstein and C. Jayaprakash, "Simple models of self-organized criticality," *Computers in Physics* **9**, 164 (1995).

G. Grinstein, Terence Hwa, and Henrik Jeldtoft Jensen, "$1/f^\alpha$ noise in dissipative transport," *Phys. Rev.* A **45**, R559 (1992).

B. Hayes, "Computer recreations," *Sci. Amer.* **250**(3), 12 (1984). An introduction to cellular automata.

Jan Hemmingsson, "Consistent results on 'Life'," *Physica* D **80**, 151 (1995). The author measures the same properties as Bak, Chen, and Creutz and finds that the power law behavior seen for smaller lattices disappears for larger lattices (1024×1024 with open boundary conditions).

John Hertz, Anders Krogh, and RIchard G. Palmer, *Introduction to the Theory of Neural Computation*, Addison-Wesley (1991).

J. J. Hopfield, "Neural networks and physical systems with emergent collective computational abilities," *Proc. Natl. Acad. Sci. USA* **79**, 2554 (1982).

W. Daniel Hillis, *The Connection Machine*, MIT Press (1985). A discussion of a new massively parallel computer architecture influenced in part by physical models of the type discussed in this chapter.

H. M. Jaeger, Chu-heng Liu, and Sidney R. Nagel, "Relaxation at the angle of repose," *Phys. Rev. Lett.* **62**, 40 (1989). The authors discuss their experiments on real sand piles.

Stuart A. Kauffman, *The Origins of Order: Self-Organization and Selection in Evolution*, Oxford University Press (1993); "Cambrian explosion and Permian quiescence: Implications of rugged fitness landscapes," *Evol. Ecol.* **3**, 274 (1989).

W. Klein, C. Ferguson, and J. B. Rundle, "Spinodals and scaling in slider block models," in *Reduction and Predictability of Natural Disasters*, J. B. Rundle, D. L. Turcotte, and W. Klein, editors, Addison-Wesley (1995).

J. A. Koza, *Genetic Programming: On the Programming of Computers by Means of Natural Selection*, MIT Press (1992).

Chris Langton, "Studying artificial life with cellular automata," *Physica* D **22**, 120 (1986). See also Christopher G. Langton, editor, *Artificial Life*, Addison-Wesley (1989); Christopher G. Langton, Charles Taylor, J. Doyne Farmer, and Steen Rasmussen, editors, *Artificial Life II*,

Addison-Wesley (1989); Christopher G. Langton, editor, *Artificial Life III*, Addison-Wesley (1994);

Roger Lewin, *Complexity: life at the edge of chaos*, MacMillan (1992). A popular exposition on complexity theory.

Elaine S. Oran and Jay P. Boris, *Numerical Simulation of Reactive Flow*, Elsevier Science Publishing (1987). Although much of this book assumes an understanding of fluid dynamics, there is much discussion of simulation methods and of the numerical solution of the differential equations of fluid flow.

Sergei Maslov, Maya Paczuski, and Per Bak, "Avalanches and 1/f noise in evolution and growth models," *Phys. Rev. Lett.* **73**, 2162 (1994).

William Poundstone, *The Recursive Universe*, Contemporary Books (1985). A book on the Game of Life that attempts to draw analogies between the patterns of Life and ideas of information theory and cosmology.

Daniel H. Rothman and Stéphane Zaleski, "Lattice-gas models of phase separation: interfaces, phase transitions, and multiphase flow," *Rev. Mod. Phys.* **66**, 1417 (1994). A comprehensive review paper.

David E. Rumelhart and James L. McClelland, *Parallel Distributed Processing: Explorations in the Microstructure of Cognition*, Vol. 1: *Foundations*, MIT Press (1986). See also Vol. 2 on applications.

L. Schulman and P. Seiden, "Statistical mechanics of a dynamical system based on Conway's Game of Life," *J. Stat. Phys.* **19**, 293 (1978).

Dietrich Stauffer, "Cellular Automata," Chapter 9 in *Fractals and Disordered Systems*, Armin Bunde and Shlomo Havlin, editors, Springer-Verlag (1991). Also see "Programming cellular automata," *Computers in Physics* **5**(1), 62 (1991).

Daniel L. Stein, editor, *Lectures in the Sciences of Complexity*, Vol. 1, Addison-Wesley (1989); Erica Jen, editor, *Lectures in Complex Systems*, Vol. 2, Addison-Wesley (1990); Daniel L. Stein and Lynn Nadel, editors, *Lectures in Complex Systems*, Vol. 3, Addison-Wesley (1991).

Patrick Sutton and Sheri Boyden, "Genetic algorithms: A general search procedure," *Amer. J. Phys.* **62**, 549 (1994). This readable paper discusses the application of genetic algorithms to Ising models and function optimization.

Tommaso Toffoli and Norman Margolus, *Cellular Automata Machines – A New Environment for Modeling*, MIT Press (1987). See also Norman Margolus and Tommaso Toffoli, "Cellular Automata Machines," in the volume edited by Doolen et al. (see above).

D. J. Tritton, *Physical Fluid Dynamics*, second edition, Oxford Science Publications (1988). An excellent introductory text that integrates theory and experiment. Although there is only a brief discussion of numerical work, the text provides the background useful for simulating fluids.

Gérard Y. Vichniac, "Cellular automata models of disorder and organization," in *Disordered Systems and Biological Organization*, E. Bienenstock, F. Fogelman Soulie, and G. Weisbuch, eds. Springer-Verlag (1986). See also Gérard Y. Vichniac, "Taking the computer seriously in teaching science (an introduction to cellular automata)," in *Microscience,* Proceedings of the UNESCO Workshop on Microcomputers in Science Education, G. Marx and P. Szucs, editors, Balaton, Hungary (1985).

M. Mitchell Waldrop, *Complexity: the emerging science at the edge of order and chaos*, Simon and Schuster (1992). A popular exposition of complexity theory.

Stephen Wolfram, editor, *Theory and Applications of Cellular Automata,* World Scientific (1986). A collection of research papers on cellular automata that range in difficulty from straightforward to specialists only. An extensive annotated bibliography also is given. Two papers in this collection that discuss the classification of one-dimensional cellular automata are S. Wolfram, "Statistical mechanics of cellular automata," *Rev. Mod. Phys.* **55**, 601 (1983), and S. Wolfram, "Universality and complexity in cellular automata," *Physica* B**10**, 1 (1984).

C H A P T E R

16
The Microcanonical Ensemble

We simulate the microcanonical ensemble and "discover" the Boltzmann distribution for systems in thermal contact with a heat bath.

16.1   INTRODUCTION

The Monte Carlo simulations of the "particles in the box" problem discussed in Chapter 7 and the molecular dynamics simulations discussed in Chapter 8 have exhibited some of the important qualitative features of macroscopic systems, e.g., the irreversible approach to equilibrium and the existence of equilibrium fluctuations in macroscopic quantities. In this chapter we apply Monte Carlo methods to the simulation of the equilibrium properties of systems with many degrees of freedom. This application will allow us to explore the methodology of statistical mechanics and to introduce the concept of temperature.

Due in part to the impact of computer simulations, the applications of statistical mechanics have expanded from the traditional areas of dense gases and liquids to the study of phase transitions, particle physics, and theories of the early universe. In fact, the algorithm introduced in this chapter was developed by a physicist interested in using computer simulations to predict experimentally verifiable quantities from lattice gauge theories, theories used to describe the interactions of fundamental particles.

16.2   THE MICROCANONICAL ENSEMBLE

We first discuss an *isolated* system for which the number of particles N, the volume V, and the total energy E are fixed and the influence of external parameters such as gravitational and magnetic fields can be ignored. In general, an isolated macroscopic system tends to a time-independent equilibrium state of maximum randomness or entropy. The *macrostate* of the system is specified by the values of E, V, and N. At the microscopic level there are a large number of different ways or *configurations* in which the macrostate (E, V, N) can be realized. A particular configuration or *microstate* is *accessible* if its properties are consistent with the specified macrostate.

All we know about the accessible microstates is that their properties are consistent with the known physical quantities of the system. Because we have no reason to prefer one microstate over another, it is reasonable to postulate that the system is *equally* likely to be in any one of its accessible microstates. To make this postulate of *equal a priori probabilities* more precise, imagine an isolated system with Ω accessible states. The probability P_s of finding the system in microstate s is

$$P_s = \begin{cases} 1/\Omega, & \text{if } s \text{ is accessible} \\ 0, & \text{otherwise .} \end{cases} \tag{16.1}$$

The sum of P_s over all Ω states is equal to unity.

The averages of physical quantities can be determined in two ways. In the usual laboratory experiment, physical quantities are measured over a time interval sufficiently long to allow the system to sample a large number of its accessible microstates. We already performed such time averages in Chapter 6, where we used the method of molecular dynamics to compute the time-averaged values of quantities such as the temperature and pressure. An interpretation of the probabilities in (16.1) that is consistent with such a time average is that during a sequence of observations, P_s yields the fraction of times that a single system is found in a given microscopic state.

Although time averages are conceptually simple, it is convenient to formulate statistical averages at a given instant of time. Instead of performing measurements on a single system, imagine a collection or *ensemble* of systems that are identical mental replicas characterized by the same macrostate. The number of systems in the ensemble equals the number of possible microstates. In this interpretation, the probabilities in (16.1) describe an ensemble of identical systems. An ensemble of systems specified by E, N, V is called a *microcanonical* ensemble. Suppose that a physical quantity A has the value A_s when the system is in the state s. Then the ensemble average of A is given by

$$<A> = \sum_{s=1}^{\Omega} A_s P_s,$$

(16.2)

where P_s is given by (16.1).

To illustrate these ideas, consider a one-dimensional model of an ideal gas in which the particles are distinguishable, noninteracting, and have only two possible velocities v_0 and $-v_0$. Because the particles are noninteracting, the size of the system and the positions of the particles are irrelevant. In Table 16.1 we show the ensemble of systems consistent with $N = 4$ and $E = 2v_0^2$. The mass of the particles is assumed to be unity.

The enumeration of the sixteen systems in the ensemble allows us to calculate ensemble averages for the physical quantities of the system. For example, inspection of Table 16.1 shows that P_n, the probability that the number of particles moving to the right is n, is given by 1/16, 4/16, 6/16, 4/16, and 1/16 for $n = 0, 1, 2, 3$, and 4, respectively. Hence, the mean number of particles moving to the right is

$$<n> = \sum n P_n = (0 \times 1 + 1 \times 4 + 2 \times 6 + 3 \times 4 + 4 \times 1)/16 = 2.$$

LLLL	LLLR	LLRR	LRRR	RRRR
	LLRL	LRLR	RLRR	
	LRLL	LRRL	RRLR	
	RLLL	RLLR	RRRL	
		RLRL		
		RRLL		

Table 16.1   The sixteen possible microstates for a one-dimensional system of $N = 4$ noninteracting particles. The letter R denotes a particle moving to the right and the letter L denotes a particle moving to the left. Each particle has speed v_0. The mass of the particles is taken to be unity and the total (kinetic) energy $E = 4(v_0^2/2)$.

16.3   THE DEMON ALGORITHM

We found in Chapter 8 that we can do a time average of a system of many particles with E, V, and N fixed by integrating Newton's equations of motion for each particle and computing the time-averaged value of the physical quantities of interest. How can we do an ensemble average at fixed E, V, and N? One way would be to enumerate all the microstates and calculate the ensemble average of the desired physical quantities as we did in the ideal gas example. However, this approach usually is not practical, because the number of microstates for even a small system is much too large to enumerate. In the spirit of Monte Carlo, we wish to develop a practical method of obtaining a representative sample of the total number of microstates. An obvious procedure is to fix V and N, change the positions and velocities of the individual particles at random, and retain the configuration if it has the desired total energy. However this procedure is very inefficient, because most configurations would not have the desired total energy and would have to be discarded.

An efficient Monte Carlo procedure has been developed by Creutz and coworkers. Suppose that we add an extra degree of freedom to the original macroscopic system of interest. For historical reasons, this extra degree of freedom is called a *demon*. The demon travels about the system transferring energy as it attempts to change the dynamical variables of the system. If a desired change lowers the energy of the system, the excess energy is given to the demon. If the desired change raises the energy of the system, the demon gives the required energy to the system if the demon has sufficient energy. The only constraint is that the demon cannot have negative energy. The demon algorithm for a classical system of particles is summarized in the following:

1. Choose a particle at random and make a trial change in its position.

2. Compute ΔE, the change in the energy of the system due to the change.

3. If $\Delta E \leq 0$, the system gives the amount $|\Delta E|$ to the demon, i.e., $E_d = E_d - \Delta E$, and the trial configuration is accepted.

4. If $\Delta E > 0$ and the demon has sufficient energy for this change ($E_d \geq \Delta E$), then the demon gives the necessary energy to the system, i.e., $E_d = E_d - \Delta E$, and the trial configuration is accepted. Otherwise, the trial configuration is rejected and the configuration is not changed.

The above steps are repeated until a representative sample of states is obtained. After a sufficient number of steps, the demon and the system will reach a compromise and agree on an average energy for each. The total energy remains constant, and because the demon is only one degree of freedom in comparison to the many degrees of freedom of the system, the energy fluctuations of the system will be order $1/N$.

How do we know that this Monte Carlo simulation of the microcanonical ensemble will yield results equivalent to the time-averaged results of molecular dynamics? The assumption that the two averages yield equivalent results is called the *ergodic* hypothesis (more accurately, the *quasi-ergodic* hypothesis). Although these two averages have not been shown to be identical in general, they have been found to yield equivalent results in all cases of interest.

16.4 ONE-DIMENSIONAL CLASSICAL IDEAL GAS

We first apply the demon algorithm to the one-dimensional classical ideal gas. Of course, we do not need to use the demon algorithm for an ideal gas because a reduction in the energy of one particle can be easily compensated by the corresponding increase in the energy of another particle. However, it is a good idea to consider a simple example first.

For an ideal gas, the energy of a configuration is independent of the positions of the particles, and the total energy is the sum of the kinetic energies of the individual particles. Hence, for an ideal gas the only coordinates of interest are the velocity coordinates. To change a configuration, we choose a particle at random and change its velocity by a random amount.

Program ideal implements a microcanonical Monte Carlo simulation of an ideal classical gas in one dimension. The variable mcs, the number of Monte Carlo steps per particle, plays an important role in Monte Carlo simulations. On the average, the demon attempts to change the velocity of each particle once during each Monte Carlo step per particle. We frequently will refer to the number of Monte Carlo steps per particle as the "time," even though this time has no obvious direct relation to a physical time.

```
PROGRAM ideal
! demon algorithm for the one-dimensional, ideal classical gas
DIM v(100)
CALL initial(N,v(),mcs,E,Ed,Ecum,Edcum,accept,dvmax)
FOR imcs = 1 to mcs
    CALL change(N,v(),E,Ed,Ecum,Edcum,accept,dvmax)
NEXT imcs
CALL averages(N,Ecum,Edcum,mcs,accept)
END

SUB initial(N,v(),mcs,E,Ed,Ecum,Edcum,accept,dvmax)
    RANDOMIZE
    INPUT prompt "number of particles = ": N
    INPUT prompt "initial energy of system = ": E
    LET Ed = 0                        ! initial demon energy
    INPUT prompt "number of MC steps per particle = ": mcs
    INPUT prompt "maximum change in velocity = ": dvmax
    ! divide energy equally among particles
    LET vinitial = sqr(2*E/N)      ! mass unity
    ! all particles have same initial velocities
    FOR i = 1 to N
        LET v(i) = vinitial
    NEXT i
    ! initialize sums
    LET Ecum = 0
    LET Edcum = 0
    LET accept = 0
END SUB
```

```
SUB change(N,v(),E,Ed,Ecum,Edcum,accept,dvmax)
    FOR i = 1 to N
        LET dv = (2*rnd - 1)*dvmax        ! trial change in velocity
        LET ipart = int(rnd*N + 1)        ! select random particle
        LET vtrial = v(ipart) + dv        ! trial velocity
        ! trial energy change
        LET de = 0.5*(vtrial*vtrial - v(ipart)*v(ipart))
        IF de <= Ed then
            LET v(ipart) = vtrial
            LET accept = accept + 1
            LET Ed = Ed - de
            LET E = E + de
        END IF
        ! accumulate data after each Monte Carlo step per particle
        LET Ecum = Ecum + E
        LET Edcum = Edcum + Ed
    NEXT i
END SUB

SUB averages(N,Ecum,Edcum,mcs,accept)
    LET norm = 1/mcs
    LET Edave = Edcum*norm           ! mean demon energy
    LET norm = norm/N
    LET accept_prob = accept*norm      ! acceptance probability
    ! system averages per particle
    LET Esave = Ecum*norm            ! mean energy per system particle
    PRINT "mean demon energy ="; Edave
    PRINT "mean system energy per particle ="; Esave
    PRINT "acceptance probability ="; accept_prob
END SUB
```

Problem 16.1   Monte Carlo simulation of an ideal gas

a.   We will use *Program ideal* to investigate some of the equilibrium properties of an ideal gas. Suppose that we assign the same initial velocity to all the particles. What is the mean value of the particle velocities after equilibrium has been reached? Choose the number of particles $N = 40$, the initial total energy $E = 10$, the initial demon energy $E_d = 0$, the maximum change in the velocity dvmax $= 2$, and the number of Monte Carlo steps per particle mcs ≥ 1000. The mass of the particles is set equal to unity.

b.   The configuration corresponding to all particles having the same velocity is not very likely, and it would be better to choose an initial configuration that is more likely to occur when the system is in equilibrium. Because this choice is not always possible, we should let the system evolve for a number of Monte Carlo steps per particle before we accumulate data for the averages. We call this number the *equilibration time*. Modify *Program ideal* so that the changes are made for nequil Monte Carlo steps per particle before averages are taken. We can estimate this time from a plot of the time average of the demon energy or other quantity of interest versus the time. Determine the mean demon energy and mean system energy per particle for the parameters in part (a).

c. Compute the mean energy of the demon and the mean system energy per particle for $E = 20$ and $E = 40$. Choose mcs = 50000 if possible. Use your result from part (b) and obtain an approximate relation between the mean demon energy and the mean system energy per particle.

d. In the microcanonical ensemble the total energy is fixed with no reference to temperature. Define the temperature by the relation $\frac{1}{2}m<v^2> = \frac{1}{2}kT_{kin}$, where $\frac{1}{2}m<v^2>$ is the mean kinetic energy per particle. Use this relation to obtain T_{kin}. How is T_{kin} related to the mean demon energy? Choose energy units such that Boltzmann's constant k is equal to unity.

e. A limitation of any Monte Carlo simulation is the finite number of particles. In part (d) we found that the mean demon energy is approximately twice the mean kinetic energy per particle. In the infinite particle limit this relation would hold exactly. Determine how close your results come to the infinite particle results for $N = 2$ and $N = 10$. If there is no statistically significant difference between your results for the two values of N, explain why finite N might not be an important limitation for the ideal gas.

16.5 THE TEMPERATURE AND THE CANONICAL ENSEMBLE

Although the microcanonical ensemble is conceptually simple, it does not represent the situation usually found in the laboratory. Most laboratory systems are not isolated, but are in thermal contact with their environment. This thermal contact allows energy to be exchanged between the laboratory system and its environment in the form of heat. The laboratory system is usually small relative to its environment. The larger system with many more degrees of freedom is referred to as the *heat reservoir* or *heat bath*.

We now consider the more realistic case for which the total energy of the *composite* system consisting of the laboratory system and the heat bath is constrained to be constant, but the energy of the laboratory system can vary. Imagine a large number of mental copies of the laboratory system and the heat bath. Considered together, the composite system is isolated and can be described by the microcanonical ensemble. However, because we are interested in the equilibrium values of the laboratory system, we need to know the probability P_s of finding the laboratory system in the microstate s with energy E_s. The ensemble that describes the probability distribution of the laboratory system in thermal equilibrium with a heat bath is known as the *canonical* ensemble.

In general, the laboratory system can be any macroscopic system that is much smaller than the heat bath. The laboratory system can be as small as an individual particle if the latter can be clearly distinguished from the particles of the heat bath. An example of such a laboratory system is the demon itself. Hence, we can consider the demon to be a system whose microstate is specified only by its energy. The demon is a model of a laboratory system in equilibrium with a heat bath.

One way of finding the form of the probability distribution of the canonical ensemble is to simulate a demon exchanging energy with an ideal gas. The ideal gas serves as the heat bath. The main quantity of interest is the probability $P(E_d)$ that the demon has energy E_d. We will find in Problem 16.2 that the form of $P(E_d)$ is given by

$$P(E_d) = \frac{1}{Z} e^{-E_d/kT}, \tag{16.3}$$

where Z is a normalization constant such that the sum over all the states of the demon is unity. The parameter T in (16.3) is called the *absolute temperature* and is measured in Kelvin (K). Boltzmann's constant k is given by $k = 1.38 \times 10^{-23}$ J/K. The probability distribution (16.3) is called the *Boltzmann* or the *canonical distribution*, and Z is called the *partition function*.

The form (16.3) of the Boltzmann distribution provides a simple way of computing T from the mean demon energy $<E_d>$ given by

$$<E_d> = \frac{\int_0^\infty E\, e^{-E/kT}\, dE}{\int_0^\infty e^{-E/kT}\, dE} = kT. \tag{16.4}$$

We see that T is proportional to the mean demon energy. Note that the result $<E_d> = kT$ in (16.4) holds only if the energy of the demon can take on a continuum of values and if the upper limit of integration can be taken to be ∞.

Problem 16.2   The Boltzmann probability distribution

a.   Add a subroutine to `Program ideal` to compute the probability distribution $P(E_d)$ of the demon. Because E_d is a continuous variable, it is necessary to place the values of E_d in appropriate bins. Plot the natural logarithm of $P(E_d)$ versus E_d and verify the form (16.3) for the Boltzmann distribution. What is the slope of this plot? Choose units such that $k = 1$ and estimate the corresponding value of T. Choose the same parameters as were used in Problem 16.1. Be sure to determine $P(E_d)$ only after thermal equilibrium has been obtained.

b.   Determine the magnitude of T from the relation (16.4). Are your two estimates of T consistent?

c.   Compare the value of T obtained in parts (a) and (b) with the value of T found in Problem 16.1 using the kinetic definition of the temperature. Is the demon in thermal equilibrium with its heat bath?

16.6   THE ISING MODEL

A popular model of a system of interacting variables in statistical physics is the *Ising* model. The model was proposed by Lenz and investigated by his graduate student, Ising, to study the phase transition from a paramagnet to a ferromagnet (cf. Brush).

Ising computed the thermodynamic properties of the model in one dimension and found that the model does not have a phase transition. However, for two and three dimensions the Ising model does exhibit a transition. The nature of the phase transition in two dimensions and the diverse applications of the Ising model are discussed in Chapter 17.

To introduce the Ising model, consider a lattice containing N sites and assume that each lattice site i has associated with it a number s_i, where $s_i = +1$ for an "up" ($\uparrow$) spin and $s_i = -1$ for a "down" ($\downarrow$) spin. A particular configuration or microstate of the lattice is specified by the set of variables $\{s_1, s_2, \ldots s_N\}$ for all lattice sites.

The macroscopic properties of a system are determined by the nature of the accessible microstates. Hence, it is necessary to know the dependence of the energy on the configuration of spins. The total energy E of the Ising model is given by

$$E = -J \sum_{i,j=\mathrm{nn}(i)}^{N} s_i s_j - H \sum_{i=1}^{N} s_i, \tag{16.5}$$

where H is proportional to a uniform external magnetic field. The first sum in (16.5) is over all nearest neighbor pairs. The *exchange constant* J is a measure of the strength of the interaction between nearest neighbor spins (see Fig. 16.1). The second sum in (16.5) represents the energy of interaction of the magnetic moments associated with the spins with an external magnetic field.

If $J > 0$, then the states $\uparrow\uparrow$ and $\downarrow\downarrow$ are energetically favored in comparison to the states $\uparrow\downarrow$ and $\downarrow\uparrow$. Hence for $J > 0$, we expect that the state of lowest total energy is *ferromagnetic*, i.e., the spins all point in the same direction. If $J < 0$, the states $\uparrow\downarrow$ and $\downarrow\uparrow$ are favored and the state of lowest energy is expected to be *antiferromagnetic*, i.e., alternate spins are aligned. If we subject the spins to an external magnetic field directed upward, the spins $\uparrow$ and $\downarrow$ possess an additional internal energy given by $-H$ and $+H$ respectively.

An important virtue of the Ising model is its simplicity. Some of its simplifying features are that the kinetic energy of the atoms associated with the lattice sites has been neglected, only nearest neighbor contributions to the interaction energy have been included, and the spins are allowed to have only two discrete values. In spite of the simplicity of the model, we will find that it exhibits very interesting behavior.

For the familiar case of classical particles with continuously varying position and velocity coordinates, the dynamics is given by Newton's laws. For the Ising model the dependence (16.5) of the energy on the spin configuration is not sufficient to determine the time-dependent properties of the system. That is, the relation (16.5) does not tell

Fig. 16.1   The interaction energy between nearest neighbor spins in the absence of an external magnetic field.

us how the system changes from one spin configuration to another and we have to introduce the dynamics separately.

In Problem 15.5 we simulated the Ising model using a cellular automata approach. The major limitation of this approach is that it is difficult for the system to sample a representative set of configurations. In addition, there is no simple measure of the temperature. The demon algorithm is much more effective at exploring the set of possible configurations, because the energy of the lattice can fluctuate slightly allowing the lattice to sample any configuration with nearly the same energy. We implement the demon algorithm by choosing a spin at random. The trial change corresponds to a flip of the spin from ↑ to ↓ or ↓ to ↑.

Because we are interested in the properties of an infinite system, we have to choose appropriate boundary conditions. The simplest boundary condition in one dimension is to choose a "free surface" so that the spins at sites 1 and N each have one nearest neighbor interaction only. In general, a better choice is periodic boundary conditions. For this choice the lattice becomes a ring and the spins at sites 1 and N interact with one another and hence have the same number of interactions as do the other spins.

What are some of the physical quantities whose averages we wish to compute? An obvious physical quantity is the *magnetization M* given by

$$M = \sum_{i=1}^{N} s_i,$$ (16.6)

and the magnetization per spin $m = M/N$. Usually we are interested in the average values $<M>$ and the fluctuations $<M^2> - <M>^2$. We can determine the temperature T as a function of the energy of the system in two ways. One way is to measure the probability that the demon has energy E_d. Because we know that this probability is proportional to $\exp(-E_d/kT)$, we can determine T from a plot of the logarithm of the probability as a function of E_d. An easier way to determine T is to measure the mean demon energy. However, because the values of E_d are not continuous for the Ising model, T not proportional to $<E_d>$ as it is for the ideal gas. We show in Appendix 16A that for $H = 0$ and the limit of an infinite system, the temperature is related to $<E_d>$ by

$$kT/J = \frac{4}{\ln(1 + 4J/<E_d>)}.$$ (16.7)

The result (16.7) comes from replacing the integrals in (16.4) by sums over the possible demon energies. Note that in the limit $|J/E_d| \ll 1$, (16.7) reduces to $kT = E_d$ as expected.

Program demon implements the microcanonical simulation of the Ising model in one dimension using spin flip dynamics and periodic boundary conditions. Once the initial configuration is chosen, the demon algorithm is similar to that described in Section 16.3. However, in contrast to the ideal gas, the spins in the one-dimensional Ising model must be chosen randomly.

```
PROGRAM demon
! demon algorithm for the d = 1 Ising model in zero magnetic field
DIM spin(1000)
LIBRARY "mygraphics"
CALL initial(N,spin(),E,Ed,M,mcs,Ecum,Edcum,Mcum,M2cum,accept)
CALL setupscreen(N,spin(),up$,down$)
FOR imcs = 1 to mcs
    CALL change(N,spin(),E,Ed,M,accept,up$,down$)
    CALL data(E,Ed,M,Ecum,Edcum,Mcum,M2cum)
NEXT imcs
CALL averages(N,Ecum,Edcum,Mcum,M2cum,mcs,accept)
END

SUB initial(N,spin(),E,Ed,M,mcs,Ecum,Edcum,Mcum,M2cum,accept)
    RANDOMIZE
    INPUT prompt "number of spins = ": N
    ! choose total energy to be multiple of 4J
    ! coupling constant J is unity
    INPUT prompt "desired total energy = ": Etot
    LET Etot = 4*int(Etot/4)
    INPUT prompt "number of Monte Carlo steps per spin = ": mcs
    ! initial configuration of spins in minimum energy state
    FOR isite = 1 to N
        LET spin(isite) = 1
    NEXT isite
    LET M = N                          ! net magnetization
    ! compute initial system energy
    LET E = -N                         ! periodic boundary conditions
    LET Ed = (Etot - E)
    PRINT "total energy = "; E + Ed
    ! initialize sums
    LET Ecum = 0
    LET Edcum = 0
    LET Mcum = 0
    LET M2cum = 0
END SUB

SUB setupscreen(N,spin(),up$,down$)
    LET r = N/(2*pi)
    CALL compute_aspect_ratio(r+2,xwin,ywin)
    SET WINDOW -xwin,xwin,-ywin,ywin
    LET dtheta = 2*pi/N
    SET COLOR "red"
    BOX AREA 1,1+0.5,1,1+0.5
    BOX KEEP 1,1+0.5,1,1+0.5 in up$
    CLEAR
    SET COLOR "blue"
    BOX AREA 1,1+0.5,1,1+0.5
    BOX KEEP 1,1+0.5,1,1+0.5 in down$
```

```
        CLEAR
        FOR i = 1 to N
            CALL showspin(N,spin(i),i,up$,down$)
        NEXT i
    END SUB

    SUB change(N,spin(),E,Ed,M,accept,up$,down$)
        FOR i = 1 to N
            LET isite = int(rnd*N + 1)        ! random spin
            ! determine neighboring spin values
            IF isite = 1 then
               LET left = spin(N)
            ELSE
               LET left = spin(isite - 1)
            END IF
            IF isite = N then
               LET right = spin(1)
            ELSE
               LET right = spin(isite + 1)
            END IF
            ! trial energy change
            LET de =  2*spin(isite)*(left + right)
            IF de <= Ed then
                ! spin flip dynamics
                LET spin(isite) = -spin(isite)
                LET M = M + 2*spin(isite)
                LET accept = accept + 1        ! number of changes accepted
                LET Ed = Ed - de
                LET E = E + de
            END IF
            CALL showspin(N,spin(isite),isite,up$,down$)
        NEXT i
    END SUB

    SUB data(E,Ed,M,Ecum,Edcum,Mcum,M2cum)
        ! accumulate data
        LET Ecum = Ecum + E
        LET Edcum = Edcum + Ed
        LET Mcum = Mcum + M
        LET M2cum = M2cum + M*M
    END SUB

    SUB averages(N,Ecum,Edcum,Mcum,M2cum,mcs,accept)
        SET COLOR "black/white"
        LET norm = 1/mcs                ! collected data after every attempt
        LET Edave = Edcum*norm
        PRINT "mean demon energy ="; Edave
        LET T = 4/log(1 + 4/Edave)
        PRINT "T ="; T
```

```
            LET Eave = Ecum*norm
            PRINT "<E> = "; Eave
            LET Mave = Mcum*norm
            PRINT "<M> ="; Mave
            LET M2ave = M2cum*norm
            PRINT "<M*M> ="; M2ave
            LET accept_prob = accept*norm/N
            PRINT "acceptance probability ="; accept_prob
        END SUB

    SUB showspin(N,dir,isite,up$,down$)
        LET r = N/(2*pi)
        LET theta = isite/r
        LET x = r*cos(theta)
        LET y = r*sin(theta)
        IF dir = 1 then
            BOX SHOW up$ at x,y
        ELSE
            BOX SHOW down$ at x,y
        END IF
    END SUB
```

Note that for $H = 0$, the change in energy due to a spin flip is either 0 or $\pm 4J$. Hence the initial energy of the system plus the demon must be an integer multiple of $4J$. Because the spins are interacting, it is difficult to choose an initial configuration of spins with precisely the desired energy. The procedure followed in SUB initial is to choose the initial configuration to be all spins up, a minimum energy configuration. The demon energy is chosen so that the total energy of the system and the demon is equal to the desired multiple of $4J$.

Problem 16.3   One-dimensional Ising model

a.   Use *Program demon* with $N = 100$ and the desired total energy Etot $= -20$. What is the initial energy assigned to the demon in SUB initial? Note that the program shows the spins as a ring. Describe the evolution of the spins. (It might be useful to insert some PAUSE statements in the program.) Change Etot and describe any qualitative changes in the evolution.

b.   Compute the time average of the demon energy and the magnetization M as a function of the time. As usual, we interpret the time as the number of Monte Carlo steps per spin. What is the approximate time for these quantities to approach their equilibrium values?

c.   Modify the program so that initial nonequilibrium configurations are not used to determine the averages of physical quantities. What are the equilibrium values of $<E_d>$, $<M>$, and $<M^2>$? The choice of mcs $= 100$ is appropriate for testing the program and yields results of approximately 20% accuracy. To obtain better than 5% results, mcs should be the order of 1000.

d. Compute T and E for $N = 100$, and the cases `Etot` $= -20, -40, -60$, and -80. Compare your results to the exact result for an infinite one-dimensional lattice, $E/N = -\tanh(J/kT)$. How do your computed results for E/N depend on N and on the number of Monte Carlo steps per spin?

e. Use the same runs as in part (d) to compute $<M^2>$ as a function of T. Does $<M^2>$ increase or decrease with T?

f. Modify *Program demon* and verify the Boltzmann form (16.3) for the energy distribution of the demon.

Problem 16.4   Additional applications

a. Modify *Program demon* so that the antiferromagnetic case $(J = -1)$ is treated. Before doing the simulation describe how you expect the spin configurations to differ from the ferromagnetic case. What is the lowest energy or ground state configuration? Run the simulation with the spins initially in their ground state, and compare your results with your expectations. Compute the mean energy per spin versus temperature and compare your results with the ferromagnetic case.

* b. Modify *Program demon* to include a nonzero magnetic field, $H \neq 0$, and compute $<E_d>$, $<M>$, and $<M^2>$ as a function of H for fixed E. Read the discussion in Appendix 16A and determine the relation of $<E_d>$ to T for your choices of H. Is the equilibrium temperature higher or lower than the $H = 0$ case for the same total energy?

* Problem 16.5   The two-dimensional Ising model

a. Simulate the two-dimensional Ising model on a square lattice with spin-flip dynamics in the microcanonical ensemble. The total number of spins $N = L^2$, where L is the length of one side of the lattice. Use periodic boundary conditions as shown in Fig. 16.2 so that spins in the left-hand column interact with spins in the right-hand column, etc. Do not include nonequilibrium configurations in your averages.

b. Compute $<E_d>$ and $<M^2>$ as a function of E. Convenient choices of parameters are $L = 10$ and `mcs` $= 500$. Assume $J = 1$ and $H = 0$. Use (16.7) to determine the dependence of T on E and plot E versus T.

c. Repeat the simulations in part (b) for $L = 20$. If necessary, increase `mcs` until your averages are accurate to within a few percent. Describe how the energy versus temperature curve changes with lattice size.

d. Modify your program to make "snapshots" of the spin configurations. Describe qualitatively the nature of the configurations at different energies or temperatures. Are they ordered or disordered? Are there domains of up or down spins?

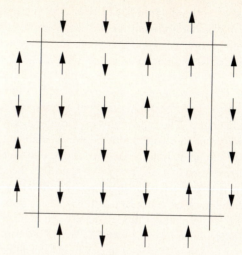

Fig. 16.2 One of the 2^N possible configurations of a system of $N = 16$ Ising spins on a square lattice. Also shown are the spins in the four nearest periodic images of the central cell that are used to calculate the energy. An up spin is denoted by ↑ and a down spin is denoted by ↓. Note that the number of nearest neighbors on a square lattice is four. The energy of this configuration is $E = -8J + 4H$ with periodic boundary conditions.

*16.7 HEAT FLOW

In our applications of the demon algorithm one demon shared its energy equally with all the spins. As a result the spins all attained the same mean energy of interaction. Many interesting questions arise when the system is not spatially uniform and is in a nonequilibrium but time-independent (steady) state.

Let us consider heat flow in a one-dimensional Ising model. Suppose that instead of all the sites sharing energy with one demon, each site has its own demon. We can study the flow of heat by requiring the demons at the boundary spins to satisfy different conditions than the demons at the other spins. The demon at spin 1 adds energy to the system by flipping this spin so that it is in its highest energy state, i.e., in the opposite direction of spin 2. The demon at spin N removes energy from the system by flipping spin N so that it is in its lowest energy state, i.e., in the same direction as spin $N - 1$. As a result, energy flows from site 1 to site N via the demons associated with the intermediate sites. In order that energy not build up at the "hot" end of the Ising chain, we require that spin 1 can only add energy to the system if spin N simultaneously removes energy from the system. Because the demons at the two ends of the lattice satisfy different conditions than the other demons, we do not use periodic boundary conditions.

The temperature is determined by the generalization of the relation (16.7), that is, the temperature at site i is related to the mean energy of the demon at site i. To control the temperature gradient, we can update the end spins at a rate different than the other spins. The maximum temperature gradient occurs if we update the end spins after every update of an internal spin. A smaller temperature gradient occurs if we update the end spins less frequently. The temperature gradient between any two spins can be determined from the temperature profile, the spatial dependence of the temperature. The energy flow can be determined by computing the magnitude of the energy per unit time that enters the lattice at site 1.

To implement this procedure we modify *Program demon* by converting the variables Ed and Edcum to arrays. We do the usual updating procedure for spins 2 through $N - 1$ and visit spins 1 and N at regular intervals denoted by nvisit.

```
PROGRAM conduct
! many demon algorithm for Ising chain
! heat added at spin 1 and subtracted at spin N
DIM spin(1000),Ed(1000),Edsum(1000),Msum(1000)
CALL initial(N,spin(),nmcs,nvisit)
FOR imcs = 1 to nmcs
    CALL change(N,spin(),Ed(),accept)
    IF mod(imcs,nvisit) = 0 then CALL heat(N,spin(),Edsum())
    CALL data(N,spin(),Ed(),Edsum(),Msum())
NEXT imcs
CALL output(N,Edsum(),Msum(),nmcs,accept)
END


SUB initial(N,spin(),nmcs,nvisit)
    RANDOMIZE
    INPUT prompt "number of spins = ": N
    INPUT prompt "number of MC steps per spin = ": nmcs
    INPUT prompt "MC steps between updates of end spins = ": nvisit
    ! initial random configuration
    FOR i = 1 to N
        IF rnd > 0.5 then
            LET spin(i) = 1
        ELSE
            LET spin(i) = -1
        END IF
    NEXT i
END SUB


SUB change(N,spin(),Ed(),accept)   ! spin flip dynamics
    ! do one Monte Carlo step
    FOR i = 2 to N - 1
```

```
                      ! pick spin at random from spins 2 to N - 1
                      LET isite = int(rnd*(N - 2) + 2)
                      ! trial energy change
                      LET de = 2*spin(isite)*(spin(isite - 1) + spin(isite + 1))
                      IF de <= Ed(isite) then
                         LET spin(isite) = - spin(isite)
                         LET accept = accept + 1
                         LET Ed(isite) = Ed(isite) - de
                      END IF
                  NEXT i
            END SUB

            SUB heat(N,spin(),Edsum())
                  ! attempt to add energy at spin 1 and remove it at spin N
                  ! possible only if spins 1 and 2 are aligned
                  ! and spins N and N - 1 are not aligned
                  IF (spin(1)*spin(2) = 1) and (spin(N)*spin(N-1) = -1) then
                     LET Edsum(1) = Edsum(1) + 2
                     LET Edsum(N) = Edsum(N) - 2
                     LET spin(1) = -spin(1)
                     LET spin(N) = -spin(N)
                  END IF
            END SUB

            SUB data(N,spin(),Ed(),Edsum(),Msum())
                  FOR i = 2 to N - 1
                     LET Edsum(i) = Edsum(i) + Ed(i)
                     LET Msum(i) = Msum(i) + spin(i)
                  NEXT i
            END SUB

            SUB output(N,Edsum(),Msum(),nmcs,accept)
                  LET norm = 1/nmcs
                  LET accept_prob = accept*norm/(N - 2)
                  PRINT "acceptance probability = "; accept_prob
                  PRINT
                  PRINT tab(2);"i";tab(16);"Ed(i)";tab(35);"T";tab(46);"M(i)"
                  PRINT
                  FOR i = 2 to N-1
                     LET edave = Edsum(i)*norm
                     LET temperature = 0
                     IF Edave <> 0 then
                        IF (1 + 4/Edave) > 0 then
                           LET temperature = 4/log(1 + 4/Edave)
                        END IF
                     END IF
                     LET M = Msum(i)*norm
                     PRINT i,Edave,temperature,M
                  NEXT i
            END SUB
```

Problem 16.6   One-dimensional heat flow

a.   As a check on *Program* conduct, modify the program so that all the demons are equivalent, i.e., impose periodic boundary conditions and do not use SUB heat. Compute the mean energy of the demon at each site and use (16.7) to define a local site temperature. Use $N \geq 22$ and mcs ≥ 1000. Is the local temperature approximately uniform? How do your results compare with the single demon case?

b.   In *Program* conduct energy is added to the system at site 1 and is removed at site N. Determine the mean demon energy for each site and obtain the corresponding local temperature and the mean energy of the system. Draw the temperature profile by plotting the temperature as a function of site number. The temperature gradient is the difference in temperature from site $N - 1$ to site 2 divided by the distance between them. (The distance between neighboring sites is unity.) Because of local temperature fluctuations and edge effects, the temperature gradient should be estimated by fitting the temperature profile in the middle of the lattice to a straight line. Reasonable choices for the parameters are $N = 22$, mcs $= 4000$, and nvisit $= 1$.

c.   The heat flux Q is the energy flow per unit length per unit time. The energy flow is the amount of energy that demon 1 adds to the system at site 1. The time is conveniently measured in terms of Monte Carlo steps per spin. Determine Q for the parameters used in part (b).

d.   If the temperature gradient $\partial T / \partial x$ is not too large, the heat flux Q is proportional to $\partial T / \partial x$. We can determine the *thermal conductivity* κ by the relation

$$Q = -\kappa \frac{\partial T}{\partial x}. \tag{16.8}$$

Use your results for $\partial T / \partial x$ and Q to estimate κ. Because of the limited number of spins and Monte Carlo steps, your results should be accurate to only about 20%. More accurate results would require at least $N = 50$ spins and 10^4 to 10^5 Monte Carlo steps per spin.

e.   Determine Q, the temperature profile, and the mean temperature for different values of nvisit. Is the temperature profile linear for all nvisit? If the temperature profile is linear, estimate $\partial T / \partial x$ and determine κ. Does κ depend on the mean temperature?

Note that in Problem 16.6 we were able to compute a temperature profile by using an algorithm that manipulated only integer numbers. The conventional approach is to solve a heat equation similar in form to the diffusion equation.

Problem 16.7   Magnetization profile

a.   Modify *Program* conduct by removing SUB heat and constraining spins 1 and N to be $+1$ and -1 respectively. Estimate the magnetization profile by plotting the mean value of the spin at each site versus the site number. Choose $N = 22$ and mcs ≥ 1000. How do your results vary as you increase N?

      b.   Compute the mean demon energy and hence the local temperature at each site. Does the system have a uniform temperature even though the magnetization is not uniform? Is the system in thermal equilibrium?

***c.**   The effect of this constraint is easier to observe in two and three dimensions than in one dimension. Write a program for a two-dimensional Ising model on a $L \times L$ square lattice. Constrain the spins at site (i, j) to be $+1$ and -1 for $i = 1$ and $i = L$ respectively. Use periodic boundary conditions in the y direction. How do your results compare with the one-dimensional case?

***d.**   Remove the periodic boundary condition in the y direction and constrain all the boundary spins from $i = 1$ to $L/2$ to be $+1$ and the other boundary spins to be -1. Choose an initial configuration where all the spins on the left half of the system are $+1$ and the others are -1. Do the simulation and draw a configuration of the spins once the system has reached equilibrium. Draw a line between each pair of spins of opposite sign. Describe the curve separating the $+1$ spins from the -1 spins. Begin with $L = 20$ and determine what happens as L is increased.

16.8   COMMENT

One advantage of doing simulations using the demon algorithm is that it is not necessary to make any demands on the random number generator. We have done a Monte Carlo simulation without random numbers! (In the one-dimensional Ising model we have to choose the trial spins at random. However, the spins can be chosen sequentially in higher dimensions.) Very fast algorithms have been developed by using one computer bit per spin and multiple demons. There also are several disadvantages associated with the microcanonical ensemble. One disadvantage is the difficulty of establishing a system at the desired value of the energy. However, the most important disadvantage for us is conceptual. That is, it is more natural for us to think of the behavior of macroscopic physical quantities as functions of the temperature rather than the total energy. Hence, we postpone further consideration of the further properties of the Ising model to Chapter 17 in the context of the canonical ensemble.

Appendix 16A   RELATION OF THE MEAN DEMON ENERGY TO THE TEMPERATURE

We know that the energy of the demon, E_d, is constrained to be positive and is given by $E_d = E_{\text{total}} - E$, where E is the energy of the system and E_{total} is the total energy. We have found in Problems 16.2 and 16.3 that the probability for the demon to have energy E_d is proportional to $e^{-E_d/kT}$. We assume that the same form of the probability distribution holds for any macroscopic system in thermodynamic equilibrium. Hence in general, $<E_d>$ is given by

$$<E_d> = \frac{\sum_{E_d} E_d\, e^{-E_d/kT}}{\sum_{E_d} e^{-E_d/kT}}, \tag{16.9}$$

where the summations in (16.9) are over the possible values of E_d. If an Ising spin is flipped in zero magnetic field, the minimum nonzero decrease in energy of the system is $4J$ (see Fig. 16.3). Hence the possible energies of the demon are $0, 4J, 8J, 12J, \ldots$ We write $x = 4J/kT$ and perform the summations in (16.9). The result is

$$<E_d/kT> = \frac{0 + xe^{-x} + 2xe^{-2x} + \ldots}{1 + e^{-x} + e^{-2x} + \ldots} = \frac{x}{e^x - 1}. \tag{16.10}$$

The form (16.7) can be obtained by solving (16.10) for T in terms of E_d. Convince yourself that the relation (16.10) is independent of dimension for lattices with an even number of nearest neighbors.

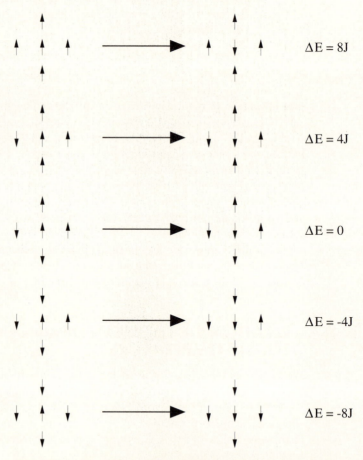

Fig. 16.3   The five possible transitions of the Ising model on the square lattice with spin flip dynamics.

If the magnetic field is nonzero, the possible values of the demon energy are $0, 2H, 4J - 2H, 4J + 2H, \cdots$. If J is a multiple of H, then the result is the same as before with $4J$ replaced by $2H$, because the possible energy values for the demon are multiples of $2H$. If the ratio $4J/2H$ is irrational, then the demon can take on a continuum of values, and thus $<E_d> = kT$. The other possibility is that $4J/2H = m/n$, where m and n are relatively prime positive integers. In this case it can be shown that (see Mak)

$$kT/J = \frac{4/m}{\ln(1 + 4J/m<E_d>)}. \tag{16.11}$$

You can test these relations for $H \neq 0$ by choosing values of J and H and computing the sums in (16.9) directly.

References and Suggestions for Further Reading

S. G. Brush, "History of the Lenz-Ising model," *Rev. Mod. Phys.* **39**, 883 (1967).

Michael Creutz, "Microcanonical Monte Carlo simulation," *Phys. Rev. Lett.* **50**, 1411 (1983). See also Gyan Bhanot, Michael Creutz, and Herbert Neuberger, "Microcanonical simulation of Ising systems," *Nuc. Phys.* B **235**, 417 (1984).

R. Harris, "Demons at work," *Computers in Physics* **4**(3), 314 (1990).

S. S. Mak, "The analytical demon of the Ising model," *Phys. Lett.* A **196**, 318 (1995).

C H A P T E R

17

Monte Carlo Simulation of the Canonical Ensemble

We discuss Monte Carlo methods for simulating equilibrium systems. Applications are made to models of magnetism and simple fluids.

17.1   THE CANONICAL ENSEMBLE

Most physical systems are not isolated, but exchange energy with their environment. Because such systems are usually small in comparison to their environment, we assume that any change in the energy of the smaller system does not have a significant effect on the temperature of the environment. We say that the environment acts as a *heat reservoir* or *heat bath* at a fixed absolute temperature T. If a small but macroscopic system is placed in thermal contact with a heat bath, the system reaches thermal equilibrium by exchanging energy with the heat bath until the system attains the temperature of the bath.

Imagine an infinitely large number of copies of a system at fixed volume V and number of particles N in equilibrium at temperature T. In Chapter 16 we verified that P_s, the probability that the system is in microstate s with energy E_s, is given by

$$P_s = \frac{1}{Z} e^{-\beta E_s}, \qquad \text{(canonical distribution)} \qquad (17.1)$$

where $\beta = 1/kT$, and Z is a normalization constant. The ensemble defined by (17.1) is known as the *canonical* ensemble. Because $\sum P_s = 1$, Z is given by

$$Z = \sum_{s=1}^{M} e^{-E_s/kT}. \qquad (17.2)$$

The summation in (17.2) is over all M accessible microstates of the system. The quantity Z is known as the *partition function* of the system.

We can use (17.1) to obtain the ensemble average of the physical quantities of interest. For example, the mean energy is given by

$$<E> = \sum_{s=1}^{M} E_s P_s = \frac{1}{Z} \sum_{s=1}^{M} E_s e^{-\beta E_s}. \qquad (17.3)$$

Note that the energy fluctuates in the canonical ensemble.

17.2   THE METROPOLIS ALGORITHM

How can we simulate a system of N particles confined in a volume V at a fixed temperature T? Because we can generate only a finite number m of the total number of M microstates, we might hope to obtain an estimate for the mean value of the physical quantity A by writing

$$<A> \approx A_m = \frac{\displaystyle\sum_{s=1}^{m} A_s e^{-\beta E_s}}{\displaystyle\sum_{s=1}^{m} e^{-\beta E_s}}. \qquad (17.4)$$

A_s is the value of the physical quantity A in microstate s. A crude Monte Carlo procedure is to generate a microstate s at random, calculate E_s, A_s, and $e^{-\beta E_s}$, and evaluate the corresponding contribution of the microstate to the sums in (17.4). However, a microstate generated in this way would likely be very improbable and hence contribute little to the sums. Instead, we use an *importance sampling* method and generate microstates according to a probability distribution function π_s.

We follow the same procedure as in Section 11.7 and rewrite (17.4) by multiplying and dividing by π_s:

$$A_m = \frac{\sum\limits_{s=1}^{m} (A_s/\pi_s)\, e^{-\beta E_s}\, \pi_s}{\sum\limits_{s=1}^{m} (1/\pi_s)\, e^{-\beta E_s}\, \pi_s}. \qquad \text{(no importance sampling)} \qquad (17.5)$$

If we generate microstates with probability π_s, then (17.5) becomes

$$A_m = \frac{\sum\limits_{s=1}^{m} (A_s/\pi_s)\, e^{-\beta E_s}}{\sum\limits_{s=1}^{m} (1/\pi_s)\, e^{-\beta E_s}}. \qquad \text{(importance sampling)} \qquad (17.6)$$

That is, if we average over a biased sample, we need to weight each microstate by $1/\pi_s$ to eliminate the bias. Although any form of π_s could be used, the form of (17.6) suggests that a reasonable choice of π_s is the Boltzmann probability itself, i.e.,

$$\pi_s = \frac{e^{-\beta E_s}}{\sum\limits_{s=1}^{m} e^{-\beta E_s}}. \qquad (17.7)$$

This choice of π_s implies that the estimate A_m of the mean value of A can be written as

$$A_m = \frac{1}{m} \sum_{s=1}^{m} A_s. \qquad (17.8)$$

The choice (17.7) for π_s is due to Metropolis et al.

Although we discussed the Metropolis sampling method in Section 11.8 in the context of the numerical evaluation of integrals, it is not necessary to read Section 11.8 to understand the Metropolis algorithm in the present context. The Metropolis algorithm can be summarized in the context of the simulation of a system of spins or particles as follows:

1. Establish an initial microstate.

2. Make a random trial change in the microstate. For example, choose a spin at random and flip it. Or choose a particle at random and displace it a random distance.

3. Compute $\Delta E \equiv E_{\text{trial}} - E_{\text{old}}$, the change in the energy of the system due to the trial change.

4. If ΔE is less than or equal to zero, accept the new microstate and go to step 8.

5. If ΔE is positive, compute the quantity $w = e^{-\beta \Delta E}$.

6. Generate a random number r in the unit interval.

7. If $r \leq w$, accept the new microstate; otherwise retain the previous microstate.

8. Determine the value of the desired physical quantities.

9. Repeat steps (2) through (8) to obtain a sufficient number of microstates.

10. Periodically compute averages over microstates.

Steps 2 through 7 give the conditional probability that the system is in microstate $\{s_j\}$ given that it was in microstate $\{s_i\}$. These steps are equivalent to the transition probability

$$W(i \rightarrow j) = \min \left(1, e^{-\beta \Delta E}\right), \qquad \text{(Metropolis algorithm)} \qquad (17.9)$$

where $\Delta E = E_j - E_i$. $W(i \rightarrow j)$ is the probability per unit time for the system to make a transition from microstate i to microstate j. Because it is necessary to evaluate only the ratio $P_j/P_i = e^{-\beta \Delta E}$, it is not necessary to normalize the probability. Note that because the microstates are generated with a probability proportional to the desired probability, all averages become arithmetic averages as in (17.8). However, because the constant of proportionally is not known, it is not possible to estimate the partition function Z in this way.

Although we choose π_s to be the Boltzmann distribution, other choices of π_s are possible and are useful in some contexts. In addition, the choice (17.9) of the transition probability is not the only one that leads to the Boltzmann distribution. It can be shown that if W satisfies the "detailed balance" condition

$$W(i \rightarrow j) \, e^{-\beta E_i} = W(j \rightarrow i) \, e^{-\beta E_j}, \qquad \text{(detailed balance)} \qquad (17.10)$$

then the corresponding Monte Carlo algorithm generates a sequence of states distributed according to the Boltzmann distribution. The derivation that the Metropolis algorithm generates states with a probability proportional to the Boltzmann probability distribution after a sufficient number of steps does not add much to our physical understanding of the algorithm. Instead, in Section 17.3 we apply the algorithm to the ideal classical gas and to a classical magnet in a magnetic field, and verify that the Metropolis algorithm yields the Boltzmann distribution after a sufficient number of trial changes have been made.

We have implicitly assumed in the above discussion that the system is ergodic. Ergodicity refers to the sampling of the important microstates of a system. In a Monte Carlo simulation, the existence of ergodicity depends on the way the trial moves are made, and on the nature of the energy barriers between microstates. For example, consider a one-dimensional lattice of Ising spins with all spins up. If the spins are updated sequentially from right to left, then if one spin is flipped, all remaining flips will be accepted regardless of the temperature because the change in energy is zero. Clearly, the system is not ergodic for this implementation of the algorithm, and we would not obtain the correct thermodynamic behavior.

17.3   VERIFICATION OF THE BOLTZMANN DISTRIBUTION

We first consider the application of the Metropolis algorithm to an ideal classical gas in one dimension. The energy of an ideal gas depends only on the velocity of the particles, and hence a microstate is completely described by a specification of the velocity (or momentum) of each particle. Because the velocity is a continuous variable, it is necessary to describe the accessible microstates so that they are countable, and hence we place the velocity into bins. Suppose we have $N = 10$ particles and divide the possible values of the velocity into twenty bins. Then the total number of microstates would be 20^{10}. Not only would it be difficult to label these 20^{10} states, it would take a prohibitively long time to obtain an accurate estimate of their relative probabilities, and it would be difficult to verify directly that the Metropolis algorithm yields the Boltzmann distribution. For this reason we consider a single classical particle in one dimension in equilibrium with a heat bath and adopt the less ambitious goal of verifying that the Metropolis algorithm generates the Boltzmann distribution for this system. The quantity of interest is the probability $P(v)\,dv$ that the system has a velocity between v and $v + dv$. The algorithm is implemented in SUB metropolis in *Program boltzmann* listed below. The array P stores the desired probability. We choose units such that Boltzmann's constant and the mass are unity.

```
PROGRAM boltzmann
! Metropolis algorithm for a particle in one dimension
DIM P(-100 to 100),accum(3)
CALL initial(v,E,beta,mcs,nequil,delta,nbin,delv)
FOR imcs = 1 to nequil               ! equilibrate system
    CALL metropolis(v,E,beta,delta,accept)
NEXT imcs
CALL initialize_sums(P(),accum(),accept,nbin)
FOR imcs = 1 to mcs
    CALL metropolis(v,E,beta,delta,accept)
    ! accumulate data after each trial change
    CALL data(P(),accum(),v,E,nbin,delv)
NEXT imcs
CALL output(P(),accum(),mcs,accept,nbin,delv)
END

SUB initial(v0,E0,beta,mcs,nequil,delta,nbin,delv)
    RANDOMIZE
    INPUT prompt "temperature = ": T
    LET beta = 1/T
    INPUT prompt "number of Monte Carlo steps = ": mcs
    LET nequil = int(0.1*mcs)
    INPUT prompt "initial speed = ": v0
    LET E0 = 0.5*v0*v0               ! initial kinetic energy
    INPUT prompt "maximum change in velocity = ": delta
    LET vmax = 10*sqr(T)
    LET nbin = 20                    ! number of bins
    LET delv = vmax/nbin             ! velocity interval
END SUB

SUB initialize_sums(P(),accum(),accept,nbin)
    FOR ibin = -nbin to nbin
        LET P(ibin) = 0
    NEXT ibin
    FOR i = 1 to 3
```

```
                LET accum(i) = 0
        NEXT i
        LET accept = 0
    END SUB

    SUB metropolis(v,E,beta,delta,accept)
        LET dv = (2*rnd - 1)*delta      ! trial velocity change
        LET vtrial = v + dv             ! trial velocity
        LET dE = 0.5*(vtrial*vtrial - v*v)      ! trial energy change
        IF dE > 0 then
            IF exp(-beta*dE) < rnd then
                EXIT SUB                ! step not accepted
            END IF
        END IF
        LET v = vtrial
        LET accept = accept + 1
        LET E = E + dE
    END SUB

    SUB data(P(),accum(),v,E,nbin,delv)
        LET accum(1) = accum(1) + E
        LET accum(2) = accum(2) + E*E
        LET accum(3) = accum(3) + v
        LET ibin = round(v/delv)
        LET P(ibin) = P(ibin) + 1
    END SUB

    SUB output(P(),accum(),mcs,accept,nbin,delv)
        LET accept = accept/mcs
        PRINT "acceptance probability ="; accept
        LET vave = accum(3)/mcs
        PRINT "mean velocity ="; vave
        LET Eave = accum(1)/mcs
        PRINT "mean energy ="; Eave
        LET E2ave = accum(2)/mcs
        LET sigma2 = E2ave - Eave*Eave
        PRINT "sigma_E = "; sqr(sigma2)
        PRINT
        PRINT " v ", "P(v)"
        PRINT
        LET v = -nbin*delv
        FOR ibin = -nbin to nbin
            IF p(ibin) > 0 then
                LET prob = p(ibin)/mcs
                PRINT v,
                PRINT using "--.###": prob
            END IF
            LET v = v + delv
        NEXT ibin
    END SUB
```

Problem 17.1   The Boltzmann distribution

a. Use *Program boltzmann* to determine the form of the probability distribution that is generated by the Metropolis algorithm. Let the temperature $T = 4$, the initial velocity $v0 = 0$, the maximum change in the particle's velocity $\delta = 4.0$, and the number of trial moves or Monte Carlo steps $mcs = 10000$. Compute the mean energy, the mean velocity, and the probability density $P(v)$.

b. Is $P(v)$ an increasing or decreasing function of the energy $E = \frac{1}{2}v^2$? Increase the number of Monte Carlo steps until the Boltzmann form of $P(v)$ is approximately verified. Verify that a plot of $\ln P(v)$ versus E yields a straight line with a slope equal to $-1/T$.

c. How do your results for the mean energy and the mean velocity compare with the corresponding exact values?

d. To insure that your results do not depend on the initial conditions, let $v0 = 2$ and compute the mean energy and velocity. How do your results compare with those found in part (a)? Explain why the computed mean particle velocity is approximately zero even though the initial particle velocities are not zero.

e. The *acceptance probability* is the fraction of trial moves that are accepted. What is the effect of changing the value of δ on the acceptance probability?

Problem 17.2   Planar spin in an external magnetic field

a. Consider a classical planar magnet with magnetic moment μ_0. The magnet can be oriented in any direction in the x-y plane, and the energy of interaction of the magnet with an external magnetic field **B** is $-\mu_0 B \cos \phi$, where ϕ is the angle between the moment and **B**. What are the possible microstates of this system? Write a Monte Carlo program to sample the microstates of this system in thermal equilibrium with a heat bath at temperature T. Compute the mean energy as a function of the ratio $\beta \mu_0 B$.

b. Do an analytical calculation of the mean energy and compare the analytical and computed results for various values of $\beta \mu_0 B$.

c. Compute the probability density $P(\phi)$ and analyze its dependence on the energy.

In Problem 17.3 we consider the Monte Carlo simulation of a classical ideal gas of N particles. It is convenient to say that one "time unit" or one "Monte Carlo step per particle" (MCS) has elapsed after N particles have had one chance each on the average to change their coordinates. If the particles are chosen at random, then during one Monte Carlo step per particle, some particles might not be chosen. Of course, all particles will be chosen equally on the average. The advantage of this definition is that the time is independent of the number of particles. However, this definition of time has no obvious relation to a physical time.

Problem 17.3   Simulation of an ideal gas in one dimension

a. Modify *Program boltzmann* to simulate an ideal gas of N particles in one dimension. Assume all particles have the same initial velocity v0 = 10. Let $N = 20$, $T = 10$, and mcs = 200. Choose the value of δ so that the acceptance probability is approximately 40%. What is the mean kinetic energy and mean velocity of the particles?

b. We might expect the total energy of an ideal gas to remain constant because the particles do not interact with one another and hence cannot exchange energy directly. What is the value of the initial total energy of the system in part (a)? Does the total energy remain constant? If not, explain how the energy changes.

c. What is the nature of the time dependence of the total energy starting from the initial condition in (a)? Estimate the number of Monte Carlo steps per particle necessary for the system to reach thermal equilibrium by computing a moving average of the total energy over a fixed time interval. Does this average change with time after a sufficient time has elapsed? What choice of the initial velocities allows the system to reach thermal equilibrium at temperature T as quickly as possible?

d. Compute the probability $P(E)\,dE$ for the system of N particles to have a total energy between E and $E + dE$. Do you expect $\ln P(E)$ to depend linearly on E? Plot $P(E)$ as a function of E and describe the qualitative behavior of $P(E)$. If the plot of $\ln P(E)$ versus E does not yield a straight line, describe the qualitative features of the plot, and determine a functional form for $P(E)$.

e. Compute the mean energy for $T = 10, 20, 40, 80$, and 120 and estimate the heat capacity from its definition $C = \partial E/\partial T$.

f. Compute the mean square energy fluctuations $<(\Delta E)^2> = <E^2> - <E>^2$ for $T = 10$ and $T = 40$. Compare the magnitude of the ratio $<(\Delta E)^2>/T^2$ with the heat capacity determined in part (e).

You might have been surprised to find in Problem 17.3d that the form of $P(E)$ is a Gaussian centered about the mean energy of the system. That is, the distribution function of a *macroscopic* quantity such as the total energy is sharply peaked about its mean value. If the microstates are distributed according to the Boltzmann probability, why is the total energy distributed according to the Gaussian distribution?

17.4   THE ISING MODEL

One of the more interesting natural phenomena in nature is magnetism. You are probably familiar with ferromagnetic materials such as iron and nickel which exhibit a spontaneous magnetization in the absence of an applied magnetic field. This nonzero magnetization occurs only if the temperature is lower than a well defined temperature known as the Curie or critical temperature T_c. For temperatures $T > T_c$, the magnetization vanishes. Hence T_c separates the disordered phase for $T > T_c$ from the ferromagnetic phase for $T < T_c$.

The origin of magnetism is quantum mechanical in nature and an area of much experimental and theoretical interest. However, the study of simple classical models of magnetism has provided much insight. The two- and three-dimensional Ising model is the most commonly studied classical model and is particularly useful in the neighborhood of the magnetic phase transition. As discussed in Chapter 16, the energy of the Ising model is given by

$$E = -J \sum_{i,j=\text{nn}(i)}^{N} s_i s_j - \mu_0 B \sum_{i=1}^{N} s_i, \tag{17.11}$$

where $s = \pm 1$, J is a measure of the strength of the interaction between spins, and the first sum is over all pairs of spins that are nearest neighbors. The second term in (17.11) is the energy of interaction of the magnetic moment with an external magnetic field. Because of the neglect of the other spin components, the Ising model does not give a complete description of ferromagnetism, especially at temperatures well below T_c.

The thermal quantities of interest for the Ising model include the mean energy $<E>$ and the heat capacity C. As we have discussed, one way to determine C at constant external magnetic field is from its definition $C = \partial <E> / \partial T$. An alternative way of determining C is to relate it to the statistical fluctuations of the total energy in the canonical ensemble (see Appendix 17A):

$$C = \frac{1}{kT^2} \left(<E^2> - <E>^2 \right). \tag{17.12}$$

Another quantity of interest is the mean magnetization $<M>$ (see (16.6)) and the corresponding thermodynamic derivative χ:

$$\chi = \lim_{H \to 0} \frac{\partial <M>}{\partial H}, \tag{17.13}$$

where H is proportional to the external magnetic field. In the following, we will refer to H as the magnetic field. The zero field magnetic susceptibility χ is an example of a linear response function, because it measures the ability of a spin to "respond" due to a change in the external magnetic field. In analogy to the heat capacity, χ is related to the fluctuations of the magnetization (see Appendix 17A):

$$\chi = \frac{1}{kT} \left(<M^2> - <M>^2 \right), \tag{17.14}$$

where $<M>$ and $<M^2>$ are evaluated in zero magnetic fields. Relations (17.12) and (17.14) are examples of the general relation between linear response functions and equilibrium fluctuations.

Now that we have specified several equilibrium quantities of interest, we implement the Metropolis algorithm for the Ising model. The possible trial change is the flip of a spin, $s_i \to -s_i$. The Metropolis algorithm was stated in Section 17.3 as a method for generating states with the desired Boltzmann probability, but the flipping of single spins also can be interpreted as a reasonable approximation to the real dynamics of an anisotropic magnet whose spins are coupled to the vibrations of the lattice. The coupling leads to random spin flips, and we expect that one Monte Carlo step per spin is proportional to the average time between single spin flips observed in the laboratory. We can regard single spin flip dynamics as a time dependent process and observe the relaxation to equilibrium after a sufficiently long time. In the following, we will frequently refer to the application of the Metropolis algorithm to the Ising model as "single spin flip dynamics."

In Problem 17.4 we use the Metropolis algorithm to simulate the one-dimensional Ising model. Note that the parameters J and kT do not appear separately, but appear together in the dimensionless ratio J/kT. Unless otherwise stated, we measure temperature in units of J/k, and set $H = 0$.

Problem 17.4   One-dimensional Ising model

a.   Write a Monte Carlo program to simulate the one-dimensional Ising model in equilibrium with a heat bath. (Modify SUB changes in *Program demon* (see Chapter 16) or see *Program ising*, listed in the following, for an example of the implementation of the Metropolis algorithm to the two-dimensional Ising model.) Use periodic boundary conditions. As a test of your program, compute the mean energy and magnetization of the lattice for $N = 20$ and $T = 1$. Draw the microscopic state (configuration) of the system after each Monte Carlo step per spin.

b.   Choose $N = 20$, $T = 1$, mcs $= 100$, and all spins up, i.e., $s_i = +1$ initially. What is the initial "temperature" of the system? Visually inspect the configuration of the system after each Monte Carlo step and estimate the time it takes for the system to reach equilibrium. Then change the initial condition so that the orientation of the spins is chosen at random. What is the initial "temperature" of the system in this case? Estimate the time it takes for the system to reach equilibrium in the same way as before.

c.   Choose $N = 20$ and equilibrate the system for mcs ≥ 100. Let mcs ≥ 1000 and determine $<E>$, $<E^2>$, $<M>$, and $<M^2>$ as a function of T in the range $0.1 \leq T \leq 5$. Plot $<E>$ as a function of T and discuss its qualitative features. Compare your computed results for $<E(T)>$ to the exact result (for $H = 0$)

$$<E> = -N \tanh \beta J. \tag{17.15}$$

Use the relation (17.12) to determine the T dependence of C.

d.   What is the qualitative dependence of $<M>$ on T? Use the relation (17.14) to estimate the T dependence of χ. One of the best laboratory realizations of a one-dimensional Ising ferromagnet is a chain of bichloride-bridged Fe^{2+} ions known as FeTAC (Greeney et al.). Measurements of χ yield a value of the exchange interaction J given by $J/k = 17.4$ K. (Experimental values of J are typically given in temperature units.) Use this value of J to plot your Monte

Carlo results for χ versus T with T given in Kelvin. At what temperature is χ a maximum for FeTAC?

e. Is the acceptance probability an increasing or decreasing function of T? Does the Metropolis algorithm become more or less efficient as the temperature is lowered?

f. Compute the probability density $P(E)$ for a system of 50 spins at $T = 1$. Choose mcs ≥ 1000. Plot $\ln P(E)$ versus $(E - <E>)^2$ and discuss its qualitative features.

We next apply the Metropolis algorithm to the two-dimensional Ising model on the square lattice. The main program is listed in the following.

```
PROGRAM ising
! Monte Carlo simulation of the Ising model on the square lattice
! using the Metropolis algorithm
DIM spin(32,32),w(-8 to 8),accum(10)
LIBRARY "csgraphics"
CALL initial(N,L,T,spin(,),mcs,nequil,w(),E,M)
FOR i = 1 to nequil          ! equilibrate system
    CALL Metropolis(N,L,spin(,),E,M,w(),accept)
NEXT i
CALL initialize(accum(),accept)
FOR pass = 1 to mcs          ! accumulate data while updating spins
    CALL Metropolis(N,L,spin(,),E,M,w(),accept)
    CALL data(E,M,accum())
NEXT pass
CALL output(T,N,mcs,accum(),accept)
END
```

In SUB initial we choose the initial directions of the spins, and compute the initial values of the energy and magnetization. To compute the total energy, we consider the interaction of a spin with its nearest neighbor spins to the north and the east. In this way we compute the energy of each interaction only once and avoid double counting. One of the most time consuming parts of the Metropolis algorithm is the calculation of the exponential function $e^{-\beta \Delta E}$. Because there are only a small number of possible values of $\beta \Delta E$ for the Ising model (see Fig. 16.3), we store the small number of different probabilities for the spin flips in the array w. The values of this array are computed in SUB initial.

```
SUB initial(N,L,T,spin(,),mcs,nequil,w(),E,M)
    RANDOMIZE
    INPUT prompt "linear dimension of lattice = ": L
    LET N = L*L                    ! number of spins
    ! temperature measured in units of J/k
    INPUT prompt "temperature = ": T
    INPUT prompt "# MC steps per spin for equilibrium = ": nequil
    INPUT prompt "# MC steps per spin for data = ": mcs
```

```
            LET M = 0
            FOR y = 1 to L                    ! random initial configuration
                FOR x = 1 to L
                    IF rnd < 0.5 then
                        LET spin(x,y) = 1   ! spin up
                    ELSE
                        LET spin(x,y) = -1
                    END IF
                    LET M = M + spin(x,y)      ! total magnetization
                NEXT x
            NEXT y
            LET E = 0
            FOR y = 1 to L                    ! compute initial energy
                IF y = L then
                    LET up = 1                ! periodic boundary conditions
                ELSE
                    LET up = y + 1
                END IF
                FOR x = 1 to L
                    IF x = L then
                        LET right = 1
                    ELSE
                        LET right = x + 1
                    END IF
                    LET sum = spin(x,up) + spin(right,y)
                    LET E = E - spin(x,y)*sum   ! total energy
                NEXT x
            NEXT y
            ! compute Boltzmann probability ratios
            FOR dE = -8 to 8 step 4
                LET w(dE) = exp(-dE/T)
            NEXT dE
        END SUB
```

One way to implement the Metropolis algorithm is to determine the change in the energy ΔE and then accept the trial flip if $\Delta E \leq 0$. If this condition is not satisfied, the second step is to generate a random number in the unit interval and compare it to $e^{-\beta \Delta E}$. Instead of this two step process, we implement the Metropolis algorithm in one step. Which method do you think is faster?

```
        SUB Metropolis(N,L,spin(,),E,M,w(),accept)
            DECLARE DEF DeltaE
            ! one Monte Carlo step per spin
            FOR ispin = 1 to N
                LET x = int(L*rnd + 1)    ! random x coordinate
                LET y = int(L*rnd + 1)    ! random y coordinate
                LET dE = DeltaE(x,y,L,spin(,))      ! compute change in energy
                IF rnd <= w(dE) then
```

```
                        LET spin(x,y) = -spin(x,y)   ! flip spin
                        LET accept = accept + 1
                        LET M = M + 2*spin(x,y)
                        LET E = E + dE
                    END IF
                NEXT ispin
            END SUB
```

A typical laboratory system has at least 10^{18} spins. In contrast, the number of spins that can be simulated typically ranges from 10^3 to 10^9. As we have discussed in other contexts, the use of periodic boundary conditions minimizes finite size effects. However, periodic boundary conditions reduce the maximum separation between spins to one half the length of the system, and more sophisticated boundary conditions are sometimes convenient. For example, we can give the surface spins extra neighbors, whose direction is related to the mean magnetization of the microstate. We adopt the simpler periodic boundary conditions in FUNCTION DeltaE in which the change in energy dE of flipping a spin is computed.

```
        FUNCTION DeltaE(x,y,L,spin(,))
            ! periodic boundary conditions
            IF x = 1 then
               LET left = spin(L,y)
            ELSE
               LET left = spin(x - 1,y)
            END IF
            IF x = L then
               LET right = spin(1,y)
            ELSE
               LET right = spin(x + 1,y)
            END IF
            IF y = 1 then
               LET down = spin(x,L)
            ELSE
               LET down = spin(x,y - 1)
            END IF
            IF y = L then
               LET up = spin(x,1)
            ELSE
               LET up = spin(x,y + 1)
            END IF
            LET DeltaE = 2*spin(x,y)*(left + right + up + down)
        END DEF
```

SUB data is called from the main program and the values of the physical observables are recorded after each Monte Carlo step per spin. The optimum time for sampling various physical quantities is explored in Problem 17.6. Note that if a flip is rejected and the old configuration is retained, thermal equilibrium is not described properly unless the old configuration is included again in computing the averages. Various variables are initialized in SUB initialize.

```
SUB initialize(accum(),accept)
    ! use array to save accumulated values of magnetization and
    ! energy. Array used to make it easier to add other quantities
    FOR i = 1 to 5
        LET accum(i) = 0
    NEXT i
    LET accept = 0
END SUB


SUB data(E,M,accum())
    ! accumulate data after every Monte Carlo step per spin
    LET accum(1) = accum(1) + E
    LET accum(2) = accum(2) + E*E
    LET accum(3) = accum(3) + M
    LET accum(4) = accum(4) + M*M
    LET accum(5) = accum(5) + abs(M)
END SUB
```

At the end of a run various averages are normalized and printed in SUB output. All averages such as the mean energy and the mean magnetization are normalized by the number of spins.

```
SUB output(T,N,mcs,accum(),accept)
    LET norm = 1/(mcs*N)                ! averages per spin
    LET accept = accept*norm
    LET eave = accum(1)*norm
    LET e2ave = accum(2)*norm
    LET mave = accum(3)*norm
    LET m2ave = accum(4)*norm
    LET abs_mave = accum(5)*norm
    CLEAR
    SET BACKGROUND COLOR "black"
    SET COLOR "yellow"
    PRINT "temperature = "; T
    PRINT "acceptance probability = "; accept
    PRINT "mean energy per spin = "; eave
    PRINT "mean squared energy per spin = "; e2ave
    PRINT "mean magnetization per spin = "; mave
    PRINT "mean of absolute magnetization per spin = "; abs_mave
    PRINT "mean squared magnetization per spin = "; m2ave
END SUB
```

Achieving thermal equilibrium can account for a substantial fraction of the total run time. The most practical choice of initial conditions is a configuration from a previous run that is at a temperature close to the desired temperature. The following subroutine saves the last configuration of a run and can be included at the end of the main loop in *Program ising*.

```
SUB save_config(N,L,T,spin(,))
    INPUT prompt "name of file for last configuration = ": file$
    OPEN #2: name file$, access output, create new
    PRINT #2: T
    FOR y = 1 to L
        FOR x = 1 to L
            PRINT #2: spin(x,y)
        NEXT x
    NEXT y
    CLOSE #2
END SUB
```

A previous configuration can be used in a later run by adding a few statements to SUB initial to allow the user to choose a previous configuration or a random configuration. A previous configuration can be read by calling the following subroutine:

```
SUB read_config(N,L,T,spin(,))
    INPUT prompt "filename?": file$
    OPEN #1: name file$, access input
    INPUT #1: T
    FOR y = 1 to L
        FOR x = 1 to L
            INPUT #1: spin(x,y)
        NEXT x
    NEXT y
    CLOSE #1
END SUB
```

Problem 17.5   Equilibration of the two-dimensional Ising model

a.   Run *Program ising* with the linear dimension of the lattice $L = 16$ and the heat bath temperature $T = 2$. Determine the time, nequil, needed to equilibrate the system, if the directions of the spins are initially chosen at random. Plot the values of E and M after each Monte Carlo step per spin. Estimate how many Monte Carlo steps per spin are necessary for the system to reach equilibrium.

b.   Write a subroutine that shows the spin configurations on the screen. One simple way to do so is to draw a solid square about each spin site and color code the orientation of the spins. Is the system "ordered" or "disordered" at $T = 2$ after equilibrium has been established?

c.   Repeat part (a) with all spins initially up. Does the equilibration time increase or decrease?

d.   Repeat parts (a)–(c) for $T = 2.5$.

Problem 17.6   Comparison with exact results

In general, a Monte Carlo simulation yields exact answers only after an infinite number of configurations have been sampled. How then can we be sure our program works correctly, and our results are statistically meaningful? One check is to ensure that our program can reproduce exact results in known limits. In the following, we test *Program ising* by considering a small system for which the mean energy and magnetization can be calculated analytically.

a.   Calculate analytically the T dependence of E, M, C and χ for the two-dimensional Ising model with $L = 2$ and periodic boundary conditions. (A summary of the calculation is given in Appendix 17B.)

b.   Use *Program ising* with $L = 2$ and estimate E, M, C, and χ for $T = 0.5$ and 0.25. Use the relations (17.12) and (17.14) to compute C and χ, respectively. Compare your estimated values to the exact results found in part (a). Approximately how many Monte Carlo steps per spin are necessary to obtain E and M to within 1%? How many Monte Carlo steps per spin are necessary to obtain C and χ to within 1%?

Now that we have checked our program and obtained typical equilibrium configurations, we consider the calculation of the mean values of the physical quantities of interest. Suppose we wish to compute the mean value of the physical quantity A. In general, the calculation of A is time consuming, and we do not want to compute its value more often than necessary. For example, we would not compute A after the flip of only one spin, because the values of A in the two configurations would almost be the same. Ideally, we wish to compute A for configurations that are statistically independent. Because we do not know *a priori* the mean number of spin flips needed to obtain configurations that are statistically independent, it is a good idea to estimate this time in our preliminary calculations.

One way to estimate the time interval over which configurations are correlated is to compute the time displaced *autocorrelation* function $C_A(t)$ defined as

$$C_A(t) = \frac{<A(t + t_0)A(t_0)> - <A>^2}{<A^2> - <A>^2}. \tag{17.16}$$

$A(t)$ is the value of the quantity A at time t. The averages in (17.16) are over all possible time origins t_0 for an equilibrium system. Because the choice of the time origin is arbitrary for an equilibrium system, C_A depends only on the time difference t rather than t and t_0 separately. For sufficiently large t, $A(t)$ and $A(0)$ will become uncorrelated, and hence $<A(t + t_0)A(t_0)> \rightarrow <A(t + t_0)><A(t_0)> = <A>^2$. Hence $C_A(t) \rightarrow 0$ as $t \rightarrow \infty$. In general, $C_A(t)$ will decay exponentially with t with a decay or correlation time τ_A whose magnitude depends on the choice of the physical quantity A as well as the physical parameters of the system, e.g., the temperature. Note that $C_A(t = 0)$ is normalized to unity.

The time dependence of the two most common correlation functions, $C_M(t)$ and $C_E(t)$ is investigated in Problem 17.7. As an example of the calculation of $C_E(t)$, consider the equilibrium time series for E for the $L = 4$ Ising model at $T = 4$: -4, -8, 0, -8, -20, -4, 0, 0, -24, -32, -24, -24, -8, -8, -16, -12. The averages of E and E^2 over these sixteen values are $<E> = -12$, $<E^2> = 240$, and $<E^2> - <E>^2 = 96$. We wish to compute $E(t)E(0)$ for all possible choices of the time origin. For example, $E(t = 4)E(0)$ is given by

$$<E(t = 4)E(0)> = \frac{1}{12}\big[(-20 \times -4) + (-4 \times -8) + (0 \times 0)$$
$$+ (0 \times -8) + (-24 \times -20) + (-32 \times -4)$$
$$+ (-24 \times 0) + (-24 \times 0) + (-8 \times -24)$$
$$+ (-8 \times -32) + (-16 \times -24) + (-12 \times -24)\big].$$

We averaged over the twelve possible choices of the origin for the time difference $t = 4$. Verify that $<E(t = 4)E(0)> = 460/3$ and $C_E(t = 4) = 7/72$.

In the above calculation of $<E(t)E(0)>$, we included all possible combinations of $E(t)E(0)$ for a given time series. To implement this procedure on a computer, we would need to store the time series in memory or in a data file. An alternative procedure is to save the last nsave values of the time series in memory and to average over fewer combinations. This procedure is implemented in SUB correl; the correlation functions are computed and printed in SUB c_output. SUB correl uses two arrays, Esave and Msave, to store the last nsave values of the energy and the magnetization at each Monte Carlo step per spin. These arrays and the arrays Ce and Cm may be initialized in a separate subroutine.

```
SUB correl(Ce(),Cm(),E,M,esave(),msave(),pass,nsave)
    ! accumulate data for time correlation functions
    ! save last nsave values of M and E
    ! index0 = array index for earliest saved time
    IF pass > nsave then
        ! compute Ce and Cm after nsave values are saved
        LET index0 = mod(pass-1,nsave) + 1
        LET index = index0
        FOR tdiff = nsave to 1 step -1
            LET Ce(tdiff) = Ce(tdiff) + E*esave(index)
            LET Cm(tdiff) = Cm(tdiff) + M*msave(index)
            LET index = index + 1
            IF index > nsave then LET index = 1
        NEXT tdiff
    END IF
    ! save latest value in array position of earliest value
    LET esave(index0) = E
    LET msave(index0) = M
END SUB


SUB c_output(N,Ce(),Cm(),accum(),mcs,nsave)
    ! compute time correlation functions
    LET ebar = accum(1)/mcs
```

```
                    LET e2bar = accum(2)/mcs
                    LET Ce(0) = e2bar - ebar*ebar
                    LET mbar = accum(3)/mcs
                    LET m2bar = accum(4)/mcs
                    LET Cm(0) = m2bar - mbar*mbar
                    LET norm = 1/(mcs - nsave)
                    PRINT
                    PRINT "t","Ce(t)","Cm(t)"
                    PRINT
                    FOR tdiff = 1 to nsave
                        ! correlation functions defined so that C(t=0) = 1
                        ! and C(infinity) = 0
                        LET Ce(tdiff) = (Ce(tdiff)*norm - ebar*ebar)/Ce(0)
                        LET Cm(tdiff) = (Cm(tdiff)*norm - mbar*mbar)/Cm(0)
                        PRINT tdiff,Ce(tdiff),Cm(tdiff)
                    NEXT tdiff
                END SUB
```

Problem 17.7   Correlation times

a.   Choose $L = 4$ and $T = 3$ and equilibrate the system. Then look at the time series of M and E after every Monte Carlo step per spin and estimate how often M changes sign. Does E change sign when M changes sign? How often does M change sign for $L = 8$ (and $T = 3$)? In equilibrium, positive and negative values of M are equally likely in the absence of an external magnetic field. Is your time series consistent with this equilibrium property? Why is it more meaningful to compute the time displaced correlation function of the absolute value of the magnetization rather than the magnetization itself if L is relatively small?

b.   Choose $L = 16$ and $T = 1$ and equilibrate the system. Then look at the time series of M. Do you find that positive and negative values of M are equally likely? Explain your results.

c.   Modify *Program ising* so that the equilibrium averaged values of $C_M(t)$ and $C_E(t)$ are computed. As a check on your program, use the time series for E given in the text to do a hand calculation of $C_E(t)$ in the way that it is computed in SUB correl and SUB c_output. Choose nsave = 10.

*d.   Estimate the correlation times from the energy and the magnetization correlation functions for $L = 8$, and $T = 3$, $T = 2.3$, and $T = 2$. Save the last nsave = 100 values of the magnetization and energy only after the system is equilibrated. Are the correlation times τ_M and τ_E comparable? One way to determine τ is to fit $C(t)$ to an assumed exponential form $C(t) \sim e^{-t/\tau}$. Another way is to define the integrated correlation time as

$$\tau = \sum_{t=1} C(t). \tag{17.17}$$

The sum is cut off at the first negative value of $C(t)$. Are the negative values of $C(t)$ physically meaningful? How does the behavior of $C(t)$ change if you average your results over longer runs? How do your estimates for the correlation times compare with your estimates of the relaxation time found in Problem 17.5? Why would the term "decorrelation time" be more appropriate than "correlation time?"

* **e.**   To describe the relaxation towards equilibrium as realistically as possible, we have randomly selected the spins to be flipped. However, if we are interested only in equilibrium properties, it might be possible to save computer time by selecting the spins sequentially. Determine if the correlation time is greater, smaller, or approximately the same if the spins are chosen sequentially rather than randomly. If the correlation time is greater, does it still save CPU time to choose spins sequentially? Why is it not desirable to choose spins sequentially in the one-dimensional Ising model?

Problem 17.8   Estimate of errors

How can we quantify the accuracy of our measurements, e.g., the accuracy of the mean energy $<E>$? As discussed in Chapter 11, the usual measure of the accuracy is the standard deviation of the mean. If we make n independent measurements of E, then the most probable error is given by

$$\sigma_m = \frac{\sigma}{\sqrt{n-1}}, \tag{17.18}$$

where the standard deviation σ is defined as

$$\sigma^2 = <E^2> - <E>^2. \tag{17.19}$$

The difficulty is that, in general, our measurements of the time series E_i are not independent, but are correlated. Hence, σ_m as given by (17.18) is an underestimate of the actual error.

How can we determine whether the measurements are independent without computing the correlation time? One way is based on the idea that the magnitude of the error should not depend on how we group the data. For example, suppose that we group every two data points to form $n/2$ new data points $E_i^{(2)}$ given by $E_i^{(g=2)} = (1/2)[E_{2i-1} + E_{2i}]$. If we replace n by $n/2$ and E by $E^{(2)}$ in (17.18) and (17.19), we would find the same value of σ_m as before provided that the original E_i are independent. If the computed σ_m is not the same, we continue this averaging process until σ_m calculated from

$$E_i^{(g)} = \frac{1}{2}[E_{2i-1}^{(g/2)} + E_{2i}^{(g/2)}] \qquad (g = 2, 4, 8, \ldots) \tag{17.20}$$

is approximately the same as that calculated from $E^{(g/2)}$.

a.   Use the above averaging method to estimate the errors in your measurements of $<E>$ and $<M>$ for the two-dimensional Ising model. Let $L = 8$, $T =$

2.269, and mcs $\geq$ 16384, and calculate averages after every Monte Carlo step per spin after the system has equilibrated. If necessary, increase the number of Monte Carlo steps for averaging. A rough measure of the correlation time is the number of terms in the time series that need to be averaged for σ_m to be approximately unchanged. What is the qualitative dependence of the correlation time on $T - T_c$?

b. Repeat for $L = 16$. Do you need more Monte Carlo steps than in part (a) to obtain statistically independent data? If so, why?

* c. The exact value of E/N for the two-dimensional Ising model on a square lattice with $L = 16$ and $T = T_c = 2/\ln(1 + \sqrt{2}) \approx 2.269$ is given by $E/N = -1.45306$ (to five decimal places). This value of T_c is exact for the infinite lattice. The exact result for E/N allows us to determine the actual error in this case. Compute $<E>$ by averaging E after each Monte Carlo step per spin for mcs $\geq 10^6$. Compare your actual error to the estimated error given by (17.18) and (17.19) and discuss their relative values.

17.5   THE ISING PHASE TRANSITION

Now that we have tested our program for the two-dimensional Ising model, we are ready to explore its properties. We first summarize some of the qualitative properties of infinite ferromagnetic systems in zero magnetic field. We know that at $T = 0$, the spins are perfectly aligned in either direction, that is, the mean magnetization per spin $m(T) = <M(T)>/N$ is given by $m(T = 0) = \pm 1$. As T is increased, the magnitude of $m(T)$ decreases continuously until $T = T_c$ at which $m(T)$ vanishes (see Fig. 17.1). Because $m(T)$ vanishes continuously rather than abruptly, the transition is termed *continuous* rather than discontinuous. (The term *first-order* describes a discontinuous transition.)

How can we characterize a continuous magnetic phase transition? Because a nonzero m implies that a net number of spins are spontaneously aligned, we designate m as the *order parameter* of the system. Near T_c, we can characterize the behavior of many physical quantities by power law behavior just as we characterized the percolation threshold (see Table 13.1). For example, we can write m near T_c as

$$m(T) \sim (T_c - T)^\beta, \tag{17.21}$$

where β is a critical exponent (not to be confused with the inverse temperature). Various thermodynamic derivatives such as the susceptibility and heat capacity diverge at T_c. We write

$$\chi \sim |T - T_c|^{-\gamma} \tag{17.22}$$

and

$$C \sim |T - T_c|^{-\alpha}. \tag{17.23}$$

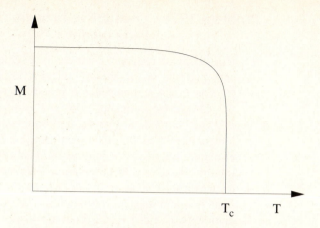

Fig. 17.1   The temperature dependence of $m(T)$, the mean
magnetization per spin, for the infinite lattice
Ising model in two dimensions.

We have assumed that χ and C are characterized by the same critical exponents above
and below T_c.

Another measure of the magnetic fluctuations is the linear dimension $\xi(T)$ of a typical magnetic domain. We expect the *correlation length* $\xi(T)$ to be the order of a lattice
spacing for $T \gg T_c$. Because the alignment of the spins becomes more correlated as T
approaches T_c from above, $\xi(T)$ increases as T approaches T_c. We can characterize the
divergent behavior of $\xi(T)$ near T_c by the critical exponent ν:

$$\xi(T) \sim |T - T_c|^{-\nu}. \tag{17.24}$$

The calculation of ξ is considered in Problem 17.9d.

As we found in our discussion of percolation in Chapter 13, a finite system cannot
exhibit a true phase transition. Nevertheless, we expect that if $\xi(T)$ is less than the
linear dimension L of the system, our simulations will yield results comparable to
an infinite system. Of course, if T is close to T_c, our simulations will be limited by
finite size effects. In the following problem, we obtain preliminary results for the T
dependence of m, $<E>$, C, and χ in the neighborhood of T_c. These results will help
us understand the qualitative nature of the ferromagnetic phase transition in the two-
dimensional Ising model.

Because we will consider the Ising model for different values of L, it will be convenient to compute intensive quantities such as the mean energy per spin, the specific
heat (per spin) and the susceptibility per spin. We will retain the same notation for both
the extensive and corresponding intensive quantities.

Problem 17.9   Qualitative behavior of the two-dimensional Ising model

a. Use *Program ising* to compute the magnetization per spin m, the mean energy per spin $<E>$, the specific heat C, and the susceptibility per spin χ. Choose $L = 4$ and consider T in the range $1.5 \leq T \leq 3.5$ in steps of $\Delta T = 0.2$. Choose the initial condition at $T = 3.5$ so that the orientation of the spins is chosen at random. Use an equilibrium configuration from a previous run at temperature T as the initial configuration for a run at temperature $T - \Delta T$. Because all the spins might overturn and the magnetization change sign during the course of your observation, estimate the mean value of $|m|$ in addition to that of m. Use at least 1000 Monte Carlo steps per spin and estimate the number of equilibrium configurations needed to obtain m and $<E>$ to 5% accuracy. Plot $<E>$, m, $|m|$, C, and χ as a function of T and describe their qualitative behavior. Do you see any evidence of a phase transition?

b. Repeat the calculations of part (a) for $L = 8$ and $L = 16$. Plot $<E>$, m, $|m|$, C, and χ as a function of T and describe their qualitative behavior. Do you see any evidence of a phase transition? For comparison, recent published Monte Carlo results for the two-dimensional Ising model are in the range $L = 10^2$ to $L = 10^3$ with order 10^6 Monte Carlo steps per spin.

c. For a given value of L, e.g., $L = 16$, choose a value of T that is well below T_c and choose the directions of the spins at random. Observe the spins evolve in time. Do you see several domains with positive and negative spontaneous magnetization? How does the magnetization evolve with time?

*d. The correlation length ξ can be obtained from the r-dependence of the spin correlation function $c(r)$. The latter is defined as:

$$c(r) = <s_i s_j> - m^2, \tag{17.25}$$

where r is the distance between sites i and j. We have assumed the system is translationally invariant so that $<s_i> = <s_j> = m$. The average is over all sites for a given configuration and over many configurations. Because the spins are not correlated for large r, we see that $c(r) \to 0$ in this limit. It is reasonable to assume that $c(r) \sim e^{-r/\xi}$ for r sufficiently large. Use this behavior to estimate ξ as a function of T. How does your estimate of ξ compare with the size of the regions of spins with the same orientation?

One of the limitations of a computer simulation study of a phase transition is the relatively small size of the systems we can study. Nevertheless, we observed in Problem 17.9 that even systems as small as $L = 4$ exhibit behavior that is reminiscent of a phase transition. In Fig. 17.2 we show our Monte Carlo data for the T dependence of the specific heat of the two-dimensional Ising model for $L = 8$ and $L = 16$. We see that C exhibits a broad maximum which becomes sharper for larger L. Does your data for C exhibit similar behavior?

Because we can simulate only finite lattices, it is difficult to obtain estimates for the critical exponents α, β, and γ by using the definitions (17.21)–(17.23) directly. We

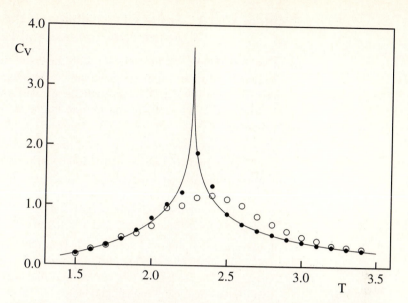

Fig. 17.2   The temperature dependence of the specific heat C (per spin) of
the Ising model on a $L = 8$ and $L = 16$ square lattice with periodic
boundary conditions. One thousand Monte Carlo steps per spin
were used for each value of the temperature. The continuous line
represents the temperature dependence of C in the limit of an
infinite lattice. (Note that C is infinite at $T = T_c$ for an infinite
lattice.)

learned in Section 13.4, we can do a *finite size scaling analysis* to extrapolate finite L
results to $L \to \infty$. For example, from Fig. 17.2 we see that the temperature at which C
exhibits a maximum becomes better defined for larger lattices. This behavior provides
a simple definition of the transition temperature $T_c(L)$ for a finite system. According to
finite size scaling theory, $T_c(L)$ scales as

$$T_c(L) - T_c(L = \infty) \sim aL^{-1/\nu}, \tag{17.26}$$

where a is a constant and ν is defined in (17.24). The finite size of the lattice is important when the correlation length

$$\xi(T) \sim L \sim |T - T_c|^{-\nu}. \tag{17.27}$$

As in Section 13.4, we can set $T = T_c$ and consider the L-dependence of M, C, and χ:

$$m(T) \sim (T_c - T)^\beta \to L^{-\beta/\nu} \tag{17.28}$$
$$C(T) \sim |T - T_c|^{-\alpha} \to L^{\alpha/\nu} \tag{17.29}$$
$$\chi(T) \sim |T - T_c|^{-\gamma} \to L^{\gamma/\nu}. \tag{17.30}$$

In Problem 17.10 we use the relations (17.28)–(17.30) to estimate the critical exponents
β, γ, and α.

Problem 17.10   Finite size scaling and the critical properties of the two-dimensional Ising model

a.   Use the relation (17.26) together with the exact result $v = 1$ to estimate the value of T_c on an infinite square lattice. Because it is difficult to obtain a precise value for T_c with small lattices, we will use the exact result $kT_c/J = 2/\ln(1 + \sqrt{2}) \approx 2.269$ for the infinite lattice in the remaining parts of this problem.

b.   Determine the specific heat C, $|m|$, and the susceptibility χ at $T = T_c$ for $L = 2, 4, 8,$ and 16. Use as many Monte Carlo steps per spin as possible. Plot the logarithm of $|m|$ and χ versus L and use the scaling relations (17.28)–(17.30) to determine the critical exponents β and γ. Assume the exact result $v = 1$. Do your log-log plots of $|m|$ and χ yield reasonably straight lines? Compare your estimates for β and γ with the exact values given in Table 13.1.

c.   Make a log-log plot of C versus L. If your data for C is sufficiently accurate, you will find that the log-log plot of C versus L is not a straight line but shows curvature. The reason for this curvature is that α in (17.23) equals zero for the two-dimensional Ising model, and hence (17.29) needs to be interpreted as

$$C \sim C_0 \ln L. \tag{17.31}$$

Is your data for C consistent with (17.31)? The constant C_0 in (17.31) is approximately 0.4995.

So far we have performed our Ising model simulations on the square lattice. How do the critical temperature and the critical exponents depend on the symmetry and the dimension of the lattice? Based on your experience with the percolation transition in Chapter 13, you might have a good idea what the answer is.

*Problem 17.11   The effects of symmetry and dimension on the critical properties of the Ising model

a.   The nature of the triangular lattice is discussed in Chapter 8 (see Fig. 8.5). The main difference between the triangular lattice and the square lattice is the number of nearest neighbors. Make the necessary modifications in your Ising program, e.g., determine the possible transitions and the values of the transition probabilities. Compute C and χ for different values of T in the interval $[1, 5]$. Assume that $v = 1$ and use finite size scaling to estimate T_c in the limit of an infinite triangular lattice. Compare your estimate of T_c to the known value $kT_c/J = 3.641$ (to three decimal places). The simulation of Ising models on the triangular lattice is relevant to the understanding of the experimentally observed phases of materials that can be absorbed on the surface of graphite.

b.   No exact results are available for the Ising model in three dimensions. Write a Monte Carlo program to simulate the Ising model on the simple cubic lattice (six nearest neighbors). Compute C and χ for T in the range $3.2 \leq T \leq 5$ in steps of 0.2 for different values of L. Estimate $T_c(L)$ from the maximum of C and χ. How do these estimates of $T_c(L)$ compare? Use the values of $T_c(L)$ that exhibit a stronger L dependence and plot $T_c(L)$ versus $L^{-1/\nu}$ for different values of ν in the range 0.5 to 1 (see (17.26)). Show that the extrapolated value of $T_c(L = \infty)$ does not depend sensitively on the value of ν. Compare your estimate for $T_c(L = \infty)$ to the known value $kT_c/J = 4.5108$ (to four decimal places).

c.   Compute $|m|$, C, and χ at $T = T_c \approx 4.5108$ for different values of L on a simple cubic lattice. Do a finite size scaling analysis to estimate β/ν, α/ν, and γ/ν. The best known values of the critical exponents for the three-dimensional Ising model are given in Table 13.1. For comparison, published Monte Carlo results in 1976 for the finite size behavior of the Ising model on the simple cubic Ising lattice are for $L = 6$ to $L = 20$; 2000–5000 Monte Carlo steps per spin were used for calculating the averages after equilibrium had been reached.

*Problem 17.12   Critical slowing down

a.   Consider the two-dimensional Ising model on a square lattice with $L = 16$. Compute $C_M(t)$ and $C_E(t)$ and determine the correlation times τ_M and τ_E for $T = 2.5, 2.4$, and 2.3. Determine the correlation times as discussed in Problem 17.17d. How do these correlation times compare with one another? Show that τ increases as the critical temperature is approached, a physical effect known as *critical slowing down*.

b.   We can define the dynamical critical exponent z by the relation

$$\tau \sim \xi^z. \tag{17.32}$$

On a finite lattice we have the relation $\tau \sim L^z$ at $T = T_c$. Compute τ for different values of L at $T = T_c$ and make a very rough estimate of z. (The value of z for the two-dimensional Ising model with spin flip dynamics is still not definitely known, but appears to be slightly greater than 2.)

The magnitude of τ found in parts (a) and (b) depends in part on our choice of dynamics. Although we have generated a trial change by the attempted flip of one spin, it is possible that other types of trial changes, e.g., the simultaneous flip of more than one spin, would be more efficient and lead to smaller correlation times and smaller values of z. A problem of much current interest is the development of more efficient algorithms near phase transitions (see Project 17.1).

17.6   OTHER APPLICATIONS OF THE ISING MODEL

Because the applications of the Ising model are so wide ranging, we can mention only a few of the applications here. In the following, we briefly describe applications of the

Ising model to first-order phase transitions, lattice gases, antiferromagnetism, and the order-disorder transition in binary alloys.

So far we have discussed the continuous phase transition in the Ising model and have found that the energy and magnetization vary continuously with the temperature, and thermodynamic derivatives such as the specific heat and the susceptibility diverge near T_c (in the limit of an infinite lattice). In Problem 17.13 we discuss a simple example of a *first-order* phase transition. Such transitions are accompanied by *discontinuous* (finite) changes in thermodynamic quantities such as the energy and the magnetization.

*Problem 17.13   The two-dimensional Ising model in an external magnetic field

a. Modify your two-dimensional Ising program so that the energy of interaction with an external magnetic field H is included. It is convenient to measure H in terms of the quantity $h = \beta H$. We wish to compute m, the mean magnetization per spin, as a function of h for $T < T_c$. Consider a square lattice with $L = 16$ and equilibrate the system at $T = 1.8$ and $h = 0$. Adopt the following procedure to obtain $m(h)$.

   i. Use an equilibrium configuration at $h = 0$ as the initial configuration for $h_1 = \Delta h = 0.2$.

   ii. Run the system for 100 Monte Carlo steps per spin before computing averages.

   iii. Average m over 80 Monte Carlo steps per spin.

   iv. Use the last configuration for h_n as the initial configuration for $h_{n+1} = h_n + \Delta h$.

   v. Repeat steps (ii)–(iv) until $m \approx 0.95$. Plot m versus h. Do the measured values of m correspond to equilibrium averages?

b. Decrease h by $\Delta h = -0.2$ in the same way as in part (a) until h passes through zero and until $m \approx -0.95$. Extend your plot of m versus h to negative h values. Does m remain positive for small negative h? Do the measured values of m for negative h correspond to equilibrium averages? Draw the spin configurations for several values of h. Do you see evidence of domains?

c. Increase h by $\Delta h = 0.2$ until the m versus h curve forms an approximately closed loop. What is the value of m at $h = 0$? This value of m is the spontaneous magnetization.

d. A first-order phase transition is characterized by a discontinuity (for an infinite lattice) in the order parameter. In the present case the transition is characterized by the behavior of m as a function of h. What is your measured value of m for $h = 0.2$? If $m(h)$ is double valued, which value of m corresponds to the equilibrium state, an absolute minima in the free energy? Which value of m corresponds to a *metastable* state, a relative minima in the free energy? What are the equilibrium and metastable values of m for $h = -0.2$? The transition from positive m to negative m is first-order because there is a discontinuous jump in the magnetization. First-order transitions exhibit *hysteresis* and the properties of the system depend on the history of the system, e.g., whether h is an increasing or decreasing function. Because of the long lifetime of metastable states near a phase transition, a system in such a state can mistakenly be interpreted as being in equilibrium. We also know that near a continuous phase transition, the relaxation to equilibrium becomes very long (see Problem 17.12), and hence a system with a continuous phase transition can behave as if it were effectively in a metastable state. For these reasons it is very difficult to distinguish the nature of a phase transition using computer simulations. This problem is discussed further in Section 17.8.

e. Repeat the above simulation for $T = 3$, a temperature above T_c. Why do your results differ from the simulations in parts (a)–(c) done for $T < T_c$?

The Ising model also describes systems that might appear to have little in common with ferromagnetism. For example, we can interpret the Ising model as a "lattice gas," where "down" represents a lattice site occupied by a molecule and "up" represents an empty site. Each lattice site can be occupied by at most one molecule, and the molecules interact with their nearest neighbors. The lattice gas is a crude model of the behavior of a real gas of molecules and is a simple lattice model of the gas-liquid transition and the critical point. What properties does the lattice gas have in common with a real gas? What properties of real gases does the lattice gas neglect?

If we wish to simulate a lattice gas, we have to decide whether to do the simulation at fixed density or at fixed chemical potential μ. The implementation of the latter is straightforward because the grand canonical ensemble for a lattice gas is equivalent to the canonical ensemble for Ising spins in an external magnetic field H, that is, the effect of the magnetic field is to fix the mean number of up spins. Hence, we can simulate a lattice gas in the grand canonical ensemble by doing spin flip dynamics. (The volume of the lattice is an irrelevant parameter.)

Another application of a lattice gas model is to the study of phase separation in a binary or A-B alloy. In this case spin up and spin down correspond to a site occupied by an A atom and B atom, respectively. As an example, the alloy β-brass has a low temperature ordered phase in which the two components (copper and zinc) have equal concentrations and form a cesium chloride structure. As the temperature is increased, some zinc atoms exchange positions with copper atoms, but the system is still ordered. However, above the critical temperature $T_c = 742$ K, the zinc and copper atoms become mixed and the system is disordered. This transition is an example of an *order-disorder* transition.

Because the number of A atoms and the number of B atoms is fixed, we cannot use spin flip dynamics to simulate a binary alloy. A dynamics that does conserve the number of down and up spins is known as *spin exchange dynamics*. In this dynamics a trial *interchange* of two nearest neighbor spins is made and the change in energy ΔE is calculated. The criterion for the acceptance or rejection of the trial change is the same as before.

*Problem 17.14   Simulation of a lattice gas

a.   Modify your Ising program so that spin exchange dynamics rather than spin flip dynamics is implemented. For example, determine the possible values of ΔE on the square lattice, determine the possible values of the transition probability, and change the way a trial change is made. If we are interested only in the mean value of quantities such as the total energy, we can reduce the computation time by not interchanging like spins. For example, we can keep a list of bonds between occupied and empty sites and make trial moves by choosing bonds at random from this list. For small lattices such a list is unnecessary and a trial move can be generated by simply choosing a spin and one of its nearest neighbors at random.

b.   Consider a square lattice with $L = 8$ and with 32 sites initially occupied. (The number of occupied sites is a conserved variable and must be specified initially.) Determine the mean energy for T in the range $1 \leq T \leq 4$. Plot the mean energy as a function of T. Does the energy appear to vary continuously?

c.   Repeat the calculations of part (b) with 44 sites initially occupied, and plot the mean energy as a function of T. Does the energy vary continuously? Do you see any evidence of a first-order phase transition?

*d.   Because the spins correspond to molecules, we can compute the single particle diffusion coefficient of the molecules. (See Problem 12.6 for a similar simulation.) Use an array to record the position of each molecule as a function of time. After equilibrium has been reached, compute $<R(t)^2>$, the mean square displacement per molecule. Is it necessary to "interchange" two like spins? If the atoms undergo a random walk, the self-diffusion constant D is defined as

$$D = \lim_{t \to \infty} \frac{1}{2dt} <R(t)^2>. \tag{17.33}$$

Estimate D for different temperatures and numbers of occupied sites.

Although you are probably familiar with ferromagnetism, e.g., a magnet on a refrigerator door, nature provides more examples of antiferromagnetism. In the language of the Ising model, antiferromagnetism means that the interaction parameter J is negative and nearest neighbor spins prefer to be aligned in opposite directions. As we will see in Problem 17.15, the properties of the antiferromagnetic Ising model on a square lattice are similar to the ferromagnetic Ising model. For example, the energy and specific heat of the ferromagnetic and antiferromagnetic Ising models are identical at all temperatures in zero magnetic field, and the system exhibits a phase transition at the Néel temperature T_N. On the other hand, the total magnetization and susceptibility of the antiferromagnetic model do not exhibit any critical behavior near T_N. Instead, we can define two sublattices for the square lattice corresponding to the red and black squares of a checkerboard and introduce the "staggered magnetization" M_s equal to the difference of the magnetization on the two sublattices. We will find in Problem 17.15 that the temperature dependence of M_s and the staggered susceptibility χ_s are identical to the analogous quantities in the ferromagnetic Ising model.

*Problem 17.15   Antiferromagnetic Ising model

a.   Modify Program ising to simulate the antiferromagnetic Ising model on the square lattice in zero magnetic field. Because J does not appear explicitly in Program ising, change the sign of the energy calculations in the appropriate places in the program. To compute the staggered magnetization on a square lattice, define one sublattice to be the sites (x, y) for which the product $\mathrm{mod}(x, 2) \times \mathrm{mod}(y, 2) = 1$; the other sublattice corresponds to the remaining sites.

b.   Choose $L = 16$ and the initial condition to be all spins up. What configuration of spins corresponds to the state of lowest energy? Compute the temperature dependence of the mean energy, specific heat, magnetization, and the susceptibility χ. Does the temperature dependence of any of these quantities show evidence of a phase transition?

c.   Compute the temperature dependence of M_s and the staggered susceptibility χ_s defined as (see (17.14))

$$\chi_s = \frac{1}{kT}\left[<{M_s}^2> - <M_s>^2\right]. \tag{17.34}$$

Verify that the temperature dependence of M_s for the antiferromagnetic Ising model is the same as the temperature dependence of M for the Ising ferromagnet. Could you have predicted this similarity without doing the simulation?

d.   In part (b) you might have noticed that χ shows a cusp. Compute χ for different values of L at $T = T_N \approx 2.269$. Do a finite size scaling analysis and verify that χ does not diverge at $T = T_N$.

e.   Consider the behavior of the antiferromagnetic Ising model on a triangular lattice. Choose $L \geq 16$ and compute the same quantities as before. Do you see any evidence of a phase transition? Draw several configurations of the system at different temperatures. Do you see evidence of many small domains at low temperatures? Is there a unique ground state? If you cannot find a unique ground state, you share the same frustration as do the individual spins in the antiferromagnetic Ising model on the triangular lattice. We say that this model exhibits *frustration* because there is no spin configuration on the triangular lattice such that all spins are able to minimize their energy (see Fig. 17.3).

The Ising model is one of many models of magnetism. The Heisenberg, Potts, and x-y models are other examples of models of magnetic materials familiar to condensed matter scientists as well as to workers in other areas. Monte Carlo simulations of these models and others have been important in the development of our understanding of phase transitions in both magnetic and nonmagnetic materials. Some of these models are discussed in Section 17.11.

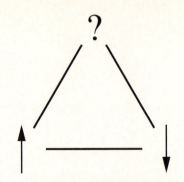

Fig. 17.3   An example of frustration
on a triangular lattice.

17.7   SIMULATION OF CLASSICAL FLUIDS

The existence of matter as a solid, liquid and gas is well known (see Fig. 17.4). Our goal in this section is to use Monte Carlo methods to gain additional insight into the qualitative differences between these three phases.

Monte Carlo simulations of classical systems are simplified considerably by the fact that the velocity (momentum) variables are decoupled from the position variables. The total energy can be written as $E = K(\{\mathbf{v}_i\}) + U(\{\mathbf{r}_i\})$, where the kinetic energy K is a function of only the particle velocities $\{\mathbf{v}_i\}$, and the potential energy U is a function of only the particle positions $\{\mathbf{r}_i\}$. Because the velocity appears quadratically in the kinetic energy, the equipartition theorem implies that the contribution of the velocity coordinates to the mean energy is $\frac{1}{2}kT$ per degree of freedom. Hence, we need to sample only the positions of the molecules, i.e., the "configurational" degrees of freedom. Is such a simplification possible for quantum systems?

The physically relevant quantities of a fluid include its mean energy, specific heat and equation of state. Another interesting quantity is the *radial distribution function* $g(r)$ which we introduced in Chapter 8. We will find in Problems 17.16–17.18 that $g(r)$ is a probe of the density fluctuations and hence a probe of the local order in the system. If only two-body forces are present, the mean potential energy per particle can be expressed as (see (8.14))

$$\frac{U}{N} = \frac{\rho}{2} \int g(r) V(r)\, d\mathbf{r}, \tag{17.35}$$

and the (virial) equation of state can be written as (see (8.15))

$$\frac{\beta P}{\rho} = 1 - \frac{\beta \rho}{2d} \int g(r)\, r \frac{dV(r)}{dr}\, d\mathbf{r}. \tag{17.36}$$

Hard core interactions. To separate the effects of the short range repulsive interaction from the longer range attractive interaction, we first investigate a model of *hard disks* with the interparticle interaction

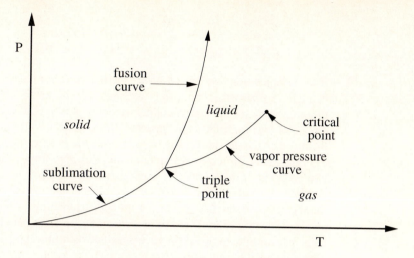

Fig. 17.4   A sketch of the phase diagram for a simple material.

$$V(r) = \begin{cases} +\infty & r < \sigma \\ 0 & r \geq \sigma . \end{cases} \tag{17.37}$$

Such an interaction has been extensively studied in one dimension (hard rods), two dimensions (hard disks), and in three dimensions (hard spheres). Hard sphere systems were the first systems studied by Metropolis and coworkers.

Because there is no attractive interaction present in (17.37), there is no transition from a gas to a liquid. Is there a phase transition between a fluid phase at low densities and a solid at high densities? Can a solid form in the absence of an attractive interaction? What are the physically relevant quantities for a system with an interaction of the form (17.37)? There are no thermal quantities such as the mean potential energy because this quantity is always zero. The major quantity of interest is $g(r)$ which yields information on the correlations of the particles and the equation of state. If the interaction is given by (17.37), it can be shown that (17.36) reduces to

$$\frac{\beta P}{\rho} = 1 + \frac{2\pi}{3}\rho\sigma^3 g(\sigma) \qquad (d = 3) \tag{17.38a}$$

$$= 1 + \frac{\pi}{2}\rho\sigma^2 g(\sigma) \qquad (d = 2) \tag{17.38b}$$

$$= 1 + \rho\sigma g(\sigma). \qquad (d = 1) \tag{17.38c}$$

We will calculate $g(r)$ for different values of r and then extrapolate our results to $r = \sigma$ (see Problem 17.16b).

Because the application of molecular dynamics and Monte Carlo methods to hard disks is similar, we discuss the latter method only briefly and do not include a program here. The idea is to choose a disk at random and move it to a trial position as implemented in the following:

```
LET itrial = int(N*rnd) + 1
LET xtrial = x(itrial) + (2*rnd - 1)*delta
LET ytrial = y(itrial) + (2*rnd - 1)*delta
```

If the new position overlaps another disk, the move is rejected and the old configuration is retained; otherwise the move is accepted. A reasonable, although not necessarily optimum, choice for the maximum displacement δ is to choose δ such that approximately one half of all trial states are accepted. We also need to fix the maximum amplitude of the move so that the moves are equally probable in all directions.

The major difficulty in implementing this algorithm is determining the overlap of two particles. If the number of particles is not too large, it is sufficient to compute the distances between the trial particle and all the other particles rather than the smaller number of particles that are in the immediate vicinity of the trial particle. For larger systems this procedure is too time consuming, and it is better to divide the system into cells and to only compute the distances between the trial particle and particles in the same and neighboring cells.

The choice of initial positions for the disks is more complicated than it might first appear. One strategy is to place each successive disk at random in the box. If a disk overlaps one that is already present, generate another pair of random numbers and attempt to place the disk again. If the desired density is low, an acceptable initial configuration can be computed fairly quickly in this way, but if the desired density is high, the probability of adding a disk will be very small (see Problem 17.17a). To reach higher densities, we might imagine beginning with the desired number of particles in a low density configuration and moving the boundaries of the central cell inward until a boundary just touches one of the disks. Then the disks are moved a number of Monte Carlo steps and the boundaries are moved inward again. This procedure also becomes more difficult as the density increases. The most efficient procedure is to start the disks on a lattice at the highest density of interest such that no overlap of disks occurs.

We first consider a one-dimensional system of hard rods for which the equation of state and $g(r)$ can be calculated exactly. The equation of state is given by

$$\frac{P}{NkT} = \frac{1}{L - N\sigma}. \tag{17.39}$$

Because hard rods cannot pass through one another, the excluded volume is $N\sigma$ and the available volume is $L - N\sigma$. Note that this form of the equation of state is the same as the van der Waals equation of state (cf. Reif) with the contribution from the attractive part of the interaction equal to zero.

Problem 17.16   Monte Carlo simulation of hard rods

a.   Write a program to do a Monte Carlo simulation of a system of hard rods. Adopt the periodic boundary condition and refer to *Program hd* in Chapter 8 for the basic structure of the program. The major difference is the nature of the "moves." Measure all lengths in terms of the hard rod diameter σ. Choose $L = 12$ and $N = 10$. How does the number density $\rho = N/L$ compare to the maximum possible density? Choose the initial positions to be on a one-dimensional grid and let the maximum displacement $\delta = 0.1$. Approximately how many Monte Carlo steps per particle are necessary to reach equilibrium? What is the equilibrium acceptance probability? Compute the pair correlation function $g(x)$.

b. Plot $g(x)$ as a function of the distance x. Why does $g(x) = 0$ for $x < 1$? Why are the values of $g(x)$ for $x > L/2$ not meaningful? What is the physical interpretation of the peaks in $g(x)$? Note that the results for $g(x)$ are for $x > 1$ because none of the hard disks are in contact in the configurations generated by the Monte Carlo algorithm. (Recall that x is measured in units of σ.) Because the mean pressure can be determined from $g(x)$ at $x = 1^{+}$ (see (17.38c)), we need to determine $g(x)$ at contact by extrapolating your results for $g(x)$ to $x = 1$. An easy way to do so is to fit the three points of $g(x)$ closest to $x = 1$ to a parabola. Use your result for $g(x = 1^{+})$ to determine the mean pressure.

c. Compute $g(x)$ at several lower densities by using an equilibrium configuration from a previous run and increasing L. How do the size and the location of the peaks in $g(x)$ change?

Problem 17.17 Monte Carlo simulation of hard disks

a. What is the maximum packing density of hard disks, that is, how many disks can be packed together in a cell of area A? The maximum packing density can be found by placing the disks on a triangular lattice with the nearest neighbor distance equal to the disk diameter. Write a simple program that adds disks at random into a rectangular box of area $A = L_x \times L_y$ with the constraint that no two disks overlap. If a disk overlaps a disk already present, generate another pair of random numbers and try to place the disk again. If the density is low, the probability of adding a disk is high, but if the desired density is high most of the disks will be rejected. For simplicity, do not worry about periodic boundary conditions and accept a disk if its center lies within the box. Choose $L_x = 6$ and $L_y = \sqrt{3}L_x/2$ and determine the maximum density $\rho = N/A$ that you can attain in a reasonable amount of CPU time. How does this density compare to the maximum packing density? What is the qualitative nature of the density dependence of the acceptance probability?

b. Adapt your Monte Carlo program for hard rods to a system of hard disks. Begin at a density ρ slightly lower than the maximum packing density ρ_0. Choose $N = 16$ with $L_x = 4.41$ and $L_y = \sqrt{3}L_x/2$. Compare the density $\rho = N/(L_x L_y)$ to the maximum packing density. Choose the initial positions of the particles to be on a triangular lattice. A reasonable first choice for the maximum displacement δ is $\delta = 0.1$. Compute $g(r)$ for $\rho/\rho_0 = 0.95, 0.92, 0.88, 0.85, 0.80, 0.70, 0.60$, and 0.30. Keep the ratio of L_x/L_y

fixed and save a configuration from the previous run to be the initial configuration of the new run at lower ρ. Allow at least 400 Monte Carlo steps per particle for the system to equilibrate and average $g(r)$ for mcs $\geq$ 400.

c. What is the qualitative behavior of $g(r)$ at high and low densities? For example, describe the number and height of the peaks of $g(r)$. If the system is crystalline, then $g(\mathbf{r})$ is not spherically symmetric. How would you compute $g(\mathbf{r})$ in this case and what would you expect to see?

d. Use your results for $g(r = 1^+)$ to compute the mean pressure P as a function of ρ (see (17.38b)). Plot the ratio PV/NkT as a function of ρ, where the "volume" V is the area of the system. How does the temperature T enter into the Monte Carlo simulation? Is the ratio PV/NkT an increasing or decreasing function of ρ? At low densities we might expect the system to act like an ideal gas with the volume replaced by $(V - N\sigma)$. Compare your low density results with this prediction.

e. Take "snapshots" of the disks at intervals of ten to twenty Monte Carlo steps per particle. Do you see any evidence of the solid becoming a fluid at lower densities?

*f. Compute an "effective diffusion coefficient" D by determining the mean square displacement $<R^2(t)>$ of the particles after equilibrium is reached. Use the relation (17.33) and identify the time t with the number of Monte Carlo steps per particle. Estimate D for the densities considered in part (a), and plot the product ρD as a function of ρ. What is the dependence of D on ρ for a dilute gas? Try to identify a range of ρ where D drops abruptly. Do you observe any evidence of a phase transition?

*g. The magnitude of the maximum displacement parameter δ is arbitrary. If δ is large and the density is high, then a high proportion of the trial moves will be rejected. On the other hand, if δ is too small, the acceptance probability will be close to unity, but the successive configurations will be strongly correlated. Hence if δ is too large or is too small, our simulation is inefficient. In practice, δ is usually chosen so that approximately half of the moves are accepted. A better criterion might be to choose δ so that the mean square displacement is a maximum for a fixed time interval. The idea is that the mean square displacement is a measure of the exploration of phase space. Fix the density and determine the value of δ that maximizes $<R^2(t)>$. What is the corresponding acceptance probability? If this probability is much less than 50%, how can you decide which criterion for δ is more appropriate?

Continuous potentials. Our simulations of hard disks have led us to conclude that there is a phase transition from a fluid at low densities to a solid at higher densities. This conclusion is consistent with molecular dynamics and Monte Carlo studies of larger systems. Although the existence of a fluid-solid transition for hard sphere and hard disk systems is well accepted, the relatively small numbers of particles used in any simulation should remind us that results of this type cannot be taken as evidence independently of any theoretical justification.

The existence of a fluid-solid transition for hard spheres implies that the transition is primarily determined by the repulsive part of the potential. We now consider a system with both a repulsive and an attractive contribution. Our primary goal will be to determine the influence of the attractive part of the potential on the structure of a liquid. We adopt as our model interaction the Lennard-Jones potential:

$$U(r) = 4\epsilon \left[\left(\frac{\sigma}{r}\right)^{12} - \left(\frac{\sigma}{r}\right)^{6} \right].$$ (17.40)

The nature of the Lennard-Jones potential and the appropriate choice of units for simulations was discussed in Chapter 8 (see Table 8.1). We consider in Problem 17.18 the application of the Metropolis algorithm to a system of N particles in a rectangular cell of fixed "volume" V interacting via the Lennard-Jones potential. Because the simulation is at fixed T, V, and N, the simulation samples configurations of the system according to the Boltzmann distribution (17.1).

Problem 17.18 Monte Carlo simulation of a Lennard-Jones system

a. The properties of a two-dimensional Lennard-Jones system with the potential energy of interaction (17.40) have been studied by many workers under a variety of conditions. Write a program to compute the total energy of a system of N particles on a triangular lattice of area $L_x \times L_y$ with periodic boundary conditions. Choose $N = 16$, $L_x = 4.6$, and $L_y = \sqrt{3}L_x/2$. Why does this energy correspond to the energy at temperature $T = 0$? Does the energy per particle change if you consider bigger systems at the same density?

b. Write a program to compute the mean energy, pressure, and radial distribution function using the Metropolis algorithm. One way of computing the change in the potential energy of the system due to a trial move of one of the particles is to use an array, pe, for the potential energy of interaction of each particle. For simplicity, compute the potential energy of particle i by considering its interaction with the other $N - 1$ particles. The total potential energy of the system is the sum of the array elements pe(i) over all N particles divided by two to account for double counting. For simplicity, accumulate data after each Monte Carlo step per particle. A reasonable choice for the bin width dr for the calculation of $g(r)$ is dr $= 0.1$.

c. Choose the same values of N, L_x, and L_y as in part (a), but give each particle an initial random displacement from its triangular lattice site of magnitude 0.2. Do the Monte Carlo simulation at a very low temperature such as $T = 0.1$. Choose the maximum trial displacement $\delta = 0.15$ and consider mcs ≥ 400. Does the system retain its hexagonal symmetry? Does the value of δ affect your results?

d. Use the same initial conditions as in part (a), but take $T = 0.5$. Choose $\delta = 0.15$ and run for a number of Monte Carlo steps per particle that is sufficient to yield a reasonable result for the mean energy. Do a similar simulation at $T = 1$ and $T = 2$. What is the best choice of the initial configuration in each case? The harmonic theory of solids predicts that the total energy of a system is due to a $T = 0$ contribution plus a term due to the harmonic oscillation of the atoms. The contribution of the latter part should be proportional to the temperature. Compare your results for $E(T) - E(0)$ with this prediction. Use the values of σ and ϵ given in Table 8.1 to determine the temperature and energy in SI units for your simulations of solid argon.

e. Describe the qualitative nature of $g(r)$ for a Lennard-Jones solid and compare it with your hard disk results for the same density.

f. Decrease the density by multiplying L_x, L_y, and all the particle coordinates by 1.07. What is the new value of ρ? Estimate the number of Monte Carlo steps per particle needed to compute E and P for $T = 0.5$ to approximately 10% accuracy. Is the total energy positive or negative? How do E and P compare to their ideal gas values? Follow the method discussed in Problem 17.17 and compute an effective diffusion constant. Is the system a liquid or a solid? Plot $g(r)$ versus r and compare $g(r)$ to your results for hard disks at the same density. What is the qualitative behavior of $g(r)$? What is the interpretation of the peaks in $g(r)$ in terms of the structure of the liquid? If time permits, consider a larger system at the same density and temperature and compute $g(r)$ for larger r.

g. Consider the same density system as in part (f) at $T = 0.6$ and $T = 1$. Look at some typical configurations of the particles. Use your results for $E(T)$, $P(T)$, $g(r)$ and the other data you have collected, and discuss whether the system is a gas, liquid, or solid at these temperatures. What criteria can you use to distinguish a gas from a liquid? If time permits, repeat these calculations for $\rho = 0.7$.

h. Compute E, P, and $g(r)$ for $N = 16$, $L_x = L_y = 10$, and $T = 3$. These conditions correspond to a dilute gas. How do your results for P compare with the ideal gas result? How does $g(r)$ compare with the results you obtained for the liquid?

*17.8 OPTIMIZED MONTE CARLO DATA ANALYSIS

As we have seen, the important physics near a phase transition occurs on long length scales. For this reason, we might expect that simulations, which for practical reasons are restricted to relatively small systems, might not be useful for simulations near a phase transition. Nevertheless, we have found that methods such as finite size scaling can yield information about how systems behave in the thermodynamic limit. We next explore some additional Monte Carlo techniques that are useful near a phase transition.

The Metropolis algorithm yields mean values of various thermodynamic quantities, e.g., the energy, at particular values of the temperature T. Near a phase transition many thermodynamic quantities change rapidly, and we need to determine these quantities at many closely spaced values of T. If we were to use standard Monte Carlo methods, we would have to do many simulations to cover the desired range of values of T. To overcome this problem, we introduce the use of *histograms* which allow us to extract more information from a single Monte Carlo simulation. The idea is to use our knowledge of the equilibrium probability distribution at one value of T (and other external parameters) to estimate the desired thermodynamic averages at neighboring values of the external parameters.

The first step of the single histogram method for the Ising model is to simulate the system at an inverse temperature β_0 which is near the values of β of interest and measure the energy of the system after every Monte Carlo step per spin (or other fixed interval). The measured probability that the system has energy E can be expressed as

$$P(E, \beta_0) = \frac{H_0(E)}{\sum_E H_0(E)}. \tag{17.41}$$

The histogram $H_0(E)$ is the number of configurations with energy E, and the denominator is the total number of measurements of E (e.g., the number of Monte Carlo steps per spin). Because the probability of a given configuration is given by the Boltzmann distribution, we have

$$P(E, \beta) = \frac{W(E)\, e^{-\beta E}}{\sum_E W(E)\, e^{-\beta E}}, \tag{17.42}$$

where $W(E)$ is the number of microstates with energy E. (This quantity is frequently called the *density of states*, and is more generally defined as the number of states per unit energy interval.) If we compare (17.41) and (17.42) and note that $W(E)$ is independent of T, we can write

$$W(E) = a_0 H_0(E) e^{\beta_0 E}, \tag{17.43}$$

where a_0 is a proportionality constant that depends on β_0. If we eliminate $W(E)$ from (17.42) by using (17.43), we obtain the desired relation

$$P(E, \beta) = \frac{H_0(E)\, e^{-(\beta - \beta_0)E}}{\sum_E H_0(E)\, e^{-(\beta - \beta_0)E}}. \tag{17.44}$$

Note that we have expressed the probability at inverse temperature β in terms of $H_0(E)$, the histogram at inverse temperature β_0.

Because β is a continuous variable, we can estimate the β dependence of the mean value of any function A that depends on E, e.g., the mean energy and the specific heat. We write the mean of $A(E)$ as

$$<A(\beta)> = \sum_E A(E) P(E, \beta). \tag{17.45}$$

If the quantity A depends on another quantity M, e.g., the magnetization, then we can generalize (17.45) to

$$<A(\beta)> = \sum_{E,M} A(E, M) P(E, M, \beta)$$

$$= \frac{\sum_{E,M} A(E, M) H_0(E, M) e^{-(\beta-\beta_0)E}}{\sum_{E,M} H_0(E, M) e^{-(\beta-\beta_0)E}}. \tag{17.46}$$

The histogram method is useful only when the configurations relevant to the range of temperatures of interest occur with reasonable probability during the simulation at temperature T_0. For example, if we simulate an Ising model at low temperatures at which only ordered configurations occur (most spins aligned in the same direction), we cannot use the histogram method to obtain meaningful thermodynamic averages at high temperatures at which most configurations are disordered.

Problem 17.19   Application of the histogram method

a.   Consider a 4×4 Ising lattice in zero magnetic field and compute the mean energy per spin, the mean magnetization per spin, the specific heat, and the susceptibility per spin for $T = 1$ to $T = 3$ in steps of $\Delta T = 0.05$. Average over at least 5000 Monte Carlo steps per spin after equilibrium has been reached for each value of T.

b.   What are the minimum and maximum values of the total energy E and magnetization M that might be observed in a simulation of a Ising model on a 4×4 lattice? Use these values to set the size of the two-dimensional array needed to accumulate data for the histogram $H(E, M)$. It is suggested that you modify *Program ising* and save $H(E, M)$ in a file. Accumulate data for $H(E, M)$ at $T = 2.27$, a value of T close to T_c, for at least 5000 Monte Carlo steps per spin after equilibration. Write a separate program to read the histogram file and compute the same thermodynamic quantities as in part (a) using (17.46). Compare your computed results with the data obtained by simulating the system directly, i.e., without using the histogram method, at the same temperatures. At what temperatures does the histogram method break down?

c.   Repeat parts (b) and (c) for a simulation centered about $T = 1.5$ and $T = 2.5$.

d.   Repeat parts (b) and (c) for an 8×8 and a 16×16 lattice at $T = 2.27$.

The histogram method can be used to do a more sophisticated finite size scaling analysis to determine the nature of a transition. Suppose that we perform a Monte Carlo simulation and observe a peak in the specific heat as a function of the temperature. What can this observation tell us about a possible phase transition? In general, we can conclude very little without doing a careful analysis of the behavior of the system at different sizes. For example, a discontinuity in the energy in an infinite system might be manifested in small systems by a peak in the specific heat. However, a phase transition in the infinite system in which the energy is continuous, but its derivative diverges at the transition, might manifest itself in the same way in a small system. Another difficulty is that the peak in the specific heat of a small system occurs at a temperature that differs from the transition temperature in the infinite system (see Project 17.3). Finally, there might be no transition at all, and the peak might simply represent a broad crossover from high to low temperature behavior (see Project 17.4).

We now discuss a method due to Lee and Kosterlitz that uses the histogram data to determine the nature of a phase transition (if it exists). To understand this method, we need to introduce the (Helmholtz) free energy F of a system. The statistical mechanics definition of F is

$$F = -kT \ln Z. \tag{17.47}$$

At low T, the factor $e^{-\beta E}$ in the partition function Z determines the dominant contributions to Z, even though there are relatively few such configurations. At high T, the factor $e^{-\beta E}$ is not very large, but the number of disordered configurations with high E is large, and hence high energy configurations dominate the contribution to Z. These considerations suggest that it is useful to define a restricted free energy $F(E)$ that includes only configurations at a particular energy E. We define

$$F(E) = -kT \ln W(E) \, e^{-\beta E}. \tag{17.48}$$

For systems with a first-order phase transition, a plot of $F(E)$ versus E will show two local minima corresponding to configurations that are characteristic of the high and low temperature phases. At low T the minimum at the lower energy will be the absolute minimum, and at high T the higher energy minimum will be the absolute minimum of F. At the transition, the two minima will have the same value of $F(E)$. For systems with no transition in the thermodynamic limit, there will only be one minimum for all T.

How will $F(E)$ behave for the relatively small lattices considered in simulations? In systems with first-order transitions, the distinction between low and high temperature phases will become more pronounced as the system size is increased. If the transition is continuous, there are domains at all sizes, and we expect that the behavior of $F(E)$ will not change significantly as the system size increases. If there is no transition, there might be a spurious double minima for small systems, but this spurious behavior should disappear for larger systems. Lee and Kosterlitz proposed the following method for categorizing phase transitions.

1. Do a simulation at a temperature close to the suspected transition temperature, and compute $H(E)$. Usually, the temperature at which the peak in the specific heat occurs is chosen as the simulation temperature.

2. Use the histogram method to compute $-\ln H_0(E) + (\beta - \beta_0)E \propto F(E)$ at neighboring values of T. If there are two minima in $F(E)$, vary β until the values of $F(E)$ at the two minima are equal. This temperature is an estimate of the possible transition temperature T_c.

3. Measure the difference ΔF at T_c between $F(E)$ at the minima and $F(E)$ at the maximum between the two minima.

4. Repeat steps (1)–(3) for larger systems. If ΔF increases with size, the transition is first-order. If ΔF remains the same, the transition is continuous. If ΔF decreases with size, there is no thermodynamic transition.

The above procedure is applicable when the phase transition occurs by varying the temperature. Transitions also can occur by varying the pressure or the magnetic field. These *field-driven transitions* can be tested by a similar method. For example, consider the Ising model in a magnetic field at low temperatures below T_c. As we vary the magnetic field from positive to negative, there is a transition from a phase with magnetization $M > 0$ to a phase with $M < 0$. Is this transition first-order or continuous? To answer this question, we can use the Lee-Kosterlitz method with a histogram $H(E, M)$ generated at zero magnetic field, and calculate $F(M)$ instead of $F(E)$. The quantity $F(M)$ is proportional to $-\ln \sum_E H(E, M) e^{-(\beta - \beta_0)E}$. Because the states with positive and negative magnetization are equally likely to occur for zero magnetic field, we should see a double minima structure for $F(M)$ with equal minima. As we increase the size of the system, ΔF should increase for a first-order transition and remain the same for a continuous transition.

Problem 17.20   Characterization of a phase transition

a. Use your modified version of `Program ising` from Problem 17.19 to determine $H(E, M)$. Read the $H(E, M)$ data from a file, and compute and plot $F(E)$ for the range of temperatures of interest. First generate data at $T = 2.27$ and use the Lee-Kosterlitz method to verify that the Ising model in two dimensions has a continuous phase transition in zero magnetic field. Consider lattices of sizes $L = 4, 8$, and 16.

b. Perform a Lee-Kosterlitz analysis of the Ising model at $T = 2$ and zero magnetic field by plotting $F(M)$. Determine if the transition from $M > 0$ to $M < 0$ is first-order or continuous. This transition is called field-driven because the transition occurs if we change the magnetic field. Make sure your simulations sample configurations with both positive and negative magnetization by using small values of L such as $L = 4, 6$ and 8.

c. Repeat part (b) at $T = 2.5$ and determine if there is a field-driven transition at $T = 2.5$.

*Problem 17.21   The Potts Model

a. In the q-state Potts model, the total energy or Hamiltonian of the lattice is given by

$$H = -J \sum_{i, j=\text{nn}(i)} \delta_{s_i, s_j}, \tag{17.49}$$

where s_i at site i can have the values $1, 2, \cdots, q$; the Kronecker delta function $\delta_{a,b}$ equals unity if $a = b$ and is zero otherwise. As before, we will measure the temperature in energy units. Convince yourself that the $q = 2$ Potts model is equivalent to the Ising model (except for a trivial difference in the energy minimum). One of the many applications of the Potts model is to helium absorbed on the surface of graphite. The graphite-helium interaction gives rise to preferred adsorption sites directly above the centers of the honeycomb graphite surface. As discussed by Plischke and Bergersen, the helium atoms can be described by a three-state Potts model.

b. The transition in the Potts model is continuous for small q and first-order for larger q. Write a Monte Carlo program to simulate the Potts model for a given value of q and store the histogram $H(E)$. Test your program by comparing the output for $q = 2$ with your Ising model program.

c. Use the Lee-Kosterlitz method to analyze the nature of the phase transition in the Potts model for $q = 3, 4, 5, 6$, and 10. First find the location of the specific heat maximum, and then collect data for $H(E)$ at the specific heat maximum. Lattice sizes of order $L \geq 50$ are required to obtain convincing results for some values of q.

*17.9 OTHER ENSEMBLES

So far, we have considered the microcanonical ensemble (fixed N, V, and E) and the canonical ensemble (fixed N, V, and T). Monte Carlo methods are very flexible and can be adapted to the calculation of averages in any ensemble. Two other ensembles of particular importance are the constant pressure and the grand canonical ensembles. The main difference in the Monte Carlo method is that there are additional "moves" corresponding to changing the volume or changing the number of particles. The constant pressure ensemble is particularly important for studying first-order phase transitions because the phase transition occurs at a fixed pressure, unlike a constant volume simulation where the system passes through a two phase coexistence region before changing phase completely as the volume is changed.

In the NPT ensemble, the probability of a microstate occurring is proportional to $e^{-\beta(E+PV)}$. For a classical system, the mean value of a physical quantity A that depends on the coordinates of the particles can be expressed as

$$<A>_{NPT} = \frac{\int_0^\infty dV e^{-\beta PV} \int d\mathbf{r}_1 d\mathbf{r}_2 \ldots d\mathbf{r}_N A(\{\mathbf{r}\}) e^{-\beta U(\{\mathbf{r}\})}}{\int_0^\infty dV e^{-\beta PV} \int d\mathbf{r}_1 d\mathbf{r}_2 \ldots d\mathbf{r}_N e^{-\beta U(\{\mathbf{r}\})}}. \qquad (17.50)$$

The potential energy $U(\{\mathbf{r}\})$ depends on the set of particle coordinates $(\{\mathbf{r}\})$. To simulate the NPT ensemble, we need to sample the coordinates $\mathbf{r}_1, \mathbf{r}_2, \cdots, \mathbf{r}_N$ of the particles and the volume V of the system. For simplicity, we assume that the central cell is a square or a cube so that $V = L^d$. It is convenient to use the set of scaled coordinates $\{\mathbf{s}\}$, where $\mathbf{s}_i$ is defined as

$$\mathbf{s}_i = \frac{\mathbf{r}_i}{L}. \qquad (17.51)$$

If we substitute (17.51) into (17.50), we can write $<A>_{\text{NPT}}$ as

$$<A>_{\text{NPT}} = \frac{\int_0^\infty dV e^{-\beta PV} V^N \int ds_1 ds_2 \ldots ds_N A(\{s\}) e^{-\beta U(\{s\})}}{\int_0^\infty dV e^{-\beta PV} V^N \int ds_1 ds_2 \ldots ds_N e^{-\beta U(\{s\})}}, \tag{17.52}$$

where the integral over $\{s\}$ is over the unit square (cube). The factor of V^N arises from the change of variables $\mathbf{r} \to \mathbf{s}$. If we let $V^N = e^{\ln V^N} = e^{N \ln V}$, we see that the quantity that is analogous to the Boltzmann factor can be written as

$$e^{-W} = e^{-\beta PV - \beta U(\{s\}) + N \ln V}. \tag{17.53}$$

Because the pressure is fixed, a trial configuration is generated from the current configuration by either randomly displacing a particle and/or making a random change in the volume, e.g., $V \to V + \delta V_{\max}(2r - 1)$, where r is a uniform random number in the unit interval. The trial configuration is accepted if the change $\Delta W \leq 0$ and with probability $e^{-\Delta W}$ if $\Delta W > 0$. It is not necessary or efficient to change the volume after every Monte Carlo step per particle.

In the grand canonical or μVT ensemble, the chemical potential μ is fixed and the number of particles fluctuates. The average of any function of the particle coordinates can be written as (in three dimensions)

$$<A>_{\mu VT} = \frac{\sum_{N=0}^\infty (1/N!) \lambda^{-3N} e^{\beta \mu N} \int d\mathbf{r}_1 d\mathbf{r}_2 \ldots d\mathbf{r}_N A(\{\mathbf{r}\}) e^{-\beta U_N(\{\mathbf{r}\})}}{\sum_{N=0}^\infty (1/N!) \lambda^{-3N} e^{\beta \mu N} \int d\mathbf{r}_1 d\mathbf{r}_2 \ldots d\mathbf{r}_N e^{-\beta U_N(\{\mathbf{r}\})}}, \tag{17.54}$$

where $\lambda = (h^2/2\pi mkT)^{1/2}$. We have made the N-dependence of the potential energy U explicit. If we write $1/N! = e^{-\ln N!}$ and $\lambda^{-3N} = e^{-N \ln \lambda^3}$, we can write the quantity that is analogous to the Boltzmann factor as

$$e^{-W} = e^{\beta \mu N - N \ln \lambda^3 - \ln N! + N \ln V - \beta U_N}. \tag{17.55}$$

If we write the chemical potential as

$$\mu = \mu^* + kT \ln(\lambda^3/V), \tag{17.56}$$

then W can be expressed as

$$e^{-W} = e^{-\beta \mu^* N - \ln N! - \beta U_N}. \tag{17.57}$$

The parameters are μ^*, V, and T. There are two possible ways of obtaining a trial configuration. The first involves the displacement of a selected particle; such a move is accepted or rejected according to the usual criteria, i.e., by the change in the potential energy U_N. In the second possible way, we choose with equal probability whether to attempt to add a particle at a randomly chosen position in the central cell or to remove a particle that is already present. In either case, the trial configuration is accepted if

W in (17.57) is increased. If W is decreased, the change is accepted with a probability equal to

$$\frac{1}{N+1} e^{\beta\left(\mu^* - (U_{N+1} - U_N)\right)} \qquad \text{(insertion)} \qquad (17.58a)$$

or

$$N e^{-\beta\left(\mu^* + (U_{N+1} - U_N)\right)}. \qquad \text{(removal)} \qquad (17.58b)$$

In this approach μ^* is an input parameter and μ is not determined until the end of the calculation when $<N>_{\mu VT}$ is obtained.

17.10   MORE APPLICATIONS

You probably do not need to be convinced that Monte Carlo methods are powerful, flexible, and applicable to a wide variety of systems. Extensions to the Monte Carlo methods that we have not discussed include multiparticle moves, biased moves where particles tend to move in the direction of the force on them, manipulation of bits for Ising-like models, the *n-fold way* algorithm for Ising-like models at low temperature, use of special processors for specific systems, and the use of parallel processing to update different parts of a large system simultaneously. We also have not described the simulation of systems with long-range potentials such as Coulombic systems and dipole-dipole interactions. For these potentials, it is necessary to include the interactions of the particles in the center cell with the infinite set of periodic images.

We conclude this chapter with a discussion of Monte Carlo methods in a context that might seem to have little in common with the types of problems we have discussed. This context is called *multivariate* or *combinatorial optimization,* a fancy way of saying, "How do you find the minimum of a function that depends on many parameters?" Problems of this type arise in many areas of scheduling and design. We explain the nature of this type of problem by an example known as the *traveling salesperson problem,* a.k.a., the *traveling salesman problem.*

Suppose that a salesperson wishes to visit N cities and follow a route such that no city is visited more than once and the end of the trip coincides with the beginning. Given these constraints, the traveling salesperson problem is to find the optimum route such that the total distance traveled is a minimum. An example of $N = 8$ cities and a possible route is shown in Fig. 17.5. All exact methods for determining the optimal route require a computing time that increases as e^N, and in practice, an exact solution can be found only for a small number of cities. The traveling salesperson problem belongs to a large class of problems known as NP-complete. (The NP refers to non-polynomial, that is, such problems cannot be done in a time proportional to some finite polynomial in N.) What is a reasonable estimate for the maximum number of cities that you can consider without the use of a computer?

To understand the nature of the different approaches to the traveling salesperson problem, consider the plot in Fig. 17.6 of the "energy" function $E(a)$. We can associate $E(a)$ with the length of the route and interpret a as a parameter that represents the order

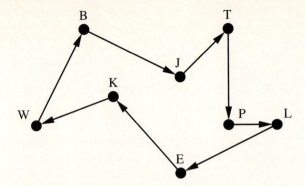

Fig. 17.5   What is the optimum route for this random
arrangement of $N = 8$ cities? The route
begins and ends at city W. A possible route
is shown.

in which the cities are visited. If $E(a)$ has several local minima, what is a good strategy
for finding the global (absolute) minimum of $E(a)$? One way is to vary a systematically
and find the value of E everywhere. This exact enumeration method corresponds to determining the length of each possible route, clearly an impossible task if the number of
cities is too large. Another way is to use a *heuristic method,* i.e., an approximate method
for finding a route that is close to the absolute minimum. One strategy is to choose a
value of a, generate a small random change δa, and accept this change if $E(a + \delta a)$ is
less than or equal to $E(a)$. This iterative improvement strategy corresponds to a search
for steps that lead downhill. Because this search usually becomes stuck in a local and

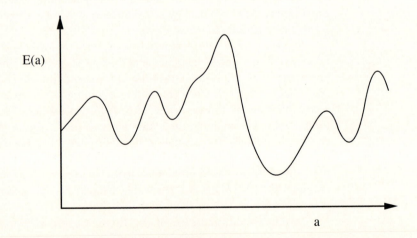

Fig. 17.6   Plot of the function $E(a)$ as a function of the parameter a.

not a global minimum, it is useful to begin from several initial choices of a and to keep the best result. What would be the application of this type of strategy to the salesperson problem?

Let us consider a seemingly unrelated problem. Suppose we wish to make a perfect single crystal. You probably know that we should first melt the material, and then lower the temperature very slowly to the desired low temperature. If we lower the temperature too quickly (a rapid "quench"), the resulting crystal would have many defects or not become a crystal at all. The gradual lowering of the temperature is known as *annealing*.

The method of annealing can be used to estimate the minimum of $E(a)$. We choose a value of a, generate a small random change δa, and calculate $E(a + \delta a)$. If $E(a + \delta a)$ is less than or equal to $E(a)$, we accept the change. However, if $\Delta E = E(a + \delta a) - E(a) > 0$, we accept the change with a probability $P = e^{-\Delta E/T}$, where T is an effective temperature. This procedure is the familiar Metropolis algorithm with the temperature playing the role of a control parameter. The *simulated annealing* process consists of first "melting" the system, that is, choosing T for which most moves are accepted, and then gradually lowering the temperature. At each temperature, the simulation should last long enough for the system to reach a steady state. The annealing schedule, i.e., the rate of temperature decrease, determines the quality of the solution.

The moral of the simulated annealing method is that sometimes it is necessary to climb a hill to reach a valley. The first application of the method of simulated annealing was to the optimal design of computers. In Problem 17.22 we apply this method to the traveling salesperson problem. Perhaps you can think of other applications.

*Problem 17.22   Simulated annealing and the traveling salesperson problem

a.   Generate a random arrangement of $N = 8$ cities in a square of linear dimension $L = \sqrt{N}$ and calculate the optimum route by hand. Then write a Monte Carlo program and apply the method of simulated annealing to this problem. For example, use two arrays to store the coordinates of each city and an array to store the distances between them. The state of the system, i.e., the route represented by a sequence of cities, can be stored in another array. The length of this route is associated with the energy of an imaginary thermal system. A reasonable choice for the initial temperature is one that is the same order as the initial energy. One way to generate random rearrangements of the route is to choose two cities at random and to interchange the order of visit. Choose this method or one that you devise and find a reasonable annealing schedule. Compare your annealing results to exact results whenever possible. Extend your results to larger N, e.g., $N = 12, 24$, and 48. For a given annealing schedule, determine the probability of finding a route of a given length. More suggestions can be found in the references.

b.   The microcanonical Monte Carlo algorithm (demon) discussed in Chapter 16 also can be used to do simulated annealing. The advantages of the demon algorithm are that it is deterministic and allows large temperature fluctuations. One way to implement the analog of simulated annealing is to impose a maximum value on the energy of the demon, $E_{d,\max}$ and then gradually lower the value of $E_{d,\max}$. Guo et al. choose the initial value of $E_{d,\max}$ to equal $\sqrt{N}/4$. Their results are comparable to the canonical simulated annealing method, but require approximately half the CPU time. Apply this method to the same routes that you considered in part (a) and compare your results.

17.11   PROJECTS

Many of the original applications of Monte Carlo methods were done for systems of approximately one hundred particles and lattices of order 32^2 spins. Most of these applications can be done with much better statistics and with larger system sizes. In the following, we discuss some recent developments, but this discussion is not complete and we have omitted other important topics such as Brownian dynamics and umbrella sampling. More ideas for projects can be found in the references.

Project 17.1   Overcoming critical slowing down

The usual limiting factor of most simulations either is the lack of computer memory or, more likely, the speed of the computer. One way to overcome the latter problem is to use a faster computer. However, near a phase transition, the most important limiting factor on even the fastest available computers is the existence of "critical slowing down" (see Problem 17.12). In this project we discuss the nature of critical slowing down and ways of overcoming it in the context of the Ising model.

  The existence of critical slowing down is related to the fact that the size of the correlated regions of spins becomes very large near the Ising critical point. The large size of the correlated regions and the corresponding divergent behavior of the correlation length ξ near T_c implies that the time τ required for a region to lose its coherence becomes very long if a *local* dynamics is used. Near T_c, we have $\tau \sim \xi^z$, which defines the dynamical critical exponent z. At $T = T_c$, we have $\tau \sim L^z$ for L sufficiently large. For the single spin flip (Metropolis) algorithm, $z \approx 2$, and τ becomes very large for $L \gg 1$. On a serial computer, the CPU time needed to obtain n configurations increases as L^2, the time needed to visit L^2 spins. This factor of L^2 is not a problem because a larger system contains proportionally more information. However, the time needed to obtain n approximately *independent* configurations is order $\tau L^2 \sim L^{2+z} \approx L^4$ for the Metropolis algorithm. We see that an increase of L by a factor of 10 requires 10^4 more computing time. Hence, the existence of critical slowing down limits the maximum value of L that can be considered.

  If we are interested only in the static properties of the Ising model, the choice of dynamics is irrelevant as long as the transition probability satisfies the detailed balance condition (17.10). As we mentioned above, the Metropolis algorithm becomes inefficient near T_c because only single spins are flipped. Hence, it is reasonable to look for a *global* algorithm for which groups or *clusters* of spins are flipped simultaneously. We already are familiar with cluster properties in the context of percolation (see Chapter 13). A naive definition of a cluster of spins might be a domain or group of parallel nearest neighbor spins. We can make this definition more explicit by introducing a bond between any two nearest neighbor spins that point in the same direction. The introduction of a bond between spins of the same sign defines a "site-bond" percolation problem. More generally, we can assume that such a bond exists with probability p and the randomness associated with the bond probability p depends on the temperature T. The dependence of p on T can be determined by requiring that the percolation transition of the clusters occurs at the Ising critical point, and by requiring that the critical exponents associated with the clusters be identical to the analogous thermal exponents. For example, we can define a critical exponent ν_p to characterize the divergence of the connectedness length of the clusters near p_c. The analogous thermal exponent ν quantifies the divergence of the thermal correlation length ξ near T_c. It is possible to show analytically that these (and other) critical exponents are identical if we define the bond probability as

$$p = 1 - e^{-2J/kT}. \qquad \text{(bond probability)} \qquad (17.59)$$

The relation (17.59) holds for any spatial dimension. What is the value of p at $T = T_c$ for the two-dimensional Ising model on the square lattice?

Now that we know how to generate clusters of spins, we use these clusters to construct a global dynamics. One way (known as the Swendsen-Wang algorithm) is to assign all bonds between parallel spins with probability p. No bonds are included between sites that have different spin values. From this configuration of bonds, we can form clusters of spins using the Hoshen-Kopelman algorithm (see Section 13.3). The smallest cluster contains a single spin. After the clusters have been identified, all the spins in each cluster are flipped with probability 1/2.

Because it is nontrivial to determine all the clusters for a given configuration, we instead explore an algorithm that flips single clusters. The idea is to "grow" a single (site-bond) percolation cluster in a way that is analogous to the single (site) percolation cluster algorithm discussed in Section 14.1. The algorithm can be implemented by the following steps:

1. Choose a seed spin at random. Its four nearest neighbor sites (on the square lattice) are the perimeter sites. Form an ordered array corresponding to the perimeter spins that are parallel to the seed spin and define a counter for the total number of perimeter spins.

2. Choose the first spin in the ordered perimeter array. Remove it from the array and replace it by the last spin in the array. Generate a random number r. If $r \leq p$, a bond exists and the spin is added to the cluster.

3. If the spin is added to the cluster, inspect its parallel perimeter spins. If such a spin is not already part of the cluster, add it to the end of the array of perimeter spins.

4. Repeat steps 2 and 3 until no perimeter spins remain.

5. Flip all the spins in the single cluster.

This algorithm is known as single cluster flip or Wolff dynamics. Note that bonds rather than sites are tested so that a spin might have more than one chance to join a cluster. In the following, we consider both the static and dynamical properties of the two-dimensional Ising model using the Wolff algorithm to generate the configurations.

a. Modify your program for the two-dimensional Ising model on a square lattice so that single cluster flip dynamics (the Wolff algorithm) is used. Compute the mean energy and magnetization for $L = 16$ as a function of T for $T = 2.0$ to 2.7 in steps of 0.1. Compare your results to those obtained using the Metropolis algorithm. How many cluster flips do you need to obtain comparable accuracy at each temperature? Is the Wolff algorithm more efficient at every temperature?

b. Fix T at the critical temperature of the infinite lattice ($T_c = 2/\ln(1 + \sqrt{2})$) and use finite size scaling to estimate the values of the various static critical exponents, e.g., γ and α. Compare your results to those obtained using the Metropolis algorithm.

c. Because we are generating site-bond percolation clusters, we can study their geometrical properties as we did for site percolation. For example, measure the distribution sn_s of cluster sizes at $p = p_c$ (see Problem 14.3a). How does n_s depend on s for large s (see Project 14.3)? What is the fractal dimension of the clusters?

d. The natural unit of time for single cluster flip dynamics is the number of cluster flips t_{cf}. Measure $C_M(t_{cf})$ and/or $C_E(t_{cf})$ and estimate the corresponding correlation time τ_{cf} for $T = 2.5, 2.4, 2.3$, and T_c for $L = 16$. As discussed in Problem 17.12, τ_{cf} can be found from the relation, $\tau_{cf} = \sum_{t_{cf}=1} C(t_{cf})$. The sum is cut off at the first negative value of $C(t_{cf})$. To compare our results for the Wolff algorithm to our results for the Metropolis algorithm, we should use the same unit of time. Because only a fraction of the spins are updated at each cluster flip, the time t_{cf} is not equal to the usual unit of time which corresponds to an update of the entire lattice or one Monte Carlo step per spin. We have that τ measured in Monte Carlo steps per spin is related to τ_{cf} by $\tau = \tau_{cf} <c>/L^2$, where $<c>$ is the mean number of spins in the single clusters, and L^2 is the number of spins in the entire lattice. Measure $<c>$ and compare your values for τ for the Wolff algorithm to the values of τ that you obtained using the Metropolis algorithm. Which values of τ are smaller?

e. Use the finite size scaling relation $\tau_{cf} \sim L^{z_{cf}}$ at $T = T_c$ to estimate z_{cf}. Because z_{cf} is small, you will find it very difficult to obtain a good estimate. Verify that the mean cluster size scales as $<c> \sim L^{\gamma/\nu}$ with $\gamma = 7/4$ and $\nu = 1$. (Note that these exponents are identical to the analogous thermal exponents.) To obtain the value of z that is directly comparable to the value found for the Metropolis algorithm, we need to rescale the time as in part (d). We have that $\tau \sim L^z \propto L^{z_{cf}} L^{\gamma/\nu} L^{-d}$. Hence, z is related to the measured value of z_{cf} by $z = z_{cf} - (d - \gamma/\nu)$. It has been suggested that $\tau \propto \ln L$, which would imply that $z = 0$, but the value of z is not known with confidence.

Project 17.2 Physical test of random number generators

In Section 12.6 we discussed various statistical tests for the quality of random number generators. In this project we will find that the usual statistical tests might not be sufficient for determining the quality of a random number generator for a particular application. The difficulty is that the quality of a random number generator for a specific application depends in part on how the subtle correlations that are intrinsic to all deterministic random number generators couple to the way that the random number sequences are used. In this project we explore the quality of two random number generators when they are used to implement single spin flip dynamics (Metropolis algorithm) and single cluster flip dynamics (Wolff algorithm) for the two-dimensional Ising model.

a. Write subroutines to generate sequences of random numbers based on the linear congruential algorithm

$$x_n = 16807 \, x_{n-1} \bmod (2^{31} - 1) \tag{17.60}$$

and the generalized feedback shift register (GFSR) algorithm

$$x_n = x_{n-103} \oplus x_{n-250}. \tag{17.61}$$

In both cases x_n is the nth random number. Both algorithms require that x_n be divided by the largest possible value of x_n to obtain numbers in the range $0 \le x_n < 1$. The GFSR algorithm requires bit manipulation and should be written in C or Fortran (see Appendices A or B). Which random number generator does a better job of passing the various statistical tests discussed in Problem 12.19?

b. Use the Metropolis algorithm and the linear congruential random number generator to determine the mean energy per spin E/N and the specific heat (per spin) C for the $L = 16$ Ising model at $T = T_c = 2/\ln(1 + \sqrt{2})$. Make ten independent runs (i.e., ten runs that use different random number seeds), and compute the standard deviation of the means σ_m from the ten values of E/N and C, respectively. Published results by Ferrenberg, Landau, and Wong are for 10^6 Monte Carlo steps per spin for each run. Calculate the differences δ_e and δ_c between the average of E/N and C over the ten runs and the exact values (to five decimal places) $E/N = -1.45306$ and $C = 1.49871$. If the ratio δ/σ_m for the two quantities is order unity, then the random number generator does not appear to be biased. Repeat your runs using the GFSR algorithm to generate the random number sequences. Do you find any evidence of statistical bias?

c. Repeat part (b) using Wolff dynamics. Do you find any evidence of statistical bias?

d. Repeat the computations in parts (b) and (c) using the random number generator supplied with your programming language.

Project 17.3   Kosterlitz-Thouless transition in the planar model

The planar model (also called the x-y model) consists of spins of unit magnitude that can point in any direction in the x-y plane. The energy or Hamiltonian function of the planar model in zero magnetic field can be written as

$$E = -J \sum_{i,j=nn(i)} \left[s_{i,x} s_{j,x} + s_{i,y} s_{j,y} \right], \tag{17.62}$$

where $s_{i,x}$ represents the x-component of the spin at the ith site, J measures the strength of the interaction, and the sum is over all nearest neighbors. We can rewrite (17.62) in a simpler form by substituting $s_{i,x} = \cos \theta_i$ and $s_{i,y} = \sin \theta_i$. The result is

$$E = -J \sum_{i,j=nn(i)} \cos(\theta_i - \theta_j), \tag{17.63}$$

where θ_i is the angle that the ith spin makes with the x axis. The most studied case is the two-dimensional model on a square lattice. In this case the mean magnetization $<\mathbf{M}> = 0$ for all temperatures $T > 0$, but nevertheless, there is a phase transition at a nonzero temperature, T_{KT}, the Kosterlitz-Thouless (KT) transition. For $T \leq T_{KT}$, the spin-spin correlation function $C(r)$ decreases as a power law for increasing r; for $T > T_{KT}$, $C(r)$ decreases exponentially. The power law decay of $C(r)$ for all $T \leq T_{KT}$ implies that every temperature below T_{KT} acts as if it were a critical point. We say that the planar model has a line of critical points. In the following, we explore some of the properties of the planar model and the mechanism that causes the transition.

a.   Write a Monte Carlo program to simulate the planar model on a square lattice using periodic boundary conditions. Because θ and hence the energy of the system is a continuous variable, it is not possible to store the previously computed values of the Boltzmann factor for each possible value of ΔE. Instead, of computing $e^{-\beta \Delta E}$ for each trial change, it is faster to set up an array w such that the array element $w(j) = e^{-\beta \Delta E}$, where j is the integer part of $1000\Delta E$. This procedure leads to an energy resolution of 0.001, which should be sufficient for most purposes.

b.   One way to show that the magnetization $<\mathbf{M}>$ vanishes for all T is to compute $<\theta^2>$, where θ is the angle that a spin makes with the magnetization $\mathbf{M}$ at any given instant. (Although the mean magnetization vanishes, $\mathbf{M} \neq 0$ at any given instant.) Compute $<\theta^2>$ as a function of the number of spins N at $T = 0.1$, and show that $<\theta^2>$ diverges as $\ln N$. Begin with a 4×4 lattice and choose the maximum change in θ_i to be $\Delta\theta_{\max} = 1.0$. If necessary, change $\theta_{\max}$ so that the acceptance probability is about 40%. If $<\theta^2>$ diverges, then the spins are not pointing along any preferred direction, and hence there is no mean magnetization.

c.   Modify your program so that an arrow is drawn at each site to show the orientation of each spin. We will look at a typical configuration and analyze it visually. Begin with a 32×32 lattice with spins pointing in random directions and do a temperature quench from $T = \infty$ to $T = 0.5$. (Simply change the value of β in the Boltzmann probability.) Such a quench should lock in some long lived, but metastable vortices. A vortex is a region of the lattice where the spins rotate by at least 2π as your eye moves around a closed path (see Fig. 17.7). To determine the center of a vortex, choose a group of four spins that are at the corners of a unit square, and determine whether the spins turn by $\pm 2\pi$ as your eye goes from one spin to the next in a counterclockwise direction around the square. Assume that the difference between the direction of two neighboring spins, $\delta\theta$, is in the range $-\pi < \delta\theta < \pi$. A total rotation of $+2\pi$ indicates the existence of a positive vortex, and a change of -2π indicates a negative vortex. Count the number of positive and negative vortices. Repeat these observations on several configurations. What can you say about the number of vortices of each sign?

Fig. 17.7   A typical configuration of the planar model on a 24×24 square lattice that has been quenched from $T = \infty$ to $T = 0$ and equilibrated for 200 Monte Carlo steps per spin after the quench. Note that there are six vortices. The circle around each vortex is a guide to the eye and is not meant to indicate the size of the vortex.

d. Write a subroutine to determine the existence of a vortex for each 1×1 square of the lattice. Represent the center of the vortices using a different symbol to distinguish between a positive and a negative vortex. Do a Monte Carlo simulation to compute the mean energy, specific heat, and number of vortices in the range from $T = 0.5$ to $T = 1.5$ in steps of 0.1. Use the last configuration at the previous temperature as the first configuration for the next temperature. Begin at $T = 0.5$ with all $\theta_i = 0$. Draw the vortex locations for the last configuration at each temperature. Use at least 1000 Monte Carlo steps per spin at each temperature to equilibrate and at least 5000 Monte Carlo steps per spin for computing the averages. Use an 8×8 or 16×16 lattice if your computer resources are limited, and larger lattices if you have sufficient resources. Describe the T dependence of the energy, specific heat, and vorticity (equal to the number of vortices per area). Plot the logarithm of the vorticity versus T for $T < 1.1$. What can you conclude about the T-dependence of the vorticity? Explain why this form is reasonable. Describe the vortex configurations. At what temperature can you find a vortex that appears to be free, i.e., a vortex that is not obviously paired up with another vortex of opposite sign?

e. The Kosterlitz-Thouless theory predicts that the susceptibility χ diverges above the transition as

$$\chi \sim A \, e^{b/\epsilon^\nu}, \tag{17.64}$$

where ϵ is the reduced temperature $\epsilon = (T - T_{KT})/T_{KT}$, $\nu = 0.5$, and A and b are nonuniversal constants. Compute χ from the relation (17.14) with $\mathbf{M} = 0$ because the mean magnetization vanishes. Assume the exponential form (17.64) for χ in the range $T = 1$ and $T = 1.2$ with $\nu = 0.7$, and find the best values of T_{KT}, A, and b. (Although the analytical theory predicts $\nu = 0.5$, simulations for small systems indicate that $\nu = 0.7$ gives a better fit.) One way to determine T_{KT}, A, and b is to assume a value of T_{KT} and then do a least squares fit of $\ln \chi$ to determine A and b. Choose the set of parameters that minimizes the variance of $\ln \chi$. How does your estimated value of T_{KT} compare with the temperature at which free vortices first appear? At what temperature does the specific heat have a peak? The Kosterlitz-Thouless theory predicts that the specific heat peak does not occur at T_{KT}. This result has been confirmed by simulations (see Tobochnik and Chester). To obtain quantitative results, you will need lattices larger than 32×32.

Project 17.4   Classical Heisenberg model in two dimensions

The energy or Hamiltonian of the classical Heisenberg model is similar to the Ising model and the planar model, except that the spins can point in any direction in three dimensions. The energy in zero external magnetic field is

$$E = -J \sum_{i,j=\mathrm{nn}(i)}^{N} \mathbf{s}_i \cdot \mathbf{s}_j, \tag{17.65}$$

where **s** is a classical vector of unit length. The spins have three components, in contrast to the spins in the Ising model which only have one component, and the spins in the planar model which have two components. We will consider the two-dimensional Heisenberg model for which the spins are located on a two-dimensional lattice.

Early simulations and approximate theories led researchers to believe that there was a continuous phase transition, similar to that found in the Ising model. The Heisenberg model received more interest after it was related to the confinement for quarks. Lattice models of the interaction between quarks, called lattice gauge theories, predict that the confinement of quarks can be explained if there are no phase transitions in these models. (The lack of a phase transition in these models implies that the attraction between quarks grows with distance.) The Heisenberg model is a two-dimensional analog of the four-dimensional models used to model quark-quark interactions. Shenker and Tobochnik used a combination of Monte Carlo and renormalization group methods to show that this model does not have a phase transition. Subsequent work on lattice gauge theories showed similar behavior.

a. Modify your Ising model program to simulate the Heisenberg model in two dimensions. One way to do so is to define three arrays, one for each of the three components of the unit spin vectors. A trial Monte Carlo move consists of randomly changing the direction of a spin, s_i. First compute a small vector $\Delta s = \Delta s_{max}(p_1, p_2, p_3)$, where $-1 \leq p_n \leq 1$ is a uniform random number, and Δs_{max} is the maximum change of any spin component. If $|\Delta s| > \Delta s_{max}$, than compute another Δs. This latter step is necessary to insure that the change in a spin direction is symmetrically distributed around the current spin direction. Next let the trial spin equal $s_i + \Delta s$ normalized to a unit vector. The standard Metropolis algorithm can now be used to determine if the trial spin is accepted. Compute the mean energy, specific heat, and susceptibility as a function of T. Choose lattice sizes of $L = 8, 16, 32$ and larger if possible and average over at least 2000 Monte Carlo steps per spin at each temperature. Is there any evidence of a phase transition? Does the susceptibility appear to diverge at a nonzero temperature? Plot the natural log of the susceptibility versus the inverse temperature, and determine the temperature dependence of the susceptibility in the limit of low temperatures.

b. Use the Lee-Kosterlitz analysis at the specific heat peak to determine if there is a phase transition.

Project 17.5 Ground state energy of the Ising spin glass

A spin glass is a magnetic system with frozen-in disorder. An example of such a system is the Ising model with the exchange constant J_{ij} between nearest neighbor spins randomly chosen to be ± 1. The disorder is said to be "frozen-in" because the set of interactions $\{J_{ij}\}$ does not change with time. Because the spins cannot arrange themselves so that every pair of spins is in its lowest energy state, the system exhibits frustration similar to the antiferromagnetic Ising model on a triangular lattice (see Problem 17.15). Is there a phase transition in the spin glass model, and if so, what is its nature? The answers to these questions are very difficult to obtain by doing simulations. One of the difficulties is that we need to do not only an average over the possible configurations of spins for a given set of $\{J_{ij}\}$, but we also need to average over different realizations of the interactions. Another difficulty is that there are many local minima in the energy (free energy at finite temperature) as a function of the configurations of spins, and it is very difficult to find the global minimum. As a result, Monte Carlo simulations typically become stuck in these local minima or metastable states. Detailed finite size scaling analyses of simulations indicate that there might be a transition in three dimensions. It is generally accepted that the transition in two dimensions is at zero temperature. In the following, we will look at some of the properties of an Ising spin glass on a square lattice at low temperatures.

a.   Write a program to apply simulated annealing to an Ising spin glass using the Metropolis algorithm with the temperature fixed at each stage of the annealing schedule (see Problem 17.22a). Search for the lowest energy configuration for a fixed set of $\{J_{ij}\}$. Use at least one other annealing schedule for the same $\{J_{ij}\}$ and compare your results. Then find the ground state energy for at least ten other sets of $\{J_{ij}\}$. Use lattice sizes of $L = 5$ and $L = 10$. Discuss the nature of the ground states you are able to find. Is there much variation in the ground state energy E_0 from one set of $\{J_{ij}\}$ to another? Theoretical calculations give an average over realizations of $\overline{E_0}/N \approx -1.4$. If you have sufficient computer resources, repeat your computations for the three-dimensional spin glass.

b.   Modify your program to do simulated annealing using the demon algorithm (see Problem 17.22b). How do your results compare to those that you found in part (a)?

Project 17.6   Zero temperature dynamics of the Ising model

We have seen that various kinetic growth models (Section 14.3) and reaction-diffusion models (Section 12.4) lead to interesting and nontrivial behavior. Similar behavior can be seen in the zero temperature dynamics of the Ising model. Consider the one-dimensional Ising model with $J > 0$ and periodic boundary conditions. The initial orientation of the spins is chosen at random. We update the configurations by choosing a spin at random and computing the change in energy ΔE. If $\Delta E < 0$, then flip the spin; else if $\Delta E = 0$, flip the spin with 50% probability. The spin is not flipped if $\Delta E > 0$. This type of Monte Carlo update is known as Glauber dynamics. How does this algorithm differ from the Metropolis algorithm at $T = 0$?

The quantity of interest is $f(t)$, the fraction of spins that flip for the first time at time t. As usual, the time is measured in terms of Monte Carlo steps per spin. Published results (Derrida, Bray, and Godrèche) for $N = 10^5$ indicate that $f(t)$

$$f(t) \sim t^{-\theta} \tag{17.66}$$

for $t \approx 3$ to $t \approx 10,000$ with $\theta \approx 0.37$. Verify this result and extend your results to the one-dimensional q-state Potts model. In the latter model each site is initially given a random integer between 1 and q. A site is chosen at random and set equal to either of its two neighbors with equal probability. The value of the exponent θ is not understood at present, but might be related to analogous behavior in reaction-diffusion models.

Project 17.7   The inverse power law potential

Consider the inverse power law potential

$$V(r) = V_0 \left(\frac{\sigma}{r}\right)^n \tag{17.67}$$

with $V_0 > 0$. One reason for interest in potentials of this form is that thermodynamic quantities such as the mean energy E do not depend on V_0 and σ separately, but depend on a single dimensionless parameter. This dimensionless parameter can be defined as

$$\Gamma = \frac{V_0}{kT} \frac{\sigma}{a}, \tag{17.68}$$

where a is defined in three and two dimensions by $4\pi a^3 \rho/3 = 1$ and $\pi a^2 \rho = 1$, respectively. The length a is proportional to the mean distance between particles. A Coulomb interaction corresponds to $n = 1$, and a hard sphere system corresponds to $n \to \infty$. What phases do you expect to occur for arbitrary n?

a. Compare the qualitative features of $g(r)$ for a "soft" potential with $n = 4$ to a system of hard disks at the same density.

b. Let $n = 12$ and compute the mean energy E as a function of T for fixed density for a three-dimensional system. Fix T and consider $N = 16, 32, 64$, and 128. Does E depend on N? Can you extrapolate your results for the N-dependence of E to $N \to \infty$? Fix N and determine E as a function of Γ. Do you see any evidence of a phase transition? If so, estimate the value of Γ at which it occurs. What is the nature of the transition if it exists?

Project 17.8   Rare gas clusters

There has been much recent interest in structures that contain many particles, but that are not macroscopic. An example is the unusual structure of sixty carbon atoms known as a "buckeyball." A less unusual structure is a cluster of argon atoms. Questions of interest include the structure of the clusters, the existence of "magic" numbers of particles for which the cluster is particularly stable, the temperature dependence of the thermodynamic quantities, and the possibility of different phases. This latter question has been subject to some controversy, because transitions between different kinds of behavior in finite systems are not nearly as sharp as they are for infinite systems.

a. Write a Monte Carlo program to simulate a three-dimensional system of particles interacting via the Lennard-Jones potential. Use open boundary conditions, i.e., do not enclose the system in a box. The number of particles N and the temperature T should be input parameters.

b. Find the ground state energy E_0 as a function of N. For each value of N begin with a random initial configuration and accept any trial displacement that lowers the energy. Repeat for at least ten different initial configurations. Plot E_0/N versus N for $N = 2$ to 20 and describe the qualitative dependence of E_0/N on N. Is there any evidence of magic numbers, that is, value(s) of N for which E_0/N is a minimum? For each value of N save the final configuration. If you have access to a three-dimensional graphics program, plot the positions of the atoms. Does the cluster look like a part of a crystalline solid?

c. Repeat part (b) using simulated annealing. The initial temperature should be sufficiently low so that the particles do not move far away from each other. Slowly lower the temperature according to some annealing schedule. Do your results for E_0/N differ from part (b)?

d. To gain more insight into the structure of the clusters, compute the mean number of neighbors per particle for each value of N. What is a reasonable criteria for two particles to be neighbors? Also compute the mean distance between each pair of particles. Plot both quantities as a function of N, and compare their dependence on N with your plot of E_0/N.

e. Is it possible to find any evidence for a "melting" transition? Begin with the configuration that has the minimum value of E_0/N and slowly increase the temperature T. Compute the energy per particle and the mean square displacement of the particles from their initial positions. Plot your results for these quantities versus T.

Project 17.9   Hard disks

Although we have mentioned (see Section 17.7) that there is reasonable evidence for a transition in a hard disk system, the nature of the transition still is a problem of current research. In this project we follow the work of Lee and Strandburg and apply the constant pressure Monte Carlo method (see Section 17.9) and the Lee-Kosterlitz method (see Section 17.8) to investigate the nature of the transition. Consider $N = L^2$ hard disks of diameter $\sigma = 1$ in a two-dimensional box of volume $V = \sqrt{3}L^2v/2$ with periodic boundary conditions. The quantity $v \geq 1$ is the reduced volume and is related to the density ρ by $\rho = N/V = 2/(\sqrt{3}v)$; $v = 1$ corresponds to maximum packing. The aspect ratio of $2/\sqrt{3}$ is used to match the perfect triangular lattice. We can perform a constant pressure (actually constant $p^* = P/kT$) Monte Carlo simulation as follows. The trial displacement of each disk is implemented as discussed in Section 17.7. Lee and Strandburg find that a maximum displacement of 0.09 gives a 45% acceptance probability. The other type of move is a random isotropic change of the volume of the system. If the change of the volume leads to an overlap of the disks, the change is rejected. Otherwise, if the trial volume $\tilde{V}$ is less than the current volume V, the change is accepted. A larger trial volume is accepted with probability

$$e^{-p^*(\tilde{V}-V)+N \ln \tilde{V}/V}. \tag{17.69}$$

Volume changes are attempted 40–200 times for each set of individual disk moves. The quantity of interest is $N(v)$, the distribution of reduced volume v. Because we need to store information about $N(v)$ in an array, it is convenient to discretize the volume in advance and choose the mesh size so that the acceptance probability for changing the volume by one unit is 40–50%. Do a Monte Carlo simulation of the hard disk system for $L = 10$ ($N = 100$) and $p^* = 7.30$. Published results are for 10^7 Monte Carlo steps. To apply the Lee-Kosterlitz method, smooth $\ln N(v)$ by fitting it to an eighth-order polynomial. Then extrapolate $\ln N(v)$ using the histogram method to determine $p_c^*(L = 10)$, the pressure at which the two peaks of $N(v)$ are of equal height. What is the value of the free energy barrier ΔF? If sufficient computer resources are available, compute ΔF for larger L (published results are for $L = 10$, 12, 14, 16, and 20) and determine if ΔF depends on L. Can you reach any conclusions about the nature of the transition?

Appendix 17A   FLUCTUATIONS IN THE CANONICAL ENSEMBLE

We first obtain the relation of the constant volume heat capacity C_V to the energy fluctuations in the canonical ensemble. We adopt the notation $U = <E>$ and write C_V as

$$C_V = \frac{\partial U}{\partial T} = -\frac{1}{kT^2} \frac{\partial U}{\partial \beta}. \tag{17.70}$$

From (17.3) we have

$$U = -\frac{\partial}{\partial \beta} \ln Z \tag{17.71}$$

and

$$\frac{\partial U}{\partial \beta} = -\frac{1}{Z^2} \frac{\partial Z}{\partial \beta} \sum_s E_s e^{-\beta E_s} - \frac{1}{Z} \sum_s E_s^2 e^{-\beta E_s}$$

$$= <E>^2 - <E^2>. \tag{17.72}$$

The relation (17.12) follows from (17.70) and (17.72). Note that the heat capacity is at constant volume because the partial derivatives were performed with the energy levels E_s kept constant. The corresponding quantity for a magnetic system is the heat capacity at constant external magnetic field.

The relation of the magnetic susceptibility χ to the fluctuations of the magnetization M can be obtained in a similar way. We assume that the energy can be written as

$$E_s = E_{0,s} - H M_s, \tag{17.73}$$

where $E_{0,s}$ is the energy in the absence of a magnetic field, H is the external applied field, and M_s is the magnetization in the s state. The mean magnetization is given by

$$<M> = \frac{1}{Z} \sum M_s \, e^{-\beta E_s}. \tag{17.74}$$

Because $\partial E_s / \partial H = -M_s$, we have

$$\frac{\partial Z}{\partial H} = \sum_s \beta M_s \, e^{-\beta E_s}. \tag{17.75}$$

Hence we obtain

$$<M> = \frac{1}{\beta} \frac{\partial}{\partial H} \ln Z. \tag{17.76}$$

If we use (17.74) and (17.76), we find

$$\frac{\partial <M>}{\partial H} = -\frac{1}{Z^2} \frac{\partial Z}{\partial H} \sum_s M_s \, e^{-\beta E_s} + \frac{1}{Z} \sum_s \beta M_s^2 \, e^{-\beta E_s}$$

$$= -\beta <M>^2 + \beta <M^2>. \tag{17.77}$$

The relation (17.14) for the zero field susceptibility follows from (17.77) and the definition (17.13).

Appendix 17B   EXACT ENUMERATION OF THE 2×2 ISING MODEL

Because the number of possible states or configurations of the Ising model increases as 2^N, we can enumerate the possible configurations only for small N. As an example, we calculate the various quantities of interest for a 2×2 Ising model on the square lattice with periodic boundary conditions. In Table 17.1 we group the sixteen states according to their total energy and magnetization.

Number spins up	Degeneracy	Energy	Magnetization
4	1	−8	4
3	4	0	2
2	4	0	0
2	2	8	0
1	4	0	−2
0	1	−8	−4

Table 17.1   The energy and magnetization of the 2^4 states of the zero field Ising model on the 2×2 square lattice. The degeneracy is the number of microstates with the same energy.

We can compute all the quantities of interest using Table 17.1. The partition function is given by

$$Z = 2\,e^{8\beta J} + 12 + 2\,e^{-8\beta J}. \tag{17.78}$$

If we use (17.71) and (17.78), we find

$$U = -\frac{\partial}{\partial \beta} \ln Z = -\frac{1}{Z}\big[2(8)e^{8\beta J} + 2(-8)e^{-8\beta J}\big]. \tag{17.79}$$

Because the other quantities of interest can be found in a similar manner, we only give the results:

$$<E^2> = \frac{1}{Z}\big[(2 \times 64)\,e^{8\beta J} + (2 \times 64)\,e^{-8\beta J}\big] \tag{17.80}$$

$$<M> = \frac{1}{Z}(0) = 0 \tag{17.81}$$

$$<|M|> = \frac{1}{Z}\big[(2 \times 4)\,e^{8\beta J} + 8 \times 2\big] \tag{17.82}$$

$$<M^2> = \frac{1}{Z}\big[(2 \times 16)\,e^{8\beta J} + 8 \times 4\big] \tag{17.83}$$

The dependence of C and χ on βJ can be found by using (17.79) and (17.80) and (17.81) and (17.83) respectively.

References and Suggestions for Further Reading

M. P. Allen and D. J. Tildesley, *Computer Simulation of Liquids*, Clarendon Press (1987). See Chapter 4 for a discussion of Monte Carlo methods.

K. Binder, editor, *Monte Carlo Methods in Statistical Physics*, second edition, Springer-Verlag (1986). Also see K. Binder, editor, *Applications of the Monte Carlo Method in Statistical Physics*, Springer-Verlag (1984) and K. Binder, editor, *The Monte Carlo Method in Condensed Matter Physics*, Springer-Verlag (1992).

Marvin Bishop and C. Bruin, "The pair correlation function: a probe of molecular order," *Amer. J. Phys.* **52**, 1106 (1984). The authors compute the pair correlation function for a two-dimensional Lennard-Jones model.

James B. Cole, "The statistical mechanics of image recovery and pattern recognition," *Amer. J. Phys.* **59**, 839 (1991). A discussion of the application of simulated annealing to the recovery of images from noisy data.

B. Derrida, A. J. Bray, and C. Godrèche, "Non-trivial exponents in the zero temperature dynamics of the 1D Ising and Potts models," *J. Phys.* A **27**, L357 (1994).

Jerome J. Erpenbeck and Marshall Luban, "Equation of state for the classical hard-disk fluid," *Phys. Rev.* A **32**, 2920 (1985). These workers use a combined molecular dynamics/Monte Carlo method and consider 1512 and 5822 disks.

Alan M. Ferrenberg, D. P. Landau, and Y. Joanna Wong, "Monte Carlo simulations: hidden errors from "good" random number generators," *Phys. Rev. Lett.* **69**, 3382 (1992).

Alan M. Ferrenberg and Robert H. Swendsen, "New Monte Carlo technique for studying phase transitions," *Phys. Rev. Lett.* **61**, 2635 (1988); "Optimized Monte Carlo data analysis," *Phys.*

Rev. Lett. **63**, 1195 (1989); "Optimized Monte Carlo data analysis," *Computers in Physics* **35**, 101 (1989). The second and third papers discuss using the multiple histogram method with data from simulations at more than one temperature.

Harvey Gould and Jan Tobochnik, "Overcoming critical slowing down," *Computers in Physics* **34**, 82 (1989).

Hong Guo, Martin Zuckermann, R. Harris, and Martin Grant, "A fast algorithm for simulated annealing," *Physica Scripta* **T38**, 40 (1991).

J. Kertész, J. Cserti and J. Szép, "Monte Carlo simulation programs for microcomputer," *Eur. J. Phys.* **6**, 232 (1985).

S. Kirkpatrick, C. D. Gelatt, and M. P. Vecchi, "Optimization by simulated annealing," *Science* **220**, 671 (1983). See also, S. Kirkpatrick and G. Toulouse, "Configuration space analysis of traveling salesman problems," *J. Physique* **46**, 1277 (1985).

J. M. Kosterlitz and D. J. Thouless, *J. Phys.* C **6**, 1181 (1973); J. M. Kosterlitz, *J. Phys.* C **7**, 1046 (1974).

D. P. Landau, "Finite-size behavior of the Ising square lattice," *Phys. Rev.* **B13**, 2997 (1976). A clearly written paper on a finite-size scaling analysis of Monte Carlo data. See also D. P. Landau, "Finite-size behavior of the simple-cubic Ising lattice," *Phys. Rev.* **B14**, 255 (1976).

D. P. Landau and R. Alben, "Monte Carlo calculations as an aid in teaching statistical mechanics," *Am. J. Phys.* **41**, 394 (1973).

Jooyoung Lee and J. M. Kosterlitz, "New numerical method to study phase transitions," *Phys. Rev. Lett.* **65**, 137 (1990); *ibid.*, "Finite-size scaling and Monte Carlo simulations of first-order phase transitions," *Phys. Rev.* B **43**, 3265 (1991).

Jooyoung Lee and Katherine J. Strandburg, "First-order melting transition of the hard-disk system," *Phys. Rev.* B **46**, 11190 (1992).

N. Metropolis, A. W. Rosenbluth, M. N. Rosenbluth, A. H. Teller, and E. Keller, "Equation of state calculations for fast computing machines," *J. Chem. Phys.* **6**, 1087 (1953).

J. Marro and R. Toral, "Microscopic observations on a kinetic Ising model," *Am. J. Phys.* **54**, 1114 (1986).

M. A. Novotny, "A new approach to an old algorithm for the simulation of Ising-like systems," *Computers in Physics* **9**1, 46 (1995). The *n*-fold way algorithm is discussed.

Ole G. Mouritsen, *Computer Studies of Phase Transitions and Critical Phenomena*, Springer-Verlag (1984).

E. P. Münger and M. A. Novotny, "Reweighting in Monte Carlo and Monte Carlo renormalization-group studies," *Phys. Rev.* B **43**, 5773 (1991). The authors discuss the histogram method and combine it with renormalization group calculations.

Michael Plischke and Birger Bergersen, *Equilibrium Statistical Physics*, second edition, Prentice Hall (1994). A graduate level text that discusses some of the more contemporary topics in statistical physics, many of which have been influenced by computer simulations.

William H. Press, Saul A. Teukolsky, William T. Vetterling, and Brian P. Flannery, *Numerical Recipes*, second edition, Cambridge University Press (1992). A Fortran program for the traveling salesman problem is given in Section 10.9.

Stephen H. Shenker and Jan Tobochnik, "Monte Carlo renormalization-group analysis of the classical Heisenberg model in two dimensions," *Phys. Rev.* B **22**, 4462 (1980).

Amihai Silverman and Joan Adler, "Animated Simulated Annealing," *Computers in Physics* **6**, 277 (1992). The authors describe a simulation of the annealing process to obtain a defect free single crystal of a model material.

Zoran Slanič, Harvey Gould, and Jan Tobochnik, "Dynamics of the classical Heisenberg chain," *Computers in Physics* **5**, 630 (1991). Unlike the Ising model, the Heisenberg model has an intrinsic dynamics and can be studied by molecular dynamics methods.

H. Eugene Stanley, *Introduction to Phase Transitions and Critical Phenomena*, Oxford University Press (1971). See Appendix B for the exact solution of the zero-field Ising model for a two-dimensional lattice.

Jan Tobochnik and G. V. Chester, "Monte Carlo study of the planar model," *Phys. Rev.* B **20**, 3761 (1979).

J. P. Valleau and S. G. Whittington, "A guide to Monte Carlo for statistical mechanics: 1. Highways," in *Statistical Mechanics, Part A*, Bruce J. Berne, editor, Plenum Press (1977). See also J. P. Valleau and G. M. Torrie, "A guide to Monte Carlo for statistical mechanics: 2. Byways," *ibid.*

I. Vattulainen, T. Ala-Nissila, and K. Kankaala, "Physical tests for random numbers in simulations," *Phys. Rev. Lett.* **73**, 2513 (1994).

18

Quantum Systems

We discuss numerical solutions of the time-independent and time-dependent Schrödinger equation and describe several Monte Carlo methods applicable to quantum systems.

18.1   INTRODUCTION

So far we have simulated the microscopic behavior of physical systems using Monte Carlo methods and molecular dynamics. In the latter method, the classical trajectory (position and momentum) of each particle is calculated as a function of time. However, in quantum systems it is impossible to use molecular dynamics methods, because the position and momentum of a particle cannot be specified simultaneously. Because a fundamental description of nature is intrinsically quantum mechanical, we cannot directly *simulate* nature on a computer (see Feynman).

Quantum mechanics does allow us to *analyze* probabilities. However, there are difficulties associated with such an analysis. Let us consider a simple probabilistic system described by the one-dimensional diffusion equation (see Section 7.3)

$$\frac{\partial P(x,t)}{\partial t} = D \frac{\partial^2 P(x,t)}{\partial x^2}. \tag{18.1}$$

$P(x,t)$ is the probability density of a particle being at position x at time t. One way to obtain a numerical solution for $P(x,t)$ is to make x and t discrete variables. Suppose we choose a mesh size for x such that the probability is given at p values of x. If we choose p to be order 10^3, we see that a straightforward calculation of $P(x,t)$ requires approximately 10^3 data points for each value of t. In contrast, the corresponding molecular dynamics calculation based on Newton's second law requires one data point.

The difficulty of the direct computational approach becomes even more apparent if there are many degrees of freedom. For example, with N particles in one dimension, we would have to calculate the probability $P(x_1, x_2, \ldots, x_N, t)$, where x_i is the coordinate of particle i. Because we need to choose a mesh of p points for each x_i, we need to specify N^p configurations at each time t. Usually we choose p to be of the same order as N, since the probability at each point in space represents useful information. Hence, we would need to compute the order of N^N configurations to obtain the desired probability at each time interval. Consequently, a doubling of the size of the system would lead to an exponential growth in the calculation time and in the memory requirements.

Although the direct computational approach is limited to systems with only a few degrees of freedom, the simplicity of this approach will aid our understanding of the behavior of quantum systems. After a summary of the general features of quantum mechanical systems in Section 18.2, we consider this approach to the time-independent Schrödinger equation in Section 18.3. In Section 18.4, we use a half-step method to generate wave packet solutions to the time-dependent Schrödinger equation.

Are there other ways of approaching probabilistic systems? Because we have already learned that the diffusion equation (18.1) can be formulated as a random walk problem, it might not surprise you that Schrödinger's equation can be analyzed in a similar way. Monte Carlo methods are introduced in Section 18.5 to obtain variational solutions of the ground state. We introduce quantum Monte Carlo methods in Section 18.6 and discuss more sophisticated Monte Carlo methods in Sections 18.7 and 18.8.

18.2   REVIEW OF QUANTUM THEORY

For simplicity, we consider a one-dimensional, nonrelativistic quantum system consisting of one particle. The state of the system is completely characterized by a *wave function* $\Psi(x, t)$, which is interpreted as a *probability amplitude*. The probability $P(x, t)\, dx$ of the particle being in a "volume" element dx centered about the position x at time t is equal to

$$P(x, t)\, dx = |\Psi(x, t)|^2 dx. \tag{18.2}$$

This interpretation of $\Psi(x, t)$ requires the use of normalized wave functions such that

$$\int_{-\infty}^{\infty} \Psi^*(x, t)\Psi(x, t)\, dx = 1, \tag{18.3}$$

where $\Psi^*(x, t)$ is the complex conjugate of $\Psi(x, t)$.

If the particle is subject to the influence of an external potential $V(x, t)$, the time evolution of $\Psi(x, t)$ is given by the time-dependent Schrödinger equation

$$i\hbar \frac{\partial \Psi(x, t)}{\partial t} = -\frac{\hbar^2}{2m} \frac{\partial^2 \Psi(x, t)}{\partial x^2} + V(x, t)\Psi(x, t), \tag{18.4}$$

where m is the mass of the particle and $\hbar$ is Planck's constant divided by 2π.

Physical quantities such as the momentum have corresponding operators. The expectation or average value of an observable A is given by

$$<A> = \int \Psi^*(x, t) A_{\text{op}} \Psi(x, t)\, dx, \tag{18.5}$$

where A_{op} is the operator corresponding to the quantity A. For example, the momentum operator corresponding to the linear momentum p is $p_{\text{op}} = -i\hbar \partial/\partial x$.

If the potential is independent of time, we can obtain solutions of (18.4) of the form

$$\Psi(x, t) = \phi(x)\, e^{-iEt/\hbar}. \tag{18.6}$$

A particle in the state (18.6) has a well-defined energy E. If we substitute (18.6) into (18.4), we obtain the time-independent Schrödinger equation

$$-\frac{\hbar^2}{2m} \frac{d^2\phi(x)}{dx^2} + V(x)\phi(x) = E\,\phi(x). \tag{18.7}$$

Note that $\phi(x)$ is an *eigenfunction* of the Hamiltonian operator

$$H_{\text{op}} = -\frac{\hbar^2}{2m} \frac{\partial^2}{\partial x^2} + V(x) \tag{18.8}$$

with the *eigenvalue* E. That is,

$$H_{\text{op}} \phi(x) = E\,\phi(x). \tag{18.9}$$

In general, there are many eigenfunctions ϕ_n, each with a particular eigenvalue E_n, which satisfy (18.9).

The general form of $\Psi(x,t)$ can be expressed as a superposition of the eigenfunctions of the operator corresponding to any physical observable. For example, if H_{op} is independent of time, we can write

$$\Psi(x,t) = \sum_n c_n \phi_n(x) e^{-iE_n t/\hbar}, \tag{18.10}$$

where Σ represents a sum over the discrete states and an integral over the continuum states. The coefficients c_n in (18.10) can be determined from the value of $\Psi(x,t)$ at any time t. For example, if we know $\Psi(x,t=0)$, we can use the orthonormality property of the eigenfunctions of any physical operator to obtain

$$c_n = \int \phi_n^*(x)\Psi(x,0)\,dx. \tag{18.11}$$

The coefficient c_n can be interpreted as the probability amplitude of a measurement of the total energy yielding the value E_n.

18.3 BOUND STATE SOLUTIONS

We consider bound state solutions of the time-independent Schrödinger equation (18.7). An important result will be that acceptable solutions to (18.7) exist only if the eigenvalues are quantized, i.e., restricted to a discrete set of energies. To be an acceptable solution, $\phi_n(x)$ must be finite for all x and bounded for large $|x|$ so that $\phi_n(x)$ can be normalized. For finite $V(x)$, $\phi_n(x)$ and $\phi_n' \equiv d\phi_n(x)/dx$ are required to be continuous, finite, and single valued for all x.

Because the time-independent Schrödinger equation is a second-order differential equation, two boundary conditions must be specified to obtain a unique solution. To simplify the analysis, we consider symmetric potentials that satisfy the condition

$$V(x) = V(-x). \tag{18.12}$$

The condition (18.12) implies that $\phi(x)$ can be chosen to have definite parity. For even parity solutions, $\phi(-x) = \phi(x)$; odd parity solutions satisfy $\phi(-x) = -\phi(x)$. The definite parity of $\phi(x)$ allows us to specify either ϕ or ϕ' at $x=0$. Hence, the parity of ϕ determines one of the boundary conditions.

To guide our choice of a suitable algorithm for the numerical solution of (18.7), recall that the solution of (18.7) with $V(x)=0$ can be expressed as a linear combination of sine and cosine functions. The oscillatory nature of this solution leads us to expect that the Euler-Cromer algorithm introduced in Chapter 3 will yield satisfactory results for $V(x) \neq 0$. The implementation of the Euler-Cromer algorithm proceeds as follows:

1. Divide the range of x into intervals of width Δx. Adopt the notation $x_s = s\Delta x$, $\phi_s = \phi(x_s)$, and $\phi_s' = \phi'(x_s)$, where s is an integer.
2. Specify the parity of $\phi(x)$. For an even parity solution choose $\phi(0)=1$ and $\phi'(0)=0$; for an odd parity solution choose $\phi(0)=0$ and $\phi'(0)=1$. The nonzero value of $\phi(0)$ or $\phi'(0)$ is arbitrary.

3. Guess a value for E.

4. Compute ϕ'_{s+1} and ϕ_{s+1} using the algorithm:

$$\phi'_{s+1} = \phi'_s + \phi''_s \, \Delta x \qquad\qquad (18.13a)$$

$$\phi_{s+1} = \phi_s + \phi'_{s+1} \, \Delta x. \qquad\qquad (18.13b)$$

5. Iterate $\phi(x)$ toward increasing x until $\phi(x)$ diverges.

6. Change E and repeat steps (4) to (5). Bracket the value of E by changing E until ϕ diverges in one direction if E is made slightly smaller and diverges in the opposite direction if E is made slightly larger.

Program eigen implements this procedure for a square well potential given by

$$V(x) = \begin{cases} 0 & \text{for } |x| \leq a \\ V_0 & \text{for } |x| > a \,. \end{cases} \qquad\qquad (18.14)$$

The input parameters are V_0 and a, the parity of the eigenfunction, the assumed value of the energy E, the step size Δx, and xmax, the maximum value of x to be plotted.

```
PROGRAM eigen
CALL parameters(V0,a,xmax,dx)
CALL plot_potential(V0,a,xmax)
CALL Euler(V0,a,dx,xmax)
END

SUB parameters(V0,a,xmax,dx)
    INPUT prompt "magnitude of well depth = ": V0
    INPUT prompt "half width of well = ": a
    INPUT prompt "step size = ": dx
    INPUT prompt "maximum value of x to be plotted = ": xmax
END SUB

SUB plot_potential(V0,a,xmax)
    ! draw potential well
    OPEN #1: screen 0,1,0,0.4
    SET COLOR "red"
    SET WINDOW -1.05*xmax,1.05*xmax,-1.1*V0,1.1*V0
    PLOT LINES: -xmax,0;-a,0;      ! horizontal line
    PLOT LINES: -a,0;-a,-V0        ! vertical line
    PLOT LINES: -a,-V0;a,-V0;a,-V0;a,0;a,0;xmax,0
END SUB

SUB Euler(V0,a,dx,xmax)
    DECLARE DEF V
    OPEN #2: screen 0,1,0.4,1
    SET WINDOW -1.01*xmax,1.01*xmax,-4,4
    INPUT prompt "even or odd parity (1 or -1) = ": parity
    CLEAR
    DO
```

```
          INPUT prompt "E = ": E
          IF E = 0 then EXIT SUB
          SET COLOR "blue"
          IF parity = -1 then
             LET phi = 0                  ! initial values at x = 0
             LET dphi = 1                 ! first derivative
          ELSE
             LET phi = 1
             LET dphi = 0
          END IF
          LET x = 0
          DO                              ! compute wave function
             LET x_old = x
             LET phi_old = phi
             LET x = x + dx
             LET d2phi = 2*(V(x,V0,a) - E)*phi       ! dimensionless units
             LET dphi = dphi + d2phi*dx   ! Euler-Cromer algorithm
             LET phi = phi + dphi*dx
             ! plot wave function
             PLOT LINES: x_old,phi_old; x,phi
             PLOT LINES: -x_old,phi_old*parity; -x,phi*parity
          LOOP until x > xmax
          SET COLOR "black"
      LOOP
END SUB

DEF V(x,V0,a)                             ! potential function V(x)
   IF abs(x) > a then
      LET V = V0
   ELSE
      LET V = 0
   END IF
END DEF
```

Problem 18.1   The infinite square well

a. Consider the infinite square well with $V(x)$ given by (18.14) with $V_0 \to \infty$.
Show analytically that the energy eigenvalues are given by $E_n = n^2\pi^2\hbar^2/8ma^2$,
where n is a positive integer. Also show that the normalized eigenfunctions
have the form

$$\phi_n(x) = \frac{1}{\sqrt{a}} \cos \frac{n\pi x}{2a} \qquad n = 1, 3, \dots \qquad \text{(even parity)} \qquad (18.15a)$$

$$\phi_n(x) = \frac{1}{\sqrt{a}} \sin \frac{n\pi x}{2a}. \qquad n = 2, 4, \dots \qquad \text{(odd parity)} \qquad (18.15b)$$

What is the parity of the ground state solution? Why does the ground state
wave function have no nodes (zeros)?

b. Use *Program eigen* with $V_0 = 150$ and $a = 1$ and find the ground state energy E_1. Choose $\Delta x = 0.01$ and xmax $= 4$ with units such that $m = \hbar = 1$. Try the initial guesses $E_1 = 1.0$ and $E_1 = 1.5$. How do you know that the ground state energy is between these two values of E_1? Obtain E_1 to two decimal places.

c. Is your numerical result for E_1 affected by your choice of Δx? Determine a value of Δx that yields E_1 to three decimal places. Is $V_0 = 150$ sufficiently large that your value for E_1 is a good approximation to an infinite square well? Does your result for E_1 depend on the magnitude of ϕ at $x = 0$?

d. Write a subroutine to normalize ϕ. What is the value of ϕ at $x = 0$? Compare your numerical result for the ground state eigenfunction with the analytical solution.

e. We can find the excited state eigenfunctions and eigenvalues in a way similar to that used for the ground state. If we count the number of nodes of ϕ, we can determine which eigenstate we have. For example, the first excited state has one node and quantum number $n = 2$. Modify *Program eigen* so that it computes the number of nodes and find the first two excited states of the infinite (i.e., $V_0 = 150$) square well.

f. Suppose you wish to find the eigenvalue corresponding to the tenth excited state of the particle in a box problem. Can you use the same value of V_0 that was used in part (e)?

It is not possible to obtain a numerical solution for $\phi(x)$ that does not diverge at sufficiently large x. The reason is that ϕ can be computed only to finite accuracy. The computed $\phi(x)$ will diverge if a sufficiently large number of iterations are performed. However, we can always calculate $\phi(x)$ to the desired accuracy by choosing an appropriate algorithm and a sufficiently small step size.

Problem 18.2   Perturbation of the ground state of an infinite square well

a. Determine the effect of a small perturbation on the eigenstates and eigenvalues of the infinite square well. Place a small rectangular bump of half-width b and height V_b symmetrically about $x = 0$ (see Fig. 18.1). Choose $V_b \ll V_0$ and $b \ll a$ and determine how the ground state energy and eigenfunction change with V_b and b. What is the relative change in E_1 for $V_b = 10$, $b = 0.1$ and $V_b = 20$, $b = 0.1$? (Set $V_0 = 150$ and $a = 1$.) Let ϕ_0 denote the ground state function for $b = 0$ and ϕ_b denote the ground state eigenfunction for $b \ne 0$. Compute the change of the overlap integral

$$\int_0^a \phi_b(x)\phi_0(x)\, dx.$$

How does this change compare to the relative change in the energy?

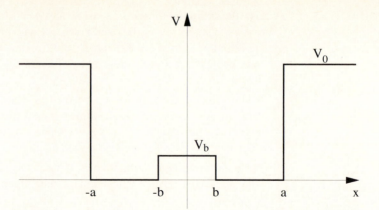

Fig. 18.1   A square well with a potential bump of height V_b in the middle.

b.   Compute the ground state energy for $V_b = 20$ and $b = 0.05$. How does the value of E_1 compare to that found in part (a) for $V_b = 10$ and $b = 0.1$?

Problem 18.3   Finite square well

a.   Consider a finite square well with $V_0 = 10$ and $a = 1$. How do you expect the value of the ground state energy to compare to its corresponding value for the infinite square well? Compute the ground state eigenvalue and eigenfunction by determining a value of E such that $\phi(x)$ has no nodes and is approximately zero for large x.

b.   Because the well depth is finite, $\phi(x)$ is nonzero in the classically forbidden region for which $E < V_0$ and $x > |a|$. Define the "penetration distance" as the distance from $x = a$ to a point where ϕ is $\sim 1/e \approx 0.37$ of its value at $x = a$. Determine the qualitative dependence of the penetration distance on the magnitude of V_0.

c.   Compute the excited eigenstates and eigenvalues for $V_0 = 10$ and $a = 1$. What is the total number of bound excited states? Why is the total number of bound states finite?

Problem 18.4   Other one-dimensional potentials

a.   Obtain numerical solutions for the first several eigenvalues and eigenfunctions of the harmonic oscillator with $V(x) = \frac{1}{2}x^2$. What value of Δx is needed for 0.1% accuracy for the ground state energy?

b.   Obtain a numerical solution of the anharmonic oscillator with $V(x) = \frac{1}{2}x^2 + bx^4$. In this case there are no analytical solutions and numerical solutions are necessary for large values of b. Compute the ground state energy for $b = 0.1$,

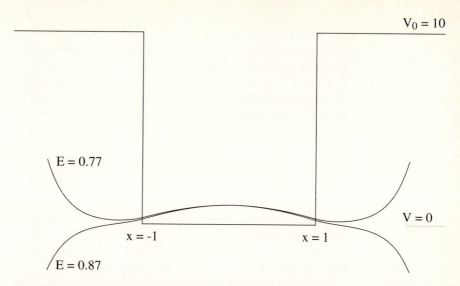

Fig. 18.2   Two numerical solutions of the time-independent Schrödinger equation for the guesses $E = 0.77$ and $E = 0.87$. The system is a particle in a square well potential with $V_0 = 10$ and $a = 1$.

0.2, and 0.5. How does the ground state energy depend on b for small b? How does the ground state eigenfunction depend on b?

c.   Find a numerical solution for the ground state of the linear potential:

$$V(x) = |x|. \tag{18.16}$$

The quantum mechanical treatment of this potential can be used to model the energy spectrum of a bound quark-antiquark system known as quarkonium.

18.4   THE TIME-DEPENDENT SCHRÖDINGER EQUATION

Although we found that the numerical solution of the time-independent Schrödinger equation (18.7) is straightforward for one particle, the numerical solution of the time-dependent Schrödinger equation (18.4) is not as straightforward. A naive approach to its numerical solution can be formulated by introducing a grid for the time coordinate and a grid for the spatial coordinate. We use the notation $t_n = t_0 + n\Delta t$, $x_s = x_0 + s\Delta x$, and $\Psi(x_s, t_n)$. The idea is to develop an algorithm that relates $\Psi(x_s, t_{n+1})$ to the value of $\Psi(x_s, t_n)$ for each value of x_s. An example of an algorithm that is first-order in Δt is given by

$$\frac{1}{\Delta t}\big[\Psi(x_s, t_{n+1}) - \Psi(x_s, t_n)\big] =$$

$$\frac{1}{(\Delta x)^2}\big[\Psi(x_{s+1}, t_n) - 2\Psi(x_s, t_n) + \Psi(x_{s-1}, t_n)\big]. \tag{18.17}$$

The right-hand side of (18.17) represents a finite difference approximation to the second derivative of Ψ with respect to x. Equation (18.17) is an example of an *explicit* scheme, since given Ψ at time t_n, we can calculate Ψ at time t_{n+1}. Unfortunately, this explicit approach leads to unstable solutions, that is, the numerical value of Ψ diverges from the exact solution as Ψ evolves in time (cf. Koonin and Meredith).

One way to avoid this difficulty is to retain the same form as (18.17), but to evaluate the spatial derivative on the right-hand side of (18.17) at time t_{n+1} rather than time t_n:

$$\frac{1}{\Delta t}\left[\Psi(x_s, t_{n+1}) - \Psi(x_s, t_n)\right] =$$
$$\frac{1}{(\Delta x)^2}\left[\Psi(x_{s+1}, t_{n+1}) - 2\Psi(x_s, t_{n+1}) + \Psi(x_{s-1}, t_{n+1})\right]. \qquad (18.18)$$

Equation (18.18) is an *implicit* method because the unknown function $\Psi(x_s, t_{n+1})$ appears on both sides. To obtain $\Psi(x_s, t_{n+1})$, it is necessary to solve a set of linear equations at each time step. More details of this approach and the demonstration that (18.18) leads to stable solutions can be found in the references.

Visscher and others (see references) have suggested an alternative approach in which the real and imaginary parts of Ψ are treated separately and defined at different times. The algorithm ensures that the total probability remains constant. If we let

$$\Psi(x, t) = R(x, t) + i\,I(x, t), \qquad (18.19)$$

then Schrödinger's equation becomes (with $\hbar = 1$)

$$\frac{dR(x, t)}{dt} = H_{op}\,I(x, t) \qquad (18.20a)$$

$$\frac{dI(x, t)}{dt} = -H_{op}\,R(x, t). \qquad (18.20b)$$

A stable method of numerically solving (18.20) is to use a form of the half-step method (see Appendix 5A). The resulting difference equations are

$$R(x, t + \Delta t) = R(x, t) + H_{op}\,I(x, t + \frac{1}{2}\Delta t)\,\Delta t \qquad (18.21a)$$

$$I(x, t + \frac{3}{2}\Delta t) = I(x, t + \frac{1}{2}\Delta t) - H_{op}\,R(x, t)\,\Delta t, \qquad (18.21b)$$

where the initial values are given by $R(x, 0)$ and $I(x, \frac{1}{2}\Delta t)$. The appropriate definition of the probability density $P = R^2 + I^2$ is not obvious, because R and I are not defined at the same time. It is easy to show that the following choice conserves the total probability:

$$P(x, t) = R(x, t)^2 + I(x, t + \frac{1}{2}\Delta t)\,I(x, t - \frac{1}{2}\Delta t) \qquad (18.22a)$$

$$P(x, t + \frac{1}{2}\Delta t) = R(t + \Delta t)\,R(x, t) + I(x, t + \frac{1}{2}\Delta t)^2. \qquad (18.22b)$$

Visscher has shown that this algorithm is stable if

$$\frac{-2\hbar}{\Delta t} \leq V \leq \frac{2\hbar}{\Delta t} - \frac{2\hbar^2}{(m\Delta x)^2},\tag{18.23}$$

where the inequality (18.23) holds for all values of the potential V.

An implementation of (18.21) is shown in SUB evolve. The real part of the wave function first is updated for all positions, and then the imaginary part is updated using the new values of the real part. The subroutine saves the imaginary part of the wave function at the previous time step so that the probability density can be computed.

```
SUB evolve(Re(),Im(),Imold(),t,V0,a,dx,dx2,dt,xmin,n)
    DECLARE DEF V
    FOR i = 1 to n
        LET x = xmin + (i-1)*dx
        LET HIm = V(x,V0,a)*Im(i) -0.5*(Im(i+1) -2*Im(i) +Im(i-1))/dx2
        ! real part defined at multiples of dt
        LET Re(i) = Re(i) + HIm*dt
    NEXT i
    FOR i = 1 to n
        LET x = xmin + (i-1)*dx
        ! dt/2 earlier than real part
        LET Imold(i) = Im(i)
        LET HRe = V(x,V0,a)*Re(i) -0.5*(Re(i+1) -2*Re(i) +Re(i-1))/dx2
        ! dt/2 later than real part
        LET Im(i) = Im(i) - HRe*dt
    NEXT i
    LET t = t + dt                      ! time of real part
END SUB
```

Before we can incorporate SUB evolve into a program, we need to choose an initial wave function. A useful form is the Gaussian wave packet with a width (variance) w centered about x_0 given by

$$\Psi(x,0) = \left(\frac{1}{2\pi w^2}\right)^{1/4} e^{ik_0(x-x_0)} \, e^{-(x-x_0)^2/4w^2}.\tag{18.24}$$

The expectation value of the initial velocity of the wave packet is $<v> = p_0/m = \hbar k_0/m$.

To begin the half-step algorithm, we need the value of $I(x, t = \frac{1}{2}\Delta t)$ and $R(x, t = 0)$. To obtain $I(x, t = \frac{1}{2}\Delta t)$, we note that a plane wave at time t evolving in a zero potential region is related to its value at $t = 0$ by a factor of $e^{-i\omega t}$, where ω is related to the kinetic energy E by

$$E = \hbar\omega = \frac{p_0^2}{2m} = \frac{\hbar^2 k_0^2}{2m}.\tag{18.25}$$

In addition, the center of the Gaussian envelope will move by an amount $<v>t$. These considerations suggest that we can approximate Ψ at time $t = \frac{1}{2}\Delta t$ by changing the phase in (18.24) by letting $ik_0(x - x_0) \rightarrow ik_0(x - x_0) - i\omega\Delta t/2 = ik_0(x - x_0) - i\hbar k_0^2 \Delta t/(4m) = ik_0(x - x_0 - \hbar k_0\Delta t/(4m))$ and changing the argument of the Gaussian by letting $x - x_0 \rightarrow x - x_0 - <v>\Delta t/2 = x - x_0 - \hbar k_0\Delta t/(2m)$. This result is not exact, because these changes require a small correction in the overall normalization factor.

For completeness, we list the main program and the other subroutines needed to make a complete program. In the program and problems we use units such that $m = 1$ and $\hbar = 1$.

```
PROGRAM tdse
! motion of a wavepacket incident on a potential
DIM Re(0:1100),Im(0:1100),Imold(0:1100)
CALL parameters(x0,k0,width,V0,a,xmin,xmax,n,dx,dx2,dt)
CALL initial_packet(Re(),Im(),x0,k0,width,xmin,n,dx,dt)
CALL set_up_windows(xmin,xmax,V0,#1,#2,#3,#4)
CALL draw_potential(V0,a,xmin,xmax,#4)
DO
    CALL evolve(Re(),Im(),Imold(),t,V0,a,dx,dx2,dt,xmin,n)
    CALL plots(Re(),Im(),Imold(),t,xmin,n,dx,#1,#2,#3,#4)
LOOP until key input
END

SUB parameters(x0,k0,width,V0,a,xmin,xmax,n,dx,dx2,dt)
    LET x0 = -15                    ! initial position of packet
    LET width = 1                   ! width of wave packet in x space
    LET k0 = 2                      ! group velocity of packet
    LET xmax = 20                   ! maximum value of x
    LET xmin = -xmax                ! minimum value of x
    LET V0 = 2
    LET a = 1                       ! half-width of potential well
    LET dx = 0.4                    ! grid size
    LET dx2 = dx*dx
    LET n = (xmax - xmin)/dx
    LET dt = 0.1                    ! time interval
END SUB

SUB initial_packet(Re(),Im(),x0,k0,width,xmin,n,dx,dt)
    ! initial Gaussian wavepacket
    LET delta2 = width*width
    LET A = (2*pi*delta2)^(-0.25)
    LET b = 0.5*k0*dt
    FOR i = 1 to n
        LET x = xmin + (i-1)*dx
        LET arg = 0.25*(x - x0)^2/delta2
        LET e = exp(-arg)
        LET Re(i) = A*cos(k0*(x - x0))*e      ! value at t = 0
        LET Im(i) = A*sin(k0*(x - x0 - 0.5*b))*e    ! value at t = dt/2
    NEXT i
END SUB
```

```
SUB set_up_windows(xmin,xmax,V0,#1,#2,#3,#4)
    OPEN #1: screen 0,1,0.75,1.0  ! probability
    SET WINDOW xmin,xmax,-0.1,0.5
    PLOT LINES: xmin,0;xmax,0
    OPEN #2: screen 0,1,0.5,0.75  ! real part
    SET WINDOW xmin,xmax,-1,1
    PLOT LINES: xmin,0;xmax,0
    OPEN #3: screen 0,1,0.25,0.5  ! imaginary part
    SET WINDOW xmin,xmax,-1,1
    PLOT LINES: xmin,0;xmax,0
    OPEN #4: screen 0,1,0,0.22    ! potential
    SET WINDOW xmin,xmax,-V0,V0
END SUB


SUB draw_potential(V0,a,xmin,xmax,#4)
    DECLARE DEF V
    WINDOW #4
    SET COLOR "red"
    PLOT LINES: xmin,0;a,0;a,0;a,V0;a,V0;xmax,V0
    SET COLOR "black/white"
    PRINT "total probability = ";
END SUB


SUB plots(Re(),Im(),Imold(),t,xmin,n,dx,#1,#2,#3,#4)
    WINDOW #2
    CLEAR
    PLOT LINES: xmin,0;xmax,0
    FOR i = 1 to n
        LET x = xmin + (i - 1)*dx
        PLOT x,Re(i);
    NEXT i
    WINDOW #3
    CLEAR
    PLOT LINES: xmin,0;xmax,0
    FOR i = 1 to n
        LET x = xmin + (i - 1)*dx
        PLOT x,Im(i);
    NEXT i
    WINDOW #1
    CLEAR
    PLOT LINES: xmin,0;xmax,0
    LET Psum = 0
    FOR i = 1 to n
        LET x = xmin + (i - 1)*dx
        LET P = Re(i)*Re(i) + Im(i)*Imold(i)
        LET Psum = Psum + P*dx
        PLOT x,P;
    NEXT i
    WINDOW #4
    SET CURSOR 1,20
```

```
    PRINT "      ";                        ! erase previous probability
    SET CURSOR 1,20
    PRINT Psum                             ! total probability should be unity
END SUB


FUNCTION V(x,V0,a)
    ! step potential
    IF x > a then
        LET V = V0
    ELSE
        LET V = 0
    END IF
END DEF
```

Problem 18.5   Evolution of a free wave packet

a. Use *Program tdse* to follow the motion of a wave packet in a potential-free region. Choose V appropriately. Let $x_0 = -15$, $k_0 = 2$, $w = 1$, dx = 0.4, and dt = 0.1. Suitable values for the minimum and maximum values of x on the grid are xmin = −20 and xmax = 20. What is the shape of the wave packet at different times? Does the shape of the wave packet depend on your choice of the parameters k_0 and w?

b. Modify *Program tdse* so that the quantities $x_0(t)$ and $w(t)$, the position and width of the wave packet as a function of time, can be measured directly. What is a reasonable definition of $w(t)$? What is the qualitative dependence of x_0 and w on t? How are your results changed if the initial width of the packet is reduced by a factor of four?

Problem 18.6   Evolution of wave packet incident on a potential step

a. Use *Program tdse* with a step potential beginning at $x = 0$ with height $V_0 = 2$. Choose $x_0 = -10$, $k_0 = 2$, $w = 1$, dx = 0.4, dt = 0.1, xmin = −20, and xmax = 20. Describe the motion of the wave packet. Does the shape of the wave packet remain a Gaussian for all t? What happens to the wave packet at $x = 0$? Determine the height and width of the reflected and transmitted wave packets, the time t_{inc} for the incident wave to hit the barrier at $x = 0$, and the time t_{ref} for the reflected wave to return to $x = -x_0$. Is $t_{ref} = t_{inc}$? If these times are not equal, explain the cause of the difference.

b. Repeat the above analysis for a step potential of height $V_0 = 10$. Is $t_{ref} \approx t_{inc}$ in this case?

c. What is the motion of a classical particle with a kinetic energy corresponding to the central wave vector $k = k_0$?

Problem 18.7   Scattering of a wave packet from a potential barrier

a.   Consider a potential barrier of the form

$$V(x) = \begin{cases} 0 & x < 0 \\ V_0 & 0 \le x \le a \\ 0 & x > a . \end{cases} \tag{18.26}$$

Generate a series of snapshots that show the wave packet approaching the barrier and then interacting with it to generate reflected and transmitted packets. Set $V_0 = 2$ and $a = 1$ and consider the behavior of the wave packet for $k_0 = 1, 1.5, 2$, and 3. Does the width of the packet increase with time? How does the width depend on k_0? For what values of k_0 is the motion of the packet in qualitative agreement with the motion of a corresponding classical particle?

b.   Consider a square well with $V_0 = -2$ and consider the same questions as in part (a).

Problem 18.8   Evolution of a wave packet inside an infinite well

a.   Consider the motion of a wave packet in an infinite square well. Approximate this well by setting $V = 0$ for $|x| \le a$ and $V = 10^4$ for $|x| > a$. Choose $a = 10$. Locate the center of the wave packet at $x = 0$. Describe the time evolution of the wave packet for $k_0 = 0$ and width $w = 1$. What is the corresponding classical motion of a particle in a box?

b.   Let $k_0 = 2$. How does the motion of the wave packet change? Is there tunneling through the well?

Problem 18.9   Evolution of two wave packets

Modify SUB `initial_packet` to include two wave packets with identical widths and speeds, with the sign of k_0 chosen so that the two wave packets approach each other. Choose their respective values of x_0 so that the two packets are well separated initially. Let $V = 0$ and describe what happens when you run the simulation. Do the two packets have any influence on each other? What do your results imply about the existence of a superposition principle?

18.5 INTRODUCTION TO VARIATIONAL METHODS

One way of obtaining a good approximation to the ground state energy is to use a variational method. This method has numerous applications, especially in atomic and molecular physics, nuclear physics, and condensed matter physics. Consider a physical system whose Hamiltonian operator H_{op} is given by (18.8). According to the variational principle, the expectation value of the Hamiltonian for an arbitrary trial wave function Ψ is greater than or equal to the ground state energy. That is,

$$<H> = E[\Psi] = \frac{\int \Psi^*(x) H_{op} \Psi(x)\, dx}{\int \Psi^*(x)\, \Psi(x)\, dx} \geq E_0, \tag{18.27}$$

where E_0 is the exact ground state energy of the system. The inequality (18.27) reduces to an equality only if Ψ is an eigenfunction of H with the eigenvalue E_0. For bound states, Ψ may be assumed to be real so that $\Psi^* = \Psi$.

The inequality (18.27) is the basis of the variational method. The procedure is to choose a physically reasonable form for the trial wave function $\Psi(x)$ that depends on one or more parameters. The expectation value $E[\Psi]$ is calculated, and the parameters are varied until a minimum of $E[\Psi]$ is obtained. This value of $E[\Psi]$ is an upper bound to the true ground state energy. Often forms of Ψ are chosen so that the integrals in (18.27) can be done analytically. To avoid this restriction we can use numerical integration methods.

One major area of application of the variational method is to atoms and molecules for which the integrals in (18.27) are multidimensional. Hence, if numerical integration methods are to be used, Monte Carlo integration methods are essential. For that reason we will use Monte Carlo integration in the following, even though we will consider only one and two body problems. Because it is inefficient to simply choose points at random to compute $E[\Psi]$, we rewrite (18.27) in a form that allows us to use importance sampling. We write

$$E[\Psi] = \frac{\int \Psi^2(x) E_L(x)\, dx}{\int \Psi^2(x)\, dx}, \tag{18.28}$$

where E_L is the *local energy*

$$E_L = \frac{H_{op} \Psi(x)}{\Psi(x)}. \tag{18.29}$$

The form of (18.28) is that of a weighted average with the weight being the normalized probability density $\Psi^2(x)/\int \Psi^2(x)\, dx$. As discussed in Section 11.7, we can sample values of x using the distribution $\Psi^2(x)$ so that the Monte Carlo estimate of $E[\Psi]$ is given by the arithmetic sum

$$E[\Psi] = \lim_{n \to \infty} \frac{1}{n} \sum_{i=1}^{n} E_L(x_i), \tag{18.30}$$

where n is the number of times that x is sampled from Ψ^2. How can we sample from Ψ^2? In general, it is not possible to use the inverse transform method (see Section 11.5) to generate a nonuniform distribution. A convenient alternative is the Metropolis method which has the advantage that only Ψ^2 for the proposed move is needed, and the unknown quantity $\int \Psi^2\, dx$ is irrelevant.

Problem 18.10   One-dimensional systems

a. It is useful to test the variational method on an exactly solvable problem. Consider the one-dimensional harmonic oscillator with $H_{op} = \frac{1}{2}p^2 + \frac{1}{2}x^2$, where we have chosen units such that the parameters m, k, and $\hbar$ are unity. Choose the trial wave function to be $\Psi(x) \sim e^{-\lambda x^2}$, with λ the variational parameter. Generate values of x chosen from a normalized $\Psi^2(x)$ using the inverse transform method, and verify that $\lambda = \frac{1}{2}$ yields the smallest upper bound. (Another way to generate a Gaussian distribution is to use the Box-Muller method discussed in Section 11.5.)

b. Repeat part (a) using the Metropolis method to generate x distributed according to $e^{-2\lambda x^2}$ and evaluate (18.30). As discussed in Section 11.8, the Metropolis method can be summarized by the following steps:

   i. Choose a trial position $x_{trial} = x_n + \delta_n$, where δ_n is a random number in the interval $[-\delta, \delta]$.
   ii. Calculate $w = p(x_{trial})/p(x_n)$, where in this case $p(x) = \Psi^2(x) = e^{-2\lambda x^2}$.
   iii. If $w \geq 1$, accept the change and let $x_{n+1} = x_{trial}$.
   iv. If $w < 1$, generate a random number r and let $x_{n+1} = x_{trial}$ if $r \leq w$.
   v. If the trial change is not accepted, then let $x_{n+1} = x_n$.

   Remember that it is necessary to wait for equilibrium (convergence to the distribution Ψ^2) before computing the average value of E_L. Look for a systematic trend in $<E_L>$ over the course of the random walk. Choose a step size that gives a reasonable value for the acceptance ratio. How many trials are necessary to obtain $<E_L>$ to 1% accuracy?

c. Instead of finding the minimum of $<E_L>$ as a function of the variational parameters, minimize the quantity

$$\sigma^2 = <E_L^2> - <E_L>^2. \tag{18.31}$$

   Verify that the exact minimum value of $\sigma^2[\Psi]$ is zero, whereas the exact minimum value of $E[\Psi]$ is unknown in general.

d. Consider the anharmonic potential $V(x) = \frac{1}{2}x^2 + bx^4$. Plot $V(x)$ as a function of x for $b = 1/8$. Use first-order perturbation theory to calculate the lowest order change in the ground state energy due to the x^4 term. Then choose a reasonable form for your trial wave function and use your Monte Carlo program to estimate the ground state energy. How does your result compare with first-order perturbation theory?

e. Consider the anharmonic potential of part (d) with $b = -1/8$. Plot $V(x)$ as a function of x. Use first-order perturbation theory to calculate the lowest order change in the ground state energy due to the x^4 term, and then use your program to estimate E_0. Do your Monte Carlo estimates for the ground state energy have a lower bound? Why or why not?

f. Modify your program so that it can be applied to the ground state of the hydrogen atom. In this case we can write $H_{op} = \frac{1}{2}p^2/\mu - e^2/r$, where μ is the reduced mass and e is the magnitude of the charge on the electron. The element of integration dx in (18.28) is replaced by $4\pi r^2\, dr$. Choose $\Psi = e^{-r/a}$, where a is the variational parameter. Measure lengths in terms of the Bohr radius $\hbar^2/me^2$ and energy in terms of the Rydberg $me^4/2\hbar^2$. In these units $\mu = e^2 = \hbar = 1$. Find the optimal value of a. What is the corresponding energy?

g. Consider the Yukawa or screened Coulomb potential for which

$$V(r) = \frac{e^2}{r} e^{-\alpha r},$$

where α is a positive constant. In this case the exact ground state and wave function can only be obtained numerically. Consider the cases $\alpha = 0.5$ and $\alpha = 1.0$ for which the most accurate numerical estimates of E_0 are -0.14808 and -0.01016 respectively. What is a good choice for the form of the trial wave function?

Problem 18.11   Ground state of Helium

Helium has long served as the testing ground for atomic trial wave functions. Consider the ground state of the helium atom with the Hamiltonian

$$H_{\mathrm{op}} = \frac{1}{2m}(p_1^2 + p_2^2) - 2e^2\Big(\frac{1}{r_1} + \frac{1}{r_2}\Big) + \frac{e^2}{r_{12}}, \tag{18.32}$$

where r_{12} is the separation between the two electrons. Assume that the nucleus is fixed and ignore relativistic effects. Choose $\Psi(\mathbf{r}_1, \mathbf{r}_2) = Ae^{-Z_{\mathrm{eff}}(r_1+r_2)/a_0}$, where Z_{eff} is a variational parameter. Estimate the upper bound to the ground state energy based on this functional form of Ψ.

The above discussion and applications of variational Monte Carlo methods has been only introductory in nature. One important application of variational Monte Carlo methods is to optimize a given trial wave function which is then used to "guide" the quantum Monte Carlo methods discussed in Sections 18.6 and 18.7. More information about variational Monte Carlo methods can be found in the references.

18.6   RANDOM WALK QUANTUM MONTE CARLO

We now introduce a Monte Carlo approach based on the relation of the Schrödinger equation to a diffusion process in imaginary time. Our approach follows that of Anderson (see references). To understand this relation, we substitute $\tau = it/\hbar$ into the time-dependent Schrödinger equation for a free particle and write (in one dimension)

$$\frac{\partial \Psi(x, \tau)}{\partial \tau} = \frac{\hbar^2}{2m} \frac{\partial^2 \Psi(x, \tau)}{\partial x^2}. \tag{18.33}$$

Note that (18.33) is similar in form to the diffusion equation (18.1), and hence we can interpret the wave function Ψ as a probability density with a diffusion constant $D = \hbar^2/2m$.

From our discussion in Chapters 7 and 12, we know that we can use a random walk algorithm to find the solution of a diffusion equation. Hence, the formal similarity between the diffusion equation and the imaginary-time free particle Schrödinger equation can be used to solve the latter by replacing it by an equivalent random walk problem. To understand how we can interpret the role of the potential energy term in the context of random walks, we write Schrödinger's equation in imaginary time as

$$\frac{\partial \Psi(x, \tau)}{\partial \tau} = \frac{\hbar^2}{2m} \frac{\partial^2 \Psi(x, \tau)}{\partial x^2} - V(x)\Psi(x, \tau). \tag{18.34}$$

If we were to ignore the first-term (the diffusion term) on the right-hand side of (18.34), then the result would be a first-order differential equation corresponding to a decay or growth process depending on the sign of V. We can obtain the solution to this first-order equation by replacing it by a random decay or growth process, e.g., radioactive decay. These considerations suggest that we can interpret (18.34) as a combination of diffusion and branching processes. In the latter, the number of walkers increases or decreases at a point x depending on the sign of $V(x)$. The walkers do not interact with each other since the Schrödinger equation (18.34) is linear in Ψ. Note that it is $\Psi \Delta x$ and *not* $\Psi^2 \Delta x$ which corresponds to the probability distribution of the random walkers. This probabilistic interpretation requires that Ψ be nonnegative.

We now use the probabilistic interpretation of (18.34) to develop a method for determining the ground state wave function and energy. The general solution of Schrödinger's equation can be written for imaginary time as (see (18.10))

$$\Psi(x, \tau) = \sum_n c_n \phi_n(x) e^{-E_n \tau}. \tag{18.35}$$

For sufficiently large τ, the dominant term in the sum in (18.35) comes from the term representing the eigenvalue of lowest energy. Hence we have

$$\Psi(x, \tau \to \infty) = c_0 \phi_0(x) e^{-E_0 \tau}. \tag{18.36}$$

From (18.36) we see that the spatial dependence $\Psi(x, \tau \to \infty)$ is proportional to the ground state eigenfunction $\phi_0(x)$. However, we also see that $\Psi(x, \tau)$ and hence the population of walkers will eventually decay to zero unless $E_0 = 0$. This problem can be avoided by measuring E_0 from an arbitrary reference energy V_{ref}, which is adjusted so that an approximate steady state distribution of random walkers is obtained.

Although we could attempt to fit the τ-dependence of the computed probability distribution of the random walkers to (18.36) and thereby extract E_0, this procedure would not yield precise values of E_0. We show in the following that E_0 also can be determined from the relation

$$E_0 = <V> = \frac{\sum n_i V(x_i)}{\sum n_i}, \tag{18.37}$$

where n_i is the number of walkers at x_i at time τ. An estimate for E_0 can be found by averaging the sum in (18.37) for several values of τ once a steady state distribution has been reached.

To derive (18.37), we rewrite (18.34) and (18.36) by explicitly introducing the reference potential V_{ref}:

$$\frac{\partial \Psi(x, \tau)}{\partial \tau} = \frac{\hbar^2}{2m} \frac{\partial^2 \Psi(x, \tau)}{\partial x^2} - \left[V(x) - V_{\text{ref}}\right] \Psi(x, \tau), \tag{18.38}$$

and

$$\Psi(x, \tau) \approx c_0 \phi_0(x) \, e^{-(E_0 - V_{\text{ref}})\tau}. \tag{18.39}$$

We first integrate (18.38) with respect to x. Because $\partial \Psi(x, \tau)/\partial x$ vanishes in the limit $|x| \to \infty$, $\int \partial^2 \Psi/\partial x^2 dx = 0$, and hence

$$\int \frac{\partial \Psi(x, \tau)}{\partial \tau} \, dx = -\int V(x)\, \Psi(x, \tau) \, dx + V_{\text{ref}} \int \Psi(x, \tau) \, dx. \tag{18.40}$$

If we differentiate (18.39) with respect to τ, we obtain the relation

$$\frac{\partial \Psi(x, \tau)}{\partial \tau} = (V_{\text{ref}} - E_0) \Psi(x, \tau). \tag{18.41}$$

We then substitute (18.41) for $\partial \Psi/\partial \tau$ into (18.40) and find

$$\int (V_{\text{ref}} - E_0) \Psi(x, \tau) \, dx = -\int V(x)\Psi(x, \tau) \, dx + V_{\text{ref}} \int \Psi(x, \tau) \, dx. \tag{18.42}$$

If we cancel the terms proportional to V_{ref} in (18.42), we find that

$$E_0 \int \Psi(x, \tau) \, dx = \int V(x)\Psi(x, \tau) \, dx, \tag{18.43}$$

or

$$E_0 = \frac{\int V(x)\Psi(x, \tau) \, dx}{\int \Psi(x, \tau) \, dx}. \tag{18.44}$$

The desired result (18.37) follows by making the connection between $\Psi(x) \, dx$ and the density of walkers at the point x.

Although the derivation of (18.37) is somewhat involved, the random walk algorithm is straightforward. One possible algorithm is as follows:

1. Place N_0 walkers at the initial set of positions x_i, where the x_i need not be on a grid.

2. Compute the reference energy, $V_{\text{ref}} = \sum_i V_i/N_0$.

3. Randomly move a walker to the right or left by a fixed step length Δs. The step length Δs is related to the time step $\Delta \tau$ by $(\Delta s)^2 = 2D\Delta \tau$. ($D = \frac{1}{2}$ in units such that $\hbar = m = 1$.)

4. Compute $\Delta V = [V(x) - V_{\text{ref}}]$ and a random number r in the unit interval. If $\Delta V > 0$ and $r < \Delta V \Delta \tau$, then remove the walker. If $\Delta V < 0$ and $r < -\Delta V \Delta \tau$, then

add another walker at x. Otherwise, just leave the walker at x. This procedure is accurate only in the limit of $\Delta\tau << 1$.

5. Repeat steps 3 and 4 for each of the N_0 walkers and compute the mean potential energy (18.44) and the actual number of random walkers. The new reference potential is given by

$$V_{\text{ref}} = <V> - \frac{a}{N_0\Delta\tau}(N - N_0),\qquad\qquad(18.45)$$

where N is the new number of random walkers and $<V>$ is their mean potential energy. The average of V is an estimate of the ground state energy. The parameter a is adjusted so that the number of random walkers N remains approximately constant.

6. Repeat steps 3–5 until the estimates of the ground state energy $<V>$ have reached a steady state value with only random fluctuations. Average $<V>$ over many Monte Carlo steps to compute the ground state energy. Do a similar calculation to estimate the distribution of random walkers.

Program qmwalk implements this algorithm for the harmonic oscillator potential. Initially, the walkers are randomly distributed within a distance w0 of the origin. Input parameters are the desired number of walkers N_0, the number of Monte Carlo steps per walker mcs, and the step size ds. The program computes the current number of walkers, the current estimate of the ground state energy, and the value of V_{ref}. The first ten percent of the samples are discarded in the averages to approximate equilibration.

```
PROGRAM qmwalk
DIM x(2000),psi(-500 to 500)
CALL initial(x(),N,NO,Vref,binsize,ds,dt,mcs,nequil,Esum)
FOR i = 1 to nequil            ! equilibration
    CALL walk(x(),N,NO,Vref,Vave,ds,dt)
NEXT i
FOR imcs = 1 to mcs
    CALL walk(x(),N,NO,Vref,Vave,ds,dt)
    CALL data(x(),psi(),N,Vref,Vave,Esum,imcs,binsize)
NEXT imcs
CALL plot_psi(psi(),binsize)
END

SUB initial(x(),N,NO,Vref,binsize,ds,dt,mcs,nequil,Esum)
    DECLARE DEF V
    RANDOMIZE
    INPUT prompt "desired number of walkers = ": NO
    INPUT prompt "step length = ": ds
    LET dt = ds*ds               ! time step
    INPUT prompt "number of Monte Carlo steps per walker = ": mcs
    LET nequil = round(0.4*mcs)
    ! choose initial number of walkers equal to desired number
    LET N = NO
    ! choose initial positions of walkers at random
```

```
                 LET width = 1                  ! initial width of region for walkers
                 LET Vref = 0
                 FOR i = 1 to N
                     LET x(i) = (2*rnd - 1)*width
                     LET Vref = Vref + V(x(i))
                 NEXT i
                 LET Vref = Vref/N
                 LET binsize = 2*ds          ! binsize for psi array
                 LET Esum = 0
                 CLEAR
                 PRINT "MC steps","N","E","Vref"
                 PRINT
             END SUB

             SUB walk(x(),N,N0,Vref,Vave,ds,dt)
                 DECLARE DEF V
                 LET Nin = N                 ! # walkers at beginning of trial
                 LET Vsum = 0
                 FOR i = Nin to 1 step -1
                     IF rnd < 0.5 then
                         LET x(i) = x(i) + ds
                     ELSE
                         LET x(i) = x(i) - ds
                     END IF
                     LET potential = V(x(i))    ! potential at x
                     LET dV = potential - Vref
                     IF dV < 0 then        ! check to add walker
                         IF rnd < -dV*dt then
                             LET N = N + 1
                             LET x(N) = x(i)       ! new walker
                             ! factor of 2 since two walkers at x
                             LET Vsum = Vsum + 2*potential
                         ELSE
                             LET Vsum = Vsum + potential    ! only do old walker
                         END IF
                     ELSE
                         IF rnd < dV*dt then     ! check to remove walker
                             LET x(i) = x(N)
                             LET N = N - 1
                         ELSE
                             LET Vsum = Vsum + potential
                         END IF
                     END IF
                 NEXT i
                 LET Vave = Vsum/N            ! mean potential
                 LET Vref = Vave - (N - N0)/(N0*dt)       ! new reference energy
             END SUB
```

```
SUB data(x(),psi(),N,Vref,Vave,Esum,imcs,binsize)
    ! accumulate data
    LET Esum = Esum + Vave
    ! bin walkers
    FOR i = 1 to N
        LET bin = round(x(i)/binsize)
        LET psi(bin) = psi(bin) + 1
    NEXT i
    IF mod(imcs,10) = 0 then
        PRINT imcs,
        PRINT using "####": N;
        PRINT,
        PRINT using "--.###": Esum/imcs;
        PRINT,
        PRINT using "--.###": Vref
    END IF
END SUB

SUB plot_psi(psi(),binsize)
    PRINT "Hit any key to see |psi|"
    DO
    LOOP until key input
    GET KEY k
    CLEAR
    LET i = 0
    LET pmax = psi(i)*psi(i)
    LET sum = pmax
    DO
        LET i = i + 1
        LET P = psi(i)*psi(i) + psi(-i)*psi(-i)
        IF P > pmax then LET pmax = P
        LET sum = sum + P
    LOOP until P = 0
    LET imax = i
    LET norm = sqr(sum*binsize)
    LET ymax = sqr(pmax)/norm
    SET WINDOW -imax,imax,-0.1*ymax,1.1*ymax
    PLOT LINES: -imax,0;imax,0
    FOR i = -imax to imax
        PLOT LINES: i-1,psi(i-1)/norm;i,psi(i)/norm
    NEXT i
END SUB

DEF V(x)
    LET V = 0.5*x*x
END DEF
```

Problem 18.12   Ground state of the harmonic and anharmonic oscillators

a. Use *Program qmwalk* to estimate the ground state energy E_0 and the corresponding eigenfunction for $V(x) = \frac{1}{2}x^2$. Choose as the desired number of walkers $N_0 = 50$, the step length ds $= 0.1$, and mcs ≥ 500. Place the walkers at random within the range $-1 \leq x \leq 1$. Compare your Monte Carlo estimate for E_0 to the exact result $E_0 = 0.5$.

b. Increase mcs by a factor of at least ten. How much improvement does this choice make for the estimate of E_0? How many Monte Carlo steps per walker are needed for 1% accuracy in E_0? Plot the spatial distribution of the random walkers and compare it to the exact result for the ground state wave function.

c. Obtain a numerical solution of the anharmonic oscillator with

$$V(x) = \frac{1}{2}x^2 + bx^3. \tag{18.46}$$

Consider $b = 0.1, 0.2$, and 0.5. A calculation of the effect of the x^3 term is necessary for the study of the anharmonicity of the vibrations of a physical system, e.g., the vibrational spectrum of diatomic molecules.

Problem 18.13   Ground state of a square well

a. Modify *Program qmwalk* to find the ground state energy and wave function for the square well potential (18.14) with $a = 1$. Choose $V_0 = 5$, $N_0 = 100$, ds $= 0.1$, and mcs $= 300$. Place the walkers at random within the range $-1.5 \leq x \leq 1.5$.

b. Increase V_0 and find the ground state energy as a function of V_0. Use your results to estimate the limiting value of the ground state energy for $V_0 \to \infty$.

Problem 18.14   Ground state of a cylindrical box

Compute the ground state energy and wave function of a two-dimensional circular potential

$$V(r) = \begin{cases} 0 & r \leq 1 \\ -V_0, & r > 1 \end{cases} \tag{18.47}$$

where $r^2 = x^2 + y^2$. Modify *Program qmwalk* by using Cartesian coordinates in two dimensions, e.g., add an array to store the positions of the y coordinates of the walkers. What happens if you begin with an initial distribution of walkers that is not cylindrically symmetric?

18.7   DIFFUSION QUANTUM MONTE CARLO

We now discuss an improvement of the random walk algorithm known as *diffusion quantum Monte Carlo*. Although some parts of the discussion might be difficult to follow initially, the algorithm is straightforward. Your understanding of the method will be enhanced by writing a program to implement the algorithm and then reading the following derivation again.

To understand the method, we introduce the concept of a Green's function or propagator defined by

$$\Psi(x', \tau) = \int G(x', x, \tau) \Psi(x, 0) \, dx. \tag{18.48}$$

From the form of (18.48) we see that $G(x', x, \tau)$ "propagates" the wave function from time zero to time τ. If we operate on both sides of (18.48) with first $(\partial/\partial \tau)$ and then with $(H_{op} - V_{ref})$, we can verify that G satisfies the equation

$$\frac{\partial G}{\partial \tau} = -(H_{op} - V_{ref})G, \tag{18.49}$$

which is the same form as the imaginary time Schrödinger equation (18.38). It is easy to verify that $G(x', x, \tau) = G(x, x', \tau)$. A formal solution of (18.49) is

$$G(\tau) = e^{-(H_{op} - V_{ref})\tau}, \tag{18.50}$$

where the meaning of the exponential of an operator is given by its Taylor series expansion.

The difficulty with (18.50) is that the kinetic and potential energy operators T_{op} and V_{op} in H_{op} do not commute. For this reason, if we want to factor (18.50), we can only approximate the exponential for short times $\Delta\tau$. To first order in $\Delta\tau$, we have

$$G(\Delta\tau) \approx G_{branch/2} \, G_{diff} \, G_{branch/2}$$
$$\equiv e^{-\frac{1}{2}(V_{op} - V_{ref})\Delta\tau} e^{-T_{op}\Delta\tau} e^{-\frac{1}{2}(V_{op} - V_{ref})\Delta\tau}, \tag{18.51}$$

where $G_{diff} \equiv e^{-T_{op}\Delta\tau}$ and $G_{branch/2} \equiv e^{-\frac{1}{2}(V_{op} - V_{ref})\Delta\tau}$ correspond to the two random processes: diffusion and branching. G_{diff} and $G_{branch/2}$ satisfy respectively the differential equations:

$$\frac{\partial G_{diff}}{\partial \tau} = -T_{op} G_{diff} = \frac{\hbar^2}{2m} \frac{\partial^2 G_{diff}}{\partial x^2} \tag{18.52}$$

and

$$\frac{\partial G_{branch/2}}{\partial \tau} = (V_{ref} - V_{op}) G_{branch/2}. \tag{18.53}$$

The solutions to (18.51)–(18.53) are

$$G_{diff}(x', x, \Delta\tau) = (4\pi D\Delta\tau)^{-1/2} e^{-(x'-x)^2/4D\Delta\tau} \tag{18.54}$$

with $D \equiv \hbar^2/2m$, and

$$G_{branch}(x', x, \Delta\tau) = e^{-\left(\frac{1}{2}[V(x)+V(x')] - V_{ref}\right)\Delta\tau}. \tag{18.55}$$

The solution (18.55) assumes that half the branching occurs before the diffusion and half afterwards as indicated in (18.51). This assumption yields a solution that is symmetric in x and x'.

From the form of (18.54) and (18.55), we can see that the diffusion quantum Monte Carlo method is similar to the random walk algorithm discussed in Section 18.6. An implementation of the diffusion quantum Monte Carlo method in one dimension can be summarized as follows:

1. Begin with a set of N_0 random walkers. There is no lattice so the positions of the walkers are continuous. It is advantageous to choose the walkers so that they are in regions of space where the wave function is known to be large.

2. Choose one of the walkers and displace it from x to x'. The new position is chosen from a Gaussian distribution with a variance $2D\Delta\tau$ and zero mean. This change corresponds to the diffusion process given by (18.54).

3. Weight the configuration x' by

$$w(x \to x', \Delta\tau) = e^{-\left(\frac{1}{2}[V(x)+V(x')]-V_{\text{ref}}\right)\Delta\tau}. \tag{18.56}$$

One way to do this weighting is to generate duplicate random walkers at x'. For example, if $w \approx 2$, we would have two walkers at x' where previously there had been one. To implement this weighting (branching) correctly, we must make an integer number of copies that is equal on the average to the number w. A simple way to do so is to take the integer part of $w + r$, where r is a uniform random number in the unit interval. The number of copies can be any nonnegative integer including zero. The latter corresponds to a termination of a walker.

4. Repeat steps 2 and 3 for all members of the ensemble, thereby creating a new ensemble at a later time $\Delta\tau$. One iteration of the ensemble is equivalent to performing the integration

$$\Psi(x', \tau) = \int G(x', x, \Delta\tau) \, \Psi(x, \tau - \Delta\tau) \, dx$$

5. The quantity of interest $\Psi(x', \tau)$ will be independent of the original ensemble $\Psi(x, 0)$ if a sufficient number of Monte Carlo steps are taken. As before, we must ensure that $N(\tau)$, the number of configurations at time τ, is kept close to the desired number N_0.

Now we can understand how the random walk algorithm discussed in Section 18.6 is an approximation to the diffusion quantum MC algorithm. First, the Gaussian distribution gives the exact distribution for the displacement of a random walker in a time $\Delta\tau$, in contrast to fixed step size in the previous algorithm which gives the average displacement of a walker. Hence, there are no systematic errors due to a finite step size. Second, if we expand the exponential in (18.55) to first-order in $\Delta\tau$ and set $V(x) = V(x')$, we obtain the branching rule used previously. (We use the fact that the uniform distribution r is the same as the distribution $1 - r$.) The diffusion quantum MC algorithm is not exact because the branching is independent of the position reached by diffusion, which is only true in the limit $\Delta\tau \to 0$. This limitation is remedied in the Green's Function Monte Carlo method where a short time approximation is not made.

One problem with the random walk methods we have discussed is that they can become very inefficient. This inefficiency is due in part to the branching process. If the potential becomes large and negative (as it is for the Coulomb potential when an electron approaches a nucleus), the number of copies of a walker will become very large. It is possible to improve the efficiency of these algorithms by introducing an importance sampling method. The idea is to use an initial guess $\Psi_T(x)$ for the wave function to guide the walkers toward the more important regions of $V(x)$. To implement this idea, we introduce the function $f(x, \tau) = \Psi(x, \tau)\Psi_T(x)$. If we calculate the quantity $\partial f/\partial t - D\partial^2 f/\partial x^2$, and use (18.38), we can show that $f(x, \tau)$ satisfies the differential equation:

$$\frac{\partial f}{\partial \tau} = D\frac{\partial^2 f}{\partial x^2} - D\frac{\partial [f F(x)]}{\partial x} - [E_L(x) - V_{\text{ref}}]f, \qquad (18.57)$$

where

$$F(x) = \frac{2}{\Psi_T}\frac{\partial \Psi_T}{\partial x}, \qquad (18.58)$$

and the local energy $E_L(x)$ is given by

$$E_L(x) = \frac{H_{\text{op}}\Psi_T}{\Psi_T} = V(x) - \frac{D}{\Psi_T}\partial^2\Psi_T/\partial x^2. \qquad (18.59)$$

The term in (18.57) containing F corresponds to a drift in the walkers away from regions where $|\Psi_T|^2$ is small (see Problem 12.18).

To incorporate the drift term into G_{diff}, we replace $(x' - x)^2$ in (18.54) by the term $(x' - x - D\Delta\tau F(x))^2$, so that the diffusion propagator becomes

$$G_{\text{diff}}(x', x, \Delta\tau) = (4\pi D\Delta\tau)^{-1/2}e^{-(x'-x-D\Delta\tau F(x))^2/4D\Delta\tau}. \qquad (18.60)$$

However, this replacement destroys the symmetry between x and x'. To restore it, we use the Metropolis algorithm for accepting the new position of a walker. The acceptance probability p is given by

$$p = \frac{|\Psi_T(x')|^2 G_{\text{diff}}(x, x', \Delta\tau)}{|\Psi_T(x)|^2 G_{\text{diff}}(x', x, \Delta\tau)}. \qquad (18.61)$$

If $p > 1$, we accept the move; otherwise, we accept the move if $r \le p$. The branching step is achieved by using (18.55) with $V(x) + V(x')$ replaced by $E_L(x) + E_L(x')$, and $\Delta\tau$ replaced by an effective time step. The reason for the use of an effective time step in (18.55) is that some diffusion steps are rejected. The effective time step to be used in (18.55) is found by multiplying $\Delta\tau$ by the average acceptance probability. It can be shown (see Hammond et al.) that the mean value of the local energy is an unbiased estimator of the ground state energy.

Another possible improvement is to periodically replace branching (which changes the number of walkers) with a weighting of the walkers. At each weighting step, each walker is weighted by G_{branch}, and the total number of walkers remains constant. After n steps, the kth walker receives a weight $W_k = \Pi_{i=1}^n G_{\text{branch}}^{(i,k)}$, where $G_{\text{branch}}^{(i,k)}$ is the branching factor of the kth walker at the ith time step. The contribution to any average quantity of the kth walker is weighted by W_k.

Problem 18.15 Diffusion Quantum Monte Carlo

a. Modify *Program qmwalk* to implement the diffusion quantum Monte Carlo method for the systems considered in Problem 18.12 or 18.13. Begin with $N_0 = 100$ walkers and $\Delta\tau = 0.01$. Use at least three values of $\Delta\tau$ and extrapolate your results to $\Delta\tau \to 0$. Reasonable results can be obtained by adjusting the reference energy every 20 Monte Carlo steps with $a = 0.1$.

* b. Write a program to apply the diffusion quantum Monte Carlo method to the hydrogen atom. In this case a configuration is represented by three coordinates.

* c. Modify your program to include weights in addition to changing walker populations. Redo part (a) and compare your results.

* Problem 18.16 Importance sampling

a. Derive the partial differential equation (18.57) for $f(x, \tau)$.

b. Modify *Program qmwalk* to implement the diffusion quantum Monte Carlo method with importance sampling. Consider the harmonic oscillator problem with the trial wave function $\Psi_T = e^{-\lambda x^2}$. Compute the statistical error associated with the ground state energy as a function of λ. How much variance reduction can you achieve relative to the naive diffusion quantum Monte Carlo method? Then consider another form of Ψ_T that does not have a form identical to the exact ground state. Try the hydrogen atom with $\Psi_T = e^{-\lambda r}$.

18.8 PATH INTEGRAL QUANTUM MONTE CARLO

The Monte Carlo methods we have discussed so far are primarily useful for the ground state, although with some effort it also is possible to find the first few excited states. In this section we discuss a Monte Carlo method that is of particular interest for computing the thermal properties of quantum systems.

We recall (see Section 7.6) that classical mechanics can be formulated in terms of the principle of least action, that is, given two points in space-time, a classical particle chooses the path that minimizes the action given by

$$S = \int_{x_0,0}^{x,t} L \, dt. \tag{18.62}$$

The Lagrangian L is given by $L = T - V$. Quantum mechanics also can be formulated in terms of the action (cf. Feynman and Hibbs). The result of this *path integral* formalism is that the real-time propagator G can be expressed as

$$G(x, x_0, t) = A \sum_{\text{paths}} e^{iS/\hbar}, \tag{18.63}$$

where A is a normalization factor. The sum in (18.63) is over all paths between $(x_0, 0)$ and (x, t), not just the path that minimizes the classical action. The presence of the imaginary number i in (18.63) leads to interference effects. As before, the propagator $G(x, x_0, t)$ can be interpreted as the probability amplitude for a particle to be at x at time t given that it was at x_0 at time zero. G satisfies the equation (see (18.48))

$$\Psi(x, t) = \int G(x, x_0, t)\Psi(x_0, 0)\, dx_0, \qquad (t > 0) \tag{18.64}$$

Because G satisfies the same differential equation as Ψ in both x and x_0, G can be expressed as

$$G(x, x_0, t) = \sum_n \phi_n(x)\phi_n(x_0)e^{-iE_n t/\hbar}, \tag{18.65}$$

where the ϕ_n are the eigenfunctions of H_{op}. For simplicity, we set $\hbar = 1$ in the following. As before, we assume that (18.65) applies for imaginary values of t, and we write

$$G(x, x_0, \tau) = \sum_n \phi_n(x)\phi_n(x_0)\, e^{-\tau E_n}. \tag{18.66}$$

First we consider the ground state. In the limit $\tau \to \infty$, we have

$$G(x_0, x_0, \tau) \to \phi_0(x_0)^2\, e^{-\tau E_0}. \qquad (\tau \to \infty) \tag{18.67}$$

From the form of (18.67) and (18.63), we see that we need to compute G and hence S to compute the properties of the ground state.

To compute S, we convert the integral in (18.62) to a sum. If we use imaginary time, the Lagrangian for a single particle of unit mass becomes

$$L = -\frac{1}{2}\left(\frac{dx}{d\tau}\right)^2 - V(x) = -E. \tag{18.68}$$

We divide the imaginary time interval τ into N equal steps of size $\Delta\tau$ and write E as

$$E(x_j, \tau_j) = \frac{1}{2}\frac{(x_{j+1} - x_j)^2}{(\Delta\tau)^2} + V(x_j), \tag{18.69}$$

where $\tau_j = j\Delta\tau$, and x_j is the corresponding displacement. If we use the rectangular approximation, the action can be written as

$$S = -i\Delta\tau \sum_{j=0}^{N-1} E(x_j, \tau_j) = -i\Delta\tau\left[\sum_{j=0}^{N-1}\frac{1}{2}\frac{(x_{j+1} - x_j)^2}{(\Delta\tau)^2} + V(x_j)\right], \tag{18.70}$$

and the probability amplitude for the path becomes

$$e^{iS} = e^{\Delta\tau\left[\sum_{j=0}^{N-1} \frac{1}{2}(x_{j+1}-x_j)^2/(\Delta\tau)^2 + V(x_j)\right]}. \tag{18.71}$$

Hence, the propagator $G(x, x_0, N\Delta\tau)$ can be expressed as

$$G(x, x_0, N\Delta\tau) = A \int dx_1 \cdots dx_{N-1} e^{\Delta\tau\left[\sum_{j=0}^{N-1} \frac{1}{2}(x_{j+1}-x_j)^2/(\Delta\tau)^2 + V(x_j)\right]}, \tag{18.72}$$

where $x \equiv x_N$, and A is an unimportant constant.

From (18.72) we see that $G(x, x_0, N\Delta\tau)$ has been expressed as a multidimensional integral with the displacement variable x_j associated with the time τ_j. The sequence $x_0, x_1, \cdots, x_N$ is a possible path and the integral in (18.72) is over all paths. Because the quantity of interest is $G(x_0, x_0, N\Delta\tau)$ (see (18.67)), we adopt the periodic boundary condition, $x_N = x_0$. The choice of x_0 in the argument of G is arbitrary for finding the ground state energy, and the use of the periodic boundary conditions implies that no point in the closed path is unique. It is thus possible (and convenient) to rewrite (18.72) by letting the sum over j go from 1 to N:

$$G(x_0, x_0, N\Delta\tau) = A \int dx_1 \cdots dx_{N-1} e^{-\Delta\tau\left[\sum_{j=1}^{N} \frac{1}{2}(x_j-x_{j-1})^2/(\Delta\tau)^2 + V(x_j)\right]}. \tag{18.73}$$

The result of the above analysis is to convert a quantum mechanical problem for a single particle into a statistical mechanics problem for N "atoms" on a ring connected by nearest neighbor "springs" with spring constant $1/(\Delta\tau)^2$. The label j denotes the order of the atoms in the ring; x_j is the displacement of atom j from its equilibrium position and not the position of the atom relative to a fixed origin.

Note that the form of (18.73) is similar to the form of the Boltzmann distribution for a single particle with $N\Delta\tau$ corresponding to the inverse temperature β. To see this relation, note that the partition function for a quantum mechanical particle contains terms of the form $e^{-\beta E_n}$, whereas (18.66) contains terms proportional to $e^{-\tau E_n}$. Hence $\beta = \tau = N\Delta\tau$. We shall see in the following how we can use this identity to simulate a quantum system at finite temperature.

We can use the Metropolis algorithm to simulate the motion of N "atoms" on a ring. Of course, these atoms are a product of our analysis just as were the random walkers we introduced in diffusion Monte Carlo and should not be confused with real particles. A summary of a possible "path integral" algorithm is as follows:

1. Choose N and $\Delta\tau$ such that $N\Delta\tau \gg 1$ (the zero temperature limit). Also choose δ, the maximum trial change in the displacement of an atom, and mcs, the total number of Monte Carlo steps per atom.

2. Choose an initial configuration for the displacements x_j which is close to the approximate shape of the ground state probability amplitude.

3. Choose an atom j at random and a trial displacement $x_{\text{trial}} \to x_j + (2r - 1)\delta$, where r is a random number uniformly distributed on the unit interval. Compute the change ΔE in the energy E, where ΔE is given by

$$\Delta E = \frac{1}{2}\left(\frac{x_{j+1} - x_{\text{trial}}}{\Delta\tau}\right)^2 + \frac{1}{2}\left(\frac{x_{\text{trial}} - x_{j-1}}{\Delta\tau}\right)^2 + V(x_{\text{trial}})$$
$$- \frac{1}{2}\left(\frac{x_{j+1} - x_j}{\Delta\tau}\right)^2 - \frac{1}{2}\left(\frac{x_j - x_{j-1}}{\Delta\tau}\right)^2 - V(x_j)$$

(18.74)

If $\Delta E < 0$, accept the change; otherwise, compute the probability $p = e^{-\Delta\tau\Delta E}$ and a random number r in the unit interval. If $r \leq p$, then accept the move; otherwise reject the trial move.

4. Update the probability density array element $P(x)$, that is, let $P(x = x_j) \to P(x = x_j) + 1$, where x is the displacement of the atom chosen in step 3 after step 3 is completed. Do this update even if the trial move was rejected. Divide the possible x values into equal size bins of width Δx.

5. Repeat steps 3 and 4 until a sufficient number of Monte Carlo steps per atom has been obtained. (Do not take data until the memory of the initial path is lost and the system has reached "equilibrium.")

Normalize the probability density $P(x)$ by dividing by the product of N and mcs. The ground state energy E_0 is given by

$$E_O = \sum_x P(x)[T(x) + V(x)],$$

(18.75)

where $T(x)$ is the kinetic energy as determined from the virial theorem:

$$2T(x) = x\frac{dV}{dx}.$$

(18.76)

It also is possible to compute T from averages over $(x_j - x_{j-1})^2$, but the virial theorem yields a smaller variance. The ground state wave function $\phi(x)$ is obtained from the normalized probability $P(x)\Delta x$ by dividing by Δx and taking the square root.

We also can find the thermodynamic properties of a particle that is connected to a heat bath at temperature $T = 1/\beta$ by not taking the $\beta = N\Delta\tau \to \infty$ limit. To obtain the ground state, which corresponds to the zero temperature limit ($\beta >> 1$), we had to make $N\Delta\tau$ as large as possible. However, we need $\Delta\tau$ to be as small as possible to approximate the continuum time limit. Hence, to obtain the ground state we need a large number of time intervals N. For the finite temperature simulation, we can use smaller values of N for the same level of accuracy as the zero temperature simulation.

The path integral method is very flexible and can be generalized to higher dimensions and many mutually interacting particles. For three dimensions, x_j is replaced by the three-dimensional displacement $\mathbf{r}_j$. Each real particle is represented by a ring of N "atoms" with a spring-like potential connecting each atom within a ring.

Problem 18.17   Path integral calculation

a. Write a program to implement the path integral algorithm for the one-dimensional harmonic oscillator. Use the structure of your Monte Carlo Lennard-Jones program from Chapter 17 as a guide. Choose $V(x) = \frac{1}{2}x^2$.

b. Let $N\Delta\tau = 15$ and consider $N = 10, 20, 40$, and 80. Equilibrate for at least 2000 Monte Carlo steps per atom and average over at least 5000 MCS. Compare your results with the exact result for the ground state energy given by $E_0 = 0.5$. Estimate the equilibration time for your calculation. What is a good initial configuration? Improve your results by using larger values of $N\Delta\tau$.

c. Find the mean energy, $<E>$, of the harmonic oscillator at temperature T given by setting $\beta = N\Delta\tau$. Find $<E>$ for $\beta = 1, 2$, and 3, and compare it with the exact result $<E> = \frac{1}{2}\coth(\beta/2)$.

d. Repeat the above calculations for the Morse potential $V(x) = 2(1 - e^{-x})^2$.

References and Suggestions for Further Reading

E. E. Anderson, *Modern Physics and Quantum Mechanics*, W. B. Saunders (1971). An undergraduate text that includes examples of wave packet motion.

J. B. Anderson, "A random walk simulation of the Schrödinger equation: H_3^+," *J. Chem. Phys.* **63**, 1499 (1975); "Quantum chemistry by random walk. H ^{2}P, H_3^+ D_{3h} $^1A'_1$, H_2 $^3\Sigma_u^+$, H_4 $^1\Sigma_g^+$, Be ^{1}S," *J. Chem. Phys.* **65**, 4121 (1976); "Quantum chemistry by random walk: Higher accuracy," *J. Chem. Phys.* **73**, 3897 (1980). These papers describe the random walk method, extensions for improved accuracy, and applications to simple molecules.

G. Baym, *Lectures on Quantum Mechanics*, W. A. Benjamin (1973). A discussion of the Schrödinger equation in imaginary time is given in Chapter 3.

H. A. Bethe, *Intermediate Quantum Mechanics*, W. A. Benjamin (1964). Applications of quantum mechanics to atomic systems are discussed.

Jay S. Bolemon, "Computer solutions to a realistic 'one-dimensional' Schrödinger equation," *Am. J. Phys.* **40**, 1511 (1972).

Siegmund Brandt and Hans Dieter Dahmen, *The Picture Book of Quantum Mechanics*, John Wiley & Sons (1985); *Quantum Mechanics on the Macintosh*, Springer-Verlag (1991). Many computer generated pictures of quantum wave functions in different contexts are shown.

David M. Ceperley and Berni J. Alder, "Quantum Monte Carlo," *Science* **231**, 555 (1986). A survey of some of the applications of quantum Monte Carlo methods to physics and chemistry.

D. F. Coker and R. O. Watts, "Quantum simulation of systems with nodal surfaces," *Mol. Phys.* **58**, 1112 (1986).

Robert M. Eisberg, *Applied Mathematical Physics with Programmable Pocket Calculators*, McGraw-Hill (1976). Chapter 8 discusses a direct numerical solution of the one-dimensional Schrödinger equation.

Robert M. Eisberg and Robert Resnick, *Quantum Physics*, second edition, John Wiley & Sons (1985). See Appendix G for a discussion of the numerical solution of Schrödinger's equation.

R. P. Feynman, "Simulating physics with computers," *Int. J. Theor. Phys.* **21**, 467 (1982). A provocative discussion of the intrinsic difficulties of simulating quantum systems.

Richard P. Feynman and A. R. Hibbs, *Quantum Mechanics and Path Integrals*, McGraw-Hill (1965).

B. L. Hammond, W. A. Lester Jr., and P. J. Reynolds, *Monte Carlo Methods in Ab Initio Quantum Chemistry*, World Scientific (1994). An excellent book on quantum Monte Carlo methods.

J. P. Killingbeck, *Microcomputer Quantum Mechanics*, second edition, Adam Hilger (1985). A book on numerical methods as well as quantum mechanics.

Steven E. Koonin and Dawn C. Meredith, *Computational Physics*, Addison-Wesley (1990). Solutions of the time-dependent Schrödinger equation are discussed in the context of parabolic partial differential equations in Chapter 7. Chapter 8 discusses Green's function Monte Carlo methods.

M. A. Lee and K. E. Schmidt, "Green's function Monte Carlo," *Computers in Physics* **6**(2), 192 (1992). A short and clear explanation of GFMC.

P. K. MacKeown, "Evaluation of Feynman path integrals by Monte Carlo methods," *Am. J. Phys.* **53**, 880 (1985). The author discusses projects suitable for an advanced undergraduate course. See also, P. K. MacKeown and D. J. Newman, *Computational Techniques in Physics*, Adam Hilger (1987).

Jean Potvin, "Computational quantum field theory. Part II: Lattice gauge theory," *Computers in Physics* **8**, 170 (1994) and "Computational Quantum Field Theory," *Computers in Physics* **7**, 149 (1993).

William H. Press, Saul A. Teukolsky, William T. Vetterling, and Brian P. Flannery, *Numerical Recipes*, second edition, Cambridge University Press (1992). The numerical solution of the time-dependent Schrödinger equation is discussed in Chapter 19.

Peter J. Reynolds, David M. Ceperley, Berni J. Alder, and William A. Lester Jr., "Fixed-node quantum Monte Carlo for molecules," *J. Chem. Phys.* **77**, 5593 (1982). This paper describes a random walk algorithm for use in molecular applications including importance sampling and the treatment of Fermi statistics.

P. J. Reynolds, J. Tobochnik, and H. Gould, "Diffusion quantum Monte Carlo," *Computers in Physics* **4**(6), 882 (1990). A short and clear explanation of DQMC.

D. Saxon, *Elementary Quantum Mechanics*, Holden-Day (1968). Numerical solutions of Schrödinger's equation are discussed in Chapter VI.

J. Tobochnik, H. Gould, and K. Mulder, "An introduction to quantum Monte Carlo," *Computers in Physics* **4**(4), 431 (1990). An explanation of the path integral method applied to one particle.

P. B. Visscher, "A fast explicit algorithm for the time-dependent Schrödinger equation," *Computers in Physics* **5**(6), 596 (1991).

19

Epilogue: The Same Algorithms Give the Same Results

We emphasize that the methods we have discussed can be applied to a wide variety of natural phenomena.

19.1   THE UNITY OF PHYSICS

Although we have discussed many topics and applications, we have covered only a small fraction of the possible computer simulations and models of natural phenomena. However, we know that the same physical principles can be applied to many kinds of phenomena. We express this point of view as *the same algorithms give the same results*. For example, the Monte Carlo methods that we applied to the simulation of classical liquids and to the analysis of quantum mechanical wave functions also were applied to the transport of neutrons and problems in chemical kinetics. Similar Monte Carlo methods are being used to analyze problems in quark confinement. Indeed, the increasing role of the computer in research is strengthening the interconnections of the various subfields of physics and the relation of physics to other disciplines.

The computer also has helped us think of natural phenomena in new ways that complement traditional methods. For example, consider a predator-prey model of the dynamics of minnows and sharks. Assume that the birth rate of the minnows is independent of the number of sharks, and that each shark kills a number of minnows proportional to their number. If we assume that $F(t)$, the number of minnows at time t, changes continuously, we can write

$$\frac{dF(t)}{dt} = [b_1 - d_1 S(t)]F(t), \tag{19.1}$$

where $S(t)$ is the number of sharks at time t, and b_1 and d_1 are constants independent of F and S. To obtain an equation for the rate of change of the sharks, we assume that the number of offspring produced by each shark is proportional to the number of minnows eaten by the shark. If we also assume that the death rate of the sharks is constant, we can write

$$\frac{dS(t)}{dt} = [b_2 F(t) - d_2]S(t). \tag{19.2}$$

Equations (19.1) and (19.2) are known as the Lotka-Volterra equations. They can be analyzed by standard methods and solved numerically using simple algorithms. Why is the dynamical behavior of (19.1) and (19.2) cyclic?

In the predator-prey model the numbers of predator and prey are assumed to change continuously and their spatial distribution is ignored. We now summarize an alternative model that can be most simply expressed as a computer algorithm. The model is a two-dimensional cellular automaton known as Wa-Tor.

1. For a desired concentration of minnows and sharks, minnows and sharks are placed at random on the sites of a lattice. The minnows and sharks are assigned random ages.

2. At time step t_n, consider each minnow sequentially. Determine the number of nearest neighbor sites that are unoccupied at time t_{n-1} and move the minnow at random to one of the unoccupied sites. If all the nearest neighbor sites are occupied, the minnow does not move.

3. If a minnow has survived for a multiple of `fbreed` iterations, the minnow has a single offspring. The new minnow is placed at the previous position of the parent minnow.

4. At time step t_n, consider each shark sequentially. If all the nearest neighbor sites of the shark at time t_{n-1} are unoccupied, move the shark at random to one of the four unoccupied sites. If one or more of the adjacent sites is occupied by a minnow, the shark moves at random to one of the occupied sites and eats the minnow.

5. If a shark moves `nstarve` times without eating, the shark dies. If a shark survives for a multiple of `sbreed` iterations, the shark has a single offspring. The new shark is placed at the previous position of the parent shark.

What is the dynamical behavior of Wa-Tor? Do Wa-Tor and the Lotka-Volterra equations exhibit similar behavior? Is the Wa-Tor model realistic? What are the advantages and disadvantages of each approach? See the references for suggestions for the numerical values of the parameters.

19.2 PERCOLATION AND GALAXIES

In addition to allowing us to investigate complex nonlinear problems and more realistic systems, the computer has reinforced one of the contemporary themes in physics, the unifying role of collective behavior. Systems composed of many individual constituents can exhibit common properties under certain conditions, even though there might be differences in the nature of the constituents and in their mutual interaction. The behavior of a system near a critical point is probably the best example of collective behavior in a familiar context. In the following, we discuss examples of collective behavior in the context of epidemiology and the structure of spiral galaxies. Our discussion follows closely the articles by Schulman and Seiden.

Consider an imaginary disease called *percolitis*. The disease conveys no immunity, and its incubation period and duration are both one day. The disease is so benign that its sufferers are able to come into contact with every member of the community so that every person comes into contact with every other person every day. At $t = 0$ one person contracts the disease from a source outside the community. Let N be the total population, t the time measured in days, p the transmission probability, and $n(t)$ the expected number of diseased individuals at time t. Convince yourself that for $N = 1000$ and $p = 0.0005$, the chance that there will be anyone suffering from the disease seven days later is vanishingly small. On the other hand, suppose that $p = 0.002$. Then $n(t = 1) = 2$, $n(t = 2) \approx 4$, and the odds are very high that after some time there will be an average number of approximately 800 victims. Do a simulation for various values of N and p and estimate the critical probability p_c such that for $p < p_c$ the average number of victims is zero, and for $p \geq p_c$ the average number of victims is nonzero. Note that no assumptions were made as to which individuals will become infected.

As its name suggests, the percolitis model has much in common with percolation. What are some of the connections? The model is sufficiently simple that an analytical solution is possible. What modifications of the model might make it more realistic for thinking about the spread of epidemics? Would you have imagined some of these modifications without thinking about how to put the problem on a computer? Do these modifications change the qualitative behavior of the model? Are analytical solutions possible in general?

The internal structure of a galaxy has traditionally been studied using Newtonian dynamics. This point of view is very useful, but can be complemented by thinking about the large scale structure of a galaxy using ideas from statistical mechanics. Because we can only briefly summarize this alternative point of view here, we encourage you to explore the properties of the percolation-based model of Schulman and Seiden by running *Program galaxy*, a simple version of their model. The basic assumption of the model is that even though a region of the galaxy might have the necessary ingredients for star formation, nothing happens if it is left alone. However, if a shock wave from a supernova passes through the gas, there is a good chance that a star will be formed. The supernova is itself the result of an earlier nearby star formation. The theory of *self-propagating star formation* is based on the importance of this mechanism. We can think of a given region of the galaxy as being like a percolitis-susceptible individual—without a source there is no percolitis. Rather than determining which regions have the necessary conditions for star formation, we summarize all the uncertainty and variability in a single parameter p, the probability that a supernova explosion in one region gives rise to star formation in a neighboring region.

The other important observation we need to make about spiral galaxies is that galaxies do not rotate rigidly (with a constant angular velocity), but to a good approximation rotate with a constant circular velocity. The properties of random self-propagating star formation and constant circular velocity are incorporated into *Program galaxy* as follows. Imagine a galaxy to be divided into concentric rings, which are divided into cells of equal size (see Fig. 19.1). Initially, a small number of cells are activated. Each cell corresponds to a region of space that is the size of a giant molecular cloud and moves with the same circular velocity v. The angular velocity is given by $\omega = v/r$, where r is the distance of the ring from the center of the galaxy. At each time step, the active cells activate neighboring cells with probability p, and then become inactive. Then the rings are rotated, and the process is repeated again in the next time step. At each time step, cells that have been active within the last 15 time steps are plotted with filled boxes, with the size of each box inversely proportional to the time since the cell become active. More details of the simulation are shown in Fig. 19.1 and in *Program galaxy*. A typical galaxy generated by *Program galaxy* is shown in Fig. 19.2.

```
PROGRAM galaxy
! model proposed by Schulman and Seiden
LIBRARY "csgraphics"
PUBLIC t,active_r(10000),active_a(10000),cell(0 to 50,300)
DECLARE PUBLIC dt
CALL parameter
CALL initial
DO
```

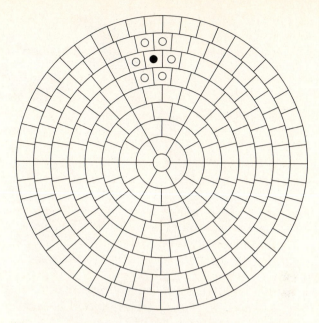

Fig. 19.1 The nature of the polar grid used in *Program galaxy*. Each cell has the same area and has on the average six nearest neighbors. The filled circle denotes an active region of star formation. At the next time step it can induce star formation in cells containing open circles. Note that as time passes, the neighbors in adjacent rings change because of differential rotation.

```
    CALL evolve
    CALL plot_spiral
    LET t = t + dt
LOOP UNTIL key input
END

SUB parameter
    PUBLIC nring,nactive,v,p,dt,s,two_pi
    LET nring = 50              ! number of rings
    LET nactive = 200          ! number of initial active cells
    LET v = 1                  ! circular velocity
    LET p = 0.18               ! star formation probability
    LET dt = 10                ! time step
    LET s = 2*pi/6             ! cell width
    LET two_pi = 2*pi
END SUB
```

Fig. 19.2   A typical structure generated by *Program galaxy*. The parameters
are the number of rings nring = 50, the initial number of active
cells nactive = 200, circular velocity $v = 1$ (200 km/s), the
probability of induced star formation $p = 0.18$, and the time step
dt = 10 (10^7 years). The structure shown is at $t = 200$ with 358
active star clusters of various ages. The diameter of the circle
representing an active star cluster is proportional to the remaining
lifetime of the cluster.

```
SUB initial
   DECLARE PUBLIC t,active_r(),active_a(),cell(,),nring,nactive
   RANDOMIZE
   MAT cell = 0
   ! randomly activate cells
   LET i = 0
   DO
     DO
         LET x = int(nring*rnd) + 1
         LET y = int(nring*rnd) + 1
         LET r = int(sqr(x*x + y*y)) + 1
```

```
                    LOOP until  r <= nring
                    LET a = int(6*r*rnd) + 1    ! array index corresponds to an angle
                    IF cell(r,a) = 0 then
                        LET i = i + 1
                        LET active_a(i) = a      ! location of active region
                        LET active_r(i) = r
                        ! activate region, stars live for 15 time steps
                        LET cell(r,a) = 15
                    END IF
                LOOP until i = nactive
                LET t = 0                        ! initial time
                CALL compute_aspect_ratio(nring,xwin,ywin)
                SET WINDOW -xwin,xwin,-ywin,ywin
                SET BACKGROUND COLOR "white"
                SET COLOR "black"
            END SUB

            SUB evolve
                DECLARE PUBLIC t,active_r(),active_a(),cell(,)
                DECLARE PUBLIC nring,nactive,v,p,s,two_pi
                DIM newactive_r(10000),newactive_a(10000)
                ! number of active star clusters for next time step
                LET newactive = 0
                FOR i = 1 to nactive
                    LET r = active_r(i)
                    LET a = active_a(i)
                    ! activate neighboring cells of same ring
                    CALL create(r,a+1,newactive,newactive_r(),newactive_a())
                    CALL create(r,a-1,newactive,newactive_r(),newactive_a())
                    LET angle = mod((a*s + v*t)/r, two_pi)
                    IF r < nring then          ! activate cells in next larger ring
                        LET wt = mod(v*t/(r+1),two_pi)
                        LET ap = int((angle - wt)*(r+1)/s)
                        LET ap = mod(ap,6*(r+1))
                        CALL create(r+1,ap,newactive,newactive_r(),newactive_a())
                        CALL create(r+1,ap+1,newactive,newactive_r(),newactive_a())
                    END IF
                    IF r > 1  then             ! activate cells in next smaller ring
                        LET wt = mod(v*t/(r-1),two_pi)
                        LET am = int((angle - wt)*(r-1)/s)
                        LET am = mod(am,6*(r-1))
                        CALL create(r-1,am,newactive,newactive_r(),newactive_a())
                        CALL create(r-1,am+1,newactive,newactive_r(),newactive_a())
                    END IF
                NEXT i
                LET nactive = newactive
                MAT active_r = newactive_r
                MAT active_a = newactive_a
            END SUB
```

```
SUB create(r,a,newactive,newactive_r(),newactive_a())
   ! create star clusters
   DECLARE PUBLIC p, cell(,)
   IF a < 1 then LET a = a + 6*r
   IF rnd < p and cell(r,a) <> 15 then
      LET newactive = newactive + 1
      LET newactive_a(newactive) = a
      LET newactive_r(newactive) = r
      LET cell(r,a) = 15          ! activate cell
   END IF
END SUB


SUB plot_spiral
   DECLARE PUBLIC t,nring,nactive,v,s, cell(,)
   CLEAR
   SET CURSOR 1,1
   PRINT "number of active star clusters ="; nactive
   FOR r = 1 to nring
      FOR a = 1 to 6*r
         IF cell(r,a) > 0 then
            LET theta = (a*s + v*t)/r
            LET x = r*cos(theta)
            LET y = r*sin(theta)
            LET plotsize = cell(r,a)/30
            BOX AREA x-plotsize,x+plotsize,y-plotsize,y+plotsize
            ! reduce star clusters lifetime
            LET cell(r,a) = cell(r,a) - 1
         END IF
      NEXT a
   NEXT r
END SUB
```

Of course our brief discussion of galaxies is not meant to convince you that the mechanism proposed by Schulman and Seiden is correct. Rather our purpose is to show how an alternative point of view can suggest new approaches in different fields. As emphasized by the authors, the dramatic images produced by computer simulations of the galaxy model show unanticipated features that can be the impetus for the development of a deeper theoretical understanding.

19.3 NUMBERS, PRETTY PICTURES, AND INSIGHT

The power of physics comes in part from its ability to give numerical agreement between theory and experiment. However, numerical agreement has little significance, unless the process of obtaining that agreement leads to insight into the phenomena of interest. For example, it is possible to design elaborate epicycle models of planetary motion which yield numerical results that are consistent with observations. Nevertheless, we prefer the Copernican approach, not for its impressive numerical success, but because it provided insight and lead to further advances by Kepler and Newton.

Computer simulations raise similar questions. The numbers produced by simulations which are consistent with experimental observations, and the pictures that are suggestive of physical phenomena are not sufficient to establish the value of a simulation. As an example, let us briefly consider a simulation of river networks. You might have seen aerial photographs of the Earth's topography and noticed the fractal-like drainage patterns formed by many rivers. A variety of random walk models can be used to generate patterns that look remarkably like river networks and even share some of their statistical properties. In these models the path of a walker represents a river, and the branching and intersections of rivers are modeled by the intersection of the paths of many walkers. However, such models have limited utility because they provide little insight into why the rivers actually have the properties they do. They do not directly incorporate the important physical processes of erosion and sedimentation.

Recently, a model has been proposed that produces realistic looking patterns of river networks whose statistical properties are consistent with observations of real networks. Unlike the random walk models, the dynamics of this lattice model reflect actual physical processes. The model consists of first creating a terrain for the network, and then defining the network on the terrain. The model can be summarized as follows:

1. The initial terrain is assumed to have a constant slope m. Each site of the lattice is given an initial height, $h(x, y) = my$.

2. Precipitation is placed on a random site of the (square) lattice.

3. Water flows from this site to a nearest neighbor site with a probability proportional to $e^{E \Delta h}$, where Δh is the difference in height between the site and a neighbor, and E is a parameter. If $\Delta h < 0$, then the probability is equal to zero, and the flow will not return to the site previously visited if there is a nonzero probability of flowing to another site.

4. Step (3) is repeated until the water flows to the bottom of the lattice, $y = 0$.

5. Each lattice point that has been visited by the flowing water has its height reduced by a constant amount D. This process represents erosion.

6. Any site at which the height difference Δh with a neighbor exceeds a critical amount M is reduced in height by an amount $\Delta h / S$, where S is another parameter in the model. This process represents mud slides.

7. Steps (2)–(6) are repeated until you wish to analyze the resulting network. The river network is defined as follows. Every site in the lattice receives one unit of precipitation. Then water flows from a site to the nearest neighbor with the smallest height. Then the water flows to the neighbor of this new site with the smallest height. This flow continues until the flow reaches the bottom of the lattice. This process is repeated for each site, and the number of times that a site receives water is recorded. The river network is defined as the set of all sites that received at least R units of water, where R is another parameter of the model.

The adjustable parameters of the model can be related to measurable quantities, and the different steps of the algorithm correspond to real dynamical processes. Hence, we can gain insight into how the results change as we vary the parameters. It would be interesting to program the above model and see what happens. You also would probably understand the algorithm better by converting the algorithm to a working program. However, significant computing resources are needed to obtain quantitative data.

19.4   WHAT ARE COMPUTERS DOING TO PHYSICS?

There is probably no need to convince you that computers are changing the way we think about the physical world. The question, "How can I formulate this problem for a computer?" has lead to new insights into old problems and is allowing us to consider new ones. The problems of galaxy formation and the evolution of river networks are just two of the many examples from the current research literature.

What will be the effect of computers in physics education? The most common use of computers has been to assist students to understand topics that have been in the curriculum for many years. So far the computer has not qualitatively changed the way we learn nor the topics we study. What will happen when computers become as common as calculators? Will computer simulation and numerical analysis make analytical methods less important? Should calculus retain its traditional importance in the curriculum? Do we understand a natural phenomenon when we are able to construct a computer model that allows us to make predictions which agree with experiment? Is it necessary to obtain at least some analytical results? What do you think should be the role of computers in education? Now that you have reached the end of this text, we expect that you already have started asking your own questions.

Computers and the visual images produced by computer models can be very seductive. However, we need to remember that the goal of science is to understand Nature. Theory and experiment have been the traditional routes to this end, and computation is now a third and complementary route. Although we have stressed the importance of computation in this text, it is important also to stress its complementary role. We must not let the rapid advances of computer technology and the easy availability of information overshadow our ultimate goal of gaining more knowledge and a deeper understanding of natural phenomena.

References and Suggestions for Further Reading

It would be impossible to list even a small subset of references to areas of physics and related disciplines that we have not discussed. Also the development of algorithms and applications in areas we have discussed is evolving rapidly. Many references to other applications and current developments can be found in archival journals. The authors edit the computer simulations column in Computers in Physics and are always looking for manuscripts that bring recent developments in research or pedagogy to a wider audience. Another place to look for recent developments is the various electronic archives available on the Internet. We encourage readers of this text and others who are using computer simulations in related contexts to browse our home page (see the preface), where we hope that many developments will be listed and discussed.

Several references relevant to this chapter are listed in the following.

A. D. Abrahams, "Channel networks: a geomorphological perspective," *Water Resour. Res.* **20**, 161 (1984). Abrahams criticizes the random walk models of river networks.

R. M. Anderson, "Population ecology of infectious disease agents," in *Theoretical Ecology: Principles and Applications*, R. M. May, editor, Blackwell (1981).

Eric Bonabeau and Laurent Dagorn, "Possible universality in the size distribution of fish schools," *Phys. Rev.* E **51**, R5220 (1995). The authors apply a model originally developed for river networks to the size distribution of schools of fish.

Silvia Burés, David P. Landau, Alan M. Ferrenberg, and Franklin A. Pokorny, "Monte Carlo computer simulation in horticulture: a model for container media characterization," *HortScience*, **28**, 1074 (1993).

Michael Creutz, Laurence Jacobs, and Claudio Rebbi, "Monte Carlo computations in lattice gauge theory," *Phys. Repts.*, **95**, 201 (1983). See also Michael Creutz, *Quarks, Gluons and Lattices*, Cambridge University Press (1983).

A. K. Dewdney, "Computer Recreations," *Sci. Amer.* **251**(#12), 14 (1984). A discussion of the Wa-Tor model.

Zvonko Fazarinc, "A viewpoint on calculus," *Hewlett-Packard Journal* **38**, 38 (March, 1987).

Richard P. Feynman, Robert B. Leighton, and Matthew Sands, *Lectures on Physics, Vol. II*, Addison-Wesley (1964). Our chapter title has been adapted from the subtitle of Section 12.1, "The same equations have the same solutions."

Robert L. Leheny, "A simple model for river network evolution," preprint (1994) and references therein.

Lawrence S. Schulman and Philip E. Seiden, "Percolation and Galaxies," *Science* **233**, 425 (1986). A discussion of percolitis is also given. The authors note that for finite N the disease always dies out eventually, but the expected time for the eradication of the disease grows exponentially with N. Also see the review paper Philip E. Seiden and Lawrence S. Schulman, *Adv. Phys.* **39**, 1 (1990).

Dietrich Stauffer, editor, *Annual Reviews of Computational Physics I and II*, World Scientific (1995).

APPENDIX

A

From BASIC to FORTRAN

Most contemporary programming languages have their roots in the 1950's and 1960's when they were developed as labor saving alternatives to writing programs in native machine language. Fortran (FORmula TRANslation) was introduced in 1957 and remains the language of choice for most scientific programming. Fortran, like all other computer languages that are widely used, is not static and continues to evolve. At present, the version of Fortran that is most widely available is compatible with Fortran 77, a standard published by the American National Standards Institute (ANSI). Almost all versions of Fortran incorporate significant extensions of Fortran 77, many of which have been incorporated into the latest standard, Fortran 90. Fortran 90 is compatible with Fortran 77 and includes extensions that will be familiar to users of C and Pascal. Some of the most important features of Fortran 90 include more control structures, recursive subroutines, dynamic storage allocation and pointers, user defined data structures, and the ability to manipulate entire arrays.

The translation from one programming language to another is usually straightforward, particularly for the relatively simple programs in this text. We will adopt a compromise between Fortran 77 and Fortran 90 that is compatible with almost all current Fortran compilers and that allows True BASIC programs to be translated more easily. The exception is graphics because of the absence of machine-independent graphics standards for C, Fortran, and Pascal.

To emphasize the similarity of True BASIC and Fortran, we first give an example of a Fortran program:

```
        PROGRAM series
*       add the first 100 terms of a simple series
        IMPLICIT NONE
        DOUBLE PRECISION sum
        INTEGER n
        sum = 0.0D0
        DO n = 1,100
```

673

```
      sum = sum + 1.0D0/DBLE(n)**2
      WRITE(*,*) n,sum
   END DO
   END
```

This program is a translation of *Program series* (see Chapter 2). The major difference between True BASIC and Fortran is that Fortran distinguishes between numerical variables which are integers and those that are not (real variables). Lower and upper case letters are treated equivalently in True BASIC and Fortran, but for convenience, we will use upper case for words that are part of the language of Fortran and variables which denote quantities that are usually written in upper case.

In the following, we discuss how to translate a True BASIC program into Fortran, and describe some of the features of Fortran 90 that make the translation easier. Of course, our discussion of Fortran is only introductory in nature and the interested reader should consult a text on Fortran (see references). The main features of Fortran are summarized below:

- *Column format*. Fortran uses a line format as does True BASIC. (In contrast, C and Pascal use punctuation to indicate the end of a statement.) Because Fortran was originally developed when card readers were the dominant input device, Fortran 77 programs have an 80 character per line fixed format. Statement numbers that label statements are placed in columns 1–5, and the actual statement may appear only in columns 7–72. If column 6 contains an alphanumeric character, then this line is a continuation of the previous line. In Fortran 90 statements can appear in any column from 1–132, and a statement is continued by appending an ampersand (&) to the line to be continued.

- *Comment statements*. A comment statement begins with an asterisk (*) (or the character C) in column 1. Fortran 77 does not allow a comment on the same line after a statement, but most implementations of Fortran use the exclamation mark (!) in the same way as True BASIC.

- *Type declarations*. If a variable is not declared in Fortran, it is assigned as a real or integer variable depending on its first letter. A variable beginning with the letter i, j, k, l, m, or n is assumed to be an integer variable; a variable beginning with any other letter is assumed to be real. Fortran does not require the type of each variable to be declared explicitly, but it is good programming practice to do so (see Appendix 3C). The explicit typing of variables can be enforced by the IMPLICIT NONE statement (similar to the OPTION TYPO statement in True BASIC).

  Variables can be INTEGER, REAL, DOUBLE PRECISION, CHARACTER, LOGICAL, or COMPLEX. Examples of such declarations are shown below.

```
      INTEGER i,j,m(10,20),counter
      REAL x,y,sum(100)
      DOUBLE PRECISION pi,w(-5:10,0:20)
      CHARACTER*12 string
      LOGICAL n(10)
      COMPLEX z
```

Fortran has two forms of real numbers, real and double precision. In general, double precision is preferred and is the precision adopted by True BASIC. Single precision variables might be useful if memory becomes a problem, because single precision variables take up half the memory of double precision variables. Note that arrays can be dimensioned in the declaration statements. Arrays also can be dimensioned in DIMENSION statements, corresponding to the DIM statement in True BASIC, but the use of the DIMENSION statement in Fortran is not recommended.

A character variable is similar to a string variable in True BASIC, but its maximum length must be declared. The above declaration of string indicates that the variable uses 12 bytes of memory and hence can hold up to 12 characters. LOGICAL variables contain the results of relational operations; only the values .TRUE. and .FALSE. are allowed. COMPLEX variables contain an ordered pair of real numbers, i.e., a real and imaginary part.

It is convenient to introduce *symbolic constants*, e.g., the value of π, that cannot be changed inadvertently. An example of the use of the PARAMETER statement to define such a constant is

```
PARAMETER (pi = 3.1415926535898D0)
```

The value of a constant defined in a parameter statement is defined only in the program unit (main program, subroutine, or function) that contains it.

- *Arithmetic expressions.* The only difference in the syntax of the arithmetic operations is exponentiation, e.g., x^3 is written in Fortran as x**3 rather than as x3̂. Expressions containing only integers are treated as integers; expressions containing both integers and reals are treated as real. For example, $1/3 = 0$, $7/3 = 2$, but $1.0/3 = 0.333333$ and $3.0 + 1/3 = 3.0$. The rule in such *mixed mode expressions* is that the weaker (or simpler) of the data types will be converted to the type of the stronger one. For example, suppose that x is a double precision variable and we make the assignment $x = 2 + 3.0$. Because 3.0 is a real constant and 2 is an integer, the integer 2 will be converted to real before addition; the result is then converted to double precision before storage in x. The accuracy of this *coerced* conversion depends on the compiler, and it is good practice to avoid mixed mode expressions and to do the conversions explicitly. For example, it is better to write $x = 3.0D0 + 2.0D0$. Constants written with the exponent letter D are of type double precision. The function REAL converts integers to real numbers, DBLE returns a double precision real value with as much precision as possible, and INT converts real numbers to integers. An example of the explicit conversion of the types of variables is shown in *Program series.f*.

- *Logical* IF. These statements are almost identical in True BASIC and Fortran with the exceptions that logical expressions are enclosed in parentheses in Fortran, and the keyword THEN must be omitted when the IF statement is on one line, e.g.,

```
IF (i .gt. 10) i = 0
```

A summary of the syntax of the relational operators is given in Table A.1.

True BASIC	Fortran
=	.eq.
<>	.ne.
>	.gt.
<	.lt.
>=	.ge.
<=	.le.
and	.and.
or	.or.

Table A.1   A comparison of the syntax of the relational operators in True BASIC and Fortran.

- DO *loops*. Simple iteration is done with a DO loop as shown in the following example:

```
      DO 100 i = 1,10
         x(i) = i**2
100   CONTINUE
```

Note the statement label 100 and the traditional use of the CONTINUE statement to mark the end of the loop. Fortran 90 and most implementations of Fortran 77 allow the DO loop to be written as

```
      DO i = 1,10
         x(i) = i**2
      END DO
```

This version of the DO loop eliminates the use of the statement number (100 in this example) and is recommended.

- DO *while*. There is no WHILE statement in standard Fortran 77, but most implementations of Fortran allow the following:

```
      DO WHILE (x .LT. 10)
         x = x + 1
      END DO
```

In Fortran 77, the equivalent logic can be written as

```
20    IF (x .GE. 10) GO TO 30
         x = x + 1
         GO TO 20
30    CONTINUE
```

In most implementations of Fortran, it is possible to write programs without using GO TO statements and statement numbers in control structures. One sign of a well written Fortran program is the limited use of such statements.

True BASIC	Fortran

`DO`
    `x = x + 1`
`LOOP UNTIL x > 10`
	`DO WHILE (x .le. 10)`
	    `x = x + 1`
	`END DO`

Table A.2   Implementation of a DO UNTIL loop in True BASIC and Fortran.

- DO *until*. There is no UNTIL statement in Fortran. In Table A.2 we show the equivalence using a DO WHILE statement. In Fortran 90 we can use an EXIT statement which takes control to the statement after the END DO statement.

```
* Fortran 90 example
DO
    x = x + 1
IF (x > 10) EXIT
END DO
```

- *Select Case*. There is no SELECT CASE statement in Fortran 77, but these statements easily can be converted to IF ... ELSE IF ... ELSE statements. Fortran 90 has a SELECT CASE statement similar to that of True BASIC.

- *Input and output*. Fortran has many ways of writing and reading data. In Table A.3 we show several examples of ways of entering data from a keyboard and showing results on a screen. Note the similarity of the PRINT and INPUT statements in True BASIC and the WRITE and READ statements in Fortran. The statements READ(*,*) and READ(5,*) allow (list-directed) data to be entered from the keyboard. The role of the *unit number* is to direct the input and output, a role that is similar to the *channel number* in True BASIC. If an asterisk is used instead of an unit number, then the input is from the keyboard (unit number 5) and the output is to the screen (unit number 6). The second argument of the READ and WRITE statements specifies the *format* of the input and output, respectively. The format can be coded as either a character string or as a separate FORMAT statement referenced by a statement label.

True BASIC		Fortran
`INPUT x,y,z`		`READ(*,*) x,y,z`
`PRINT x,y,z`		`WRITE(*,*) x,y,z`
`INPUT astring$`		`READ(*,100) astring`
	`100`	`FORMAT(A10)`
`INPUT PROMPT "v = ": v`		`WRITE(*,*) 'v = '`
		`READ(*,*) v`
`PRINT USING "--#.####": w`		`WRITE(*,'(F8.4)') w`

Table A.3   Comparison of simple input and output statements in True BASIC and Fortran.

The following program illustrates the use of format statements. The format statements may appear anywhere within the program unit in which the read and write statements appear. The most common *edit descriptors* are A (alphanumeric), F (floating point), E (exponential), I (integer), and X (space). For example, F8.4 prints a real variable with eight spaces, four of which are reserved for the fractional part of the number, A10 prints a character variable in a width of ten spaces, and 2X skips two spaces. 10F12.3 is an example of a repetitive edit descriptor, where 10 is the repeat count.

```
        PROGRAM output
        IMPLICIT NONE
        DOUBLE PRECISION x,y,z
        x = 3.14159D0
        y = 2.03D0
        z = -17.5D0
        WRITE(*,10) x,y,z
        WRITE(*,20) x,y,z
        WRITE(*,30) x,y,z
10      FORMAT(F10.3,2E13.4)
20      FORMAT(F10.3,2(2x,E13.4))
30      FORMAT(1X,'x = ',F10.4,2X,'y = ', F10.4,2X,'z =',F10.3)
        END
```

The output of the program is

```
------------------------------------------
    3.142   0.2030E+01  -0.1750E+02
    3.142      0.2030E+01     -0.1750E+02
x =     3.1416  y =     2.0300  z =    -17.500
------------------------------------------
```

Each dash represents one column of output.

- *Files.* Data files can be organized sequentially (the default) or by record. We will consider only sequential organization. The conversions between some simple True BASIC and Fortran statements are as follows:

```
! True BASIC
OPEN #1: NAME "data.out", CREATE NEW, ACCESS OUTPUT
OPEN #2: NAME "data.in", CREATE OLD, ACCESS INPUT
OPEN #3: NAME "test.dat", CREATE OLD, ACCESS INPUT
INPUT #2: s,t,u
PRINT #1: x,y,z
CLOSE #1
CLOSE #2
CLOSE #3

*       FORTRAN
        OPEN(1,FILE='data.out',STATUS='NEW')
```

```
OPEN(2,FILE='data.in',STATUS='OLD')
OPEN(3,FILE='test.dat',STATUS='UNKNOWN')

READ(2,*) s,t,u
WRITE(1,*) x,y,z

CLOSE(1)
CLOSE(2)
CLOSE(3)
```

- *Program units*. A typical Fortran program consists of subprograms such as subroutines and functions that are similar to True BASIC. The main program has an optional PROGRAM header and a required END statement. The way that subroutines are called and information is shared between subprograms is similar to True BASIC. An example of the passing of shared information via arguments in a subroutine call follows:

```
DOUBLE PRECISION w(-8:8)
INTEGER spin(30,30)
CALL Metropolis(spin,w)
.
.
END

SUBROUTINE Metropolis(spin,w)
DOUBLE PRECISION w(-8:8)
INTEGER spin(30,30)
.
.
END
```

Note that the declaration statements for all the variables passed to a subroutine must be repeated in the subroutine itself. Unexpected results may occur when the declarations in the calling program are not identical to those in the subprogram. The keyword RETURN in Fortran is used instead of EXIT SUB in True BASIC.

- *Function declaration*. The DECLARE DEF statement in True BASIC does not exist in Fortran. An example of the use of the FUNCTION statement is given in the following:

```
REAL ran2
INTEGER seed
seed = 1427
x = ran2(seed)
.
.
END
```

```
REAL FUNCTION ran2(idum)
INTEGER idum
.
.
ran2 = ...
END
```

- COMMON *statements*. An alternative to passing variables in subprogram argument lists is to use COMMON statements. This use is similar to the use of the DECLARE PUBLIC statement in True BASIC. This way is convenient when the number of arguments becomes too large or when many subroutines require access to the same information. In the following are several examples of the format of the *labeled* COMMON statement.

```
PROGRAM example
DOUBLE PRECISION x(100),y(100)
COMMON/position/x,y
INTEGER N,spin(100,100)
COMMON/lattice/N,spin
CALL flip
.
.
END

SUBROUTINE flip
INTEGER N,spin(100,100)
COMMON/lattice/N,spin
.
.
END
```

The labels position and lattice are arbitrary and allow us to include more than one common statement.

Module statements in Fortran 90 make the use of the common statement obsolete. An example of such a module construction follows:

```
MODULE position
    DOUBLE PRECISION x(100),y(100)
END MODULE position
```

A program unit that needs the variables in MODULE position does so by including the statement USE position.

- DATA *statements*. True BASIC uses a combination of a DATA statement and a READ statement to assign data to variables. In Fortran this assignment is done in one statement as shown in Table A.4.
- *Intrinsic functions*. The Fortran intrinsic function names given in Table A.5 are called *generic* names meaning that they work for all relevant data types of the

True BASIC	Fortran
| ```
DATA 1,2,3
READ x,y,z
DATA 1,2,3,3,3
FOR i = 1 TO 5
   READ x(i)
NEXT i
``` | ```
DATA x,y,z/1.0,2.0,3.0/

REAL x(5)/1.0,2.0,3.0,3.0,3.0/
``` |

Table A.4   Comparison of use of DATA statements in True BASIC and Fortran.

arguments, e.g., integers, reals, and double precision numbers. The data type of the value returned by the function depends on the data type of its arguments.

The following program is a translation of *Program free_fall* in Chapter 3. All variables are declared in the main program and in each subroutine. For convenience, we have used upper case to distinguish the syntax of Fortran.

```
      PROGRAM free_fall
*     no air resistance
      IMPLICIT NONE
      DOUBLE PRECISION y,v,a,g,t,dt
      INTEGER counter,nshow
*     initial conditions and parameters
      CALL initial(y,v,a,g,t,dt)
```

True BASIC	Fortran
sqr(x)	sqrt(x)
mod(x,y)	mod(x,y)
abs(x)	abs(x)
sin(x)	sin(x)
cos(x)	cos(x)
tan(x)	tan(x)
atn(x)	atan(x)
exp(x)	exp(x)
log(x)	log(x)
log10(x)	log10(x)
truncate(x,0)	int(x)
max(x,y)	max(x,y)
min(x,y)	min(x,y)
sgn(x)	sign(1,x)

Table A.5   Listing of the more common intrinsic functions in True BASIC and Fortran. Note that ran(iseed) is a common name for a random number generator; iseed is an integer seed for the generator.

```
*       print initial conditions
        CALL print_table(y,v,a,t,nshow)
        counter = 0
        DO WHILE (y .GE. 0.0D0)
           CALL Euler(y,v,a,g,t,dt)
           counter = counter + 1
           IF (mod(counter,nshow) .EQ. 0) THEN
              CALL  print_table(y,v,a,t,nshow)
           END IF
        END DO
*       print values at surface
        CALL  print_table(y,v,a,t,nshow)
        END

        SUBROUTINE initial(y,v,a,g,t,dt)
        IMPLICIT NONE
        DOUBLE PRECISION y,v,a,g,t,dt
*       initial time (sec)
        t = 0.0D0
*       initial height (m)
        y = 10.0D0
        v = 0.0D0
*       (magnitude) of accel due to gravity
        g = 9.8D0
        a = -g
        WRITE(*,*)'time step dt = '
        READ(*,*) dt
        END

        SUBROUTINE Euler(y,v,a,g,t,dt)
        IMPLICIT NONE
        DOUBLE PRECISION y,v,a,g,t,dt
*       use velocity at beginning of interval
        y = y + v*dt
*       following included to remind us that acceleration constant
*       y positive upward
        a = -g
        v = v + a*dt
        t = t + dt
        END

        SUBROUTINE print_table(y,v,a,t,nshow)
        IMPLICIT NONE
        DOUBLE PRECISION y,v,a,t
        INTEGER nshow
        IF (t .EQ. 0.0D0) THEN
           WRITE(*,*) 'number of time steps between output = '
           READ(*,*) nshow
           WRITE(*,*)
           WRITE(*,10)
```

```
          WRITE(*,*)
      END IF
      WRITE(*,20) t,y,v,a
10    FORMAT(5x,'time',7x,'y',6x,'velocity',6x,'acceleration')
20    FORMAT(3f10.3,5x,f10.3)
      END
```

The following program is a translation of *Program Ising* in Chapter 17. The program uses a linear congruential random number generator to generate the nth random number, $r_n = ar_{n-1} \bmod m$, with $m = 2^{31} - 1$ and $a = 7^5 = 16807$. Because ar_{n-1} is typically larger than a 32 bit integer, the modulus function is performed by the program in a special way to avoid overflows. First we specify two constants p and q such that $m = aq + p$. Then we compute k equal to the integer part of r_{n-1}/q. The modulus $r_{n-1} \bmod q$ is equivalent to $r_{n-1} - kq$. It can be shown that $r_n = a(r_{n-1} - kq) - kp$ if $r_n \geq 0$; otherwise, $r_n = a(r_{n-1} - kq) - kp + m$.

```
      PROGRAM ising
*     Metropolis algorithm for the Ising model on a square lattice
      IMPLICIT NONE
      INTEGER pass,mcs,nequil,N,L,tsave,spin(32,32)
      DOUBLE PRECISION w(-8:8),Ce(0:20),Cm(0:20),esave(100),msave(100)
      DOUBLE PRECISION E,M,T,accept,cum(10)
      CALL initial(N,L,T,mcs,nequil,tsave,E,M,w,spin)
*     equilibrate system
      DO pass = 1,nequil
         CALL Metropolis(N,L,spin,E,M,w,accept)
      END DO
      CALL initialize(cum,Ce,Cm,esave,msave,accept)
*     accumulate data while updating spins
      DO pass = 1,mcs
         CALL Metropolis(N,L,spin,E,M,w,accept)
         CALL data(E,M,cum)
         CALL correl(Ce,Cm,E,M,esave,msave,pass,tsave)
      END DO
      CALL output(T,N,mcs,cum,accept)
      CALL c_output(Ce,Cm,cum,mcs,tsave)
      END

      SUBROUTINE initial(N,L,T,mcs,nequil,tsave,E,M,w,spin)
      IMPLICIT NONE
      INTEGER mcs,nequil,tsave,N,L,x,y,up,right,dE,seed,spin(32,32)
      DOUBLE PRECISION w(-8:8),E,M,T,sum
      REAL dummy,rnd
      WRITE(*,*)'linear dimension of lattice = '
      READ(*,*) L
      WRITE(*,*)'reduced temperature = '
      READ(*,*) T
      N = L*L
      WRITE(*,*) '# MC steps per spin for equilibrium = '
```

```
          READ(*,*) nequil
          WRITE(*,*) '# MC steps per spin = '
          READ(*,*) mcs
          WRITE(*,*) 'random number seed = '
          READ(*,*) seed
*         seed must not equal 0 to initialize rnd
          dummy = rnd(seed)
*         random initial configuration
          DO y = 1,L
             DO x = 1,L
                IF (rnd(0) .LT. 0.5) THEN
                   spin(x,y) = 1
                ELSE
                   spin(x,y) = -1
                END IF
                M = M + spin(x,y)
             END DO
          END DO
*         compute initial energy
          DO y = 1,L
*            periodic boundary conditions
             IF (y .EQ. L) THEN
                up = 1
             ELSE
                up = y + 1
             END IF
             DO x = 1,L
                IF (x .EQ. L) THEN
                   right = 1
                ELSE
                   right = x + 1
                END IF
                sum = spin(x,up) + spin(right,y)
                E = E - spin(x,y)*sum
             END DO
          END DO
*         compute Boltzmann probability ratios
          DO dE = -8,8,4
             w(dE) = exp(-dE/T)
          END DO
          tsave = 10
          END

          SUBROUTINE initialize(cum,Ce,Cm,esave,msave,accept)
          IMPLICIT NONE
          INTEGER i
          DOUBLE PRECISION Ce(0:20),Cm(0:20),esave(100),msave(100)
          DOUBLE PRECISION accept,cum(10)
          DO i = 1,100
             esave(i) = 0
```

```
                  msave(i) = 0
            END DO
            DO i = 1,20
                  Ce(i) = 0
                  Cm(i) = 0
            END DO
            DO i = 1,5
                  cum(i) = 0
            END DO
            accept = 0
            END

            SUBROUTINE Metropolis(N,L,spin,E,M,w,accept)
*           one Monte Carlo step per spin
            IMPLICIT NONE
            INTEGER ispin,N,L,x,y,dE,DeltaE,spin(32,32)
            DOUBLE PRECISION w(-8:8),E,M,accept
            REAL rnd
            DO ispin = 1,N
*                 random x and y coordinates for trial spin
                  x = int(L*rnd(0)) + 1
                  y = int(L*rnd(0)) + 1
                  dE = DeltaE(x,y,L,spin)
                  IF (rnd(0) .LE. w(dE)) THEN
                        spin(x,y) = -spin(x,y)
                        accept = accept + 1
                        M = M + 2*spin(x,y)
                        E = E + dE
                  END IF
            END DO
            END

            INTEGER FUNCTION DeltaE(x,y,L,spin)
*           periodic boundary conditions
            IMPLICIT NONE
            INTEGER L,x,y,left,right,up,down,spin(32,32)
            IF (x .EQ. 1) THEN
                  left = spin(L,y)
                  right = spin(2,y)
            ELSE IF (x .EQ. L) THEN
                  left = spin(L-1,y)
                  right = spin(1,y)
            ELSE
                  left = spin(x-1,y)
                  right = spin(x+1,y)
            END IF
            IF (y .EQ. 1) THEN
                  up = spin(x,2)
                  down = spin(x,L)
            ELSE IF (y .EQ. L) THEN
```

```
            up = spin(x,1)
            down = spin(x,L-1)
         ELSE
            up = spin(x,y+1)
            down = spin(x,y-1)
         END IF
         DeltaE = 2*spin(x,y)*(left + right + up + down)
         END


         SUBROUTINE data(E,M,cum)
*        accumulate data after every Monte Carlo step per spin
         IMPLICIT NONE
         DOUBLE PRECISION E,M,cum(10)
         cum(1) = cum(1) + E
         cum(2) = cum(2) + E*E
         cum(3) = cum(3) + M
         cum(4) = cum(4) + M*M
         cum(5) = cum(5) + abs(M)
         END


         SUBROUTINE output(T,N,mcs,cum,accept)
*        average per spin
         IMPLICIT NONE
         INTEGER N,mcs
         DOUBLE PRECISION T,accept,norm,eave,e2ave,mave,m2ave
         DOUBLE PRECISION abs_mave,cum(10)
         norm = 1.0/(mcs*N)
         accept = accept*norm
         eave = cum(1)*norm
         e2ave = cum(2)*norm
         mave = cum(3)*norm
         m2ave = cum(4)*norm
         abs_mave = cum(5)*norm
         WRITE(*,*)
         WRITE(*,'(A14,F8.4)') 'Temperature = ',T
         WRITE(*,'(A25,F8.4)') 'acceptance probability = ',accept
         WRITE(*,'(A23,F8.4)') 'mean energy per spin = ', eave
         WRITE(*,'(A30,F10.4)') 'mean squared energy per spin = ', e2ave
         WRITE(*,'(A29,F8.4)') 'mean magnetization per spin = ', mave
         WRITE(*,'(A25,F8.4)') 'mean abs(mag) per spin = ', abs_mave
         WRITE(*,'(A28,F10.4)') 'mean squared mag. per spin = ', m2ave
         END


         SUBROUTINE save_config(N,L,T,spin)
         IMPLICIT NONE
         DOUBLE PRECISION T
         INTEGER N,L,x,y,spin(32,32)
         CHARACTER*10 filename
         WRITE(*,*)'name of file for last configuration = '
         READ(*,*) filename
```

```
                    OPEN(2,FILE=filename,STATUS='NEW')
                    WRITE(2,*) T
                    DO y = 1,L
                       DO x = 1,L
                          WRITE(2,*) spin(x,y)
                       END DO
                    END DO
                    CLOSE(2)
                    END


                    SUBROUTINE read_config(N,L,T,spin)
                    IMPLICIT NONE
                    DOUBLE PRECISION T
                    INTEGER N,L,x,y,spin(32,32)
                    CHARACTER*10 filename
                    WRITE(*,*)'filename ? '
                    READ(*,*) filename
                    OPEN(1,FILE=filename,STATUS='OLD')
                    READ(1,*)T
                    DO y = 1,L
                       DO x = 1,L
                          READ(1,*) spin(x,y)
                       END DO
                    END DO
                    CLOSE(2)
                    END


                    SUBROUTINE correl(Ce,Cm,E,M,esave,msave,pass,tsave)
          *         accumulate data for time correlation functions
          *         save last tsave values of M and E
                    IMPLICIT NONE
                    INTEGER tdiff,tsave,pass,index0,index
                    DOUBLE PRECISION E,M,Ce(0:20),Cm(0:20),esave(100),msave(100)
          *         index0 = array index for earliest saved time
                    index0 = mod(pass-1,tsave) + 1
                    IF (pass .GT. tsave) THEN
          *            compute Ce and Cm after tsave values are saved
                       index = index0
                       DO tdiff = tsave,1,-1
                          Ce(tdiff) = Ce(tdiff) + E*esave(index)
                          Cm(tdiff) = Cm(tdiff) + M*msave(index)
                          index = index + 1
                          IF (index .GT. tsave) index = 1
                       END DO
                    END IF
          *         save latest value in array position of earliest value
                    esave(index0) = E
                    msave(index0) = M
                    END
```

```
              SUBROUTINE c_output(Ce,Cm,cum,mcs,tsave)
*      compute time correlation functions
              IMPLICIT NONE
              INTEGER tdiff,tsave,mcs
              DOUBLE PRECISION Ce(0:20),Cm(0:20),cum(10)
              DOUBLE PRECISION ebar,e2bar,mbar,m2bar,norm
              ebar = cum(1)/mcs
              e2bar = cum(2)/mcs
              Ce(0) = e2bar - ebar*ebar
              mbar = cum(3)/mcs
              m2bar = cum(4)/mcs
              Cm(0) = m2bar - mbar*mbar
              norm = 1.0D0/(mcs-tsave)
              WRITE(*,*)
              WRITE(*,*) '    t ','  Ce(t) ','   Cm(t)'
              WRITE(*,*)
              DO tdiff = 1,tsave
*          define correlation function so that C(0) = 1
*          and C(infinity) = 0
                  Ce(tdiff) = (Ce(tdiff)*norm - ebar*ebar)/Ce(0)
                  Cm(tdiff) = (Cm(tdiff)*norm - mbar*mbar)/Cm(0)
                  WRITE(*,15) tdiff,Ce(tdiff), Cm(tdiff)
15                FORMAT(I6,2(F10.5))
              END DO
              END


              REAL FUNCTION rnd(seed)
*      linear congruential random number generator with shuffling
*      based on ran1 in "Numerical Recipes" second edition
              IMPLICIT NONE
              INTEGER a,m,q,p,n,ndiv,j,k,seed
              REAL rm,rmax
*      m = 2**31 - 1  and  m = a*q + p
              PARAMETER (a = 16807, m = 2147483647, rm = 1.0/m)
              PARAMETER (q = 127773, p = 2836, n = 32, ndiv = 1 + (m-1)/n)
              PARAMETER (rmax = 1.0 - 1.2e-7)
              INTEGER r(n),r0,r1
              SAVE r,r0,r1
              DATA r/n*0/
*      initialize table of random numbers
              IF(seed .NE. 0) THEN
                  r1 = abs(seed)
                      DO j = n+8,1,-1
                          k = r1/q
                          r1 = a*(r1-k*q) - p*k
                          IF (r1 .LT. 0) r1 = r1 + m
                          IF (j .LE. n) r(j) = r1
                      ENDDO
                  r0 = r(1)
              END IF
```

```
*       beginning when not initializing
*       compute r1 = mod(a*r1,m) without overflows
        k = r1/q
        r1 = a*(r1 - k*q) - p*k
        IF (r1 .LT. 0) r1 = r1 + m
        j = 1 + r0/ndiv
        r0 = r(j)
        r(j) = r1
        rnd = min(rm*r0,rmax)
        END
```

The following program implements the fluid lattice gas model described in Chapter 15. We have used the bit manipulation functions, btest and ibset, found in Fortran 90 and on Digital Equipment Corporation and IBM compilers.

```
        PROGRAM latgas
*       simulate lattice gas model with periodic boundary conditions
*       must be modifed slightly to include barrier
*       graphics routines not listed
        IMPLICIT NONE
        INTEGER lat(10000),nn(10000,0:5)
        INTEGER Lx,Ly,nstep,nplot,istep
        INTEGER rule(0:1024)
        CALL initial(lat,Lx,Ly,nstep,nplot)
*       CALL initial_graphics(Lx,Ly)
        CALL nntable(nn,Lx,Ly)
        CALL ruletable(rule)
*       CALL plot(lat,Lx,Ly)
        DO istep = 1,nstep
           CALL update(lat,rule,nn,Lx,Ly)
*          IF (MOD(istep,nplot) .EQ. 0) THEN
*             CALL plot(lat,Lx,Ly)
*          END IF
        END DO
*       CALL end_graphics
        END

        SUBROUTINE initial(lat,Lx,Ly,nstep,nplot)
        IMPLICIT NONE
        INTEGER lat(10000)
        INTEGER Lx,Ly,nstep,nplot
        INTEGER i,j,n
        Lx = 50
        Ly = 20
        nstep = 100
        nplot = 1
*       begin with no particles
        DO n = 1,Lx*Ly
           lat(n) = 0
        END DO
```

```
*       fill block in center of lattice with 6 particles per site
        DO j = Ly/2-1,Ly/2+1
           DO i = Lx/2-1,Lx/2+1
              lat((j-1)*Lx + i) = 63
           END DO
        END DO
        END

        SUBROUTINE update(lat,rule,nn,Lx,Ly)
        IMPLICIT NONE
        INTEGER Lx,Ly,i,j,n,dir,barrier
        PARAMETER (barrier = 7)
        INTEGER latn(10000),lat(10000),nv(0:5)
        INTEGER rule(0:1024),nn(10000,0:5)
*       bounce back boundary conditions v goes to -v
        DATA nv/3,4,5,0,1,2/
        DATA latn/10000*0/
        save latn
*       move particles
        DO j = 1,Ly
           DO i = 1,Lx
              n = (j-1)*Lx + i
                 DO dir = 0,5
                 IF (BTEST(lat(nn(n,dir)),dir)) THEN
                    IF (BTEST(lat(n),barrier)) THEN
*                      reflection
                       latn(nn(n,dir)) = IBSET(latn(nn(n,dir)),nv(dir))
                    else
*                      particle moves from nearest neighbor
                       latn(n) = IBSET(latn(n),dir)
                    END IF
                 END IF
                 END DO
*              WRITE(9,*)n,lat(n),latn(n)
           END DO
        END DO
*       collisions
        DO i = 1,Lx
           DO j = 1,Ly
              n = (j-1)*Lx + i
              IF (.NOT.BTEST(lat(n),barrier)) THEN
                 lat(n) = rule(latn(n))
                 latn(n) = 0
              END IF
           END DO
        END DO
        END

        SUBROUTINE ruletable(rule)
        IMPLICIT NONE
```

```
*       6-bit saturated deterministic rule
        INTEGER rule(0:1024)
        INTEGER i
        DO i = 0,256
            rule(i) = i
        END DO
        rule(21) = 42
        rule(42) = 21
        rule(9) = 36
        rule(18) = 9
        rule(36) = 18
        rule(27) = 45
        rule(45) = 54
        rule(54) = 27
        rule(19) = 37
        rule(37) = 19
        rule(50) = 41
        rule(41) = 50
        rule(22) = 13
        rule(13) = 22
        rule(26) = 44
        rule(44) = 26
        rule(11) = 38
        rule(38) = 11
        rule(25) = 52
        rule(52) = 25
        END

        SUBROUTINE nntable(nn,Lx,Ly)
*       nn(n,dir) gives the neighbor whose particle moving in
*       direction dir will move to site n.
        IMPLICIT NONE
        INTEGER nn(10000,0:5)
        INTEGER Lx,Ly,i,j,ip,im,jp,jm,n
*       odd values of y
        DO j = 1,Ly,2
            DO i = 1,Lx
                n = (j-1)*Lx + i
                ip = i + 1
                IF (ip .GT. Lx) ip = 1
                im = i - 1
                IF (im .LT. 1) im = Lx
                jp = j + 1
                IF (jp .GT. Ly) jp = 1
                jm = j - 1
                IF (jm .LT. 1) jm = Ly
                nn(n,0) = (j - 1)*Lx + im
                nn(n,1) = (jp - 1)*Lx + im
                nn(n,2) = (jp - 1)*Lx + i
                nn(n,3) = (j - 1)*Lx + ip
```

```
                   nn(n,4) = (jm - 1)*Lx + i
                   nn(n,5) = (jm - 1)*Lx + im
              END DO
          END DO
*     even values of y
      DO j = 2,Ly,2
         DO i = 1,Lx
            n = (j - 1)*Lx + i
            ip = i + 1
            IF (ip .GT. Lx) ip = 1
            im = i - 1
            IF (im .LT. 1) im = Lx
            jp = j + 1
            IF (jp .GT. Ly) jp = 1
            jm = j - 1
            IF (jm .LT. 1) jm = Ly
            nn(n,0) = (j - 1)*Lx + im
            nn(n,1) = (jp - 1)*Lx + i
            nn(n,2) = (jp - 1)*Lx + ip
            nn(n,3) = (j - 1)*Lx + ip
            nn(n,4) = (jm - 1)*Lx + ip
            nn(n,5) = (jm - 1)*Lx + i
         END DO
      END DO
      END
```

References and Suggestions for Further Reading

Gary D. Doolen, editor *Lattice Gas Methods for Partial Differential Equations*, Addison-Wesley (1990). The article "Density and Velocity Dependence of Reynolds Numbers for Several Lattice Gas Models," by K. Diemer, K. Hunt, S. Chen, T. Shimomura, and G. D. Doolen describes the bit representation for many lattice models.

Ellis Horowitz, *Fundamentals of Programming Languages*, Computer Science Press (1984).

Michael Metcalf, *Effective FORTRAN 77*, Clarendon Press (1985) and Michael Metcalf and John Reid, *Fortran 90 Explained*, Oxford Science Publications (1990).

Larry Nyhoff and Sanford Leestma, *Fortran 77 for Engineers and Scientists*, third edition, Macmillan (1992).

William H. Press, Saul A. Teukolsky, William T. Vetterling, and Brian P. Flannery, *Numerical Recipes in Fortran*, second edition, Cambridge University Press (1992).

Richard L. Wexelblat, editor, *History of Programming Languages*, Academic Press (1981).

B

From BASIC to C

C was developed in 1972 by Dennis Ritchie at Bell Laboratories. The language is a general purpose and a system programming language. It features an economy of expression, a rich set of operators, and modern data structures and control statements. Although the primary goal of C++ is to introduce an object oriented extension to C, some features of C++ considerably improve the C language, and we include a discussion of some of these features. C and C++ are the languages of choice for most commercial software developers, and are replacing Pascal as the first languages taught in introductory computer science courses. If you do not know C programming, then the following summary will give you an idea of the nature of C and hopefully encourage you to learn more. If you already know C, you might be surprised to learn how much True BASIC and C have in common. Of course, our discussion of C and C++ is only introductory, and the interested reader should consult one of the many popular texts available.

We first give an example of a C program that is a simple translation of *program series* (see Chapter 2):

```
/* program series */
/* the # sign must appear in column 1 for many C preprocessors */
#include <stdio.h>     /* library of input/output functions */
#include <math.h>      /* library of math functions */

void main()            /* void optional for main */
/*    add the first 100 terms of a simple series */
  {
      double sum;
      int n;
      sum = 0.0;
      for (n = 1; n <= 100; n++)
        {
          sum =  sum + 1.0/pow((double) n,2.0);
```

```
/*        real C programmers would write
          sum + = 1.0/pow((double) n,2.0);    */
          printf("%d %f\n",n,sum);
          }
    }
```

Although *series.c* is simple, it allows us to discuss many of the important features of C:

- *Comment statements*. Any characters between /∗ and ∗/ are ignored by the compiler and are used for comments. Comments can span more than one line. C does not have a PROGRAM statement.

- *Headers*. C is a very concise language, and most input and output functions and many mathematical functions are declared in separate header files or libraries. The statement #include <stdio.h> at the beginning of *program series* allows standard input and output functions to be used; the statement #include <math.h> allows the function pow and other mathematical functions to be used.

- *Program structure*. A C program is a collection of functions, one of which must be called main. C does not distinguish between functions and subroutines. The analog of a subroutine is a function that does not return a value. If a function does not return a value, then the keyword void is used before the function. It is good practice to declare the names of all functions before they are used. This declaration is required in C++ and we adopt this convention.

- *Format*. C uses a free style format. Statements end with a semicolon, and groups of statements, called compound statements or blocks, are enclosed by braces { }. There is no semicolon after the brace }. C is case sensitive, and sum refers to a different variable than Sum.

- *Assignment*. C uses the equals sign to indicate assignment.

- *Arithmetic expressions*. C has the same binary arithmetic operators (+, -, *, /) as True BASIC. C does not have an operator for raising a number to a power. Instead, C uses the function pow(x,y) to express x^y.

- *Data types*. All variables must be declared along with their type. There are four built-in types from which complex data structures (such as arrays) can be constructed. These four types are int (integer), float (single precision real), double (double precision real), and char (character). In C, as in Fortran, mixed type expressions are allowed. The only case where care is necessary is integer division. For example, 1/3 evaluates to 0. It is good practice to write floating point constants with explicit decimal points even if they have integral values.

- *Loops*. The FOR NEXT loop in True BASIC is implemented with a for loop in C. For example, FOR n = 1 to 100 is written in C as for (n = 1; n <= 100; n++). The statements included in the for loop are contained within the braces { }. (In *program series* there are two such statements in the for loop.) The expression n++ means replace n by n+1 and is almost universally used in for loops. Note that the expression n++ is evaluated at the end of each pass through the loop and not at the beginning. Similarly n-- means replace n by n-1. The `for` loop in C can be extended in various ways that cannot be implemented in True BASIC or Fortran.

- *Type conversion.* We need to pay careful attention to the type of variable used in a function. For example, the function pow(x,y) expects x and y to be double precision variables. Conversion operators, or *casts* can be used for explicit conversions. For example, we need to write (double) n to convert the integer variable n to double precision.

- *If statements.* C and True BASIC share many of the same decision structures. Two examples of an `if` construct follow:

Example 1

```
if (x < 0) x = -x;
```

Note that the parenthesis must be included, and the keyword `then` is not used in C.

Example 2

```
if (x == 5)
   {
      y = 6;
      z = 10;
   }
else
   {
      y = 8;
      z = 12;
   }
```

A summary of the relational and logical operators in C is given in Table B.1. C distinguishes between assignment and the relational operator that tests for equality and uses == for expressing the latter. Note that = is syntactically correct in an if statement, but usually leads to undesired consequences and is a frequent source of error. Note also how groups of statements are collected into a compound statement by the braces {}.

- *While statements.* The following example demonstrates the `while` loop. The boolean test can be at the beginning or at the end of the loop.

Example 3

```
while (x < 10)
   {
      x = x + 1;
      y = y + x;
/*    real C programmers would combine above two lines and write
      y += ++x;        */
   }
```

True BASIC	C
=	==
<>	!=
>	>
<	<
>=	>=
<=	<=
and	&&
or	||

Table B.1   A comparison of the relational operators in True BASIC and C.

Example 4

```
do
  {
     x = x + 1;
     y = x*y;
/*   real C programmers would combine above two lines and write
     y* = ++x;           */
  }
while (x < 10);
```

An important feature of C is delayed evaluation of boolean expressions. For example, the statement A or B (written in C as (A || B) is evaluated from left to right. If A evaluates to true, then B is not evaluated. This feature not only speeds up the computation, but is very convenient because B might be meaningless if A is true. For example, consider the expression (A == 0) || (C == B/A). If A is equivalent to 0, then the entire expression is true. However, if we evaluated the second half of the expression, we would obtain an overflow error.

- *Simple output.* As with most languages, input and output in C is complicated and is best explained by examples. The letter f in the function name printf in *program series* stands for "formatted." The first argument is a format string that specifies how the external representation of the data is to be written. For the string "%d %f\n", the format %d causes the value of an integer to be printed in a decimal representation in as many positions as needed. The conversion character f can be used for both float and double data variables. Note the effect of the blank space included in the format string. The newline character \n must be included explicitly.

As mentioned, all C programs consist of a set of functions with the function main as the starting point. In the following, we list a program that illustrates how variables are passed. In C the value of a variable is passed, not its address or location in memory. For example, in *program pass* the value of x in the main program is unchanged even though its value in example has changed. The reason is that the variable x in function example is stored in a different memory location than the variable x in the main program. To allow functions to have access to the values of variables changed in the functions they call, we must pass the address of the variables. In program pass &y refers to the address of the variable y, and *y refers to the contents or value of the variable y. As discussed below, arrays such as z in program pass are pointers to the memory location of the first element of an array. Hence, the address of an array argument in a function is automatically passed.

```c
/* program pass */
#include <stdio.h>      /* library of input/output functions */

/* declare function type before using it */
int example(int x, int *y, int z[]);
void main()
    {
        int x = 5, y = 6;
        int z[2],w;
        z[0] = 10;
        z[1] = 20;
        w = example(x,&y,z);
        printf("%d %d %d %d %d\n",x,y,z[0],z[1],w);
    }

int example(int x, int *y, int z[])
    {
        int product;
        x = 50;
        *y = 60;
        z[0] = 100;
        z[1] = 200;
        product = x*(*y);
        return product;
    }
```

Verify that the output of *program pass* is 5 60 100 200 3000. The value of the function example (in this case 3000), is returned by the statement return, not by assigning to the function name a value as done in True BASIC. Some additional features of C are discussed in the following.

- *Initialization.* The analog in C to using the DATA and READ statements in True BASIC is to assign initial values to variables in the declaration statements. An example is given in *Program pass*.

  The (default) variables we have used so far have been declared inside a function and are called *internal* or *local* variables. In C, they are known as *automatic* variables, because their memory space is automatically allocated when the function is entered and released when it is exited. The initial value of an automatic variable for each function call is unknown and typically is garbage. Examples of *static* variables are given in *program ising* below. Static variables maintain their values between function calls. In addition there are external or global variables which are declared near the top of a file outside of any functions including `main`. When the program is compiled, `static` and `external` variables are initialized to zero.

- *Arrays*. Arrays are much different in C than in True BASIC or Fortran. To understand arrays in C, it is helpful to think of an array name as a pointer to the first element of the array. The declaration of an array declares the size s of the array. The index of an array always ranges from 0 to $s - 1$. For example, the declaration

```
int sum[10];
```

reserves memory for 10 integer variables, `sum[0]`, `sum[1]`, ..., `sum[9]`. In C there is no way of defining an array such as `sum(10 to 20)` in True BASIC. Examples of the declaration of one-, two-, and three-dimensional arrays are given in the following.

```
float sum[100];
char str[32];
int m[10][20];
double w[10][30][20];
```

One of the remarkable properties of C to a BASIC programmer is the equivalence of arrays and pointers. After the declaration of `int sum[10]`, the name `sum` denotes a pointer to the array element `sum[0]`. That is, the array name already refers to the address of a variable. Hence passing arrays to a function is relatively straightforward (see how z is passed in *program pass*). Because arrays are pointers, the size of the first dimension of an array need not be specified when declaring array arguments in functions. However, the sizes of all other dimensions must be specified.

A string can be represented as an array of `char` with an extra element needed for the terminating null character, which signifies the end of the string. One string variable can be assigned to another using the `strcpy` function shown below. To use string functions such as `strcpy` the library `string.h` must be included in the file. Other string functions also are available.

```
char str1[15], str2[15];
strcpy(str1,"hello");   /* assign "hello" to str1 */
strcpy(str2,str1);      /* assign str1 to str2 */
```

- *Standard functions*. A summary of the more common mathematical functions is given in Table B.2. These functions can be used when `#include <math.h>` is included at the beginning of the source file.

- *Input and Output*. In Table B.3 we show several examples of ways of entering data from a keyboard and showing the results on a screen. Note that it is necessary to include the format for the variables in the output function `printf` and the input function `scanf`. For example, `%i` or `%d`, `%f`, `%c` denote integer, float, and char data types, respectively. For input it is necessary to use `%lf` for double type variables. Note that the address of a variable is passed to the `scanf` function so that `scanf` can change the contents of the variable.

True BASIC	C
sqr(x)	sqrt(x)
truncate(x,0)	floor(x)
mod(x,y)	fmod(x,y)
abs(x)	fabs(x)
sin(x)	sin(x)
cos(x)	cos(x)
tan(x)	tan(x)
atn(x)	atan(x)
exp(x)	exp(x)
log(x)	log(x)
log10(x)	log10(x)

Table B.2 Summary of the more common mathematical functions in True BASIC and C.

The notation %4e prints float or double variables in exponential notation with four decimal places. The character \n causes control to go to the next line. The following program illustrates how to format output.

```
/* output */
#include <stdio.h>
main()
   {
      double x,y,z;
      x = 3.14159;
      y = 2.03;
      z = -17.5;
      printf("%10.3f%13.4e%13.4e\n",x,y,z);
      printf("%10.3f  %13.4e  %13.4e\n",x,y,z);
      printf("x = %10.4f y = %10.4f z = %10.3f\n",x,y,z);
   }
```

True BASIC	C
	double x,y,z,v,w; char astring[12];
INPUT x,y,z	scanf("%lf %lf %lf",&x,&y,&z);
PRINT x,y,z	printf("%f %f %f \n",x,y,z);
INPUT astring$	scanf("%s",astring);
INPUT PROMPT "v = ": v	printf("v = ");
	scanf("%lf",&v);
PRINT USING "--#.####": w	printf("%8.4f \n",w);

Table B.3 Comparison of simple input and output statements in True BASIC and C.

The output of the program is

```
---------------------------------------------
     3.142   2.0300e+00 -17.5000e+00
     3.142      2.0300e+00 -17.5000e+00
x =      3.1416  y =      2.0300  z =     -17.500
---------------------------------------------
```

Each dash represents one column of output.

- *Files*. In the following we compare how True BASIC and C handle reading a file and writing to a file. The first argument of a fprintf or fscanf function is the file pointer.

```
! True BASIC
OPEN #1: NAME "data.out", CREATE NEW, ACCESS OUTPUT
OPEN #2: NAME "data.in", CREATE OLD, ACCESS INPUT
OPEN #3: NAME "test.dat", CREATE OLD, ACCESS INPUT
INPUT #2: s,t,u
PRINT #1: x,y,z
CLOSE #1
CLOSE #2
CLOSE #3
```

```
/* C example */
#include <stdio.h>
FILE *fp1, *fp2, *fp3;
fp1 = fopen("data.out","w");
fp2 = fopen("data.in","r");
fp3 = fopen("data.app","a");
fscanf(fp2,"%lf %lf %lf\n",&s,&t,&u);
fprintf(fp1,"%f %f %f\n",x,y,z);
fclose(fp1);
fclose(fp2);
fclose(fp3);
```

C++. For our purposes there are two main improvements in C++ in comparison to C. First, it is much easier to pass the address of the variable. We discuss this difference using a C++ version of *program pass* shown below. Note that in the declaration of the function example, we pass the address of the variable &y. However, inside the function and in all calls to the function, we use y without the address operator. This way of passing the address is analogous to using var in argument lists in Pascal. C++ also has a simpler way of performing input and output functions using *stream io*. In C++ cin and cout are the input and output stream, respectively. In *program pass* we use cout to write output to the screen without specifying the format. Also, note that C++ has another way of adding comments on a single line. Any characters appearing after // are ignored by the compiler.

```
/* program pass C++ version */
#include <iostream.h> // library of C++ input/output functions

int example(int x, int &y, int z[]);
void main()
  {
      int x = 5, y = 6;
      int z[2], w;
      z[0] = 10;
      z[1] = 20;
      w = example(x,y,z);
      cout << x << " " << y << " " << z[0] << " " << z[1] << " " << w;
  }

int example(int x, int &y, int z[])
   {
     int product;
     x = 50;
     y = 60;
     z[0] = 100;
     z[1] = 200;
     product = x*y;
     return product;
   }
```

Program listings. The following program is a translation of `Program free_fall` (see Chapter 3). Note that we have retained the flavor of the original programs and this program and the other translations do not look like programs written by experienced C programmers.

```
/* program free_fall C version */
#include <stdio.h>

void initial(double *y,double *v,double *a,double *g,double *t,
             double *dt);
void Euler(double *y,double *v,double *a,double g,double *t,
           double dt);
void print_table(double y,double v,double a,double t,int *nshow);

main()
/* no air resistance */
 {
     double y,v,a,g,t,dt;
     int counter,nshow;
/*   initial conditions and parameters */
     initial(&y,&v,&a,&g,&t,&dt);
/*   print initial conditions */
```

```
        print_table(y,v,a,t,&nshow);
        counter = 0;
        while (y >= 0)
          {
            Euler(&y,&v,&a,g,&t,dt);
/* next 2 lines usually written as   if (++counter % nshow == 0)    */
            counter = counter + 1;
            if (counter % nshow == 0)   /* % is modulus operator */
                print_table(y,v,a,t,&nshow);
          }
/*   print values at surface */
        print_table(y,v,a,t,&nshow);
  }


void initial(double *y,double *v,double *a,double *g,double *t,
             double *dt)
  {
      *t = 0;          /* initial time (sec)  */
      *y = 10;         /* initial height (m)  */
      *v = 0;          /* initial velocity    */
      *g = 9.8;        /* (magnitude) of accel due to gravity */
      *a = -(*g);
      printf("time step dt = ");
      scanf("%lf",dt);
  }


void Euler(double *y,double *v,double *a,double g,double *t,double dt)
  {
/*   use velocity at beginning of interval */
      *y = *y + (*v)*dt;
/*   reminder that acceleration is constant */
/*   y positive upward  */
      *a = -g;
      *v = *v + (*a)*dt;
      *t = *t + dt;
  }


void print_table(double y,double v,double a,double t,int *nshow)
  {
    if (t == 0.0)
      {
        printf("number of time steps between output = ");
        scanf("%d",nshow);
        printf(" \n");
        printf(" time (s)     y (m)    velocity (m/s)   accel (m/s^2) \n");
        printf(" \n");
      }
    printf("%10.3f,%10.3f,%10.3f,%12.3f\n",t,y,v,a);
  }
```

The next program is a C++ translation of *Program free_fall* (see Chapter 3). The program illustrates formatting using stream io. The libraries iostream.h and iomanip.h (for formatting) must be included.

```c
/* program free_fall  C++ version */
#include <stdio.h>
#include <iostream.h>
#include <iomanip.h>
/* need iostream.h and iomanip.h for stream output using cin and cout
*/

void initial(double &y,double &v,double &a,double &g,double &t,
             double &dt);
void Euler(double &y,double &v,double &a,double g,double &t,
           double dt);
void print_table(double y,double v,double a,double t,int &nshow);

main()
/* no air resistance */
 {
     double y,v,a,g,t,dt;
     int counter,nshow;
/*   initial conditions and parameters */
     initial(y,v,a,g,t,dt);
/*   print initial conditions */
     print_table(y,v,a,t,nshow);
     counter = 0;
     while (y >= 0)
       {
         Euler(y,v,a,g,t,dt);
         counter = counter + 1;
         if (counter % nshow == 0)
          print_table(y,v,a,t,nshow);
       }
       print_table(y,v,a,t,nshow);   // print values at surface
 }

void initial(double &y,double &v,double &a,double &g,double &t,
             double &dt)
  {
     t = 0;        // initial time (sec)
     y = 10;       // initial height (m)
     v = 0;        // initial velocity
     g = 9.8;      // (magnitude) of accel due to gravity
     a = -g;
     cout << "time step dt = ";
     cin >> dt;    // read in value for dt
  }
```

```
void Euler(double &y,double &v,double &a,double g,double &t,double dt)
  {
/*     use velocity at beginning of interval */
      y = y + v*dt;
      a = -g;      // y positive upward
      v = v + a*dt;
      t = t + dt;
  }


void print_table(double y,double v,double a,double t,int &nshow)
   {
    if (t == 0)
      {
       cout << "number of time steps between output = ";
       cin >> nshow;
       cout << "\n  time (s) y (m)  velocity (m/s)   accel (m/s^2) "
            << "\n\n";
      }
      /* setw(n) allows n spaces for a number
         setprecision(m) provides m decimal places   */
      cout << setw(10) << setprecision(3) << t;
      cout << setw(10) << setprecision(3) << y;
      cout << setw(10) << setprecision(3) << v;
      cout << setw(12) << setprecision(3) << a << "\n";
   }
```

The following program is a translation of `Program Ising` (see Chapter 17). The program uses a linear congruential random number generator to generate the nth random number $r_n = a r_{n-1}$ modulo m, with $m = 2^{31} - 1$ and $a = 7^5 = 16807$. Because $a r_{n-1}$ is typically larger than a 32 bit integer, the modulus function is implemented such that overflows are avoided. First we specify two constants p and q such that $m = aq + p$. Then we compute k equal to the integer part of r_{n-1}/q. The modulus $r_{n-1} \bmod q$ is equivalent to $r_{n-1} - kq$. It can be shown that $r_n = a(r_{n-1} - kq) - kp$ if the right-hand side is greater than or equal to 0, and $r_n = a(r_{n-1} - kq) - kp + m$ otherwise. Note the use of the #define construction in the definition of the function rnd. Such a construction is used to define a symbolic name or symbolic constant. The "min function" in rnd is an example of a macro.

```
/* program ising */
#include <stdio.h>
#include <math.h>

void initial(int *N,int *L,double *T,int *mcs,int *nequil,int *tsave,
          double *E,double *M,double w[],int spin[][33]);
void Metropolis(int N,int L,int spin[][33],double *E,double *M,
          double w[],double *accept);
int DeltaE(int x,int y,int L,int spin[][33]);
void data(double E,double M,double cum[]);
```

```
           void output(double T,int N,int mcs,double cum[],double accept);
           void save_config(int N,int L,double T,int spin[][33]);
           void read_config(int N,int L,double *T,int spin[][33]);
           void correl(double Ce[],double Cm[],double E,double M,
                       double esave[],double msave[],int pass,int tsave);
           void c_output(double Ce[],double Cm[],double cum[],int mcs,int tsave);
           float rnd(int seed);

           main()
             {
/*     Metropolis algorithm for the Ising model on a square lattice */
             int pass,mcs,nequil,N,L,tsave,spin[33][33];
             double w[17],Ce[21],Cm[21],esave[101],msave[101];
             double E,M,T,accept,cum[11];
             initial(&N,&L,&T,&mcs,&nequil,&tsave,&E,&M,w,spin);
/*     equilibrate system */
             for (pass = 1;pass <= nequil;pass++)
                Metropolis(N,L,spin,&E,&M,w,&accept);
             accept = 0;
/*     accumulate data while updating spins */
             for (pass = 1; pass <= mcs; pass++)
               {
                 Metropolis(N,L,spin,&E,&M,w,&accept);
                 data(E,M,cum);
                 correl(Ce,Cm,E,M,esave,msave,pass,tsave);
               }
             output(T,N,mcs,cum,accept);
             c_output(Ce,Cm,cum,mcs,tsave);
           }

       void initial(int *N,int *L,double *T,int *mcs,int *nequil,int *tsave,
                    double *E,double *M,double w[],int spin[][33])
         {
         int x,y,up,right,dE,seed;
         double sum,dummy;
         printf("linear dimension of lattice = ");
         scanf("%d",L);
         printf("reduced temperature = ");
         scanf("%lf",T);
         *N = (*L)*(*L);
         printf("# MC steps per spin for equilibrium = ");
         scanf("%d",nequil);
         printf("# MC steps per spin = ");
         scanf("%d",mcs);
         printf("random number seed = ");
         scanf("%d",&seed);
/*     seed must not equal 0 to initialize rnd */
         dummy = rnd(seed);
/*     random initial configuration */
```

```
              for (y = 1; y <= *L; y++)
                for (x = 1; x <= *L; x++)
                  {
                    if (rnd(0) < 0.5)
                      spin[x][y] = 1;
                    else
                      spin[x][y] = -1;
                    *M = *M + spin[x][y];
                  }
/*      compute initial energy */
              for (y = 1; y <= *L; y++)
                {
/*        periodic boundary conditions */
                  if (y==(*L))
                    up = 1;
                  else
                    up = y + 1;
                  for (x = 1; x <= *L; x++)
                    {
/*            conditional expression */
                      if (x==(*L))
                        right = 1;
                      else
                        right = y + 1;
                      sum = spin[x][up] + spin[right][y];
                      *E = *E - spin[x][y]*sum;
                    }
                }
/*      compute Boltzmann probability ratios  */
              for (dE = -8; dE <= 8; dE=dE+4)
                w[dE+8] = exp(-dE/(*T));
              *tsave = 10;
          }


void Metropolis(int N,int L,int spin[][33],double *E,double *M,
                double w[],double *accept)
      {
/*      one Monte Carlo step per spin  */
        int ispin,x,y,dE;
        for (ispin=1; ispin <= N; ispin++)
          {
/*          random x and y coordinates for trial spin  */
            x = L*rnd(0) + 1;
            y = L*rnd(0) + 1;
            dE = DeltaE(x,y,L,spin);
            if (rnd(0) <= w[dE+8])
              {
                spin[x][y] = -spin[x][y];
```

```
                        *accept = *accept + 1;
                        *M = *M + 2*spin[x][y];
                        *E = *E + dE;
                       }
                     }
            }

        int DeltaE(int x,int y,int L,int spin[][33])
           {
/*      periodic boundary conditions */
           int dE,left,right,up,down;
           if (x == 1)
              {
              left = spin[L][y];
              right = spin[2][y];
              }
           else if (x == L)
              {
              left = spin[L-1][y];
              right = spin[1][y];
              }
           else
              {
              left = spin[x-1][y];
              right = spin[x+1][y];
              }
           if (y == 1)
              {
              up = spin[x][2];
              down = spin[x][L];
              }
           else if (y == L)
              {
              up = spin[x][1];
              down = spin[x][L-1];
              }
           else
              {
              up = spin[x][y+1];
              down = spin[x][y-1];
              }
           dE = 2*spin[x][y]*(left + right + up + down);
           return dE;
          }

        void data(double E,double M,double cum[])
           {
/*      accumulate data after every Monte Carlo step per spin */
           cum[1] = cum[1] + E;
```

```
                    cum[2] = cum[2] + E*E;
                    cum[3] = cum[3] + M;
                    cum[4] = cum[4] + M*M;
                    cum[5] = cum[5] + fabs(M);
                }

       void output(double T,int N,int mcs,double cum[],double accept)
           {
/*     average per spin */
           double norm,eave,e2ave,mave,m2ave,abs_mave;
           norm = 1.0/(mcs*N);
           accept = accept*norm;
           eave = cum[1]*norm;
           e2ave = cum[2]*norm;
           mave = cum[3]*norm;
           m2ave = cum[4]*norm;
           abs_mave = cum[5]*norm;
           printf(" \n");
           printf("Temperature = %8.4f\n",T);
           printf("acceptance probability = %8.4f\n",accept);
           printf("mean energy per spin = %8.4f\n", eave);
           printf("mean squared energy per spin = %10.4f\n", e2ave);
           printf("mean magnetization per spin = %8.4f\n", mave);
           printf("mean abs(mag) per spin = %8.4f\n", abs_mave);
           printf("mean squared mag. per spin = %10.4f\n", m2ave);
           }


       void save_config(int N,int L,double T,int spin[][33])
           {
           int x,y;
           FILE *fp;    /* note the use of the variable type FILE */
           char filename[20];
           printf("name of file for last configuration = ");
           scanf("%s",filename);
           fp = fopen(filename,"w");
           fprintf(fp,"%f\n",T);
           for (y=1; y <= L; y++)
              for (x=1; x <= L; x++)
                 fprintf(fp,"%d\n",spin[x][y]);
           fclose(fp);
           }


       void read_config(int N,int L,double *T,int spin[][33])
           {
           int x,y;
           FILE *fp;    /* note the use of the variable type FILE */
           char filename[20];
           printf("name of file for last configuration = ");
           scanf("%s",filename);
```

```
            fp = fopen(filename,"r");
            fscanf(fp,"%f\n",T);
            for (y=1; y <= L; y++)
               for (x=1; x <= L; x++)
                  fscanf(fp,"%d\n",spin[x][y]);
            fclose(fp);
     }

   void correl(double Ce[],double Cm[],double E,double M,
               double esave[],double msave[],int pass,int tsave)
      {
/*     accumulate data for time correlation functions */
/*     save last tsave values of M and E */
       int tdiff,index0,index;
/*     index0 = array index for earliest saved time */
       index0 = ((pass - 1) % tsave) + 1;
       if (pass > tsave)
/*       compute Ce and Cm after tsave values are saved */
         {
          index = index0;
          for (tdiff=tsave; tdiff >= 1; tdiff--)
            {
             Ce[tdiff] = Ce[tdiff] + E*esave[index];
             Cm[tdiff] = Cm[tdiff] + M*msave[index];
             index = index + 1;
             if(index > tsave) index = 1;
             }
          }
/*     save latest value in array position of earliest value */
       esave[index0] = E;
       msave[index0] = M;
     }

   void c_output(double Ce[],double Cm[],double cum[],int mcs,int tsave)
      {
/*     compute time correlation functions */
       int tdiff;
       double ebar,e2bar,mbar,m2bar,norm;
       ebar = cum[1]/mcs;
       e2bar = cum[2]/mcs;
       Ce[0] = e2bar - ebar*ebar;
       mbar = cum[3]/mcs;
       m2bar = cum[4]/mcs;
       Cm[0] = m2bar - mbar*mbar;
       norm = 1.0/(mcs-tsave);
       printf("\n    t       Ce(t)   Cm(t)        \n\n");
       for (tdiff = 1;tdiff <= tsave;tdiff++)
         {
/*         define correlation function so that C(0) = 1 and
```

```
                            C(infinity) = 0   */
                            Ce[tdiff] = (Ce[tdiff]*norm - ebar*ebar)/Ce[0];
                            Cm[tdiff] = (Cm[tdiff]*norm - mbar*mbar)/Cm[0];
                            printf("%6d %10.5f %10.5f  \n",tdiff,Ce[tdiff],Cm[tdiff]);
                        }
                }

        float rnd(int seed)
            {
/*      linear congruential random number generator with shuffling */
/*      based on ran1 of second edition of "Numerical Recipes" */
#define min(a,b)   ((a) < (b) ? (a) : (b))
#define a 16807
#define m 2147483647  /* 2^31 - 1  */
#define rm 1.0/m
#define q 127773   /*  m = a*q + p  */
#define p 2836
#define n 32
#define ndiv (1 + (m-1)/n)
#define rmax (1.0 - 1.2e-7)
            static int r[n+1],r0,r1;
            int j,k;
            if (seed != 0)
/*      initialize table of random numbers  */
                {
                  r1 = abs(seed);
                for (j = n+8;j>=1;j--)
                  {
                  k = r1/q;
                  r1 = a*(r1-k*q) - p*k;
                  if (r1 < 0) r1 = r1 + m;
                  if (j < n) r[j] = r1;
                    }
                  r0 = r[1];
                  }
/*      beginning when not initializing    */
/*      compute r1 = mod(a*r1,m) without overflows  */
            k = r1/q;
            r1 = a*(r1 - k*q) - p*k;
            if (r1 < 0) r1 = r1 + m;
            j = 1 + r0/ndiv;
            r0 = r[j];
            r[j] = r1;
            return min(rm*r0,rmax);
            }
```

The following C program listing simulates the fluid lattice gas described in Chapter 15. Note the use of the bitwise and operator, &, and the bitwise or operator, |.

```
/* program latgas */
#include <stdio.h>

void initial(int lat[],int *lx,int *ly,int *nstep,int *nplot);
/*    void initial_graphics(int lx,int ly);  */
void update(int lat[],int rule[],int nn[][6],int lx,int ly);
void nntable(int nn[][6],int lx,int ly);
void ruletable(int rule[]);
/*    void plot(int lat[],int lx,int ly);   */

/* note how arrays are initialized */
int nv[] = {3,4,5,0,1,2};        /* global variables available */
int mask[] = {1,2,4,8,16,32,64,128}; /* to all functions */

main()
/* simulates lattice gas fluid model
   graphics routines not listed */
{
  int lat[10000],nn[10000][6];
  int lx,ly,nstep,nplot,istep;
  int rule[1024];
  initial(lat,&lx,&ly,&nstep,&nplot);
/*    initial_graphics(lx,ly); */
  nntable(nn,lx,ly);
  ruletable(rule);
/*    plot(lat,lx,ly);     */
  for (istep = 1; istep <= nstep; istep++)
    {
      update(lat,rule,nn,lx,ly);
/*        if (istep % nplot==0)
             plot(lat,lx,ly);  */
    }
}

void initial(int lat[],int *lx,int *ly,int *nstep,int *nplot)
{
  int i,j,n;
  /* system size */
  *lx = 50;
  *ly = 20;
  /* simulation parameters */
  *nstep = 100;
  *nplot = 1;
  /* begin with no particles */
  for (n = 0; n < (*lx)*(*ly); n++)
    lat[n] = 0;
  /* fill block in center of lattice with 6 particles per site */
  for (j = -1; j <= 1; j++)
    for (i = -1; i <= 1; i++)
```

```
            {
               n = (j+(*ly)/2)*(*lx) + i + (*lx)/2;
               lat[n] = 63;
            }
   }


   void update(int lat[],int rule[],int nn[][6],int lx,int ly)
   {
     int i,j,n,dir;
     int latn[10000];
     /* bounce back boundary conditions v goes to -v */
     extern int nv[];
     extern int mask[];
     /* move particles */
     for (j=0; j<ly; j++)
       for (i=0; i<lx; i++)
         {
            n = j*lx + i;
            for (dir=0; dir<6 ;dir++)
            if ((lat[nn[n][dir]] & mask[dir]) != 0)
              {
                 if ((lat[n] & mask[7]) != 0)
                 /* reflection */
                 latn[nn[n][dir]] = latn[nn[n][dir]] | mask[nv[dir]];
              else
                 /* particle moves from nearest neighbor */
                 latn[n] = latn[n] | mask[dir];
              }
         }/* end for */

     /* collisions */
     for (j=0; j<ly; j++)
       for (i=0; i<lx; i++)
         {
            n = j*lx + i;
            if ((lat[n] & mask[7]) == 0)
              {
                 lat[n] = rule[latn[n]];
                 latn[n] = 0;
              }
         } /* end for */
   }


   void ruletable(int rule[])
   /* 6-bit saturated deterministic rule */
   {
     int i;
     for (i=0; i<256; i++)
       rule[i] = i;
```

```
            rule[21] = 42;
            rule[42] = 21;
            rule[9] = 36;
            rule[18] = 9;
            rule[36] = 18;
            rule[27] = 45;
            rule[45] = 54;
            rule[54] = 27;
            rule[19] = 37;
            rule[37] = 19;
            rule[50] = 41;
            rule[41] = 50;
            rule[22] = 13;
            rule[13] = 22;
            rule[26] = 44;
            rule[44] = 26;
            rule[11] = 38;
            rule[38] = 11;
            rule[25] = 52;
            rule[52] = 25;
}

void nntable(int nn[][6],int lx,int ly)
     /* nn[n][dir] gives the neighbor whose particle moving in
        direction dir will move to site n. */
{
  int i,j,ip,im,jp,jm,n;
  for (j=0; j < ly; j = j+2)
    for (i=0; i < lx; i++)
      {
        n = j*lx + i;
        ip = i + 1;
        if (ip > lx-1) ip = 0;
          im = i-1;
        if (im < 0) im = lx - 1;
          jp = j + 1;
        if (jp > ly-1) jp = 0;
          jm = j - 1;
        if (jm < 0) jm = ly - 1;
        nn[n][0] = j*lx + im;
        nn[n][1] = jp*lx + im;
        nn[n][2] = jp*lx + i;
        nn[n][3] = j*lx + ip;
        nn[n][4] = jm*lx + i;
        nn[n][5] = jm*lx + im;
      }
    for (j=1; j < ly;j = j+2)
      for (i=0; i < lx;i++)
        {
```

```
        n = j*lx + i;
        ip = i + 1;
        if (ip > lx-1) ip = 0;
        im = i - 1;
        if (im < 0) im = lx - 1;
        jp = j + 1;
        if (jp > ly-1) jp = 0;
        jm = j - 1;
        if (jm < 0) jm = ly - 1;
        nn[n][0] = j*lx + im;
        nn[n][1] = jp*lx + i;
        nn[n][2] = jp*lx + ip;
        nn[n][3] = j*lx + ip;
        nn[n][4] = jm*lx + ip;
        nn[n][5] = jm*lx + i;
    }
}
```

References and Suggestions for Further Reading

Leendert Ammeraal, *C++ for Programmers*, John Wiley & Sons (1991). The text assumes a background in programming and introduces C++ very efficiently.

John J. Barton and lee R. Nackman, *Scientific and Engineering C++*, Addison-Wesley (1994). The text is particularly useful for someone with a background in Fortran or C.

Stephen C. Dewhurst and Kathy T. Stark, *Programming in C++*, Prentice Hall (1989).

Gary D. Doolen, editor, *Lattice Gas Methods for Partial Differential Equations,* Addison-Wesley (1990). The article "Density and Velocity Dependence of Reynolds Numbers for Several Lattice Gas Models," by K. Diemer, K. Hunt, S. Chen, T. Shimomura, and G. D. Doolen describes the bit representation for many lattice models.

Brian W. Kernighan and Dennis M. Ritchie, *The C Programming Language*, second edition, Prentice Hall (1988). An excellent reference.

William H. Press, Saul A. Teukolsky, William T. Vetterling, and Brian P. Flannery, *Numerical Recipes in Fortran*, second edition, Cambridge University Press (1992).

Robert J. Traister, *Leaping from BASIC to C++*, Harcourt-Brace (1994).

I N D E X